Cosmic Dust from the Laboratory to the Stars

Rafael Rodrigo • Jürgen Blum • Hsiang-Wen Hsu •
Detlef V. Koschny •
Anny-Chantal Levasseur-Regourd •
Jesús Martín-Pintado • Veerle J. Sterken •
Andrew Westphal
Editors

Cosmic Dust from the Laboratory to the Stars

Originally published in *Space Science Reviews* in the Topical Collection "Cosmic Dust from the Laboratory to the Stars"

Springer

Editors
Rafael Rodrigo
International Space Science Institute
Bern, Switzerland

Jürgen Blum
Institut für Geophysik und extraterrestrische Physik
Technische Universität Braunschweig
Braunschweig, Germany

Hsiang-Wen Hsu
LASP
University of Colorado Boulder
Boulder, CO, USA

Detlef V. Koschny
SCI-S
European Space Agency
Noordwijk, The Netherlands

Anny-Chantal Levasseur-Regourd
LATMOS
Sorbonne Université
Paris, France

Jesús Martín-Pintado
Astrophysics, Centro de Astrobiología (CSIC/INTA)
Torrejón de Ardoz, Spain

Veerle J. Sterken
Physics Department
ETH Zurich IPPA
Zürich, Switzerland

Andrew Westphal
Space Sciences Laboratory
University of California at Berkeley
Berkeley, CA, USA

ISSN 1385-7525 Space Sciences Series of ISSI
ISBN 978-94-024-2009-8

Cover Image: This new NASA/ESA Hubble Space Telescope image shows the Lagoon Nebula, an object with a deceptively tranquil name. The region is filled with intense winds from hot stars, churning funnels of gas, and energetic star formation, all embedded within an intricate haze of gas and pitch-dark dust. Credit: NASA, ESA, J. Trauger (Jet Propulsion Laboratory)

Printed on acid-free paper

This Springer imprint is published by the registered company Springer Nature B.V.
The registered company address is: Van Godewijckstraat 30, 3311 GX Dordrecht, The Netherlands

Preface

This topical volume entitled *Cosmic Dust form the Laboratory to the Stars* has been edited by: Rafael Rodrigo, Jürgen Blum, Hsiang-Wen Hsu, Detlef Koschny, Anny-Chantal Levasseur-Regourd, Jesús Martín-Pintado, Veerle Sterken and Andrew Westphal. The papers are the result of a collaborative effort started with an ISSI Workshop organized in Bern, Switzerland. The workshop took place between in October 31 and November 4, 2016. It was attended by 49 participants (including nine conveners and six young scientists) from a broad range of disciplines and backgrounds (all of them specially invited to come to ISSI for this purpose), resulting in an intensive and stimulating experience. This volume is not intended to be the proceedings of the Workshop, but a collection of in-depth review papers informed by the contributions and discussions at the Workshop.

The main purpose was to stimulate discussion on the state of the art of the science devoted to Cosmic Dust. Dust is ubiquitous – it is found from the Earth's atmosphere to the edges of our Universe. It is studied in many ways including in-situ measurement, remote observation (in our Solar System through to distant galaxies) and by analysis of returned samples. Progress in the last decade has been significant with sample return missions, in-situ measurements and ground- and space-based measurements, as well as progress in laboratory experiments and modeling. One of the challenges in the study of dust has been to harmonize the different techniques used in the study of dust and the various environments in which dust is found. It is now nearly 18 years since the last major book publication in the field (Interplanetary Dust, Grün et al., 2001). The time gap and the pace of development merited the next such initiative – now with an emphasis where possible on interconnections, similarities, differences and on synthesizing results from different techniques into one coherent view. A new aspect is that astrobiological connections are also considered. Moreover, the dust hazard, and future technology and space mission requirements and scenarios was also addressed. During the Workshop, 46 communications were presented (including four from young scientists). An important outcome of the workshop is this book published in the Topical Collection in Space Science Reviews and as a book in the Space Sciences Series of ISSI (SSSI) by Springer Verlag. The outcome is on this way a compendium with a coherent view on dust from the different disciplines involved in cosmic dust science and recommendations for future dust research and space missions. The main aim is to provide a coherent picture of the current state of the subject. The readers can find a complete review of the development of dust science, as well as a comprehensive summary of the different chapters in the first paper of this book by Eberhard Grün, Harald Krüger and Ralf Srama.

The editors warmly compliment the conveners and organizers of the Workshop who tirelessly conducted the whole workshop and the edition of this book to a successful and fast fruition with enthusiasm and dedication. Let us also thank all those participated in this memorable event, especially all ISSI staff, and in the writing of this excellent book. Congratulations to all.

Rafael Rodrigo, former Executive Director of International Space Science Institute, Bern, Switzerland, actually at Spanish Ministry of Science and Innovation.

Group picture of the workshop held in November 2016 in Bern

Contents

Space Sci Rev (2019) 215:46
https://doi.org/10.1007/s11214-019-0610-1

The Dawn of Dust Astronomy

Eberhard Grün[1,2] · Harald Krüger[3] · Ralf Srama[4]

Received: 20 November 2018 / Accepted: 16 September 2019 / Published online: 14 October 2019

Abstract We review the development of dust science from the first ground-based astronomical observations of dust in space to compositional analysis of individual dust particles and their source objects. A multitude of observational techniques is available for the scientific study of space dust: from meteors and interplanetary dust particles collected in the upper atmosphere to dust analyzed in situ or returned to Earth. In situ dust detectors have been developed from simple dust impact detectors determining the dust hazard in Earth orbit to dust telescopes capable of providing compositional analysis and accurate trajectory determination of individual dust particles in space. The concept of Dust Astronomy has been developed, recognizing that dust particles, like photons, carry information from remote sites in space and time. From knowledge of the dust particles' birthplace and their bulk properties, we learn about the remote environment out of which the particles were formed. Dust Observatory missions like Cassini, Stardust, and Rosetta study Saturn's satellites and rings and the dust environments of comet Wild 2 and comet Churyumov-Gerasimenko, respectively. Supplemented by simulations of dusty processes in the laboratory we are beginning to understand the dusty environments in space.

Keywords Dust · Interstellar medium · Dust detection · Zodiacal cloud · Interplanetary dust

Cosmic Dust from the Laboratory to the Stars
Edited by Rafael Rodrigo, Jürgen Blum, Hsiang-Wen Hsu, Detlef Koschny, Anny-Chantal Levasseur-Regourd, Jesús Martín-Pintado, Veerle Sterken and Andrew Westphal

✉ H. Krüger
harald.krueger@mps.mpg.de

1 Max-Planck-Institut für Kernphysik, Heidelberg, Germany

2 LASP, University of Colorado, Boulder, CO, USA

3 Max-Planck-Institut für Sonnensystemforschung, Göttingen, Germany

4 Institut für Raumfahrtsysteme, Universität Stuttgart, Stuttgart, Germany

1 Introduction

Dusty phenomena in space have been studied by astronomers for centuries: the zodiacal light, comets, and meteors. The zodiacal light is a prominent phenomenon observable by the human eye in the morning and evening sky in non-polluted areas on Earth. Already in 1683 the Italian astronomer Giovanni Domenico Cassini presented the correct explanation: it is sunlight scattered by dust particles orbiting the Sun. Comets shed large amounts of dust during their passages through the inner solar system that is mostly visible in the cometary coma and dust tail. The third dusty phenomenon visible with the naked eye, meteors, can be observed during any clear night, but there are special periods, so called meteor showers or meteor storms, when the rate is thousand times enhanced. It was the coincidence between such meteor showers and the simultaneous apparitions of comets that suggested their genetic relation. Triangulation of meteor trails in the atmosphere, especially during meteor showers, confirmed the relation of some meteor streams to comets. The physical interrelation of the three phenomena was illuminated by the work of the American astronomer Fred Whipple in the early 1950s (Whipple 1949, 1951, 1955).

Interstellar dust became a topic of astrophysical research in the early 1930s when the existence of extinction, weakening, and scattering of starlight in the interstellar medium (ISM) was realized (cf. Trümpler 1930). Since then astronomical observations provided information about the properties of the dust in the ISM. Dense clouds of interstellar dust are visible as dark nebulae against the brighter background of the Milky Way. The distance dependent attenuation of blue star light is much stronger than that of red light; therefore, extinction causes distant stars to appear redder than expected, a phenomenon referred to as interstellar reddening. Broad absorption features at infrared (IR) wavelengths are clues to the composition of interstellar dust grains, e.g. the water ice feature at 3.1 μm, and silicate features at 10 and 18 μm. Dark interstellar clouds are frequently related to star formation regions. Here, dust plays an important role by radiative cooling of collapsing clouds and eventually providing seeds for condensation and accretion of larger bodies that are building blocks of planetary objects (Dorschner 2001). Even from expanding shells of supernova explosions dust became introduced into primitive solar system material (Leitner and Hoppe 2019).

The ultimate sources of dust are stars breeding heavy elements by nuclear fusion. Dust particles bear information on their formation process, and any subsequent processes which they became part of. In interstellar space, heavy elements, i.e. elements heavier than helium, often reside in dust grains. Together with their corresponding atomic and ionized species they trace stellar evolution in our galaxy.

In 2016, during the final phases of the Rosetta mission around comet 67P/Churyumov-Gerasimenko, the spacecraft flew sufficiently close to the nucleus so that several instruments simultaneously observed outbursts of gas and dust (Grün et al. 2016; Agarwal et al. 2017, and references therein). Monitoring instruments, like the star sensors STR, the dust impact detector GIADA, and the gas sensor ROSINA/COPS, were the first to detect outbursts at the cometary surface. The STR cameras observed stray light from the emitted dust cloud in their fields-of-view, GIADA detected impacts of individual sub-millimetre sized dust particles on its sensors and determined their mass and speed, and COPS recorded the enhanced gas flux from the outburst but also the gas puffs from sublimating icy dust particles entering the pressure sensor. The Alice UV spectrometer recorded signs of gas and dust emissions from the nucleus. In some cases the OSIRIS cameras could identify the outburst location on the nucleus and observe the expanding dust cloud and streaks of centimetre-sized particles passing near the camera. The atomic force microscope instrument MIDAS provided microscopic and morphologic information on collected submicrometer-sized grains, and the

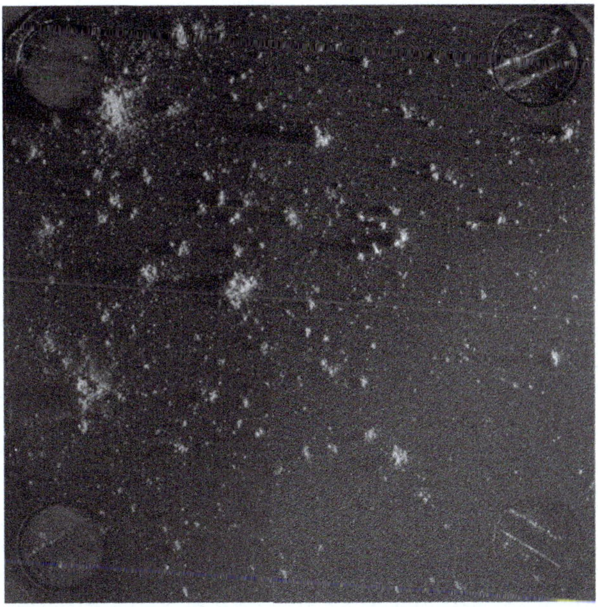

Fig. 1 Comet particles collected and analyzed by the COSIMA instrument on board Rosetta. The particles were collected between 10 April 2015 and 27 May 2015 at distances between 91 km and 321 km from the nucleus of comet 67P (at heliocentric distance between 1.89 and 1.55 AU), at a collection speed of ~10 m/s. For scale the screw head has 1.5 mm diameter (Langevin et al. 2016; Hilchenbach et al. 2016, ESA/ROSETTA/MPS FOR COSIMA TEAM MPS/CSNSM/UNIBW/ TUORLA/IWF/IAS/ESA/BUW/ MPE/LPC2E/LCM/FMI/ UTU/LISA/UOFC/VH&S)

COSIMA dust analyzer obtained structural and compositional information on collected submillimetre sized grains (Fig. 1). Also, the plasma instruments RPC recorded an effect of the expanding gas and dust cloud. By the combination of this wealth of information on the solid component of the nucleus, we can obtain a better understanding of this comet and the processes occurring on its surface. This described episode demonstrates the many facets that the study of dust and its interrelation with its environment has. Rosetta was a dust observatory mission that provided one of the most recent highlights in dust astronomy.

Since the beginning of space flight in 1957 the interest in space dust has enormously expanded. The first questions asked were: How much dust is in near-Earth space and how big is it? The reasons were obvious: Travelling with a typical speed of 20 km s^{-1}, even submillimetre-sized meteoroids pose a serious threat to space vehicles. Most early satellites carried dust detectors in order to characterize the meteoroid hazard for space flight. The long path from these early days of simple in situ dust detection to the present days of dust astronomy is portrayed in this article.

Since the early 1960s the community of astronomers and dust scientists met and exchanged their data and ideas at big international conferences, like COSPAR (Committee on Space Research) and IAU (International Astronomical Union) meetings. Soon it was recognized that it was mutually advantageous to bring together these partially overlapping communities of scientists. Therefore, specialized international dust meetings regularly took place at varying locations which produced significant proceedings volumes. In later years dust science spread to broader topics and was discussed at meetings of the Meteoritical Society, at geophysical (AGU, EGS) and planetary science meetings (DPS, EPSC, ACM).

In 1978 Tony McDonnell took up the task to assemble a comprehensive review of the "Cosmic Dust" field. The book (McDonnell 1978) covered a wide range of topics; from interstellar dust to comets, meteors, and zodiacal light, over lunar craters to dust particles collected in the atmosphere, from dust dynamics to laboratory simulation and impact physics. About 20 years later, in 2001, a new version of a comprehensive review of the field was assembled by Eberhard Grün, Bo Gustafson, Stanley Dermott and Hugo Fechtig (Grün et al.

2001). Besides the classical topics of cometary, interplanetary and interstellar dust, instrumentation and dust dynamics, it covered new subjects like near-Earth dust and space debris, dusty rings, thermal properties of dust and physical processes on interplanetary dust, and a review of comprehensive models of interplanetary dust. The book also contained an extensive historical perspective section by Hugo Fechtig et al. (2001) which described the early attempts and failures of dust detections until the grandiose achievements of the space missions to comet Halley and the beginning of the success story of infrared dust astronomy.

Progress in the last decade has been significant with sample return missions, in situ measurements and ground- and space-based measurements, as well as in laboratory experiments and modelling. Therefore, the time gap from the last comprehensive review and the pace of development merits a new coherent view in a conference and a book, now with an emphasis on interconnections, similarities, differences and on synthesizing results from different techniques. Astrobiological connections are a new aspect considered.

The first section of this chapter describes the turbulent early days of dust research that were focussed on the meteoroid hazard when the community painfully learned how to fly reliable in situ dust instruments. The second section characterizes the multitude of observational techniques contributing to our knowledge about cosmic dust. This section also describes the development of in situ instruments to complete dust telescopes and dust observatories in space by which dust astronomy is performed (Grün et al. 2005). Important laboratory experiments are discussed in section three. In section four we synthesize our understanding of the zodiacal cloud and give reference to modelling of other dust environments.

2 Dangerous Dust

Meteoroids are part of the space environment. The danger from millimetre and centimetre sized meteoroids to manned and unmanned activities was obvious from the beginning of the space age. Smaller particles could be dangerous to not well shielded satellites and to manned extravehicular activity as well. Therefore, at the beginning, all space-faring agencies started a vigorous program to characterize the meteoroid hazard as part of their space activity.

A set of simple microphones were the first dust detectors flown in 1950 in the US on board a V2 rocket. Parallel to the development of space dust detectors were appropriate simulation techniques. In order to simulate impacts and collision phenomena in space, accelerators were necessary to provide projectiles in the speed range at 10 $\mathrm{km\,s^{-1}}$ and beyond. The work horse of accelerators for millimetre and bigger projectiles was the light gas gun. This gun consisted of a conventional powder gun that pushes a piston into a barrel containing hydrogen gas which itself accelerates the projectile. By this method a projectile speed up to 12 $\mathrm{km\,s^{-1}}$ was reached, however the wear on the gun was excessive and, therefore, only speeds up to 5 $\mathrm{km\,s^{-1}}$ were routinely achieved. A wide range of projectile materials can be accelerated with this method.

An electrostatic accelerator for micrometer sized and smaller particles is similar to nuclear physics devices that accelerate charged particles by high electric voltages (Friichtenicht 1962). Potentials of several million Volts are obtained by van-de-Graaff generators. High charges on electrically conducting particles are obtained by bringing these particles in contact with the fine tip of a tungsten needle which is at a high electric potential. In a 2 MV accelerator micrometer sized iron particles can reach speeds of 12 $\mathrm{km\,s^{-1}}$, or 0.1 μm sized particles reach up to 35 $\mathrm{km\,s^{-1}}$. Speeds as high as 100 $\mathrm{km\,s^{-1}}$ have been observed for sub-micrometer sized grains. Electrically conducting particles consisting of iron, carbon,

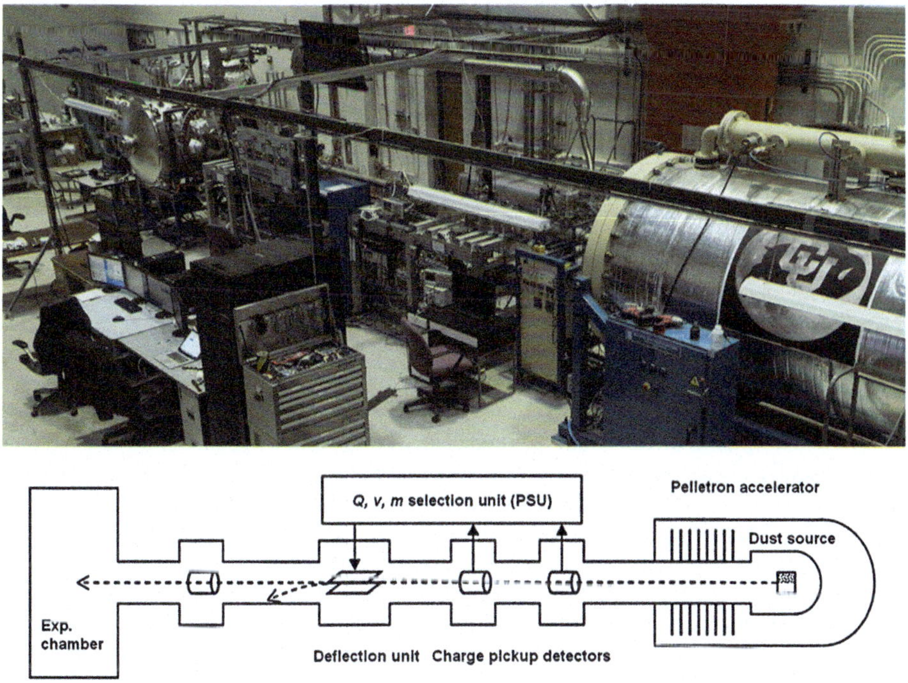

Fig. 2 Dust accelerator at the University of Colorado/LASP with 3 MV acceleration voltage. The schematics are shown at the bottom (after Horanyi et al. 2016)

aluminum, and non-conducting glass, mineral, or plastic particles coated by conducting materials have been used in these accelerators. Currently, only the dust accelerator at the Colorado Center for Lunar Dust and Atmospheric Studies (CCLDAS, Thomas et al. 2013) is in routine operation (Fig. 2).

Many of the earliest dust detectors were single shot detectors responding to a single signal upon impact of a dust particle on to the sensor. These detectors had no means to directly distinguish dust impacts from any noise they responded to. Most popular were momentum sensors, consisting of a single piezoelectric crystal attached to an impact plate. Any signal exceeding a threshold was interpreted as an impacting meteoroid. Often such detector systems were calibrated by simply dropping glass beads on to them because they were not sensitive enough for projectiles from an electrostatic dust accelerator (Auer 2001).

In the 1960s, early microphone data from NASA's Explorer and some Russian Cosmos and Salyut satellites were interpreted as near-Earth dust enhancements by factors of 1000 and higher than the dust density in interplanetary space. Dust collections by primitive collectors on sounding rockets above the Earth's dense atmosphere seemed to confirm the satellite observations.

However, the early investigators did not appreciate that there were a variety of interfering effects that microphones respond to as well. These include external effects like cosmic rays hitting sensitive first stages of the amplifiers, or time variable ambient plasma effects that can interfere with sensitive dust detectors or temperature variations during a satellite's passage through the Earth's shadow causing thermal cracking. But also interferences generated inside the satellite like micro-vibrations caused by other spacecraft systems, or switching high currents or voltages could generate electromagnetic interferences in the dust detector.

Likewise, pre- and post-flight contaminations of the collector surfaces invalidated the earlier in-flight rocket dust collections.

Measurements of the zodiacal light brightness set stringent constraints on the total dust density in interplanetary space. Theoretical analysis by Shapiro et al. (1966) and Colombo et al. (1966) discussed various aspects of 'The Earth dust belt: fact or fiction?' and concluded that a near-Earth enhancement of at most a factor of 10 could be reasonably explained but no enhancement factors of 100 or more. Indeed, the current best estimate of the near-Earth gravitational enhancement is only a factor of 2. At about the same time, Nilsson (1966) expressed doubts about the Earth's dust cloud based on experimental grounds. He used witness microphone detectors that were shielded form dust impacts but otherwise identical to active ones. They recorded about the same event rate as the active ones and hence, contradicted the dust belt hypothesis. Therefore, the initial high fluxes reported were erroneous and were caused by the combined effects of immature dust detectors and the harsh near-Earth environment.

The most reliable of the single parameter detectors were the penetration detectors on board several early satellites and space probes (e.g. Explorers 16 and 23, Pioneers 10 and 11, D'Aiutolo et al. 1967). These detectors consisted of a large number of pressurized cells (termed 'beer cans') that recorded the decrease in gas pressure when a wall was punctured by a meteoroid. The walls were metal sheets (stainless steel or copper beryllium) of 25 or 50 μm thickness and had penetration limits of 10^{-9} and 10^{-8} g, respectively at 20 $\mathrm{km\,s^{-1}}$ impact speed (Humes et al. 1974). The Pegasus detectors were large area (about 200 m^2) detectors that recorded penetrations of up to 0.4 mm-thick aluminium sheets by the discharge of a capacitor. Detectors like these determined the flux of meteoroids in near-Earth space in the 10 micrometer to millimetre size range. This range was important for the assessment of the meteoroid hazard for typical satellites and the manned missions ahead. Cour-Palais (1969) published the official NASA meteoroid model that described the Earth's natural meteoroid environment. It showed that no special and demanding precautions had to be taken to protect space systems against the meteoroid hazard and that the Apollo missions to the Moon were safe to be conducted. It also showed that a large space station protected by an effective 1 cm thick aluminium shield could survive in the natural near-Earth meteoroid environment.

As a consequence, the interest in meteoroids shifted from technological to astrophysical questions and the amount of money to support dust studies dramatically declined. From early major groups at the NASA centres, only a few individual scientists continued to work in the meteoroid field, in Russia almost all meteoroid work subsided. Only a few universities and research laboratories, primarily the University of Kent at Canterbury and the Max-Planck-Institute for Nuclear Physics at Heidelberg, kept on advancing the meteoroid field.

Previous experience has shown that in situ measurements of space dust are low-level measurements, they require highest sensitivity for extremely scarce impact events in a hostile time-varying space environment. Only one dust impact per month is not uncommon, therefore, the foremost challenge was to develop means to distinguish dust impacts from noise. Careful characterization of the impact signal at a dust accelerator is a fundamental requirement. Multiple separate dust sensors, with one or more sensors shielded from dust impacts, provide some means of statistical distinction of impact events from noise events. Amplitude or even waveform measurements of the recorded signals can help directly to identify noise and dust signals. The most reliable method, however, is to record several signals from a single dust impact in coincidence. Multi-coincidence detectors based on the impact ionization process were successfully flown throughout the solar system and, together with appropriate data analysis, provided significant results about various dust environments (see below). A well-proven approach is to spend as much effort to understand the noise behaviour of a dust instrument in a specific environment as to analyze dust impact data.

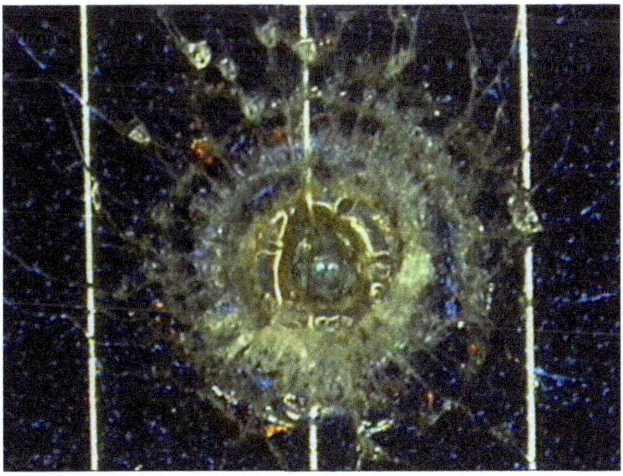

Fig. 3 Space debris impact on to Hubble Space Telescope solar cells that were returned to ground (NASA image)

Recently, the near-Earth dust environment has been getting back into the focus of space agencies. The ever increasing human space activity has led to a hazardous condition that could jeopardize future space activities in low Earth-orbit (below approx. 1000 km altitude). Abandoned upper rocket stages exploded in space which had not completely used up their liquid fuel. Collisions of satellites over the Earth poles where many orbits of polar satellites cross each other shed large amounts of long-living debris into Earth orbit. Already in 1978, Kessler and Cour-Palais (1978) noted that the increased population of satellites will eventually lead to increased amounts of debris by accidental collisions. Shielding satellites by massive shields will eventually worsen the situation because even more mass needs to be transported to Earth orbit which will finally become debris. Kessler predicted that after 50 to 100 years a run-away process will be started and shatter all matter in low-Earth-orbit, leading to the formation of an artificial collisional ring until atmospheric drag will clear this region of space again.

Current measurements of the near-Earth debris by radar, lidar, and other methods (Fig. 3) indicate that it is already in or near the self-destruction mode, also evidenced by the frequent collision avoidance manoeuvres of the International Space Station (ISS) with trackable debris. Micrometeoroid or space debris impacts damaged some focal plane detectors of the XMM-Newton and Swift-XRT X-ray observatories (Carpenter et al. 2008). Clustering in the times of impacts is indicative of dust generation events. Such clusters have been reported by NASA's Long Duration Exposure Facility (LDEF) IDE instrument (Oliver et al. 1993; Cooke et al. 1993) and by ESA's DEBIE instrument (Schwanenthal et al. 2005). Currently, ESA estimates that almost one million space debris objects bigger than 1 cm orbit the Earth. Because of the disastrous effects of a collision of such a projectile with an active spacecraft, most space agencies adopted measures to reduce the detrimental effects of space flight. These measures include to empty liquid fuel from used rocket stages, to bring in-operational satellites to orbits where the risk of collisions is low or to "de-orbit" them into the Earth atmosphere. However, an ultimate solution to the space debris problem has not been found.

From 1984 to 1990, LDEF, was exposed to the space environment at about 450 km altitude. Studies of impact craters on returned LDEF surfaces, and a similar study by the European Eureca satellite, showed that the near-Earth meteoroid flux is about a factor of 2 higher than the flux in deep space because of gravitational concentration by the Earth (McDonnell and Gardner 1998). Depending on the particle size, natural meteoroids are outnumbered by man-made space debris: below about 10 µm and above about 100 µm in size debris particles are more abundant than natural meteoroids. Therefore, monitoring the near-Earth dust environment is presently a routine activity of all major space agencies.

The meteoroid hazard in interplanetary space is currently not well understood, and human space travel beyond the Earth-Moon system will require higher attention in the future. A space station in Earth vicinity like the planned Deep Space Gateway in lunar orbit would be an excellent platform for future active or passive dust detection techniques. Pegasus-type in situ dust detectors may be needed to monitor the dust flux in interplanetary space between Venus and Mars. In addition, during the future human colonization of Mars, dust from the Martian satellites may play a similar hazardous role as space debris in the Earth environment. Interstellar travel by automatic probes to α Centauri and other near-by stars at 30,000 km s^{-1} speed (0.1 speed of light) pose an unpredictable hazard: at such a high speed micrometer-sized interstellar particles carry the energy of rifle bullets!

3 Multifaceted Scientific Dust Observations

According to the 2017 IAU definition, meteoroids are solid natural objects of size smaller than roughly 1 meter moving in, or coming from, interplanetary space. Interplanetary dust (or micrometeoroids) is the small end ($<\sim$30 micrometer) of the meteoroid size distribution. However, there is a continuous distribution of objects up to the size of asteroids, comets and planetary satellites that are in intimate relation with and constitute the interplanetary meteoroid complex. There are several scientific methods to observe and analyze meteoroids and interplanetary dust (Koschny et al. 2019).

3.1 Large-Scale Distribution and Composition of Interplanetary and Interstellar Dust by Remote Observations

Since the 17th century, when Giovanni Cassini explained the relation between zodiacal light and dust in interplanetary space, zodiacal light observations have been the main tool to study space dust. However, observations from the ground are handicapped by weather effects, atmospheric dust and light pollution, and airglow. Such observations have the additional difficulty that the brightness along the line of sight (LOS) is generated from particles located at different heliocentric distances and observed at different scattering angles. Because the light intensity scattered at a certain angle strongly depends on particle size, such observations cannot separate the radial dust density from the size distribution. Only in the special case when observations are performed along the same LOS from two different positions in space can the spatial density of particles along the section between the two observation positions directly be derived (Dumont, 1983). With the photometers on board the Helios space probes Leinert et al. (1981) succeeded in the measurement of zodiacal light between 0.3 and 1 AU with the same scattering geometry. The authors determined the radial brightness profile ($\sim r^{-1.3}$) and found an inclination of 2° of the symmetry plane of the zodiacal cloud with respect to the ecliptic plane. From these observations Leinert and Grün (1990) concluded that the zodiacal cloud has a flattened lenticular shape with an axial ratio of 1:7.

Outside the Earth's orbit, zodiacal light was observed by photometers on board the Pioneer 10 and 11 spacecraft (Weinberg et al. 1974). The zodiacal light brightness was found to exceed the background out to 3.3 AU distance from the Sun (Hanner and Weinberg 1973). On board Pioneer 10 and 11 there were two more dust instruments besides the zodiacal light photometer: the beer can penetration detectors (see below) and the Asteroid Meteoroid

Detector (AMD, Soberman et al. 1974). The latter instrument consisted of four telescopes designed to detect scattered sunlight from individual meteoroids passing through their field-of-view. From the timing and the amplitudes of the received light signals, the trajectory of the meteoroid could in principle be reconstructed. Because of too low preset threshold sensitivity, however, the instrument was very noisy and no single trajectory could be uniquely identified (Auer 1974). Due to this inability to determine meteoroid trajectories of the Pioneer AMD instrument, the proposed twin AMD instrument was removed from the preliminary Voyager payload. Given the unique launch window of Voyager (1976–1978) for the planetary grand tour to the four giant gas planets, no replacement was found in time and, hence, no dedicated dust instrument was onboard the Voyager spacecraft.

An interplanetary dust particle scatters only less than 10% of the incident sunlight and contributes to the zodiacal light. The rest of the absorbed energy (>90%) is re-emitted as thermal infrared radiation, mostly in the 10 to 50 μm mid-IR wavelength range. Because of this effect, zodiacal infrared emission is a much more prominent astronomical phenomenon than emission at visible wavelengths; however, infrared observations have to be performed from above the Earth's atmosphere.

A milestone in the observations of the local zodiacal dust disc was the first all-sky survey at infrared wavelengths, conducted by the Infrared Astronomical Satellite (IRAS) in 1983. It mapped the sky at 12, 25, 60 and 100 μm wavelengths. Besides the dominant foreground zodiacal emission, several hundred thousand sources were discovered including stars, galaxies, and solar system objects. Many stars with dust discs were identified through their infrared excess radiation; among them dust discs around Vega, β Pictoris, Fomalhaut, and ϵ Eridani (Blum 2019). A surprise was the observed structure in the zodiacal cloud: IRAS discovered a complex system of asteroid bands and many comet trails (Eaton et al. 1984; Sykes et al. 1986). Later on, the COBE satellite obtained unprecedented information on the overall structure of the zodiacal cloud to an accuracy of better than 3% (Kelsall et al. 1998a, 1998b). Using the Spitzer infrared Space Telescope, Reach et al. (2007) found 80% of the Jupiter family comets to possess dust trails. An extreme example is comet 73P/Schwassmann-Wachmann (Fig. 4) that in 1995 started to break up into many large pieces which spread along the comet's original orbit. Recently, the Planck mission extended the observations of the zodiacal cloud to millimetre wavelengths. Given that IR observations from near the Earth are handicapped by the dense zodiacal foreground, a 'bird's eye view' from an IR observatory above the ecliptic plane (e.g. at 40° latitude) would be able to examine the extent and fine structure of the 'warm' zodiacal cloud, and, finally, may observe the 'cold' outer Trans-Neptunian dust disc. It may also provide an unobscured view into interstellar space at mid-IR wavelengths.

IR observations also provide compositional information about the dust particles that emit or absorb IR radiation. IR features are often the only way to obtain direct information on the composition of dust; this is especially true for interstellar and circumstellar space. Beside these compositionally identified features there are many prominent unidentified features in the dusty interstellar environment.

Amorphous silicates were the first solids discovered in circumstellar space. They were identified by their prominent broad features at 9.7 and 18 μm. Crystalline silicates such as pyroxenes and olivines have sharper features and were also commonly detected in the circumstellar environment (cf. Fig. 5). These features are very common in oxygen-rich asymptotic giant branch (AGB) stars. The most common carbon bearing solid in carbon-rich stars is SiC which has a feature at 11.3 μm. Finally, water ice has a feature at 2.05 μm.

Carbon bearing molecules display a number of features in the 3 to 11 μm range. They were identified in the spectra of planetary nebulae and came as a complete surprise. These

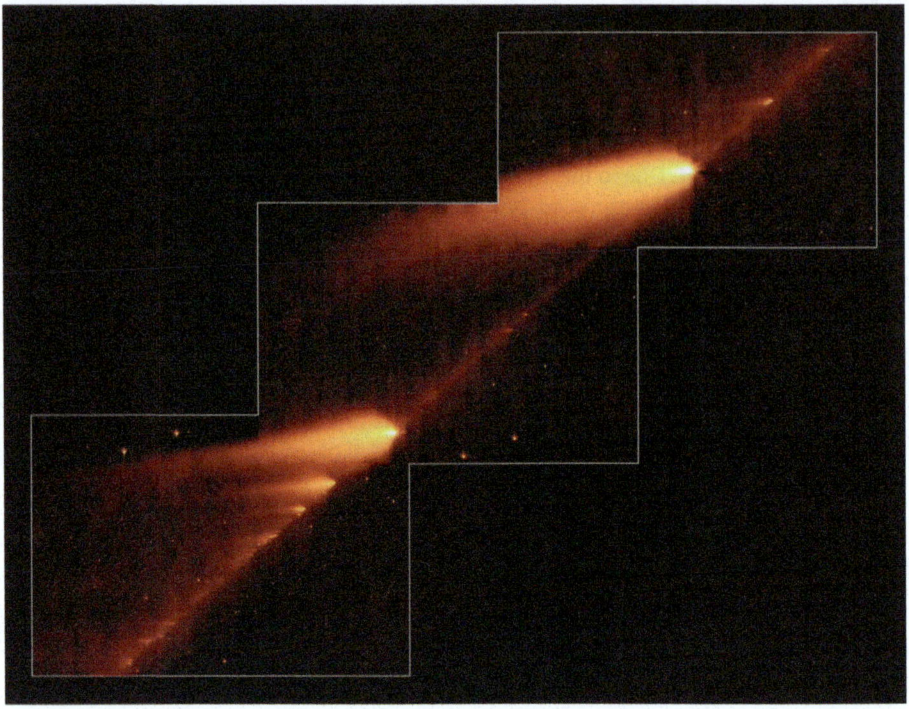

Fig. 4 Breakup of comet 73P/Schwassmann-Wachmann (NASA Spitzer Space Telescope). Individual fragments are spread along the original orbit and display their own dusty tails (Reach et al. 2009). Comet 73P/Schwassman-Wachmann 3 images taken from May 4 to 6 show at least 36 distinct fragments. Credit: NASA/JPL-Caltech

features are generally identified as originating from stretching and bending modes of polycyclic aromatic hydrocarbons (PAHs), although it has recently been shown that they may also originate from amorphous hydrocarbons. Natural terrestrial counterparts of these compounds may be coal, kerogen and petroleum.

More complex hydrocarbons were identified in interstellar space. Interstellar dust grains with typical sizes in the sub-micrometer range are believed to consist of a silicate and/or carbonaceous nucleus with an ice mantle if the temperature is low enough.

In our solar system, since the exploration of comet 1P/Halley in the 1980s, it has been known that a significant fraction of the cometary organic matter exists in the solid phase and is ejected from the nucleus as dust particles. For example, phosphorus, a key element in living organisms, was recently identified in the gas phase of comet 67P/Churymov-Gerasimenko by the ROSINA instrument on board Rosetta (Altwegg et al. 2016). The presence of organic molecules which may be relevant for the formation of more complex biogenic compounds in living organisms raises the question whether space dust may harbour the precursors of life (cf. Pintado et al. 2019).

Most planetary missions carried cameras on board to observe the target planet(s), satellites and rings. The TV cameras on board both Voyager spacecraft discovered new dusty phenomena around all giant planets (Smith et al. 1979a, 1979b, 1981, 1982, 1986, 1989). The Jovian system harbours a complex ring system inside about 3 Jupiter radii (R_J) from Jupiter (Showalter et al. 2008; Hamilton and Krüger 2008) and powerful volcanoes on Io that spew plumes of ash as high as 300 km above the surface (McEwen et al. 2004). At

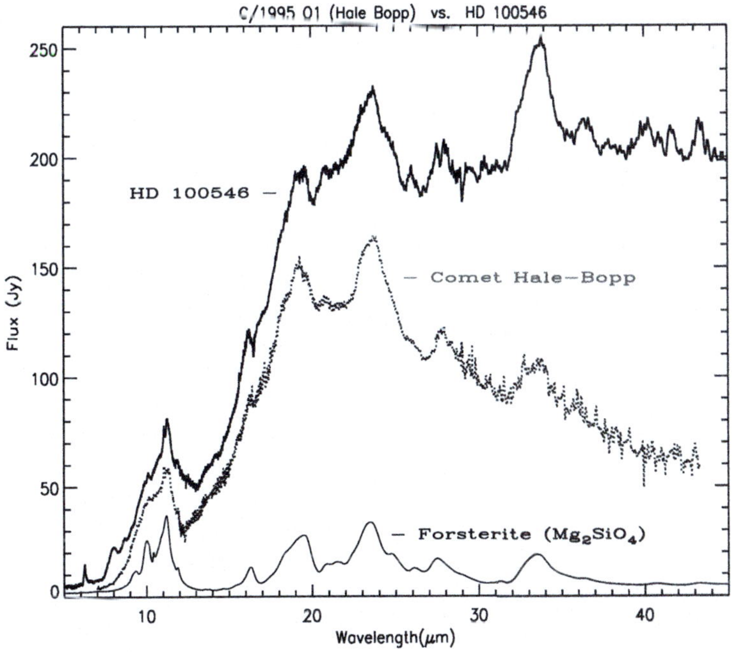

Fig. 5 Infrared spectra of the Young Stellar Object HD 100546 (top curve) compared to those of comet Hale-Bopp (middle), both taken with the Infrared Space Observatory (ISO). The bottom curve shows a laboratory spectrum of forsterite (van den Ancker 1999)

Saturn the Voyager cameras observed small inner satellites, interacting gravitationally with one another and with particles in the dense rings (A, B, C, and F rings). In the B ring the mysterious spokes phenomenon was discovered (Grün et al. 1983, Morfill et al. 1983, Goertz and Morfill 1983) that is still awaiting full explanation. Beyond the F ring two tenuous rings (G and E rings) were observed. At Uranus nine rings are described, along with the locations and sizes of 10 shepherd satellites. Like the rings of Uranus, those of Neptune are composed of very dark material. On the biggest geologically evolved satellite, Triton, at least two active plumes driven by sublimating nitrogen ice were found. Later the cameras on Galileo and Cassini followed up on further investigating the dusty phenomena in the Jovian and Saturnian systems. It was found that geysers on the small icy Saturnian moon Enceladus feed the broad dusty E ring.

3.2 Meteor Observations

On a clear night, about 1 to 10 faint meteors can be observed by the naked eye per hour. Meteors are caused by centimetre-sized and bigger meteoroids that enter the Earth's atmosphere at speeds greater than 10 km s^{-1}. Once in a while, when a meter-sized meteoroid hits the atmosphere, a bright fireball appears. More sensitive methods to observe and measure the trajectories of meteors are photographic and TV cameras and telescopes. Spectroscopy of meteors shows mostly lines of meteoritic metals (Na, Mg, Ca, and Fe, Vojacek et al. 2015), besides strong emission lines of atmospheric origin (O, N, and N_2).

While the speed of meteors can be accurately determined, the size or mass of the meteoroid can only be estimated. Calibration measurements of the luminous efficiency and ionization coefficient have been performed only with micrometer-sized particles at relevant entry speeds exceeding 10 km s^{-1} (Friichtenicht and Becker 1971; Thomas et al. 2016). For centimetre sized meteoroids only theories are available to obtain the meteoroid mass from the observed measurement (Koschny et al. 2019). Effects of meteors on planetary atmospheres like the formation of layers of smoke and ice particles and atmospheric gas-phase chemistry are discussed in Plane et al. (2018).

Even though our collections contain more than 10,000 meteorites, only sixteen falls of meteorites have been observed with adequate equipment to derive their interplanetary orbits. These meteorites are of the ordinary chondrite and achondrite types and their heliocentric orbits prior to entering the Earth's atmosphere had aphelia in the asteroid belt, directly indicating a genetic relationship between these meteorites and asteroids. A comprehensive review of meteor research was given by Ceplecha et al. (1998). However, all observational methods require a dark and clear sky and, therefore, they cannot provide a complete record of the Earth meteoroid environment.

About 80 years ago, radar techniques were developed to observe faint meteor trails even during daylight. Radar meteors can be detected using either high power and large aperture radars (like Arecibo or other big radar dishes) or specular backscattering meteor radars such as the Advanced Meteor Orbit Radar (AMOR) in New Zealand and the Canadian Meteor Orbit Radar (CMOR). The first type of radar observes the head-echo of the meteor which is the signal reflected from the plasma region travelling with the meteoroid. The detection of the head-echo allows for the determination of instantaneous meteor altitude, in-beam speed and deceleration. However, it suffers from the low head echo rates and from the low precision of the true velocity measurement. A specular meteor radar detects echoes from the ionized trail produced by meteors ablating in the Earth atmosphere. From the measurements of the same meteor trail by several transmitting and receiving stations on the ground, the velocity of the heliocentric orbit can be inferred.

Three major complete meteor surveys have been performed by averaging over daily and seasonal variations. The first study of apparent radiant concentrations used the Harvard Meteor Radar Project (HRMP, cf. Taylor and Elford 1998). More recently, Galligan and Baggaley (2004, 2005) reported a meteor orbit study from AMOR. Highly sensitive AMOR observations indicate even an interstellar origin of a small percentage ($<1\%$) of the observed meteors (Baggaley et al. 1993, cf. Hajdukova 2016). In the Northern hemisphere the lower sensitivity CMOR (Brown et al. 2008) is operated. Besides the study of meteor streams, radar meteor data is used for the comprehensive analysis of the sporadic meteor background. While meteor streams show up in narrow (i.e. few degrees wide) concentrations of apparent radiants (not corrected for Earth motion), also sporadic meteors display wide and diffuse concentrations of radiants (Fig. 6). The symmetric north and south apex radiant concentrations correspond to meteoroids at the descending and ascending nodes of their inclined orbits. Similarly, the helion and antihelion radiants represent meteoroids on the outgoing and ingoing legs of their eccentric orbits. The radiant distributions are corrected for atmosphere observing biases which include correction for the radar's collecting area, the altitude dependent initial trail radius effect, Faraday rotation effects, the finite velocity effect and the pulse repetition factor (Campbell-Brown 2008). The different radiant concentrations have different mean observed speeds and have been tentatively associated with different parent objects: helion and antihelion meteors with Jupiter family comets ($\sim$35 km s^{-1}), toroidal meteors with Halley group comets ($\sim$45 km s^{-1}), and apex meteors with long-period comets ($\sim$55 km s^{-1}, Jones et al. 2005).

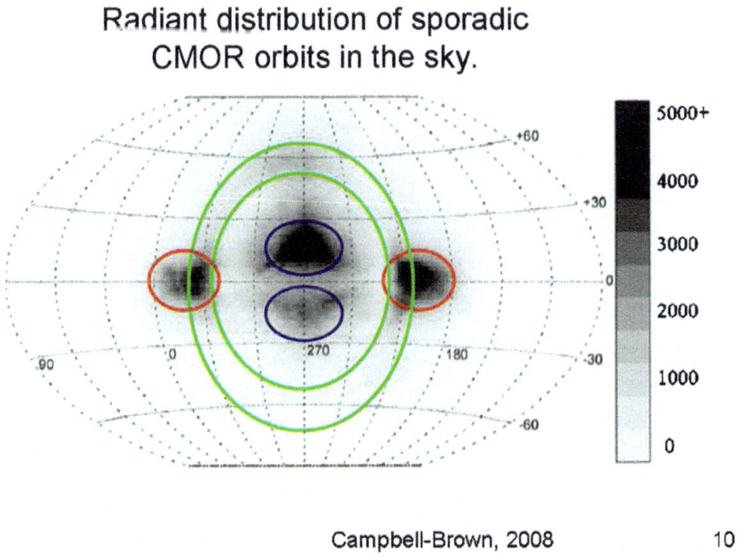

Fig. 6 Apparent radiant distribution of sporadic CMOR orbits in the sky, corrected for in-atmosphere observing biases (after Campbell-Brown 2008) with individual types of radiant concentrations indicated by colored ellipses. Red: helion (left) and antihelion (right), blue: north and south apex, and green: toroidal type. Because CMOR is located at 43° North latitude the southern components are reduced or not visible

3.3 Composition of Meteorites

Football-sized and bigger meteoroids may survive the entry into the atmosphere, and some residual material may fall to the ground, where such meteorites can be picked up and examined. On the Antarctic continent almost any rock one finds on the blue ice surface is a meteorite fallen some time ago and transported to its finding site by ice motion.

Because of their exotic extraterrestrial nature, meteorites belong to the best analyzed rock samples on Earth. The most abundant meteorites are chondrites because they contain up to millimetre-sized round particles (chondrules) embedded in a fine-grained matrix (Fig. 7). These chodrules are composed mostly of silicate minerals, apparently being melted and solidified in space. Chondrites, especially carbonaceous chondrites (the matrix of which contains abundant carbonaceous material) consist of a mixture of elements believed to comprise the mean composition of the entire planetary system (called "cosmic abundances"), except for some volatile elements, like hydrogen and helium. This primitive material has not been modified by thermal processes contrary to the interior of planets. Less than 10% of the observed meteorite falls are stony meteorites (achondrites). They consist of material similar to terrestrial basalts or plutonic rocks that have been differentiated and reprocessed to a lesser or greater degree due to melting and re-crystallization on or within meteorite parent bodies. About 5% of the meteorites are iron meteorites consisting of almost pure iron-nickel. They are believed to originate from planetary cores. One percent of the meteorites are stony-iron meteorites (Pallasites) consisting of centimetre-sized olivine crystals in an iron-nickel matrix. While the chondrites represent the elemental composition of the terrestrial planets, the other meteorite classes (achondrites, pallasites, and iron meteorites) have been used as analogue materials for different compositional shells inside the Earth.

Mineralogical, chemical and isotopic analyses of meteorites have provided deep insights into the physical and chemical conditions during their formation, i.e. into the history of the

Fig. 7 Meteorite types in clockwise order: carbonaceous chondrite, stony meteorite, stony-iron meteorite, iron meteorite (after Lipschutz and Schultz, Encyclopedia of the Solar System 2nd. Edition)

planetary system as a whole. The asteroid belt has always been suspected to be a source of meteoroids. Most asteroids have orbits between Mars and Jupiter. Hirayama suggested that asteroids with very similar orbits form families that have been generated by the break-up of a larger asteroid in a catastrophic collision about 100 million years ago (Hirayama 1923). Nesvorny et al. (2003) identified the IR asteroid bands with asteroid families that were generated less than 10 million years ago. Also large impacts on Mars and on the Moon shed material into interplanetary space, several lunar and Martian meteorites were found on Earth.

Organic compounds identified in carbonaceous chondrites of aqueously altered petrographic types 1 and 2 include amines and amides; alcohols, aldehydes, and ketones; aliphatic and aromatic hydrocarbons; sulfonic and phosphonic acids; amino, hydroxy-carboxylic, and carboxylic acids; purines and pyrimidines; and kerogen-type material (Sephton 2002). An extensive range of amino acids was found in the Murchison CM2 chondrite. Organic solid materials in carbonaceous chondrites are dominantly an insoluble macromolecular component resembling terrestrial kerogen, which is the most common type of organic matter on Earth. The latter results from the decomposition of microbial and higher living tissue at the bottom of oceans or lakes, and it is used as a nutrient by bacteria. All known life is based on organic compounds and water; both of these are present in carbonaceous chondrites. Analyses of the organic matter in meteorites can provide insights into the types of chemical reactions and organic compounds which could have been significant on the prebiotic Earth (Pintado et al. 2019).

Fig. 8 Chondritic dust particle collected in the Earth's stratosphere (left, NASA image). Single CAI particle collected from comet Wild 2 by NASA Stardust mission (right, cf. Zolensky et al. 2006)

Up to now no meteorite has been linked to a cometary parent body. However, isotopic analyses of meteorites found that about 10^{-5} of the meteorite material has non-solar system isotopic abundance and is of pre-solar (interstellar) origin.

3.4 Structure and Composition of Collected Cosmic Dust Particles

It is obvious that analyses of micrometeorites collected in the stratosphere could give similar information as meteorites. A high flying aircraft like the U2 spy plane can cruise at 20 km altitude for many hours. On its wings it carries a several 100 cm^2 flat plate dust collector which sweeps huge amounts of air providing a strong concentration effect on the collector.

Extraterrestrial grains (conveniently termed IDPs, which stands for Interplanetary Dust Particles) of about 5 to 50 μm in diameter are collected that way (Fig. 8). The lower size limit is due to contamination by smaller terrestrial particles. Micrometer and sub-micrometer-sized particles from volcanic eruptions can reach these altitudes in significant amounts. The upper limit is caused by the scarcity of bigger particles, e.g. only about ten IDPs of more than 10 μm in size are collected during one hour of aircraft flight. The extraterrestrial nature of the collected particles was demonstrated by their chemical composition which often resembles chondritic meteorites and the finding in IDPs of traces of solar wind helium and tracks from the exposure to high energetic ions in space (Brownlee et al. 1976). NASA has routinely performed cosmic dust collections by airplanes since 1981.

Like meteorites, IDPs do not represent an unbiased sample of meteoroids in space. Only small interplanetary dust particles of a few to 10 μm diameter are decelerated in the tenuous atmosphere above 100 km altitude, especially if their entry speed is low. At this altitude the deceleration is gentle, and the grains do not reach the temperature of substantial evaporation ($T \approx 800\ ^{\circ}C$), because they effectively radiate away excessive heat. These decelerated dust particles subsequently sediment through the atmosphere and become accessible to collection in the atmosphere, on the Earth's surface and on the deep-sea floor. At 20 km altitude the concentration of interplanetary particles of 10 μm diameter is several thousand times higher than in space. On the ground, especially the polar ice caps and mid-ocean sediments provide concentrations of ablated meteoritic material. Most of the micrometeorites collected in antarctic ice have primitive carbonaceous composition. Nesvorny et al. (2010) suggest that Jupiter family comets are the main source of zodiacal dust inside 5 AU and, thus, the most likely source of these micrometeorites.

In deep-sea sediments, cosmic dust deposits were already recognized in the 19th century (Murray et al. 1884). Recently, supernova-generated dust was identified on the ocean-floor, indicating that our solar system was hit by a nearby supernova shock wave that exploded approximately 50 pc away from our solar system 2–3 Myr ago (Knie et al. 2004; Fry et al. 2016). Extensive studies of the particle composition and variations in dust influx have also been performed (Prasad et al. 2017). Cosmic dust in the Earth atmosphere is discussed in detail in Plane et al. (2018). An influence of cosmic dust on the Earth climate is also being debated.

The first samples of extraterrestrial origin were the lunar samples returned by the American Apollo astronauts (380 kg) and by the automatic Russian Luna probes (ca. 0.3 kg). Analyses of the samples revealed their similarity with terrestrial material. Age determinations of lunar samples suggest that the Moon formed at approximately the same time as the Earth, perhaps by a giant impact on to the young Earth (Canup and Asphaug 2001).

Recently, dust samples were collected and brought back to Earth from comet Wild 2 by the Stardust mission (Brownlee 2014) and from asteroid Itokawa by the Hayabusa mission (Nakamura et al. 2011). In 2004 during a close (240 km) flyby of comet Wild 2 at a speed of $6\ \mathrm{km\,s^{-1}}$ Stardust collected more than 10,000 cometary dust particles on its $0.1\ \mathrm{m^2}$ aerogel collector and returned them to Earth (Brownlee et al. 2006). Among the Stardust collection of cometary particles there were those containing large crystalline Calcium Aluminum Inclusions (CAI, Fig. 8) formed close to the Sun at temperatures above 1300 °C. They gave proof that large scale mixing occurred in the protoplanetary disc (Brownlee 2014; Levasseur-Regourd et al. 2019).

In addition to the cometary particle collector, the Stardust spacecraft was equipped with a second aerogel collector specially designed for the collection of interstellar particles. It was exposed to the interstellar dust stream for about 200 days during the cruise phase of the spacecraft (Sterken et al. 2014). A total of seven particles were identified which are of potential interstellar origin (Westphal et al. 2014). The two largest particles contained crystalline silicates and, thus, may have been circumstellar condensates (Sterken et al. 2019). Alternative origins of the particles have not yet been finally eliminated, and definitive tests through isotopic analyses are necessary to prove their interstellar origin.

In 2005 the Japanese Hayabusa mission (Nakamura et al. 2011) touched down on asteroid Itokawa and picked up about 1500 dust particles from the surface and returned them to Earth. Mineral chemistry of the samples suggests that this ca. 0.5 km sized asteroid consists of pieces of the interior portions of a once larger asteroid. Dust from asteroid Itokawa proved to be the link between S-type asteroids and ordinary chondrites.

Future realization of current plans for sample return missions to asteroids, comets, Martian satellites, Trojans, and Galilean satellites will eventually provide further glimpses into the nature of early solar system materials.

3.5 Size Distribution of Meteoroids Obtained by Microcrater Studies

In 1969, when NASA's astronauts returned samples from the Moon, it was immediately recognized that the lunar rocks were peppered by micrometeoroid impacts (Fig. 9). Impact craters were identified on individual rocks from millimetre down to sub-micrometer in size. Several effects, however, made the analysis difficult. Since the exposure time of a given surface on a lunar rock could not reliably be determined, no absolute flux values could

Fig. 9 Microcraters on the glassy surface of a lunar rock sample (NASA image)

be derived but only relative values for different crater sizes. Morrison and Zinner (1977) concluded that only the steepest slope measured on a single surface was least vulnerable to effects from variations in exposure conditions and possible shielding by thin coatings of dust on the rocks. Calibration of microcrater dimensions with respect to meteoroid sizes was provided by Hörz et al. (1975) who found a size-dependent crater-to-projectile diameter ratio range of 2 to 9 for meteoroids of 10^{-18} to 1 g masses at an assumed average impact speed of 20 km s^{-1}. Grün et al. 1985 used these results and combined them with in situ measurements in order to derive the present microcrater production flux on the lunar surface. Studies of the effects of secondary microcraters produced by ejecta from primary impact craters on lunar samples showed that the number of microcraters is significantly increased for crater diameters below 10 μm. It was concluded that the interplanetary flux of meteoroids below 10^{-9} g is up to 2 orders of magnitude smaller than the ejecta flux on the lunar surface.

Since the return of lunar surface samples by the Apollo and Luna missions, the study of impact craters on material exposed to space (e.g. on the LDEF mission) has been used to characterize the flux of interplanetary micrometeoroids and man-made space debris particles (e.g. Love and Brownlee 1995, and McDonnell and Gardner 1998).

3.6 Dust Detections by in Situ Detectors Throughout the Planetary System

In situ dust detectors are carried by space probes throughout the solar system from within Mercury's orbit (Helios) to beyond Pluto's orbit (New Horizons) where no other method is capable to detect and analyze dust (Table 1). Based on the early negative experience to reliably detect dust impacts in near-Earth space, extra effort was spent to make dust detectors more reliable. Highly reliable impact momentum sensors were realized with Giotto's Dust Impact Detection System (DIDSY, McDonnell et al. 1987; McDonnell 1987) in the dusty environment of Halley's comet. DIDSY employed various impact detection principles, most notably five piezoelectric crystals mounted at various positions of Giotto's 3 m^2 front and rear shields. Careful laboratory and theoretical calibrations (McDonnell et al. 1989) led to sensitivity maps of the front shield for coincident recordings by the impact sensors. For hypervelocity impacts of Halley particles on to the Giotto shield at 69 km s^{-1} impact speed, the resultant crater ejecta enhance the effective impulse by factors of up to 10 (McDonnell et al.

Table 1 In situ dust detectors in interplanetary space: distance of operation, mass sensitivity and sensitive area

Spacecraft	Launch Year	Distances (AU)	Mass threshold (g)	Area (m^2)
Pioneer 8	1967	0.97–1.09	2×10^{-13}	0.0094
Pioneer 9	1968	0.75–0.99	2×10^{-13}	0.0074
HEOS 2	1972	1	2×10^{-16}	0.01
Pioneer 10	1972	1–18	2×10^{-9}	0.26
Pioneer 11	1973	1–10	10^{-8}	0.26
Helios 1/2	1974/76	0.3–1	10^{-14}	0.012
Galileo	1989	0.7–5.3	10^{-15}	0.1
Hiten	1990	1	10^{-15}	0.01
Ulysses	1990	1–5.4	10^{-15}	0.1
GORID	1996	1	10^{-15}	0.1
Cassini CDA	1997	0.7–10	2×10^{-16}	0.09
Cassini HRD	1997	0.7–10	3×10^{-13}	0.006
Nozomi	1998	1–1.5	10^{-15}	0.01
New Horizons	2006	2.6–>45[a]	2×10^{-12}	0.1
IKAROS-ALADDIN[b]	2010	0.72–1.1	10^{-9}	0.54
LADEE[c]	2013	1	2×10^{-16}	0.01
BepiColombo MMO-MDC[d]	2018	0.31-0.47	1×10^{13}	0.006
DESTINY$^+$-DDA[e]	~2023	0.75–1.0	10^{-16}	0.035
MMX CMDM[f]	~2024	1–1.5	$\sim 10^{-10}$	~1
Europa Clipper SUDA[g]	~2025	5	10^{-16}	0.025

[a] Jan. 2019

[b] Interplanaetary Kite-craft Accelerated by Radiation of the Sun, Arrayed Large-Area Dust Detectors in Interplanetray Space

[c] Lunar Atmosphere and Dust Environment Explorer

[d] Mercury Magnetospheric Orbiter, Mercury Dust Monitor

[e] Demonstration and Experiment of Space Technology for Interplanetary Voyage Phaethon Flyby and Dust Science Dust Analyzer

[f] Martian Moons Exploration Circum-Martian Dust Monitor

[g] Surface Dust Mass Analyzer

1984). Piezo-electric sensors were also employed on the Dust Impact Monitor (DIM) onboard the lander Rosetta/Philae which landed on comet 67P/Churymov-Gerasimenko (Seidensticker et al. 2007; Krüger et al. 2015b). Modern microphone detectors like the Mercury dust monitor (MDM) onboard the BepiColombo/Mercury Magnetosphere Orbiter mission (MMO) use four highly sensitive piezoelectric ceramic sensor plates of 16 cm^2 each. The full waveform of any signal will be recorded and transmitted to Earth (Nogami et al. 2010). Calibration of electrostatic dust accelerators provides means to identify dust impacts, to measure the momentum transferred, and to estimate coarsely the impact speed from the signal form.

The penetration detectors on board the interplanetary Pioneer 10 and 11 spacecraft reliably recorded very low impact rates in the outer planetary system (Humes 1980). During the passages through the dusty environments of Jupiter and Saturn a handful of penetra-

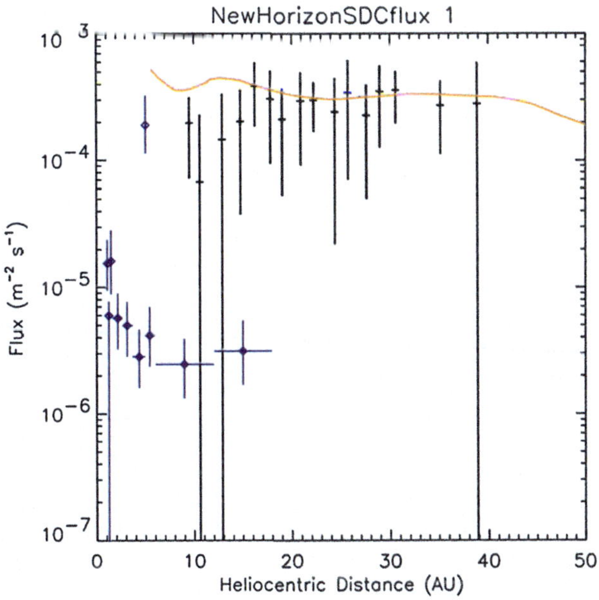

Fig. 10 Interplanetary dust flux in the outer planetary system. Black: flux onto New Horizons SDC for grains with radii >0.6 μm (Poppe et al. 2010; Piquette et al. 2019). The red curves represents a model of the predicted SDC flux (Poppe 2016). Blue: Pioneer 10 flux (Humes 1980)

tions were recorded but because of the instrumental dead-time of 87 minutes it was not clear whether more undetected penetrations had occurred. Actually, indeed, Pioneer 11 had passed close to Jupiter's Gossamer ring which was discovered 5 years later in 1979 by the Voyager 1 space probe. A principal way to check the total number of hits received by the Pioneer 10 and 11 penetration detectors was to see at which number of recorded hits (initial number of 234 pressurized cells) no further counts were eventually recorded. However, this check became impossible because beyond approximately 20 AU the detectors responded to the very low ambient temperatures by freezing of the fill gas (argon and nitrogen, Humes, personal communication). The pressure in the cells dropped and the electrical discharge (which is used to detect the pressure drop) triggered. This discharge heated the gas and the pressure increased again, and after some time the whole process repeated. Because of this effect no useful data were obtained from the Pioneer 10 and 11 detectors beyond 20 AU. Nevertheless, these instruments characterized the meteoroid environment out to about 20 AU (Fig. 10) and confirmed that the outer solar system is safe to unmanned missions.

Despite the existence of reliable multi-coincidence dust detectors there was the desire to fly simple and inexpensive dust detectors on some space missions. On board the VEGA-1 and 2 missions to comet Halley, Simpson et al. (1986) used new light weight and low power consumption dust detectors. A PVDF detector consisted of a metal-coated and permanently polarized polyvinylidene fluoride film. A micrometeorite impact on to such a detector removes a portion of the metal surface layer and excavates some of the permanently polarized PVDF dielectric material underneath, thus generating a charge signal to determine the magnitude of the impact (Simpson and Tuzzolino 1985; Shu et al. 2013).

Because of their low demand on spacecraft resources PVDF films were used as dust detectors on a number of missions. PVDF detectors provided useful dust information in the dusty environments of comets (Dust Flux Monitor Instrument on the Stardust mission to comet 81P/Wild 2, Tuzzolino et al. 2004) and planetary rings (High Rate Detector (HRD) of CDA on Cassini, Kempf 2008). In the Earth environment, the SPADUS instrument (Tuzzolino et al. 2005) on the Advanced Research and Global Observation Satellite (ARGOS) used two arrays of PVDF dust sensors in a time-of-flight (TOF) arrangement that provided measurements of flux, velocity, and trajectory of big ($\sim 10^{-9}$ kg) natural meteoroids as well as space debris particles. On the other hand, the measurements of small ($\sim 10^{-12}$ kg) particles on the Cosmic Dust Experiment (CDE, Poppe et al. 2011) on the Aeronomy of Ice in the Mesosphere (AIM) mission suffered from high noise rates because PVDF displays both pyroelectric and piezoelectric properties and is affected by temperature variations and mechanical vibrations. Contrary to the SPADUS instrument, CDE could not rely on coincident signals from the same dust particle. The strong temperature contrast in sunlight and in the Earth's shadow caused both interferences from thermal cracking and from mismatch of electronic components.

On the New Horizons mission to Pluto and the Kuiper belt (Stern 2008) the light-weight Venetia Burney Student Dust Counter (SDC, Horanyi et al. 2008) was added at a late stage of spacecraft development. The instrument consists of fourteen independent PVDF film impact sensors of approximately 0.1 m^2 sensitive area each, twelve of which are exposed to space with the remaining two reference sensors shielded and placed on the underside of the instrument. The measured signal-to-noise ratio is approximately 0.5, hence, the measured meteoroid flux has at least a factor of two uncertainty (Fig. 10, Piquette et al. 2019). The large error bars are partially a result of the small number of events recorded, but are mostly the effects of a too small number of reference detectors. The number of counts and their errors on these detectors have six times the weight in comparison to the counts on the open detectors. This SDC data analysis demonstrates the importance for simple impact instruments like the PVDF detectors of having an independent characterization of noise. Without reference detectors the observed signals could be anything (noise event or dust impact), therefore such simple impact detectors require another independent means like reference detectors if applied in uncharted territory.

PVDF-type detectors were also used for the Arrayed Large-Area Dust Detectors in INterplanetary space (ALADDIN, sensor area 0.54 m^2) on board the Japanese Space Agency's (JAXA) IKAROS mission. This instrument measured interplanetary dust above 10 μm in size between Venus and Earth orbits (Yano et al. 2013; Hirai et al. 2014). Of 4427 events recorded within 300 days 60% had to be ruled out as noise (Hirai et al. 2017). The remaining 1773 events were classified as most promising dust impact events because they resulted in a dust flux compatible with the flux at 1 AU (Grün et al. 1985).

In the future, the Circum-Martian Dust Monitor (CMDM) will be a 1 m^2 thin-film sensor on board the Martian Moons Exploration (MMX) sample return mission to the Phobos and Deimos. CMDM will employ the outermost layer of the multi-layer thermal spacecraft insulation (MLI) together with piezo-electric sensors (Kobayashi et al. 2018a) to search for putative Martian dust rings (Soter 1971) which are still waiting for discovery (Showalter et al. 2006). An even larger (~ 3 m^2) PVDF detector will measure dust beyond the asteroid belt onboard the planned OKEANOS mission to the Jupiter trojans (Okada et al. 2018).

The only dedicated instruments on board the Voyager spacecraft to observe dust were the cameras which discovered and characterized the extensive ring systems around all four giant planets. However, there was a surprise dust instrument on board no one had foreseen. On August 26 1981, during the Voyager 2 Saturn flyby the plasma wave instrument recorded a very intense burst of noise spikes more than 100 times above the background in the few-minute interval around the ring plane crossing (Gurnett et al. 1983). The maximum intensity was almost exactly centred on the ring plane crossing. The signals consisted of many brief impulses occurring at a rate of several hundred per second close to the newly discovered G-ring. The investigators interpreted the enhanced rate of spikes as impacts from a cloud of dust particles onto the spacecraft skin. During all subsequent flybys of the outer planets Uranus and Neptune this instrument recorded also concentrations of dust impact spikes in the corresponding ring systems. Similarly, in 1985 during a passage through the tail of Comet Giacobini-Zinner a plasma wave instrument onboard the NASA ICE spacecraft observed numerous dust impact spikes within a distance of 30,000 km from the comet nucleus (Gurnett et al. 1986).

More recently, the plasma WAVES instruments on board the interplanetary STEREO A and B spacecraft recorded two types of spikes on their electric field antennas (Meyer-Vernet et al. 2009). The investigators suggested the voltage spikes on a single antenna to be due to a strongly varying flux of nanometer-sized dust particles arriving from the solar direction. These particles would be more than 10 times smaller than the earlier detected beta-meteoroids (see below). A second type of signals in the wideband waveform analyzer was based on coincident signals from all three antennas; these where identified to be due to impacts of submicrometer and micrometer-sized dust particles, roughly compatible with classical interplanetary and interstellar grains. Laboratory dust accelerator calibration of plasma wave antennas has quantified the charge signals associated with hypervelocity dust impacts on materials specific to STEREO (Collette et al. 2015). Since 2009 the rate of single spikes has ceased on STEREO A but not on STEREO B. No such anomaly was observed on both spacecraft for the triple signals referring to the flux of micrometer sized grains (Malaspina et al. 2015), and no explanation for the disappearance of nanometer dust spikes on STEREO-A (but not STEREO-B) was found either. Recently, the interpretation of the single hits as dust signals was put into question (Kellogg et al. 2018). An independent confirmation of the interplanetary flux of nanometer-sized dust particles from the solar direction is still missing.

Dust instruments in combination with solar wind spectrometers and magnetometers may give new information on how plasma-dust interactions affect solar wind dynamics and composition. A nano-dust analyzer (O'Brien et al. 2014) positioned at 1 AU and observing the flow of dust particles will be a suitable instrument to determine the formation, dynamics, and composition of nano-dust particles in the inner solar system (Juhasz and Horanyi 2013). Unfortunately, the Parker Solar Probe mission does not carry such a dedicated dust instrument when it will pass through the densest region of the zodiacal cloud in the solar F-corona within 10 solar radii (0.05 AU) from the Sun with a speed of almost 200 $\mathrm{km\,s^{-1}}$ (Mann et al. 2004). The plasma wave antennae as well as many other instruments and spacecraft systems on board may be affected by impinging high-speed dust particles.

The Cassini spacecraft orbiting Saturn from 2004 to 2017 carried both a dedicated Cosmic Dust Analyzer (CDA, Srama et al. 2004) and a Radio and Plasma Wave Science instrument (RPWS Gurnett et al. 2004). In this case both methods could be directly compared and the relative sensitivities were determined (Ye et al. 2014).

The most successful and versatile impact phenomenon used in dust detection and analysis is impact ionization. This phenomenon had been theoretically predicted by Yu. P. Raizer (1960). Shortly after the first dust accelerator was developed at the TRW Company Friichtenicht and Slattery (1963) reported the experimental verification of impact ionization. An impact of a fast ($>1\ \mathrm{km\,s^{-1}}$) particle on to a solid target generates an expanding impact plasma cloud. In an impact ionization detector ions and electrons are separated by an external electric field and two coincident and complementary charge signals are recorded that provide a very sensitive and reliable method of dust detection. The early impact ionization detectors flown on the OGO satellites and on the lunar Explorer 35 (Alexander et al. 1971) were time-of-flight (TOF) systems consisting of a thin film front sensor and a rear sensor 0.1 m apart. Both the dust penetration of the front film and the impact on to the solid rear sensor generated coincident impact charge signals. However, misled by the early false reports of a very high micro-particle flux in the Earth' environment, Alexander selected only 0.0005 $\mathrm{m^2}$ as sensitive area of a single detector. As a consequence no impacts were recorded.

Berg and Richardson (1969) extended this idea by combining 16 TOF tubes into single 0.01 m^2 dust detectors that successfully operated for more that 7 years on the interplanetary space probes Pioneer 8 and 9. These detectors made the first important observations of interplanetary dust and discovered a flow of particles on hyperbolic orbits arriving roughly from the solar direction (Berg and Grün 1973). This outward dust flow was interpreted by Zook and Berg (1975) as small grains being generated by collisions of meteoroids near the Sun and being expelled by the prevailing action of solar radiation pressure. Because the ratio of solar radiation pressure force F^{rad} and solar gravity F_{grav} is often called $\beta = F^{rad}/F_{grav}$, the authors coined the name beta-meteoroids for these particles. For particles to be affected by radiation pressure β has to be large: A particle with $\beta \geq 0.5$ released from a parent body on a circular heliocentric orbit will leave the solar system on a hyperbolic trajectory. Since their discovery β-meteoroids have been observed by dust instruments on Helios 1 (Grün et al. 1980) and Ulysses (Baguhl et al. 1995; Wehry et al. 2004).

In 1968 the European Space Research Organization ESRO—the precursor organization of ESA—offered the opportunity to fly an instrument on the HEOS 2 satellite. The Max-Planck-Institute for Nuclear Physics proposed a new dust detector and was selected. From laboratory experiments (Grün and Rauser 1969) it was known that a film has a strong effect on projectiles that penetrate it at high speeds even if the film is much thinner than the diameter of the projectile. Therefore, a film-less impact ionization dust detector was flown on the HEOS-2 satellite (Dietzel et al. 1973, Hoffmann et al. 1975). It consisted of a hemispherical impact target of about 0.01 m^2 sensitive area and a central ion collector. A voltage of 350 V between the target and the central electrode separated the impact charges, and two coincident signals were recorded upon impact of a fast dust particle on to the target. Within 2.5 years more than 400 particles were recorded on the highly eccentric orbit between 10,000 and 240,000 km above the Earth's surface (Fechtig et al. 1979). At large distances (>60,000 km) the observed flux of interplanetary dust was mostly randomly distributed in time. Groups of several particles within a few hours were interpreted as members of ejecta clouds generated by large meteoroid impacts on the Moon. Swarms of impacts were recorded in the auroral zone of the Earth's magnetic field and in the perigee region of the spacecraft orbit. Electrostatic disruption of large porous meteoroids has been suggested as a mechanism of these bursts. These latter phenomena await confirmation by modern dust instruments in the Earth-Moon system.

A major improvement of the simple HEOS dust detector configuration was a ten times increase of sensitive area (0.1 m^2). Based on the high reliability of impact detection, a channeltron ion detector was introduced inside the central ion collector, thereby providing a third coincident signal upon dust impact. Additionally, variable commandable detection thresholds and a variable coincidence scheme controlled by a programmable processor were implemented. Such a detector was successfully operated on board Galileo for 6 years en route to Jupiter and for about 8 years in the harsh Jovian environment (Grün et al. 1992a; Krüger et al. 2010). The twin detector on the Ulysses mission was operated in interplanetary space for 17 years (Grün et al. 1992b; Krüger et al. 2015a). In a noisy environment, multi-coincidence dust detectors like the ones on Galileo, Ulysses and recently the LDEX detector on LADEE (Horanyi et al. 2014) have the advantage that they provide means of identification of even unexpected noise sources (cf. Baguhl et al. 1993). Thereby, data analysis methods can be developed that reliably distinguish noise from dust impact events. By reprogramming Galileo's onboard data processing computer, this detector reached the same sensitivity and reliability as the Ulysses detector despite the fact that Galileo's data transmission rate to the ground was lower by more than a factor of hundred due to a spacecraft antenna failure (Grün et al. 1995).

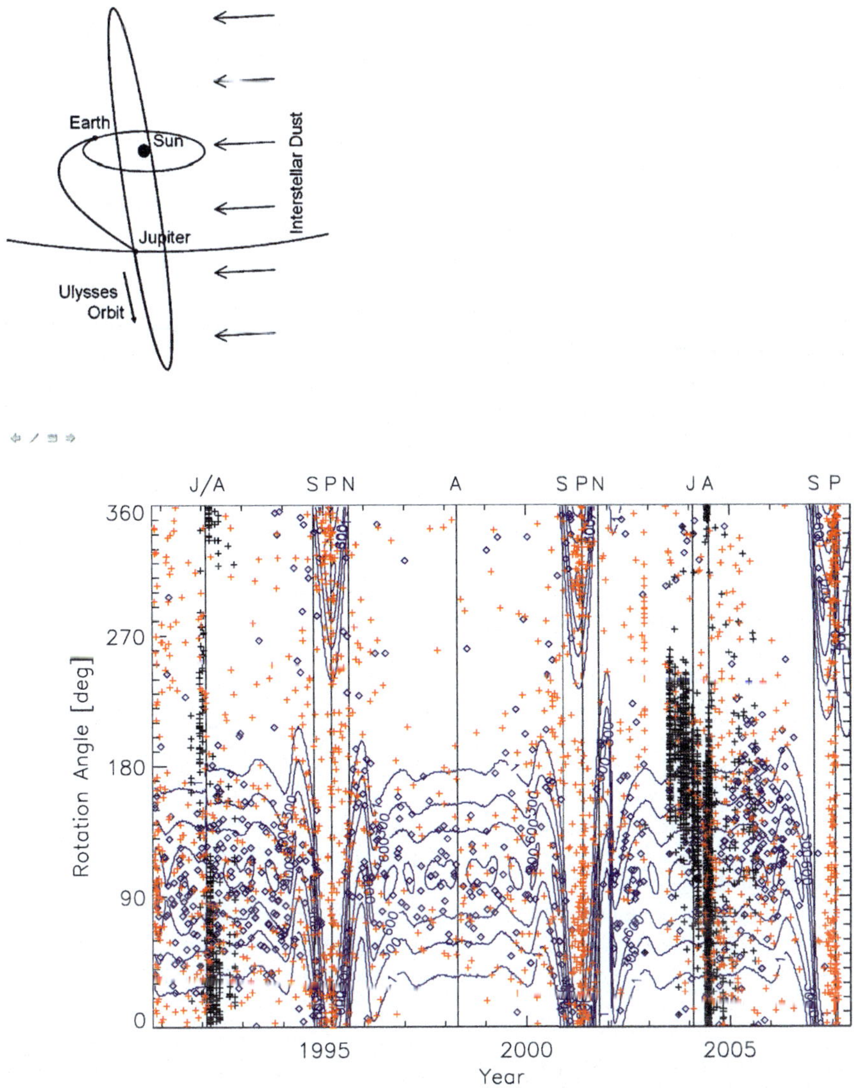

Fig. 11 Top panel: Ulysses orbit. Positions of Earth at launch and of Jupiter at flyby. Interstellar dust flow is almost perpendicular to the Ulysses orbit plane. Bottom panel: Impact direction (rotation angle) verus time for all dust impacts detected by Ulysses between launch in October 1992 and the end of dust instrument operation in November 2007. Each symbol indicates an individual dust particle impact. Red crosses: Interplanetary dust; black crosses: Jupiter stream particles; blue diamonds: interstellar dust particles, blue contour lines show the effective sensor area for dust particles approaching from the upstream direction of interstellar helium. Vertical dashed lines and labels at the top indicate Ulysses's Jupiter flybys (J), perihelion passages (P), aphelion passages (A), south polar passes (S), and north polar passes of Ulysses (N) (adapted from Krüger et al. 2015a)

Among the significant discoveries by the Galileo and Ulysses dust instruments were: the detection of streams of nanometer sized dust particles from the volcanoes of Jupiter's moon Io (Grün et al. 1993a,b; Horanyi et al. 1993; Zook et al. 1996; Graps et al. 2000; Krüger et al. 2006; Hillier et al. 2019), impact-generated dust clouds surrounding the Galilean moons and the formation of a tenuous dust ring (Krüger et al. 2003; Krivov et al. 2002), and dust in and beyond Jupiter's gossamer rings (Krüger et al. 2009). By a Jupiter flyby, the Ulysses spacecraft was brought on to an orbit almost perpendicular to the ecliptic plane (Fig. 11). This provided the basis to probe the three-dimensional structure of the zodiacal dust cloud (Grün et al. 1997). The delay of the Ulysses launch by about 4 years and the subsequent Jupiter flyby caused the final orbit plane to rotate by about 90° compared to the originally planned launch date, which enabled the detection of an interstellar dust flow through the planetary system (Grün et al. 1994, Krüger et al. 2007, 2015a; Strub et al. 2015, 2019). With the original launch date interstellar dust particles would not have entered the Ulysses dust sensor.

Through the 8-year long measurements by Galileo the dust environment in the Jovian system is better understood than the natural meteoroid environment in the Earth-Moon system. The spare unit of the Ulysses dust instrument GORID (Geostationary Orbit Impact Detector) was flown on the Russian EXPRESS spacecraft and characterized the natural and space debris environment at ~36,000 km altitude (Drolshagen et al. 1997). Only recently, measurements by the dust detector on the lunar orbiter LADEE (Lunar Atmosphere and Dust Environment Explorer, 2013) identified also an impact induced dust cloud around our Moon. The Lunar Dust Experiment (LDEX) instrument observed particles ejected by impacts of sporadic meteoroids and from meteoroid streams on to the lunar surface (Szalay and Horanyi 2016a,b).

3.7 Composition Measurements by Dust Analyzers

Dust instruments that characterize more than merely dust impacts and their magnitudes are called dust analyzers. The chemical and isotopic compositions and physical properties of dust particles can provide insights into the physical and chemical conditions during solar system formation and into the history of the planetary system as a whole.

The first dust analyzers used in space were compositional analyzers. Early laboratory studies of impact ionization already used a TOF impact mass spectrometer to analyze the ions released during a hypervelocity impact (Auer and Sitte 1968; Hansen 1968). The mass resolutions of the spectra obtained in the laboratory were low ($M/\Delta M \sim 30$), mostly because of electronic limitations. The first low mass-resolution dust analyzer was flown on Helios 1, 2 in 1974/76 (Dietzel et al. 1973). This dust analyzer consisted of a venetian blind type impact target behind which the ion collector was situated at the end of a one meter long tube. This TOF ion mass analyzer had only a marginal mass resolution of 5–10. In interplanetary space it identified three types of spectra dominated by low mass (<30 amu), medium mass (30–50 amu) and high mass (>50 amu) species (Altobelli et al. 2006). However, the main value of this mass analyzer was that each recorded ion spectrum reliably marked a dust impact. This feature was especially important since the noise rate was high. Helios 1 recorded 235 impacts during its 10 orbits about the Sun between 0.3 and 1 AU (Grün et al. 1980). Each of the spinning Helios spacecraft carried two dust analyzers of 0.006 m^2 sensitive areas each: the ecliptic sensor that scanned through the Sun direction was covered by a thin film, and the open south sensor on Helios 1 was shielded from direct sunlight by the spacecraft frame. Although, both sensors had widely overlapping fields-of-view the south sensor recorded significantly higher impact rates at otherwise similar impact parameters (impact direction and impact charge). A comparison with penetration studies showed that particles which did not penetrate the entrance film must have had bulk densities below 1000 $kg\,m^{-3}$. Approximately 30% of the particles on high eccentricity orbits (e > 0.4) had such low densities (Grün et al. 1980; Pailer and Gruen 1980).

A breakthrough was reached with the dust analyzers on the Halley missions, Giotto, VeGa 1, and 2 (Kissel et al. 1986a, 1986b). By inclusion of an electrostatic reflectron

(Mamyrin et al. 1973) and faster electronics in the TOF spectrometer the mass resolution was improved to $M/\Delta M \sim 100$. Because of the dusty cometary environment these instruments needed only 5×10^{-4} m^2 sensitive area. The dust impact mass spectrometers PIA and PUMA on the GIOTTO and VEGA spacecraft, respectively, allowed the compositional analysis of individual particles (Kissel et al. 1986a,b). The measurements showed that each particle is an intimate mixture of a mineral core, ices, and organic molecules. Since the impact velocity was large (>60 km s^{-1}) only atomic ions were identified in the Halley case. Nevertheless, Kissel and Krueger (1987) found evidence for the organic nature of cometary material. Important results were the discovery of a prominent carbonaceous component in Halley dust, and a wide scatter in some isotopic ratios. Cometary particulates are an intimate mixture of two end-member components, namely refractory carbonaceous (rich in the elements H, C, N, and O) and stony material (rich in rock-forming elements as Si, Mg, Fe), respectively. The stony component comprises silicates, metals, oxides, sulfides and others. Both end-member components do not occur as pure components but are mixed to the finest scale. Magnesium isotope ratios showed only a slight variation around the nominal solar value, whereas the isotopic ratio of $^{12}C/^{13}C$ showed large variations from grain to grain, but on average it was also solar like (Jessberger and Kissel 1991). The average elemental composition was found to be solar like, but significantly enriched in volatile elements H, C, N, and O compared to CI chondrites (Jessberger et al. 1987).

A 20 times enlarged version (10^{-2} m^2 sensitive area) of the Halley mass analyzer with mass resolution $M/\Delta M \sim 100$ was flown on the Stardust mission to comet Wild 2 (Kissel et al. 2004). For the first time a series of positive and negative ion mass spectra from the impact of (apparently) interstellar and cometary dust particles has been collected. In the spectra of 45 presumably interstellar particles, quinone derivates were identified as constituents in the organic component. 29 spectra obtained during the flyby of Comet 81P/Wild 2 on 2 January 2004 suggest the predominance of organic matter (Kissel et al. 2005). The authors found that organic material in cometary dust seems to have lost most of its hydrogen and oxygen as water and carbon monoxide. These are now present in the comet as gas phases, whereas the dust is rich in nitrogen-containing species.

The Cassini CDA instrument which was developed by Srama et al. (2004) is a combination of the Galileo type dust detector with a linear impact ionization mass analyzer of mass resolution $M/\Delta M \sim 30$ and 0.016 m^2 sensitive area (Fig. 12). The impact ionization detector and compositional analyzer is combined with a PVDF High Rate Detector (HRD) in order to time-resolve the flux of micrometer-sized particles during the fast passage through denser ring regions of Saturn. In contrast to the Galileo and Ulysses dust detectors which were mounted on a spinning spacecraft (Ulysses) or the spinning spacecraft section in the case of Galileo, CDA was on board the 3-axis stabilized Cassini spacecraft. In order to provide some control of the pointing directions, CDA was mounted on an instrument controlled turn-table. Another significant feature during the initial dust measurements was reduced count rate data transmitted to ground through the continuous spacecraft engineering data stream while high resolution raw data and spectral data was included in the sporadic science data stream. This way CDA took dust measurements during Cassini's 7 year long trip through the planetary system to Saturn. Handicapped by operational constraints, CDA was able to obtain only two spectra of interplanetary particles inside Jupiter's orbit (Hillier et al. 2007). Their iron-rich composition with the lack of silicates and magnesium is in strong contrast to the silicate-rich minerals in interplanetary dust particles collected in the Earth's stratosphere. From hundreds of mass spectra obtained during Jupiter flyby CDA confirmed the compositional relation of dust stream particles with the volcanic plumes on Io. During 13 years in orbit about Saturn, CDA measured thousands of mass spectra of particles in Saturn's E-ring and from the interior of Enceladus (Fig. 13). Postberg et al. (2011) conclude

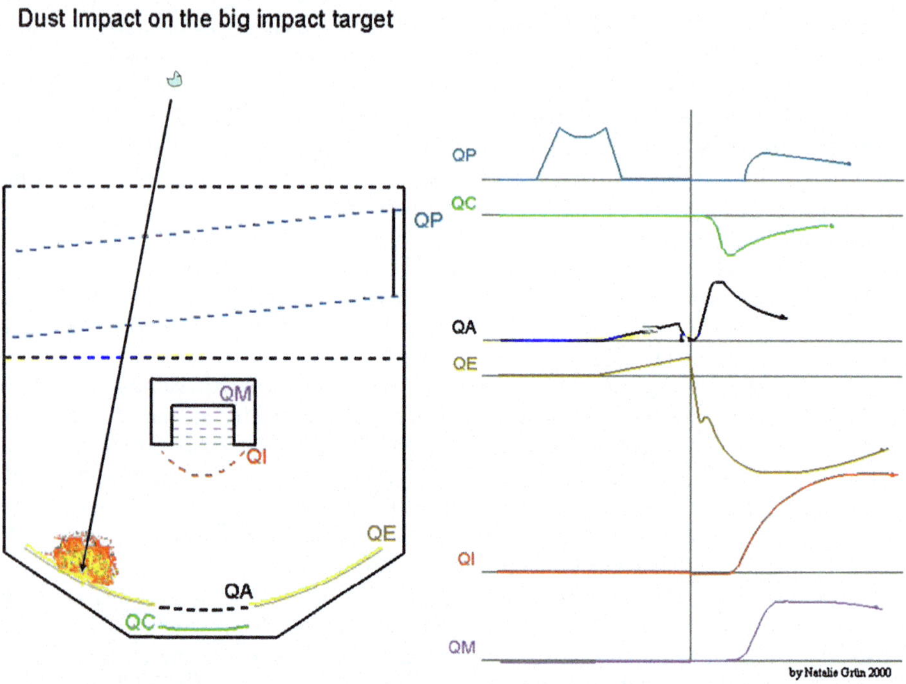

Fig. 12 Cassini Cosmic Dust Analyzer, CDA, left. Instrument schematics with signals from an impact onto the large hemispherical target (right). Impacts onto the centre target generate a mass spectrum of the released ions (Srama et al. 2004)

Fig. 13 Co-added spectra of salt-rich water-ice particles measured by Cassini CDA in the E-ring of Saturn. Although the particles are still predominantly water, these spectra typically show very few pure water and Na-hydrate clusters, if any. They are characterized by an abundant Na^+ mass line followed by a peak sequence of hydroxyl-cluster-ions $(NaOH)_n$ Na^+ (cf. Postberg et al. 2009)

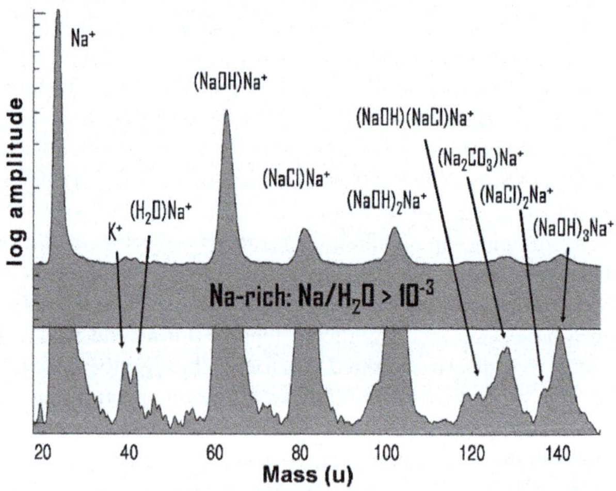

that a salt-water reservoir is the source of a compositionally stratified plume on Enceladus (see also Hsu et al. 2015; Hillier et al. 2019). CDA also identified particles above the main rings (Hsu et al. 2018), dust from the Kuiper belt and interstellar dust grains (Altobelli et al. 2016).

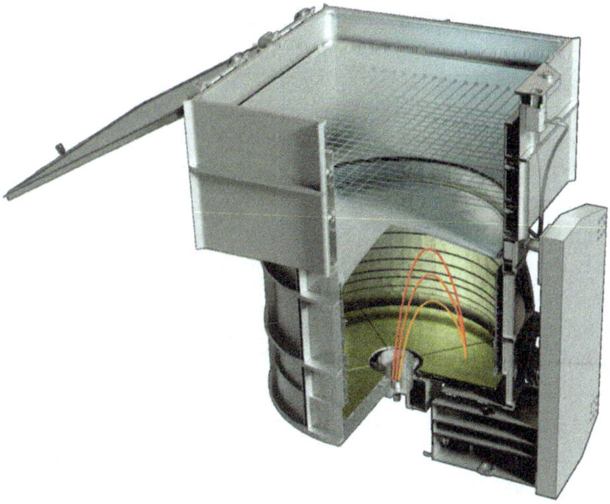

Fig. 14 Dust Telescope consisting of a trajectory sensor (on top) in combination with a large-area mass analyzer (below). In the cut-out ion paths (red) from the impact on the target to the central ion collector are visible (from the FOSSIL proposal, Horanyi et al. courtesy Tibor Balint, JPL)

The Rosetta high resolution time-of-flight ion mass spectrometer for the analysis of cometary particles, COSIMA, included a dust collector, a manipulator unit for target handling, and a dust analysis station. More than 10,000 particles were collected which range from 50 to more than 500 μm in size. Most collected particles originate from the disruption of large aggregates (>1 mm in size) during the collection process (Langevin et al. 2016; Merouane et al. 2016). The particles are composed of various minerals (silicates, Fe sulfides, etc., Hilchenbach et al. 2016). Fray et al. (2016) report the detection by COSIMA of solid organic matter in the dust particles emitted by comet 67P/Churyumov-Gerasimenko; the carbon in the organic material is bound in very large macromolecular compounds, analogous to the insoluble organic matter found in carbonaceous chondrite meteorites.

The cometary sampling and composition (COSAC) experiment on board the Rosetta lander Philae reported the detection of volatile organic molecules that were most likely released from dust particles collected in the instrument's entrance tube when Philae bounced at the cometary surface. The identified compounds are consistent with the first steps of photochemical and/or radiolytic evolution in ices—steps that could have taken place before the comet formed, in the protosolar molecular cloud or on the edge of the solar nebula (Goesmann et al. 2015). One of the key objectives of the Philae mission was the in situ analysis of a fresh sample of cometary material drilled from underneath the surface. This experiment failed, however, due to Philae's unfavourable landing conditions at the Abydos landing site. Keeping in mind that comets contain the most pristine material from the early solar system we know of, in situ sample analysis of fresh sub-surface cometary material in combination with cryogenic sample return should be given the highest priority for future cometary missions.

Recently, noncontact methods to analyze dust particles became available and provide means to derive the trajectories of dust particles in space. The combination of a trajectory sensor with a large area compositional analyser is called a Dust Telescope (Fig. 14, Srama et al. 2005, Sternovsky et al. 2011). It measures the trajectory together with the composition of dust particles entering the instrument. In the near future the Japanese German collaboration in the DESTINY^{+} mission will employ such a dust telescope (Kobayashi et al. 2018b). This mission will focus on analyses of dust populations at 1 AU and the contributions from asteroid (3200) Phaethon which is the parent object of the Geminid meteor stream. The dust telescope will be able to distinguish lunar ejecta particles from cometary and asteroidal interplanetary as well as interstellar dust particles and to characterize their compositional similarities and differences. The measurements will provide a direct comparison of the composition of interstellar raw material with the more processed material from comets and asteroids. Such analyses will complement currently planned asteroid sample return missions and pave the way for future interstellar and interplanetary dust collection missions from near-Earth space (Strub et al. 2019).

A dust telescope of similar design will be on board NASA's Europa Clipper mission to Jupiter's moon Europa. The SUrface Dust Mass Analyzer (SUDA) will measure the composition of ballistic dust particles populating the thin exospheres that were detected around each of the Galilean moons. Since these particles are direct samples from the moons' icy surfaces, unique composition data will be obtained that will help to define and constrain the geological activities on and below the moons' surface. SUDA will characterise the surface composition, habitability, the icy crust, and exchange processes with the deeper interior of the Jovian icy moon Europa (Kempf 2018).

3.8 Dust Observatories in Space

Comets have long been recognized as carriers of pristine, unheated and unaltered material from the early formation stages of our planetary system and at the same time to be significant sources of dust in the solar system. Therefore, the interest in studying these solid, icy, and gaseous objects has been longstanding. Cometary dust observatory missions, generally, have dust science as one of their prime mission objectives. Therefore, they carry advanced dust analyzers together with a full set of other environmental monitors. Examples are the Halley missions, Giotto and VeGa, Stardust to comet Wild 2, and Rosetta to comet Churyumov-Gerasimenko (Levasseur-Regourd et al. 2019). Each of these missions carried several dedicated instruments to pursue various aspects of dust science (Table 2).

Because of the high eccentricity of their orbits, comets are difficult to reach and to rendezvous with from Earth. Cheaper ways are flyby missions that just cross the orbit of a comet close to the nucleus. In 1986 the Russian VeGa 1 and 2 and ESA's Giotto missions were the first missions to make close observations of Halley's comet. The VeGa missions passed by the nucleus at about 9000 km and Giotto at 600 km distance. During their high-speed flybys at about 70 $km\,s^{-1}$ relative to the comet they took images of the nucleus, determined its optical properties, and characterized the dust emission, the neutral gas and plasma environments.

The three space probes carried a complement of specifically designed dust instruments close to the comet. Most of these instruments were impact ionization detectors and mass spectrometers, but also piezo-electric microphones, and thin film detectors were flown. Important results were the discovery of (1) both, smaller and bigger grains in the coma than the 1 to 10 μm sized grains that had been anticipated from astronomical observations (McDonnell et al. 1987, 1989, Mazets et al. 1987), (2) the existence of a significant carbonaceous component in Halley dust (Kissel and Krueger 1987), and (3) a much wider scatter in some isotopic ratios (Solc et al. 1987) than has been found in any other extraterrestrial material.

Analysis of observations by the Halley Multicolor Camera of the dust distribution near the nucleus of the Halley comet (Keller et al. 1990) revealed the occurrence of particle fragmentation close to the comet's nucleus. This suggested a sublimation process in which only a part of the gas comes directly from the surface, while a fraction is liberated during the fragmentation of dust particles. Such a distributed source of CO has been directly observed by the Giotto neutral gas mass spectrometer (NMS, Eberhardt et al. 1987) in the inner coma less than 20,000 km from the nucleus. It was proposed that the CO or a very short-lived

Table 2 Comet dust analyzers, mass sensitivity and sensitive area, and compositional resolution

Mission Instrument	Encounter Year	Comet	Mass threshold (g)	Area (m^2)	Compositional resolution $M/\Delta M$
Giotto Dust Impact Detection System (DIDSY)	1986 1992	Halley, Grigg Skjellerup	10^{-8}	1	–
Giotto Particle Impact Analyzer (PIA)	1986	Halley	2×10^{-15}	0.0005	100
VeGa 1/2 Dust Counter and Mass Analyzer (DUCMA)	1986	Halley	10^{-11}	0.0075	–
VeGa 1/2 Particle Impact Mass Analyzer (PUMA)	1986	Halley	2×10^{-15}	0.0005	100
VeGa 1/2 Solid Particle Experiment (SP)-1	1986	Halley	2×10^{-15}	0.0081	–
VeGa 1/2 SP-2	1986	Halley	10^{-11}	0.05	–
Stardust Aerogel Collector	2004	Wild 2	$\sim 10^{-12}$	0.1	>1000 (in laboratory)
Stardust Cometary and Interstellar Dust Analyzer (CIDA)	2004 2011	Wild 2 Tempel 1	2×10^{-15}	0.009	200
Stardust Dust Flux Monitor Instrument (DFMI)	2004 2011	Wild 2 Tempel 1	10^{-5}, 10^{-12} 10^{-8}, 3×10^{-12}	0.7, 0.002	–
Rosetta Grain Impact Analyzer and Dust Accumulator (GIADA)	2014–2016	Churyumov-Gerasimenko	10^{-7}	0.01	–
Rosetta Cometary Secondary Ion Mass Analyzer (COSIMA)	2014–2016	Churyumov-Gerasimenko	10^{-7}	0.0003	2000
Rosetta Micro-Imaging Dust Analysis System (MIDAS)	2014–2016	Churyumov-Gerasimenko	10^{-16}	10^{-5}	–
Rosetta/Philae Surface Electric Sounding and Acoustic Monitoring Experiment-Dust Impact Monitor (SESAME-DIM)	2014	Churyumov-Gerasimenko	10^{-4}	0.007	–

parent of this molecule was released in the coma from cometary dust grains, such as the carbon, hydrogen, oxygen, and nitrogen rich "CHON" particles.

In 1992 during Giotto's extended mission the spacecraft flew by comet Grigg-Skjellerup at a distance of about 200 km from the nucleus. During the encounter, several meteoroid impacts were detected on Giotto's front shield (McDonnell et al. 1993). The particle masses were found to exceed 10^{-8} kg, suggesting that the mass distribution of the cometary dust was dominated by relatively large particles. The results indicate a higher rate of mass loss from the nucleus than previously thought, and hence a higher dust-to-gas mass ratio.

The cosmochemical relation of cometary dust to meteorites and IDPs was in the focus of the Stardust mission. In 2004 it passed by comet Wild 2 within 230 km at a relative speed of 6 km s^{-1} (Brownlee et al. 2006). The spacecraft carried the 0.1 m^2 Aerogel Dust Collectors, a dust spectrometer (Cometary and Interstellar Dust Analyzer, CIDA), and a PVDF Dust Flux Monitor Instrument (DFMI). During the flyby, DFMI observed a highly varying dust flux on small spatial scales, which was explained by the presence of jets and by particle fragmentation (Green et al. 2004). The mass of the dust coma was dominated by larger particles (up to millimetre sizes), as was found for comets Halley and Grigg-Skjellerup.

The Stardust spacecraft collected thousands of particles from comet 81P/Wild 2 and returned them to Earth for laboratory study. Upon collection, these particles produced hypervelocity impact features on the aerogel collector surfaces. The morphologies of the deceleration tracks were created by particles varying from dense mineral grains to loosely bound, polymineralic aggregates, ranging from tens of nanometres to hundreds of micrometers in size (Hörz et al. 2006). Some particles larger than a few micrometers were seemingly undamaged, whereas smaller or finer-grained components were severely altered. A major portion of the collected particles larger than a micrometer was composed of the silicate minerals olivine and pyroxene (Brownlee et al. 2006). For example, x-ray spectral analysis of a single particle showed three major components: sulfide pyrrhotite, enstatite, and fine-grained porous aggregate material with approximately chondritic elemental composition. Another particle as well as many of the other fragments are isotopically and mineralogically linked to CAIs which are exotic refractory components well-known from primitive meteorites that appear to have formed in the inner regions of the solar nebula. In addition to silicates and abundant sulfides, the collected comet samples contain organic materials. However, potential organic contaminants in the aerogel collector have greatly complicated the interpretation of the organic portions of the samples returned by the Stardust spacecraft (Sandford et al. 2010).

The comet samples returned by the Stardust mission showed that the cold regions of the early solar system were not isolated but that the comets formed from mixed materials, including familiar high-temperature meteoritic materials, such as chondrule fragments that were transported to cold nebular regions (Brownlee 2014). Interstellar dust collections by Stardust have been discussed above (Sterken et al. 2019).

After ten years of interplanetary cruise and three flybys at Earth and at Mars the Rosetta spacecraft reached comet Churyumov-Gerasimenko in August 2014. For more than 2 years Rosetta followed the comet from 3.6 AU inbound, through perihelion at 1.24 AU, to 3.6 AU outbound. During that time it orbited the approximately 2 km radius nucleus from about 10 km to several hundred kilometres distance until September 2016 when the mission was terminated by disposing the spacecraft on to the comet's surface. In November 2014 Rosetta delivered the Philae Lander to the nucleus surface where it took measurements of the nucleus properties at the Agilkia and Abydos landing sites. A synthesis of the morphology of dust observations at comet 67P was recently given by Güttler et al. (2019), and a comprehensive review of the science results from the observations of the various Rosetta instruments is found in Levasseur-Regourd et al. (2019). Here, we stress the complementary nature of the dust observations by various instruments.

The GIADA dust instrument on board Rosetta (Colangeli et al. 2007) was a combination of the Grain Detection System (GDS) and the Impact Sensor (IS). GDS consisted of a light curtain (of 100 cm^2 area) generated by four laser diodes to illuminate passing dust particles so that they could be sensed by a series of photodiodes. The Impact Sensor was located 10 cm below GDS and consisted of a square aluminium plate (sensitive area 100 cm^2) equipped with five piezoelectric sensors. From the combination of signals from both systems the speed and the momentum of cometary grains were determined (Della Corte et al.

Fig. 15 Image of Comet 67P/Churyumov–Gerasimenko captured by Rosetta's OSIRIS narrow-angle camera on 12 August 2015 (17:35 GMT) at a distance of 332 km from the comet nucleus. Dust emission and an outburst (cf. Vincent et al. 2016) are visible on the Sun-lit (top) side of the nucleus (ESA/Rosetta/MPS for OSIRIS Team MPS/UPD/LAM/IAA/SSO/INTA/UPM/DASP/IDA)

2015). Recordings by the IS alone provided just the momentum of the grains. GDS-only signals were obtained and were interpreted as very low density (∼1 kg/m^3) dust particle impacts which did not trigger the more sensitive IS detector (Fulle et al. 2015). Independent confirmation of the existence of such particles is still pending.

One of the prime objectives of the Rosetta mission was to develop an understanding of cometary activity. Early in the mission, a plan was developed to implement an activity campaign involving most orbiter instruments in coordinated observations of an active area at the surface and the resulting dust jet above. The most critical aspect was a fly-through the jet by Rosetta to measure in situ gas (by the ROSINA instrument) and dust flux (GIADA) and to determine the chemical and physical dust properties (COSIMA, MIDAS) in and outside the jet. These measurements were to be compared with long-term (several nucleus rotations) remote sensing observations of optical, near-infrared, and thermal properties (OSIRIS, VIRTIS, MIRO) of the active region and its surroundings. Long-term limb observations (ALICE, MIRO, OSIRIS, VIRTIS) just above the nucleus' surface of the associated jet were planned to complement the campaign. Such observations were foreseen for mid 2015; however, an unexpected phenomenon terminated all plans for an activity campaign: the star sensors onboard Rosetta became confused by thousands of big dust particles in their field-of-view and could no longer support the attitude determination of the spacecraft. As a consequence the spacecraft had to retreat to several 100 km distance during perihelion passage of the comet where in situ observations of gas and dust became impossible and only the camera OSIRIS could observe a firework of outbursts during 3 months around perihelion (Vincent et al. 2016; Fig. 15).

Fortunately, in the last year of the mission, when the Rosetta spacecraft could approach the nucleus again within 30 km distance, three outbursts were serendipitously observed by several in situ and remote sensing instruments. On 19 February 2016, at a distance of 2.4 AU, nine Rosetta instruments observed an outburst of gas and dust from the nucleus of comet 67P/Churyumov-Gerasimenko. Among these instruments were cameras and spectrometers ranging from UV over visible to microwave wavelengths, in situ gas, dust and plasma instruments, and one dust collector (Grün et al. 2016). Furthermore, on 3 July 2016, six Rosetta instruments detected an outburst of gas, dust and ice particles at 3.3 AU from the Sun and took images of the outburst site (Agarwal et al. 2017). The largest number of dust particles in an outburst was recorded by GIADA on 5 September 2016 at 3.6 AU (Della Corte et al. 2017). Gas, dust, and ice particles were simultaneously recorded by ROSINA and the star sensors, and an image of the plume was obtained by OSIRIS. The combination of observations from various instruments bears great potential for unravelling the secrets of the comet.

Voyager fly-bys at the outer planets Jupiter, Saturn, Uranus, and Neptune demonstrated that dust is a major constituent of these planets' environments. There is an intimate interrelation between the satellites, rings, and magnetospheres. The Galileo mission visited the Jupiter environment with a simple dust detector, cameras, plasma and fields instruments. However, Saturn was the first planetary system that was explored by a dust observatory. The central dust instrument was the Cassini Dust Analyzer (CDA) that had capabilities to measure the composition and charge state of detected particles (Hsu et al. 2018). Together with the cameras and the full range of particles and fields instruments, Cassini characterized the interrelations between Saturn's satellites, magnetosphere and rings. Using Cassini nano-dust stream measurements, it was possible to derive the interplanetary magnetic field structure during the 2013 Saturn aurora campaign (Hsu et al. 2013, 2016). Due to the complex dynamical interactions with the interplanetary magnetic field, a fraction of fast nanodust particles emerging from the Saturnian system was sent back into the magnetosphere and could be detected by a spacecraft located within. It was demonstrated by CDA compositional measurements that Enceladus' particles probe the deep interior conditions of this satellite (Postberg et al. 2011, Hsu et al. 2015, 2018).

4 Laboratory Experiments with Dust

Important elements of dust science are laboratory experiments for the study of dust properties and processes. Such experiments address the formation, release from parent bodies, and destruction of dust and its interaction with its environment. We summarize here the most important simulation experiments that had an impact on the development of the field in the past. Recently, many more laboratory experiments have been developed to deepen our understanding of the interrelation between dust and its environment in space.

4.1 High Speed Collisions

The importance of dust accelerator experiments for the development and calibration of dust detectors has been already mentioned. Because of the limited number of projectile compositions used in the past, the need for an increased number of different projectile materials was recognized (cf. Hillier et al. 2014). Similarly, the study of craters on planetary and man-made collector surfaces benefits from impact experiments (cf. Hörz et al. 1975). Impacts on to ices (Timmermann and Grün 1991; Koschny and Grün 2001) support studies of planetary processes in the outer planetary system where ices are abundant. Dust accelerator experiments simulating the entry of meteors into the atmosphere provide important calibration of the ionization yield for radar meteor studies (Thomas et al. 2016).

An important parameter to estimate the life time of a meteoroid against mutual collisions is $\Gamma(u) = M_T/m_{pc}$, where m_{pc} is the mass of the smallest projectile that at impact speed u is able to catastrophically shatter the target object of mass M_T. The parameter Γ is related to the threshold energy density Q_s ($\mathrm{J\,kg^{-1}}$) required to shatter the target: $Q_s = u^2/(2\Gamma)$. Q_s is defined as the ratio of projectile kinetic energy to target mass needed to produce the largest intact fragment of mass M_L that contains half of the target mass M_T (Holsapple et al. 2002). Q_s is obtained by impact experiments. A few sample values of Γ for 10^{-3} kg targets at impact speed of 20 $\mathrm{km\,s^{-1}}$ are $\Gamma = 2.2 \times 10^5$ for crystalline rock (Hörz et al. 1975) and 9×10^5 for gypsum and pyrophyllite (Nakamura et al. 2015). For porous target bodies Love et al. (1993) found that $\Gamma \sim (1 - f)^{3.6}$, where f is the porosity of the target material. For example a porous aggregate particle of 50% porosity (half the density of the solid building block material) has a factor $\sim$10 smaller Γ value than the solid material.

4.2 Low Speed Collisions

An important step of dust evolution from its formation in the atmospheres of cool red giant stars to meteorites and interplanetary dust is the accretion of interstellar dust in the early solar system. For several decades important experimental studies have been performed in order to better understand the build-up of planetary bodies from dust and gas during the formation of the planetary system (cf. Blum et al. 2000). Two major phases in the formation of solid planetary bodies can be distinguished: first, the aggregation of dust particles and clusters in the solar nebula caused by low-velocity mutual collisions to kilometre-sized planetesimals and second, the accretion of planetary cores and planets due to intermediate and high-velocity planetesimal collisions and gravitational attraction. A comprehensive review of the dust growth in protoplanetary discs is given in Blum (2019). Here we only want to stress the importance of laboratory investigations for the first step of accumulation of interstellar dust to larger aggregates.

Laboratory studies were performed to empirically determine sticking and fragmentation efficiencies in low velocity collisions of single dust grains and aggregates, and in intermediate to higher-velocity impact processes. The evolution of morphological structures of growing dust aggregates is an important aspect of these studies. It turned out that, after a period of rapid collisional growth of porous dust aggregates to sizes of a few centimetres, the protoplanetary dust particles are subject to bouncing collisions, in which their porosity is considerably decreased (Blum 2010, 2019). According to these studies, direct formation of kilometre-sized planetesimals by collisional sticking is unlikely, implying that collective effects, such as the streaming instability and the gravitational instability in dust-enhanced regions of the protoplanetary disc, are the best candidates for the processes leading to planetesimals. Studies of freshly released dust from comet 67P/Churyumov-Gerasimenko are underway to compare them with the predicted structures of aggregates during comet formation (Ellerbroek et al. 2017)

4.3 Environmental Effects

Meteoroid bombardment of airless planetary surfaces comminutes the surface material and causes the build-up of a regolith layer. Regolith consists of impact-generated fragments which did not escape the gravity of the planetary body. Further impacts on to the regolith cause gardening and redistribution of material on the surface. Also UV and plasma exposure of regolith surfaces causes mobilization, transport, and ejection of dust particles from the surface. These processes are subject to extensive theoretical and laboratory studies (Szalay and Poppe 2019).

Electrostatic dust transport was first suggested five decades ago to explain the images of the Surveyor 5, 6 and 7 missions showing a lunar horizon glow: on the Moon a bright hovering cloud was observed shortly after sunset (Criswell 1973; Rennilson and Criswell 1974).

Fig. 16 Dust hopping from a central reservoir during exposure to plasma and UV radiadtion (Wang et al. 2016)

Other examples showing evidence for dust transport across vast regions without winds or flowing water are the intermittently appearing radial spoke features first seen by the Voyagers above the rings of Saturn (Smith et al. 1981, 1982), and the accumulation of fine dust in "ponds" on the surface of asteroid Eros imaged by the NEAR-Shoemaker mission (Robinson et al. 2001).

In all these examples, both dust charges and electric fields predicted from current surface-plasma interaction models cannot give an explanation. However, Wang et al. (2016) recently observed jumping dust particles of about 100 μm diameter above a dusty surface when it was exposed to UV radiation or to plasmas in vacuum in the laboratory (Fig. 16). These experiments explained that the interaction of a dusty surface with ultraviolet radiation and/or plasmas is a volume effect. The emission and re-absorption of photo- and/or secondary electrons from the walls of micro-cavities formed between insulating dust particles below the surface are responsible for generating unexpectedly large charges and intense electric fields. The experimenters observed particles jumping to several millimetres height and dust ejection speeds up to $0.6\ \mathrm{m\,s^{-1}}$. In the low gravity environment of asteroids and comets this effect could lead to dust jumps of several 100 m height and similar distances. The results indicate that electrostatic dust transport may be efficient in shaping the surfaces of airless bodies, such as surface morphology and porosity, and it may lead to space weathering.

Space weathering is another regolith process, which describes the darkening and reddening of airless planetary surface materials with time, together with changes to the depths of absorption bands in their optical spectra. This process has been invoked to explain the mismatched optical spectra of lunar rocks and regolith, and between those of asteroids and meteorites. The formation of nanophase iron particles on regolith grains as a result of micrometeorite impacts or irradiation by the solar wind has been proposed as the main cause of the change in the optical properties. Simulations of dust impacts by nano-second-pulse laser irradiation of olivine grains reproduced the optical changes. By observations with a transmission electron microscope Sasaki et al. (2001) found within the vapor-deposited rims of olivine grains nano-phase iron particles similar to those observed in the rims of space-weathered lunar regolith grains.

The Hayabusa 2 sample return mission arrived at the carbonaceous asteroid Ryugu in 2018. Images revealed a rough asteroid surface with abundant boulders which unexpectedly looks fine-grained regolith (Watanabe et al. 2019; Sugita et al. 2019; Jaumann et al. 2019). The spacecraft successfully deployed three landers to the asteroid surface and collected samples from two sites which will be returned to Earth in December 2020. The OSIRIS-REx sample return mission is currently orbiting the asteroid Bennu and mapping at centimetre resolution its surface with cameras, visible, infrared, and X-ray spectrometer (Enos and Lauretta 2019). In July 2020 the spacecraft will take regolith samples from the surface and return them to Earth in 2023.

4.4 Dust Release from Dirty Snow

Already at the time when Giotto reached comet Halley in 1986, ESA mentioned the planetary Cornerstone mission Comet Nucleus Sample Return (CNSR) in its Horizon 2000 plan. CNSR should have followed NASA's Comet Rendezvous Asteroid Flyby mission (CRAF) as the next step in comet research. A joint ESA-NASA study group was set up to define the science of CNSR and to identify steps necessary to technologically implement such a mission. At the same time some members of the science team proposed to the German Science Foundation (DFG) a project to study how an ice-dust mixture reacts to solar irradiation in vacuum and to use the big solar simulation chamber left over from the Helios mission at DLR Cologne (Grün 1991). The comet simulation experiment (KOSI) was selected and became part of a special DFG project "Small bodies in the solar system". About 50 experimental and theoretical scientists and students and 10 technicians from 8 science institutes in Germany and 12 participating scientists from Austria, France, Israel, Netherlands, SU/Russia, and USA took part in the project (Grün et al. 1992c; Sears et al. 1999). Related experimental investigations took place in Tel Aviv (Bar-Nun and Laufer 2003), Graz (Kömle et al. 1996), and Dushanbe (Ibadinov et al. 1991). Eleven KOSI experiments were performed from May 1987 to May 1993. During insolation of the dirty ice, gas and dust emissions were determined. The initially homogeneous loose ice dust mixture decomposed into a stratified composition. At the surface a hot coherent but very fluffy dust mantle ($\sim$100 kg/m^3 density) developed over a solidified porous water ice layer (Stöffler et al. 1991). Below the hard ice layer more volatile ices like CO_2 ice were found which overlaid the cold original sample material. The emitted dust particles were mostly fragments of the dust mantle consisting of aggregates of the original building blocks mixed into the ices (Fig. 17, Kochan et al. 1990, 1998, Grün et al. 1993a). These particles resembled the fluffy IDPs collected in the Earth's stratosphere (cf. Fig. 8). Recently, Rosetta observations stimulated a renewed interest in the results from laboratory studies of cometary processes (Gudipati et al. 2015; Poch et al. 2016, Jost et al. 2017).

5 Understanding the Zodiacal Cloud

Understanding the meteoroid environment is the subject of dust modelling. The models describe the life of meteoroids from their formation to their disappearance. Scientists already tried to understand the observed phenomena in the era when only ground based optical observations of dust and meteoroids were available. Foremost, Fred Whipple influenced the dust field with his ground-breaking work on comets (Whipple 1950), their relationship to meteors (Whipple 1951), and through his studies of the meteoritic complex as a whole (Whipple 1967).

Understanding the dynamics of meteoroids is an important aspect of any modelling the meteoroid environment. Based on the fact that dust in space is subject to gravitational interaction, Öpik described the gravitational scattering of an interplanetary object by a planet

Fig. 17 Dust release from an icy dust mixture during the KOSI Experiments (top) and collected dust particles (bottom, Kochan et al. 1998)

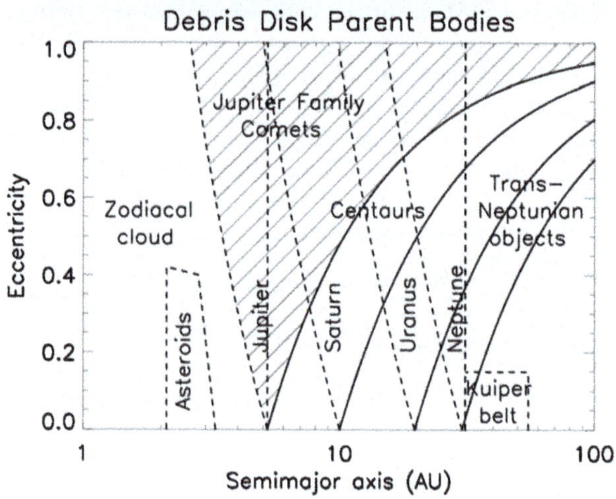

Fig. 18 Relationship between small bodies in the solar system and the giant planets. The triangular-shaped regions pointing at the semi-major axes of the giant planets delineate the scattering zone of the respective planet

(Öpik 1951). While his work aimed at the distribution of interplanetary matter, it became also the basis of interplanetary travel to other planets using gravity assist. Figure 18 shows the eccentricity vs. semi-major axis diagram of the small body reservoirs (Kuiper belt objects and asteroids) and their relation to the planetary scattering zones. Jupiter family comets are related to Trans-Neptunian objects via the Centaurs.

In addition to gravity, small meteoroids are affected by radiation pressure (cf. Burns et al. 1979). Absorbing particles with sizes comparable to the wavelength of visible sunlight (~0.5 μm) are affected most strongly, and the radiation pressure force may exceed solar gravity ($\beta > 1$). In this case particles leave the solar system on hyperbolic trajectories. Such β meteoroids were first identified in the data from the Pioneer 8 and 9 space probes (Zook and Berg 1975). The effect of radiation pressure was earlier applied to comet tails by Finson and Probstein (1968a,b). A comet tail can be described by an array of syndynes and synchrones where particles are sorted according to their repulsive force from solar radiation pressure and their ejection time from the nucleus, respectively. The secular effect of radiation pressure on meteoroid orbits, the Poynting-Robertson effect, was quantified by Wyatt and Whipple (1950). Through the Poynting-Robertson effect, interplanetary particles lose angular momentum and orbital energy and eventually spiral into the Sun. In most space environments, dust particles are electrically charged (Horanyi 1996), and, hence, small, submicrometer-sized dust particles are strongly affected by their interaction with the interplanetary magnetic field as well (e.g. Morfill and Grün 1979a,b).

Finally, the life times of meteoroids are limited by mutual collisions. Dohnanyi (1970) demonstrated that catastrophic collisions among meteoroids (i.e. the impact of a small meteoroid completely shatters the bigger meteoroid) are more effective in limiting the life time of sporadic meteoroids than erosive collisions (i.e. the impact of a small meteoroid generates just a crater on the bigger meteoroid). Dohnanyi (1969) also showed that the mass distribution of smaller asteroids (<100 km in size) is in a collisional quasi-steady-state: in any given size interval the number of destroyed asteroids equals the number of fragments generated by collisions from bigger asteroids. In this case the cumulative mass distribution has the form $m^{-\alpha}$ with $\alpha \approx 5/6$. The steeper mass distribution ($\alpha > 5/6$) of sporadic meteoroids is unstable and changes with time unless there is a sufficient source for such particles.

The first quantitative empirical dust model was deveoped by Cour-Palais (1969), aimed at describing to space engineers the meteoroid environment in near-Earth space. It was based on results from the Harvard Radio Meteor Project (summarized by Southword and Sekanina 1973) and from the penetration detectors on board the Explorer 16, 23, and the Pegasus satellites. The model provided the flux of meteoroids in interplanetary space and in the near-Earth environment, taking into account gravitational enhancement and Earth shielding effects. Grün et al. (1985) published an interplanetary flux model on the basis of lunar micro-crater statistics and previous satellite data. It covered meteoroid masses from 10^{-18} to 1 g and assumed an average collision speed of 20 km s^{-1} at 1 AU. It was demonstrated that the life time of meteoroids with masses below 10^{-9} kg (~100 μm) is dominated by the Poynting-Robertson effect whereas bigger particles are destroyed by mutual collisions. Results from impact experiments into basalt by Fujiwara et al. (1977) were used to quantify the collisional destruction of meteoroids. At 1 AU the collisional life time of centimetre-sized meteoroids was calculated to be approximately 10^4 years.

In 1993 Divine published his "Five populations of interplanetary meteoroids" model (Divine 1993). It was an empirical model based on data from radar meteor surveys, the earlier interplanetary flux model by Grün et al. (1985), and space probe impact detector measurements ranging from 0.3 AU (Helios) to 20 AU (Pioneer 10). The model consistently described the interplanetary micrometeoroid environment via five distinct dust populations defined by their distributions separable in mass and orbital elements. Meteoroid fluxes, densities, and directional properties were calculated throughout the solar system using Keplerian dynamics. This model was later updated by Staubach et al. (1997) in order to include radiation pressure effects and thereby better represent the small particle flux measurements by Ulysses, Galileo, and other in situ dust measurements. A comprehensive review of the earlier dust models was given by Staubach et al. (2001).

NASA's Meteoroid Engineering Model, MEM, was developed by Jones (2004) and McNamara et al. (2004) on the basis of sporadic meteor observations by the Canadian Meteor Orbit Radar (CMOR, Jones and Brown 1993) together with zodiacal light observations from Helios and lunar crater statistics. The model used orbital element distributions from meteor populations that were tied to actual comet families and asteroids (Jones et al. 2001). Since meteors must intersect the Earth orbit, this model can only be used to predict fluxes, speeds, and directions from approximately 0.5 to 2.0 AU.

A first evolutionary model of the sporadic meteoroid background inside Jupiter's orbit was published by Dikarev et al. (2004) and was implemented in ESA's Interplanetary Meteoroid Environment Model, IMEM. Contrary to earlier attempts, this model starts from the orbital elements of the dominant known sources of interplanetary dust: Jupiter family comets and asteroids. With this model the era of descriptions of the interplanetary dust cloud by just a handful of dust populations is left, and now even small computers can evaluate the dust densities and fluxes at any desired position in space. Six populations of dust from asteroids, 72 populations of dust from comets and other parent objects on Jupiter-crossing orbits, and one stream of interstellar dust, were introduced in the IMEM model (for a full description see Dikarev et al. 2019). The model assumes that big meteoroids ($\geq 10^{-5}$ g) stay in the orbits of their parent objects (because collisions destroy them before they are removed by dynamical effects) while the orbits of smaller meteoroids evolve under planetary gravity and the Poynting-Robertson effect. Thermal radiation measurements by the COBE DIRBE instrument (Kelsall et al. 1998a, 1998b), in situ data from the dust instruments onboard Galileo and Ulysses (Grün et al. 1997), and lunar microcrater distributions were used to calibrate the contributions from the known sources. However, due to an inconsistency between the COBE DIRBE observations and the Galileo and Ulysses in situ detections, the Galileo and Ulysses data were adjusted such that the grain sizes were increased by a factor of 2.5 with respect to the usual instrument calibration (Dikarev et al. 2019). Attempts to include meteor orbits from the Advanced Meteor Orbit Radar AMOR in the model (Galligan and Baggaley 2004) failed since the AMOR orbital distributions were incompatible with the COBE latitudinal IR brightness profile under the assumptions of the model.

Recently, Nesvorny et al. (2010, 2011a, 2011b) published a dynamical model of the zodiacal cloud and sporadic meteors. The sources of this evolutionary model are asteroids, Jupiter-family comets, Halley-type comets, and Oort-cloud comets. The model describes zodiacal infrared observations by the Infrared Astronomical Satellite (IRAS, Hauser et al. 1984), distributions of radar meteors from the Canadian Meteor Orbit Radar (CMOR, Campbell-Brown 2008), and properties of micrometeorites recovered in Antarctica (Engrand and Maurette 1998). The authors found that Jupiter family comets are the main source of meteor concentrations arriving at the Earth from the helion and antihelion directions (cf. Fig. 6), as well as the sources of micrometeorites in our collections (micrometeorites collected in Antarctic ice have primitive carbonaceous composition). At Earth, asteroidal dust contributes less than 10%. However, the authors rather crudely extrapolated from the collision life time given by Grün et al. (1985) and applied a fitting parameter which multiplies the collision life times. They found that the orbits of some particle populations must be much further evolved by the Poynting-Robertson effect than the original collisional life times would allow. The fitting parameter must be 10 to 30 in order to be compatible with infrared latitudinal brightness profiles and to match the speed distribution measured by the radar meteor observatories. They even required that about 1% of particles from Oort cloud comets orbitally evolve by Poynting–Robertson drag to reach orbits with semi-major axis a $\sim$1 AU. Such meteoroids are expected to produce meteors with radiants near the apex

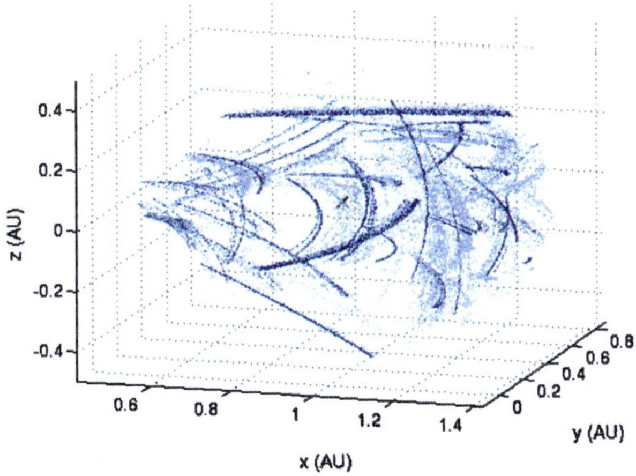

Fig. 19 Prediction of the dust trail structure within 0.5 AU from the Earth on 1st October 2049 (ESA IMEX study, Soja et al. 2014). Dust positions are given in heliocentric ecliptic coordinates. Dark blue particles have mass 1 × 10–6 kg and light blue particles have mass 1 × 10–9 kg. Earth is the black cross

of Earth's orbital motion. Nesvorny et al. (2011b) also include a set of Jupiter family and Oort cloud comets that are similar, but dynamically evolved beyond their normal activity life times. This is to account for a dust component resulting from the spontaneous disruption of such dormant comets. This model has recently been applied in the inner solar system to develop a comprehensive model of the meteoroid environment around Mercury and to explain structural and temporal variability in Mercury's exosphere (Pokorny et al. 2018).

Motivated by the need to understand the meteoroid hazard to human exploration beyond Earth vicinity, ESA supported the IMEX study (Interplanetary Meteoroid Environment for eXploration) at the Institut für Raumfahrtsysteme in Stuttgart (Soja et al. 2014, 2015a, 2015b). The study was based on the observations by Reach et al. (2007) with the Spitzer Space Telescope. The latter authors found that 30 out of 34 Jupiter-family comets had known debris trails. Thus, the detection rate exceeded 80%, indicating that debris trails are a generic feature of short-period comets. In the IMEX study the evolution of dust streams from 362 Jupiter family comets, 40 Halley-type comets and 18 Encke-type comets were calculated between previous five apparitions and the year 2080. A test of the model with the Leonid meteor storm demonstrated that it predicts within 20 minutes the maximum and the duration of the November 1999 storm. The amplitude of the profile, however, was not matched well because of uncertainties of the cometary dust production rate and mass distribution. The application to the dust streams of comet 67P/Churyumov-Gerasimenko results in the prediction of the ejection velocities, the dust size index, and a dust production rate of this comet. Also a meteor stream from this comet is predicted at Mars for the year 2062. The meteoroid stream environment of the Earth in September 2049 is shown in Fig. 19.

Currently, the space agencies ESA and NASA are in the process of updating their interplanetary meteoroid models. The ESA Interplanetary Meteoroid Environment Model, IMEM2, has been developed at the Institut für Raumfahrtsysteme in Stuttgart (Soja et al. 2019). The orbit evolution of dust emitted from various types of sources (Jupiter family, and Halley type comets and asteroids) is followed for a million years and the collision history is determined (Soja et al. 2016). The model is fitted to the lunar micro-crater distribution (Morrison and Zinner 1977; Morrison and Clanton 1979), and meteor observational data (Campbell-Brown 2008, Galligan and Baggaley 2005). Calibration of dust densities and fluxes is provided by a fit to the absolute infrared brightnesses observed by COBE (Kelsall et al. 1998a, 1998b). In this model it was found that the collisional life times are significantly extended as compared to the ones assumed by Grün et al. (1985) for rocky meteoroid material. It was also determined that about 80% of the dust observable by infrared methods originates from Jupiter family comets, 20% from asteroids, and only 0.5% from Halley type comets. Both findings support the hypothesis that interplanetary dust particles have mostly a porous fluffy structure just like cometary particles have (cf. Fig. 1 and Fig. 5).

In the outer solar system, beyond Jupiter's orbit, the sources of dust are different, including a dominant Kuiper Belt component, but the particles are subject to the same perturbing and modifying effects as in the inner solar system. An understanding of the dust cloud in this region is important for quantifying the interplanetary dust input in the Jovian and Saturnian systems: A comprehensive model has been built by Poppe 2016 (see also Szalay and Poppe 2019).

In addition to the flow of interplanetary matter mostly released from comets and asteroids, there is a flow of interstellar grains through the planetary system. Following the earlier predictions (Bertaux and Blamont 1976; Gustafson and Misconi 1979; Morfill and Grün 1979b) of the flow of electrically charged interstellar dust through the planetary system such flow was successfully detected by the Ulysses dust instrument (Grün et al. 1994). The models by Landgraf (2000), Sterken et al. (2012, 2013, 2015) and Strub et al. (2019) show strong spatial and temporal variations of the interstellar dust density due to its interactions with solar radiation pressure and the interplanetary magnetic field (Sterken et al. 2019; Krüger et al. 2019).

The dynamical dust processes in circumplanetary environments of the outer planets are even more complex than in interplanetary space because of the proximity to the central body, the multitude of satellites some of which are active dust/ice emitters, the existence of rings, and the strong magnetospheric effects. Nevertheless, a detailed understanding of most of the observed phenomena has been developed. For a comprehensive review see Spahn et al. (2019).

Despite the great advances made in the last years in understanding the heliospheric dust environment, there remain many important questions to be answered. The overall goal is to collect a compositional dust inventory for the whole planetary system from the solar F-corona to the Kuiper belt and to characterize its interrelation with the planetary, interplanetary, and interstellar environment.

We have summarized the long history of dust research starting from the perception by many scientists that dust is dangerous to encounter, dirty and impure in its properties and difficult to quantify it individually and collectively. In our present understanding dust is a treasure chest of information on its sources and the processes that shape its properties throughout space and time. This book gives testimony that dust bridges astrophysics, planetary science, and life science.

Acknowledgements Open access funding provided by Max Planck Society.

References

J. Agarwal et al., Evidence of sub-surface energy storage in comet 67P from the outburst of 2016 July 03. Mon. Not. R. Astron. Soc. **469**, 606–625 (2017)

W.M. Alexander, C.W. Arthur, J.L. Bohn, Lunar Explorer 35 and OGO 3: dust particle measurements in selenocentric and cislunar space from 1967 to 1969, in *Space Research XI*, ed. by K.Y.A. Kondratyev, M.J. Rycroft, C. Sagan (1971, Akademie-Verlag, Berlin) pp. 279–285

N. Altobelli, E. Grün, M. Landgraf, A new look into the Helios dust experiment data: presence of interstellar dust inside the Earth's orbit. Astron. Astrophys. **448**, 243–252 (2006)

N. Altobelli et al., Flux and composition of interstellar dust at Saturn from Cassini's Cosmic Dust Analyzer. Science **352**, 312–318 (2016)

K. Altwegg et al., Prebiotic chemicals-amino acid and phosphorus-in the coma of comet 67P/Churyumov-Gerasimenko. Sci. Adv. **2**(5), e1600285–e1600285 (2016)

S. Auer, The asteroid belt: doubts about the particle concentration measured with the Asteroid/Meteoroid detector on Pioneer 10. Science **186**, 650–652 (1974)

S. Auer, Instrumentation, in *Interplanetary Dust*, ed. by E. Grün, B.A.S. Gustafson, H. Fechtig, S.F. Dermott (Springer, Heidelberg, 2001)

S. Auer, K. Sitte, Detection technique for micrometeoroids, using impact ionization. Earth Planet. Sci. Lett. **4**, 178–183 (1968)

W.J. Baggaley, A.D. Taylor, D.I. Steel, The influx of meteoroids with hyperbolic heliocentric orbits. Meteoroids and their Parent Bodies, pp. 53–56 (1993)

M. Baguhl, E. Grün, G. Linkert, D. Linkert, N. Siddique, Identification of 'small' dust impacts in the Ulysses dust detector data. Planet. Space Sci. **41**, 1085–1098 (1993)

M. Baguhl, D.P. Hamilton, E. Grün, S.F. Dermott, H. Fechtig, M.S. Hanner, J. Kissel, B.A. Lindblad, D. Linkert, G. Linkert, I. Mann, J.A.M. McDonnell, G.E. Morfill, C. Polanskey, R. Riemann, G. Schwehm, P. Staubach, H.A. Zook, Dust measurements at high ecliptic latitudes. Science **268**, 1016–1019 (1995)

A. Bar-Nun, D. Laufer, First experimental studies of large samples of gas laden amorphous "cometary" ices. Icarus **161**, 157–163 (2003)

O.E. Berg, E. Grün, Evidence of hyperbolic cosmic dust particles, in *Space Research XIII*, ed. by M.J. Rycroft, S.K. Runcorn (Akademie-Verlag, Berlin, 1973), pp. 1047–1055

O.E. Berg, F.F. Richardson, The Pioneer 8 cosmic dust experiment. Rev. Sci. Instrum. **40**, 1333–1337 (1969)

J.L. Bertaux, J.E. Blamont, Possible evidence for penetration of interstellar dust into the solar system. Nature **262**, 263–266 (1976)

J. Blum, Dust growth in protoplanetary disks—a comprehensive experimental/theoretical approach. Res. Astron. Astrophys. **10**, 1199–1214 (2010)

J. Blum, *Dust Evolution in Protoplanetary Discs and the Formation of Planetesimals*. Space Science Reviews Book, vol. 69 (2019, in press)

J. Blum et al., Growth and form of planetary seedlings: results from a microgravity aggregation experiment. Phys. Rev. Lett. **85**, 2426–2429 (2000)

P. Brown, R.J. Weryk, D.K. Wong, J. Jones, The Canadian meteor orbit radar meteor stream catalogue. Earth Moon Planets **102**, 209–219 (2008)

D. Brownlee, The stardust mission: analyzing samples from the edge of the solar system. Annu. Rev. Earth Planet. Sci. **42**, 179–205 (2014)

D.E. Brownlee, D. Tomandl, M.B. Blanchard, Ferry, G.V. Kyte F, An Atlas of extraterrestrial particles collected with NASA U-2 aircraft, 1974–1976, NAS TMX (1976)

D.E. Brownlee et al., Comet 81P/Wild 2 under a microscope. Science **314**, 1711 (2006)

J.A. Burns, P.L. Lamy, S. Soter, Radiation forces on small particles in the solar system. Icarus **40**, 1–48 (1979). https://doi.org/10.1016/0019-1035(79)90050-2

M.D. Campbell-Brown, High resolution radiant distribution and orbits of sporadic radar meteoroids. Icarus **196**, 144–163 (2008)

R.M. Canup, E. Asphaug, Origin of the Moon in a giant impact near the end of the Earth's formation. Nature **412**, 708–712 (2001). https://doi.org/10.1038/35089010

J.D. Carpenter, A. Wells, A.F. Abbey, R.M. Ambrosi, Meteoroid and space debris impacts in grazing-incidence telescopes. Astron. Astrophys. **483**, 941–947 (2008)

Z. Ceplecha, J. Borovicka, W.G. Elford, D.O. Revelle, R.L. Hawkes, V. Porubcan, M. Simek, Meteor phenomena and bodies. Space Sci. Rev. **84**, 327–471 (1998)

L. Colangeli et al., The Grain Impact Analyser and Dust Accumulator (GIADA) experiment for the Rosetta mission: design, performances and first results. Space Sci. Rev. **128**, 803–821 (2007)

A. Collette, G. Meyer, D. Malaspina, Z. Sternovsky, Laboratory investigation of antenna signals from dust impacts on spacecraft. J. Geophys. Res. Space Phys. **120**, 5298–5305 (2015)

G. Colombo, D.A. Lautmann, I.I. Shapiro, The Earth's dust belt: fact or fiction? 2. Gravitational focussing and Jacobi capture. J. Geophys. Res. **71**, 5705 (1966)
W.J. Cooke, J.D. Mulholland, J.P. Oliver, IDE constraints on the beta meteoroid population. Adv. Space Res. **13**, 119–122 (1993)
B.G. Cour-Palais (ed.), *Meteoroid Environment Model-1969*. Near-Earth to Lunar Surface, NASA SP, vol. 8013 (1969).
D.R. Criswell, Horizon-glow and the motion of lunar dust, in *Photon and Particle Interactions with Surfaces in Space*, ed. by R.J.L. Grard (Springer, New York, 1973), pp. 545–556
C.T. D'Aiutolo, W.H. Kinard, R.J. Naumann, Recent NASA meteoroid penetration results from satellites. Smithson. Contrib. Astrophys. **11**, 239 (1967)
V. Della Corte et al., GIADA: shining a light on the monitoring of the comet dust production from the nucleus of 67P/Churyumov-Gerasimenko. Astron. Astrophys. **583**, A13 (2015)
V. Della Corte, A. Rotundi, M. Fulle, S. Ivanovski, J. Agarwal, E. Gruen, 67P/Churyumov-Gerasimenko: the 2016 September 5 dust-rich event characterized by GIADA onboard Rosetta, in *European Planetary Science Congress EPSC2017-889* (2017)
H. Dietzel, G. Eichhorn, H. Fechtig, E. Grün, H.J. Hoffmann, J. Kissel, The HEOS A-2 and Helios micrometeoroid experiments. J. Phys. E, Sci. Instrum. **6**, 209–217 (1973)
V. Dikarev, E. Grün, J. Baggaley, D. Galligan*, M. Landgraf, R. Jehn, Modeling the sporadic meteoroid background cloud. Earth Moon Planets **95**, 109–122 (2004)
V. Dikarev, E. Grün, J. Baggaley, D. Galligan, M. Landgraf, R. Jehn, A single physical model for diverse meteoroid data sets (2019). arXiv:1902.02977
N. Divine, Five populations of interplanetary meteoroids. J. Geophys. Res. **98**, 17029–17048 (1993)
J.S. Dohnanyi, Collisional model of asteroids and their debris. J. Geophys. Res. **74**, 2531–2554 (1969)
J.S. Dohnanyi, On the origin and distribution of meteoroids. J. Geophys. Res. **75**, 3468–3493 (1970)
J. Dorschner, Interstellar dust and circumstellar dust disks, in *Interplanetary Dust*. ed. by E. Grün, B.A.S. Gustafson, S. Dermott, H. Fechtig. Astronomy and Astrophysics Library, vol. 727 (Springer, Berlin, 2001), p. 727. 804 p., ISBN 3-540-42067-3
G. Drolshagen, H. Svedhem, E. Grün, O. Grafodatsky, V. Verhoturov, U. Prokopiev, V. Gusyelnikov, Situ measurement of cosmic dust and space debris in the geostationary orbit, in *Proceedings of 2nd European Conf. on Space Debris*, Darmstadt, March 17–19, 1997 (1997)
N. Eaton, J.K. Davies, S.F. Green, The anomalous dust tail of comet P/Tempel 2. Mon. Not. R. Astron. Soc. **211**, 15P–19P (1984)
P. Eberhardt et al., To CO and N_2 abundance in comet P/Halley. Astron. Astrophys. **187**, 481 (1987)
L.E. Ellerbroek et al., The footprint of cometary dust analogues—I. Laboratory experiments of low-velocity impacts and comparison with Rosetta data. Mon. Not. R. Astron. Soc. **469**, S204–S216 (2017)
C. Engrand, M. Maurette, Carbonaceous micrometeorites from Antarctica. Meteorit. Planet. Sci. **33**, 565–580 (1998)
H.L. Enos, D.S. Lauretta, A rendezvous with asteroid Bennu. Nat. Astron. **3**, 363 (2019)
H. Fechtig, E. Gruen, G. Morfill, Micrometeoroids within ten Earth radii. Planet. Space Sci. **27**, 511–531 (1979)
H. Fechtig, C. Leinert, O.E. Berg, Historical perspectives, in *Interplanetary Dust* ed. by E. Grün, B.A.S. Gustafson, S. Dermott, H. Fechtig. Astronomy and Astrophysics Library (Springer, Berlin, 2001), 804 p., ISBN 3-540-42067-3
M.J. Finson, R.F. Probstein, A theory of dust comets. I. Model and equations. Astrophys. J. **154**, 327–352 (1968a)
M.L. Finson, R.F. Probstein, A theory of dust comets. II. Results for comet Arend-Roland. Astrophys. J. **154**, 353–380 (1968b)
N. Fray et al., High-molecular-weight organic matter in the particles of comet 67P/Churyumov-Gerasimenko. Nature **538**, 72–74 (2016)
J.F. Friichtenicht, Two-million-volt electrostatic accelerator for hypervelocity research. Rev. Sci. Instrum. **33**, 209–212 (1962)
J.F. Friichtenicht, D.G. Becker, Measurements of the ionization probability of CU and LaB^6 simulated micrometeors. Astrophys. J. **166**, 717 (1971)
J.F. Friichtenicht, J.C. Slattery, Ionization associated with hypervelocity impact. NASA Technical Note, D-2091 (1963)
B.J. Fry, B.D. Fields, J.R. Ellis, Radioactive iron rain: transporting ^{60}Fe in supernova dust to the ocean floor. Astrophys. J. **827**, 48 (2016). https://doi.org/10.3847/0004-637X/827/1/48
A. Fujiwara, G. Kamimoto, A. Tsukamoto, Destruction of basaltic bodies by high-velocity impact. Icarus **31**, 277–288 (1977)
M. Fulle et al., Density and charge of pristine fluffy particles from comet 67P/Churyumov-Gerasimenko. Astrophys. J. **802**, L12 (2015)

D.P. Galligan, W.J. Baggaley, The orbital distribution of radar-detected meteoroids of the solar system dust cloud. Mon. Not. R. Astron. Soc. **353**, 422–446 (2004)

D.P. Galligan, W.J. Baggaley, The radiant distribution of AMOR radar meteors. Mon. Not. R. Astron. Soc. **359**, 551–560 (2005)

C.K. Goertz, G.E. Morfill, A model for the formation of spokes in Saturn's ring. Icarus **53**, 219–229 (1983)

F. Goesmann et al., Organic compounds on comet 67P/Churyumov-Gerasimenko revealed by COSAC mass spectrometry. Science **349**, aab0689 (2015)

A.L. Graps, E. Grün, H. Svedhem, H. Krüger, M. Horanyi, A. Heck, S. Lammers, Io as a source of the jovian dust streams. Nature **405**, 48–50 (2000)

S.F. Green, J.A.M. McDonnell, N. McBride, M.T.S.H. Colwell, A.J. Tuzzolino, T.E. Economou, P. Tsou, B.C. Clark, D.E. Brownlee, The dust mass distribution of comet 81P/Wild 2. J. Geophys. Res., Planets **109**, E12S04 (2004)

E. Grün, Interrelation between KOSI experiments and comet nucleus sampling. Space Sci. Rev. **56**, 105–108 (1991)

E. Grün, Interplanetary Dust and the Zodiacal light, in *Encyclopedia of the Solar System*, ed. by P. Weisman, L. McFadden, T. Johnson (Academic Press, New York, 1999), pp. 673–695

E. Grün, P. Rauser, Penetration studies of iron dust particles in thin foils, in *Space Research IX*, ed. by K.S.W. Champion, P.A. Smith, R.L. Smith-Rose (North-Holland, Amsterdam, 1969), pp. 147–154

E. Grün, N. Pailer, H. Fechtig, J. Kissel, Orbital and physical characteristics of micrometeoroids in the inner solar system as observed by Helios 1. Planet. Space Sci. **28**, 333–349 (1980)

E. Grün, G.E. Morfill, R.J. Terrile, T.V. Johnson, G.H. Schwehm, The evolution of spokes in Saturn's B ring. Icarus **54**, 227–252 (1983)

E. Grün, H.A. Zook, H. Fechtig, R.H. Giese, Collisional balance of the meteoritic complex. Icarus **62**, 244–272 (1985)

E. Grün, H. Fechtig, M.S. Hanner, J. Kissel, B.A. Lindblad, D. Linkert, G.E. Morfill, H.A. Zook, The Galileo dust detector. Space Sci. Rev. **60**, 317–340 (1992a)

E. Grün, H. Fechtig, R.H. Giese, J. Kissel, D. Linkert, D. Maas, J.A.M. McDonnell, G.E. Morfill, G. Schwehm, H.A. Zook, The Ulysses dust experiment. Astron. Astrophys. Suppl. Ser. **92**, 411–423 (1992b)

E. Grün, J. Benkhoff, J. Gebhard, Past, present and future KOSI comet simulation experiments. Ann. Geophys. **10**, 190–197 (1992c)

E. Grün, J. Gebhard, A. Bar-Nun, J. Benkhoff, H. Dueren, G. Eich, R. Hische, W.F. Huebner, H.U. Keller, G. Klees, Development of a dust mantle on the surface of an insolated ice-dust mixture—results from the KOSI-9 experiment. J. Geophys. Res. **98**, 15 (1993a)

E. Grün, H.A. Zook, M. Baguhl, A. Balogh, S.J. Bame, H. Fechtig, R. Forsyth, M.S. Hanner, M. Horanyi, J. Kissel, B.-A. Lindblad, D. Linkert, G. Linkert, I. Mann, J.A.M. McDonnell, G.E. Morfill, J.L. Phillips, C. Polanskey, G. Schwehm, N. Siddique, P. Staubach, J. Svestka, A. Taylor, Discovery of jovian dust streams and interstellar grains by the Ulysses spacecraft. Nature **362**, 428–430 (1993b)

E. Grün, B.A.S. Gustafson, I. Mann, M. Baguhl, G.E. Morfill, P. Staubach, A. Taylor, H.A. Zook, Interstellar dust in the heliosphere. Astron. Astrophys. **286**, 915–924 (1994)

E. Grün, B.A.S. Gustafson, H. Fechtig, S.F. Dermott (eds.), *Interplanetary Dust* (Springer, Heidelberg, 2001)

E. Grün, M. Baguhl, N. Divine, H. Fechtig, D.P. Hamilton, M.S. Hanner, J. Kissel, B.-A. Lindblad, D. Linkert, G. Linkert, I. Mann, J.A.M. McDonnell, G.E. Morfill, C. Polanskey, R. Riemann, G. Schwehm, N. Siddique, P. Staubach, H.A. Zook, Three years of Galileo dust data. Planet. Space Sci. **43**(8), 953–969 (1995). https://doi.org/10.1016/0032-0633(94)00234-I

E. Grün, P. Staubach, M. Baguhl, D.P. Hamilton, H.A. Zook, S. Dermott, B.A. Gustafson, H. Fechtig, J. Kissel, D. Linkert, G. Linkert, R. Srama, M.S. Hanner, C. Polanskey, M. Horanyi, B.A. Lindblad, I. Mann, J.A.M. McDonnell, G.E. Morfill, G. Schwehm, South-North and radial traverses through the interplanetary dust cloud. Icarus **129**(2), 270–288 (1997). https://doi.org/10.1006/icar.1997.5789

E. Grün, R. Srama, H. Krüger, S. Kempf, V. Dikarev, S. Helfert, G. Moragas-Klostermeyer, 2002 Kuiper prize lecture: dust astronomy. Icarus **174**, 1–14 (2005)

E. Grün et al., The 2016 Feb 19 outburst of comet 67P/CG: an ESA Rosetta multi-instrument study. Mon. Not. R. Astron. Soc. **462**, 220–234 (2016)

M.S. Gudipati et al., Laboratory studies towards understanding comets. Space Sci. Rev. **197**, 101–150 (2015)

D.A. Gurnett, E. Grün, D. Gallagher, W.S. Kurth, F.L. Scarf, Micron-sized particles detected near Saturn by the Voyager plasma wave instrument. Icarus **53**, 236–254 (1983)

D.A. Gurnett, T.F. Averkamp, F.L. Scarf, E. Grun, Dust particles detected near Giacobini-Zinner by the ICE plasma wave instrument. Geophys. Res. Lett. **13**, 291–294 (1986)

D.A. Gurnett et al., The Cassini radio and plasma wave science investigation. Space Sci. Rev. **114**, 395–463 (2004)

B.A.S. Gustafson, N.Y. Misconi, Streaming of interstellar grains in the solar system. Nature **282**, 276–278 (1979)
C. Güttler, T. Mannel, A. Rotundi, S. Merouane, M. Fulle, D. Bockelee-Morvan, J. Lasue, A.-C. Levasseur-Regourd, J. Blum, G. Naletto, H. Sierks, M. Hilchenbach, C. Tubiana, F. Capaccioni, J.A. Paquette, A. Flandes, F. Moreno, J. Agarwal, D. Bodewits, I. Bertini, G.P. Tozzi, K. Hornung, Y. Langevin, H. Krüger, A. Longobardo, V. Della Corte, I. Toth, G. Filachione, S.L. Ivanovski, S. Mottola, G. Rinaldi, Synthesis of the morphological description of cometary dust at comet 67P. Astron. Astrophys. **630**, A24 (2019)
M. Hajdukova Jr., The occurrence of interstellar particles in the vicinity of the Sun an overview—25 years of research, in *International Meteor Conference Egmond*, The Netherlands, 2–5 June 2016 (2016), pp. 105–110.
D.P. Hamilton, H. Krüger, Jupiter's shadow sculpts its gossamer rings. Nature **453**, 72–75 (2008). https://doi.org/10.1038/nature06886
M.S. Hanner, J.L. Weinberg, Gegenschein observations from Pioneer 10. Sky Telesc. **45**, 217–218 (1973)
D.O. Hansen, Mass analysis of ions produced by hypervelocity impact. Appl. Phys. Lett. **13**, 89 (1968)
M.G. Hauser, F.C. Gillett, F.J. Low, T.N. Gautier, C.A. Beichman, H.H. Aumann, G. Neugebauer, B. Baud, N. Boggess, J.P. Emerson, IRAS observations of the diffuse infrared background. Astrophys. J. **278**, L15–L18 (1984)
M. Hilchenbach et al., Comet 67P/Churyumov-Gerasimenko: close-up on dust particle fragments. Astrophys. J. **816**, L32 (2016)
J.K. Hillier, S.F. Green, N. McBride, N. Altobelli, F. Postberg, S. Kempf, J. Schwanethal, R. Srama, J.A.M. McDonnell, E. Grün, Interplanetary dust detected by the Cassini CDA chemical analyser. Icarus **190**, 643–654 (2007)
J.K. Hillier, Z. Sternovsky, S.P. Armes, L.A. Fielding, F. Postberg, S. Bugiel, K. Drake, R. Srama, A.T. Kearsley, M. Trieloff, Impact ionisation mass spectrometry of polypyrrole-coated pyrrhotite microparticles. Planet. Space Sci. **97**, 9–22 (2014)
J.K. Hillier, F. Postberg, S. Hsu, *Dust Emission by Active Moons*. Space Science Reviews Book, vol. 69 (2019, in press)
T. Hirai, M.J. Cole, M. Fujii, S. Hasegawa, T. Iwai, M. Kobayashi, R. Srama, H. Yano, Microparticle impact calibration of the arrayed large-area dust detectors in INterplanetary space (ALADDIN) onboard the solar power sail demonstrator IKAROS. Planet. Space Sci. **100**, 87–97 (2014). https://doi.org/10.1016/j.pss.2014.05.009
T. Hirai, H. Yano, M. Fujii, S. Hasegawa, N. Moriyama, C. Okamoto, M. Tanaka, Data screening and reduction in interplanetary dust measurement by IKAROS-ALADDIN. Adv. Space Res. **59**, 1450–1459 (2017)
K. Hirayama, Families of asteroids, in *Annales de'Observatoire de Tokyo. Appendice* vol. 11, (1923), pp. 55–93
H.-J. Hoffmann, H. Fechtig, E. Grün, J. Kissel, Temporal fluctuation and anisotropy of the micrometeoroid flux in the Earth-Moon system. Planet. Space Sci. **23**, 985–991 (1975)
K. Holsapple, I. Giblin, K. Housen, A. Nakamura, E. Ryan, Asteroid impacts: laboratory experiments and scaling laws, in *Proceedings of the ASTEROIDS III* (University of Arizona Press, Tucson, 2002)
M. Horanyi, Charged dust dynamics in the solar system. Annu. Rev. Astron. Astrophys. **34**, 383–418 (1996)
M. Horanyi, G. Morfill, E. Grün, The dusty ballerina skirt of Jupiter. J. Geophys. Res. **98**, A12, (1993)
M. Horanyi et al., The student dust counter on the New Horizons Mission. Space Sci. Rev. **140**, 387–402 (2008)
M. Horányi, Z. Sternovsky, M. Lankton, C. Dumont, S. Gagnard, D. Gathright, E. Grün, D. Hansen, D. James, S. Kempf, B. Lamprecht, R. Srama, J.R. Szalay, G. Wright, The lunar dust experiment (LDEX) onboard the lunar atmosphere and dust environment explorer (LADEE) mission. Space Sci. Rev. **185**(1–4), 93–113 (2014). https://doi.org/10.1007/s11214-014-0118-7
M. Horanyi, D. James, S. Kempf, T. Munsat, Z. Sternovsky, The SSERVI-IMPACT dust accelerator facility at the University of Colorado, in *Lunar and Planetary Science Conference*, vol. 47 (2016), p. 1653
F. Hörz, D.E. Brownlee, H. Fechtig, J.B. Hartung, D.A. Morrison, G. Neukum, E. Schneider, J.F. Vedder, D.E. Gault, Lunar microcraters: implications for the micrometeoroid complex. Planet. Space Sci. **23**, 151–172 (1975)
F. Hörz et al., Impact features on stardust: implications for comet 81P/Wild 2 dust. Science **314**, 1716 (2006)
H.-W. Hsu, K.C. Hansen, M. Horanyi, S. Kempf, A. Mocker, G. Moragas-Klostermeyer, F. Postberg, R. Srama, B. Zieger, Probing IMF using nanodust measurements from inside Saturn's magnetosphere. Geophys. Res. Lett. **40**, 2902–2906 (2013)
H.-W. Hsu et al., Ongoing hydrothermal activities within Enceladus. Nature **519**, 207–210 (2015)

H.-W. Hsu, S. Kempf, S.V. Badman, W.S. Kurth, F. Postberg, R. Srama, Interplanetary magnetic field structure at Saturn inferred from nanodust measurements during the 2013 aurora campaign. Icarus **263**, 10–16 (2016)

H.-W. Hsu et al., In situ collection of dust grains falling from Saturn's rings into its atmosphere. Science **362**, 3185 (2018)

D.H. Humes, Results of Pioneer 10 and 11 meteoroid experiments: interplanetary and near-Saturn. J. Geophys. Res. **85**, 5841–5852 (1980)

D.H. Humes, J.M. Alvarez, R.L. O'Neal, W.H. Kinard, The interplanetary and near-Jupiter meteoroid environments. J. Geophys. Res. **79**(25), 3677–3684 (1974)

K.I. Ibadinov, A.A. Rahmonov, A.S. Bjasso, Laboratory simulation of cometary structures, in *IAU Colloq. 116: Comets in the Post-Halley Era*, vol. 167 (1991), p. 299

R. Jaumann et al., Images from the surface of asteroid Ryugu show rocks similar to carbonaceous chondrite meteorites. Science **365**, 817–820 (2019)

E.K. Jessberger, J. Kissel, Chemical properties of cometary dust and a note on carbon isotopes, in *Comets in the Post-Halley Era* 2, ed. by R.L. Newburn Jr., M. Neugebauer, J. Rahe (Kluwer Academic, Dordrecht, 1991), pp. 1075–1092

E.K. Jessberger, J. Kissel, H. Fechtig, F.R. Krueger, On the average chemical composition of cometary dust, in *Physical Processes in Comets, Stars, and Active Galaxies*, ed. by W. Hillebrandt, E. Meyer-Hofmeister, H.-C. Thomas (Springer, Heidelberg, 1987), pp. 26–33

J. Jones, Meteoroid Engineering Model—Final Report. Space Environments and Effects Program SEE/CR-2004-400, NASA Marshall Space Flight Center (2004)

J. Jones, P. Brown, Sporadic meteor radiant distributions: orbital survey results. Mon. Not. R. Astron. Soc. **265**, 524–532 (1993)

J. Jones, M. Campbell, S. Nikolova, Modelling of the sporadic meteoroid sources, in *Meteoroids 2001 Conference*, vol. 495 (2001), pp. 575–580.

J. Jones, P. Brown, K.J. Ellis, A.R. Webster, M. Campbell-Brown, Z. Krzemenski, R.J. Weryk, The Canadian Meteor Orbit Radar: system overview and preliminary results. Planet. Space Sci. **53**, 413–421 (2005)

B. Jost, A. Pommerol, O. Poch, Z. Yoldi, S. Fornasier, P.H. Hasselmann, C. Feller, N. Carrasco, C. Szopa, N. Thomas, Bidirectional reflectance and VIS-NIR spectroscopy of cometary analogues under simulated space conditions. Planet. Space Sci. **145**, 14–27 (2017)

A. Juhasz, M. Horanyi, Dynamics and distribution of nano-dust particles in the inner solar system. Geophys. Res. Lett. **40**, 2500–2504 (2013)

H.U. Keller, M.L. Marconi, N. Thomas, Hydrodynamic implications of particle fragmentation near cometary nuclei. Astron. Astrophys. **227**, L1–L4 (1990)

P.J. Kellogg, K. Goetz, S.J. Monson, Are STEREO single hits dust impacts? J. Geophys. Res. **123**, 7211–7219 (2018). https://doi.org/10.1029/2018JA025554

T. Kelsall, J.L. Weiland, B.A. Franz, W.T. Reach, R.G. Arendt, E. Dwek, H.T. Freudenreich, M.G. Hauser, S.H. Moseley, N.P. Odegard, R.F. Silverberg, E.L. Wright, The COBE diffuse infrared background experiment search for the cosmic infrared background. II. Model of the interplanetary dust cloud. Astrophys. J. **508**, 44–73 (1998a)

T. Kelsall, J.L. Weiland, B.A. Franz, W.T. Reach, R.G. Arendt, E. Dwek, H.T. Freudenreich, M.G. Hauser, S.H. Moseley, N.P. Odegard, R.F. Silverberg, E.L. Wright, The COBE diffuse infrared background experiment search for the cosmic infrared background. II. Model of the interplanetary dust cloud. Astrophys. J. **508**, 44–73 (1998b)

S. Kempf, Interpretation of high rate dust measurements with the Cassini dust detector CDA. Planet. Space Sci. **56**, 378–385 (2008)

S. Kempf, The Surface Dust Analyzer (SUDA) on Europa Clipper, in *European Planetary Science Congress*, vol. 12 (2018), p. 462

D.J. Kessler, B.G. Cour-Palais, Collision frequency of artificial satellites: the creation of a debris belt. J. Geophys. Res. **83**, 2637–2646 (1978)

J. Kissel, F.R. Krueger, The organic component in dust from comet Halley as measured by the PUMA mass spectrometer on board Vega 1. Nature **326**, 755–760 (1987)

J. Kissel, D.E. Brownlee, K. Büchler, C.B. Clark, H. Fechtig, E. Grün, K. Hornung, E.B. Igenbergs, F.R. Jessberger Kruger, H. Kuczera, E.K Mc Donnell, G.E. Morfill, J. Rahe, G.H. Schwehm, Z. Zekanina, N.G. Utterback, H. Völk, H.A. Zook, Composition of comet Halley dust particles from Giotto observations Nature **321**, 336–338 (1986a)

J. Kissel, R.Z. Sagdeev, J.L. Bertaux, V.N. Angarov, J. Audouze, J.E. Blamont, K. Büchler, H. von Hoerner, N.A. Inogamov, V.N. Khromov, W. Knabe, F.R. Krueger, Y. Langevin, A.C. Levasseur-Regourd, G.G. Managadze, S.N. Podkolzin, V.D. Sharipo, S.R. Tabaldyev, B.V. Zubkov, Composition of comet Halley dust particles from VEGA observations. NATURE **321**(6067), 280–282 (1986b)

J. Kissel, F.R. Krueger, J. Silen, B.C. Clark, The cometary and interstellar dust analyzer at comet 81P/Wild 2. Science **304**, 1774–1776 (2004)

J. Kissel, F.R. Krueger, J. Silen, Analysis of cosmic dust by the 'Cometary and Interstellar Dust Analyser' (CIDA) onboard the stardust spacecraft, in *Dust in Planetary Systems*, vol. 12080 (2005), p. 95

K. Knie, G. Korschinek, T. Faestermann, E.A. Dorfi, G. Rugel, A. Wallner, ^{60}Fe anomaly in a deep-sea manganese crust and implications for a nearby supernova source. Phys. Rev. Lett. **93**, 171103 (2004). https://doi.org/10.1103/PhysRevLett.93.171103

M. Kobayashi, H. Krüger, H. Senshu, K. Wada, O. Okudaira, S. Sasaki, H. Kimura, In situ observations of dust particles in Martian dust belts using a large-sensitive-area dust sensor. Planet. Space Sci. **10**, 1 (2018a). https://doi.org/10.1016/j.pss.2017.12.011

M. Kobayashi, R. Srama, H. Krüger, T. Arai, H. Kimura, Destiny + dust analyser, in *Lunar and Planetary Institute Science Conference Abstracts*, vol. 49 (2018b). 2050

H. Kochan, L. Ratke, K. Thiel, E. Grün, Particle emission from artificial cometary surfaces: material science aspects. Proc. Lunar Planet. Sci. Conf. **20**, 401 (1990)

H. Kochan, W.F. Huebner, D.W.G. Sears, Simulation experiments with cometary analogous material. Earth Moon Planets **80**, 369 (1998)

N.I. Kömle, G. Kargl, K. Thiel, K. Seiferlin, Thermal properties of cometary ices and sublimation residua including organics. Planet. Space Sci. **44**, 675–689 (1996)

D. Koschny, E. Grün, Impacts into ice-silicate mixtures: ejecta mass and size distributions. Icarus **154**, 402–411 (2001)

D. Koschny, R.H. Soja, C. Engrand, G.J. Flynn, J. Lasue, T. Nakamura, D. Malaspina, A.C. Levasseur-Regourd, A.R. Poppe, V.J. Sterken, J.M. Trigo-Rodriguez, *Interplanetary Dust, Meteoroids, Meteors and Meteorites*. Space Science Reviews Book, vol. 69 (2019, in press)

A.V. Krivov, H. Krüger, E. Grün, K.-U. Thiessenhusen, D.P. Hamilton, A tenuous dust ring of Jupiter formed by escaping ejecta from the Galilean satellites. J. Geophys. Res., Planets **107**, 5002-1 (2002)

H. Krüger, A.V. Krivov, M. Sremcevic, E. Grün, Impact-generated dust clouds surrounding the Galilean moons. Icarus **164**, 170–187 (2003)

H. Krüger, A.L. Graps, D.P. Hamilton, A. Flandes, R.J. Forsyth, M. Horanyi, E. Grün, Ulysses jovian latitude scan of high-velocity dust streams originating from the jovian system. Planet. Space Sci. **54**, 919–931 (2006)

H. Krüger, M. Landgraf, N. Altobelli, E. Grün, Interstellar dust in the solar system. Space Sci. Rev. **130**, 401–408 (2007)

H. Krüger, D.P. Hamilton, R. Moissl, E. Grün, Galileo in-situ dust measurements in Jupiter's gossamer rings. Icarus **203**, 198–213 (2009)

H. Krüger et al., Galileo dust data from the jovian system: 2000 to 2003. Planet. Space Sci. **58**, 965–993 (2010)

H. Krüger, P. Strub, E. Grün, V.J. Sterken, Sixteen years of Ulysses interstellar dust measurements in the solar system. I. Mass distribution and gas-to-dust mass ratio. Astrophys. J. **812**, 139 (2015a)

H. Krüger, K.J. Seidensticker, H.-H. Fischer, T. Albin, I. Apathy, W. Arnold, A. Flandes, A. Hirn, M. Kobayashi, A. Loose, A. Peter, M. Podolak, Dust impact monitor (SESAME-DIM) measurements at comet 67P/Churyumov-Gerasimenko. Astron. Astrophys. **583**, A15 (2015b). https://doi.org/10.1051/0004-6361/201526400

H. Krüger, P. Strub, N. Altobelli, V.J. Sterken, R. Srama, E. Grün, Interstellar dust in the solar system: model versus in-situ spacecraft data. Astron. Astrophys. **626**, A37 (2019). https://doi.org/10.1051/0004-6361/201834316

M. Landgraf, Modeling the motion and distribution of interstellar dust inside the heliosphere. J. Geophys. Res. **105**, 10303–10316 (2000)

Y. Langevin, M. Hilchenbach, N. Ligier, S. Merouane, K. Hornung, C. Engrand, R. Schulz, J. Kissel, J. Rynö, P. Eng, Typology of dust particles collected by the COSIMA mass spectrometer in the inner coma of 67P/Churyumov-Gerasimenko. Icarus **271**, 76–97 (2016)

Ch. Leinert, E. Grün, in *Interplanetary Dust, in Physics of the Inner Heliosphere I*, ed. by R. Schwenn, E. Marsch (Springer, Berlin, 1990), pp. 207–275

Ch. Leinert, I. Richter, E. Pitz, B. Planck, The zodiacal light from 1.0 to 0.3 AU as observed by the Helios space probes. Astron. Astrophys. **103**, 177–188 (1981)

J. Leitner, P. Hoppe, A new population of dust from stellar explosions among meteoritic stardust. Nat. Astron. (2019). https://doi.org/10.1038/s41550-019-0788-x

A.-C. Levasseur-Regourd, J. Agarwal, H. Cottin, C. Engrand, G. Flynn, M. Fulle, T. Gombosi, Y. Langevin, J. Lasue, T. Mannel, S. Merouane, O. Poch, N. Thomas, A. Westphal, *Cometary Dust*. Space Science Reviews Book, vol. 69 (2019, in press)

S.G. Love, D.E. Brownlee, A direct measurement of the terrestrial mass accretion rate of cosmic dust. Science **262**, 550–553 (1995)

S.G. Love, F. Hörz, D.E. Brownlee, Target porosity effects in impact cratering and collisional disruption. Icarus **105**, 216–224 (1993)
D.M. Malaspina, L.E. O'Brien, F. Thayer, Z. Sternovsky, A. Collette, Revisiting STEREO interplanetary and interstellar dust flux and mass estimates. J. Geophys. Res. Space Phys. **120**, 6085–6100 (2015)
B.A. Mamyrin, V.I. Karatyev, D.V. Shmikk, V.A. Zagulin, Mass-reflectron—a new high-resolution nonmagnetic time-of-flight mass-spectrometer. Pis'ma Zh. Eksp. Teor. Fiz. **64**, 82 (1973), or: Sov. Phys. JETP **37**(1)
I. Mann et al., Dust near the Sun. Space Sci. Rev. **110**, 269–305 (2004)
E.P. Mazets et al., Dust in comet P/Halley from VEGA observations. Astron. Astrophys. **187**, 699 (1987)
J.A.M. McDonnell (ed.), *Cosmic Dust* (Wiley, Chichester, 1978)
J.A.M. McDonnell, The Giotto dust impact detection system. J. Phys. E, Sci. Instrum. **20**, 741–758 (1987)
J.A.M. McDonnell, D.J. Gardner, Meteoroid morphology and densities: decoding satellite impact data. Icarus **133**, 25–35 (1998)
J.A.M. McDonnell et al., The impact of dust grains on fast flyby spacecraft: momentum multiplication measurements and theory. Adv. Space Res. **4**(9), 297–301 (1984)
J.A.M. McDonnell, G.C. Evans, S.T. Evans, W.M. Alexander, W.M. Burton, J.G. Firth, E. Bussoletti, R.J.L. Grard, M.S. Hanner, Z. Sekanina, The dust distribution within the inner coma of comet P/Halley 1982i—encounter by Giotto's impact detectors. Astron. Astrophys. **187**, 719–741 (1987)
J.A.M. McDonnell, S.F. Green, E. Grün, J. Kissel, S. Nappo, G.S. Pankiewicz, C.H. Perry, In situ exploration of the dusty coma of comet P/Halley at Giotto's encounter—flux rates and time profiles from 10 to the −19th KG to 10 to the −5th KG. Adv. Space Res. **9**, 277–280 (1989)
J.A.M. McDonnell, N. McBride, R. Beard, E. Bussoletti, L. Colangeli, P. Eberhardt, J.G. Firth, R. Grard, S.F. Green, J.M. Greenberg, Dust particle impacts during the Giotto encounter with comet Grigg-Skjellerup. Nature **362**, 6422 (1993). https://doi.org/10.1038/362732a0
A. McEwen, L. Keszthelyi, R. Lopes, P. Schenk, J. Spencer, The lithosphere and surface of Io, in *Jupiter: Planet, Satellites & Magnetosphere*, ed. by F. Bagenal, B. McKinnon, T. Dowling (Cambridge University Press, Cambridge, 2004), pp. 307–328
H. McNamara, J. Jones, B. Kauffman, R. Suggs, W. Cooke, S. Smith, Meteoroid engineering model (MEM): a meteoroid model for the inner solar system. Earth Moon Planets **95**(1–4), 123–139 (2004). https://doi.org/10.1007/s11038-005-9044-8
S. Merouane et al., Dust particle flux and size distribution in the coma of 67P/Churyumov-Gerasimenko measured in situ by the COSIMA instrument on board Rosetta. Astron. Astrophys. **596**, A87 (2016)
N. Meyer-Vernet, M. Maksimovic, A. Czechowski, I. Mann, I. Zouganelis, K. Goetz, M.L. Kaiser, O.C. St. Cyr, J.-L. Bougeret, S.D. Bale, Dust detection by the wave instrument on STEREO: nanoparticles picked up by the solar wind? Sol. Phys. **256**, 463–474 (2009)
G.E. Morfill, E. Grün, The motion of charged dust particles in interplanetary space—I. The zodiacal dust cloud. Planet. Space Sci. **27**, 1269–1282 (1979a)
G.E. Morfill, E. Grün, The motion of charged dust particles in interplanetary space—II. Interstellar grains. Planet. Space Sci. **27**, 1283–1292 (1979b)
G.E. Morfill, E. Grün, C.K. Goertz, T.V. Johnson, On the evolution of Saturn's 'spokes'—theory. Icarus **53**, 230–235 (1983)
D.A. Morrison, U.S. Clanton, Properties of microcraters and cosmic dust of less than 100 Ådimensions, in *Proc. Lunar Sci. Conf. 10th* (1979), p. 1649
D.A. Morrison, E. Zinner, 12054 and 76215: new measurements of interplanetary dust and solar flare fluxes, in *Proc. Lunar Sci. Conf. 8th*, (1977), pp. 841–863
J. Murray, M. Renard, J. Murray, On the microscopic characters of volcanic ashes and cosmic dust, and their distribution in the deep sea deposits. Proc. R. Soc. Edinb. **12**, 474–495 (1884). https://doi.org/10.1017/S0370164600000924
T. Nakamura et al., Itokawa dust particles: a direct link between S-type asteroids and ordinary chondrites. Science **333**, 1113 (2011)
A.M. Nakamura, F. Yamane, T. Okamoto, S. Takasawa, Size dependence of the disruption threshold: laboratory examination of millimeter-centimeter porous targets. Planet. Space Sci. **107**, 45–52 (2015)
D. Nesvorny, W.F. Bottke, H.F. Levison, L. Dones, Recent origin of the solar system dust bands. Astrophys. J. **591**, 486–497 (2003)
D. Nesvorny, P. Jenniskens, H.F. Levison, W.F. Bottke, D. Vokrouhlicky, M. Gounelle, Cometary origin of the zodiacal cloud and carbonaceous micrometeorites. Implications for hot debris disks. Astrophys. J. **713**, 816–836 (2010)
D. Nesvorny, D. Janches, D. Vokrouhlicky, P. Pokorny, W.F. Bottke, P. Jenniskens, Dynamical model for the zodiacal cloud and sporadic meteors. Astrophys. J. **743**, 129 (2011a)
D. Nesvorny, D. Vokrouhlicky, P. Pokorny, D. Janches, Dynamics of dust particles released from oort cloud comets and their contribution to radar meteors. Astrophys. J. **743**, 37 (2011b)

C. Nilsson, Some doubts about the Earth's dust cloud. Science **153**, 1242–1246 (1966)
K. Nogami et al., Development of the Mercury dust monitor (MDM) onboard the BepiColombo mission. Planet. Space Sci. **58**, 108–115 (2010)
L. O'Brien, S. Auer, A. Gemer, E. Grün, M. Horanyi, A. Juhasz, S. Kempf, D. Malaspina, A. Mocker, E. Moebius, R. Srama, Z. Sternovsky, Development of the nano-dust analyzer (NDA) for detection and compositional analysis of nanometer-size dust particles originating in the inner heliosphere. Rev. Sci. Instrum. **85**, 035113 (2014). https://doi.org/10.1063/1.4868506
T. Okada, Y. Kebukawa, J. Aoki, J. Matsumoto, H. Yano, T. Iwata, O. Mori, J.-P. Bibring, S. Ulamec, R. Jaumann, Science exploration and instrumentation of the OKEANOS mission to a Jupiter Trojan asteroid using the solar power sail. Planet. Space Sci. **161**, 99–106 (2018)
H.W. Oliver, J.G. Moore, R.F. Sikora, Internal structure of the Sierra Nevada batholith based on specific gravity and gravity measurements. Geophys. Res. Lett. **20**, 2179–2182 (1993)
E.J. Öpik, Collision probabilities with the planets and the distribution of interplanetary matter. Proc. R. Ir. Acad., A Math. Phys. Sci. **54**, 165–199 (1951)
N. Pailer, E. Gruen, The penetration limit of thin films. Planet. Space Sci. **28**(3), 321–331 (1980). https://doi.org/10.1016/0032-0633(80)90021-5
J.M. Pintado, J. Burchell, H. Cottin, E. Dartois, I. Loes ten Kate, S. Kwok, P. Lavvas, G. Munoz Caro, F. Postberg, *Astrobiology and Dust*. Space Science Reviews Book, vol. 69 (2019, in press)
M. Piquette, A. Poppe, E. Bernardoni, J. Szalay, D. James, M. Horanyi, A. Stern, H. Weaver, J. Spencer, C. Olkin, Student dust counter: status report at 38 AU. Icarus **321**, 116–125 (2019)
J.M.C. Plane, G.J. Flynn, A. Määttänen et al., Space Sci. Rev. **214**, 23 (2018). https://doi.org/10.1007/s11214-017-0458-1
O. Poch, A. Pommerol, B. Jost, N. Carrasco, C. Szopa, N. Thomas, Sublimation of water ice mixed with silicates and tholins: evolution of surface texture and reflectance spectra, with implications for comets. Icarus **267**, 154–173 (2016)
P. Pokorny, M. Sarantos, D. Janches, A comprehensive model of the meteoroid environment around Mercury. Astrophys. J. **863**, doi.org/10.3847/1538–4357/aad051 (2018)
A.R. Poppe, An improved model for interplanetary dust fluxes in the outer Solar System. Icarus **264**, 369–386 (2016)
A. Poppe, D. James, B. Jacobsmeyer, M. Horanyi, First results from the Venetia Burney Student Dust Counter on the New Horizons mission. Geophys. Res. Lett. **37**, L11101 (2010)
A. Poppe, D. James, M. Horanyi, Measurements of the terrestrial dust influx variability by the cosmic dust experiment. Planet. Space Sci. **59**, 319–326 (2011)
F. Postberg, S. Kempf, J. Schmidt, N. Brilliantov, A. Beinsen, B. Abel, U. Buck, R. Srama, Sodium salts in E-ring ice grains from an ocean below the surface of Enceladus. Nature **459**, 1098 (2009)
F. Postberg, E. Grün, M. Horanyi, S. Kempf, H. Krüger, J. Schmidt, F. Spahn, R. Srama, Z. Sternovsky, M. Trieloff, Compositional mapping of planetary moons by mass spectrometry of dust ejecta. Planet. Space Sci. **59**, 1815–1825 (2011)
M.S. Prasad, N.G. Rudraswami, A.A. de Araujo, V.D. Khedekar, Unmelted cosmic metal particles in the Indian Ocean. Meteorit. Planet. Sci. **52**, 1060–1081 (2017)
Yu.P. Raizer, Residual ionization of a gas expanding in vacuum. Sov. Phys. JETP **10**, 411–416 (1960)
W.T. Reach, M.S. Kelley, M.V. Sykes, A survey of debris trails from short-period comets. Icarus **191**, 298–322 (2007)
W.T. Reach, J. Vaubaillon, M.S. Kelley, C.M. Lisse, M.V. Sykes, Distribution and properties of fragments and debris from the split comet 73P/Schwassmann-Wachmann 3 as revealed by Spitzer Space Telescope. Icarus **203**, 571–588 (2009)
J.J. Rennilson, D.R. Criswell, Surveyor observations of lunar horizon-glow. Moon **10**, 121–142 (1974)
M.S. Robinson, P.C. Thomas, J. Veverka, S. Murchie, B. Carcich, The nature of ponded deposits on Eros. Nature **413**, 396–400 (2001)
S.A. Sandford et al., Assessment and control of organic and other contaminants associated with the stardust sample return from comet 81P/Wild 2. Meteorit. Planet. Sci. **45**, 406–433 (2010)
S. Sasaki et al., Production of iron nanoparticles by laser irradiation in a simulation of lunar-like space weathering. Nature **410**(6828), 555-7 (2001)
J.P. Schwanenthal et al., in *Proceedings of the 4th European Conference on Space Debris (ESA SP-587)*, ESA/ESOC, Darmstadt, Germany, 18–20 April 2005, ed. by D. Danesy (2005), p. 177
D.W.G. Sears, H.W. Kochan, W.F. Huebner, Invited review: laboratory simulation of the physical processes occurring on and near the surfaces of comet nuclei. Meteorit. Planet. Sci. **34**, 497–525 (1999)
K.J. Seidensticker et al., Sesame—an experiment of the Rosetta Lander Philae: objectives and general design. Space Sci. Rev. **128**, 301–337 (2007)
M.A. Sephton, Organic compounds in carbonaceous meteorites. Nat. Prod. Rep. **19**, 292–311 (2002)

I.C. Shapiro, D.A. Lautman, G. Colombo, The Earth's dust belt: fact or fiction? 1. Forces perturbing dust particle motion. J. Geophys. Res. **71**, 5695–5704 (1966)
M.R. Showalter, D.P. Hamilton, P.D. Nicholson, A deep search for Martian dust rings and inner moons using the Hubble Space Telescope. Planet. Space Sci. **54**, 844–854 (2006)
M.R. Showalter I. de Pater, G. Verbanac, D.P. Hamilton, J.A. Burns, Properties and dynamics of Jupiter's Gossamer Rings from Galileo, Voyager, Hubble and Keck images. Icarus **195**, 361–377 (2008). https://doi.org/10.1016/j.icarus.2007.12.012
A. Shu, S. Bugiel, E. Grün, J. Hillier, M. Horányi, T. Munsat, R. Srama, Cratering studies in polyvinylidene fluoride (PVDF) thin films. Planet. Space Sci. **89**, 29–35 (2013)
J.A. Simpson, A.J. Tuzzolino, Polarized polymer films as electronic pulse detectors of cosmic dust particles. Nucl. Instrum. Methods Phys. Res. A **236**, 187–202 (1985)
J.A. Simpson et al., Dust counter and mass analyser (DUCMA) measurements of comet Halley's coma from VEGA spacecraft. Nature **321**, 278–280 (1986)
B.A. Smith et al., The Jupiter system through the eyes of Voyager 1. Science **204**, 951–957 (1979a)
B.A. Smith et al., The Galilean satellites and Jupiter—Voyager 2 imaging science results. Science **206**, 927–950 (1979b)
B.A. Smith et al., Encounter with Saturn—Voyager 1 imaging science results. Science **212**, 163–191 (1981)
B.A. Smith et al., A new look at the Saturn system—the Voyager 2 images. Science **215**, 504–537 (1982)
B.A. Smith et al., Voyager 2 in the Uranian system—imaging science results. Science **233**, 43–64 (1986)
B.A. Smith et al., Voyager 2 at Neptune: imaging science results. Science **246**, 1422–1449 (1989)
R.K. Soberman, S.L. Neste, K. Lichtenfeld, Particle concentration in the asteroid belt from Pioneer 10. Science **183**, 320–321 (1974)
R.H. Soja, M. Sommer, J. Herzog, R. Srama, E. Grün, J. Rodmann, P. Strub, J. Vaubaillon, A. Hornig, L. Bausch, The interplanetary meteoroid environment for eXploration—(IMEX) project, in *Proceedings of the International Meteor Conference*, Giron, France, 18–21 September 2014 (2014), pp. 146–149
R.H. Soja et al., Meteor storms and showers with the IMEX model, in *Proceedings of the International Meteor Conference*, Mistelbach, Austria (2015b), pp. 66–69
R.H. Soja, M. Sommer, J. Herzog, J. Agarwal, J. Rodmann, R. Srama, J. Vaubaillon, P. Strub, A. Hornig, L. Bausch, E. Grün, Characteristics of the dust trail of 67P/Churyumov-Gerasimenko: an application of the IMEX model. Astron. Astrophys. **583**, A18 (2015a). https://doi.org/10.1051/0004-6361/201526184
R.H. Soja, G.J. Schwarzkopf, M. Sommer, J. Vaubaillon, T. Albin, J. Rodmann, E. Grün, R. Srama, Collisional lifetimes of meteoroids, in *International Meteor Conference*, Egmond, The Netherlands, 2–5 June 2016 (2016), pp. 284–286.
R.H. Soja, E. Grün, P. Strub, T. Albin, M. Millinger, J. Vaubaillon, W. Alius, G. Camodeca, F. Hein, J. Laskar, M. Gastineau, M. Sommer, G.J. Schwarzkopf, J. Herzog, K. Gutsche, N. Skuppin, R. Srama, IMEM2: a meteoroid environment model for the inner solar system. Astron. Astrophys. **628**, A109 (2019). https://doi.org/10.1051/0004-6361/201834892
M. Solc, J. Kissel, V. Vanysek, Carbon-isotope ratio in PUMA 1 spectra of P/Halley dust. Astron. Astrophys. **187**, 385–387 (1987)
S. Soter, Dust Belts of Mars. Ph.D. Thesis, Cornell University (1971)
R.B. Southworth, Z. Sekanina, Physical and dynamical studies of meteors. NASA CR-2316 (1973)
F. Spahn, M. Sachse, H.-W. Hsu, S. Kempf, M. Horanyi, *Circumplanetary Dust Populations*. Space Science Reviews Book, vol. 69 (2019, in press)
R. Srama et al., The Cassini cosmic dust analyzer. Space Sci. Rev. **114**, 465–518 (2004)
R. Srama, M. Rachev, A. Srowig, V. Dikarev, S. Helfert, S. Kempf, D. Linkert, G. Moragas-Klostermeyer, E. Grün, Performance of an advanced dust telescope, in *4th European Conference on Space Debris*, vol. 587 (2005), pp. 171–178
P. Staubach, E. Grün, R. Jehn, The meteoroid environment near Earth. Adv. Space Res. **19**, 301–308 (1997)
P. Staubach, E. Grün, M. Matney, in *Interplanetary Dust* ed. by E. Grün, B.A.S. Gustafson, S. Dermott, H. Fechtig (Springer, Heidelberg, 2001), pp. 347–384
V.J. Sterken, N. Altobelli, S. Kempf, G. Schwehm, R. Srama, E. Grün, The flow of interstellar dust into the solar system. Astron. Astrophys. **538**, A102 (2012)
V.J. Sterken, N. Altobelli, S. Kempf, H. Krüger, R. Srama, P. Strub, E. Grün, The filtering of interstellar dust in the solar system. Astron. Astrophys. **552**, A130 (2013)
V.J. Sterken et al., Stardust interstellar preliminary examination X: impact speeds and directions of interstellar grains on the stardust dust collector. Meteorit. Planet. Sci. **49**, 1680–1697 (2014)
V.J. Sterken, P. Strub, H. Krüger, R. von Steiger, P. Frisch, Sixteen years of Ulysses interstellar dust measurements in the solar system. III. Simulations and data unveil new insights into local interstellar dust. Astrophys. J. **812**, 141 (2015)
V.J. Sterken, A.J. Westphal, N. Altobelli, D. Malaspina, F. Postberg, *Interstellar Dust in the Solar System*. Space Science Reviews Book, vol. 69 (2019, in press)

S.A. Stern, The New Horizons Pluto Kuiper Belt mission: an overview with historical context. Space Sci. Rev. **140**, 3–21 (2008)

Z. Sternovsky, E. Grün, K. Drake, J. Xie, M. Horanyi, R. Srama, S. Kempf, F. Postberg, A. Mocker, S. Auer, H. Krüger, Novel instrument for dust astronomy: dust telescope, in *Aerospace Conference*, 2011 IEEE, 5–12 March 2011 (2011), pp. 1–8. https://doi.org/10.1109/AERO.2011.5747300.

D. Stöffler, H. Düren, J. Knölker, R. Hische, A. Bischoff, Cometary analogue material: preparation, composition, and thin section petrography. Geophys. Res. Lett. **18**, 285 (1991)

P. Strub, H. Krüger, V.J. Sterken, Sixteen years of Ulysses interstellar dust measurements in the solar system. II. Fluctuations in the dust flow from the data. Astrophys. J. **812**, 140 (2015)

P. Strub, V.J. Sterken, R. Soja, H. Krüger, E. Grün, R. Srama, Heliospheric modulation of the interstellar dust flow on to Earth. Astron. Astrophys. **621**, A54 (2019). https://doi.org/10.1051/0004-6361/201832644

S. Sugita et al., The geomorphology, color, and thermal properties of Ryugu: implications for parent-body process. Science **364**, eaaw0422 (2019)

M.V. Sykes, L.A. Lebovsky, D.M. Hunten, F. Low, The discovery of dust trails in the orbits of periodic comets. Science **232**, 1115–1117 (1986)

J.R. Szalay, M. Horanyi, Lunar meteoritic gardening rate derived from in situ LADEE/LDEX measurements. Geophys. Res. Lett. **43**, 4893–4898 (2016a)

J.R. Szalay, M. Horanyi, Detecting meteoroid streams with an in-situ dust detector above an airless body. Icarus **275**, 221–231 (2016b)

J.R. Szalay, A. Poppe, *Dust Phenomena Relating to Airless Bodies*. Space Science Reviews Book, vol. 69 (2019, in press)

A.D. Taylor, W.G. Elford, Meteoroid orbital element distributions at 1 AU deduced from the Harvard Radio Meteor Project observations. Earth Planets Space **50**, 569–575 (1998). https://doi.org/10.1186/BF03352150

E. Thomas, S. Auer, K. Drake, M. Horányi, T. Munsat, A. Shu, FPGA cross-correlation filters for real-time dust detection and selection. Planet. Space Sci. **89**, 71–76 (2013)

E. Thomas, M. Horanyi, D. Janches, T. Munsat, J. Simolka, Z. Sternovsky, Measurements of the ionization coefficient of simulated iron micrometeoroids. Geophys. Res. Lett. **43**, 3645–3652 (2016)

T. Timmermann, E. Grün, Plasma emission from high velocity impacts of microparticles onto water ice, in *IAU Colloq. 126: Origin and Evolution of Interplanetary Dust*, vol. 173 (1991), p. 375

R.J. Trümpler, Absorption of light in the galactic system. Publ. Astron. Soc. Pac. **42**, 214 (1930)

A.J. Tuzzolino, T.E. Economou, B.C. Clark, P. Tsou, D.E. Brownlee, S.F. Green, J.A.M. McDonnell, N. McBride, M.T.S.H. Colwell, Dust measurements in the coma of comet 81P/Wild 2 by the dust flux monitor instrument. Science **304**, 1776–1780 (2004)

A.J. Tuzzolino, T.E. Economou, R.B. McKibben, J.A. Simpson, S. BenZvi, L. Blackburn, H.D. Voss, H. Gursky, Final results from the space dust (SPADUS) instrument flown aboard the Earth-orbiting ARGOS spacecraft. Planet. Space Sci. **53**, 903–923 (2005)

M.E. van den Ancker, Circumstellar material in young stellar objects. PhD thesis, University of Amsterdam, The Netherlands (1999)

J.-B. Vincent et al., Summer fireworks on comet 67P. Mon. Not. R. Astron. Soc. **462**, S184–S194 (2016)

V. Vojacek, J. Borovicka, P. Koten, P. Spurný, R. Štork, Catalogue of representative meteor spectra. Astron. Astrophys. **580**, A67 (2015)

X. Wang, J. Schwan, H.-W. Hsu, E. Grün, M. Horanyi, Dust charging and transport on airless planetary bodies. Geophys. Res. Lett. **43**, 6103–6110 (2016)

S. Watanabe et al., Hayabusa 2 arrives at the carbonaceous asteroid 162173 Ryugu—a spinning top-shaped rubble pile. Science **364**, 268–272 (2019)

A. Wehry, H. Krüger, E. Grün, Analysis of Ulysses data: radiation pressure effects of dust particles. Astron. Astrophys. **419**, 1169–1174 (2004). https://doi.org/10.1051/0004-6361:20035613

J.L. Weinberg, M.S. Hanner, D.E. Beeson, L.M. DeShields, B.A. Green, Background starlight observed from Pioneer 10. J. Geophys. Res. **76**, 3665–3670 (1974)

A.J. Westphal et al., Evidence for interstellar origin of seven dust particles collected by the stardust spacecraft. Science **345**, 786–791 (2014)

F.L. Whipple, Comets, meteors and the interplanetary complex. Astron. J. **54**, 179 (1949)

F.L. Whipple, A comet model. I. The acceleration of comet Encke. Astrophys. J. **111**, 375–394 (1950)

F.L. Whipple, A comet model. II. Physical relations for comets and meteors. Astrophys. J. **113**, 464 (1951)

F.L. Whipple, A comet model. III. The zodiacal light. Astrophys. J. **121**, 750 (1955)

F.L. Whipple, On maintaining the meteoritic complex, in *Zodiacal Light and the Interplanetary Medium*. NASA-SP, vol. 150 (1967), pp. 409–426

S.P. Wyatt, F.L. Whipple, The Poynting-Robertson effect on meteor orbits. Astrophys. J. **111**, 134–141 (1950)

H. Yano, T. Hirai, C. Okamoto, M. Fujii, M. Tanaka, N. Moriyama, M. Burchell (IKAROS-ALADDIN Team), Heliocentric variation of cosmic dust flux measured by the IKAROS-ALADDIN between the Earth and Venus, in *Lunar Planet. Sci. Conf.*, The Woodlands, TX, vol. 44 (2013), #2743

S.-Y. Ye, D.A. Gurnett, W.S. Kurth, T.F. Averkamp, S. Kempf, H.-W. Hsu, R. Srama, E. Grün, Properties of dust particles near Saturn inferred from voltage pulses induced by dust impacts on Cassini spacecraft. J. Geophys. Res. Space Phys. **119**, 6294–6312 (2014)

M.E. Zolensky et al., Mineralogy and petrology of comet 81P/Wild 2 nucleus samples. Science **314**, 1735 (2006)

H.A. Zook, O.E. Berg, A source for hyperbolic cosmic dust particles. Planet. Space Sci. **23**, 183–203 (1975)

H.A. Zook, E. Grün, M. Baguhl, D. Hamilton, G. Linkert, J.C. Liou, R. Forsyth, J.L. Phillips, Solar wind magnetic field bending of jovian dust trajectories. Science **274**, 1501–1503 (1996)

Space Sci Rev (2019) 215:43
https://doi.org/10.1007/s11214-019-0607-9

Interstellar Dust in the Solar System

Veerle J. Sterken[1,2] · Andrew J. Westphal[3] · Nicolas Altobelli[4] · David Malaspina[5] · Frank Postberg[6,7]

Received: 18 April 2018 / Accepted: 31 July 2019 / Published online: 9 October 2019

Abstract Interstellar dust from the Local Interstellar Cloud was detected unambiguously for the first time in 1992 (Grün et al. in Nature 362:428–430, 1993). Since then, great progress has been made in observing local interstellar dust in the Solar System using a variety of methods that, all together, provide complementary views of the dust particles from our local galactic neighborhood. The complementary methods discussed in this paper are: (1) *in situ* observations with dust detectors, (2) sample return, (3) observations of dust in the infrared, and (4) detections using spacecraft antennae. We review the current state of the art of *local interstellar dust* research, with a special focus on the advances made in the last ∼10 years of interstellar dust research. We introduce this paper with an overview of the definitions of interstellar dust. We describe the dynamics of the dust particles moving through the heliosphere and report on the progress made in the modelling efforts especially in the last decade. We also review the currently available *in situ* measurements of interstellar dust flux, speed, direction and size distribution from various missions, in specific from Ulysses and Cassini, and their interpretation in context of the dust dynamics studies. Interstellar dust composition is also reviewed from Cassini in situ time of flight measurements and from the Stardust sample return mission that both took place in the last decade. Finally,

Cosmic Dust from the Laboratory to the Stars
Edited by Rafael Rodrigo, Jürgen Blum, Hsiang-Wen Hsu, Detlef Koschny, Anny-Chantal Levasseur-Regourd, Jesús Martín-Pintado, Veerle Sterken and Andrew Westphal

✉ V.J. Sterken
info@veerlesterken.ch

1 Astronomisches Institut Universität Bern (AIUB), Bern, Switzerland

2 Institute of Applied Physics (IAP), University of Bern, Bern, Switzerland

3 Space Sciences Laboratory, University of California at Berkeley, Berkeley, USA

4 European Space Agency, ESAC, Madrid, Spain

5 LASP, University of Colorado, Boulder, USA

6 Institute of Earth Sciences, University of Heidelberg, Heidelberg, Germany

7 Institute of Geological Sciences, Freie Universität Berlin, Berlin, Germany

also new dust measurements from spacecraft antennae are reviewed. The paper concludes with a discussion on currently still open questions, and an outlook for the future.

Keywords Dust · ISM · LIC · Meteoroids · Zodiacal dust · Interplanetary medium · Interstellar dust

1 Introduction

Because of ambiguity in the literature, we begin this review with a definition of terms. A *rock* is an object composed of one or more minerals or organic phases—this is the usual geological definition. *Interstellar dust* (ISD) consists of tiny building blocks that formed anywhere outside the solar system. We make no attempt to define tiny in this context, but perhaps a loose definition is an upper limit in diameter of ~1 mm. *Interstellar Medium (ISM)-formed* dust consists of rocks that formed primarily in the interstellar medium. *Circumstellar* dust consists of rocks that formed primarily in circumstellar environments, generally in circumstellar outflows. Thus, interstellar dust consists of a mixture of circumstellar and ISM-formed dust. The definitions are not entirely orthogonal: ISM-formed dust may contain circumstellar components, or may consist of heavily modified circumstellar precursors, or both. These definitions are also not entirely satisfying—for example, ISM-formed dust could more properly be called "interstellar" dust, and interstellar dust could be called "extrasolar" dust—but we retain these definitions to maintain consistency with the nomenclature in the literature, which developed historically.

The motion of the solar system through the Local Interstellar Cloud (LIC[1]) of warm and partially-ionized gas and dust causes part of this dust to pass through the solar system, which enables us to study such "extraordinary" dust in situ, thus *nearby*. Observations of interstellar dust in the solar system have been made, so far, in five contexts.

1. Dust detectors on board the Galileo (Baguhl et al. 1996; Altobelli et al. 2005), Ulysses (Grün et al. 1993, 1994; Baguhl et al. 1996; Landgraf et al. 2000; Krüger et al. 2015; Strub et al. 2015; Frisch et al. 1999) and Cassini (Altobelli et al. 2003, 2016) spacecraft have made unambiguous detections of interstellar dust inside the heliosphere.[2] These are collectively the *in-situ ISD* observations (Sects. 3 and 4).
2. Circumstellar dust has been preserved in primitive extraterrestrial materials, in the form of so-called *pre-solar grains*. These have been identified in the matrices of primitive meteorites, in so-called *Interplanetary Dust Particles* (IDPs) collected in the stratosphere, and in micrometeorites collected from ice in Greenland and Antarctica. Because they are recognized by their highly non-solar isotopic compositions, pre-solar grains are samples of circumstellar dust: we do not yet know of an observational signature that allows us to recognize ISM-formed dust in solar-system materials. However, several lines of evidence (oxygen isotopes, crystalline fraction of silicates, D/H ratio in chondritic water, the concentration of refractory organics) are converging and point to a fraction of unaltered interstellar material in comets and undifferentiated asteroids of order 5–10% (Alexander et al. 2017). The discovery of pre-solar grains in primitive, ancient solar-system materials opened a new window on astrophysics, and demonstrated conclusively that some

[1] The Local Interstellar Cloud is a warm, low-density cloud surrounding the solar system that is located itself in a hot and even lower-density region called the "Local Bubble" (Frisch et al. 2011).

[2] The heliosphere is the region of space around the Sun that is dominated by the solar wind plasma, with respect to the ISM plasma.

interstellar materials survived the violent, hot formation of the solar system. The story is a familiar one in science, with confusion following progress following confusion, and on and on (Pais 1986, Inward bound). This paper focuses on contemporary[3] extrasolar dust only. For a review of the extensive literature on pre-solar grains, see Zinner (2014).
3. Observations of ISD in the solar system have come from a $\sim 0.1\ \mathrm{m}^2$ aerogel and Al foil collector on the Stardust mission that was exposed to the interstellar dust stream for 200 days. A massively-distributed search of this collector resulted in a constraint on the ISD flux independent of in-situ ISD observations, and in the identification of seven particles whose properties are consistent with an interstellar origin. Ongoing laboratory work will confirm (or not) the interstellar origin of these candidates, through measurements of the oxygen isotopic compositions (Sect. 5).
4. Isotropic foreground radiation has been detected due to the ISD, using IRAS and COBE (DIRBE) observations in the Infrared (Rowan-Robinson and May 2013) (Sect. 6.1).
5. Finally, the presence of interstellar dust has been inferred from serendipitous observations of impacts on spacecraft antennae (Sect. 6.2).

Studies of the dynamics of interstellar dust in the heliosphere for different particle properties (Sect. 2) support the understanding and interpretation of these various types of observations.

This paper reviews the current status of interstellar dust research in the solar system with an emphasis on the progress in the last decade. It focuses on new insights in the dust dynamics, in situ measurements of the interstellar dust flux and composition, interstellar dust sample return results from the Stardust mission, and indirect measurements using spacecraft antennae. Because of our somewhat arbitrary size cut, here we exclude from discussion the interstellar "asteroid" A/2017 U1 (Meech et al. 2017; Feng and Jones 2018; Zhang 2018), and the interstellar meteors (Baggaley 2000) that are currently questioned (Hajdukova et al. 2019). These subjects are reviewed in Hajdukova et al. (2019) and Koschny et al. (2019).

This book review paper is an outcome from a workshop "*Cosmic dust from the laboratory to the stars*", that was held at the International Space Sciences Institute (ISSI) in Bern (Switzerland). The topic of this paper follows up on the results from an ISSI working group on interstellar dust in the heliosphere that was held in October 1997 and whose outputs are summarized in Frisch et al. (1999) and papers in the JGR volume containing Frisch (2000).

2 Dynamics and Simulations of Interstellar Dust in the Solar System

The Sun and the planets currently move through the LIC at a velocity of $\sim 26\ \mathrm{km\,s^{-1}}$ (Witte 2004; McComas et al. 2015) in the direction of the neighbouring "G-cloud" that is less than 0.2 pc away from us and we will cross into the G-cloud within a few thousand years from now (Redfield and Linsky 2000; Redfield et al. 2004). The motion of interstellar dust passing through the solar system as a consequence of this (relative) motion has been studied through computer simulations in the context of in situ observations. We first review in Sect. 2.1 the forces acting on the ISD particles and their effects on the dust flow patterns. Then we discuss the ISD dynamics as the particles pass through the boundary regions of the heliosphere (Sect. 2.2). In both sections, we highlight the state-of-the-art modeling.

[3]This contemporary extrasolar dust may be modified by its journey through the solar system and near the Sun though.

2.1 Dynamics of Interstellar Dust in the Solar System

The three most important forces that influence the trajectories of interstellar dust moving through the solar system are solar gravity, solar radiation pressure force (Sect. 2.1.1) and the Lorentz force due to the charged particles' motion through the interplanetary magnetic field (IMF) (Sect. 2.1.2). Other forces, like planetary perturbations, Poynting-Robertson effect, Yarkovsky effect and solar wind corpuscular drag can be generally neglected (Altobelli 2004; Gustafson 1994).

2.1.1 Solar Gravity and Solar Radiation Pressure Force

The solar gravitational force on a dust particle is described by:

$$\mathbf{F}_G = -\frac{GM_\odot m_p}{|\mathbf{r}|^3}\mathbf{r} \tag{1}$$

with G the gravitational constant, $M_\odot$ the mass of the Sun, m_p the dust particle mass, and $\mathbf{r}$ the position vector of the dust particle with respect to the Sun. The solar radiation pressure force is expressed as:

$$\mathbf{F}_{SRP} = \frac{A_p Q_{pr} S_0}{c}\frac{\mathbf{r}}{|\mathbf{r}|^3} \tag{2}$$

with A_p the projected surface of the dust particle ($A_p = \pi a^2$ for a spherical particle with radius a), Q_{pr} the radiation pressure efficiency weighted by the solar spectrum, S_0 the solar flux density at 1 AU from the Sun, and c the speed of light. Q_{pr} depends on the particle optical properties.[4]

Because the solar gravity and the solar radiation pressure force are both radial and decline quadratically with increasing distance from the Sun, they can be combined in a single dimensionless ratio:

$$\beta = \frac{|\mathbf{F}_{SRP}|}{|\mathbf{F}_G|} = \frac{A_p Q_{pr} S_0}{cGM_\odot m_p} \tag{3}$$

β depends on the particle optical properties, thus on composition, morphology, and size. It is a constant value if the particles are assumed not to be altered on their path through the solar system (e.g. through evaporation) and if we neglect variations in the solar flux. The β versus mass relationship is expressed in a so-called *β-curve*, specific for each dust material, morphology, and density. Examples of such β-curves are the coloured curves in Fig. 5 (Sect. 4.3 of this paper). Silicates have lower maximum β-values than carbonaceous materials, and aggregates tend to have a "flattened" β-curve with respect to their compact counterparts. More examples of β-curves can be found in Schwehm (1976), Gustafson (1994), Kimura and Mann (2000), Kimura et al. (2003a) and Sterken et al. (2015).

The β-value of the ISD is important for its dynamics. In the absence of Lorentz forces, particles with $\beta = 1$ move through the solar system on straight trajectories because the solar radiation pressure compensates for the effects of solar gravity. Particles with $\beta < 1$ are deflected towards the Sun on hyperbolic trajectories, by the dominance of the solar gravitational over solar radiation pressure force. This is illustrated in Fig. 1 (left, from Sterken

[4]See Schwehm (1976) and Burns et al. (1979) for calculating radiation pressure efficiencies, and Silsbee and Draine (2016) and Kimura (2017) for a discussion on the radiation pressure efficiencies for ISD, applied in specific to the Stardust mission (Sect. 5).

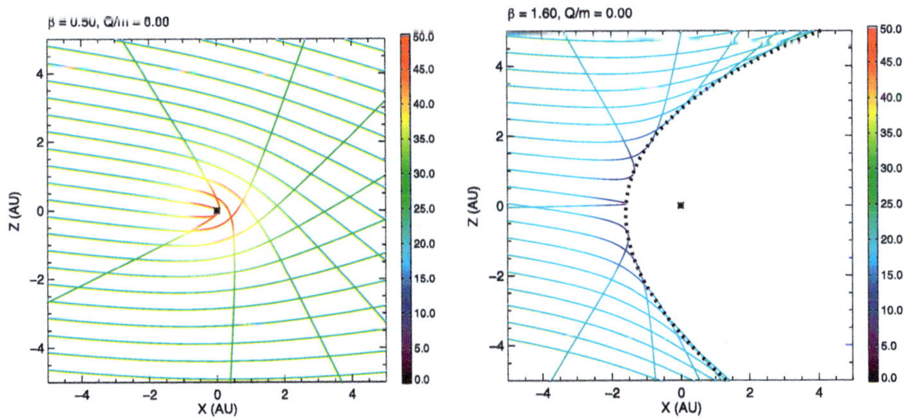

Fig. 1 Simulated dust trajectories for solar gravity and solar radiation pressure force only, for $\beta < 1$ (left) and $\beta > 1$ (right). The colors indicate the velocity of the particles with respect to the Sun (the black cross) in $\mathrm{km\,s^{-1}}$ and the dust moves from the left to the right in the simulations. From Sterken et al. (2012). Reproduced with permission from Astronomy & Astrophysics, ©ESO

et al. 2012) for $\beta = 0.5$. As the particles move out of the solar system, they accelerate again to their original speed of about 26 $\mathrm{km\,s^{-1}}$ for ISD particles from the LIC. As a result of the solar gravitational attraction, these particles form a region of *gravitational focusing* downstream of the Sun with respect to the inflow vector of the ISD (see Fig. 1, left). This gravitational focusing region is located at approximately $+79°$ and $-8°$ ecliptic longitude and latitude (Frisch et al. 1999; Landgraf 1998) as was determined from Ulysses and Galileo ISD data.[5] Because of this gravitational focusing and thus for particles with $\beta < 1$, the ISD number densities are a factor of 2–5 times larger than the undisturbed number density in the ecliptic plane, and up to a factor of 10 larger in the focusing region itself (Landgraf et al. 1999b; Strub et al. 2019). This factor depends on the distance to the Sun and on the β-value of the particles.[6] The number density enhancements can be calculated numerically or analytically. For the equations we refer to Danby and Camm (1957), Wallis (1987), Landgraf et al. (1999b). Very small and very large particles (sizes approximately below 0.1 μm or above 0.4 μm radius for silicates) belong to the size category of particles with β-values < 1, that cause such a region of gravitational focusing. However, very small particles are also deflected by Lorentz force (see Sect. 2.1.2).

Particles with $\beta > 1$ decelerate as they approach the Sun and are diverted on hyperbolic trajectories in a direction *away* from the Sun, see Fig. 1 (right), from Sterken et al. (2012) for $\beta = 1.6$. Such particles do not enter a paraboloid-shaped region around the Sun of which

[5] A newer analysis of the complete Ulysses dataset, but excluding the period where the dust flow was shifted in direction (2005, see Sect. 3.2), has resulted in a derived flow direction of $+75° \pm 30$ ecliptic longitude and $-13° \pm 4$ ecliptic latitude (Strub et al. 2015). This corresponds roughly to the flow direction of the interstellar helium from IBEX He data between 2009 and 2013: $+75.6° \pm 1.4$ ecliptic longitude, and $-5.12° \pm 0.27$ ecliptic latitude (Schwadron et al. 2015) and earlier ISD directions derived from Ulysses data by Landgraf (1998), Frisch et al. (1999), and Kimura et al. (2003b). The average speed of the ISD particles measured by Ulysses was $24 \pm 12\ \mathrm{km\,s^{-1}}$ (Krüger et al. 2015), compatible with the Helium inflow speed of $26.3 \pm 0.4\ \mathrm{km\,s^{-1}}$ (Witte 2004) and with earlier determination of the heliocentric ISD *inflow* speed $V_{\infty,ISD} = 25.7 \pm 0.5\ \mathrm{km\,s^{-1}}$ from the then available Ulysses ISD data by Kimura et al. (2003b).

[6] See Sterken et al. (2013) for examples of gravitational focusing effects for $\beta = 0.5$ at Asteroid, Jupiter and Saturn distance from the Sun, in the ecliptic plane.

its size depends on the β-value of the particles. This void region in space is called the *β-cone*: its size depends on the particle properties (mass, surface optical properties) while its direction depends, like the gravitational focusing cone, on the inflow vector of the ISD.

Comparison of numerical simulations of ISD trajectories with spacecraft observations of ISD mass and velocity can constrain the particle optical properties and thus composition, provided that sufficient data are available. This method of constraining the ISD particle material properties using spacecraft data and numerical simulations, is called *β-spectroscopy* (Altobelli 2004). Landgraf et al. (1999a) used this method to conclude that the maximum β-value of the ISD particles from Ulysses data between 1992 and 1999 is between 1.4 and 1.8 for particles of mass 1×10^{-17} kg to 3×10^{-16} kg.

2.1.2 Lorentz Force

The solar gravity and radiation pressure force alone cause a static, location-dependent flow of ISD through the heliosphere. However, dust particles moving through a plasma in a radiation field like from the Sun, acquire an equilibrium charge as a result of competing currents (Mukai 1981; Horanyi 1996). They are ionized by the radiation from the Sun (*photoionization*), they lose electrons by secondary electron emission, they capture protons, and they collect electrons from the solar wind. Since the solar wind and the ionizing radiation from the Sun both decline quadratically with the distance to the Sun, these particles acquire a relatively constant nett equilibrium charge corresponding to a surface potential of about +5 V as they travel through the solar system in interplanetary space (Kempf et al. 2004).[7]

Assuming compact spheres, the net charge on the particle surface is $q = 4\pi\epsilon_0 a U$ and the charge-to-mass ratio is related to the particle size and density as:

$$\frac{Q}{m} = \frac{3\epsilon_0 U}{\rho a^2} \tag{4}$$

with ρ the particle bulk density, a its radius, U the equilibrium potential and ϵ_0 the permittivity of free space. The larger the ISD particle is, the smaller its charge-to-mass ratio Q/m is, and opposite. The smallest particles (i.e. < 0.1 μm) acquire a larger equilibrium potential than their larger counterparts due to the so-called "small-particle effect" (they have larger photoelectric yields). Kimura (2016) reviews this effect in detail and Slavin et al. (2012, Fig. 2) and Kimura and Mann (1998, Fig. 3) illustrate it in their Figures. Aggregate particles also have larger equilibrium potentials than their compact equivalents due to the small particle effect (see Ma et al. 2013). When dust particles move through different plasma environments, they also obtain different charges: sometimes largely negative (e.g. in some cases in planetary magnetospheres down to −30 V; Horanyi 1996), sometimes positive and larger than in the IMF, like in the boundary regions of the heliosphere (Sect. 2.2).

The IMF originates from the solar magnetic field that is convected outward, "frozen" in the solar wind. Because the Sun rotates, a spiral-type pattern emerges in the IMF, referred to as the "Parker spiral" (Parker 1958). The solar magnetic field can be approximated for dust trajectory computer simulations by modeling a dipole during solar minimum, a mix of magnetic field polarities at solar maximum, and a dipole with opposite polarity at the next solar minimum approximately 11 years later (Grün et al. 1994; Landgraf 2000; Sterken et al. 2012). The 22-year solar magnetic cycle dominates the IMF and because a charged

[7] Mukai (1981) calculated an equilibrium potential of +0.5 to +6 V for graphite at 1 AU and +4 to +14 V for silicate grains depending on solar wind conditions. For charging in other plasma conditions (e.g. in the heliosphere boundary regions), we refer to Sect. 2.2.

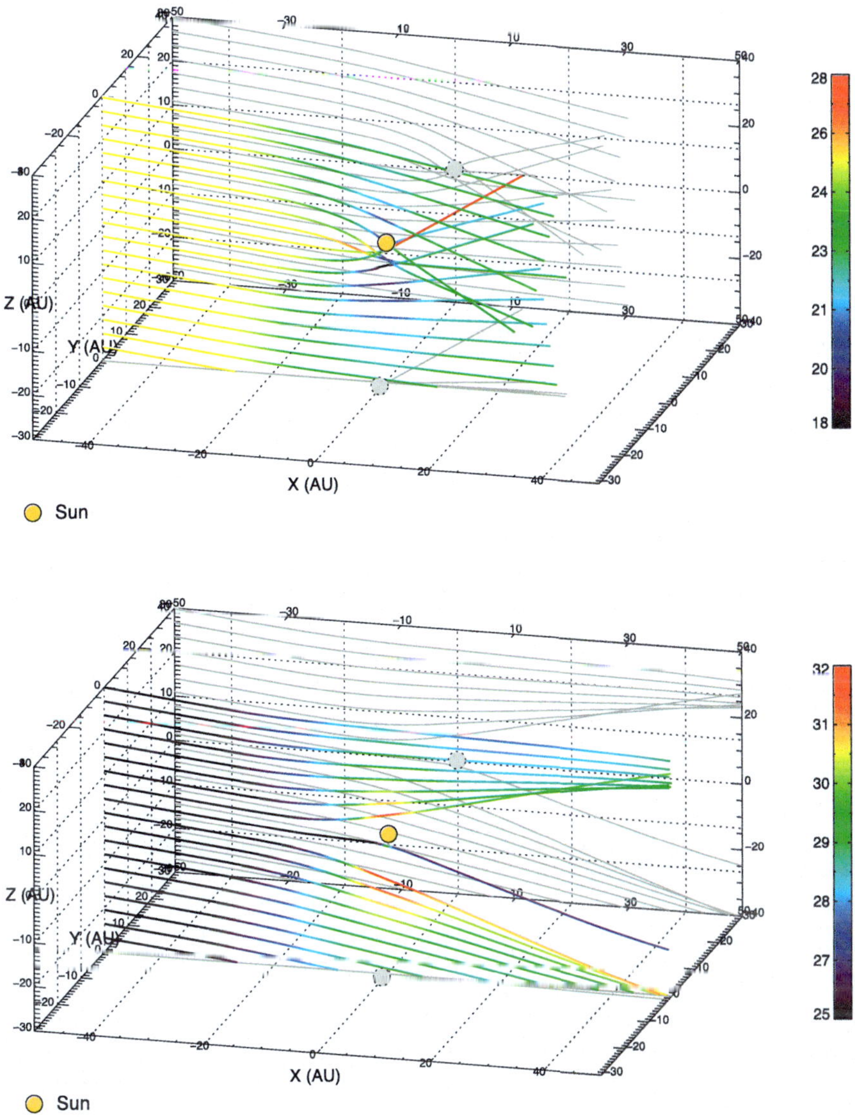

Fig. 2 Simulated dust trajectories for $\beta = 1$ and $Q/m = 1\ \mathrm{C\,kg^{-1}}$ in the focusing (upper graph) and defocusing (lower graph) phase of the 22-year solar magnetic cycle. The colors indicate the velocity of the particles with respect to the Sun in $\mathrm{km\,s^{-1}}$. From Sterken et al. (2012). Reproduced with permission from Astronomy & Astrophysics, ©ESO

particle that moves through a magnetic field experiences a Lorentz force, the 22-year solar magnetic cycle 'forces' the ISD trajectories into a *focusing* phase (during solar minimum) where the particles are focused *towards* the solar equatorial plane (see Fig. 2, upper graph), and into a *defocusing* phase at the next solar minimum, where the particles are defocused *away* from the solar equatorial plane (Fig. 2, lower graph). The magnitude and direction of the Lorentz force is determined by the particle charge-to-mass ratio (larger for small particles), the IMF direction and strength (smaller and more azimuthal at larger distances from the Sun), and the solar wind speed which is much faster (several hundreds of $\mathrm{km\,s^{-1}}$) than the ISD particle speed ($\sim 26\ \mathrm{km\,s^{-1}}$) (Witte 2004; McComas et al. 2015) with respect to the Sun. The Lorentz force is described as:

$$\mathbf{F}_L = \frac{Q}{m}(\dot{\mathbf{r}}_{p,sw} \times \mathbf{B}_{SW}) \tag{5}$$

with Q/m the charge-to-mass ratio of the dust particle, $\dot{\mathbf{r}}_{p,sw}$ the velocity vector of the dust with respect to the solar wind, and $\mathbf{B}_{SW}$ is the magnetic field vector of the IMF at the location of the dust particle. The solar equatorial plane is tilted with approximately 7° with respect to the ecliptic plane. For large charge-to-mass ratios (e.g. $Q/m \approx 10\ \mathrm{C\,kg^{-1}}$), the dust trajectories become more complicated than a mere *focusing* or *defocusing*, and density enhancements of such small ISD particles can occur at locations and during times that are not quite intuitive at a first glance. For this reason, Monte Carlo computer simulations of ISD trajectories are very helpful to understand or predict the interstellar dust fluxes measured in the solar system by in situ dust instruments (see Sects. 2.1.3 and 3.2).

The combination of gravity, solar radiation pressure force and Lorentz force leads to a modulation of the β-cones and of the zone of gravitational focusing, with time.[8] Worth to note is that during the defocusing phase of the solar cycle, the ISD particles with large enough Q/m are not only deflected away from the ecliptic plane, but they are also concentrated, as a result of Lorenz force, at higher ecliptic latitudes of ca. 10-20 AU above the ecliptic plane (see Fig. 46 in Sterken et al. 2012).

The forces acting on the ISD lead to a variable ISD size distribution that depends on space, time and particle properties like mass, morphology and optical properties (thus particle composition). For the size distribution in the ISM, generally a power-law distribution is assumed as is derived from many astronomical observations like observations of the gas phase and of absorption of starlight by the dust (e.g., see Mathis et al. 1977; Draine and Lee 1984; Weingartner and Draine 2001; Zubko et al. 2004; Wang et al. 2015). These derived size distributions are representative for long (kpc-scale) distances and the *Local* Interstellar Cloud may thus have different properties (like size distribution) from these. The filtering effect of the discussed forces in space and time and with respect to a reference size distribution is described in detail for Saturn, Jupiter and Asteroid distances in Sterken et al. (2013) and for Earth distance in Strub et al. (2019). Also for the Ulysses mission, computer simulations demonstrate the time- and space-dependent filtering of the interstellar dust size distribution, e.g. as illustrated in Fig. 12 in Sterken et al. (2015). In these studies, the filtering for small particles at the heliosphere boundary was not taken into account (we refer to Sect. 2.2 for this).

2.1.3 *A Brief Review of ISD Trajectory Simulation Studies Inside the Solar System*

The motion of interstellar dust in the solar system was studied and predicted long before the first *in-situ* detections in the solar system took place, with the above-mentioned physics as a basis: Bertaux and Blamont (1976) predicted that these particles should be concentrated due to gravitational focusing of the Sun (Sect. 2.1.1). Levy and Jokipii (1976) followed up on this work and postulated that the net charge that the interstellar particles would carry determines whether they can enter the heliosphere or not (see Sect. 2.2 for further discussion). Subsequently, Gustafson and Misconi (1979) and Morfill and Gruen (1979) performed the first

[8] Figures 45 and 46 in Sterken et al. (2012) visualize such "Lorentz-force-modulated β-cones".

numerical Monte Carlo computer simulations of interstellar dust trajectories that include all three the Lorentz force, solar gravity, and solar radiation pressure force and they demonstrated the focusing and defocusing effect due to the 22-year IMF cycle (Sect. 2.1.2). In 1992, the first interstellar dust particles were detected *in situ* with the Ulysses dust detector (see Sect. 3.2) by Grün et al. (1993) which prompted further numerical simulation studies of the interstellar dust flow through the solar system. Landgraf (1998, 2000) and Landgraf et al. (2000) simulated the ISD trajectories for an assumed hypothetical composition (silicate with inclusions of other chemical compounds) that is consistent with the spectral signature as observed by astronomical means (Draine and Lee 1984; Greenberg and Li 1996). Computer simulations of ISD dynamics were used to interpret the variable Ulysses ISD flux measurements from 1992 until 2000 (Landgraf 2000; Landgraf et al. 2000) and Landgraf (1998) and Frisch et al. (1999) derived the gas-to-dust mass ratio in the LIC from ISD measurements by Ulysses and Galileo to be $R_{g/d} = 40$ and $R_{g/d} = 94^{+46}_{-38}$. Czechowski and Mann (2003) studied the penetration of small interstellar dust through the heliosphere including boundary regions (inside of the heliopause). They found that a wavy current sheet can facilitate interstellar dust to enter the solar system, and suggested that beyond 40 AU, there is a stream of small interstellar dust that can enter the heliosphere at low ecliptic latitudes (see also Sect. 2.2).

Sterken et al. (2012, 2013) followed up on the dust dynamics simulation work of Landgraf (1998) by studying the flow and filtering of ISD for a broad range of dust composition assumptions, and tested a few variations of the Interplanetary Magnetic Field (IMF) models. These simulations were used for interpreting and understanding the full Ulysses dataset from 1992–2008 (Sect. 3.2). Sterken et al. (2015) also tested different descriptions of the configuration of the IMF, for which a Parker model is a sufficient approximation, but they concluded that the filtering in the heliosphere boundary regions still is a necessary additional component to be able to explain in the future the full Ulysses dataset (see Sect. 3.2). Finally, Strub et al. (2019) studied the flow of ISD onto the Earth using computer simulations, which is useful for interpreting data from missions near the Earth's orbit. Dynamical studies and the above mentioned simulations were used to predict and interpret ISD data from missions like Cassini (Altobelli et al. 2003; Sterken et al. 2012; Altobelli et al. 2016), Stardust (Landgraf et al. 1999b; Sterken et al. 2014), and Ulysses (Landgraf et al. 1999a; Landgraf 2000; Landgraf et al. 2000, 2003; Sterken et al. 2015).

2.2 Interstellar Dust Transport into the Heliosphere

Interstellar dust grains in the local interstellar medium are charged to a surface electrostatic potential of typically +0.5 to +1 V: Grün and Svestka (1996) give a detailed discussion on the grain charging in the interstellar medium. When these particles enter the heliosphere boundary regions, which have different plasma conditions from the ISM or inner heliosphere, their equilibrium potential can increase to about +8 V or +10 V (Linde and Gombosi 2000; Slavin et al. 2012).[9] Kimura and Mann (1999) elaborate on the importance of secondary electron emission for the dust transport into the heliosphere. The particles are also subject to increased magnetic field strengths and slowed down solar wind speeds in the transition regions from ISM to inner heliosphere. Linde and Gombosi (2000) calculated the filtering of small interstellar dust particles as a result of Lorentz forces in the transitional

[9] Kimura and Mann (1998) found +12 V for Silicate particles and +6 V for Carbon for particles with radius about 0.3 μm, while Alexashov et al. (2016) estimated a potential of +2 to +3 V. Ma et al. (2013) calculated higher charges for aggregates than for compact spheres (see Sect. 2.1.2)

region between the external environment and the inner heliosphere[10] during the 1996 solar minimum. They find a strong increase in filtering for grain sizes between 0.2 and 0.1 μm and below: large particles can penetrate this region freely, while small ones are rather deflected around the heliosphere instead. Note that 1996 is in the defocusing phase of the solar cycle for the outer heliosphere region where particle deflection varies significantly between solar cycle magnetic phases, see Fig. 19 in Sterken et al. (2015).[11] Based on a static MHD-model for describing the IMF and plasma conditions for calculating particle charging, Slavin et al. (2012) simulated that particles smaller than 0.01 μm are excluded completely while micron-sized particles flow freely into the solar system and particle sizes in-between are partially filtered. They calculated the global distribution of ISD inside the heliosphere as a result thereof. Time-dependence of the interplanetary magnetic field throughout the particle travel time is important for comparison with in situ data and was not yet taken into account in the model, but Slavin et al. (2012) demonstrated the variable particle charges in the boundary regions of the heliosphere and close to the Sun, and for smaller particles ("small particle effect", see Sect. 2.1.2). Alexashov et al. (2016) also modeled the ISD distribution near the interface region of the heliosphere for a time-independent magnetic field configuration based on a (static) 3D MHD model used for modelling the IMF and plasma conditions, with the aim to describe the filtering and directional deviations of the ISD near the heliosphere-LIC interface region. Their results roughly confirm earlier studies, they report on features in the heliosphere boundary regions due to the Lorentz force, and they conclude that the dust distribution near the heliopause[12] depends on the direction of the Interstellar Magnetic Field. They also calculated one time-dependent distribution of ISD in the solar system, but do not go in detail for analyzing the results, which is beyond the scope of the paper.

Recapitulating, models from Landgraf (2000), Sterken et al. (2012), Strub et al. (2019) assume a time-variable IMF polarity based on the solar wind magnetic field (through the Parker spiral) for calculating the particle trajectories during their journey through the solar system because these are moving at roughly 5.5 AU/year. Particle trajectories in these models are calculated at distances up to 50 AU from the Sun only, ignoring the effects of the heliosphere boundary regions. Particle surface potentials are assumed to be static (+5 V), as roughly justified for these distances (see Fig. 2 in Slavin et al. (2012) and Kimura and Mann (1998)). In contrast, Slavin et al. (2012) utilizes a 3D solar wind model that includes the modification of the solar wind plasma by pickup-ion mass-loading and a variable particle potential as function of the 3D solar wind plasma thermal properties. This is important in the heliosphere boundary regions that are included in the model. In contrast to the former models, this model considers two static solar magnetic polarities during the particles' journeys through and around the heliosphere.

Note that regions like the heliosphere boundary not only act as a *filter* for the size of the particles, but also for the morphology of the particles: because of their higher charges, aggregates are filtered out more easily than compact particles of the same size (Ma et al. 2013).

Once in the heliosphere, there is the second *filtering* of the dust particles near the solar system due to the higher values of the IMF strength. The Monte-Carlo based, time-dependent simulations described in Sect. 2.1.3 only take into account this second filtering

[10] The inner heliosphere is the part of the heliosphere where the solar wind dominates the plasma and is still supersonic.

[11] Figure 19 in Sterken et al. (2015) illustrates how the incoming particles move through a different phase of the solar cycle in the solar system than they did earlier when crossing the heliosphere boundary regions.

[12] The heliopause is the boundary between the solar wind dominated inner heliosphere, and the region around the heliosphere where interstellar medium plasma dominates.

near the solar system, while the (rather static) models from Slavin et al. (2012), Alexashov et al. (2016) describe the filtering near the heliosphere interface. A combination of both, including full time-dependence of all magnetic fields, is crucial for the study of the particle dynamics and for comparison with spacecraft data. Sterken et al. (2015) suggested that this may be the reason why only two distinct periods of Ulysses ISD data can be understood with the current simulations, but not the whole time-series of data at once: a full model of the heliosphere boundary combined with the solar system filtering is needed for further study.

3 In-situ Measurements and Interpretation of Interstellar Dust

3.1 Cassini, Helios, and Galileo Measurements

A fleet of interplanetary probes, carrying in-situ dust instruments, have returned measurements of the ISD flux in the Solar System, either on their way to their main destination or as secondary mission science objectives. In particular, the Helios, Galileo, Ulysses and Cassini missions have provided insights into the dynamics and composition of interstellar grains found in our Solar System at different heliocentric distances. Each of those missions was equipped with dust detectors allowing an equivalent detection sensitivity in the submicrometer size range and therefore provided comparable data. All dust detectors carried by these probes have a dust detection system based on the collection and separation of the charges (positive ions and electrons) resulting from the vaporization of the grains upon impact at hypervelocity (typically from a few up to a few tens of $\mathrm{km\,s^{-1}}$) on the detector's target. Only the Cosmic Dust Analyser (CDA) on-board Cassini and the Helios instrument were equipped with sub-systems to retrieve the elemental composition of the grains using time-of-flight (TOF) mass spectroscopy. We refer the reader to Dietzel et al. (1973), Grün et al. (1984, 1992) and Srama et al. (2004) for a detailed description of the different instruments, and in particular to Table 1 in Srama et al. (2004) for a comparison of the instruments' performances.

ISD measurements have been performed at different locations and time in interplanetary space, as well as from within the Saturnian system. In all cases, special care has been taken to discriminate impacts of interstellar grains from other dust populations. Often, a direct determination of the grain's origin based on a measurement of its impact velocity has not been possible, because of the error bar of the impact speed determination (typically, about a factor of two). Instead, the ISD identification relies on the total amount of positive ions collected upon impact combined with directionality criteria, as described in Altobelli et al. (2003, 2006); Altobelli (2004).

In the case of the Ulysses spacecraft, the ISD grain identification was somewhat easier due to the measurements performed at high ecliptic latitudes, where the abundance of IDP grains is reduced compared to the ecliptic plane, combined with the favorable spinning configuration of the instrument with respect to the direction of the ISD flow. In contrast, the analysis of the Helios, Galileo and Cassini data have been more challenging. The Helios ISD measurements were performed with the dust instrument on board the Helios-A spacecraft between 1974, day of year 353 and 1979, day of year 002, covering heliocentric distances ranging from 0.3 AU to 0.98 AU. The spacecraft trajectory was within the ecliptic plane for an heliocentric orbital period of 190 days. The ISD measurements performed by the Galileo spacecraft were acquired between January 1990 and mid-1993 as the probe was traversing the interplanetary space within the Venus and Mars orbits (the spacecraft had to perform multiple planetary flybys in order to gain the energy to reach Jupiter). While the

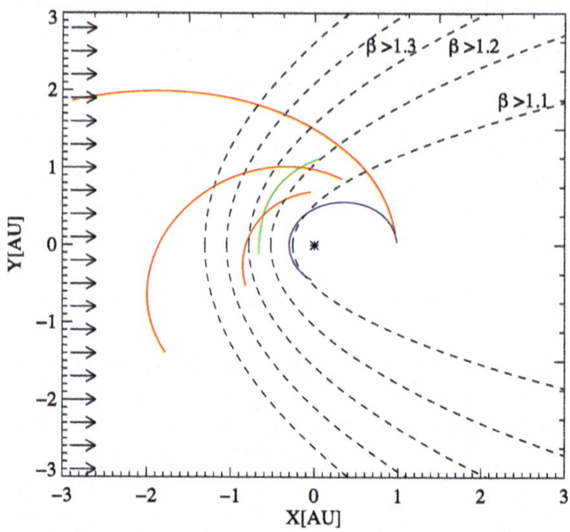

Fig. 3 Trajectory segments of the Galileo (red), Cassini (green) and Helios (blue) missions where ISD grains were detected, in the ecliptic plane. The coordinate system has the Sun as origin. The $+X$ axis is aligned with the projection on the ecliptic of the ISD downstream direction. The dotted lines indicate the ISD exclusion boundaries shaped by radiation pressure for different values of the β parameter. From Krüger et al. (2019a). Reproduced with permission from Astronomy & Astrophysics, ©ESO

dust instrument was on all the time, only three segments of the interplanetary trajectory were geometrically favorable to ISD stream monitoring. The CDA instrument was able to perform ISD measurements during the interplanetary cruise phase from 0.7 AU and 1.2 AU and during the nominal science phase around Saturn. While the Helios and Galileo dust detectors were attached to the rotating structure of their respective spacecraft, the CDA instrument can be pointed toward fixed directions as decided during the science planning of the mission (Srama et al. 2004).

Overall, a major achievement of those ISD measurements at different locations (from 0.3 to 10 AU) and time within the solar system was a confirmation of the ISD grain dynamics, as a proxy to some of their physical properties (mass distribution, radiation pressure efficiency, charge-to-mass ratio). The different segments of trajectories where ISD grain detection was achieved are shown in Fig. 3. The overall decrease of the ISD flux with decreasing heliocentric distance (see Fig. 4) is expected due to the size dependent filtering effect of radiation pressure. ISD grain models considering various compositions and structure predict β values below 2, using Mie scattering theory and Ulysses observations (Kimura et al. 2003a). At around 1.5 AU, ISD grains must have $\beta > 1.5$ in order to be filtered out by radiation pressure—beyond 5 AU, where the Ulysses measurements were performed, and *a-fortiori* at Saturn's heliocentric distance, where Cassini measurements were performed, the filtering by radiation pressure does not reduce significantly the ISD flux, which is then measured unperturbed (however, within the fluctuations due to the Solar Activity cycle on the smaller grains sensitive to the Lorentz forces, see Sect. 2.1.2).

3.2 Ulysses ISD Flux and Direction Measurements and Interpretation

The first detection of ISD in the solar system with in situ measurements by Ulysses (Grün et al. 1993) was the beginning of a 16-year long survey of the ISD flux, mass and flow direction by one and the same mission throughout three quarters of the 22-year Hale cycle. The mission was very suited for studying the ISD dynamics because of the long observation span and because of the peculiar orbit of Ulysses with respect to the ISD flow: not only Ulysses flew out of the ecliptic plane, but it also had its orbit serendipitously almost perpendicular

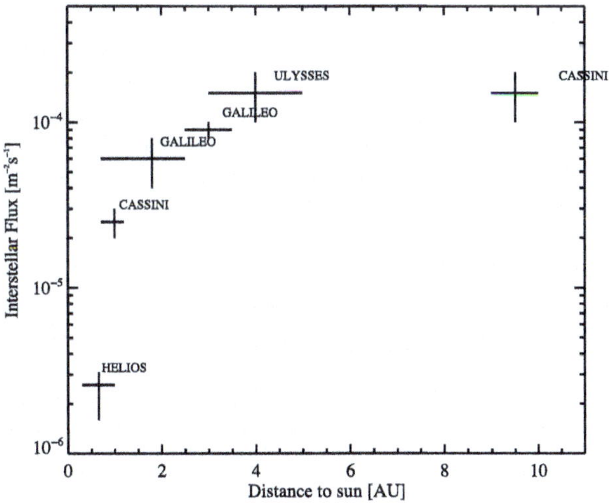

Fig. 4 ISD flux measured by the different instruments on-board Cassini (in interplanetary space and at Saturn), Galileo, Ulysses and Helios. The Helios data were taken between December 1974 and January 1980, the Cassini data were taken in 1999, the Galileo data during 3 orbit segments between January 1990 and early 1993, Ulysses data from ~ 1992–2000 and Cassini at Saturn (~ 10 AU) represents the average flux between mid-2004 and 2013

to the ISD inflow vector because the mission launch date was delayed with ca. 4 years from May 1986 until October 1990. Only at perihelion, the IDP particles and ISD particles came from the same (prograde) direction. During the largest part of Ulysses' orbit, ISD particles came from the retrograde direction while IDP (generally) came from the prograde direction, which facilitated their mutual distinction on a statistical basis, together with the fact that out of the ecliptic plane, IDPs are much less present than in the ecliptic.

Landgraf et al. (2003) simulated the distribution of the ISD in the solar system and compared the simulation results to Ulysses and Galileo data as far as available: the bulk ISD population had a size of 0.3 μm ($Q/m = 0.59$ C kg^{-1}, $\beta = 1.1$), 0.4 μm and 0.2 μm in radius, assuming the astronomical silicates β-curve from Gustafson (1994). The focusing and defocusing phase of the solar cycle were represented in the data. Krüger et al. (2007) reported on a southward shift in dust flow direction in ecliptic latitude of at least 30° as measured by Ulysses, that was unexplained until then. The full Ulysses dataset from 1992 until 2007 was analyzed in 3 subsequent papers: Krüger et al. (2015) analyzed the mass distribution of the complete data set, Strub et al. (2015) focused on the variability of the ISD flux, flow direction, mass index and flow width, and Sterken et al. (2015) interpreted these data using Monte Carlo computer simulations of the dust trajectories. Details of the ISD candidate selection criteria were slightly different, and are described in the two first papers.

Krüger et al. (2015) calculated from the Ulysses in situ data a gas-to-dust mass ratio in the Local Interstellar Cloud (LIC) of ca. $R_{g/d} \approx 231^{+102}_{-68}$ (for an assumed inflow velocity $V_\infty = 26$ km s^{-1}) with a local dust density in the LIC of $(1.68 \pm 0.48) \times 10^{-24}$ kg m^{-3} for the same assumptions. Also, micron sized interstellar particles were confirmed in the Ulysses data. The derived gas-to-dust mass ratio is higher than the value of $R_{g/d} \approx 94^{+46}_{-38}$ presented in Frisch et al. (1999). The latter was based on the earliest Galileo and Ulysses measurements, while the former was based on Ulysses data between 1992 and the end of 2007. A comparison of the full mission datasets is given in Krüger et al. (2019a). Krüger et al. (2015) presents a mass distribution and cumulative mass distribution for the full data set from 1992 until 2007. The total cumulative flux is ca. 7×10^{-4} m^{-2} s^{-1} (Krüger et al. 2015). Strub et al. (2015) concluded that the Ulysses ISD data indeed show the effect of the focusing and defocusing phase of the solar cycle in the flux, as was also reported in Landgraf et al. (2003) for the first subset of the data. A total shift in dust direction of $50° \pm 7°$ was confirmed that coincided with an increase of the flux by a factor of 4 in only 8 months (Strub et al. 2015). Finally, Sterken et al. (2015) studied the full β–Q/m parameter space with simulations for interpreting the Ulysses data. They concluded that the fluxes and flow directions of the bulk of the 'big' ($a > 0.2$ μm) dust particles between 1992 and 2003 in the data were compatible with the simulations for dust with $\beta > 1$ and $Q/m \approx 0.5\ \mathrm{C\,kg^{-1}}$. This corresponds well with the dominating particle population from Landgraf et al. (2003) with $\beta = 1.1$ and $Q/m = 0.59\ \mathrm{C\,kg^{-1}}$. However, such (big) particles did not explain the second half of the data provided by Ulysses: the measured shift in dust direction and the increase in dust flux were compatible to simulations of ISD with $Q/m \approx 4\ \mathrm{C\,kg^{-1}}$ instead, implying that these 'big' ISD particles are of lower density than originally assumed. The flux of the smaller particles ($a < 0.2$ μm) could not be fitted with simulations before 2003, but the shift in dust flow direction and increase in flux in 2005 were reproducible features by the Lorentz force (the small particles not being of particular low density). The authors conclude that a time-dependent filtering of the small particles at the heliosphere boundary regions may be responsible for the fact that parts of the data set can be explained, but not all the data in one go.

4 Interstellar Dust Flux and Composition from Helios and Cassini

In contrast to the Ulysses instrument, the Helios dust instrument and Cassini's CDA were equipped with a subsystem able to record time of flight mass spectra of cations of the plasma cloud generated by high-velocity impacts (Dietzel et al. 1973; Srama et al. 2004; Postberg et al. 2009a). This allows compositional in situ measurements of individual dust grains impacting the detector's target.

4.1 Helios Measurements

The mass spectra acquired by the Helios detector had a low resolution compared to detectors of later generation. A mass resolution of $5 < \frac{\Delta M}{M} < 10$ was reached with the best performances obtained between 16 and 75 a.m.u. The ISD grains were identified on the basis of their dynamical properties and an average spectrum from all ISD grains was built. Despite the poor mass resolution, two main peaks could be identified: cations with masses around 25–30 u. compatible with Mg and Al bearing silicates dominate the spectrum while a second and lower peak between 50 and 60 u. would be compatible with iron. However, a more detailed elemental composition, with relative abundances of the grain constituents could not be derived from the Helios data.

4.2 Cassini CDA Measurements

Since 2010, CDA has carried out systematic observations of interstellar dust from the Local Interstellar Cloud (LIC). Altobelli et al. (2016) showed that over 30 dust impacts recorded between 2006 and 2013 could be dynamically separated from the background flow of interplanetary dust (IDP) and the overwhelming number of icy dust originating from the Saturnian system (Hillier et al. 2007; Postberg et al. 2008, 2009b). Each of these 36 impacts provided a mass spectrum very likely showing the composition of dust from our interstellar neighborhood. Together with the few ISD candidate grains brought back to earth by the Stardust mission (Westphal et al. 2014c, and Sect. 5 of this paper), the CDA ISD dataset provided the first quantitative compositional information of contemporary interstellar dust crossing the solar system. These measurements at Saturn allowed to examine LIC dust at the largest heliocentric distance ever achieved in situ, reducing the amount of dynamical filtering as compared to former detections performed closer to the Sun. In addition, the relative position of Saturn and the Sun, with respect to the ISD stream over the sample period, permitted the detection of grains with a wide range of β-values (Sterken et al. 2012).

4.3 Cassini CDA Measurements: ISD Identification and Dynamical Properties

The identification of the faint ISD population compared to the dominant background particles native to the Saturnian system was a challenge. In particular, detection criteria based on the velocity determination from the CDA Ion Channel did not provide sufficient accuracy to distinguish Saturn bound from unbound particles. Hence, a specific detection method was implemented, utilizing an estimate of the minimum impact velocity from the presence of the oxygen and carbon signals in the TOF spectra, because the cation formation of these two species requires a certain minimum impact velocity (Postberg et al. 2009a; Fiege et al. 2014). All TOF mass spectra from dust impacts on the Cassini CAT subsystem between 2004 and 2013 were analyzed, when CDA was pointed within 50° of the reference ISD upstream direction as defined by the Ulysses and Galileo measurements (Sect. 3). This selection criterion results in over 30,000 TOF mass spectra to be investigated (Altobelli et al. 2016). From this primary dataset, most particles showed an obvious water composition signature in the spectra and thus were rejected as endogenic to the Saturnian system. For the other particles the minimum velocity with respect to Saturn was inferred and compared to the expected velocities of Saturn bound particles as well as interplanetary dust particles. A population of a few tens of particles with sufficient velocities as expected for ISD grains entering Saturn's system stood out clearly. The analysis was repeated with a control group consisting of all particles with TOF spectra coming from the opposite direction to the reference ISD upstream direction. As expected, no such high velocity dust impacts were found there.

The flux of the ISD grains was determined by fitting simultaneously the cumulative distribution of the detected particles as function of time and of the distance to Saturn. The flux value of $F = (1.5 \pm 0.1) \times 10^{-4}\ \mathrm{m^{-2}\,s^{-1}}$ derived following this approach is the heliocentric value, corrected for Saturn's motion (Altobelli et al. 2016). The directional distribution of the ISD grains suggests a well collimated flux ($< 20°$) around the reference ISD direction at Saturn. Note that the ISD time collection period benefited in 2010 of a focusing configuration of the IMF for the smaller particles (Sterken et al. 2012, 2015).

The derived mass distribution, in agreement with the Ulysses measurements, confirm the existence of grains with $\beta > 1$ with a maximum reached between 10^{-17} kg and 10^{-16} kg (Altobelli et al. 2016). By deriving the β ratio and the mass, an independent consistency check for the bulk composition was obtained. Synthetic β-mass curves were computed using Mie theory for different grain models and composition and were compared with the β-mass curve inferred from the grain dynamics (Fig. 5).

4.4 Cassini CDA Measurements: ISD Composition

Despite the relatively low mass resolution of CDA spectra ($\frac{M}{\Delta M} \approx 30$) the abundances of Mg, Fe, Ca, O, Si, and C could be determined. The dynamical separation flux coincided well with a compositional separation from the IDP background. Whereas the IDP background exhibited a wide distribution of elemental abundances, spectra of ISD grains provided specific Mg/Fe and Mg/Ca ratios indicating a surprisingly homogeneous abundance of the major rock forming metals. Each LIC-ISD grain detected by CDA contains the major rock forming elements (Mg, Si, Fe, Ca) in roughly cosmic abundances, with only small grain-to-grain variations (Altobelli et al. 2016). Hence, CDA shows there is compositional

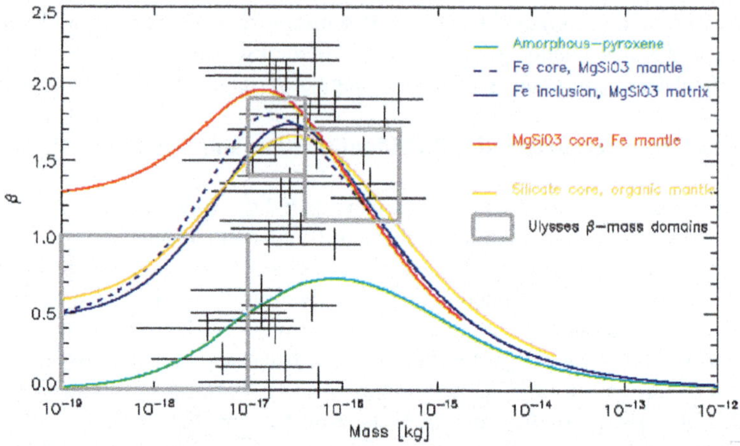

Fig. 5 This diagram shows the β-mass distribution of measured ISD grains by the Ulysses and Cassini spacecraft, as well as the expected β-mass values for different constitutive grain models. The area delimited by grey rectangles show the β-mass values as inferred from the Ulysses data, while the black crosses correspond to measurements performed by Cassini-CDA. The coloured curves show the theoretical β-mass values expected for the different grain models as listed in the plot legends. From Altobelli et al. (2016). Reprinted with permission from AAAS

homogeneity for particles of size 200 nm (Altobelli et al. 2016), as condensation products of the LIC, rather than being heterogenous samples of grains of the first generation formed around dying stars. They are homogenous because they are made of heavily processed and mixed material from different star populations. The composition observed by Altobelli et al. (2016) is different from what is observed in circumstellar dust (Leitner et al. 2012), most obviously the variation of Mg/Si, Mg/Fe is significantly smaller in the spectra obtained at Saturn (Fig. 6).

In the observed size regime, neither carbon rich grains (graphite, SiC) or pure metal or oxide grains were detected by CDA, which sets an upper limit of 8% at the 2σ confidence level for these grain species (Altobelli et al. 2016). This finding is in contrast with the isotopically and compositionally diverse populations of circumstellar dust inherited from AGB Stars and supernovae, which typically consist of silicates (e.g., olivine, pyroxene), with minor contributions (few %) of oxides (e.g., corundum, hibonite) and carbonaceous grains, mainly silicon carbide with an abundance of $> 20\%$, possibly up to 50% (Leitner et al. 2012). The upper limits for sulfur and carbon from CDA LIC-ISD grains observation indicate a substantial depletion of carbon and sulfur compared to the abundances inferred from the gas-absorption spectra of the LIC (Altobelli et al. 2016).

Although the CDA results indicate roughly cosmic element abundances of ISD grains, there seems to be a slight (20%) Si deficit (Fig. 7). A Si deficit is also supported by astronomical observations (Frisch and Slavin 2013) and can be explained by a higher abundance of either oxides and/or metals, similar to grains returned by the Stardust mission (Westphal et al. 2014c, see Sect. 5).

The dynamical analysis of the Cassini ISD grains based on plausible ranges of β-values for the 36 detections (Altobelli et al. 2016) is in good agreement with silicate particles enriched with metallic Fe. In contrast, the β-ratio of pure Fe-Mg silicates without metallic Fe enrichment cannot exceed unity, which is in disagreement with the high β-ratio of most grains above 5×10^{-17} kg. As no indications for abundant organic material is found in CDA

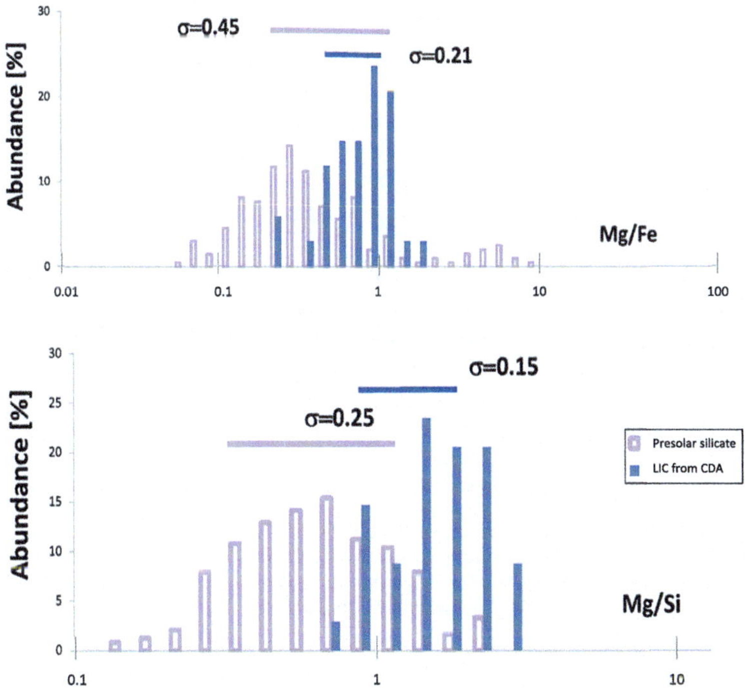

Fig. 6 Comparison of element wt% ratios in CDA ISD grains with pre-solar material. From Altobelli et al. (2016). Reprinted with permission from AAAS

ISD mass spectra it is suggested that at least some of the iron identified by CDA has to be in metallic form (Altobelli et al. 2016). However, the inferred upper limit for carbon does not exclude the presence of thin organic mantles on the ISD grains causing an increase in the β-ratio.

5 The Stardust Interstellar Dust Sample Return

In addition to a collector that returned cometary dust (Levasseur-Regourd et al. 2018, this book), the Stardust spacecraft returned a 0.1 m^2 aerogel and Al foil collector that was exposed to the interstellar dust stream for ~200 days in 2000 and 2002. The Stardust Interstellar Preliminary Examination (ISPE) (Westphal et al. 2014b), a consortium of > 50 professional scientists and >30,000 amateur scientists worldwide (Westphal et al. 2014a), consisted of coordinated projects to identify impacts in aerogel tiles and Al foils by optical (Westphal et al. 2014a) and SEM imaging (Stroud et al. 2014), development and implementation of novel sample preparation techniques (Frank et al. 2014), synchrotron Fourier-Transform Infrared Spectroscopy (FTIR) (Bechtel et al. 2014) and X ray microprobeanalyses (Butterworth et al. 2014; Brenker et al. 2014; Simionovici et al. 2014; Flynn et al. 2014; Gainsforth et al. 2014), laboratory hypervelocity capture experiments (Postberg et al. 2014), numerical modeling of dust transport in the heliosphere (Sterken et al. 2014), and elemental and isotopic composition analyses of residues in Al foils (Stroud et al. 2014), concluded in 2014 with a special issue of Meteoritics and Planetary Science (MAPS) (ISPE 2014), and a Research Article in *Science* (Westphal et al. 2014c) announcing the discovery of seven

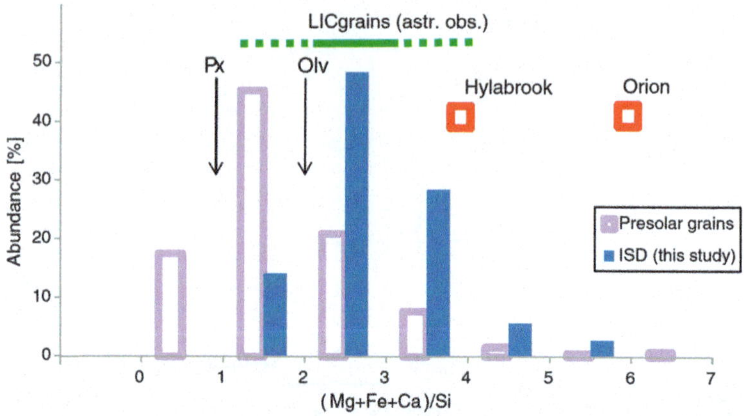

Fig. 7 Comparison of (Mg + Fe + Ca)/Si ratio histograms (in atom %) of CDA ISD grains, presolar circumstellar material, LIC ISD determined by astronomical observations and ISD returned by the Stardust mission. Intermediate between pyroxene (Px) and olivine (Olv) lies the maximum of presolar grains. An overabundance of refractory oxides in presolar grains causes a *tail* to higher values. Values for the LIC center here between 2 and 3 and are either also caused by oxides or by the presence of metallic Fe phases. The latter has also been observed in the Stardust particles Orion and Hylabrook which therefore plot at the far right. From Altobelli et al. (2016). Reprinted with permission from AAAS

particles of probable interstellar origin, as well as a measurement of the interstellar dust flux. The 2σ upper limit on the interstellar dust flux is lower than expected from dust modeling predictions, but is consistent with a large fraction of ISD consisting of particles with sufficiently large β to significantly suppress the flux at the heliocentric distance of particle collection by the Stardust spacecraft, about 2 AU.

At the conclusion of the ISPE, three impacts were identified in the aerogel collectors, and four were identified in the Al foil collectors. These particles are inconsistent in trajectory, capture speed, flux or composition with an origin as spacecraft secondaries or interplanetary dust (Westphal et al. 2014c). The two largest particles, identified in aerogel, contained crystalline silicates and so were likely to be circumstellar condensates (Westphal et al. 2014c). Both contained forsteritic olivine cores. One, Orion, also contained spinel and an iron phase that may have been consistent with nanophase iron. The other, Hylabrook, contained an amorphous Mg-bearing silicate, and also was surrounded by an Fe-bearing phase. Their measured densities were less than 1 $\mathrm{g\,cm^{-3}}$. These particles were of sufficiently low speed that they survived practically intact. Assuming that they had an interstellar origin, these particles were probably fluffy aggregates, consistent with their low overall densities, so that the ratio of light pressure force to gravitation force (β) was comparable to unity (between 1 and 1.6; Sterken et al. 2014). The third particle identified in aerogel was of sufficiently high speed that no terminal particle survived, so its composition is unknown. It was probably compact since it does not appear to have slowed significantly in the heliosphere. The four impacts in the Al foils were diverse in mineralogy and chemistry.

The crystallinity of Orion and Hylabrook was a surprise, but is not necessarily inconsistent with the astronomical constraint on the crystalline fraction of interstellar silicates (Kemper et al. 2004). If silicates are amorphized by low-energy ion bombardment in the ISM, large particles such as Orion and Hylabrook, which are much larger than typical interstellar particles, might retain crystalline cores. Indeed, their sizes alone may imply that Orion and Hylabrook are members of an entirely different population of particles than that which constitutes the bulk of interstellar dust.

Confirmation of interstellar origin of these candidates and others in the future—the search for new impacts is ongoing—depends on the ongoing development of ultrareliable techniques for measuring oxygen isotopic composition in these extremely challenging samples. Recently, Silsbee and Draine (2016) modeled the optical properties of interstellar dust, and concluded that the two large interstellar dust candidates, Orion and Hylabrook, had to have very unusual optical properties if they were interstellar. However, in a separate modeling study Kimura (2017) concluded that radiation pressure on Mg-silicate grains with nanophase Fe inclusions is sufficient high to be consistent with the Stardust observations.

6 Other Interstellar Dust Measurements

Several additional alternative techniques exist with which interstellar dust in the solar system have been measured. These are infrared observations, in situ measurements, and meteoroid observations. Of these, the in situ antenna measurements are the most promising. Other methods like PVDF detectors, or accelerometers (Thorpe et al. 2016, 2017) are too insensitive to measure interstellar dust. Interstellar meteor detections by ground-based radar systems (e.g. Baggaley 2000) are reviewed in Hajdukova et al. (2019) and Koschny et al. (2019).

6.1 Infrared Measurements

Infrared (IR) emission from dust far out in the galaxy, or in other galaxies has been used to investigate the dust properties and amounts, since the era of space infrared astronomy (e.g. the Planck and Herschel telescopes). Several attempts were also made to detect a signature of interstellar dust in, or near the solar system with infrared space telescope data. Grogan et al. (1996) modeled the infrared emission of interstellar dust in the solar system including effects like gravitational focusing for big interstellar dust, and electromagnetic forces for the smaller particles. They concluded that the prospects to find interstellar dust in the infrared data from COBE and IRAS are high enough to justify a search in the data. Indeed, a signature of an isotropic component of ISD (infrared) foreground radiation was seen in the data of the IRAS and COBE satellites by Rowan-Robinson and May (2013), by comparing models of cometary, asteroidal and interstellar dust in the solar system with infrared data from these spacecraft. The derived spatial mass-density of the interstellar dust is in agreement with values from Kimura et al. (2003a) at 1.5 AU distance from the Sun (approx. $10^{-25}\ \mathrm{kg\,m^{-3}}$), but is an order of magnitude higher than as measured by Ulysses at 4 AU. The authors argue that a study including dependence on radial distance, and a dynamical study are the next important steps to take.

6.2 Interstellar Dust Detections with Spacecraft Antennae

6.2.1 Antenna Measurements

Analysis of Voyager 1 and Voyager 2 data by Scarf et al. (1982) and Gurnett et al. (1983) determined that observed "intense impulsive noise bursts" were caused by dust grains impacting spacecraft surfaces at high velocities, thereby vaporizing and ionizing themselves and a fraction of the spacecraft surface material. Such impacts generate short-lived plasma clouds local to the spacecraft, which perturb the electric potential measured by plasma wave instruments.

While plasma wave instruments are, by their nature, not designed to detect dust grains, information concerning the mass, speed, and direction of the impacting dust can, under some circumstances, be estimated (Gurnett et al. 1983; Kurth et al. 2006; Andersson et al. 2015; Kellogg et al. 2016; Ye et al. 2016; Thayer et al. 2016; Oberc 1996; Zaslavsky 2015; Pantellini et al. 2012; McBride and Jam 1999; Collette et al. 2014, 2015, 2016). When plasma wave instruments are used as dust detectors, the entire spacecraft body surface area is sensitive to dust, resulting in count rates that can be orders of magnitude higher than those achieved by conventional dust detectors.

Dust has been detected by plasma wave instruments at the outer planets (Gurnett et al. 1987, 1991, 2004; Meyer-Vernet et al. 1986, 2009a; Pedersen et al. 1991; Tsintikidis et al. 1996; Kurth et al. 2006), near comets (Gurnett et al. 1986; Laakso et al. 1989; Neubauer et al. 1990; Tsurutani et al. 2003), and in the solar wind (Meyer-Vernet et al. 2009b; St. Cyr et al. 2009; Malaspina et al. 2014; Kellogg et al. 2016).

6.2.2 Interstellar Dust Detections and Database

Interstellar dust was identified in the STEREO and Wind plasma wave instrument data sets through a yearly modulation of the dust impact count rates (Zaslavsky et al. 2012; Malaspina et al. 2014; Wood et al. 2015; Kellogg et al. 2016).

The extensive Wind plasma wave dataset contains an observational history of the dust environment near 1 AU spanning at least two full solar cycles (Malaspina and Wilson 2016). This data set is uniquely valuable for interstellar dust studies in that: (i) it provides continuous observations over > 22 years, (ii) it includes a large number of dust impacts ($> 107{,}000$), and (iii) these data allow the direction that each dust grain was traveling when it impacted the spacecraft to be estimated. The Wind dust database is now publicly available on NASA's Space Science Data Facility Coordinated Data Analysis Web (Malaspina 2017).

Preliminary analysis of the Wind dust database indicates that significant changes in the flux ($6\times$) and arrival direction ($\sim 20°$) of interstellar micron dust at 1 AU occurred between 2005 and 2006 (Wood et al. 2015). Analysis of Ulysses interstellar dust observations using dedicated dust detection instrumentation also reveals shifts in interstellar dust flux and direction between 2005 and 2006 (Strub et al. 2011, 2015). These observations are consistent with recent simulation results predicting a change in near-micron interstellar dust flow direction through the inner heliosphere between 2005 and 2006 due primarily to solar cycle variation (Sterken et al. 2015). The Wind database is now ready to be used with detailed modeling (Strub et al. 2019) to analyze the ISD flow near the Earth in more depth.

7 Discussion

Great progress has been made in interstellar dust research in the last decade but nevertheless many open questions remain to be solved.

7.1 Dust Dynamics and Size Distribution

The flux and flow direction of the interstellar dust has been quite well understood in its relation to the interplanetary magnetic field configurations (Sects. 2, 3.2). Also the observed shift in dust flow direction in 2006 has been identified as likely to be caused by the Lorentz force in the vicinity of the solar system. However, a complete fit from the Ulysses data (1992–2008) for one particle size and one set of material properties has not been found and a more extensive modeling, including the outskirts of the heliosphere, is required to fully understand the variations herein (Sterken et al. 2015) and to fully grasp the filtering of the dust and hence, it's original mass distribution in the LIC. New datasets from e.g. the Wind mission (Malaspina et al. 2014) are still to be understood in the context of dynamics modeling. For doing so, the required interstellar dust models up to 50 AU from the Sun, with a high enough resolution, are available already (Strub et al. 2019). Using these new data and models as a starting point will help to understand the flux and flow of interstellar dust, and in specific the relatively unexplored role of the outer parts of the heliosphere herein. Interstellar dust may as such even be used as an extra *observable* for constraining the different heliosphere models of these boundary regions of the heliosphere (Sterken et al. 2015).

7.2 Dust Composition

Analyses of the interstellar dust composition from Cassini (small particles up to 0.1 μm) and the Stardust mission (μm-sized particles for the intact captured dust, and sub-μm particles for the crater residues) conclude about a homogeneous population with the major rock forming elements (Mg, Fr, Ca, Si) in about cosmic abundances (Cassini) versus a rather diverse collection of particles (Stardust). However, Stardust observed a different ISD population than Cassini since Cassini could only detect grains with radii up to $a < 0.2$ μm whereas the Stardust ISD grains where over 50 time more massive ($a > 1$ μm). None show conclusive evidence of carbon-rich particles so far, in contrast to astronomical observations of ISD in the ISM. More research into dust composition is required to get a full image, especially concerning the presence of Carbon (or why it is not present), and the question whether Sulfide is contained in it. While Slavin and Frisch (2008) suggest a depletion of carbonaceous particles in the ISM around the heliosphere based on an overabundance of carbon in the gas phase with respect to solar photospheric abundances, Kimura (2015) proposes that carbon-rich dust particles may be altered on their path through the solar system. Sulfur in the diffuse interstellar medium is generally undepleted (Jenkins 2009) and is thus not expected to be found condensed in the form of LIC dust.

During the formation of our solar system, most presolar dust grains were destroyed or heavily processed. A minor population of circumstellar presolar grains survived processing in the protosolar disk and can be recognized by their extremely diverse isotopic composition (Zinner et al. 2007), but it is unknown whether those grains are representative of the grain populations in the interstellar medium (Frisch et al. 1999; Kimura et al. 2003a; Zhukovska et al. 2008). The homogeneous composition observed by the CDA can be explained by destruction, recondensation and equilibration processes in the ISM that homogenise an initially diverse population that started as circumstellar dust. This is supported by astronomical observations of the diffuse ISM demonstrating that the most condensable elements (atomic mass > 23) are depleted in the gas phase and hence are bound in solids, while the lifetime of interstellar grains against destruction by supernova shocks of about 0.5 Ga is much shorter than the average residence time of matter in the ISM of 2.5 Ga (Zhukovska et al. 2008). During this residence time, ISD grains frequently cycle between the hot interstellar medium (low density regions carved by supernovae), the warm diffuse medium (accessible by spectroscopic observations) and cold molecular clouds, which are star formation regions. Altobelli et al. (2016) suggest that searches for presolar interstellar grains in meteorites lead by isotopic anomalies could miss a population of isotopically inconspicuous presolar grains that recondensed in the ISM (e.g. Bradley et al. 1999; Keller and Messenger 2011).

Presolar grains have a maximum at a silicate composition intermediate between pyroxene and olivine and a tail to high values caused by a bias (overabundance) of refractory oxides. The CDA LIC ISD values center between 2 and 3 (Fig. 7). Such high values are caused either by oxides or metallic Fe, phases that have also been observed in the Stardust particles "Orion" and "Hylabrook". It should be noted, that the presolar circumstellar grain population is biased; for example there is possibly selective loss of amorphous grains or a higher fraction of resistant oxides due to metamorphic destruction on asteroids or comets. Nevertheless, it is highly unlikely that the presolar grain population initially had a distribution similar to the CDA ISD data: In this case the presolar population should have lost primarily the oxide or metal component, which is, however, unlikely, as oxides are rather more resistant against metamorphic destruction.

7.3 Dust Morphology

Also the morphology and density of the particles is not yet fully understood: the Cassini mission with its smaller particles concluded on compact interstellar dust (Altobelli et al. 2016) while the large micron-sized Stardust particles showed surprisingly low densities (Westphal et al. 2014c). Dust dynamics simulations in context of the complete Ulysses data analysis concluded on rather compact dust for the small particles ($\lesssim 0.2\ \mu m$) and low densities for the larger particles ($\gtrsim 0.2\ \mu m$) (Sterken et al. 2015). This is consistent with the Cassini and Stardust observations. Calibration of impact ionization instruments for low-density micron-sized dust particles has to be undertaken for obtaining a reliable and updated mass distribution, and hence, is important for the dust-to-gas mass ratio in the LIC (Sterken et al. 2016).

8 Outlook

Interstellar dust inside the solar system has been studied since 25 years, after its first unambiguous in situ detection by the Ulysses Dust Detector by Grün et al. (1993). Ulysses flew out of the ecliptic plan, but also in the ecliptic and closer to Earth there was ISD detected by in situ missions like Galileo, Helios[13] and Cassini. In the last decade, much progress has been made in the understanding the dust dynamics, its interaction with the heliosphere, and especially in its compositional analysis, by using a wide variety of complementary methods.

The full dataset of Ulysses (1992–2008) has been analysed and interpreted (Krüger et al. 2015; Strub et al. 2015; Sterken et al. 2015), new datasets from spacecraft antenna measurements became available (Malaspina et al. 2014), for the first time, contemporary interstellar dust particles were returned to Earth for further study in the laboratory (Westphal et al. 2014c), and their composition was measured in space using a time-of-flight mass spectrometer (Altobelli et al. 2016). Also, infrared measurements have confirmed the presence of interstellar dust in the solar system (Rowan-Robinson and May 2013), and although interstellar meteors are still heavily under debate (Hajdukova et al. 2019), the discovery of the first interstellar asteroid in 2017 (Feng and Jones 2018; Zhang 2018), indicates that indeed more "big" interstellar dust particles may be discovered or measured in the future with refined techniques (e.g. radar, see Koschny et al. 2019; Hajdukova et al. 2019).

The past decade, several mission concepts for *dust observatories* that would improve our knowledge of interstellar and interplanetary dust in the solar system have been proposed. These have built upon improved instrumentation. Their main asset is the combined use of the different technologies used so far. For instance, a *Dust Telescope* is a dust instrument,

[13]ISD data from Helios were sampled between 1974 and 1980, before Ulysses, but were recognized in the data and analysed only after the first Ulysses ISD detections (Altobelli 2004; Altobelli et al. 2006).

including time-of-flight mass spectrometer, combined with a trajectory detector. An *Active Dust Collector* (Grün et al. 2012) combines the dust telescope with a collection system providing two main advantages: easier identification of the particles in the aerogel tracks thanks to the trajectory sensor, and better tracking of the source of the particle and hereby a better distinction between interstellar dust and interplanetary dust. The SARIM (*SAmple Return of Interstellar Matter*) mission was proposed for ESA's Cosmic Vision 2015–2025 Programme in 2007 (Srama et al. 2009). This mission concept contained such an active dust collector and was intended to fly at the Sun-Earth Lagrange point L_2 to avoid the Earth's space debris, and would be positioned in the free solar wind. More missions for dust astronomy were proposed, like Cosmic DUNE (*DUst Near the Earth*, NASA proposal) in 2008 (Grün and Srama 2006), SARIM+ (ESA proposal, 2012), and the *Solar System Debris Disk* (S2D2) mission concept (proposal for science themes to ESA, 2013; Srama et al. 2013). NASA's Europa Mission will fly to Jupiter and has a dedicated dust detector on board (SUDA; Kempf et al. 2014) that potentially may detect and analyze the composition of ISD. Finally, the Japanese Destiny+ mission will visit the asteroid Phaeton and will carry on board an in situ dust analyzer with a mass resolution of $m/\Delta m \approx 150$ (Krüger et al. 2017). Interstellar dust measurements are foreseen (Krüger et al. 2019b). This instrument will be able to measure both cations and anions using two sensor heads in parallel, allowing O and C to be measured with a much higher sensitivity than before (using only cations during former missions), as well as S, P, SO_4, PO_4 and CO_3 (Masanori et al. 2018). With the developments in instrument capabilities and the upcoming missions (Europa Mission, Destiny+), the future looks good for further interstellar dust research and discoveries.

Acknowledgements FP received financial support from the German Research Foundation (DFG) projects PO 1015/3-1, /4-1, and ERC Consolidator Grant 724908-Habitat OASIS.

Publisher's Note Springer Nature remains neutral with regard to jurisdictional claims in published maps and institutional affiliations.

References

C.M.O. Alexander, L.R. Nittler, J. Davidson, F.J. Ciesla, Measuring the level of interstellar inheritance in the solar protoplanetary disk. Meteorit. Planet. Sci. **52**, 1797–1821 (2017). https://doi.org/10.1111/maps.12891

D.B. Alexashov, O.A. Katushkina, V.V. Izmodenov, P.S. Akaev, Interstellar dust distribution outside the heliopause: deflection at the heliospheric interface. Mon. Not. R. Astron. Soc. **458**, 2553–2564 (2016). https://doi.org/10.1093/mnras/stw514

N. Altobelli, Monitoring of the interstellar dust stream in the inner solar system using data of different spacecraft. PhD thesis, Ruprecht-Karls-Universität Heidelberg (2004)

N. Altobelli, S. Kempf, M. Landgraf, R. Srama, V. Dikarev, H. Krüger, G. Moragas-Klostermeyer, E. Grün, Cassini between Venus and Earth: detection of interstellar dust. J. Geophys. Res. Space Phys. **108**, 8032 (2003). https://doi.org/10.1029/2003JA009874

N. Altobelli, S. Kempf, H. Krüger, M. Landgraf, M. Roy, E. Grün, Interstellar dust flux measurements by the Galileo dust instrument between the orbits of Venus and Mars. J. Geophys. Res. Space Phys. **110**, A07102 (2005). https://doi.org/10.1029/2004JA010772

N. Altobelli, E. Grün, M. Landgraf, A new look into the Helios dust experiment data: presence of interstellar dust inside the Earth's orbit. Astron. Astrophys. **448**, 243–252 (2006). https://doi.org/10.1051/0004-6361:20053909

N. Altobelli, F. Postberg, K. Fiege, M. Trieloff, H. Kimura, V.J. Sterken, H.W. Hsu, J. Hillier, N. Khawaja, G. Moragas-Klostermeyer, J. Blum, M. Burton, R. Srama, S. Kempf, E. Gruen, Flux and composition of interstellar dust at Saturn from Cassini's Cosmic Dust Analyzer. Science **352**, 312–318 (2016). https://doi.org/10.1126/science.aac6397

L. Andersson, T.D. Weber, D. Malaspina, F. Crary, R.E. Ergun, G.T. Delory, C.M. Fowler, M.W. Morooka, T. McEnulty, A.I. Eriksson, D.J. Andrews, M. Horanyi, A. Collette, R. Yelle, B.M. Jakosky, Dust observations at orbital altitudes surrounding Mars. Science **350**, 0398 (2015). https://doi.org/10.1126/science.aad0398

W.J. Baggaley, Advanced Meteor Orbit Radar observations of interstellar meteoroids. J. Geophys. Res. **105**(10), 10353–10362 (2000). https://doi.org/10.1029/1999JA900383

M. Baguhl, E. Grün, M. Landgraf, In situ measurements of interstellar dust with the ULYSSES and Galileo spaceprobes. Space Sci. Rev. **78**, 165–172 (1996). https://doi.org/10.1007/BF00170803

H.A. Bechtel, G.J. Flynn, C. Allen, D. Anderson, A. Ansari, S. Bajt, R.K. Bastien, N. Bassim, J. Borg, F.E. Brenker, J. Bridges, D.E. Brownlee, M. Burchell, M. Burghammer, A.L. Butterworth, H. Changela, P. Cloetens, A.M. Davis, R. Doll, C. Floss, D.R. Frank, Z. Gainsforth, E. Grün, P.R. Heck, J.K. Hillier, P. Hoppe, B. Hudson, J. Huth, B. Hvide, A. Kearsley, A.J. King, B. Lai, J. Leitner, L. Lemelle, H. Leroux, A. Leonard, R. Lettieri, W. Marchant, L.R. Nittler, R. Ogliore, W.J. Ong, F. Postberg, M.C. Price, S.A. Sandford, J.A.S. Tresseras, S. Schmitz, T. Schoonjans, G. Silversmit, A.S. Simionovici, V.A. Solé, R. Srama, F.J. Stadermann, T. Stephan, V.J. Sterken, J. Stodolna, R.M. Stroud, S. Sutton, M. Trieloff, P. Tsou, A. Tsuchiyama, T. Tyliszczak, B. Vekemans, L. Vincze, J. von Korff, A.J. Westphal, N. Wordsworth, D. Zevin, M.E. Zolensky, Stardust Interstellar Preliminary Examination III: infrared spectroscopic analysis of interstellar dust candidates. Meteorit. Planet. Sci. **49**, 1548–1561 (2014). https://doi.org/10.1111/maps.12125

J.L. Bertaux, J.E. Blamont, Possible evidence for penetration of interstellar dust into the solar system. Nature **262**, 263–266 (1976). https://doi.org/10.1038/262263a0

J.P. Bradley, L.P. Keller, T.P. Snow, M.S. Hanner, G.J. Flynn, J.C. Gezo, S.J. Clemett, D.E. Brownlee, J.E. Bowey, An infrared spectral match between GEMS and interstellar grains. Science **285**, 1716–1718 (1999). https://doi.org/10.1126/science.285.5434.1716

F.E. Brenker, A.J. Westphal, L. Vincze, M. Burghammer, S. Schmitz, T. Schoonjans, G. Silversmit, B. Vekemans, C. Allen, D. Anderson, A. Ansari, S. Bajt, R.K. Bastien, N. Bassim, H.A. Bechtel, J. Borg, J. Bridges, D.E. Brownlee, M. Burchell, A.L. Butterworth, H. Changela, P. Cloetens, A.M. Davis, R. Doll, C. Floss, G. Flynn, P. Fougeray, D.R. Frank, Z. Gainsforth, E. Grün, P.R. Heck, J.K. Hillier, P. Hoppe, B. Hudson, J. Huth, B. Hvide, A. Kearsley, A.J. King, B. Lai, J. Leitner, L. Lemelle, H. Leroux, A. Leonard, R. Lettieri, W. Marchant, L.R. Nittler, R. Ogliore, W.J. Ong, F. Postberg, M.C. Price, S.A. Sandford, J.A.S. Tresseras, A.S. Simionovici, V.A. Solé, R. Srama, F. Stadermann, T. Stephan, V.J. Sterken, J. Stodolna, R.M. Stroud, S. Sutton, M. Trieloff, P. Tsou, A. Tsuchiyama, T. Tyliszczak, J. Korff, N. Wordsworth, D. Zevin, M.E. Zolensky, Stardust Interstellar Preliminary Examination V: XRF analyses of interstellar dust candidates at ESRF ID13. Meteorit. Planet. Sci. **49**, 1594–1611 (2014). https://doi.org/10.1111/maps.12206

J.A. Burns, P.L. Lamy, S. Soter, Radiation forces on small particles in the solar system. Icarus **40**, 1–48 (1979). https://doi.org/10.1016/0019-1035(79)90050-2

A.L. Butterworth, A.J. Westphal, T. Tyliszczak, Z. Gainsforth, J. Stodolna, D.R. Frank, C. Allen, D. Anderson, A. Ansari, S. Bajt, R.K. Bastien, N. Bassim, H.A. Bechtel, J. Borg, F.E. Brenker, J. Bridges, D.E. Brownlee, M. Burchell, M. Burghammer, H. Changela, P. Cloetens, A.M. Davis, R. Doll, C. Floss, G. Flynn, E. Grün, P.R. Heck, J.K. Hillier, P. Hoppe, B. Hudson, J. Huth, B. Hvide, A. Kearsley, A.J. King, B. Lai, J. Leitner, L. Lemelle, H. Leroux, A. Leonard, R. Lettieri, W. Marchant, L.R. Nittler, R. Ogliore, W.J. Ong, F. Postberg, M.C. Price, S.A. Sandford, J.A.S. Tresseras, S. Schmitz, T. Schoonjans, G. Silversmit, A.S. Simionovici, V.A. Solé, R. Srama, F.J. Stadermann, T. Stephan, V.J. Sterken, R.M. Stroud, S. Sutton, M. Trieloff, P. Tsou, A. Tsuchiyama, B. Vekemans, L. Vincze, J. von Korff, N. Wordsworth, D. Zevin, M.E. Zolensky, Stardust Interstellar Preliminary Examination IV: Scanning transmission X-ray microscopy analyses of impact features in the Stardust Interstellar Dust Collector. Meteorit. Planet. Sci. **49**, 1562–1593 (2014). https://doi.org/10.1111/maps.12220

A. Collette, E. Grün, D. Malaspina, Z. Sternovsky, Micrometeoroid impact charge yield for common spacecraft materials. J. Geophys. Res. Space Phys. **119**, 6019–6026 (2014). https://doi.org/10.1002/2014JA020042

A. Collette, G. Meyer, D. Malaspina, Z. Sternovsky, Laboratory investigation of antenna signals from dust impacts on spacecraft. J. Geophys. Res. Space Phys. **120**, 5298–5305 (2015). https://doi.org/10.1002/2015JA021198

A. Collette, D.M. Malaspina, Z. Sternovsky, Characteristic temperatures of hypervelocity dust impact plasmas. J. Geophys. Res. Space Phys. **121**, 8182–8187 (2016). https://doi.org/10.1002/2015JA022220

A. Czechowski, I. Mann, Penetration of interstellar dust grains into the heliosphere. J. Geophys. Res. Space Phys. **108**, 8038 (2003). https://doi.org/10.1029/2003JA009917

J.M.A. Danby, G.L. Camm, Statistical dynamics and accretion. Mon. Not. R. Astron. Soc. **117**, 50 (1957). https://doi.org/10.1093/mnras/117.1.50

H. Dietzel, G. Eichhorn, H. Fechtig, E. Grun, H.J. Hoffmann, J. Kissel, The HEOS 2 and HELIOS micrometeoroid experiments. J. Phys. E, Sci. Instrum. **6**, 209–217 (1973). https://doi.org/10.1088/0022-3735/6/3/008

B.T. Draine, H.M. Lee, Optical properties of interstellar graphite and silicate grains. Astrophys. J. **285**, 89–108 (1984). https://doi.org/10.1086/162480

F. Feng, H.R.A. Jones, Oumuamua as a messenger from the Local Association. Astrophys. J. Lett. **852**, L27 (2018). https://doi.org/10.3847/2041-8213/aaa404. 1711.08800

K. Fiege, M. Trieloff, J.K. Hillier, M. Guglielmino, F. Postberg, R. Srama, S. Kempf, J. Blum, Calibration of relative sensitivity factors for impact ionization detectors with high-velocity silicate microparticles. Icarus **241**, 336–345 (2014). https://doi.org/10.1016/j.icarus.2014.07.015

G.J. Flynn, S.R. Sutton, B. Lai, S. Wirick, C. Allen, D. Anderson, A. Ansari, S. Bajt, R.K. Bastien, N. Bassim, H.A. Bechtel, J. Borg, F.E. Brenker, J. Bridges, D.E. Brownlee, M. Burchell, M. Burghammer, A.L. Butterworth, H. Changela, P. Cloetens, A.M. Davis, R. Doll, C. Floss, D. Frank, Z. Gainsforth, E. Grün, P.R. Heck, J.K. Hillier, P. Hoppe, B. Hudson, J. Huth, B. Hvide, A. Kearsley, A.J. King, J. Leitner, L. Lemelle, H. Leroux, A. Leonard, R. Lettieri, W. Marchant, L.R. Nittler, R. Ogliore, W.J. Ong, F. Postberg, M.C. Price, S.A. Sandford, J.A.S. Tresseras, S. Schmitz, T. Schoonjans, G. Silversmit, A. Simionovici, V.A. Sol, R. Srama, F.J. Stadermann, T. Stephan, V. Sterken, J. Stodolna, R.M. Stroud, M. Trieloff, P. Tsou, A. Tsuchiyama, T. Tyliszczak, B. Vekemans, L. Vincze, J. von Korff, A.J. Westphal, N. Wordsworth, D. Zevin, M.E. Zolensky, Stardust Interstellar Preliminary Examination VII: synchrotron X-ray fluorescence analysis of six Stardust interstellar candidates measured with the Advanced Photon Source 2-ID-D microprobe. Meteorit. Planet. Sci. **49**, 1626–1644 (2014). https://doi.org/10.1111/maps.12144

D.R. Frank, A.J. Westphal, M.E. Zolensky, Z. Gainsforth, A.L. Butterworth, R.K. Bastien, C. Allen, D. Anderson, A. Ansari, S. Bajt, N. Bassim, H.A. Bechtel, J. Borg, F.E. Brenker, J. Bridges, D.E. Brownlee, M. Burchell, M. Burghammer, H. Changela, P. Cloetens, A.M. Davis, R. Doll, C. Floss, G. Flynn, E. Grün, P.R. Heck, J.K. Hillier, P. Hoppe, B. Hudson, J. Huth, B. Hvide, A. Kearsley, A.J. King, B. Lai, J. Leitner, L. Lemelle, H. Leroux, A. Leonard, R. Lettieri, W. Marchant, L.R. Nittler, R. Ogliore, W.J. Ong, F. Postberg, M.C. Price, S.A. Sandford, J.A.S. Tresseras, S. Schmitz, T. Schoonjans, G. Silversmit, A.S. Simionovici, V.A. Solé, R. Srama, T. Stephan, V.J. Sterken, J. Stodolna, R.M. Stroud, S. Sutton, M. Trieloff, P. Tsou, A. Tsuchiyama, T. Tyliszczak, B. Vekemans, L. Vincze, J. von Korff, N. Wordsworth, D. Zevin, Stardust Interstellar Preliminary Examination II: curating the interstellar dust collector, picokeystones, and sources of impact tracks. Meteorit. Planet. Sci. **49**, 1522–1547 (2014). https://doi.org/10.1111/maps.12147

P.C. Frisch, Foreword. J. Geophys. Res. **105**, 10237–10238 (2000). https://doi.org/10.1029/1999JA900349

P.C. Frisch, J.D. Slavin, Interstellar dust close to the Sun. Earth Planets Space **65**, 175 (2013)

P.C. Frisch, J.M. Dorschner, J. Geiss, J.M. Greenberg, E. Grün, M. Landgraf, P. Hoppe, A.P. Jones, W. Krätschmer, T.J. Linde, G.E. Morfill, W. Reach, J.D. Slavin, J. Svestka, A.N. Witt, G.P. Zank, Dust in the local interstellar wind. Astrophys. J. **525**, 492–516 (1999). https://doi.org/10.1086/307869. astro-ph/9905108

P.C. Frisch, S. Redfield, J.D. Slavin, The interstellar medium surrounding the Sun. Annu. Rev. Astron. Astrophys. **49**, 237–279 (2011). https://doi.org/10.1146/annurev-astro-081710-102613

Z. Gainsforth, F.E. Brenker, A.S. Simionovici, S. Schmitz, M. Burghammer, A.L. Butterworth, P. Cloetens, L. Lemelle, J.A.S. Tresserras, T. Schoonjans, G. Silversmit, V.A. Solé, B. Vekemans, L. Vincze, A.J. Westphal, C. Allen, D. Anderson, A. Ansari, S. Bajt, R.K. Bastien, N. Bassim, H.A. Bechtel, J. Borg, J. Bridges, D.E. Brownlee, M. Burchell, H. Changela, A.M. Davis, R. Doll, C. Floss, G. Flynn, P. Fougeray, D. Frank, E. Grün, P.R. Heck, J.K. Hillier, P. Hoppe, B. Hudson, J. Huth, B. Hvide, A. Kearsley, A.J. King, B. Lai, J. Leitner, H. Leroux, A. Leonard, R. Lettieri, W. Marchant, L.R. Nittler, R. Ogliore, W.J. Ong, F. Postberg, M.C. Price, S.A. Sandford, R. Srama, T. Stephan, V. Sterken, J. Stodolna, R.M. Stroud, S. Sutton, M. Trieloff, P. Tsou, A. Tsuchiyama, T. Tyliszczak, J. von Korff, D. Zevin, M.E. Zolensky, Stardust Interstellar Preliminary Examination VIII: identification of crystalline material in two interstellar candidates. Meteorit. Planet. Sci. **49**, 1645–1665 (2014). https://doi.org/10.1111/maps.12148

J.M. Greenberg, A. Li, What are the true astronomical silicates? Astron. Astrophys. **309**, 258–266 (1996)

K. Grogan, S.F. Dermott, B.A.S. Gustafson, An estimation of the interstellar contribution to the zodiacal thermal emission. Astrophys. J. **472**, 812 (1996). https://doi.org/10.1086/178110

E. Grün, R. Srama (Cosmic Dune Team), The Cosmic DUNE dust astronomy mission, in *European Planetary Science Congress 2006* (2006), p. 292

E. Grün, J. Svestka, Physics of interplanetary and interstellar dust. Space Sci. Rev. **78**, 347–360 (1996). https://doi.org/10.1007/BF00170821

E. Grün, H. Fechtig, J. Kissel, The micrometeorite experiment on HELIOS. Geochim. Cosmochim. Acta, Suppl. (1984)

E. Grün, H. Fechtig, M.S. Hanner, J. Kissel, B.A. Lindblad, D. Linkert, D. Maas, G.E. Morfill, H.A. Zook, The Galileo dust detector. Space Sci. Rev. **60**, 317–340 (1992). https://doi.org/10.1007/BF00216860

E. Grün, H. Zook, M. Baguhl, A. Balogh, S. Bame, H. Fechtig, R. Forsyth, M. Hanner, M. Horanyi, J. Kissel, B.A. Lindblad, D. Linkert, G. Linkert, I. Mann, J. McDonnel, G. Morfill, J. Phillips, C. Polanskey, G. Schwehm, N. Siddique, P. Staubach, J. Svestka, A. Taylor, Discovery of Jovian dust streams and interstellar grains by the Ulysses spacecraft. Nature **362**, 428–430 (1993)

E. Grün, B. Gustafson, I. Mann, M. Baguhl, G.E. Morfill, P. Staubach, A. Taylor, H.A. Zook, Interstellar dust in the heliosphere. Astron. Astrophys. **286**, 915–924 (1994)

E. Grün, Z. Sternovsky, M. Horanyi, V. Hoxie, S. Robertson, J. Xi, S. Auer, M. Landgraf, F. Postberg, M.C. Price, R. Srama, N.A. Starkey, J.K. Hillier, I.A. Franchi, P. Tsou, A. Westphal, Z. Gainsforth, Active cosmic dust collector. Planet. Space Sci. **60**, 261–273 (2012). https://doi.org/10.1016/j.pss.2011.09.006

D.A. Gurnett, E. Grun, D. Gallagher, W.S. Kurth, F.L. Scarf, Micron-sized particles detected near Saturn by the Voyager plasma wave instrument. Icarus **53**, 236–254 (1983). https://doi.org/10.1016/0019-1035(83)90145-8

D.A. Gurnett, T.F. Averkamp, F.L. Scarf, E. Grun, Dust particles detected near Giacobini-Zinner by the ICE plasma wave instrument. Geophys. Res. Lett. **13**, 291–294 (1986). https://doi.org/10.1029/GL013i003p00291

D.A. Gurnett, W.S. Kurth, K.L. Scarf, J.A. Burns, J.N. Cuzzi, Micron-sized particle impacts detected near Uranus by the Voyager 2 plasma wave instrument. J. Geophys. Res. **92**(14), 14959–14968 (1987). https://doi.org/10.1029/JA092iA13p14959

D.A. Gurnett, W.S. Kurth, L.J. Granroth, S.C. Allendorf, R.L. Poynter, Micron-sized particles detected near Neptune by the Voyager 2 plasma wave instrument. J. Geophys. Res. **96**, 19 (1991). https://doi.org/10.1029/91JA01270

D.A. Gurnett, W.S. Kurth, D.L. Kirchner, G.B. Hospodarsky, T.F. Averkamp, P. Zarka, A. Lecacheux, R. Manning, A. Roux, P. Canu, N. Cornilleau-Wehrlin, P. Galopeau, A. Meyer,hajdukova:2018iaubook A. Meyer, R. Boström, G. Gustafsson, J.E. Wahlund, L. Åhlen, H.O. Rucker, H.P. Ladreiter, W. Macher, L.J.C. Woolliscroft, H. Alleyne, M.L. Kaiser, M.D. Desch, W.M. Farrell, C.C. Harvey, P. Louarn, P.J. Kellogg, K. Goetz, A. Pedersen, The Cassini radio and plasma wave investigation. Space Sci. Rev. **114**, 395–463 (2004). https://doi.org/10.1007/s11214-004-1434-0

B.A.S. Gustafson, Physics of zodiacal dust. Annu. Rev. Earth Planet. Sci. **22**, 553–595 (1994). https://doi.org/10.1146/annurev.ea.22.050194.003005

B.S. Gustafson, N. Misconi, Streaming of interstellar grains in the solar system. Nature **282**, 276–278 (1979)

M. Hajdukova, V.J. Sterken, P. Wiegert, Interstellar meteoroids (2019)

J.K. Hillier, S.F. Green, N. McBride, J.P. Schwanethal, F. Postberg, R. Srama, S. Kempf, G. Moragas-Klostermeyer, J.A.M. McDonnell, E. Grün, The composition of Saturn's E ring. Mon. Not. R. Astron. Soc. **377**, 1588–1596 (2007). https://doi.org/10.1111/j.1365-2966.2007.11710.x

M. Horanyi, Charged dust dynamics in the solar system. Annu. Rev. Astron. Astrophys. **34**, 383–418 (1996). https://doi.org/10.1146/annurev.astro.34.1.383

ISPE, Meteorit. Planet. Sci. **49**(9), 1509–1733 (2014). http://onlinelibrary.wiley.com/doi/10.1111/maps.2014.49.issue-9/issuetoc

E.B. Jenkins, A unified representation of gas-phase element depletions in the interstellar medium. Astrophys. J. **700**, 1299–1348 (2009). https://doi.org/10.1088/0004-637X/700/2/1299

L.P. Keller, S. Messenger, On the origins of GEMS grains. Geochim. Cosmochim. Acta **75**, 5336–5365 (2011). https://doi.org/10.1016/j.gca.2011.06.040

P.J. Kellogg, K. Goetz, S.J. Monson, Dust impact signals on the wind spacecraft. J. Geophys. Res. Space Phys. **121**, 966–991 (2016). https://doi.org/10.1002/2015JA021124

F. Kemper, W.J. Vriend, A.G.G.M. Tielens, The absence of crystalline silicates in the diffuse interstellar medium. Astrophys. J. **609**, 826–837 (2004). https://doi.org/10.1086/421339. astro-ph/0403609

S. Kempf, R. Srama, N. Altobelli, S. Auer, V. Tschernjawski, J. Bradley, M.E. Burton, S. Helfert, T.V. Johnson, H. Krüger, G. Moragas-Klostermeyer, E. Grün, Cassini between Earth and asteroid belt: first in-situ charge measurements of interplanetary grains. Icarus **171**, 317–335 (2004). https://doi.org/10.1016/j.icarus.2004.05.017

S. Kempf, N. Altobelli, C. Briois, E. Grün, M. Horanyi, F. Postberg, J. Schmidt, R. Srama, Z. Sternovsky, G. Tobie, M. Zolotov, SUDA: a dust mass spectrometer for compositional surface mapping for a mission to Europa, in *European Planetary Science Congress*, vol. 9 (2014), EPSC2014-229

H. Kimura, Interstellar dust in the Local Cloud surrounding the Sun. Mon. Not. R. Astron. Soc. **449**, 2250–2258 (2015). https://doi.org/10.1093/mnras/stv427

H. Kimura, On the photoelectric quantum yield of small dust particles. Mon. Not. R. Astron. Soc. **459**, 2751–2761 (2016). https://doi.org/10.1093/mnras/stw820. 1604.03664

H. Kimura, High radiation pressure on interstellar dust computed by light-scattering simulation on fluffy agglomerates of magnesium-silicate grains with metallic-iron inclusions. Astrophys. J. Lett. **839**, L23 (2017). https://doi.org/10.3847/2041-8213/aa6c2d. 1704.02066
H. Kimura, I. Mann, The electric charging of interstellar dust in the solar system and consequences for its dynamics. Astrophys. J. **499**, 454–462 (1998). https://doi.org/10.1086/305613
H. Kimura, I. Mann, Filtering of the interstellar dust flow near the heliopause: the importance of secondary electron emission for the grain charging. Earth Planets Space **51**, 1223–1232 (1999). https://doi.org/10.1186/BF03351596
H. Kimura, I. Mann, Selection effects on interstellar dust in heliosphere. Adv. Space Res. **25**, 299–302 (2000). https://doi.org/10.1016/S0273-1177(99)00952-7
H. Kimura, I. Mann, E.K. Jessberger, Composition, structure, and size distribution of dust in the local interstellar cloud. Astrophys. J. **583**, 314–321 (2003a). https://doi.org/10.1086/345102
H. Kimura, I. Mann, E.K. Jessberger, Elemental abundances and mass densities of dust and gas in the local interstellar cloud. Astrophys. J. **582**, 846–858 (2003b). https://doi.org/10.1086/344691
D. Koschny, R.H. Soja, C. Engrand, G.J. Flynn, J. Lasue, A.C. Levasseur-Regourd, T. Nakamura, D. Malaspina, A.R. Poppe, V.J. Sterken, J.M. Trigo-Rodríguez, Interplanetary dust, meteoroids, meteors and meteorites. Space Sci. Rev. **215**(4), 1–62 (2019)
H. Krüger, M. Landgraf, N. Altobelli, E. Grün, Interstellar dust in the solar system. Space Sci. Rev. **130**, 401–408 (2007). https://doi.org/10.1007/s11214-007-9181-7. 0706.3110
H. Krüger, P. Strub, E. Grün, V.J. Sterken, Sixteen years of Ulysses interstellar dust measurements in the solar system. I. Mass distribution and gas-to-dust mass ratio. Astrophys. J. **812**, 139 (2015). https://doi.org/10.1088/0004-637X/812/2/139. 1510.06180
H. Krüger, M. Kobayashi, T. Arai, R. Srama, B.V. Sarli, H. Kimura, G. Moragas-Klostermeyer, R. Soja, N. Altobelli, E. Grün, Dust analysis on board the Destiny+ mission to 3200 Phaethon, in *European Planetary Science Congress*, vol. 11 (2017), EPSC2017-204
H. Krüger, N. Altobelli, P. Strub, V. Sterken, R. Srama, E. Grün, Interstellar dust in the inner solar system: model versus in-situ spacecraft data. Astron. Astrophys. **626**, A37 (2019a)
H. Krüger, P. Strub, R. Srama, M. Kobayashi, T. Arai, H. Kimura, T. Hirai, G. Moragas-Klostermeyer, N. Altobelli, V. Sterken, J. Agarwal, E. Grün, Modelling DESTINY+ interplanetary and interstellar dust measurements en route to the active asteroid (3200) Phaethon. Planet. Space Sci. **172**, 22–42 (2019b)
W.S. Kurth, T.F. Averkamp, D.A. Gurnett, Z. Wang, Cassini RPWS observations of dust in Saturn's E ring. Planet. Space Sci. **54**, 988–998 (2006). https://doi.org/10.1016/j.pss.2006.05.011
H. Laakso, R. Grard, A. Pedersen, G. Schwehm, Impacts of large dust particles on the VEGA spacecraft. Adv. Space Res. **9**, 269–272 (1989). https://doi.org/10.1016/0273-1177(89)90273-1
M. Landgraf, PhD thesis. Ruprecht-Karls-Univ, Heidelberg (1998)
M. Landgraf, Modeling the motion and distribution of interstellar dust inside the heliosphere. J. Geophys. Res. **105**(10), 10303–10316 (2000)
M. Landgraf, K. Augustsson, E. Grün, B.A.S. Gustafson, Deflection of the local interstellar dust flow by solar radiation pressure. Science **286**, 2319–2322 (1999a)
M. Landgraf, M. Müller, E. Grün, Prediction of the in-situ dust measurements of the stardust mission to comet 81P/Wild 2. Planet. Space Sci. **47**, 1029–1050 (1999b). https://doi.org/10.1016/S0032-0633(99)00031-8. astro-ph/9904204
M. Landgraf, W.J. Baggaley, E. Grün, H. Krüger, G. Linkert, Aspects of the mass distribution of interstellar dust grains in the solar system from in situ measurements. J. Geophys. Res. **105**(10), 10343–10352 (2000)
M. Landgraf, H. Krüger, N. Altobelli, E. Grün, Penetration of the heliosphere by the interstellar dust stream during solar maximum. J. Geophys. Res. Space Phys. **108**, 8030 (2003). https://doi.org/10.1029/2003JA009872
J. Leitner, C. Vollmer, P. Hoppe, J. Zipfel, Characterization of presolar material in the CR chondrite Northwest Africa 852. Astrophys. J. **745**, 38 (2012). https://doi.org/10.1088/0004-637X/745/1/38
A.C. Levasseur-Regourd, J. Agarwal, H. Cottin, C. Engrand, G. Flynn, M. Fulle, T. Gombosi, Y. Langevin, J. Lasue, T. Mannel, S. Merouane, O. Poch, N. Thomas, A. Westphal, Cometary dust. Space Sci. Rev. **214**(3), 64 (2018). https://doi.org/10.1007/s11214-018-0496-3
E.H. Levy, J.R. Jokipii, Penetration of interstellar dust into the Solar System. Nature **264**, 423–424 (1976)
T.J. Linde, T.I. Gombosi, Interstellar dust filtration at the heliospheric interface. J. Geophys. Res. **105**(10), 10411–10418 (2000). https://doi.org/10.1029/1999JA900149
Q. Ma, L.S. Matthews, V. Land, T.W. Hyde, Charging of aggregate grains in astrophysical environments. Astrophys. J. **763**, 77 (2013). https://doi.org/10.1088/0004-637X/763/2/77. 1210.0459
D. Malaspina, Coordinated Data Analysis Web: The Wind ISD Database (2017). wi_l3-dustimpact_waves. https://cdaweb.sci.gsfc.nasa.gov/index.html/

D.M. Malaspina, L.B. Wilson, A database of interplanetary and interstellar dust detected by the Wind spacecraft. J. Geophys. Res. Space Phys. **121**, 9369–9377 (2016). https://doi.org/10.1002/2016JA023209

D.M. Malaspina, M. Horányi, A. Zaslavsky, K. Goetz, L.B. Wilson, K. Kersten, Interplanetary and interstellar dust observed by the Wind/WAVES electric field instrument. Geophys. Res. Lett. **41**, 266–272 (2014). https://doi.org/10.1002/2013GL058786

M. Masanori, R. Srama, H. Krüger, T. Arai, H. Kimura, DESTINY+ Dust Analyzer, in *49th Lunar and Planetary Science Conference 2018* (2018) (LPI Contrib. No. 2083)

J.S. Mathis, W. Rumpl, K.H. Nordsieck, The size distribution of interstellar grains. Astrophys. J. **217**, 425–433 (1977). https://doi.org/10.1086/155591

N. McBride, M. Jam, Meteoroid impacts on spacecraft: sporadics, streams, and the 1999 Leonids. Planet. Space Sci. **47**, 1005–1013 (1999). https://doi.org/10.1016/S0032-0633(99)00023-9

D.J. McComas, M. Bzowski, P. Frisch, S.A. Fuselier, M.A. Kubiak, H. Kucharek, T. Leonard, E. Möbius, N.A. Schwadron, J.M. Sokół, P. Swaczyna, M. Witte, Warmer local interstellar medium: a possible resolution of the Ulysses-IBEX enigma. Astrophys. J. **801**, 28 (2015). https://doi.org/10.1088/0004-637X/801/1/28

K.J. Meech, R. Weryk, M. Micheli, J.T. Kleyna, O.R. Hainaut, R. Jedicke, R.J. Wainscoat, K.C. Chambers, J.V. Keane, A. Petric, L. Denneau, E. Magnier, T. Berger, M.E. Huber, H. Flewelling, C. Waters, E. Schunova-Lilly, S. Chastel, A brief visit from a red and extremely elongated interstellar asteroid. Nature **552**, 378–381 (2017). https://doi.org/10.1038/nature25020

N. Meyer-Vernet, M.G. Aubier, B.M. Pedersen, Voyager 2 at Uranus—grain impacts in the ring plane. Geophys. Res. Lett. **13**, 617–620 (1986). https://doi.org/10.1029/GL013i007p00617

N. Meyer-Vernet, A. Lecacheux, M.L. Kaiser, D.A. Gurnett, Detecting nanoparticles at radio frequencies: Jovian dust stream impacts on Cassini/RPWS. Geophys. Res. Lett. **36**, L03103 (2009a). https://doi.org/10.1029/2008GL036752

N. Meyer-Vernet, M. Maksimovic, A. Czechowski, I. Mann, I. Zouganelis, K. Goetz, M.L. Kaiser, O.C. St. Cyr, J.L. Bougeret, S.D. Bale, Dust detection by the Wave Instrument on STEREO: nanoparticles picked up by the solar wind? Sol. Phys. **256**, 463 (2009b)

G.E. Morfill, E. Gruen, The motion of charged dust particles in interplanetary space. I—The zodiacal dust cloud. II—Interstellar grains. Planet. Space Sci. **27**, 1269–1292 (1979). https://doi.org/10.1016/0032-0633(79)90105-3

T. Mukai, On the charge distribution of interplanetary grains. Astron. Astrophys. **99**, 1–6 (1981)

F.M. Neubauer, K.H. Glassmeier, A.J. Coates, R. Goldstein, M.H. Acuna, Hypervelocity dust particle impacts observed by the Giotto magnetometer and plasma experiments. Geophys. Res. Lett. **17**, 1809–1812 (1990). https://doi.org/10.1029/GL017i011p01809

P. Oberc, Electric antenna as a dust detector. Adv. Space Res. **17**, 105–110 (1996). https://doi.org/10.1016/0273-1177(95)00766-8

A. Pais, *Inward Bound: Of Matter and Forces in the Physical World* (Oxford University Press, London, 1986)

F. Pantellini, S. Belheouane, N. Meyer-Vernet, A. Zaslavsky, Nano dust impacts on spacecraft and boom antenna charging. Astrophys. Space Sci. **341**, 309–314 (2012). https://doi.org/10.1007/s10509-012-1108-4. 1205.1430

E.N. Parker, Dynamics of the interplanetary gas and magnetic fields. Astrophys. J. **128**, 664 (1958). https://doi.org/10.1086/146579

B.M. Pedersen, N. Meyer-Vernet, M.G. Aubier, P. Zarka, Dust distribution around Neptune—grain impacts near the ring plane measured by the Voyager planetary radio astronomy experiment. J. Geophys. Res. **96**, 19 (1991). https://doi.org/10.1029/91JA01601

F. Postberg, S. Kempf, J.K. Hillier, R. Srama, S.F. Green, N. McBride, E. Grün, The E-ring in the vicinity of Enceladus. II. Probing the moon's interior—the composition of E-ring particles. Icarus **193**, 438–454 (2008). https://doi.org/10.1016/j.icarus.2007.09.001

F. Postberg, S. Kempf, D. Rost, T. Stephan, R. Srama, M. Trieloff, A. Mocker, M. Goerlich, Discriminating contamination from particle components in spectra of Cassini's dust detector CDA. Planet. Space Sci. **57**, 1359–1374 (2009a). https://doi.org/10.1016/j.pss.2009.06.027

F. Postberg, S. Kempf, J. Schmidt, N. Brilliantov, A. Beinsen, B. Abel, U. Buck, R. Srama, Sodium salts in E-ring ice grains from an ocean below the surface of Enceladus. Nature **459**, 1098–1101 (2009b). https://doi.org/10.1038/nature08046

F. Postberg, J.K. Hillier, S.P. Armes, S. Bugiel, A. Butterworth, D. Dupin, L.A. Fielding, S. Fujii, Z. Gainsforth, E. Grün, Y.W. Li, R. Srama, V. Sterken, J. Stodolna, M. Trieloff, A. Westphal, C. Achilles, C. Allen, A. Ansari, S. Bajt, N. Bassim, R.K. Bastien, H.A. Bechtel, J. Borg, F. Brenker, J. Bridges, D.E. Brownlee, M. Burchell, M. Burghammer, H. Changela, P. Cloetens, A. Davis, R. Doll, C. Floss, G. Flynn, D. Frank, P.R. Heck, P. Hoppe, G. Huss, J. Huth, A. Kearsley, A.J. King, B. Lai, J. Leitner, L. Lemelle, A. Leonard, H. Leroux, R. Lettieri, W. Marchant, L.R. Nittler, R. Ogliore, W.J. Ong, M.C. Price, S.A. Sandford, J.A.S. Tressaras, S. Schmitz, T. Schoonjans, K. Schreiber, G. Silversmit, A.

Simionovici, V.A. Solé, F. Stadermann, T. Stephan, R.M. Stroud, S. Sutton, P. Tsou, A. Tsuchiyama, T. Tyliczszak, B. Vekemans, L. Vincze, D. Zevin, M.E. Zolensky, Stardust Interstellar Preliminary Examination IX: high-speed interstellar dust analog capture in Stardust flight-spare aerogel. Meteorit. Planet. Sci. **49**, 1666–1679 (2014). https://doi.org/10.1111/maps.12173

S. Redfield, J.L. Linsky, The three-dimensional structure of the warm local interstellar medium. II. The Colorado model of the local interstellar cloud. Astrophys. J. **534**, 825–837 (2000). https://doi.org/10.1086/308769

S. Redfield, B.E. Wood, J.L. Linsky, Physical structure of the local interstellar medium. Adv. Space Res. **34**, 41–45 (2004). https://doi.org/10.1016/j.asr.2003.02.053

M. Rowan-Robinson, B. May, An improved model for the infrared emission from the zodiacal dust cloud: cometary, asteroidal and interstellar dust. Mon. Not. R. Astron. Soc. **429**, 2894–2902 (2013). https://doi.org/10.1093/mnras/sts471. 1212.4759

F.L. Scarf, D.A. Gurnett, W.S. Kurth, R.L. Poynter, Voyager 2 plasma wave observations at Saturn. Science **215**, 587–594 (1982). https://doi.org/10.1126/science.215.4532.587

N.A. Schwadron, E. Möbius, T. Leonard, S.A. Fuselier, D.J. McComas, D. Heirtzler, H. Kucharek, F. Rahmanifard, M. Bzowski, M.A. Kubiak, J.M. Sokół, P. Swaczyna, P. Frisch, Determination of interstellar He parameters using five years of data from the IBEX: beyond closed form approximations. Astrophys. J. Suppl. Ser. **220**, 25 (2015). https://doi.org/10.1088/0067-0049/220/2/25

G. Schwehm, Radiation pressure on interplanetary dust particles, in *Interplanetary Dust and Zodiacal Light*, ed. by H. Elsaesser, H. Fechtig. Lecture Notes in Physics, vol. 48 (Springer, Berlin, 1976), pp. 459–463. https://doi.org/10.1007/3-540-07615-8_526

K. Silsbee, B.T. Draine, Radiation pressure on fluffy submicron-sized grains. Astrophys. J. **818**, 133 (2016). https://doi.org/10.3847/0004-637X/818/2/133. 1508.00646

A.S. Simionovici, L. Lemelle, P. Cloetens, V.A. Solé, J.A.S. Tresseras, A.L. Butterworth, A.J. Westphal, Z. Gainsforth, J. Stodolna, C. Allen, D. Anderson, A. Ansari, S. Bajt, N. Bassim, R.K. Bastien, H.A. Bechtel, J. Borg, F.E. Brenker, J. Bridges, D.E. Brownlee, M. Burchell, M. Burghammer, H. Changela, A.M. Davis, R. Doll, C. Floss, G. Flynn, D.R. Frank, E. Grün, P.R. Heck, J.K. Hillier, P. Hoppe, B. Hudson, J. Huth, B. Hvide, A. Kearsley, A.J. King, B. Lai, J. Leitner, A. Leonard, H. Leroux, R. Lettieri, W. Marchant, L.R. Nittler, R. Ogliore, W.J. Ong, F. Postberg, M.C. Price, S.A. Sandford, S. Schmitz, T. Schoonjans, G. Silversmit, R. Srama, F.J. Stadermann, T. Stephan, V.J. Sterken, R.M. Stroud, S. Sutton, M. Trieloff, P. Tsou, A. Tsuchiyama, T. Tyliszczak, B. Vekemans, L. Vincze, J. Korff, N. Wordsworth, D. Zevin, M.E. Zolensky, Stardust Interstellar Preliminary Examination VI: quantitative elemental analysis by synchrotron X-ray fluorescence nanoimaging of eight impact features in aerogel. Meteorit. Planet. Sci. **49**, 1612–1625 (2014). https://doi.org/10.1111/maps.12208

J.D. Slavin, P.C. Frisch, The boundary conditions of the heliosphere: photoionization models constrained by interstellar and in situ data. Astron. Astrophys. **491**, 53–68 (2008). https://doi.org/10.1051/0004-6361:20078101

J.D. Slavin, P.C. Frisch, H.R. Müller, J. Heerikhuisen, N.V. Pogorelov, W.T. Reach, G. Zank, Trajectories and distribution of interstellar dust grains in the heliosphere. Astrophys. J. **760**, 46 (2012). https://doi.org/10.1088/0004-637X/760/1/46. 1210.1127

R. Srama, T.J. Ahrens, N. Altobelli, S. Auer, J.G. Bradley, M. Burton, V.V. Dikarev, T. Economou, H. Fechtig, M. Görlich, M. Grande, A. Graps, E. Grün, O. Havnes, S. Helfert, M. Horanyi, E. Igenbergs, E.K. Jessberger, T.V. Johnson, S. Kempf, A.V. Krivov, H. Krüger, A. Mocker-Ahlreep, G. Moragas-Klostermeyer, P. Lamy, M. Landgraf, D. Linkert, G. Linkert, F. Lura, J.A.M. McDonnell, D. Möhlmann, G.E. Morfill, M. Müller, M. Roy, G. Schäfer, G. Schlotzhauer, G.H. Schwehm, F. Spahn, M. Stübig, J. Svestka, V. Tschernjawski, A.J. Tuzzolino, R. Wäsch, H.A. Zook, The Cassini cosmic dust analyzer. Space Sci. Rev. **114**, 465–518 (2004). https://doi.org/10.1007/s11214-004-1435-z

R. Srama, T. Stephan, E. Grün, N. Pailer, A. Kearsley, A. Graps, R. Laufer, P. Ehrenfreund, N. Altobelli, K. Altwegg, S. Auer, J. Baggaley, M.J. Burchell, J. Carpenter, L. Colangeli, F. Esposito, S.F. Green, H. Henkel, M. Horanyi, A. Jäckel, S. Kempf, N. McBride, G. Moragas-Klostermeyer, H. Krüger, P. Palumbo, A. Srowig, M. Trieloff, P. Tsou, Z. Sternovsky, O. Zeile, H.P. Röser, Sample return of interstellar matter (SARIM). Exp. Astron. **23**, 303–328 (2009). https://doi.org/10.1007/s10686-008-9088-7

R. Srama, E. Gruün, A. Krivov, R. Soja, V. Sterken, Z. Sternovsky, S2d2: solar system debris disk (2013). http://www.irs.uni-stuttgart.de/cosmicdust/missions/debrisdisk/

O.C. St. Cyr, M.L. Kaiser, N. Meyer-Vernet, R.A. Howard, R.A. Harrison, S.D. Bale, W.T. Thompson, K. Goetz, M. Maksimovic, J.L. Bougeret, D. Wang, S. Crothers, STEREO SECCHI and S/WAVES observations of spacecraft debris caused by micron-size interplanetary dust impacts. Sol. Phys. **256**, 475–488 (2009). https://doi.org/10.1007/s11207-009-9362-5

V.J. Sterken, N. Altobelli, S. Kempf, G. Schwehm, R. Srama, E. Grün, The flow of interstellar dust into the solar system. Astron. Astrophys. **538**, A102 (2012). https://doi.org/10.1051/0004-6361/201117119

V.J. Sterken, N. Altobelli, S. Kempf, H. Krüger, R. Srama, P. Strub, E. Grün, The filtering of interstellar dust in the solar system. Astron. Astrophys. **552**, A130 (2013). https://doi.org/10.1051/0004-6361/201219609

V.J. Sterken, A.J. Westphal, N. Altobelli, E. Grün, J.K. Hillier, F. Postberg, R. Srama, C. Allen, D. Anderson, A. Ansari, S. Bajt, R.S. Bastien, N. Bassim, H.A. Bechtel, J. Borg, F.E. Brenker, J. Bridges, D.E. Brownlee, M. Burchell, M. Burghammer, A.L. Butterworth, H. Changela, P. Cloetens, A.M. Davis, R. Doll, C. Floss, G. Flynn, D. Frank, Z. Gainsforth, P.R. Heck, P. Hoppe, B. Hudson, J. Huth, B. Hvide, A. Kearsley, A.J. King, B. Lai, J. Leitner, L. Lemelle, H. Leroux, A. Leonard, R. Lettieri, W. Marchant, L.R. Nittler, R. Ogliore, W.J. Ong, M.C. Price, S.A. Sandford, J.A.S. Tresseras, S. Schmitz, T. Schoonjans, G. Silversmit, A. Simionovici, V.A. Solé, T. Stephan, J. Stodolna, R.M. Stroud, S. Sutton, M. Trieloff, P. Tsou, A. Tsuchiyama, T. Tyliszczak, B. Vekemans, L. Vincze, J. von Korff, N. Wordsworth, D. Zevin, M.E. Zolensky, Stardust Interstellar Preliminary Examination X: impact speeds and directions of interstellar grains on the Stardust dust collector. Meteorit. Planet. Sci. **49**, 1680–1697 (2014). https://doi.org/10.1111/maps.12219

V.J. Sterken, P. Strub, H. Krüger, R. von Steiger, P. Frisch, Sixteen years of Ulysses interstellar dust measurements in the solar system. III. Simulations and data unveil new insights into local interstellar dust. Astrophys. J. **812**, 141 (2015). https://doi.org/10.1088/0004-637X/812/2/141

V. Sterken, G. Moragas-Klostermeyer, J. Hillier, L. Fielding, J. Lovett, S. Armes, N. Fechler, R. Srama, S. Bugiel, K. Hornung, Impact ionization experiments with porous cosmic dust particle analogs, in *EGU General Assembly Conference Abstracts, EGU General Assembly Conference Abstracts*, vol. 18 (2016), EPSC2016-16018

R.M. Stroud, C. Allen, A. Ansari, D. Anderson, S. Bajt, N. Bassim, R.S. Bastien, H.A. Bechtel, J. Borg, F.E. Brenker, J. Bridges, D.E. Brownlee, M. Burchell, M. Burghammer, A.L. Butterworth, H. Changela, P. Cloetens, A.M. Davis, R. Doll, C. Floss, G. Flynn, D.R. Frank, Z. Gainsforth, E. Grün, P.R. Heck, J.K. Hillier, P. Hoppe, J. Huth, B. Hvide, A. Kearsley, A.J. King, P. Kotula, B. Lai, J. Leitner, L. Lemelle, H. Leroux, A. Leonard, R. Lettieri, W. Marchant, L.R. Nittler, R. Ogliore, W.J. Ong, F. Postberg, M.C. Price, S.A. Sandford, J.A.S. Tresseras, S. Schmitz, T. Schoonjans, K. Schreiber, G. Silversmit, A.S. Simionovici, V.A. Solé, R. Srama, T. Stephan, V.J. Sterken, J. Stodolna, S. Sutton, M. Trieloff, P. Tsou, A. Tsuchiyama, T. Tyliszczak, B. Vekemans, L. Vincze, A.J. Westphal, J. von Korff, D. Zevin, M.E. Zolensky, Stardust Interstellar Preliminary Examination XI: identification and elemental analysis of impact craters on Al foils from the Stardust Interstellar Dust Collector. Meteorit. Planet. Sci. **49**, 1698–1719 (2014). https://doi.org/10.1111/maps.12136

P. Strub, V.J. Sterken, H. Krüger, E. Grün, M. Horanyi, Interstellar dust flow through the solar system, in *American Institute of Physics Conference Series*, ed. by V.Y. Nosenko, P.K. Shukla, M.H. Thoma, H.M. Thomas. American Institute of Physics Conference Series, vol. 1397 (2011), pp. 385–386. https://doi.org/10.1063/1.3659855

P. Strub, H. Krüger, V.J. Sterken, Sixteen years of Ulysses interstellar dust measurements in the solar system. II. Fluctuations in the dust flow from the data. Astrophys. J. **812**, 140 (2015). https://doi.org/10.1088/0004-637X/812/2/140. 1508.03242

P. Strub, V.J. Sterken, R. Soja, H. Krüger, E. Grün, R. Srama, Heliospheric modulation of the interstellar dust flow on to Earth. Astron. Astrophys. **621**, A54 (2019). https://doi.org/10.1051/0004-6361/201832644

F.M. Thayer, D.M. Malaspina, A. Collette, Z. Sternovsky, Variation in relative dust impact charge recollection with antenna to spacecraft potential on STEREO. J. Geophys. Res. Space Phys. **121**, 4998–5004 (2016). https://doi.org/10.1002/2015JA021983

J.I. Thorpe, C. Parvini, J.M. Trigo-Rodríguez, Detection and measurement of micrometeoroids with LISA Pathfinder. Astron. Astrophys. **586**, A107 (2016). https://doi.org/10.1051/0004-6361/201527658

J.I. Thorpe, T.B. Littenberg, J. Baker, J. Slutsky (The LISA Pathfinder Team) LISA Pathfinder as a micrometeoroid instrument. J. Phys. Conf. Ser. **840**, 012007 (2017). https://doi.org/10.1088/1742-6596/840/1/012007

D. Tsintikidis, D.A. Gurnett, W.S. Kurth, L.J. Granroth, Micron-sized particles detected in the vicinity of Jupiter by the Voyager plasma wave instruments. Geophys. Res. Lett. **23**, 997–1000 (1996). https://doi.org/10.1029/96GL00961

B.T. Tsurutani, D.R. Clay, L.D. Zhang, B. Dasgupta, D. Brinza, M. Henry, A. Mendis, S. Moses, K.H. Glassmeier, G. Musmann, I. Richter, Dust impacts at comet P/Borrelly. Geophys. Res. Lett. **30**, 2134 (2003). https://doi.org/10.1029/2003GL017580

M.K. Wallis, Penetration of charged interstellar dust into the solar system. Mon. Not. R. Astron. Soc. **227**, 331–339 (1987). https://doi.org/10.1093/mnras/227.2.331

S. Wang, A. Li, B.W. Jiang, Very large interstellar grains as evidenced by the mid-infrared extinction. Astrophys. J. **811**, 38 (2015). https://doi.org/10.1088/0004-637X/811/1/38. 1508.03403

J.C. Weingartner, B.T. Draine, Dust grain-size distributions and extinction in the Milky Way, Large Magellanic Cloud, and Small Magellanic Cloud. Astrophys. J. **548**, 296–309 (2001). https://doi.org/10.1086/318651. astro-ph/0008146

A.J. Westphal, D. Anderson, A.L. Butterworth, D.R. Frank, R. Lettieri, W. Marchant, J. von Korff, D. Zevin, A. Ardizzone, A. Campanile, M. Capraro, K. Courtney, M.N. Criswell, D. Crumpler, R. Cwik, F.J. Gray, B. Hudson, G. Imada, J. Karr, L.L.W. Wah, M. Mazzucato, P.G. Motta, C. Rigamonti, R.C. Spencer, S.B. Woodrough, I.C. Santoni, G. Sperry, J.N. Terry, N. Wordsworth, T. Yahnke, C. Allen, A. Ansari, S. Bajt, R.K. Bastien, N. Bassim, H.A. Bechtel, J. Borg, F.E. Brenker, J. Bridges, D.E. Brownlee, M. Burchell, M. Burghammer, H. Changela, P. Cloetens, A.M. Davis, R. Doll, C. Floss, G. Flynn, Z. Gainsforth, E. Grün, P.R. Heck, J.K. Hillier, P. Hoppe, J. Huth, B. Hvide, A. Kearsley, A.J. King, B. Lai, J. Leitner, L. Lemelle, H. Leroux, A. Leonard, L.R. Nittler, R. Ogliore, W.J. Ong, F. Postberg, M.C. Price, S.A. Sandford, J.A.S. Tresseras, S. Schmitz, T. Schoonjans, G. Silversmit, A.S. Simionovici, V.A. Solé, R. Srama, T. Stephan, V.J. Sterken, J. Stodolna, R.M. Stroud, S. Sutton, M. Trieloff, P. Tsou, A. Tsuchiyama, T. Tyliszczak, B. Vekemans, L. Vincze, M.E. Zolensky, Stardust interstellar preliminary examination I: identification of tracks in aerogel. Meteorit. Planet. Sci. **49**, 1509–1521 (2014a). https://doi.org/10.1111/maps.12168

A.J. Westphal, H.A. Bechtel, F.E. Brenker, A.L. Butterworth, G. Flynn, D.R. Frank, Z. Gainsforth, J.K. Hillier, F. Postberg, A.S. Simionovici, V.J. Sterken, R.M. Stroud, C. Allen, D. Anderson, A. Ansari, S. Bajt, R.K. Bastien, N. Bassim, J. Borg, J. Bridges, D.E. Brownlee, M. Burchell, M. Burghammer, H. Changela, P. Cloetens, A.M. Davis, R. Doll, C. Floss, E. Grün, P.R. Heck, P. Hoppe, B. Hudson, J. Huth, B. Hvide, A. Kearsley, A.J. King, B. Lai, J. Leitner, L. Lemelle, H. Leroux, A. Leonard, R. Lettieri, W. Marchant, L.R. Nittler, R. Ogliore, W.J. Ong, M.C. Price, S.A. Sandford, J.A.S. Tresseras, S. Schmitz, T. Schoonjans, G. Silversmit, V.A. Solé, R. Srama, F. Stadermann, T. Stephan, J. Stodolna, S. Sutton, M. Trieloff, P. Tsou, A. Tsuchiyama, T. Tyliszczak, B. Vekemans, L. Vincze, J. Korff, N. Wordsworth, D. Zevin, M.E. Zolensky, Final reports of the Stardust Interstellar Preliminary Examination. Meteorit. Planet. Sci. **49**, 1720–1733 (2014b). https://doi.org/10.1111/maps.12221

A.J. Westphal, R.M. Stroud, H.A. Bechtel, F.E. Brenker, A.L. Butterworth, G.J. Flynn, D.R. Frank, Z. Gainsforth, J.K. Hillier, F. Postberg, A.S. Simionovici, V.J. Sterken, L.R. Nittler, C. Allen, D. Anderson, A. Ansari, S. Bajt, R.K. Bastien, N. Bassim, J. Bridges, D.E. Brownlee, M. Burchell, M. Burghammer, H. Changela, P. Cloetens, A.M. Davis, R. Doll, C. Floss, E. Grün, P.R. Heck, P. Hoppe, B. Hudson, J. Huth, A. Kearsley, A.J. King, B. Lai, J. Leitner, L. Lemelle, A. Leonard, H. Leroux, R. Lettieri, W. Marchant, R. Ogliore, W.J. Ong, M.C. Price, S.A. Sandford, J.A.S. Tresseras, S. Schmitz, T. Schoonjans, K. Schreiber, G. Silversmit, V.A. Solé, R. Srama, F. Stadermann, T. Stephan, J. Stodolna, S. Sutton, M. Trieloff, P. Tsou, T. Tyliszczak, B. Vekemans, L. Vincze, J. Von Korff, N. Wordsworth, D. Zevin, M.E. Zolensky, Evidence for interstellar origin of seven dust particles collected by the Stardust spacecraft. Science **345**, 786–791 (2014c). https://doi.org/10.1126/science.1252496

M. Witte, Kinetic parameters of interstellar neutral helium. Review of results obtained during one solar cycle with the Ulysses/GAS-instrument. Astron. Astrophys. **426**, 835–844 (2004). https://doi.org/10.1051/0004-6361:20035956

S.R. Wood, D.M. Malaspina, L. Andersson, M. Horanyi, Hypervelocity dust impacts on the Wind spacecraft: correlations between Ulysses and Wind interstellar dust detections. J. Geophys. Res. Space Phys. **120**, 7121–7129 (2015). https://doi.org/10.1002/2015JA021463

S.Y. Ye, W.S. Kurth, G.B. Hospodarsky, T.F. Averkamp, D.A. Gurnett, Dust detection in space using the monopole and dipole electric field antennas. J. Geophys. Res. Space Phys. **121**, 11 (2016). https://doi.org/10.1002/2016JA023266

A. Zaslavsky, Floating potential perturbations due to micrometeoroid impacts: theory and application to S/WAVES data. J. Geophys. Res. Space Phys. **120**, 855–867 (2015). https://doi.org/10.1002/2014JA020635

A. Zaslavsky, N. Meyer-Vernet, I. Mann, A. Czechowski, K. Issautier, G. Le Chat, F. Pantellini, K. Goetz, M. Maksimovic, S.D. Bale, J.C. Kasper, Interplanetary dust detection by radio antennas: mass calibration and fluxes measured by STEREO/WAVES. J. Geophys. Res. Space Phys. **117**, A05102 (2012). https://doi.org/10.1029/2011JA017480

Q. Zhang, Prospects for backtracing 1I/'Oumuamua and future interstellar objects. Astrophys. J. Lett. **852**, L13 (2018). https://doi.org/10.3847/2041-8213/aaa2f7. 1712.08059

S. Zhukovska, H.P. Gail, M. Trieloff, Evolution of interstellar dust and stardust in the solar neighbourhood. Astron. Astrophys. **479**, 453–480 (2008). https://doi.org/10.1051/0004-6361:20077789. 0706.1155

E. Zinner, Presolar grains, in *Treatise on Geochemistry, Vol. 1. Meteorites and Cosmochemical Processes*, 2nd edn. (2014), pp. 181–213

E. Zinner, S. Amari, R. Guinness, C. Jennings, A.F. Mertz, A.N. Nguyen, R. Gallino, P. Hoppe, M. Lugaro, L.R. Nittler, R.S. Lewis, NanoSIMS isotopic analysis of small presolar grains: search for Si_3N_4 grains from AGB stars and Al and Ti isotopic compositions of rare presolar SiC grains. Geochim. Cosmochim. Acta **71**, 4786–4813 (2007). https://doi.org/10.1016/j.gca.2007.07.012

V. Zubko, E. Dwek, R.G. Arendt, Interstellar dust models consistent with extinction, emission, and abundance constraints. Astrophys. J. Suppl. Ser. **152**, 211–249 (2004). https://doi.org/10.1086/382351. astro-ph/0312641

Space Sci Rev (2018) 214:52
https://doi.org/10.1007/s11214-018-0486-5

Dust Evolution in Protoplanetary Discs and the Formation of Planetesimals

What Have We Learned from Laboratory Experiments?

Jürgen Blum[1]

Received: 24 November 2017 / Accepted: 2 February 2018 / Published online: 28 February 2018

Abstract After 25 years of laboratory research on protoplanetary dust agglomeration, a consistent picture of the various processes that involve colliding dust aggregates has emerged. Besides sticking, bouncing and fragmentation, other effects, like, e.g., erosion or mass transfer, have now been extensively studied. Coagulation simulations consistently show that µm-sized dust grains can grow to mm- to cm-sized aggregates before they encounter the bouncing barrier, whereas sub-µm-sized water-ice particles can directly grow to planetesimal sizes. For siliceous materials, other processes have to be responsible for turning the dust aggregates into planetesimals. In this article, these processes are discussed, the physical properties of the emerging dusty or icy planetesimals are presented and compared to empirical evidence from within and without the Solar System. In conclusion, the formation of planetesimals by a gravitational collapse of dust "pebbles" seems the most likely.

Keywords Protoplanetary dust · Planetesimals · Planet formation

1 Introduction

It is generally assumed that dust in protoplanetary discs (PPDs) grows from initial sizes of ~ 0.1–1 µm to planetesimals. The latter are objects of sizes in the range ~ 1–1000 km and possess sufficient gravitational attraction to allow further accretion of more material in subsequent collisions so that they finally end up in planetary embryos and planets. For bodies smaller than planetesimals, gravitational accretion is not possible in two-body collisions, because typical collision velocities exceed by far their gravitational escape speed. Thus, planetesimals play an essential intermediate role in planet formation. Mutual collisions among

Cosmic Dust from the Laboratory to the Stars
Edited by Rafael Rodrigo, Jürgen Blum, Hsiang-Wen Hsu, Detlef Koschny, Anny-Chantal Levasseur-Regourd, Jesús Martín-Pintado, Veerle Sterken and Andrew Westphal

✉ J. Blum
j.blum@tu-bs.de

[1] Institut für Geophysik und extraterrestrische Physik, Technische Universität Braunschweig, Mendelssohnstr. 3, 38106 Braunschweig, Germany

protoplanetary dust particles are important for our understanding of the build-up of larger bodies.

Based upon the pioneering work of Weidenschilling (1977), Weidenschilling and Cuzzi (1993) present a comprehensive overview of relative particle motion in the disc. For individual dust grains or small aggregates with diameters smaller than $\sim$ 10–100 μm, Brownian motion is the dominating source for collisions. Typical collision velocities in this regime are $\lesssim 1\ \mathrm{mm\,s^{-1}}$. Aggregates larger than $\sim$ 10–100 μm are subject to systematic drift motions relative to the gas, which causes relative velocities, and, thus, collisions, among particles of different sizes. One cause for the relative drift between dust and gas lies in the sub-Keplerian orbital velocity of the pressure-supported gas, which typically rotates around the central star slower than with Keplerian velocity by $\Delta v \sim c^2/v_K$, with c and v_K being the local sound speed of the gas and the Keplerian velocity, respectively. As the dust aggregates do not feel the pressure support, their equilibrium orbital velocity would be Keplerian in the absence of a gaseous nebula. However, in the presence of the gas, small dust aggregates with gas-dust coupling times smaller than the orbital time scale are forced by the gas to move with sub-Keplerian velocity and, thus, feel a net inward-directed force, which causes a radial drift velocity that increases with increasing aggregate size. Following Weidenschilling and Cuzzi (1993), these radial drift velocities reach a maximum value of $\sim 60\ \mathrm{m\,s^{-1}}$ for bodies with $\sim$ 1 m diameter, which limits the lifetimes of these bodies to $\sim$ 100 years at 1 au. In the particular nebula model used by Weidenschilling and Cuzzi (1993), the gas-grain coupling time of m-sized aggregates equals the orbital time scale at 1 au. Aggregates larger than about 1 m travel essentially with Keplerian orbital speeds so that they feel a steady headwind of the slower-orbiting gas. As the frictional effect of this headwind depends on the surface-to-mass ratio of the dust aggregates, the resulting inward drift velocity decreases with increasing aggregate size in this size regime. Aggregates with different dust-gas coupling times (i.e. different sizes for a given porosity) thus differentially drift in the radial direction, which causes collisions among these particles. For the same reason, also in the azimuthal (i.e. orbital-velocity) direction, dust aggregates possess relative velocities to one another, which also leads to collisions. An example of the resulting collision velocities can be seen in Fig. 1.

Another cause for systematic drift of dust particles relative to the gas is the sedimentary motion of the particles towards the disc midplane. Also here, dust aggregates with larger surface-to-mass ratios sediment slower than those with smaller ones so that collisions among aggregates with different sizes result. This effect is greater for higher elevations above the midplane, due to a higher downward-directed force component and lower gas densities. Naturally, dust particles in the nebula midplane are not affected by sedimentation so that a quantification requires knowledge about the position above the midplane.

Finally, gas turbulence plays also an important role in the evolution of dust aggregates, because it also causes dust particles to collide. Due to the nature of the turbulence, it causes stochastic particle motion, particularly (but not exclusively) for larger dust aggregates, depending also on the strength of the turbulence. Thus, also dust aggregates of identical aerodynamic properties (or sizes) can collide.

Dust particles condense as refractory materials (oxides, metals, silicates) in the hot inner disc, semi-volatile materials (carbonaceous and organic matter) in the middle disc, and ices (H_2O, CO_2, CO, NH_3, CH_4) in the cold outer disc, respectively. Due to the findings by the *Rosetta* mission, particularly that the cometary nucleus possesses a high D/H ratio (Altwegg et al. 2015) and contains a high abundance of molecular O_2 (Bieler et al. 2015), it is also possible that the water ice in the outer parts of the solar nebula never reached temperatures high enough to vaporise it so that cometary water ice might be a pre-solar component of the condensed matter in the Solar System. Formation models of protoplanetary discs predicted this (Visser et al. 2009).

Over the past 25 years, a number of experimental investigations have tried to shed light on the outcomes of protoplanetary dust collisions. This paper summarizes the knowledge gained by these laboratory and microgravity experiments. Moreover, the possible growth mechanisms towards planetesimals, resulting from the collision models, will be presented. Finally, predictions about planetesimal properties from the corresponding growth models will be compared with empirical evidence.

2 Outcomes of Laboratory Experiments

The systematic search for a collision model of protoplanetary dust started with the work by Blum and Münch (1993) who showed that collisions among mm-sized dust aggregates consisting of μm- or nm-sized silicate grains do not result in sticking, but rather in bouncing (for impact velocities between 0.15 and $\sim 1\ \mathrm{m\,s^{-1}}$) or fragmentation (for impact velocities $\gtrsim 1\ \mathrm{m\,s^{-1}}$). Since then, a considerable number of papers have been published that expanded the parameter space to dust (aggregate) sizes between ~ 1 μm and ~ 10 cm and the whole range of size ratios between projectile and target, and impact velocities between $\sim 10^{-3}\ \mathrm{m\,s^{-1}}$ and $\sim 100\ \mathrm{m\,s^{-1}}$, for silicate monomer grains of ~ 1 μm diameter. The review by Blum and Wurm (2008) summarizes the state of the art about 10 years ago. Güttler et al. (2010) then used all available information at that time to derive the first comprehensive collision model for silicate aggregates. Since then, numerous new experiments have been performed, expanding, for instance, the parameter space to water ice (Gundlach and Blum 2015) as well as CO_2 ice (Musiolik et al. 2016a) and CO_2-H_2O ice mixtures (Musiolik et al. 2016b). In the following, I will summarize our knowledge about the possible collisional outcomes and their dependence on projectile and target size as well as grain material and monomer size.

2.1 General Collisional Outcomes

I will start with the results for aggregates consisting of μm-sized SiO_2 grains, because for this material, the body of empirical evidence is satisfactory.

Outcomes for Similar-Sized Collision Partners When the size ratio between the colliding dust aggregates is not too different from unity, the following general collision outcomes have been identified:

- **Sticking.**
  When the collision energy is small compared to the van-der-Waals binding energy of the colliding dust aggregates, sticking occurs inevitably if the collisions are at least partially inelastic (Dominik and Tielens 1997). For higher impact energies, the degree of inelasticity determines the fate of the colliding dust aggregates. Güttler et al. (2010) identified three processes that lead to the complete transfer of both colliding bodies into a more massive dust aggregate, i.e., hit-and-stick behaviour for very small impact velocities, sticking with deformation/compaction of the aggregates, and deep penetration of a somewhat smaller projectile into a larger target aggregate. The supporting laboratory and microgravity experiments were performed by Blum et al. (1998), Wurm and Blum (1998), Blum et al. (2000), Blum and Wurm (2000), Krause and Blum (2004), Langkowski et al.

(2008), Weidling et al. (2012), Kothe et al. (2013), Weidling and Blum (2015), Brisset et al. (2016, 2017) and Whizin et al. (2017).

When colliding dust particles or aggregates are in the hit-and-stick regime (for which the impact energy is insufficient to cause rolling of the dust grains upon impact, see Dominik and Tielens 1997), the corresponding aggregates develop a fractal morphology (Dominik and Tielens 1997; Kempf et al. 1999; Blum et al. 2000; Krause and Blum 2004; Blum 2006), with a fractal dimension in the range $D_f \approx 1.1 \ldots 1.9$, depending on the gas pressure if the collisions are caused by Brownian motion (Paszun and Dominik 2006). For higher impact speeds, the growing aggregates become more compact (Dominik and Tielens 1997; Blum and Wurm 2000; Paszun and Dominik 2009; Wada et al. 2008), but possibly remain fractal with fractal dimensions of $D_f \approx 2.5$ (Wada et al. 2008).

A general description of the outcomes in hierarchical coagulation is provided by Dominik et al. (2016).

- **Bouncing.**
  When the amount of energy dissipation during the collision is insufficient to allow sticking, but still not large enough to disrupt the colliding bodies (see below), the collisions result in bouncing. Bouncing was found in a variety of experimental investigations (Blum and Münch 1993; Heißelmann et al. 2007; Weidling et al. 2012; Kothe et al. 2013; Landeck 2016; Brisset et al. 2016, 2017). Although bouncing might seem to be important only for halting growth or stretching growth time scales, it turned out that bouncing collisions lead to the gradual compaction of dust aggregates. Weidling et al. (2009) experimentally found that the equilibrium filling factor (or packing density) of bouncing aggregates is ~ 0.36.
- **Fragmentation.**
  Experiments have also shown that collisions at high speeds lead to the fragmentation of the colliding dust aggregates (Blum and Münch 1993; Beitz et al. 2011; Schräpler et al. 2012; Deckers and Teiser 2013; Bukhari Syed et al. 2017). Recently, Bukhari Syed et al. (2017) analysed the influence of the aggregate size, size ratio and impact velocity on the outcome of fragmenting collisions. They could show that the mass ratio of the largest fragment to the original aggregate mass decreases with increasing impact velocity and that also the slope of the power law of the fragment size distribution is velocity dependent. On top of that, the onset of fragmentation, i.e. the transition velocity from bouncing to fragmentation, increases for decreasing aggregate mass.
- **Abrasion.**
  In recent microgravity experiments on low-velocity collisions in a many-particle system, it was discovered that cm-sized dust aggregates suffer a gradual mass loss, although their collision velocities were smaller than the threshold velocity for fragmentation (Kothe 2016). Abrasion was not observed for velocities below $\sim 0.1\ \mathrm{m\,s^{-1}}$ and then gradually increased in strength with increasing velocity. The preliminary results by Kothe (2016) indicate that the abrasion efficiency is relatively weak so that aggregates need on the order of 1000 collisions before they are completely destroyed.

Outcomes When Small Projectiles Hit Large Targets When a rather small projectile dust aggregate hits a much larger target aggregate, the following additional collision outcomes may occur:

- **Mass transfer.**
  Above the fragmentation limit of the small projectile aggregate, part of its mass is permanently transferred to the target aggregate so that the latter gains mass. In the past years, many experimental investigations have studied mass transfer (Wurm et al. 2005b; Teiser

and Wurm 2009a,b; Güttler et al. 2010; Teiser et al. 2011; Beitz et al. 2011; Meisner et al. 2013; Deckers and Teiser 2014; Bukhari Syed et al. 2017). Typical efficiencies of the mass-transfer process are between a few and ~ 50 per cent and the deposited mass is compressed to a volume filling factor of typically 0.3–0.4 (Teiser et al. 2011; Meisner et al. 2013).

- **Cratering.**
  Experiments have shown that in the same impact-velocity range as mass transfer, but for larger projectile aggregate, the target agglomerates lose mass by cratering (Wurm et al. 2005a; Paraskov et al. 2007). Similar to the classical high-velocity-impact cratering effect, the impinging projectile excavates more mass than it transfers to the target, so that the target's mass budget is negative. Cratering can be regarded as a transition process to fragmentation of the target. Bukhari Syed et al. (2017) have shown that the (relative) mass loss of the target with increasing impact energy of the projectile is a continuous function of the impact energy. The amount of excavated crater mass per impact can be substantial, with up to 35 times the projectile mass, depending on the strength of the target material (Wurm et al. 2005a; Paraskov et al. 2007).
- **Erosion.**
  Again for similar impact velocities, but much smaller projectile aggregates or dust monomers, the effect of erosion has been identified in the laboratory (Schräpler and Blum 2011; Schräpler et al. 2018) and by numerical simulations (Seizinger et al. 2013; Krijt et al. 2015). The erosion efficiency increases with increasing impact velocity and decreases with increasing impactor mass, so that there is a smooth transition to mass transfer. Seizinger et al. (2013) argue that the erosion efficiency of monomers and small aggregates is higher than that of large aggregate projectiles, because some of the initially eroded mass gets pushed into the target aggregate again and is, thus, re-accreted by the target in impacts with large aggregates. Erosion has also been observed for impacts of µm-sized ice particles into ice agglomerates (Gundlach and Blum 2015).

The detailed physics of most of the before-mentioned collisional outcomes, i.e., their dependence on collision velocity, size and size ratio of the dust aggregates, their porosity or monomer-particle size have been compiled in the reviews by Blum and Wurm (2008) and Güttler et al. (2010). Güttler et al. (2010) presented the first complete collision model, including all effects known at the time. They distinguished in their model collisions between compact and highly porous dust aggregates and also differentiated between collisions among similar-sized dust aggregates and those between aggregates of very different sizes, because some of the outcomes only occur above or below a certain size ratio between projectile and target aggregate. Short-versions of this collision model have been developed by, e.g., Windmark et al. (2012b) and Birnstiel et al. (2016).

The latest experimental findings since the Güttler et al. (2010) model will soon be incorporated into a forthcoming dust-aggregate collisions model (see Kothe (2016) for a preliminary version). An example of this model is shown in Fig. 1 where on the axes the sizes of the colliding dust aggregates are shown and the coloured regions denote the collisional outcomes. It must be mentioned that the contours and their boundaries can considerably vary with (i) distance to the central star, (ii) PPD model, (iii) turbulence strength, (iv) monomer material, and (v) monomer size (distribution), respectively (see Sect. 2.2).

2.2 Influence of Material and Monomer Size

The parameter-space coverage for dust aggregates consisting of micrometre-sized silica monomer grains is quite advanced so that the resulting collision model may be considered

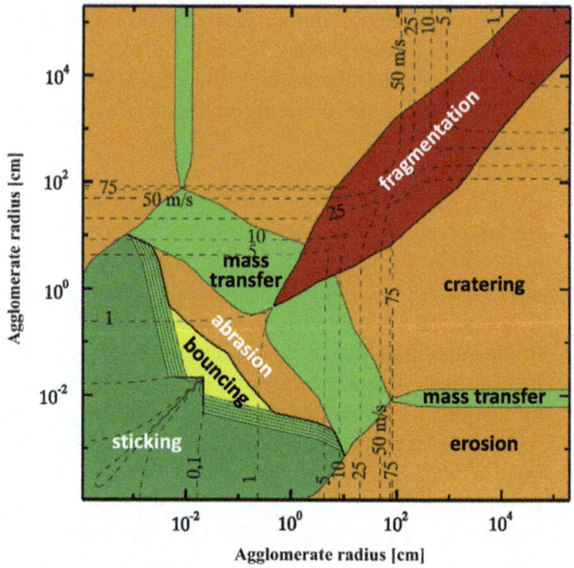

Fig. 1 Example of the latest collision model under development at TU Braunschweig for SiO_2 monomer particles with 0.75 μm radius, a minimum mass solar nebula model at 1 AU in the midplane, and a turbulence strength of $\alpha = 10^{-3}$, respectively. Collisional outcomes are marked in *colour* and labelled. *Dashed contours* mark the mean collision velocities in units of $\mathrm{m\,s^{-1}}$. Image credit: Kothe and Blum (in prep.)

relatively reliable. However, the situation is much worse if we consider other abundant materials or nanometre-size constituents of the aggregates. While the latter case can be scaled using dust-collision models, e.g. by Dominik and Tielens (1997), Wada et al. (2008) and Lorek et al. (2018), other grain materials require new experimental results at realistic temperatures. Some progress in this field has been made for micrometre-sized water-ice grains and aggregates thereof. Experiments by Gundlach and Blum (2015) indicate that the sticking-to-bouncing threshold of icy monomer grains is a factor ~ 10 higher than the corresponding value for silica grain. The same factor of ~ 10 can be found for the bouncing-erosion threshold (Gundlach and Blum 2015). It must be mentioned that Gundlach and Blum (2015) also found that the stickiness of small water-ice grains heavily depends on temperature. While for temperatures below ~ 200 K the sticking threshold is constant at $\sim 10\ \mathrm{m\,s^{-1}}$, its value increases steadily for temperatures above ~ 200 K. It has recently be shown by Gärtner et al. (2017) that the reason for this behaviour is the increase in thickness of a diffuse surface layer around the ice particles for temperatures exceeding ~ 200 K. Thus, laboratory experiments on the study of protoplanetary ice aggregation have to be performed at low ambient temperatures and vacuum conditions (Gärtner et al. 2017).

That the material properties may have an enormous impact on the growth behaviour of protoplanetary dust has long been speculated. Kouchi et al. (2002) and Kudo et al. (2002) performed low-velocity impact experiments with "interstellar organic matter analogs". They found that in the right temperature regime (around ~ 250 K), this material possesses an enhanced stickiness. However, for lower and higher temperatures, the material was either too hard or not viscous enough to allow for sufficient sticking. Flynn et al. (2013) found that individual monomer grains of a chondritic porous interplanetary dust particle of possible cometary origin were coated with a ~ 100 nm thick organic mantle. They estimated the tensile strength of the coated μm-sized grains to be at least several hundred Pa, comparable to or even higher than the values measured for silica particles. Thus, we may expect that organic coatings can aid planetesimal formation through sticking collisions. However, much more work in this field is required before the role of organic matter has been solved.

Besides water ice, there is only little experimental data on the collision outcome of micrometre-sized particles and their aggregates consisting of other materials. Recently, first laboratory experiments on the collision properties of CO_2 aggregates have been reported by Musiolik et al. (2016a). It seems that their sticking threshold does not differ from that of SiO_2.

3 Pathways to Planetesimals

Based on the above described collision model for protoplanetary dust aggregates, three pathways to planetesimals seem to be feasible, which will be described, including their pros and cons, in the following.

3.1 Formation of Planetesimals by the Gentle Gravitational Collapse of a Concentrated Cloud of Dust "Pebbles"

If the majority of the dust grains consist of siliceous material, Monte-Carlo simulations of the protoplanetary aggregation process have shown that after an initial process of fractal growth, sticking and bouncing collisions lead to the formation of rather compact mm- to cm-sized aggregates with volume filling factors of ~ 0.36 within $\sim 10^4$ orbital time scales (Zsom et al. 2010). Further growth is inhibited by the bouncing barrier and no other growth process (i.e., mass transfer) can be reached. It turned out that the maximum aggregate size depends on (i) the PPD model, with a minimum mass solar nebula model resulting in the biggest aggregates (Zsom et al. 2010), (ii) the distance to the central star and (iii) the composition of the dust (Lorek et al. 2018). Increasing distances to the central star and increasing dust-to-ice ratios result is smaller maximum aggregate sizes (Lorek et al. 2018). Moreover, the aggregates are fluffier further away from the central star, for smaller monomer grains and if the dust-to-ice ratio is low (Lorek et al. 2018). In the literature, the dust aggregates in this stage of growth are termed "pebbles". The presence of mm- to dm-sized "pebbles" (and partially the absence of larger particles) was confirmed in a number of PPD observations (van Boekel et al. 2004; D'Alessio et al. 2006; Natta et al. 2007; Birnstiel et al. 2010; Ricci et al. 2010; Pérez et al. 2012; Trotta et al. 2013; Testi et al. 2014; Pérez et al. 2015; Tazzari et al. 2016; Liu et al. 2017).

Under the conditions that

1. the Stokes numbers St of the "pebbles" are in the range $\mathrm{St} \sim 10^{-3} \ldots 5$ (Yang et al. 2017),
2. the "metallicity" of the PPD is larger than a Stokes-number-dependent threshold value, and
3. the local dust-to-gas mass ratio is above unity,

the streaming instability (SI, see below and Youdin and Goodman (2005) for the original work) can further concentrate the "pebble" cloud until a gravitational collapse leads to the formation of planetesimals. Here, the Stokes number is defined by $\mathrm{St} = \tau_f \Omega$, with τ_f and Ω being the gas-dust friction time and the orbital frequency, respectively. The "metallicity" is defined by the vertically-integrated dust-to-gas ratio. A relation between the minimum required "metallicity" and the Stokes number of the aggregates can be found in Yang et al. (2017). An absolute minimum "metallicity" of ~ 0.015 is required for $\mathrm{St} \sim 0.1$ to allow the SI to work.

The SI is a collective-particle effect in which the frictional feedback from the solid to the gaseous component of the disc is taken into account. In contrast to a single "pebble", a highly-concentrated region of "pebbles" with a local dust-to-gas mass ratio above unity has two major effects: (i) Due to the mutual aerodynamic shielding, the effective cross section of the region is much smaller than for all dust "pebbles" combined. This means that the SI region experiences less headwind and, thus, radially drifts much slower and azimuthally much faster (see Sect. 1). By this, it can catch up with all individual "pebbles" on its orbit (and with those that reach its orbit during their radial inward drift), which are then incorporated into the SI region by which its mass grows. (ii) Owing to the large mass concentration, the SI region accelerates the ambient gas, which further reduces friction and yields a local minimum in radial drift. Numerical simulations have shown that this can lead to the gravitational collapse of the "pebble" cloud (Johansen et al. 2007).

Lorek et al. (2018) showed that the known collision properties of dust aggregates indeed lead to the formation of "pebbles" with sufficiently large Stokes numbers. Numerical work on the SI has resulted in predictions of the mass-frequency distribution function of planetesimals, which can be fitted by a power law with exponent -1.6 ± 0.1 (Simon et al. 2016, 2017; Schäfer et al. 2017).

3.1.1 Benefits

The formation of planetesimals by the SI avoids any problems that arise for dust aggregates larger than the "pebble" size, i.e., the bouncing barrier for aggregate sizes $\gtrsim$ mm-cm (Zsom et al. 2010), the fragmentation barrier for impact velocities $\gtrsim 1\ \mathrm{m\,s^{-1}}$ (Bukhari Syed et al. 2017), the erosion barrier (Schräpler et al. 2018), or the drift barrier (Weidenschilling 1977). On top of that, the SI is a fast process. As soon as the conditions (see above) are met, the gravitational collapse occurs on the order of tens to a few thousand orbital time scales (Johansen et al. 2007; Yang et al. 2017).

3.1.2 Problems

As already stated above, the SI requires a minimum "metallicity" and relatively high spatial pre-concentration of the "pebbles". Although it is difficult for the coagulation models to achieve the "optimal" Stokes number for the SI ($\mathrm{St} \sim 0.1$), they can reach a "minimal" Stokes number of $\mathrm{St} \sim 1.5 \times 10^{-3}$, which requires a "metallicity" of 0.03 or higher (Lorek et al. 2018). In turn, such a high "metallicity" requires a partial dissipation of the gaseous PPD of initial solar "metallicity" so that planetesimal formation can only occur at later stages of the disc phase. To achieve the required pre-concentration of the "pebbles", various ideas have been suggested, including concentration inside or between turbulence eddies or concentration in pressure bumps (see, e.g., Johansen et al. 2014).

3.2 Formation of Planetesimals by Collisional Growth

An alternative approach to planetesimal formation has been described by Windmark et al. (2012a,b), Garaud et al. (2013) and Booth et al. (2018). After reaching the bouncing barrier, the gap to fragmentation, mass transfer or cratering (see Fig. 1) is overcome by, e.g., assuming a velocity distribution of the dust aggregates. Once new small projectile aggregates are formed in the destructive processes, mass transfer allows some of the dust aggregates to grow to planetesimal sizes.

3.2.1 *Benefits*

The processes of fragmentation, cratering, and mass transfer have been empirically proven to exist for dust aggregates so that they can be regarded as robust. Assumptions of velocity distributions seem reasonable, because turbulence-induced velocities have a stochastic nature and variations in the mass density of aggregates will lead to variations in drift velocities for aggregates of the same mass. At 1 au, 100-m-sized aggregates may form within $1 \dots 5 \times 10^4$ years (Windmark et al. 2012a; Garaud et al. 2013).

3.2.2 *Problems*

Although the time scales for growth to the 100-metre level are reasonably short at 1 au, they are much longer further out in the disc, and the maximum aggregate sizes are thus considerably smaller. Garaud et al. (2013) showed that at 30 au, even after 6×10^5 years, the maximum aggregate size is only a few metres. Typically, the growth time scales are so long that radial drift (outside dust traps) limits the maximum size achievable (Booth et al. 2018). Recently, Schräpler et al. (2018) showed that even under the most favourable conditions of growth everywhere in the parameter space shown in Fig. 1, except for the part marked "erosion", no aggregates larger than ~ 0.1 m can form, due to the overwhelming and self-supporting effect of erosion.

3.3 Formation of Planetesimals Consisting of Sub-Micrometre-Sized Water-Ice Particles

The biggest obstacles to direct collisional growth to planetesimals are (i) the low velocity for the transition from sticking to bouncing, which limits the "pebble" size before the bouncing barrier is hit, and (ii) the low collision energies required for compaction, which results in fast radial drift time scales. To overcome both, smaller and stickier monomer grains can be used (Wada et al. 2009; Okuzumi et al. 2012; Kataoka et al. 2013). In their planetesimal-formation model, Kataoka et al. (2013) assume that the material is dominated by the relatively sticky water ice (Gundlach and Blum 2015) and that the ice grains possess radii of 0.1 μm, for which compaction is harder to achieve than for 1 μm-sized grains. They calculate the collisional growth path of the resulting ice aggregates at 5 au and 8 au. Following an initial fractal growth phase, which leads to minimal mass densities of $\sim 10^{-5}\ \mathrm{g\,cm^{-3}}$, the aggregates then are compacted in mutual collisions, by the ram pressure of the gas and due to self-gravity. At the end of the simulation, the icy aggregates possess radii of ~ 10 km and mass densities of $\sim 0.1\ \mathrm{g\,cm^{-3}}$.

3.3.1 *Benefits*

Due to the choice of very small ice particles with high restructuring impedance and high sticking threshold, the bouncing barrier is never reached. The growth time scales to planetesimals are very short ($\sim 10^4$ years), due to the high porosity and, thus, high capture cross section of the fluffy aggregates (Krijt et al. 2015). As a consequence, radial drift is negligible. Results have been confirmed by Krijt et al. (2015) and Lorek et al. (2018).

3.3.2 Problems

Relaxing one of the assumptions (high stickiness, high resistance to compaction) does not lead to planetesimal sizes, and the resulting (smaller) aggregates suffer considerable radial drifts. The role of erosion for the survival of icy agglomerates was studied by Krijt et al. (2015), who found that erosion can limit growth if the erosion threshold is smaller than typical relative velocities between small and large aggregates. Empirical data for sub-micron-sized ice particles are still missing. On top of that, the collisional physics of highly fluffy (ice) aggregates has not been studied so far so that judging the collisional outcomes is difficult.

4 Properties of Planetesimal Formed in the Different Scenarios

Based on our knowledge about the collision physics of dust and ice aggregates, all of the three pathways to planetesimals (see Sect. 3) are in principle feasible (but mind their individual problems, as stated in Sect. 3). Obviously, we need more laboratory experiments and refined collision models to assess whether or not any of the three formation models is really capable of predicting the formation of planetesimals.

Alternatively, we may consider the predictions of the diverse models about the physical properties of the resulting planetesimals to better assess the likeliness of the respective planetesimal-formation scenario. Several distinct physical quantities can be estimated with which the three models can be distinguished. These comprise the size of the resulting planetesimals, the volume filling factor (i.e., the fraction of total planetesimal volume actually occupied by matter), the tensile strength (i.e., the internal cohesion of the material), the collisional strength (i.e. the energy required to fragment the colliding bodies such that the biggest surviving mass equals half the original mass), the Knudsen diffusivity (i.e., the resistance to gas flow), and the thermal conductivity, respectively. Table 1 compiles estimates of these values for the three formation models and expands on earlier approaches by Blum et al. (2006, 2014). Please mind that the values shown in Table 1 for the gravitational-collapse model are only valid for small planetesimals, because only then the "pebbles" survive the gravitational collapse intact (Wahlberg Jansson and Johansen 2014; Wahlberg Jansson et al. 2017). However, the approach used by Wahlberg Jansson and Johansen (2014) and Wahlberg Jansson et al. (2017) ignores feedback of the collapsing dust particles to the gas cloud. A more rigorous treatment of the collapsing two-phase-flow problem was performed by Shariff and Cuzzi (2015), but the implications to the fate of the "pebbles" have not been considered yet. In any event, the average lithostatic pressure of a body with $R = 50$ km in radius and an average mass density $\rho = 1000\ \mathrm{kg\,m^{-3}}$ is $p = \frac{4}{15}\pi G\rho^2 R^2 = 1.4 \times 10^5$ Pa, with G being the gravitational constant, which exceeds the crushing strength of the pebbles. Thus, for planetesimals larger than ~ 10 km in this formation model, the physical values should approach those of the mass-transfer model.

A word of caution is necessary before we use Table 1 for a comparison between the various formation scenarios. While the gravitational-collapse and the mass-transfer model are physically distinct, this is only partly the case for the mass-transfer and the icy-agglomerates model. Both scenarios rely on the intrinsic "stickiness" of the grains, but vary only in the material and size of the dust/ice grains and, thus, potentially in the region of the protoplanetary disc where they can be applied.

It is evident from Table 1 that in principle some of the listed parameters are sensitive enough to be used to distinguish between the various formation models. Exceptions might

Table 1 Comparison between the three formation scenarios of planetesimals described in Sect. 3

	Gravitational collapse (Sect. 3.1)	Mass transfer (Sect. 3.2)	Icy agglomerates (Sect. 3.3)
Size of planetesimals [km]	$\lesssim 1000$ [1]	$\lesssim 1$ [2–4]	~ 10 [5]
Volume filling factor	$0.36 \times 0.6 \approx 0.2$ [6–7] $\sim 0.4^*$	~ 0.4 [8]	~ 0.1 [5]
Tensile strength of interior [Pa]	~ 1–10 [9–10]	$\sim 10^3$–10^4 [8, 11]	$\sim 10^3$–10^4 (guess)
Critical fragmentation energy for 1 m-sized body [$\mathrm{J\,kg^{-1}}$]	$\sim 10^{-5}$ [12]	$\sim 10^2$ [12]	$\sim 10^2$ [12]
Normalised Knudsen diffusivity	$\equiv 1$	$\sim 10^{-4} \ldots 10^{-3}$ [13]	$\sim 10^{-5} \ldots 10^{-4}$ [13]
Thermal conductivity [$\mathrm{W\,m^{-1}\,K^{-1}}$]	10^{-3}–1 [14] (conduction/radiation)	10^{-2}–10^{-1} [14] (conduction)	10^{-2}–10^{-1} [14] (conduction)

References: [1] Schäfer et al. (2017), [2] Windmark et al. (2012b), [3] Windmark et al. (2012a), [4] Garaud et al. (2013), [5] Kataoka et al. (2013), [6] Weidling et al. (2009), [7] Zsom et al. (2010), [8] Kothe et al. (2010), [9] Skorov and Blum (2012), [10] Blum et al. (2014), [11] Blum et al. (2006), [12] Krivov et al. (2018), [13] Gundlach et al. (2011), [14] Gundlach and Blum (2012).

*For planetesimals with $R \gtrsim 10$–50 km.

be the thermal conductivity, for which the uncertainties and overlaps of the different models are too large, and the volume filling factor for large planetesimals, which is affected by lithostatic compression. Uncertainties in the quantification of the thermal conductivity, the tensile strength and the fragmentation energy arise from their strong dependencies on material properties and temperature. The latter becomes of utmost importance for the thermal conductivity of planetesimals formed through a gravitational collapse, because here radiative heat transport, which is intrinsically strongly temperature dependent, may dominate over conduction (Blum et al. 2017).

5 Empirical Evidence

We have seen above that the three formation models of planetesimals allow predictions, which can be compared to empirical evidence in our own Solar System or in extrasolar planetary system. Below, several criteria are listed, which allow to draw conclusions on the formation process of planetesimals.

1. Size-frequency distribution in the asteroid and Kuiper belt.
   Both, the asteroid belt and the Kuiper belt exhibit a "knee" in the size-frequency distribution around ~ 100 km. Asteroids smaller than that size are collisional fragments, whereas larger bodies are assumed to be primordial planetesimals. This may indicate that planetesimals were (on average) born big (Morbidelli et al. 2009), but Weidenschilling (2011) pointed out that the current size distribution in the asteroid belt can also be reproduced with sub-km-sized planetesimals.
2. Debris discs.
   Debris discs are thought to represent the dusty end of a collision cascade among planetesimals. Thus, modelling the collision processes and fitting them to the observed debris-

disc brightnesses may provide information about the planetesimal properties in these extrasolar pre-planetary systems. Recently, Krivov et al. (2018) modelled the collisional cascades within debris discs for both, the gravitational-collapse and the mass-transfer scenario of planetesimal formation. As a matter of fact, Krivov et al. (2018) found agreement between observed and predicted debris-disc brightnesses for both models. However, due to the different collisional strengths (see Table 1), the number of small (sub-km-sized) bodies in the two scenarios is very different. Although these bodies are unobservable in debris discs, our own Solar System might provide a clue to whether the mass-transfer or the gravitational-collapse model more likely represent the true formation scenario. Krivov et al. (2018) point out that only the latter is in agreement with the observed low number of sub-km-sized bodies in the Solar System.

3. Fractal particles in comet 67P.
   The discovery of fractal particles in comet 67P-Churyumov-Gerasimenko (hereafter 67P) by two *Rosetta* instruments (Fulle et al. 2015, 2016b; Mannel et al. 2016) can only be explained if the comet consists of larger entities between which the primordial fractal dust aggregates are captured (Fulle and Blum 2017). In the mass-transfer model, the relatively high impact velocities would certainly destroy the pebbles and render them into a compact configuration. The same is true for the icy-planetesimal formation model, which does not provide sufficiently large (cm-sized) void spaces to store the fractal particles until today.
4. Physical properties of comet 67P.
   It was argued earlier that comets can only be formed through the gravitational-collapse process, because their dust activity requires a gas pressure below the dry dust layer that exceeds the sum of cohesion and gravitational force of the dust (Kührt and Keller 1994, 1996; Skorov and Blum 2012; Blum et al. 2014; Gundlach et al. 2015; Gundlach and Blum 2016). With the *Rosetta* mission, a much more detailed investigation on the make-up of cometary nuclei became possible. Recently, Blum et al. (2017) showed that observations of the cometary properties by various *Rosetta* and *Philae* instruments (sub-surface and surface temperatures, size-frequency distribution of surface and coma dust, tensile strength) point towards comet 67P consisting of "pebbles" with 3–6 mm radii.
5. General properties of comets.
   Blum et al. (2017) also showed that a formation of comet 67P by the gravitational collapse of a "pebble" cloud correctly predicts the porosity and continued dust activity of the comet.

As already discussed before (see Table 3 in Blum et al. 2006), comets are highly porous objects. Experiments revealed that bouncing dust "pebbles" possess a volume filling factor of ~ 0.36 (Weidling et al. 2009; Zsom et al. 2010). Concentrating these dust aggregates by a (gentle) gravitational collapse results in a packing density of ~ 0.6 (Onoda and Liniger 1990). Combining these two filling factors results in the overall packing density of ~ 0.22, which is compatible with estimates for comet 67P (Kofman et al. 2015; Pätzold et al. 2016; Fulle et al. 2016a). The two alternative models fail to predict this value (see Table 1).

To sustain gas and dust activity of comets over many apparitions, a process is required that quasi-continuously emits gas and dust in the mass-proportion present inside the comet nucleus. To achieve this, the sub-surface gas pressure of the sublimating ices has to overcome the tensile strength of the surface material (Kührt and Keller 1994, 1996; Skorov and Blum 2012; Blum et al. 2014; Gundlach et al. 2015; Gundlach and Blum 2016). Only the hierarchical architecture of the comet following its formation through the gravitational collapse of a cloud of "pebbles" predicts sufficiently small tensile strengths

(see Table 1) to allow dust activity for realistic temperatures of the sub-surface ices (Skorov and Blum 2012; Blum et al. 2014; Gundlach et al. 2015; Gundlach and Blum 2016). As soon as the gas pressure exceeds the tensile strength of the overlaying dust layer, the latter is torn off and transported away. Due to the now free access to the ambient vacuum, the increased local outgassing rate decreases the ice temperature so that ice sublimation slows down and the (local) pressure becomes insufficient to continue dust emission, until a sufficiently thick dust layer has formed again, whereupon the process repeats.
6. Evidence for "pebbles" in other comets.
The flyby of the EPOXI spacecraft by comet 103P/Hartley revealed "pebbles" of the right size to have formed the comet by the streaming instability and a subsequent gravitational collapse (Kretke and Levison 2015). Moreover, comets leave trails of dust along their orbits with particle sizes large enough to not be subjected to considerable radiation pressure. A recent analysis of the trail profiles of comet 67P indicates dust "pebbles" of millimetre sizes to be the dominant constituents of the trail (Soja et al. 2015), in agreement with earlier estimates by Agarwal et al. (2010) and Fulle et al. (2010).

6 Conclusion

In this paper, I showed in Sect. 2 that the current state-of-the-art of pre-planetary dust-aggregate collisions is such that the initial dust growth by sticking collisions leads to mm- to cm-sized aggregates ("pebbles") for siliceous materials and monomer sizes of $\sim 1\ \mu\text{m}$. At this stage, inter-aggregate bouncing becomes and important effect, which decreases the porosity of the "pebbles" until they reach filling factors of ~ 0.36.

Two scenarios can explain the formation of planetesimals out of these "pebbles": (i) the gravitational-collapse scenario, which requires a high spatial concentration of the "pebbles" (e.g., by the streaming instability) and a subsequent gravitational collapse, or (ii) the mass-transfer scenario, in which the aggregates further grow by high-velocity collisions between small projectile aggregates and larger target aggregates; here, the projectiles disintegrate upon impact and transfer part of their mass to the target. Pros and cons of the two models are discussed in Sect. 3.

For sub-micrometre sized water-ice monomers, a third scenario for the formation of planetesimals is possible, which relies on the high sticking and restructuring thresholds of very small ice grains. Also for this model, pros and cons are discussed in Sect. 3.

The planetesimals resulting from the three formation models possess very different properties, which were shown in Table 1 in Sect. 4. These properties in principle allow a comparison to "real" planetesimals. Such a comparison was made in Sect. 5 where I compiled pieces of evidence that suggest that planetesimals consist of macroscopic dust "pebbles". The paucity of sub-km-sized bodies in the Solar System, the discovery of primordial fractal particles on comet 67P as well as many general physical properties of comets support the conjecture that planetesimals formed by a smooth gravitational collapse of dust "pebbles", following a spatial concentration of dust aggregates by the streaming instability.

Although the *Rosetta* mission to comet 67P has provided us with many new insights into cometary nuclei, we are far away from a comprehensive and closed picture of their formation, although, as stated above, the gravitational collapse of a "pebble" cloud, seems to be a viable assumption. First of all, it is unclear whether the gravitational collapse can lead to km-sized objects at all or produces initially much larger bodies. In the latter case, cometary nuclei must be collisional fragments, which seem to have morphologically and compositionally survived the break-up of their parent bodies quite intact. However, even if planetesimals form small (and by whatever process), it is unclear how they could have survived without undergoing catastrophic collisions (Morbidelli and Rickman 2015). Clearly, more work on the formation and (collisional) evolution of planetesimals needs to be done.

Acknowledgements I thank the Deutsche Forschungsgemeinschaft and the Deutsches Zentrum für Luft- und Raumfahrt for continuous support. I thank Stefan Kothe for providing me with Fig. 1. I also thank Stu Weidenschilling for his constructive suggestions during the review process. Finally, I thank ISSI for inviting me to a wonderful and fruitful workshop.

References

J. Agarwal, M. Müller, W.T. Reach, M.V. Sykes, H. Boehnhardt, E. Grün, The dust trail of Comet 67P/Churyumov-Gerasimenko between 2004 and 2006. Icarus **207**, 992–1012 (2010). https://doi.org/10.1016/j.icarus.2010.01.003

K. Altwegg, H. Balsiger, A. Bar-Nun, J.J. Berthelier, A. Bieler, P. Bochsler, C. Briois, U. Calmonte, M. Combi, J. De Keyser, P. Eberhardt, B. Fiethe, S. Fuselier, S. Gasc, T.I. Gombosi, K.C. Hansen, M. Hässig, A. Jäckel, E. Kopp, A. Korth, L. LeRoy, U. Mall, B. Marty, O. Mousis, E. Neefs, T. Owen, H. Rème, M. Rubin, T. Sémon, C.-Y. Tzou, H. Waite, P. Wurz, 67P/Churyumov-Gerasimenko, a Jupiter family comet with a high D/H ratio. Science **347**(27), 1261952 (2015). https://doi.org/10.1126/science.1261952

E. Beitz, C. Güttler, J. Blum, T. Meisner, J. Teiser, G. Wurm, Low-velocity collisions of centimeter-sized dust aggregates. Astrophys. J. **736**, 34 (2011). https://doi.org/10.1088/0004-637X/736/1/34

A. Bieler, K. Altwegg, H. Balsiger, A. Bar-Nun, J.-J. Berthelier, P. Bochsler, C. Briois, U. Calmonte, M. Combi, J. de Keyser, E.F. van Dishoeck, B. Fiethe, S.A. Fuselier, S. Gasc, T.I. Gombosi, K.C. Hansen, M. Hässig, A. Jäckel, E. Kopp, A. Korth, L. Le Roy, U. Mall, R. Maggiolo, B. Marty, O. Mousis, T. Owen, H. Rème, M. Rubin, T. Sémon, C.-Y. Tzou, J.H. Waite, C. Walsh, P. Wurz, Abundant molecular oxygen in the coma of comet 67P/Churyumov-Gerasimenko. Nature **526**, 678–681 (2015). https://doi.org/10.1038/nature15707

T. Birnstiel, L. Ricci, F. Trotta, C.P. Dullemond, A. Natta, L. Testi, C. Dominik, T. Henning, C.W. Ormel, A. Zsom, Testing the theory of grain growth and fragmentation by millimeter observations of protoplanetary disks. Astron. Astrophys. **516**, 14 (2010). https://doi.org/10.1051/0004-6361/201014893

T. Birnstiel, M. Fang, A. Johansen, Dust evolution and the formation of planetesimals. Space Sci. Rev. **205**, 41–75 (2016). https://doi.org/10.1007/s11214-016-0256-1

J. Blum, Dust agglomeration. Adv. Phys. **55**, 881–947 (2006). https://doi.org/10.1080/00018730601095039

J. Blum, M. Münch, Experimental investigations on aggregate-aggregate collisions in the early solar nebula. Icarus **106**, 151 (1993). https://doi.org/10.1006/icar.1993.1163

J. Blum, G. Wurm, Experiments on sticking, restructuring, and fragmentation of preplanetary dust aggregates. Icarus **143**, 138–146 (2000). https://doi.org/10.1006/icar.1999.6234

J. Blum, G. Wurm, The growth mechanisms of macroscopic bodies in protoplanetary disks. Annu. Rev. Astron. Astrophys. **46**, 21–56 (2008). https://doi.org/10.1146/annurev.astro.46.060407.145152

J. Blum, G. Wurm, T. Poppe, L.-O. Heim, Aspects of laboratory dust aggregation with relevance to the formation of planetesimals. Earth Moon Planets **80**, 285–309 (1998). https://doi.org/10.1023/A:1006386417473

J. Blum, G. Wurm, S. Kempf, T. Poppe, H. Klahr, T. Kozasa, M. Rott, T. Henning, J. Dorschner, R. Schräpler, H.U. Keller, W.J. Markiewicz, I. Mann, B.A. Gustafson, F. Giovane, D. Neuhaus, H. Fechtig, E. Grün, B. Feuerbacher, H. Kochan, L. Ratke, A. El Goresy, G. Morfill, S.J. Weidenschilling, G. Schwehm, K. Metzler, W.-H. Ip, Growth and form of planetary seedlings: results from a microgravity aggregation experiment. Phys. Rev. Lett. **85**, 2426–2429 (2000). https://doi.org/10.1103/PhysRevLett.85.2426

J. Blum, R. Schräpler, B.J.R. Davidsson, J.M. Trigo-Rodríguez, The physics of protoplanetesimal dust agglomerates. I. Mechanical properties and relations to primitive bodies in the solar system. Astrophys. J. **652**, 1768–1781 (2006). https://doi.org/10.1086/508017

J. Blum, B. Gundlach, S. Mühle, J.M. Trigo-Rodriguez, Comets formed in solar-nebula instabilities!—An experimental and modeling attempt to relate the activity of comets to their formation process. Icarus **235**, 156–169 (2014). https://doi.org/10.1016/j.icarus.2014.03.016

J. Blum, B. Gundlach, M. Krause, M. Fulle, A. Johansen, J. Agarwal, I. von Borstel, X. Shi, X. Hu, M.S. Bentley, F. Capaccioni, L. Colangeli, V. Della Corte, N. Fougere, S.F. Green, S. Ivanovski, T. Mannel, S. Merouane, A. Migliorini, A. Rotundi, R. Schmied, C. Snodgrass, Evidence for the formation of comet 67P/Churyumov-Gerasimenko through gravitational collapse of a bound clump of pebbles. Mon. Not. R. Astron. Soc. **469**, 755–773 (2017). https://doi.org/10.1093/mnras/stx2741

R.A. Booth, F. Meru, M.H. Lee, C.J. Clarke, Breakthrough revisited: investigating the requirements for growth of dust beyond the bouncing barrier. Mon. Not. R. Astron. Soc. **475**, 167–180 (2018). https://doi.org/10.1093/mnras/stx3084

J. Brisset, D. Heißelmann, S. Kothe, R. Weidling, J. Blum, Submillimetre-sized dust aggregate collision and growth properties. Experimental study of a multi-particle system on a suborbital rocket. Astron. Astrophys. **593**, 3 (2016). https://doi.org/10.1051/0004-6361/201527288

J. Brisset, D. Heißelmann, S. Kothe, R. Weidling, J. Blum, Low-velocity collision behaviour of clusters composed of sub-millimetre sized dust aggregates. Astron. Astrophys. **603**, 66 (2017). https://doi.org/10.1051/0004-6361/201630345

M. Bukhari Syed, J. Blum, K. Wahlberg Jansson, A. Johansen, The role of pebble fragmentation in planetesimal formation. I. Experimental study. Astrophys. J. **834**, 145 (2017). https://doi.org/10.3847/1538-4357/834/2/145

P. D'Alessio, N. Calvet, L. Hartmann, R. Franco-Hernández, H. Servín, Effects of dust growth and settling in T Tauri disks. Astrophys. J. **638**, 314–335 (2006). https://doi.org/10.1086/498861

J. Deckers, J. Teiser, Colliding decimeter dust. Astrophys. J. **769**, 151 (2013). https://doi.org/10.1088/0004-637X/769/2/151

J. Deckers, J. Teiser, Macroscopic dust in protoplanetary disks—from growth to destruction. Astrophys. J. **796**, 99 (2014). https://doi.org/10.1088/0004-637X/796/2/99

C. Dominik, A.G.G.M. Tielens, The physics of dust coagulation and the structure of dust aggregates in space. Astrophys. J. **480**, 647–673 (1997). https://doi.org/10.1086/303996

C. Dominik, D. Paszun, H. Borel, The structure of dust aggregates in hierarchical coagulation. ArXiv e-prints (2016)

G.J. Flynn, S. Wirick, L.P. Keller, Organic grain coatings in primitive interplanetary dust particles: implications for grain sticking in the Solar Nebula. Earth Planets Space **65**, 1159–1166 (2013). https://doi.org/10.5047/eps.2013.05.007

M. Fulle, J. Blum, Fractal dust constrains the collisional history of comets. Mon. Not. R. Astron. Soc. **469**, 39–44 (2017). https://doi.org/10.1093/mnras/stx971

M. Fulle, L. Colangeli, J. Agarwal, A. Aronica, V. Della Corte, F. Esposito, E. Grün, M. Ishiguro, R. Ligustri, J.J. Lopez Moreno, E. Mazzotta Epifani, G. Milani, F. Moreno, P. Palumbo, J. Rodríguez Gómez, A. Rotundi, Comet 67P/Churyumov-Gerasimenko: the GIADA dust environment model of the Rosetta mission target. Astron. Astrophys. **522**, 63 (2010). https://doi.org/10.1051/0004-6361/201014928

M. Fulle, V. Della Corte, A. Rotundi, P. Weissman, A. Juhasz, K. Szego, R. Sordini, M. Ferrari, S. Ivanovski, F. Lucarelli, M. Accolla, S. Merouane, V. Zakharov, E. Mazzotta Epifani, J.J. López-Moreno, J. Rodríguez, L. Colangeli, P. Palumbo, E. Grün, M. Hilchenbach, E. Bussoletti, F. Esposito, S.F. Green, P.L. Lamy, J.A.M. McDonnell, V. Mennella, A. Molina, R. Morales, F. Moreno, J.L. Ortiz, E. Palomba, R. Rodrigo, J.C. Zarnecki, M. Cosi, F. Giovane, B. Gustafson, M.L. Herranz, J.M. Jerónimo, M.R. Leese, A.C. López-Jiménez, N. Altobelli, Density and charge of pristine fluffy particles from comet 67P/Churyumov-Gerasimenko. Astrophys. J. Lett. **802**, 12 (2015). https://doi.org/10.1088/2041-8205/802/1/L12

M. Fulle, V. Della Corte, A. Rotundi, F.J.M. Rietmeijer, S.F. Green, P. Weissman, M. Accolla, L. Colangeli, M. Ferrari, S. Ivanovski, J.J. Lopez-Moreno, E.M. Epifani, R. Morales, J.L. Ortiz, E. Palomba, P. Palumbo, J. Rodriguez, R. Sordini, V. Zakharov, Comet 67P/Churyumov-Gerasimenko preserved the pebbles that formed planetesimals. Mon. Not. R. Astron. Soc. **462**, 132–137 (2016a). https://doi.org/10.1093/mnras/stw2299

M. Fulle, N. Altobelli, B. Buratti, M. Choukroun, M. Fulchignoni, E. Grün, M.G.G.T. Taylor, P. Weissman, Unexpected and significant findings in comet 67P/Churyumov-Gerasimenko: an interdisciplinary view. Mon. Not. R. Astron. Soc. **462**, 2–8 (2016b). https://doi.org/10.1093/mnras/stw1663

P. Garaud, F. Meru, M. Galvagni, C. Olczak, From dust to planetesimals: an improved model for collisional growth in protoplanetary disks. Astrophys. J. **764**, 146 (2013). https://doi.org/10.1088/0004-637X/764/2/146

S. Gärtner, B. Gundlach, T.F. Headen, J. Ratte, J. Oesert, S.N. Gorb, T.G.A. Youngs, D.T. Bowron, J. Blum, H.J. Fraser, Micrometer-sized water ice particles for planetary science experiments: influence of surface structure on collisional properties. Astrophys. J. **848**, 96 (2017). https://doi.org/10.3847/1538-4357/aa8c7f

B. Gundlach, J. Blum, Outgassing of icy bodies in the Solar System—II: Heat transport in dry, porous surface dust layers. Icarus **219**, 618–629 (2012). https://doi.org/10.1016/j.icarus.2012.03.013

B. Gundlach, J. Blum, The stickiness of micrometer-sized water-ice particles. Astrophys. J. **798**, 34 (2015). https://doi.org/10.1088/0004-637X/798/1/34

B. Gundlach, J. Blum, Why are Jupiter family comets active and asteroids in cometary-like orbits inactive? How hydrostatic compression leads to inactivity. Astron. Astrophys. **589**, 111 (2016). https://doi.org/10.1051/0004-6361/201527260

B. Gundlach, Y.V. Skorov, J. Blum, Outgassing of icy bodies in the Solar System—I. The sublimation of hexagonal water ice through dust layers. Icarus **213**, 710–719 (2011). https://doi.org/10.1016/j.icarus.2011.03.022

B. Gundlach, J. Blum, H.U. Keller, Y.V. Skorov, What drives the dust activity of comet 67P/Churyumov-Gerasimenko? Astron. Astrophys. **583**, 12 (2015). https://doi.org/10.1051/0004-6361/201525828

C. Güttler, J. Blum, A. Zsom, C.W. Ormel, C.P. Dullemond, The outcome of protoplanetary dust growth: pebbles, boulders, or planetesimals? I. Mapping the zoo of laboratory collision experiments. Astron. Astrophys. **513**, 56 (2010). https://doi.org/10.1051/0004-6361/200912852

D. Heißelmann, J. Blum, H.J. Fraser, Experimental studies on the aggregation properties of ice and dust in planet-forming regions, in *Proc. 58th Int. Astronautical Congress 2007* (Int. Astronautical Federation, Paris, 2007), IAC-07-A2.1.02

A. Johansen, J.S. Oishi, M.-M. Mac Low, H. Klahr, T. Henning, A. Youdin, Rapid planetesimal formation in turbulent circumstellar disks. Nature **448**, 1022–1025 (2007). https://doi.org/10.1038/nature06086

A. Johansen, J. Blum, H. Tanaka, C. Ormel, M. Bizzarro, H. Rickman, The multifaceted planetesimal formation process, in *Protostars and Planets VI* (2014), pp. 547–570

A. Kataoka, H. Tanaka, S. Okuzumi, K. Wada, Fluffy dust forms icy planetesimals by static compression. Astron. Astrophys. **557**, 4 (2013). https://doi.org/10.1051/0004-6361/201322151

S. Kempf, S. Pfalzner, T.K. Henning, N-particle-simulations of dust growth. I. Growth driven by Brownian motion. Icarus **141**, 388–398 (1999). https://doi.org/10.1006/icar.1999.6171

W. Kofman, A. Herique, Y. Barbin, J.-P. Barriot, V. Ciarletti, S. Clifford, P. Edenhofer, C. Elachi, C. Eyraud, J.-P. Goutail, E. Heggy, L. Jorda, J. Lasue, A.-C. Levasseur-Regourd, E. Nielsen, P. Pasquero, F. Preusker, P. Puget, D. Plettemeier, Y. Rogez, H. Sierks, C. Statz, H. Svedhem, I. Williams, S. Zine, J. Van Zyl, Properties of the 67P/Churyumov-Gerasimenko interior revealed by CONSERT radar. Science **349**(2), 0639 (2015). https://doi.org/10.1126/science.aab0639

S. Kothe, Mikrogravitationsexperimente zur Entwicklung eines empirischen Stoßmodells für protoplanetare Staubagglomerate. Ph.D. thesis, TU Braunschweig, 2016

S. Kothe, C. Güttler, J. Blum, The physics of protoplanetesimal dust agglomerates. V. Multiple impacts of dusty agglomerates at velocities above the fragmentation threshold. Astrophys. J. **725**, 1242–1251 (2010). https://doi.org/10.1088/0004-637X/725/1/1242

S. Kothe, J. Blum, R. Weidling, C. Güttler, Free collisions in a microgravity many-particle experiment. III. The collision behavior of sub-millimeter-sized dust aggregates. Icarus **225**, 75–85 (2013). https://doi.org/10.1016/j.icarus.2013.02.034

A. Kouchi, T. Kudo, H. Nakano, M. Arakawa, N. Watanabe, S.-i. Sirono, M. Higa, N. Maeno, Rapid growth of asteroids owing to very sticky interstellar organic grains. Astrophys. J. Lett. **566**, 121–124 (2002). https://doi.org/10.1086/339618

M. Krause, J. Blum, Growth and form of planetary seedlings: results from a sounding rocket microgravity aggregation experiment. Phys. Rev. Lett. **93**(2), 021103 (2004). https://doi.org/10.1103/PhysRevLett.93.021103

K.A. Kretke, H.F. Levison, Evidence for pebbles in comets. Icarus **262**, 9–13 (2015). https://doi.org/10.1016/j.icarus.2015.08.017

S. Krijt, C.W. Ormel, C. Dominik, A.G.G.M. Tielens, Erosion and the limits to planetesimal growth. Astron. Astrophys. **574**, 83 (2015). https://doi.org/10.1051/0004-6361/201425222

A.V. Krivov, A. Ide, T. Löhne, A. Johansen, J. Blum, Debris disc constraints on planetesimal formation. Mon. Not. R. Astron. Soc. **474**, 2564–2575 (2018). https://doi.org/10.1093/mnras/stx2932

T. Kudo, A. Kouchi, M. Arakawa, H. Nakano, The role of sticky interstellar organic material in the formation of asteroids. Meteorit. Planet. Sci. **37**, 1975–1983 (2002). https://doi.org/10.1111/j.1945-5100.2002.tb01178.x

E. Kührt, H.U. Keller, The formation of cometary surface crusts. Icarus **109**, 121–132 (1994). https://doi.org/10.1006/icar.1994.1080

E.K. Kührt, H.U. Keller, On the importance of dust in cometary nuclei. Earth Moon Planets **72**, 79–89 (1996). https://doi.org/10.1007/BF00117506

A. Landeck, Stöße zwischen hochporösen Staubagglomeraten im Laborfallturm. Bachelor Thesis, TU Braunschweig, 2016

D. Langkowski, J. Teiser, J. Blum, The physics of protoplanetesimal dust agglomerates. II. Low-velocity collision properties. Astrophys. J. **675**, 764–776 (2008). https://doi.org/10.1086/525841

Y. Liu, T. Henning, C. Carrasco-González, C.J. Chandler, H. Linz, T. Birnstiel, R. van Boekel, L.M. Pérez, M. Flock, L. Testi, L.F. Rodríguez, R. Galván-Madrid, The properties of the inner disk around HL Tau: multi-wavelength modeling of the dust emission. Astron. Astrophys. **607**, 74 (2017). https://doi.org/10.1051/0004-6361/201629786

S. Lorek, P. Lacerda, J. Blum, Local growth of dust- and ice-mixed aggregates as cometary building blocks in the solar nebula. Astron. Astrophys. (2018, in press). https://doi.org/10.1051/0004-6361/201630175

T. Mannel, M.S. Bentley, R. Schmied, H. Jeszenszky, A.C. Levasseur-Regourd, J. Romstedt, K. Torkar, Fractal cometary dust—a window into the early Solar system. Mon. Not. R. Astron. Soc. **462**, 304–311 (2016). https://doi.org/10.1093/mnras/stw2898

T. Meisner, G. Wurm, J. Teiser, M. Schywek, Preplanetary scavengers: growing tall in dust collisions. Astron. Astrophys. **559**, 123 (2013). https://doi.org/10.1051/0004-6361/201322083

A. Morbidelli, H. Rickman, Comets as collisional fragments of a primordial planetesimal disk. Astron. Astrophys. **583**, 43 (2015). https://doi.org/10.1051/0004-6361/201526116

A. Morbidelli, W.F. Bottke, D. Nesvorný, H.F. Levison, Asteroids were born big. Icarus **204**, 558–573 (2009). https://doi.org/10.1016/j.icarus.2009.07.011

G. Musiolik, J. Teiser, T. Jankowski, G. Wurm, Collisions of CO_2 ice grains in planet formation. Astrophys. J. **818**, 16 (2016a). https://doi.org/10.3847/0004-637X/818/1/16

G. Musiolik, J. Teiser, T. Jankowski, G. Wurm, Ice grain collisions in comparison: CO_2, H_2O, and their mixtures. Astrophys. J. **827**, 63 (2016b). https://doi.org/10.3847/0004-637X/827/1/63

A. Natta, L. Testi, N. Calvet, T. Henning, R. Waters, D. Wilner, Dust in protoplanetary disks: properties and evolution, in *Protostars and Planets V* (2007), pp. 767–781

S. Okuzumi, H. Tanaka, H. Kobayashi, K. Wada, Rapid coagulation of porous dust aggregates outside the snow line: a pathway to successful icy planetesimal formation. Astrophys. J. **752**, 106 (2012). https://doi.org/10.1088/0004-637X/752/2/106

G.Y. Onoda, E.G. Liniger, Random loose packings of uniform spheres and the dilatancy onset. Phys. Rev. Lett. **64**, 2727–2730 (1990). https://doi.org/10.1103/PhysRevLett.64.2727

G.B. Paraskov, G. Wurm, O. Krauss, Impacts into weak dust targets under microgravity and the formation of planetesimals. Icarus **191**, 779–789 (2007). https://doi.org/10.1016/j.icarus.2007.05.008

D. Paszun, C. Dominik, The influence of grain rotation on the structure of dust aggregates. Icarus **182**, 274–280 (2006). https://doi.org/10.1016/j.icarus.2005.12.018

D. Paszun, C. Dominik, Collisional evolution of dust aggregates. From compaction to catastrophic destruction. Astron. Astrophys. **507**, 1023–1040 (2009). https://doi.org/10.1051/0004-6361/200810682

M. Pätzold, T. Andert, M. Hahn, S.W. Asmar, J.-P. Barriot, M.K. Bird, B. Häusler, K. Peter, S. Tellmann, E. Grün, P.R. Weissman, H. Sierks, L. Jorda, R. Gaskell, F. Preusker, F. Scholten, A homogeneous nucleus for comet 67P/Churyumov-Gerasimenko from its gravity field. Nature **530**, 63–65 (2016). https://doi.org/10.1038/nature16535

L.M. Pérez, J.M. Carpenter, C.J. Chandler, A. Isella, S.M. Andrews, L. Ricci, N. Calvet, S.A. Corder, A.T. Deller, C.P. Dullemond, J.S. Greaves, R.J. Harris, T. Henning, W. Kwon, J. Lazio, H. Linz, L.G. Mundy, A.I. Sargent, S. Storm, L. Testi, D.J. Wilner, Constraints on the radial variation of grain growth in the AS 209 circumstellar disk. Astrophys. J. Lett. **760**, 17 (2012). https://doi.org/10.1088/2041-8205/760/1/L17

L.M. Pérez, C.J. Chandler, A. Isella, J.M. Carpenter, S.M. Andrews, N. Calvet, S.A. Corder, A.T. Deller, C.P. Dullemond, J.S. Greaves, R.J. Harris, T. Henning, W. Kwon, J. Lazio, H. Linz, L.G. Mundy, L. Ricci, A.I. Sargent, S. Storm, M. Tazzari, L. Testi, D.J. Wilner, Grain growth in the circumstellar disks of the young stars CY Tau and DoAr 25. Astrophys. J. **813**, 41 (2015). https://doi.org/10.1088/0004-637X/813/1/41

L. Ricci, L. Testi, A. Natta, K.J. Brooks, Dust grain growth in ρ-Ophiuchi protoplanetary disks. Astron. Astrophys. **521**, 66 (2010). https://doi.org/10.1051/0004-6361/201015039

U. Schäfer, C.-C. Yang, A. Johansen, Initial mass function of planetesimals formed by the streaming instability. Astron. Astrophys. **597**, 69 (2017). https://doi.org/10.1051/0004-6361/201629561

R. Schräpler, J. Blum, The physics of protoplanetesimal dust agglomerates. VI. Erosion of large aggregates as a source of micrometer-sized particles. Astrophys. J. **734**, 108 (2011). https://doi.org/10.1088/0004-637X/734/2/108

R. Schräpler, J. Blum, A. Seizinger, W. Kley, The physics of protoplanetesimal dust agglomerates. VII. The low-velocity collision behavior of large dust agglomerates. Astrophys. J. **758**, 35 (2012). https://doi.org/10.1088/0004-637X/758/1/35

R. Schräpler, J. Blum, S. Krijt, J.-H. Raabe, The physics of protoplanetesimal dust agglomerates. X. High-velocity collisions between small and large dust agglomerates as growth barrier. Astrophys. J. **853**, 74 (2018). https://doi.org/10.3847/1538-4357/aaa0d2

A. Seizinger, S. Krijt, W. Kley, Erosion of dust aggregates. Astron. Astrophys. **560**, 45 (2013). https://doi.org/10.1051/0004-6361/201322773

K. Shariff, J.N. Cuzzi, The spherically symmetric gravitational collapse of a clump of solids in a gas. Astrophys. J. **805**, 42 (2015). https://doi.org/10.1088/0004-637X/805/1/42

J.B. Simon, P.J. Armitage, R. Li, A.N. Youdin, The mass and size distribution of planetesimals formed by the streaming instability. I. The role of self-gravity. Astrophys. J. **822**, 55 (2016). https://doi.org/10.3847/0004-637X/822/1/55

J.B. Simon, P.J. Armitage, A.N. Youdin, R. Li, Evidence for universality in the initial planetesimal mass function. Astrophys. J. Lett. **847**, 12 (2017). https://doi.org/10.3847/2041-8213/aa8c79

Y. Skorov, J. Blum, Dust release and tensile strength of the non-volatile layer of cometary nuclei. Icarus **221**, 1–11 (2012). https://doi.org/10.1016/j.icarus.2012.01.012

R.H. Soja, M. Sommer, J. Herzog, J. Agarwal, J. Rodmann, R. Srama, J. Vaubaillon, P. Strub, A. Hornig, L. Bausch, E. Grün, Characteristics of the dust trail of 67P/Churyumov-Gerasimenko: an application of the IMEX model. Astron. Astrophys. **583**, 18 (2015). https://doi.org/10.1051/0004-6361/201526184

M. Tazzari, L. Testi, B. Ercolano, A. Natta, A. Isella, C.J. Chandler, L.M. Pérez, S. Andrews, D.J. Wilner, L. Ricci, T. Henning, H. Linz, W. Kwon, S.A. Corder, C.P. Dullemond, J.M. Carpenter, A.I. Sargent, L. Mundy, S. Storm, N. Calvet, J.A. Greaves, J. Lazio, A.T. Deller, Multiwavelength analysis for interferometric (sub-)mm observations of protoplanetary disks. Radial constraints on the dust properties and the disk structure. Astron. Astrophys. **588**, 53 (2016). https://doi.org/10.1051/0004-6361/201527423

J. Teiser, G. Wurm, Decimetre dust aggregates in protoplanetary discs. Astron. Astrophys. **505**, 351–359 (2009a). https://doi.org/10.1051/0004-6361/200912027

J. Teiser, G. Wurm, High-velocity dust collisions: forming planetesimals in a fragmentation cascade with final accretion. Mon. Not. R. Astron. Soc. **393**, 1584–1594 (2009b). https://doi.org/10.1111/j.1365-2966.2008.14289.x

J. Teiser, M. Küpper, G. Wurm, Impact angle influence in high velocity dust collisions during planetesimal formation. Icarus **215**, 596–598 (2011). https://doi.org/10.1016/j.icarus.2011.07.036

L. Testi, T. Birnstiel, L. Ricci, S. Andrews, J. Blum, J. Carpenter, C. Dominik, A. Isella, A. Natta, J.P. Williams, D.J. Wilner, Dust evolution in protoplanetary disks, in *Protostars and Planets VI* (2014), pp. 339–361

F. Trotta, L. Testi, A. Natta, A. Isella, L. Ricci, Constraints on the radial distribution of the dust properties in the CQ Tauri protoplanetary disk. Astron. Astrophys. **558**, 64 (2013). https://doi.org/10.1051/0004-6361/201321896

R. van Boekel, M. Min, C. Leinert, L.B.F.M. Waters, A. Richichi, O. Chesneau, C. Dominik, W. Jaffe, A. Dutrey, U. Graser, T. Henning, J. de Jong, R. Köhler, A. de Koter, B. Lopez, F. Malbet, S. Morel, F. Paresce, G. Perrin, T. Preibisch, F. Przygodda, M. Schöller, M. Wittkowski, The building blocks of planets within the 'terrestrial' region of protoplanetary disks. Nature **432**, 479–482 (2004). https://doi.org/10.1038/nature03088

R. Visser, E.F. van Dishoeck, S.D. Doty, C.P. Dullemond, The chemical history of molecules in circumstellar disks. I. Ices. Astron. Astrophys. **495**, 881–897 (2009). https://doi.org/10.1051/0004-6361/200810846

K. Wada, H. Tanaka, T. Suyama, H. Kimura, T. Yamamoto, Numerical simulation of dust aggregate collisions. II. Compression and disruption of three-dimensional aggregates in head-on collisions. Astrophys. J. **677**, 1296–1308 (2008). https://doi.org/10.1086/529511

K. Wada, H. Tanaka, T. Suyama, H. Kimura, T. Yamamoto, Collisional growth conditions for dust aggregates. Astrophys. J. **702**, 1490–1501 (2009). https://doi.org/10.1088/0004-637X/702/2/1490

K. Wahlberg Jansson, A. Johansen, Formation of pebble-pile planetesimals. Astron. Astrophys. **570**, 47 (2014). https://doi.org/10.1051/0004-6361/201424369

K. Wahlberg Jansson, A. Johansen, M. Bukhari Syed, J. Blum, The role of pebble fragmentation in planetesimal formation. II. Numerical simulations. Astrophys. J. **835**, 109 (2017). https://doi.org/10.3847/1538-4357/835/1/109

S.J. Weidenschilling, Aerodynamics of solid bodies in the solar nebula. Mon. Not. R. Astron. Soc. **180**, 57–70 (1977). https://doi.org/10.1093/mnras/180.1.57

S.J. Weidenschilling, Initial sizes of planetesimals and accretion of the asteroids. Icarus **214**, 671–684 (2011). https://doi.org/10.1016/j.icarus.2011.05.024

S.J. Weidenschilling, J.N. Cuzzi, Formation of planetesimals in the solar nebula, in *Protostars and Planets III*, ed. by E.H. Levy, J.I. Lunine, 1993, pp. 1031–1060

R. Weidling, J. Blum, Free collisions in a microgravity many-particle experiment. IV.—Three-dimensional analysis of collision properties. Icarus **253**, 31–39 (2015). https://doi.org/10.1016/j.icarus.2014.12.010

R. Weidling, C. Güttler, J. Blum, F. Brauer, The physics of protoplanetesimal dust agglomerates. III. Compaction in multiple collisions. Astrophys. J. **696**, 2036–2043 (2009). https://doi.org/10.1088/0004-637X/696/2/2036

R. Weidling, C. Güttler, J. Blum, Free collisions in a microgravity many-particle experiment. I. Dust aggregate sticking at low velocities. Icarus **218**, 688–700 (2012). https://doi.org/10.1016/j.icarus.2011.10.002

A.D. Whizin, J. Blum, J.E. Colwell, The physics of protoplanetesimal dust agglomerates. VIII. Microgravity collisions between porous SiO_2 aggregates and loosely bound agglomerates. Astrophys. J. **836**, 94 (2017). https://doi.org/10.3847/1538-4357/836/1/94

F. Windmark, T. Birnstiel, C.W. Ormel, C.P. Dullemond, Breaking through: the effects of a velocity distribution on barriers to dust growth. Astron. Astrophys. **544**, 16 (2012a). https://doi.org/10.1051/0004-6361/201220004

F. Windmark, T. Birnstiel, C. Güttler, J. Blum, C.P. Dullemond, T. Henning, Planetesimal formation by sweep-up: how the bouncing barrier can be beneficial to growth. Astron. Astrophys. **540**, 73 (2012b). https://doi.org/10.1051/0004-6361/201118475

G. Wurm, J. Blum, Experiments on preplanetary dust aggregation. Icarus **132**, 125–136 (1998). https://doi.org/10.1006/icar.1998.5891

G. Wurm, G. Paraskov, O. Krauss, Ejection of dust by elastic waves in collisions between millimeter- and centimeter-sized dust aggregates at 16.5 to 37.5 m/s impact velocities. Phys. Rev. E **71**(2), 021304 (2005a). https://doi.org/10.1103/PhysRevE.71.021304

G. Wurm, G. Paraskov, O. Krauss, Growth of planetesimals by impacts at ~ 25 m/s. Icarus **178**, 253–263 (2005b). https://doi.org/10.1016/j.icarus.2005.04.002

C.-C. Yang, A. Johansen, D. Carrera, Concentrating small particles in protoplanetary disks through the streaming instability. Astron. Astrophys. **606**, 80 (2017). https://doi.org/10.1051/0004-6361/201630106

A.N. Youdin, J. Goodman, Streaming instabilities in protoplanetary disks. Astrophys. J. **620**, 459–469 (2005). https://doi.org/10.1086/426895

A. Zsom, C.W. Ormel, C. Güttler, J. Blum, C.P. Dullemond, The outcome of protoplanetary dust growth: pebbles, boulders, or planetesimals? II. Introducing the bouncing barrier. Astron. Astrophys. **513**, 57 (2010). https://doi.org/10.1051/0004-6361/200912976

Space Sci Rev (2018) 214:64
https://doi.org/10.1007/s11214-018-0496-3

Cometary Dust

Anny-Chantal Levasseur-Regourd[1] · Jessica Agarwal[2] · Hervé Cottin[3] · Cécile Engrand[4] · George Flynn[5] · Marco Fulle[6] · Tamas Gombosi[7] · Yves Langevin[8] · Jérémie Lasue[9] · Thurid Mannel[10,11] · Sihane Merouane[2] · Olivier Poch[12] · Nicolas Thomas[13] · Andrew Westphal[14]

Received: 13 December 2017 / Accepted: 16 March 2018 / Published online: 28 March 2018

Abstract This review presents our understanding of cometary dust at the end of 2017. For decades, insight about the dust ejected by nuclei of comets had stemmed from remote observations from Earth or Earth's orbit, and from flybys, including the samples of dust returned to Earth for laboratory studies by the Stardust return capsule. The long-duration Rosetta mission has recently provided a huge and unique amount of data, obtained using numerous instruments, including innovative dust instruments, over a wide range of distances from the

Cosmic Dust from the Laboratory to the Stars
Edited by Rafael Rodrigo, Jürgen Blum, Hsiang-Wen Hsu, Detlef Koschny, Anny-Chantal Levasseur-Regourd, Jesús Martín-Pintado, Veerle Sterken and Andrew Westphal

✉ M. Fulle
fulle@oats.inaf.it

A.C. Levasseur-Regourd
aclr@latmos.ipsl.fr

1 LATMOS, Sorbonne Université, CNRS, IPSL, UVSQ, Campus Pierre et Marie Curie, BC 102, 4 place Jussieu, 75005 Paris, France

2 Max-Planck-Institut für Sonnensystemforschung, Justus-von-Liebig-Weg, 3, 37077, Göttingen, Germany

3 Laboratoire Interuniversitaire des Systèmes Atmosphériques (LISA), UMR CNRS 7583, Université Paris-Est Créteil et Université Paris Diderot, Institut Pierre Simon Laplace, 94000 Créteil, Paris, France

4 Centre de Sciences Nucléaires et de Sciences de la Matière (CSNSM), CNRS/IN2P3 – Université Paris Sud – UMR 8609, Université Paris-Saclay, 91405, Bâtiment 104, Orsay Campus, Orsay, France

5 SUNY-Plattsburgh, 101 Broad St, Plattsburgh, NY 12901, USA

6 INAF – Osservatorio Astronomico, Via Tiepolo 11, 34143 Trieste, Italy

7 Department of Climate and Space Sciences and Engineering, University of Michigan, Ann Arbor, MI, 48109, USA

8 Institut d'Astrophysique Spatiale (IAS), CNRS/Université Paris Saclay, Bâtiment 121, Orsay Campus, Orsay, 91405, France

9 IRAP, Université de Toulouse, CNRS, UPS, CNES, Toulouse, France

Sun and from the nucleus. The diverse approaches available to study dust in comets, together with the related theoretical and experimental studies, provide evidence of the composition and physical properties of dust particles, e.g., the presence of a large fraction of carbon in macromolecules, and of aggregates on a wide range of scales. The results have opened vivid discussions on the variety of dust-release processes and on the diversity of dust properties in comets, as well as on the formation of cometary dust, and on its presence in the near-Earth interplanetary medium. These discussions stress the significance of future explorations as a way to decipher the formation and evolution of our Solar System.

Keywords Dust · Comets · Jupiter-family comets · Organics · Aggregates · Rosetta · Stardust · Comets: coma, nucleus, trail · Comets: individual: 1P/Halley, 9P/Tempel 1, 67P/Churyumov-Gerasimenko, 81P/Wild 2, C/1995 O1 Hale-Bopp · Solar System formation · Comet formation · Origin of life · Cosmic dust · Solar nebula

1 Introduction

By the middle of the XXth century, Whipple (1950) inferred, through his analysis of the non-keplerian motion of comets, the existence of cometary nuclei consisting of ices embedded with dust, i.e. micron-sized non-volatile particles. Dust ejected with gases from the nuclei, in combination with solar gravity and radiation pressure when comets approach the Sun, was recognized as the means by which the long-yellowish cometary dust tails are produced (e.g., Brandt and Chapman 1981). Significant progress in our understanding of cometary dust has been achieved, since the 1980's, by a series of space missions performing flyby of nuclei and by remote observations of dust in cometary environments, as summarized below.

1.1 Past Flyby Missions to Comets

The choice made in 1980 by the European Space Agency (ESA), to perform a flyby of comet 1P/Halley soon after its perihelion passage, triggered in situ space exploration of cometary dust. The soviet Vega missions complemented the European Giotto mission, allowing, through the pathfinder concept, to flyby the nucleus with a miss-distance of about 600 km in March 1986. Vega also provided unique results on dust elemental composition within 1P/Halley's coma; the PUMA instrument (Dust Impact Mass Analyser) on board Vega-1 discovered dust particles to consist of rock-forming minerals, and also of carbon-hydrogen-oxygen-nitrogen compounds, so-called CHONs (Kissel et al. 1986; Krueger et al. 1991).

On board Giotto, DID (Dust Impact Detector) and OPE (Optical Probe Experiment) monitored dust impacts and dust scattering properties along the trajectory of the spacecraft within the coma. OPE, through measurements of the solar light locally scattered by

[10] Space Research Institute, Austrian Academy of Sciences, Schmiedlstrasse 6, 8042 Graz, Austria

[11] Physics Institute, University of Graz, Universitätsplatz 5, 8010 Graz, Austria

[12] Institut de Planétologie et d'Astrophysique de Grenoble (IPAG), Univ. Grenoble Alpes, CNRS, IPAG, 38000 Grenoble, France

[13] Physikalisches Institut, Universität Bern, Sidlerstrasse 5, 3012, Bern, Switzerland

[14] Space Sciences Laboratory, U.C. Berkeley, Berkeley, CA, 94720-7450, USA

dust, determined changes in the dust properties (Levasseur-Regourd et al. 1999). In addition, a comparison between DID and OPE dust data suggested, through a non-isotropic dust dynamical model, the geometric albedo and the density of the dust particles to be extremely low, ranging between 3% and 10% and between 50 and 500 $kg\,m^{-3}$, respectively (Fulle et al. 2000). No significant distributed gas sources were identified, implying no significant sublimation of dust (Rubin et al. 2011). After its successful mission to 1P/Halley, Giotto in 1992 flew-by a Jupiter Family Comet (JFC), 26P/Grigg-Skjellerup. The presence of a possible fragment of the nucleus was suspected within its coma (Levasseur-Regourd et al. 1993). Although the data were consistent with impacts of large particles on the spacecraft generating a transient dust cloud (Le Duin et al. 1996), the presence of a fragment of the main nucleus was also inferred from fluctuations in the energetic particle data recorded by EPONA experiment on board Giotto (McKenna-Lawlor and Afonin 1999).

In subsequent years four more JFC comets were visited by NASA spacecraft, from Deep-Space 1 at 19P/Borrelly in 2001 to Stardust-NExT at 9P/Tempel 1 in 2011. The densities of the nuclei of 9P/Tempel 1 and of 103P/Hartley 2, previously estimated from non-gravitational forces effects, were confirmed to be significantly smaller than 10^3 $kg\,m^{-3}$, implying that they are very porous objects (e.g., Barucci et al. 2011). Stardust mission to 81P/Wild 2 in 2004 confirmed the presence of organic compounds (Brownlee et al. 2004). Clusters of dust within 81P/Wild 2 and 9P/Tempel 1 comae were interpreted in terms of dust fragmentation (Tuzzolino et al. 2004; Economou et al. 2013). The Deep Space 1 mission to comet 19P/Borrelly in 2001 granted further evidence of spatially complex coma dust distributions.

In addition, samples of cometary dust collected in the Stardust return capsule provided exceptional results on cometary dust collected during the 6.1 $km\,s^{-1}$ flyby in the coma of 81P/Wild 2. Analyses of dust impacts on aluminum foils or tracks within aerogel cells established the presence of particles of different internal tensile strength (Brownlee et al. 2006; Hörz et al. 2006; Burchell et al. 2008) and of complex organics (most likely including glycine) with heterogeneous and un-equilibrated distributions (Matrajt et al. 2008; Elsila et al. 2009).

To summarize, studies during cometary flybys of dust particles ejected by the dark and highly non-spherical nuclei of comet 1P/Halley and of five JFCs, suggest that cometary dust is not only rich in organic compounds, but also diverse, dark, and quite porous (see e.g. for reviews Kolokolova et al. 2004; Cochran et al. 2015).

1.2 Past Remote Observations of Dust in Comae, Tails and Trails

Remote observations of cometary dust have been mostly triggered by space missions to comets, by the passage of new spectacular comets, such as C/1995 O1 Hale-Bopp, by new observational techniques, such as infrared surveys, and by the development of various numerical models and experimental simulations.

The spectroscopic observations of Hale-Bopp in the near infrared from the ISO space telescope showed that particles emitted from the nucleus contained silicates, particularly magnesium-rich crystalline olivine (forsterite) and crystalline magnesium-rich pyroxenes, along with a large proportion of amorphous silicates (Crovisier 1997; Wooden et al. 1999). The detection in spectra of crystalline material provided a first hint that some minerals in cometary dust had encountered high temperatures during their history before being stored in the comet nucleus, implying that they are not only consisting of amorphous grains directly inherited from the presolar cloud (see e.g. for reviews Cochran et al. 2015; Wooden et al. 2017).

The brightness and partial linear polarization (thereafter referred to as polarization) of solar light scattered by dust in the comae of bright comets (including the spectacular C/1995 O1 Hale-Bopp) and of targets of space missions have long been monitored from Earth-based observatories. Of major interest are brightness images, which have made it possible to point out anisotropies in dust density within comae, and to infer from them constraints on the spin axis orientation of various comets, and the variations of a given set of observations with the phase angle (α), possibly accessible from the backscattering region near 0 deg to about 100 deg, and the wavelength (λ).

The main interest of the polarization is that, except very close to the nucleus, it does not vary with the dust spatial density, nor with the distances to the Sun and to the observer. It just varies with α and λ, and it depends on the bulk properties of dust along the line of sight, such as the size, the structure and the complex refractive indices, and thus the geometric albedo (Levasseur-Regourd and Hadamcik 2003; Kolokolova et al. 2004). The first combination of polarimetric measurements obtained by various teams on Earth and from a space probe took place during 1P/Halley return in 1985–1986 (Dollfus et al. 1988); OPE on Giotto provided clues (Levasseur-Regourd et al. 1999) to changes in dust properties from measurements in three visible spectral domains at $\alpha = 73$ deg and from about 10^3 to 10^5 km from the nucleus.

Local and remote measurements of the polarization of cometary dust are most delicate, since they need to be performed through narrow interference filters. Such filters, which were already developed for 1P/Halley observations (typically through the international Halley Watch) to avoid the contamination by less-polarized molecular bands, reduce the intensity of the collected light (Levasseur-Regourd 1999; Kiselev et al. 2004). Developments in remote imaging polarimetry made it possible to distinguish between anisotropies in dust density and changes in intrinsic dust properties, and have confirmed the existence of such changes within jet-like coma features and inner comae, below a few thousands of km from the nucleus (Renard et al. 1996; Hadamcik et al. 1997). Data on whole coma polarization have been used to build up polarimetric phase curves for comets extensively observed by various teams, at different times and thus different phase angles. Within a given wavelength range, free of gaseous emissions, phase curves are well defined; their smoothness may be indicative of scattering by irregular particles (Levasseur-Regourd et al. 1996; Kiselev et al. 2015). For $\alpha \geq 25$ deg, the polarization usually increases with increasing wavelength, at least up to 2 μm, where infrared emission is still negligible. Different types of dust properties have been suggested, from the positive branches (at least from $\alpha \geq 25$ deg) of the phase curves within a given wavelength range, with high-polarization comets that usually also present a strong silicate emission feature near 10 μm (Levasseur-Regourd 1999).

Outbursts in activity of cometary nuclei (seldom observed remotely), as well as partial fragmentation and total disintegration of nuclei, and even induced impacts (from Deep Impact mission), have been noticed to lead often to an increase in polarization within the coma (Hadamcik and Levasseur-Regourd 2003; Furusho et al. 2007). It may be speculated that such events triggered the ejection of dust particles stored deep enough inside the nucleus to present slightly different dust properties. Cometary polarization curves are generally consistent with light scattering by a population mixing irregular dust particles with more compact large particles, and with optical indices similar to silicates and absorbing organics (Lasue et al. 2009; Kolokolova and Kimura 2010), as detailed and compared with recent in situ results in Sect. 5.2.

Dust is accelerated in the coma by the gas pressure up to a terminal velocity which also depends on the dust size. Dust with such an ejection velocity leaves the coma, and is swept into the dust tail by the solar radiation pressure. Monte-Carlo models of dust tails have provided information on the time evolution of the dust loss rate, size distribution and ejection velocity for many long and short period comets (Fulle 2004), showing that the dust mass, and sometimes also the optical cross-section, is dominated by the largest ejected chunks.

Finally, many short-period comets are followed in their orbital motion by a trail of debris with sizes typically above 100 μm and up to meters. Due to their low cross-section-to-mass ratio, such dust particles are only little affected by solar radiation pressure and remain close to the parent body's orbit for many revolutions around the Sun, before they are eventually dispersed by planetary encounters (Soja et al. 2015). Infrared observations from space led to the discovery of debris trails as narrow and elongated structures along the orbits of periodic comets (Davies et al. 1984; Sykes et al. 1986; Reach et al. 2007). Trails represent an intermediate step between the cometary dust and the meteoritic streams (Kresak 1993; Koschny et al. 2017).

Following their discovery with the Infra Red Astronomical Satellite (IRAS) (Sykes et al. 1986; Sykes and Walker 1992), surveys at thermal infrared and visible wavelengths have been conducted (Reach et al. 2007; Ishiguro et al. 2009), and trails of some individual comets have been the subject of targeted observations. The width of a trail and its surface brightness as a function of the nucleus distance have been used to constrain the dust ejection velocities, past production rates and size distribution through numerical modelling of the dust motion under the influence of solar gravity and radiation pressure (Reach et al. 2000; Ishiguro et al. 2003; Sarugaku et al. 2007; Ishiguro 2008; Kelley et al. 2008; Stevenson et al. 2014). Ejection velocities are of the order of $m\,s^{-1}$ (Sykes et al. 1990; Davies et al. 1997).

The combination of observations in scattered visible and thermal infrared light yields a dark geometric albedo of a few percent (Ishiguro et al. 2002; Agarwal et al. 2010; Sarugaku et al. 2015). Trails are associated with a large fraction of the observed short-period comets and can be considered one of their generic features (Reach et al. 2007). The derived production rates of trail material suggest that cm-sized particles dominate the optical cross-section of dust released from comets (Reach et al. 2007; Agarwal et al. 2010; Vaubaillon and Reach 2010; Arendt 2014). Dust production rates derived from trail formation models are therefore a good proxy for the total loss rate of refractory material from a comet (Reach et al. 2000).

The next section presents the context of the Rosetta rendezvous mission, which has recently allowed significant progress in our understanding of cometary dust. Key results on the composition and on the physical properties of dust particles are then developed in Sects. 3 and 4, with emphasis on Stardust samples and on Rosetta in situ analyses. Finally, various topics related to cometary dust release to comparisons with other approaches, and to implications for topics related to origins are discussed in Sect. 5.

2 Context of the Rosetta Mission

One of the main objectives of the mission was to characterize the properties of the dust released by the nucleus of a Jupiter Family Comet, together with its evolution through a long-duration rendezvous (Glassmeier et al. 2007). The Rosetta cometary probe accompanied 67P/Churyumov-Gerasimenko comet (thereafter 67P/C-G) from August 2014 to September 2016. During its 26-month investigation, the solar distance decreased from about 3.6 au (astronomical units) down to 1.2 au at perihelion (13 Aug. 2015), and then increased again up to 3.8 au. Meanwhile, the distance of Rosetta to the nucleus decreased to about 10 km before the release of the Philae lander (12 Nov. 2014) and to a few meters, when the signal was cut off immediately before the landing of Rosetta (30 Sept. 2016), while increasing by several hundred kilometres whenever the comet was very active.

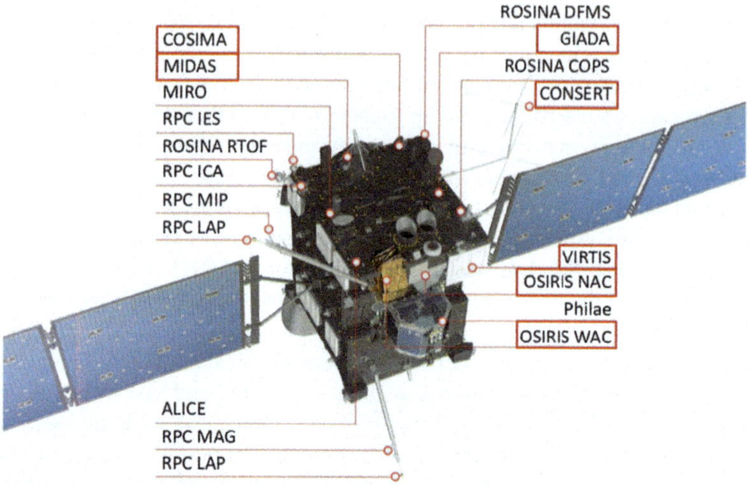

Fig. 1 Rosetta with labelled instruments, including the lander before its release. The dedicated dust analysing instruments COSIMA, GIADA and MIDAS, as well as the OSIRIS camera system, the VIRTIS VIS/NIR imaging spectrometer and the CONSERT antenna are located on the panel of Rosetta that is mainly oriented towards the cometary nucleus. After ESA/ATG medialab

2.1 Rosetta Dust Instruments

Among the 11 instruments on board Rosetta, 3 were specifically dedicated to the study of cometary dust (Fig. 1). COSIMA (COmetary Secondary Ion Mass Analyser) collected dust particles, imaged them with the CosiScope optical microscope with a resolution of 14 μm per pixel, and analysed their composition with a mass spectrometer. GIADA (Grain Impact Analyser and Dust Accumulator) measured the optical cross-section, speed and momentum for particle sizes of about 0.1 to 1 mm, and the cumulative flux of dust particles smaller than 5 μm. MIDAS (Micro-Imaging Dust Analysis System), an atomic force microscope, imaged the 3D surface in the tens of nanometers to micrometres size range. Other instruments could detect dust additionally to their main science goals. For instance, the scientific camera system on board, OSIRIS (Optical, Spectroscopic, and Infrared Remote Imaging System) detected dust and debris over a wide range of sizes and distances from the spacecraft, which allowed the dust phase function to be retrieved, while the bistatic radar on board Rosetta and Philae, CONSERT (Comet Nucleus Sounding Experiment by Radiowave Transmission), provided information about the porosity and the sizes of the voids within the small lobe of the nucleus.

The COSIMA instrument collected, imaged and finally analysed the dust with a time of flight secondary ion mass spectrometer (TOF-SIMS) (Kissel et al. 2007). Dust collection was achieved by exposing three 1 cm^2 targets to the cometary dust stream for durations of hours to days. Afterwards, a robotic unit brought the targets to one of the analytical stations. The first step was to image the targets using the CosiScope with a resolution of 14 $\mu m\,pixel^{-1}$. Selected particles were then probed by TOF-SIMS: The surface layers of the dust were irradiated with a pulsed indium ion beam (of $35 \times 50\ \mu m^2$) to generate secondary ions representative of the composition of the sampled area. The masses and numbers of either the negative or the positive ions were then determined, allowing the identification of the relative composition of the surface of the dust particles. With a mass resolution ($m/\Delta m$) of about 1400 (full-width half maximum of 100) the separation of elements from organic ions containing hydrogen from the same integer mass is in most cases possible. COSIMA also carried a chemical station to warm the targets prior to exposure to either clean them or study the evolution of samples. To summarize, COSIMA determines the elemental and isotopic composition, sheds light on the molecular composition, and characterizes the morphology for particles between 14 μm and a few hundred μm.

The GIADA instrument determined the physical properties of the dust in real time, including momentum, speed and cross-section of the individual particles (Colangeli et al. 2007; Della Corte et al. 2014). These parameters enabled the determination of the dust flux and of the mass of the individual particles, the estimation of the shape and density of the particles, and the place of origin of the dust on the nucleus. GIADA had three subsystems: Grain Detection System (GDS), Impact Sensor (IS) and Micro-Balances System (MBS). GDS consisted of a 3 mm thin light curtain created with 4 laser diodes; if a particle passed the curtain it scattered the light, which was then detected by 4 photodiodes; calibration indicates that the sizes range from about 0.1 to 1 mm (Ferrari et al. 2014). The IS is a 0.5 mm thick and (0.1×0.1 m^2) sized aluminium diaphragm equipped with 5 piezoelectric sensors; if a particle impacted the IS, the momentum and the time difference between GDS and IS signal were measured. Finally, the MBS consisted of 5 quartz microbalances sampling all the 2π solid angle of space outside the Rosetta payload and dedicated to the measurement of the cumulative flux of particles below 5 μm in size.

The MIDAS instrument was the first atomic force microscope (AFM) to be flown in space (Riedler et al. 2007). It collected dust particles and imaged their surfaces in 3D with resolutions of a few nanometres to a few μm. MIDAS had 64 targets of about 3.5 mm^2 cross-section, mounted on a wheel. For particle collection the wheel could drive a target to the position behind a focussing funnel with a shutter that could be opened for dust exposure. To image the collected dust, the target was moved to the AFM part of the instrument. It consisted of 16 tips on a piezoelectric XYZ-stage that allowed accurate movement across the target. The height measurement was performed by amplitude-modulated atomic force microscopy. A typical scientific sequence of MIDAS was to pre-scan an area of a target, expose it to the dust flux of the comet, and re-scan the same area with low resolution. Possibly detected dust particles were then scanned with optimised parameters. The results are three dimensional surface images of cometary dust that allow the determination of size, shape and morphology.

OSIRIS was the scientific camera system on board Rosetta, consisting of a Narrow Angle Camera (NAC) and a Wide Angle Camera (WAC) (Keller et al. 2007). During the Rosetta mission, both cameras regularly observed the nucleus and the coma in different filter bands, monitoring the surface and the gas and dust environment of the comet. In addition to the diffuse background of unresolved dust, OSIRIS regularly provided images of individual dust particles. Depending on the circumstances, it was possible to constrain the sizes of the dust particles, their velocities, rotation rates, and/or colour and to retrieve the dust phase function.

VIRTIS (Visible and Infrared Thermal Imaging Spectrometer) obtained information on the composition and physical properties of dust both on the surface of the nucleus and in the coma (Coradini et al. 2007).

Finally, CONSERT (COmet Nucleus Sounding Experiment by Radiowave Transmission) was a bistatic radar, designed to provide unique information about the interior of the nucleus, by propagating electromagnetic waves (at 90 MHz, i.e. by 3 m wavelength) between the Philae lander and the Rosetta probe (Kofman et al. 2007). The CONSERT radar probed the small lobe of 67P/C-G to depths of a couple of hundred meters, providing constraints on its internal structure (Ciarletti et al. 2017).

2.2 Remote Observations of Dust in 67P/C-G

Once 67P/Churyumov-Gerasimenko was chosen as the target of the Rosetta rendezvous mission, numerous remote observations were performed during the 2008–2009 return and carefully analyzed in preparation of the Rosetta mission. Polarimetric observations provided clues to a seasonal effect, as well as to the presence of both rather large particles and small grains, likely in the form of fluffy aggregates (Hadamcik et al. 2010, 2016). Dust coma structures have constrained properties of the dust and of the nucleus of 67P/C-G (Vincent et al. 2013), although the links between the coma structures observed at large scale (Snodgrass et al. 2017) and those detected by Rosetta still need to be clarified. Tail modelling of 67P/C-G has provided information useful to plan the Rosetta mission, such as the expected dust flux and the contamination to star-trackers frames by individual dust particles (Fulle et al. 2010). The dust size distribution showed a strong time-evolution, with the optical cross-section dominated by the largest ejected dust before perihelion, and by the smallest ejected dust after (Fulle et al. 2004; Moreno et al. 2010; Fulle et al. 2010).

The debris trail of comet 67P/C-G was observed multiple times prior to the Rosetta mission. It was discovered by IRAS (Sykes and Walker 1992), and subsequently observed in both visible and thermal infrared light (Ishiguro 2008; Kelley et al. 2008; Agarwal et al. 2010). Numerical models aiming to reproduce the observed surface brightness find that the production rate of about 0.1 mm-sized particles around perihelion must have been sufficiently high to also dominate the surface brightness of the coma at that time, without a strong additional contribution by smaller particles that would not subsequently populate the debris trail; sizes are in the 0.1 to 100 mm range, with an albedo of a few percent (Agarwal et al. 2010; Soja et al. 2015).

A unique long-term campaign (Snodgrass et al. 2016a,b, 2017) of remote coordinated observations of 67P/C-G, from ultraviolet to radio wavelengths, was organized for the 2013–2016 return, in order to contribute to the support the mission through early characterization of the comet (Snodgrass et al. 2016a) and to provide numerous remote observations, including observation of dust in 67P/C-G coma, tail and trail (Snodgrass et al. 2016b; Moreno et al. 2016, 2017). Although the observing conditions from Earth or its environment were difficult due to the comet's low surface brightness and its low solar elongation around its perihelion passage, numerous observations took place, from a wide range of telescopes distributed on a broad geographical spread, including robotic instruments (Snodgrass et al. 2016b), large ground based telescopes, such as the Russian 6-m (Rosenbush et al. 2017) and space telescopes, such as HST (Hadamcik et al. 2016). The level of activity and the peak in activity dust brightness (about 2 weeks after perihelion passage) were consistent with 67P/C-G observations at previous returns; an extensive review of all these observations is provided in Snodgrass et al. (2017).

Such remote observations should provide a link between the ground-truth provided by Rosetta, to be presented in the next two sections, and future remote observations of other JFCs, including observations of cometary dust.

3 Composition of dust particles

Deciphering the composition of cometary dust particles is a key to understanding where and when they formed, and what were the physico-chemical conditions of their formation environment. Composition can be addressed from four main points of views: the elemental composition, the mineralogy of the dust particles, their organic content and the isotopic fractionation of some of their elements. Being building blocks of cometary nuclei, these dust particles provide valuable information about comets, which may be more easily accessible than direct measurements of the nucleus and its interior.

3.1 Elemental Composition

The relative abundance of chemical elements in an object of the Solar System is a useful tool to assess the degree of "pristinity" of the object, i.e. the extent to which it was submitted to transformation (mainly differentiation) since its formation 4.5 Ga ago. The closer to the solar photosphere composition, the more the object has kept a composition close to the one of the solar nebula where it accreted. Carbonaceous chondrites of the CI type (such as the Orgueil meteorite) are the most chemically primitive meteorites, their elemental composition being very similar to that of the solar photosphere. CI carbonaceous chondrites are commonly used as a reference for comparison with other objects in the Solar System. Most of their elements have a "solar abundance" to within 10% (Lodders 2010). Light elements H, C, O, N and noble gases are exceptions. If these elements display a rather low abundance in CI meteorites relative to the solar photosphere, they may have not been accreted in the parent body of the meteorite since they were not present in the solid phase at that time and parent bodies of the chondrites formed in the warm inner Solar System. Comets, however, are expected to have formed in the outer Solar System, at low temperature, hence they are expected to contain more light elements than CI chondrites.

As briefly mentioned in Sect. 1.1, the elemental composition of some of the main constituents have been measured in dust particles emitted from comets 1P/Halley, 81P/Wild 2 and 67P/C-G. The two spacecraft flying by 1P/Halley in March 1986 (Vega 1 and Giotto) revealed that the dust particles were a mixture of mineral and organic matter (Lawler and Brownlee 1992). A mean elemental composition was measured in the particles that impacted the mass spectrometers at very high velocities (about 77 $km\,s^{-1}$ for Vega 1 and 2, and 68 $km\,s^{-1}$ for Giotto). Results showed that comet 1P/Halley dust particles have a chondritic composition within a factor of two for most of the elements detected. However, they are enriched in three elements: carbon (×11), nitrogen (×8) and hydrogen (×4). The conclusion was that comets would be therefore less altered and more primitive than carbonaceous chondrites CI (Jessberger et al. 1988).

The Stardust mass spectrometer (Kissel et al. 2004) revealed the importance of organic matter in 81P/Wild 2 coma. The results on light elements were inconclusive during the flyby and could not be confirmed in samples returned by Stardust from 81P/Wild 2 (Brownlee et al. 2006; Hörz et al. 2006). Because the materials were captured at 6.1 $km\,s^{-1}$ in aerogel and aluminum foil collectors, organics and other fragile materials were severely affected (Brownlee 2014; Keller et al. 2006; Sandford et al. 2010). Elements such as C and N could not be measured, and volatile elements such as S were apparently vaporized and redistributed as a halo around the tracks in aerogel (Ishii et al. 2008a) or were lost from residues in craters in the Al foil. The other elements that could be quantified displayed chondritic abundance (Flynn et al. 2006; Ishii et al. 2008b; Lanzirotti et al. 2008; Leroux et al. 2008; Stephan et al. 2008) but with large error bars because of the strong heterogeneity among particles. Much more about the mineralogy of cometary particles was learned from Stardust and will be discussed in the next subsection.

The composition of dust particles emitted from the Jupiter family comet 67P/C-G was measured with the COSIMA instrument on board the Rosetta spacecraft. Analyses with COSIMA differ from those of previous missions in two aspects: (1) the Rosetta spacecraft followed the nucleus for about 26 months through a large portion of its orbit, as opposed to

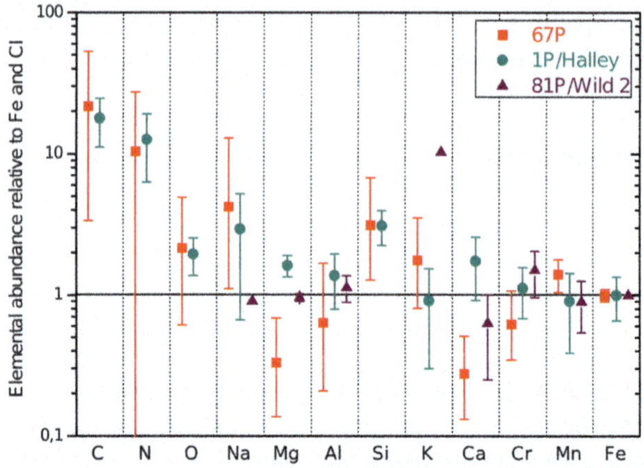

Fig. 2 Comet 67P/C-G dust elemental ratios relative to Fe and to the CI chondrite composition (Lodders 2010) compared to 1P/Halley's dust and 81P/Wild 2's dust collected in Stardust aerogel. The values correspond to the relative ratios (E/Fe)comet/(E/Fe)CI. The N/C atomic ratio has been measured by Fray et al. (2017). The error bars for 67P/C-G data come from the uncertainties on the determination of Relative Sensitivity Factors' (RSFs) uncertainties and can be amplified by successive normalizations. For instance, the uncertainty displayed for C is the addition of uncertainty on the RSF for calculation of C/Si, plus uncertainty for Si/Fe, plus uncertainty on C/Fe in CI-type chondrite. Therefore, error bars can be "artificially" enhanced due to successive normalizations to compare all elements on a same plot (this explains why error bars on N are so large in this plot, compared to the actual value measured for $N/C = 0.035 \pm 0.011$ (Bardyn et al. 2017)

previous flyby spacecraft that spent only a few hours within the coma; COSIMA collected more than 35,000 cometary particles and fragments (Merouane et al. 2017), with sizes ranging from about 10 to 1000 μm, and about 250 of them could be analysed; (2) The low impact velocity of the collected particles, $<10\ \mathrm{m\,s^{-1}}$ (Rotundi et al. 2015), on the COSIMA targets preserved the dust chemical properties and part of the physical structure (e.g. the porosity of the particles) (Hornung et al. 2016; Langevin et al. 2016).

The relative elemental composition of 67P/C-G particles have been measured in about 30 particles for the following elements: C, N, O, Na, Mg, Al, Si, K, Ca, Cr, Mn and Fe (Bardyn et al. 2017; Fray et al. 2017). As shown in Fig. 2, relative abundances in 67P/C-G are roughly chondritic, except for C and N. This is broadly similar to the results from 1P/Halley and 81P/Wild 2. In 67P/C-G, $C/Si = 5.5^{+1.4}_{-1.2}$, which is close to the protosolar abundance of 7.19 ± 0.83 (to be compared to 0.76 ± 0.10 found in CI carbonaceous chondritic meteorites) (Lodders 2010). Regarding N abundance ($N/C = 0.035 \pm 0.011$ Fray et al. 2017), which is strongly depleted compared to protosolar value of $N/C = 0.3 \pm 0.1$ (Lodders 2010) and quite similar to values found in CI chondrites ($N/C = 0.04$ Alexander et al. 2012). Some caution has to be taken in comparing data in representations such as Fig. 2, which shows a double normalization: to Fe and then CI abundances. No absolute value could be measured, only elemental ratios. For instance, in comet 67P/C-G particles, C/Si ratio is calculated in the particles on the basis of a standard material with a known composition (silicon carbide) measured with the ground COSIMA instrument. For some other elements, for instance Si, this is Si/Fe, which is measured (on the basis of a San Carlos olivine standard). Hence, for the purpose of comparison, every elemental ratio has to be normalized to the same element, here Fe was chosen (from which other normalizations such as C/Si and N/C follow, which increases error bars in Fig. 2). Those ratios are subsequently normalized to the same ratio measured in CI chondrites. The selection of Fe for normalization leads to the conclusion that, for instance, Si appears to be in excess in both comets 1P/Halley and 67P/C-G. If normalization would have been performed relative to Si, then Fe would appear to be depleted. Nonetheless, the abundance of Si and Na show a similar trend in comets 1P/Halley and 67P/C-G. Na was also observed in the upper atmosphere of Mars by the MAVEN spacecraft, as the most abundant element detected during the ablation of particles from the passage of comet C/2013 A1 (Benna et al. 2015). Mg and Ca are quite low in 67P/C-G, which is not understood. The high K in Fig. 2 is based on TOF-SIMS analyses of craters in the Al foil of Stardust and of slices of tracks in the aerogel (Stephan et al. 2008). K was detected in only 2 craters out of 7 analysed, and in 5 tracks out of 21 analysed, thus providing only an estimate but not a bulk value (Flynn et al. 2006).

One important feature of 67P/C-G data is that the particles have been collected throughout the 26 months of the mission, at various heliocentric distances and most probably from different locations on the nucleus, and they show different morphologies. However, they display a quite similar composition, which might indicate a rather homogeneous composition of the dust in the nucleus. The data for the three comets points toward rather primitive bodies, globally close to solar abundance, except for nitrogen, which, although enriched with regard to the CI composition, is always strongly depleted with regard to the solar value.

3.2 Mineralogy

Mineralogy of the cometary dust particles has long been quite poorly constrained. However, several measurements are now available to get a better picture of the mineralogy of cometary dust. Historically, remote sensing observations in the infrared range have first been used to get a grip on the composition.

The Spitzer Space Telescope imaging spectrometer characterized the ejecta from comet 9P/Tempel 1 produced by the Deep Impact encounter (Lisse et al. 2006). 5- to 35-micrometer emission spectra of the ejecta were obtained, allowing to identify specific minerals and infer their relative proportions. This emission is inhibited at dust grain sizes below the wavelength of the emission, so the technique is much less sensitive to sub-micron grains, and for larger grains the emission is proportional to the surface area of the grain, so surface area weighted abundances were reported rather than mass weighted abundances. Conversion to mass or molar fractions requires the assumption that all minerals have the same size-frequency distribution, but hypervelocity disruption experiments on different types of meteorites show different size-frequency distributions, suggesting the distribution may be different for different minerals (Flynn et al. 2005). In addition, the best fit to the spectral data depends on the basis set of minerals selected for inclusion in the fitting routine. With these limitations, the spectrum was best fit by 18.2% of the emission from forsterite, 10% from amorphous silicate, 18.8% from Mg–Fe-sulfide, 8.2% from clay, 6.8% from diopside, 5.9% from orthoenstatite, 5.0% from fayalite, 4.7% from carbonates, 4.0% from amorphous carbon, 2.3% from ionized PAHs, and only 2.9% from water ice crystals, with the remainder from gaseous water and minor phases (Lisse et al. 2006). No assessment was done of the possibility of either mineral destruction or formation from the shock and heating of the high speed impact which deposited about 2×10^{10} J of kinetic energy into a localized region on 9P/Tempel 1.

Samples returned by the Stardust mission contain thousands of particles from less than 1 μm to about 100 μm (Brownlee 2014). They enabled a detailed analysis of the mineralogy of cometary dust particles returned from comet 81P/Wild 2. Relatively refractory materials—silicates, oxides, sulfides and metals—were captured in the aerogel collectors with little alteration, and, over the decade since recovery, have been the subject of intense scrutiny in terrestrial laboratories. Key observations from the analysis of about 5% of the total collection studied so far include:

1. Although interstellar silicates are known from astronomical observations to be almost entirely amorphous, approximately half of the silicates in 81P/Wild 2 samples are crystalline (Westphal et al. 2009; Stodolna et al. 2014).
2. Among crystalline minerals, olivines, pyroxenes and Fe sulfides are the most abundant minerals present in the 81P/Wild 2 samples.
3. Approximately 1% of the particles recovered from 81P/Wild 2 are assemblages of highly refractory minerals, reminiscent of Calcium-Aluminum-rich Inclusions (CAIs) found in chondritic meteorites, which are thought to be high-temperature condensates and perhaps the first solids to have formed in the cooling inner solar nebula (Brownlee et al. 2006; Zolensky et al. 2006; Joswiak et al. 2017).
4. The collection also contains numerous small igneous "rocks"—assemblages of minerals that apparently cooled from a melt. These are reminiscent of chondrules, found in chondritic meteorites, which are thought to be cooled melt droplets. The origin of chondrules is unclear (and is controversial), but it is certain that they do not share a common origin with CAIs, and generally appear to be a few Myrs younger than CAIs, based on ^{26}Al dating, although ^{26}Al ages are controversial (Ogliore et al. 2015; Nakamura et al. 2008; Kööp et al. 2016; Gainsforth et al. 2015).
5. 81P/Wild 2 samples contained large amount of fine-grained matrix (individual size of minerals <1 μm) that did not survive well the impact. This fine-grained component got mixed with melted aerogel in the walls of the impact tracks and contains olivine, pyroxenes and Fe sulfides. The composition of this fine-grained component is close to the chondritic CI composition (Leroux et al. 2008; Leroux 2012; Leroux and Jacob 2013).
6. Although not common, Fe–Ni metal particles at least up to 20 μm in size are present in 81P/Wild 2. One particle that was studied in detail has an unusual composition reminiscent of metals found in ureilites, a family of meteorites that is poorly understood but may have a composition consistent with smelting of silicates in a carbon-rich environment, with Fe originally in silicates reduced to metal (Westphal et al. 2012; Humayun et al. 2015).
7. Small amounts of minerals possibly resulting from aqueous alteration have been identified in 81P/Wild 2 samples, like magnetite grains (Stodolna et al. 2012), carbonates (Flynn 2008; Flynn et al. 2009a; Mikouchi et al. 2007; Wirick et al. 2007; Leroux 2012), and a copper iron sulfide (cubanite) in association with pyrrothite and pentlandite (Berger et al. 2011). No hydrated silicates have been identified in 81P/Wild 2 samples.

These observations, combined with observations of oxidation state (Westphal et al. 2017; Butterworth et al. 2017; Keller and Berger 2014) are pointing to a dramatic and complicated process, perhaps including complete vaporization and recondensation of a substantial fraction of the material in the inner nebula, that must have been responsible for converting interstellar dust to nebular and cometary dust. Despite the evidence for the presence of inner Solar System material in 81P/Wild 2, the distribution of types and compositions of materials in 81P/Wild 2 differs from materials observed in meteorites.

1. The sizes of crystalline aggregates in Stardust samples are significantly smaller that those in meteorites (e.g. presence of micro-CAIS, micro-chondrules...).
2. Micro-CAIs found in 81P/Wild 2 do not span the entire range of volatility of CAIs observed in meteorites but are concentrated at the most volatile end of the CAI volatility distribution (Joswiak et al. 2017).

3. Microchondrules in 81P/Wild 2 are Fe–O rich, similar to type II chondrules, while chondrules in meteorites are dominated by those of type I (Fe–O poor) (Nakamura et al. 2008; Gainsforth et al. 2015).
4. Olivines in 81P/Wild 2 are unequilibrated and show a wide range of Fe content, implying a wide sampling of materials from the protosolar disk (Frank et al. 2014).
5. Pyroxenes minerals are about as abundant as olivine in 81P/Wild 2, which is in sharp contrast with carbonaceous chondrites where the abundance of olivine mineral dominates over that of pyroxenes (except for CR chondrites).
6. Only a few presolar grains have been found in 81P/Wild 2, but it was shown that the collection procedure in the aerogel or on the aluminium foils potentially resulted in the destruction of a substantial amount of presolar grains (Floss et al. 2013; Croat et al. 2015). Once this collection effect is taken into account, the corrected abundance of presolar grains in 81P/Wild 2 samples is on the order of 600–800 ppm, rather comparable to that of the most primitive Solar System material (McKeegan et al. 2006; Stadermann et al. 2008; Messenger et al. 2009; Floss et al. 2013; Leitner et al. 2010, 2012; Floss and Haenecour 2016).
7. The oxidation state of Fe in 81P/Wild 2 is also not observed in any meteorite family, although the effects of thermal metamorphism and especially aqueous alteration in asteroids complicate this comparison (Westphal et al. 2009; Leroux et al. 2009; Ogliore et al. 2010; Westphal et al. 2017).

In the case of the Rosetta mission at 67P/C-G, the mineralogy is not straightforward to assess from the analyses of the dust instruments (COSIMA, GIADA and MIDAS). The analysis of reflectance spectroscopy measurements indicated the presence of opaque minerals at the surface of the comet, that could be Fe sulfides (Quirico et al. 2016). The analysis of 67P/C-G by COSIMA, as well as the identification of size and density of the particles by GIADA are compatible with a composition including silicates, Fe sulfides and carbon (Bardyn et al. 2017; Fulle et al. 2016b). The ratio of amorphous to crystalline silicates could not be measured in situ. There is however no evidence from the COSIMA mass spectra analyzed so far for the presence of hydrated silicates in 67P/C-G dust particles (Bardyn et al. 2017; Stenzel et al. 2017). From these different characterisations, it appears that the mineralogy of cometary dust is highly diverse and shows sometimes unexpected components (Wooden et al. 2017).

There are clear differences between the inferred mineralogical content of the 9P/Tempel 1 ejecta from the Deep Impact mission, and the 81P/Wild 2 particles collected by Stardust despite both being focused in the 10 to 50 μm size range. The major difference is the detectable presence of hydrated silicates in 9P/Tempel 1 ejecta, but their absence in the 81P/Wild 2 material. This could be explained in 3 ways: (i) the spectrum fit in 9P/Tempel 1 could be explained without the need for hydrated silicates, as those fits rely on mineral reference spectra that might not represent all spectral features related to shapes and minerals sizes in particular; (ii) it could reflect the destruction of micrometer-sized hydrated silicates by hypervelocity aerogel capture in the 81P/Wild 2 samples, (iii) it could be an indigenous characteristic of 81P/Wild 2. No evidence for hydrated silicate was found in the analysis of 67P/C-G dust particles so far (Bardyn et al. 2017). Another clear difference is the olivine to pyroxene abundance ratio, which is about 2 to 1 in the 9P/Tempel 1 ejecta compared to about equal abundances in 81P/Wild 2 dust (e.g. Dobrica et al. 2009 and references therein).

3.3 Organic Composition

Organic matter in comets can be divided into two components. The first is volatile, stored in an icy phase dominated by H_2O, CO_2 and CO, and is released from the nucleus into the cometary atmosphere when the comet gets close to the Sun on its elliptical orbit. Part of this volatile fraction is released directly from the nucleus when ices sublimate, while other compounds are ejected on dust particles that are ejected from the nucleus during the active phase of the comet and sublimate in the atmosphere. The second component is a refractory phase that remains solid either on the nucleus and on the particles.

Regarding the dust organic composition, though not operating as planned at the surface of the nucleus of 67P/C-G, two of the instruments of the Philae lander were able to detect a few new molecules from dust particles. During the first bounce of the lander above the surface of 67P/C-G, dust particles were lifted at first touchdown and serendipitously entered the instruments COSAC (Cometary Sampling and Composition experiment) and Ptolemy. Some volatile compounds were outgassed from the particles and were analysed by direct mass spectrometry. Detections of nitrogen bearing molecules, and for the first time, CH_3COCH_3 (acetone), CH_3CONH_2 (acetamide), C_2H_5CHO (propionaldehyde) and CH_3NCO (methyl isocyanate) were reported with the instrument COSAC (Goesmann et al. 2015), while a sequence of compounds that could be a signature of small chains of formaldehyde polymers were detected by the instrument Ptolemy (Wright et al. 2015). Those results have recently been reconsidered and detection of CH_3NCO, CH_3CONH_2, C_2H_5CHO and formaldehyde polymers questioned in the light of other measurements performed by the ROSINA (Rosetta Orbiter Spectrometer for Ion and Neutral Analysis) instrument after an impact of dust particles in the instrument less than 10 minutes after their ejection from the nucleus, which means that some volatile compounds were not yet outgassed from the dust (Altwegg et al. 2017). The mass spectra of all three instruments show very similar patterns of mainly CHO-bearing molecules that sublimate from particles at temperatures of 275 K. Detection of toluene ($C_6H_5CH_3$) is also proposed. Glycine (NH_2–CH_2–CO_2H), the simplest of the amino acids, which are the building blocks of proteins, was also detected with ROSINA. Its distribution in the atmosphere of the comet 67P/C-G is consistent with a release from the dust particles heated in the cometary atmosphere, rather than a direct sublimation from the nucleus (Altwegg et al. 2016).

Since the exploration of comet 1P/Halley, it has been observed that a significant fraction of the cometary organic matter is also present in a phase remaining solid whatever the distance and temperature reached by the dust particles ejected from the nucleus (Fomenkova et al. 1994; Kissel and Krueger 1987). The nature of this refractory component has eluded a proper characterization for a very long time. Indeed, the measurements performed on the dust samples returned from comet 81P/Wild 2, regarding the organic matter, were hampered by the presence of a significant amount of carbon in the collecting aerogel, which made it rather difficult to unravel the actual cometary organic matter from the one already present in the aerogel, and a mixture of the two reacting together at the time of the impact of the particles in the aerogel (Sandford et al. 2010). Nevertheless, the cometary organic matter detected can be split into three groups: one is similar to Insoluble Organic Matter (IOM) observed in carbonaceous meteorites, the second is highly aromatic refractory matter contained in nanoglobule-like features (de Gregorio et al. 2011), and the third a volatile aliphatic organic found as a halo (Sandford et al. 2006), similar to the S halo; detection of glycine originating from the dust particles was also reported (Elsila et al. 2009). A quantification of the carbon content could not be achieved in Stardust samples, but it seems that 81P/Wild 2 particles are rather low in carbonaceous content compared to what was anticipated in comets. It was proposed that this low amount could be due to degradation during capture at high velocity, at about 6 $km\,s^{-1}$ (Brownlee 2014). It could however also be due to the sampling of an organic poor region of the coma, which is not representative of the whole comet (Westphal et al. 2017; Zubko et al. 2012).

As discussed earlier, the elemental composition of dust particles captured in 67P/C-G coma at low velocity (about 10 $m\,s^{-1}$, possibly preserving the nature and structure of the particles) has been probed with the COSIMA instrument in the Rosetta spacecraft, throughout the 26 months of the mission. It has been shown that the carbon in dust particles is present in the form of macromolecules that bear mass spectral similarities with the Insoluble Organic Matter found in carbonaceous meteorites (Fray et al. 2016). A recent quantification of this macromolecular component points to an abundance of roughly 45% in mass of the dust particles, regardless of the size, shape and period of collect of the particles (Bardyn et al. 2017). It appears that in 67P/C-G particles, only this kind of complex macromolecular form could be detected at the spatial resolution of the instrument (≈40 μm). This rather large amount of organic matter is consistent with values that were indirectly deduced from the density of the particles at $(52 \pm 8)\%$ in volume of organic matter (Fulle et al. 2016b) and from the radar CONSERT permittivity characterization of the nucleus inferring a C/Si of at least 5.7 (Herique et al. 2016).

3.4 Isotopic Composition

Access to the isotopic composition of cometary dust particles requires direct measurements on the particles. Therefore they have only been extensively made on samples returned from comet 81P/Wild 2. At the time of writing, only very limited information has been derived from the Rosetta mission, but it is expected that more results will be published in the coming years.

Hydrogen, Carbon, Nitrogen, Sulfur Isotopes Measurements for H, C and N isotopes have only been made in Stardust samples so far, and in the organic phase. Unfortunately, 81P/Wild 2 samples are not very rich in organics, either due to the roughness of the hypervelocity capture that destroyed most of the organics, or because 81P/Wild 2 was not an organic-rich comet (Brownlee 2014). The H, C or N isotopic composition of Stardust samples was reported (McKeegan et al. 2006; Stadermann et al. 2008; de Gregorio et al. 2011; Matrajt et al. 2008, 2013). The D/H ratios vary between the terrestrial (V-SMOW) value and up to a factor of three that of V-SMOW. The $^{13}C/^{12}C$ ratio shows moderate variations with regard to the terrestrial value ($\delta^{13}C$ from −20 to −50 per mil, here δ refers to relative differences). The $^{15}N/^{14}N$ ratios cluster around the terrestrial value and are also consistent with values found in most meteorites, with the exception of a few analyses showing excesses in ^{15}N ($\delta^{15}N$ from +100 to +500 per mil). At the submicron level, ^{15}N-rich hotspots are found, with a maximum of 1300 ± 400 per mil in $\delta^{15}N$ are observed. Such elevated values are also found in IDPs and in some carbonaceous chondrites (Busemann et al. 2006; Briani et al. 2009). The sulfur isotopic composition of several 81P/Wild 2 impact residues and one sulfide captured in the aerogel was found compatible with solar isotopic composition (Heck et al. 2012; Ogliore et al. 2012), showing that these S-rich components of 81P/Wild 2 were formed in the Solar System. A ^{33}S enrichment in a cosmic symplectite in Stardust samples was found, that could result from photochemical irradiation of solar nebular gas (Nguyen et al. 2015). The sulfur isotopic composition of 67P/C-G dust is also compatible with the normal value (Paquette et al. 2017). Unfortunately, only the ^{34}S and ^{32}S isotopes could be measured, so a potential ^{33}S excess could not be assessed for 67P/C-G dust.

Oxygen and Silicon Isotopes The origin of the peculiar distribution of oxygen isotopic composition of Solar System materials has been controversial since the discovery of ^{16}O-rich materials (Clayton 1993; Clayton and Mayeda 1999). In a three-isotope plot

($^{17}O/^{16}O$ vs. $^{18}O/^{16}O$), ^{16}O-rich phases, particularly CAIs, form a line with a slope close to 1. This line is unexpected in a system controlled only by mass-dependent fractionation effects, which gives a slope close to 1/2. Three leading models aim to explain the data falling on the slope 1 line: (i) mixing between the initial ^{16}O-rich solar value (McKeegan et al. 2011) and an ^{16}O-depleted reservoir resulting from the self-shielding of the CO molecule (Lyons and Young 2005); (ii) mass-independent fractionation could explain the data without the need of mixing several reservoirs (Thiemens and Heidenreich 1983; Thiemens 1999; Chakraborty 2008), (iii) and presolar heterogeneous chemistry and later mixing between two presolar reservoirs with different O isotopic compositions (Dominguez 2010). Oxygen isotopes of cometary dust could only be measured with precision on the returned Stardust samples, using micro- and nanoanalytical techniques. The coarse-grained particles from 81P/Wild 2 exhibit O isotopic composition that is broadly consistent with that of chondritic meteorites and their components, with olivine and pyroxenes plotting in the carbonaceous chondrite range, while refractory minerals are ^{16}O-rich (McKeegan et al. 2006; Nakamura et al. 2008; Simon et al. 2008; Bridges et al. 2012; Nakashima et al. 2012; Ogliore et al. 2015; Defouilloy et al. 2017). There are however differences, like the recent measurement of Mg-rich pyroxenes that are ^{16}O-rich, suggesting a condensation origin for these minerals (Defouilloy et al. 2017). The statistics are not yet sufficient to distinguish between them, but the correlation of oxygen isotopic composition of olivine and pyroxene with iron content shows evidence for an affinity with silicates in CR meteorites and Tagish Lake-like chondrites (Defouilloy et al. 2017). Some fine-grained materials in 81P/Wild 2, however, preserved in the bulbs of tracks in aerogel, show a much larger dispersion in oxygen isotopes, extending the CCAM line from the terrestrial value to very ^{16}O-poor materials (Ogliore et al. 2015), up to 200 per mil, similar to cosmic symplectite (Sakamoto et al. 2007). The significance of this is unclear, but may point toward the two-component mixing model to explain the O isotopic composition of Solar System solids. The silicon isotopic composition of dust particles (likely sputtered by the solar wind) was measured by ROSINA for 67P/C-G. The Si isotopic composition shows a depletion of the heavy silicon isotopes ^{29}Si and ^{30}Si with respect to ^{28}Si and solar abundances (Rubin et al. 2017). The origin of the heavy Si isotope depletion, still debated, could suggest a presolar signature in silicates present in dust particles.

Short Lived Radionuclides: ^{26}Al The short-lived radionuclide ^{26}Al was abundant in the early Solar System, with $^{26}Al/^{27}Al \approx 5.2 \times 10^{-5}$ (Kita et al. 2013). Its β-decay to ^{26}Mg (half-life 0.72 Myr, energy 4.0 MeV) was a significant source of heating, and probably played an important role in the internal heating and differentiation of early-forming planetesimals. Its presence as a live radioactivity in the nebula also allows dating of the formation of individual phases in meteorites with respect to the formation of CAIs, the earliest solids to have formed in the Solar System, although this dating technique requires that ^{26}Al was homogeneously distributed in the early protoplanetary disk (Kita et al. 2013). Another limitation is also that suitable phases with high Al/Mg ratios, such as plagioclase, are required, so that the contribution of ^{26}Mg originating from the decay of ^{26}Al can be discriminated from the naturally occurring stable ^{26}Mg isotope. Unfortunately, more common phases such as olivine or enstatite are not suitable because of their low Al/Mg ratios. Al–Mg systematics have been measured in four 81P/Wild 2 particles (Matzel et al. 2010; Ogliore et al. 2012; Nakashima et al. 2015), and in all four cases no evidence of live ^{26}Al at the time of their crystallisation was found. If these formed after injection of ^{26}Al into the nebula, they must have formed >3 Myr after meteoritic CAIs formed. An alternative explanation for this observation is that they formed before ^{26}Al injection into the nebula, or that ^{26}Al was not homogeneously distributed in the early protoplanetary disk. A late Mg isotopic exchange through metamorphism of chemical alteration also remain in the range of possibilities.

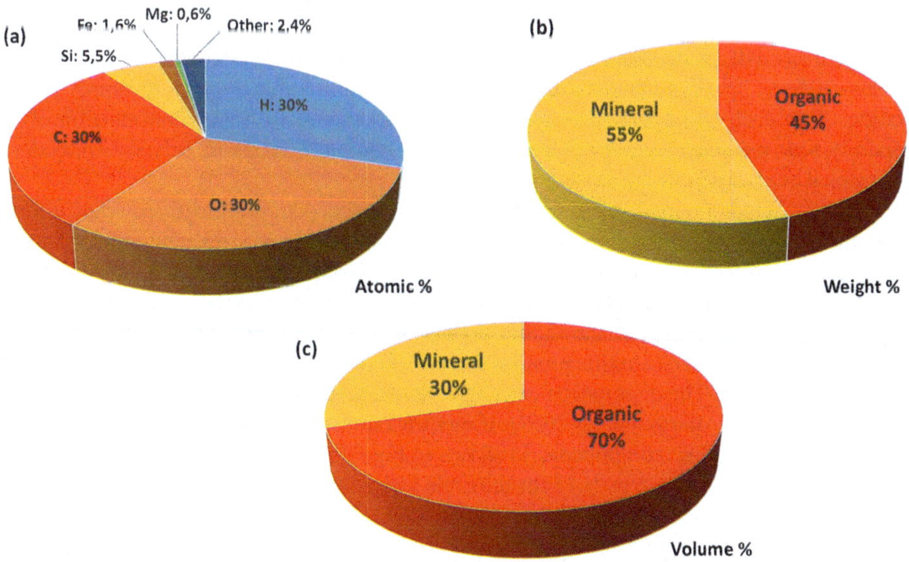

Fig. 3 Composition of 67P/C-G dust particles. (**a**) Relative abundance of elements in atom numbers for 67P/C-G dust particles. In this chart, H abundance is not directly measured and the *atomic ratio* H/C = 1 is assumed. Elements under the label "other" include mainly N, Na, Si, and Mn, which relative abundances have been measured with COSIMA, and elements such as S and Ni, not yet quantified with COSIMA, but assumed to have a Solar abundance. (**b**) Tentative repartition between a mineral and an organic component assuming that elements such as C, N, H and some of the O are the constitutive elements of the organic phase, while the remaining O and other elements are included in the mineral phase. For the oxygen, it is assumed as an upper limit that minerals can be at most with a SiO_4 stoechiometry, the remaining O being incorporated in the organic phase (Bardyn et al. 2017). (**c**) Relative abundance in volume of mineral and organic components in 67P/C-G dust particles. This is calculated considering that a relative density for minerals of 3, and a relative density of 1 for organic molecules

Noble Gases Stardust samples contain cometary light noble gases: He, Ne (Marty et al. 2008; Palma et al. 2007, 2013, 2015). Most of the noble gases were found in the wall of the Stardust tracks, probably carried by the fine-grained fraction of the particles. The concentrations of noble gases can be quite high, which may be consistent with formation in the inner Solar System and consequent exposure to high fluences of solar energetic particles. The $^3He/^4He$ isotopic ratio lies between the solar wind value (post-D burning helium) and the protosolar nebula ratio. Hence the 81P/Wild 2 matter possibly was therefore exposed to the irradiation of the proto-Sun. The Ne isotopic composition of 81P/Wild 2 matter is compatible with that of Ne in chondrites, rather than being close to the solar composition.

3.5 Summary

Most information in this composition section derives either from samples returned from comet 81P/Wild 2 by the Stardust mission or 67P/C-G dust particles analyzed in situ by the Rosetta spacecraft. Despite the small size of the sample collection, the principal limitation in the analysis of 81P/Wild 2 samples returned by Stardust is instrumental. Some instruments now used routinely to analyze these samples did not exist at launch of the mission, e.g. nanoSIMS (Stadermann et al. 2008). New laboratory analytical techniques that are just now being brought to bear on cometary dust particles include nanoFTIR (Dominguez et al. 2014),

Atom Probe (Heck et al. 2014) and microbeam laser resonance ionization mass spectrometry such as CHILI (Chicago Instrument for Laser Ionization Liu et al. 2017). Focused-Ion Beam (FIB) is opening new windows also through dramatically improved sample preparation. With each improvement in analytical technique, and with each introduction of an entirely new technique, there is the potential for new discoveries in existing collections (Westphal et al. 2016).

The same kind of remark can be made regarding Rosetta data. Although providing unique information in the direct vicinity of the comet since 2014, one has to keep in mind that the instrumentation was selected in the mid 90's on the basis of technology available at that time. Nonetheless a global composition of cometary particles can be derived from these data as is shown in Fig. 3. It shows that organic matter in an important component of the dust particles, however quantification of some elements (e.g. H and S) is still missing at the time of writing. The distribution of O between the mineral and the organic phase is still poorly constrained (Bardyn et al. 2017). If one tries to estimate a distribution in volume of the mineral and organic components, as shown in Fig. 3(c), one has to rely on a very rough estimation of the relative density (3 for minerals and 1 for the organic molecules Robertson 2002) which is subject to debate.

4 Physical Properties of Dust Particles

4.1 Morphology

The morphology of dust particles describes their structure and shape and how these emerge from the characteristics of the particle's components. It is crucial to investigate as it is a key property to understand basic parameters of comets, such as porosity, strength and thermal properties. Polarimetric observations of comets suggested that the basic morphology of cometary dust is twofold, containing more compact entities, either particles or aggregates, and more porous aggregate particles (see Sect. 1.2, Sect. 4.2). Stardust (Brownlee 2014) brought the only available samples of cometary dust to Earth; although the majority of the particles were structurally altered during collection, it became evident that cometary dust consists of aggregates. The most accurate morphological analysis of cometary dust to date stems from Rosetta.

MIDAS provides topographic information on dust particles from 1 to 50 μm in size (Bentley et al. 2016), while COSIMA obtained resolved images of dust particles from 30 μm to 1 mm (Hilchenbach et al. 2016). The dust particles investigated by MIDAS (Fig. 4) and COSIMA (Fig. 5) provide the least altered sampling of cometary dust over scales from 1 μm to 1 mm, as they were collected on their own target with relative velocities in the range from 1 to 15 $m\,s^{-1}$ (Fulle et al. 2015a) at distances from the comet nucleus lower than 500 km (Bentley et al. 2016; Merouane et al. 2016). COSIMA provided evidence for break-up events of large parent particles. Indeed, large numbers of particles were collected during a single exposure (down to 1 day) by specific target areas with no collection on other target areas exposed at the same time (Merouane et al. 2016; Langevin et al. 2017). Particles are also likely to break-up and spread over the MIDAS targets upon collection (Bentley et al. 2016; Ellerbroek et al. 2017).

The information obtained by COSIMA and MIDAS using two different approaches led to overall similar conclusions applying to a range of sizes of cometary dust extending over 3 orders of magnitudes. Both MIDAS (Bentley et al. 2016) and COSIMA (Langevin et al. 2016) identified particles with well-defined boundaries. Of these, the smallest individual

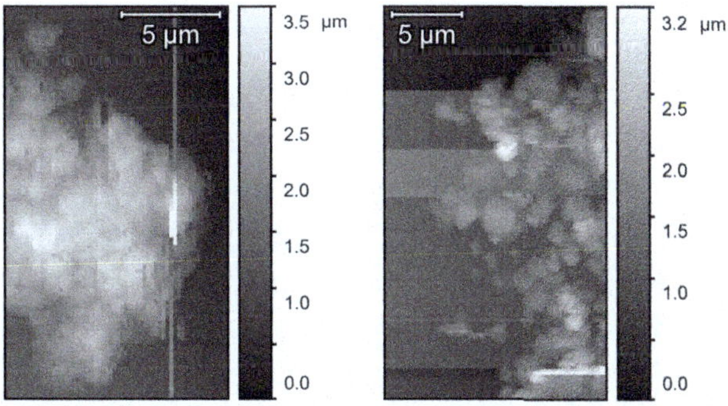

Fig. 4 Left panel: A typical compact particle collected with MIDAS, extending beyond the field of view and exhibiting a well-defined boundary. The surface features at smaller scales are interpreted as individual units composing the dust. Right panel: A large fluffy particle (also extending beyond the field of view of MIDAS) that is interpreted to represent the remains of a fractal aggregate with fractal dimension $D_f = 1.7 \pm 0.1$ that compacted on collection (Mannel et al. 2016). Another possibility, the rebound of a larger dust particle bouncing on the target and depositing a surface layer, is less likely due to the structural differences found in experimental tests (Ellerbroek et al. 2017). ESA/Rosetta/IWF for the MIDAS team IWF/ESA/LATMOS/Universiteit Leiden/Universität Wien

particles detected with MIDAS are about 1 μm in size (Bentley et al. 2016) while the largest one identified by COSIMA is about 350 μm in size. Each of these particles exhibit textural substructure (Mannel et al. 2016; Langevin et al. 2016), down to about 25 nm for MIDAS. Thus, they can be defined "compact aggregates" or eventually "compact particles" (left-hand panel of Fig. 4 and Fig. 5a). Many compact aggregates collected by COSIMA broke up into smaller components when analysed because of electrostatic forces (Hilchenbach et al. 2017). The reason why no particle smaller than 1 μm has been detected by MIDAS could be due to a collection bias or to a dearth of such particles in the cometary environment (Bockelée-Morvan et al. 2017).

Most particles impacting COSIMA collection plates are destroyed into clusters of fragments, which exhibit morphologies ranging from sub-components connected by a matrix (Fig. 5b) to rubble piles (Fig. 5c) and shattered clusters (Fig. 5d) with separated sub-components and a low height to size ratio (1:10). This range of morphologies can be attributed to a large extent to the relationship between the incoming velocity and the tensile strength of the particle (Hornung et al. 2016; Ellerbroek et al. 2017). There is continuity with compact aggregates, corresponding to a higher tensile strength to velocity ratio, as demonstrated by the break-up of such particles into rubble piles observed under the ion gun of COSIMA. Specular reflections observed by COSIMA may indicate that monocrystalline grains are likely to have been collected as part of large aggregates (Langevin et al. 2017).

The collected particles observed by MIDAS and COSIMA therefore provide evidence for textural complexity over a range of scales from 1 μm to 1 mm, of main importance for comparison with interplanetary dust particles and micrometeorites (as discussed in Sect. 5.3). Porosity at such scales may be compared with the porosity of the nucleus, in the 60% to 80% range as deduced by the CONSERT experiment from the permittivity of the small lobe (Kofman et al. 2015; Herique et al. 2016), thus contributing to the overall low density of 67P/C-G of about 530 kg m^{-3} (Pätzold et al. 2016; Jorda et al. 2016). The results of dust collection experiments MIDAS and COSIMA provide the first view of the hierarchical structure of dust in comets.

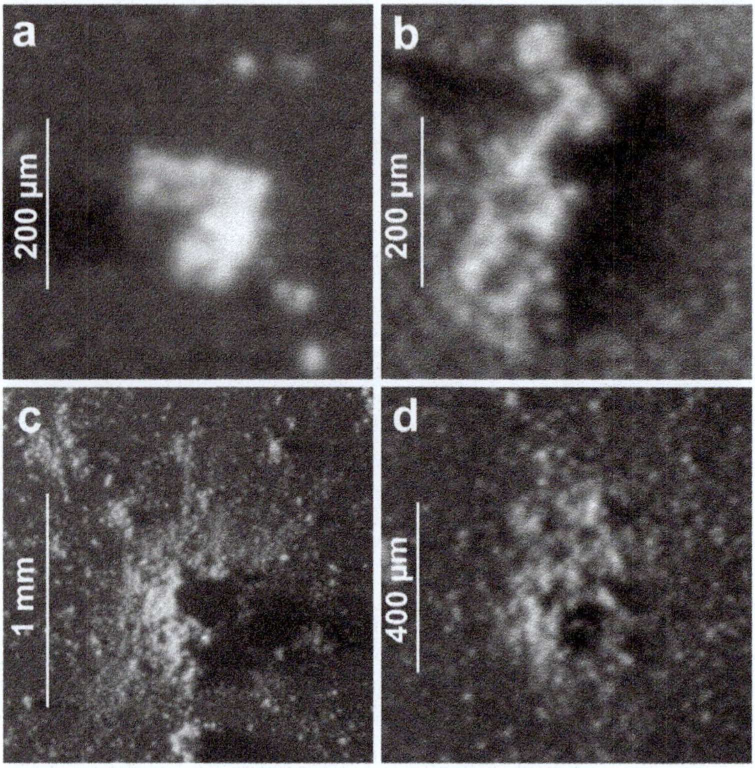

Fig. 5 Morphology of particles collected by COSIMA, using the classification of Langevin et al. (2016); (**a**) "compact" particle exhibiting well-defined boundary; (**b**) "glued cluster", with sub-components apparently linked together by a matrix; (**c**) "rubble pile", with a central mound surrounded by an apron of smaller sub-components; (**d**) "shattered cluster", with widely distributed components and a size to height ratio smaller than 10. Such particles present morphological similarities with the largest particles collected by MIDAS (Fig. 4), which are 10 times smaller

4.2 Dust Bulk Density and Microporosity

The interaction of photons with compact aggregates as observed by COSIMA is dominated by surface scattering. Conversely, the mean free path of photons in the clusters of fragments observed by COSIMA (Figs. 5b, c, d) is in the range of 25 µm, indicating a microporosity >50% (Langevin et al. 2017). The particle collected by MIDAS that is presented in the right-hand panel of Fig. 4 has a even larger porosity: the underlying surface can be observed between sub-components (Mannel et al. 2016). The fractal dimension of the associated parent particle was determined via a density-density correlation function to be $D_f = 1.7 \pm 0.1$ (Mannel et al. 2016), which can be related to a volume filling factor ϕ (the fraction of volume occupied by matter) by $\phi = (R/r)^{D_f - 3}$, where R and r are the volume-equivalent radii of the aggregate and of the grains composing the aggregate, respectively (Fulle and Blum 2017). The microporosity, defined as $1 - \phi$, is 99% for this MIDAS particle (Fulle and Blum 2017). Here we define fluffy particles as those with a microporosity >90%.

In the 67P/C-G coma, compact and fluffy particles are consistent with the morphology of aerogel tracks produced by high-speed impacts of 81P/Wild 2 particles observed by Stardust (Brownlee 2014): carrot-like aerogel tracks (type A, 65% of the total number of tracks)

produced by impacts of compact aggregates of minerals (Burchell et al. 2008), and bulbous tracks consistent with the explosion of fluffy aggregates in the aerogel (Trigo-Rodríguez et al. 2008). Bulbous type-B tracks (33% of the tracks) show terminal particles, which are aggregates of minerals, lacking in bulbous type-C tracks, 2% of the tracks (Burchell et al. 2008). GIADA confirms this scenario (Della Corte et al. 2015), observing dust belonging to two main families, compact (a) and fluffy (b) particles. (a) All compact particles provide a signal to the IS (Rotundi et al. 2015) and are consistent with Stardust type-A tracks. Their internal tensile strength T is consistent with $\log(T/1\ \mathrm{kPa}) = 1.48\phi - 0.2$ (Güttler et al. 2009), in agreement with the models of dust impacts on the COSIMA collection plate (Hornung et al. 2016). Typical strengths of compact particles are $1 < T < 6$ kPa for $0.15 < \phi < 0.66$, i.e. a microporosity between 34% and 85%. COSIMA has mostly observed compact particles (according to the definition above), which are porous aggregates fragmented on the instrument funnel and collection plates only (Merouane et al. 2016). (b) The dust showers observed by GDS only are explained in terms of fractal aggregates of $D_f \leq 2$ fragmented at a few meters from the spacecraft (Fulle et al. 2015a). About 30% of the observed GDS dust flux is observed in showers (Fulle and Blum 2017), and their parent fractal particles are consistent with type B and C tracks. The showers associated with a compact particle are consistent with type-B tracks (Fulle et al. 2016b). A comparison with the relative occurrence of B and C tracks is impossible, because GIADA showers have a cross-section larger than GDS (Fulle and Blum 2017), so that the compact particles associated to a shower may not enter GDS and may have a momentum below the IS detection limit. Hugoniots (which provide the states on both sides of a shock wave in a one-dimensional deformation in solids) of hypervelocity impacts by projectiles of density lower than the Stardust aerogel predict explosions inside the aerogel (as observed in tracks B and C) instead of surface craters (Knudson and Lemke 2013).

The GIADA measurements of both cross-section and mass for 271 compact particles (Fig. 6) has provided the average bulk density $\rho_d = 785^{+520}_{-115}\ \mathrm{kg\,m^{-3}}$, 1σ (Fulle et al. 2017). The lower value estimated by COSIMA in the same size range covered by Fig. 6 is about 200 $\mathrm{kg\,m^{-3}}$ (Hornung et al. 2016). This value has been inferred by impact models of particles porous enough to fragment on the COSIMA collection plate at the observed impact speed of a few $\mathrm{m\,s^{-1}}$, so that this lower density refers to a subset of compact particles only, missing all densest ones (Fulle et al. 2018). GIADA estimates an aspect ratio ≤ 5 of $\approx 98\%$ of the compact particles (Fig. 6), consistent with the largest observed aspect ratio of 3.2 observed over 24 COSIMA compact particles. The low aspect ratio and rotating frequency of the particles make improbable dust fragmentation by centrifugal forces (Fulle et al. 2015b). Coupled to the bulk density of the 67P/C-G nucleus (Pätzold et al. 2016; Jorda et al. 2016), the average dust bulk density provides an average dust microporosity of 60%, i.e. $\phi = 0.4$ (Fulle et al. 2017), affected (as well as the density) by a wide dispersion. The microporosity of compact aggregates does not reach anyway that of fluffy aggregates, always >90% (Fulle and Blum 2017). The average bulk density of 67P/C-G compacted and dried dust is $1925^{+2030}_{-560}\ \mathrm{kg\,m^{-3}}$, 1σ (Fulle et al. 2017). Many particles have a bulk density $> 4000\ \mathrm{kg\,m^{-3}}$, i.e. are sub-mm aggregates of minerals with almost no porosity, which are consistent with the Stardust type-A tracks producing a single terminal particle (Burchell et al. 2008). The minerals composing these sub-mm aggregates necessarily condensed in the inner solar nebula, thus confirming its complete mixing up to 30 au in the first Myr (Ciesla 2011), as already inferred from Stardust data (Brownlee 2014).

In contrast to Stardust, which supports a CI-chondritic composition (Brownlee 2014), the composition of 1P/Halley and 67P/C-G dust matches the carbon solar abundance (Lodders 2003; Bardyn et al. 2017; Jessberger 1989). Then, the dust and nucleus bulk densities constrain the dust/ice mass ratio of the 67P/C-G nucleus and its average composition: $(4 \pm 1)\%$

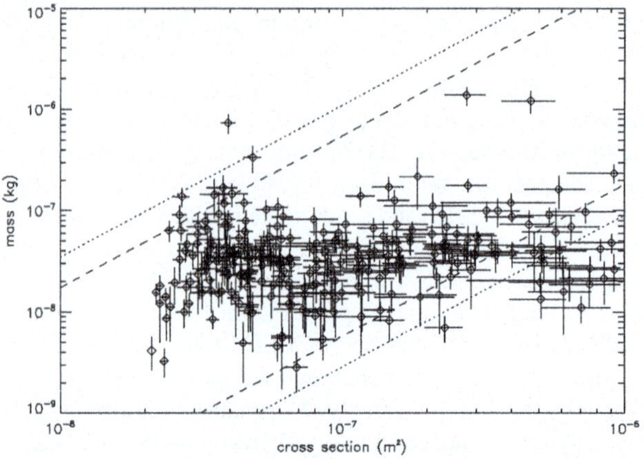

Fig. 6 Mass and cross-section measurements of compact particles detected by GIADA from August 2014 to September 2016 (the error bars refer to 1σ standard error of the 271 GDS+IS measurements). The data are compared with the trends of prolate and oblate ellipsoids of aspect ratio of 10 (dotted lines) and 5 (dashed lines), respectively, and with dust bulk densities of Fe sulphides, $\rho_1 = 4600\ \mathrm{kg\,m^{-3}}$ (upper lines), and of hydrocarbons, $\rho_2 = 1200\ \mathrm{kg\,m^{-3}}$ (lower lines). The GDS signal saturates at a cross section $>10^{-6}\ \mathrm{m}^2$. Particles with cross-sections $<2 \times 10^{-8}\ \mathrm{m}^2$ (and most of those with mass $<10^{-8}$ kg) were too small and fast to be detected by GDS. The flux at masses $>2 \times 10^{-7}$ kg was very low during the entire mission due to the spacecraft safety constraints. All these facts explain why all data are distributed in a mostly horizontal cloud. The particles close to the upper dotted and dashed lines have an almost zero porosity, and are compact aggregates of minerals condensed in the inner protoplanetary disc (they are too large to have an interstellar origin Brownlee 2014). The particles at right of the lower dotted and dashed lines have both a large porosity and a composition dominated by hydrocarbons. The particles in between are a porous mixture of minerals and hydrocarbons

of sulfides, $(20 \pm 8)\%$ of ices, $(22 \pm 2)\%$ of silicates, and $(54 \pm 5)\%$ of hydrocarbons, in volume abundances (Fulle et al. 2017). This dust composition matches that measured by COSIMA, taking into account the uncertainties affecting the elemental abundances (Fulle et al. 2018). The refractory/ices mass ratio in the nucleus is close to 7.5, implying a 67P/C-G water content lower than in CI-chondrites (Fulle 2017). The average refractory/ices mass ratio in 67P/C-G dust is larger than the nucleus one, being about 20 (Fulle et al. 2017). Since the dust cross-section in a coma is much larger than that of the sunlit nucleus, such a low ice content in dust is still consistent with the dominant water loss from dust observed e.g. in 103P/Hartley 2 (Fulle et al. 2016a).

4.3 Size Distributions

The range of sizes of dust particles detected by Rosetta in the coma was from 1 µm (Bentley et al. 2016) to about 1 m (Fulle et al. 2016c). Evidence that individual sub-µm particles are present in comae is weak. For instance, the low-phase negative polarization and backscattering of dust comae (Hanner and Newburn 1989), are well fit by large (>3 µm) fractal aggregates (Mann et al. 2004; Kolokolova et al. 2004; Levasseur-Regourd et al. 2008; Lasue et al. 2009), as further discussed in Sect. 5.2. The submicron-sized grains composing the dust aggregates, well sampled by Stardust in 81P/Wild 2 (Brownlee 2014) and MIDAS in 67P/C-G (Bentley et al. 2016), are linked by sticking strengths orders of magnitude larger than gas pressures in comae (Skorov and Blum 2012; Gundlach et al. 2015). Processes or mechanisms that might overcome these sticking forces in compact aggregates of size >1 µm have not been proposed (see Sect. 4.5). Most micron-sized MIDAS particles are interpreted as fragments of larger parent particles (Bentley et al. 2016), with a cut-off which seems to exclude any electrostatic bias (as detailed in Sect. 4.5). The only measurements supporting the presence of single submicron-sized particles in comae are (i) the smallest non-clustered impact craters on the Al-foils of the Stardust mission (Hörz et al. 2006)—clusters of craters are another evidence of fluffy aggregates in 81P/Wild 2—and (ii) two PIA data-point in 1P/Halley at mass $<10^{-15}$ kg (McDonnell et al. 1989). The cumulative sub-micron size distribution inferred from the craters on the Stardust Al foils has a power index <1.5 (Price et al. 2010), and therefore there is consistency between the observed flux of sub-micron particles and the fragmentation of incoming fractal parent particles of $D_f \leq 2$, as detailed in Sect. 4.5. Fractal aggregates with $D_f < 2$ are optically thin, so that their optical, thermal and dynamical properties mostly mimic that of single submicron-sized grains (Bertini et al. 2007), explaining e.g. long dust tails and dust temperatures up to 500 K at Sun distances >1 au. The composition of the fractal aggregates affects the internal multi-scattering, i.e. the albedo of the aggregates (Bertini et al. 2007).

Dust fluences measured at comets 1P/Halley, 26P/Grigg-Skjellerup, 81P/Wild 2 and 9P/Tempel 1 were sampling dust masses $>10^{-15}$ kg, i.e. dust with $R > 0.7$ μm if $\rho_d \approx 800\ \mathrm{kg\,m^{-3}}$. They were converted to a dust mass distribution at the nucleus surface by means of isotropic models, but they provided unphysical results (Fulle et al. 1995, 2000) as confirmed by Rosetta data, which show that the dust subsolar ejection is an order of magnitude larger than at terminator (Della Corte et al. 2015). Since dust in comets has a volume filling factor covering many orders of magnitude, unrealistic assumptions are needed to convert the momentum measured by flyby fluences into sizes. Rosetta was the first space mission to measure a reliable dust size distribution in a comet, because the Rosetta dust instruments (COSIMA, GIADA and MIDAS) have provided, for the first time, information on the microporosity of the collected particles. GIADA has observed showers of fluffy dust fragmented at a few meters from Rosetta (Fulle et al. 2015a). Far from Rosetta, any significant (i.e. involving $> 5\%$ of the total ejected mass) dust fragmentation and sublimation in 67P/C-G coma are excluded by all available Rosetta data (Fulle et al. 2016a; Moreno et al. 2017).

A summary of all the power indexes collected by GIADA, COSIMA, OSIRIS, ROLIS and ground-based observations is shown in Fig. 7. Before 2015 equinox, most dust is coming from Hapi, a dust deposit close to the northern pole. After the 2015 equinox, dust is coming from the erosion of meters around the southern pole, showing that the pristine dust has a power index close to -4 over all dust sizes. The size distribution characterises the relative abundance of particles of different sizes in an ensemble. If approximated by a power-law, the characteristic quantity is its exponent. The smaller the (negative) size distribution exponent, the stronger is the relative contribution of small particles. For exponents < -4, not only the cross-section, but also the mass is dominated by the smallest particles present. The mass is concentrated in the largest particles for exponents > -4, and for exponents > -3, also the cross-section is predominantly in the largest particles. This occurred before the 2015 equinox at sizes <1 mm (Rotundi et al. 2015; Fulle et al. 2016c; Marschall et al. 2016; Merouane et al. 2016), implying that most 67P/C-G dust coma light was scattered by mm-sized particles. Dust data collected after the 2016 equinox are too scarce to provide significant statistics. In most comets and in 67P/C-G, the differential dust size distribution has a power index shallower than -4 (Fulle 2004), so that the integrated dust mass loss depends always on the largest ejected chunk, which however remains unknown in most comets. At 67P/C-G, OSIRIS has observed the largest pebbles and chunks escaping the nucleus, of sizes ranging from about 1 cm at 3.5 au (Rotundi et al. 2015), up to about 1 meter at perihelion (Fulle et al. 2016c; Ott et al. 2017). The measurement of the ejected dust mass is complicated by

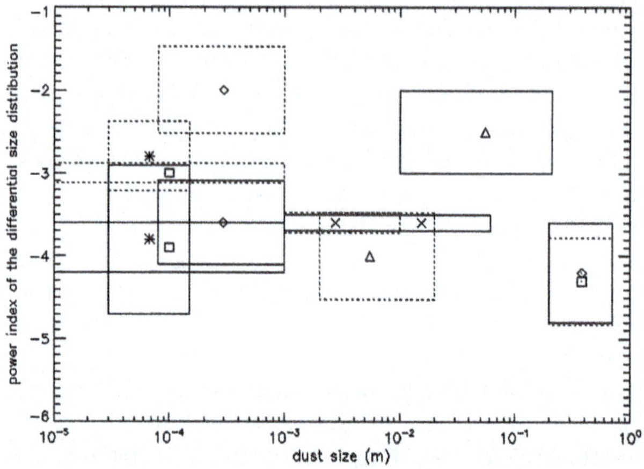

Fig. 7 Power index of the differential size distribution of 67P/C-G. Dotted lines: power index before the 2015 equinox. Continuous lines: power index after the 2015 equinox, up to just after the 2015 perihelion. The ranges along the x axis show the instrument sensitivity, the ranges along the y axis are given by the uncertainty of the power index. Stars: COSIMA data (Merouane et al. 2017). Diamonds: GIADA data (Rotundi et al. 2015; Fulle et al. 2016c). Triangles: OSIRIS observations of single particles in the coma (Rotundi et al. 2015; Fulle et al. 2016c; Ott et al. 2017). Dotted diamond: OSIRIS observations of pebbles in Sais (Pajola et al. 2017a). Dotted square: ROLIS observations of pebbles in Agilkia (Pajola et al. 2017a). Squares: Ground-based observations, tail models (Moreno et al. 2017). Crosses: Ground-based observations, trail models (Moreno et al. 2017)

the fact that some largest ejected chunks remain in metastable orbits around the nucleus up to the following perihelion passage (Fulle 1997; Rotundi et al. 2015), while a significant fraction of the ejected mass falls back on the nucleus surface (Thomas et al. 2015a; Pajola et al. 2017a; Fulle et al. 2017). The relationship between the size distribution of the ejected dust and that of the dust deposits is complex, and depends on the details of the fall-back processes and of the gas release from the deposits (Pajola et al. 2017a).

4.4 Optical and Thermal Properties

The phase curve of the diffuse coma observed by OSIRIS during observations covering phase angles from 10 to 155 deg within 2.5 hours seems inconsistent with the commonly adopted average cometary dust phase function (Kolokolova et al. 2004). Such an OSIRIS dust phase function shows a stronger back-scattering than assumed so far (Bertini et al. 2017). The OSIRIS dust phase function is consistent with most individual Earth-based or Earth-orbiting observations of comets. Observations by OSIRIS of dust bursts within 100 m from Rosetta suggest that the OSIRIS phase function is valid for dust sizes up to 2.5 mm (Fulle et al. 2018).

The diffuse coma has a color consistent with the average nucleus one at phase angles <30 deg, and is significantly bluer above (Bertini et al. 2017). The dust has a negligible phase reddening at phase angles <90 deg, indicating a coma dominated by single scattering (Bertini et al. 2017). The spectra of individual particles measured in the 535–882 nm range cover all the spectral slopes observed in the reddest and bluest nucleus areas (Frattin et al. 2017).

Dust in 67P/C-G coma shows a temperature a few per cents above the equilibrium one outside outbursts, but it increases up to factors 3–4 during outbursts, suggesting the ejection of smaller particles during outbursts (Bockelée-Morvan et al. 2017). During steady activity, the inferred minor flux of fractal aggregates of $D_f < 2$ explains all the features of the IR coma spectra observed by VIRTIS (Visible and Infrared Thermal Imaging Spectrometer) (Bockelée-Morvan et al. 2017). The dust composition and physical properties discussed above seem consistent with these data, although scattering models fitting all the observations are still unavailable.

4.5 Charged Dust

The amount of electric charge collected by fluffy and sub-μ m dust particles affects their motion in comae and tails and their lifetime against electrostatic fragmentation. The Rosetta mission provided the first direct evidence for the presence of electrically charged dust particles in a coma. These observations were possible as a result of the interaction between dust and the spacecraft electric potential, which was oscillating around −10 V (Nilsson et al. 2015) within 100 km from the 67P/C-G nucleus.

The fundamental physics of charged cometary dust has been studied for a long time (Mendis and Horányi 2013). Dust charging is a delicate interplay between several currents, including electron and ion collection currents from the surrounding plasma, ultraviolet induced photoelectron currents, secondary electron emission (due to energetic ion and/or electron bombardment), thermo-ionic emission, field emission (due to large surface fields), etc. As a rule of thumb, a dust particle collects about 700 extra electrons per unit particle volume-equivalent radius (measured in μm) and unit electric potential difference between the particle surface and the surrounding plasma (measured in V). This means that in the solar wind a particle of 1 μm size can have about 10^4 extra electrons. Based on our experience with mesospheric dust charging at Earth it can be assumed that most of the extra electrons are deep inside the dust particle and they can electrostatically disrupt fluffy, highly friable dust particles (Hill and Mendis 1981). An updated version (Mendis and Horányi 2013) predicts that a 10^{-19} kg particle can be electrostatically disrupted if its tensile strength is less than about 0.5 MPa. In general, the tensile strength of a particle charged at the potential U is $T = \epsilon_0/2\ (U/R)^2$ (Boehnhardt and Fechtig 1987). Regarding dust electrical resistance, COSIMA estimates a specific resistance of 10^{16} ohm mm^2 m^{-1} after the interaction of an Indium ion beam with the particles (Hilchenbach et al. 2017).

The GIADA dust showers have been modelled in terms of cm-sized fluffy dust aggregates charged by the flux of secondary electrons from the spacecraft, so that they decrease their potential from $-2 < U < +2$ V in the coma to $-17 < U < -5$ V close to the spacecraft (Fulle et al. 2015a). This potential decrease is sufficient to disrupt fractal particles (independent of their size if $D_f \approx 5/3$), because the tensile strength linking the grains of a dust aggregate is $T = 160\ [R/(1\ \mu\text{m})]^{D_f - 11/3}$ Pa (Skorov and Blum 2012; Fulle and Blum 2017). The fragments are then decelerated by the electrostatic interaction between negatively charged fractal fragments and the spacecraft (Fulle et al. 2015a). This model predicts a fractal dimension $D_f < 1.9$ (Fulle et al. 2016a) consistent with the measurements on the fluffy MIDAS particle, $D_f = 1.7 \pm 0.1$ (Mannel et al. 2016). The dust is deflected away from Rosetta at a charge-to-mass ratio of about 1 C kg^{-1} (Fulle et al. 2015a). This may explain the few MIDAS detections between 1 and 10 μm, with no submicron-sized samples, well above MIDAS resolution. However, most MIDAS detections are clustered, suggesting fragmentation of parent particles of size >10 μm at the impact on the collection plate (Bentley et al. 2016). Only compact particles with $R < 0.5$ μm have $q/m = 3\epsilon_0 U/(\rho_d R^2) > 1$ C kg^{-1}, implying that sub-micron compact particles only could have been pushed away by the spacecraft potential. The GIADA and MIDAS sub-mm fractal fragments of $D_f < 2$, decelerated at speeds of about 1 $cm\,s^{-1}$ by the spacecraft electric field, have a kinetic energy of 0.2–20 keV (Fulle et al. 2016a), so that the bursts of 0.2–20 keV electrons coming from the nucleus (direct dust) or from the Sun (dust reflected by the solar radiation pressure) measured by RPC/IES (Radio Plasma Consortium—Ion and Electron Sensor) on Rosetta (Burch et al. 2015) are not evidence of charged nano-dust.

These facts show that the electrostatic fragmentation of fractal aggregates of $D_f < 2$ is independent of their size and requires a dust charging much larger than occurring in the coma. Electrostatic forces may fragment only sub-micron sized compact particles. Fluffy particles carry always an important amount of charge vs. their mass, so that their motion is significantly affected by electric and magnetic fields, confirming the interpretations of striae in dust tails in terms of charged dust (Notni and Tiersch 1987).

4.6 Summary

Stardust and Rosetta have revealed extreme complexity in the physical properties of cometary dust. Dust particles are aggregates of grains, with volume filling factors covering many orders of magnitude. Cometary dust aggregates range from extremely compact to extremely fluffy. The most compact ones have shapes sometimes very far from spheres, and sub-mm thermally annealed aggregates of minerals observed both by Stardust and Rosetta suggest an origin in the inner protoplanetary disc. On the opposite end, Rosetta data suggest the presence of extremely fluffy aggregates of equivalent bulk density $< 1\ kg\,m^{-3}$. Between these two extremes, the aggregates cover a wide range of porosities. Their hierarchical structure, clearly imaged by Rosetta for the first time, provides constraints to the processes of dust condensation in the protosolar nebula and the protoplanetary disc.

The complexity of the physical parameters describing cometary dust challenges the interpretation of data from flyby missions and ground-based observations, always based on multi-parametric models which are hard to constrain, because averaging is not a suitable way to define shape, equivalent bulk density and sizes, and because their distributions evolve in time according to nucleus seasons, as seen at Rosetta.

5 Discussion

5.1 Dust Release, Continuous and Abrupt Processes

The nucleus of a comet is the origin of dust seen in the inner coma and, ultimately, in the cometary dust tail and trail. However, the details of the processes that initially lift the dust are, at best, uncertain.

The brightness distributions of inner comae are characterized by maximum intensities in a direction approximately sunward (Keller et al. 1994). This observed quasi-continuous emission is occasionally augmented by outburst phenomena, which are abrupt, short, and irregular events of mass loss. During the approach of the Deep Impact spacecraft to comet 9P/Tempel 1, outbursts were repeatedly observed (A'Hearn et al. 2005) while the rendezvous of Rosetta with comet 67P/C-G provided 26 months of monitoring during which an outburst was observed once in April 2014, followed by observations at increased frequency after May 2015 at all observed heliocentric distances. Outbursts were seen on a daily basis around the perihelion passage (Vincent et al. 2016a).

The largest liftable particle radius is usually given as $R = 3C_D Z u_0 R_n^2/(8GM\rho_d)$, where R_n is the nucleus radius, GM is the standard gravitational parameter (using the mass of the comet). Z is the surface gas mass flux, u_0 is the velocity of the gas at the surface, C_D is the gas drag efficiency, and ρ_d the dust bulk density (Gombosi and Horanyi 1986; Harmon et al. 2004). u_0 is usually taken as the thermal expansion velocity at the surface multiplied a correction factor of order of 9/4 (Finson and Probstein 1968). Typical values for R are of the order of a centimetre. This equation, coming from the balance between gas drag and gravity at the surface, has been widely used to explain particle mass loss from the sub-micron to the millimetre range with H_2O being the driving volatile at heliocentric distances of less than about 2.7 AU. On the other hand, the observations of 67P/C-G suggest ejection of particles much larger than could be explained by this equation (Fulle et al. 2016c; Ott et al. 2017).

Outbursts are indicative of a rapid release of large amounts of gas, which presumably would allow larger chunks to be ejected. It is important to note that the driving volatile initiating the outburst may not be H_2O but possibly more volatile species close to the surface and heated via conduction through the surface layer. On the other hand, with the exception of very specific cases (Montalto et al. 2008), even the larger outbursts are not the dominant source of dust in the coma. Most of the comae observed during flybys and that of 67P/C-G show stable dust coma brightness distributions on timescales of 1/10th to 1/20th of a rotation period. Hence, it remains appropriate to treat the continuous, more stable, emission separately from the more abrupt, short duration, outbursts and to assess their source processes. We look at the continuous emission first.

It might be assumed that a more uniform, apparently insolation-driven emission from much of the nucleus surface would be a straightforward assumption on the basis of observations of the inner comae of several comets. This would be H_2O driven and controlled through the largest liftable mass. However, this assumption is surprisingly simple to challenge. In the largest liftable mass calculation, van der Waals (vdW) forces are ignored but simple calculations (Kührt and Keller 1996) show that for micron-sized particles the cohesive forces can exceed vapour pressure forces by several orders of magnitude (Skorov and Blum 2012; Gundlach et al. 2015). Hence, there is inconsistency in a simple model of the dust ejection process with a size distribution extending from sub-micron particles (as seen at 1P/Halley, for example) to decimetre-sized chunks (as indicated at 67P/C-G).

It is also highly questionable whether dust emission is solely insolation-driven. More localized emission may give the appearance of being uniform and insolation-driven as a result of gas drag and lateral gas flow. This has been tested by fits to the gas densities measured at 67P/C-G.

Initial fits (Bieler et al. 2015; Marschall et al. 2016; Fougere et al. 2016) were consistent in showing the dominance of one particular area (Hapi) during the early phase of the Rosetta mission. Other work has gone further by testing even more localized emission distributions. For example, an insolation-driven model and a model where "cliffs" (topographic slopes of greater than 30 degrees) were active together with emission from the Hapi region (between the two lobes) of the nucleus were compared by Marschall et al. (2017). An active "cliff" hypothesis had been favoured on the basis of OSIRIS observations (Vincent et al. 2016b). The fits were statistically identical. The cliff hypothesis removes issues with cohesive forces where gravity is sufficient to breach the vdW forces but it requires steady activity from these specific sites if it is a dominant mechanism. It is also inconsistent with dominant dust ejection from a dust deposit (Hapi), and with observations of surface changes in flat dusty regions (i.e., Imhotep, Ash, and Anubis) of the nucleus that we must assume are activity driven (Auger et al. 2015; El-Maarry et al. 2017), although even here it is not proven that dust has actually been ejected from some of these sites. Hence, there is still no indisputable conclusion on the source distribution on the nucleus and the geomorphology of the source. This dramatically reduces our ability to constrain source processes and this is further hindered by the apparent absence of ices directly at the surface.

Sunshine et al. (2006) demonstrated that the surface area of H_2O ice on a cometary nucleus is insufficient to provide the measured gas production rates even in the case of free sublimation. Hence, a concept where H_2O ice just below a (partially) desiccated, weakly bound, layer of refractory material needs to be conceived that also resolves the problem with vdW forces. Some mechanisms may plausibly resolve this issue but are difficult to prove. Some examples include i) refractory particles being separated from each other by an icy "glue" (Houpis et al. 1985) that sublimes to release the particles without a vdW interaction, ii) small scale trapping of ices to produces smaller "explosions" on unresolved scales, (iii) electrostatic lifting to break vdW forces (see Sect. 4.5). Neither computer nor laboratory simulations of these processes have been presented. However, even in the case of the electrostatic mechanism, some circumstantial evidence can be found. For example, 67P/C-G shows ponded deposits (Thomas et al. 2015b) that are also seen on asteroids and have been attributed to the effect of electrostatic forces (Hartzell et al. 2013).

The causes of cometary outbursts have been the subject of research for many decades (Hughes 1990), but remain unresolved. Outbursts observed at 67P/C-G near perihelion cluster around local sunrise and early afternoon (Vincent et al. 2016a). These authors infer that at least two different outburst mechanisms occur in 67P/C-G, attributing the morning events to near-surface thermal stress during the fast temperature rise, and the afternoon events to a process in a deeper layer that would be reached by the diurnal heatwave around that time. The typical mass ejected during a bright, spacecraft-detected outburst is of order 10^3 to 10^5 kg (Knollenberg et al. 2016; Vincent et al. 2016a; Grün et al. 2016; Agarwal et al. 2017). Assuming that, during the 2016 perihelion of 67P/C-G, an average 10^4 kg per day were emitted during outbursts, this would only be a negligible fraction of the total dust production rate of $>1500\ \mathrm{kg\,s^{-1}}$ (Fulle et al. 2016c; Ott et al. 2017).

The high nucleus porosity and its high dust-to-ice mass ratio might appear inconsistent with models involving the presence of sealed, pressurised cavities close to the surface that will release a large amount of gas and dust in a short time when opened (Belton et al. 2008; Ipatov and A'Hearn 2011; Belton et al. 2013). However, combinations of different volatiles with differing sublimation temperatures offer a large number of possible alternatives. The crystallisation of amorphous water ice has been proposed as an energy source for cometary outbursts (Prialnik and Bar-Nun 1992; Belton et al. 2008), and as a source of super-volatiles trapped in the amorphous matrix before. However, only the phase transition of pure amorphous ice is exothermic, any pollution by super-volatiles makes it endothermic (Kouchi and Sirono 2001). Also the destabilisation of clathrates and release of trapped volatiles has been suggested to fuel outbursts from sealed sub-surface cavities (Mousis et al. 2015). An alternative scenario ascribes dust outbursts to the deepening of pre-existing cracks under thermal stress to depths where pristine, super-volatiles are present that violently sublimate for a short time when first exposed to sunlight (Skorov et al. 2016).

In summary, the simplistic model of dust production resulting from gas drag overcoming gravity is no longer tenable. However, the dominant mechanism is still unclear. While activity from "cliffs" has been observed by Rosetta (Vincent et al. 2016a; Pajola et al. 2017b), it remains difficult to prove that it dominates the emission. The relationship of the uppermost (optically visible) surface to icy material just below the surface and the means of releasing the dust component remains a subject of speculation. Outburst phenomena are of interest as a means of lifting large chunks of material but it appears proven that the outbursts observed and quantified at 67P/C-G do not dominate the total visible particle emission of the comet. The exact mechanism is unknown and multiple mechanisms cannot be ruled out.

5.2 Comparison Between Dust Properties and Clues from Light Scattering Observations

While unique space missions to a few comets are now revealing the properties of dust particles, a large amount of information on cometary dust is in the form of remote sensing data, collected by telescopes from the ground or in Earth orbit. The interpretation of these datasets to retrieve the physico-chemical properties of cometary dust is not straightforward and requires experimental and numerical simulations.

Laboratory Studies Measurements performed in the laboratory on analogues are mandatory to interpret properly the observational data and understand the chemical composition (content and nature of organic molecules, minerals and ices) and the physical properties (particle size, shape, structure etc.) of the dust emitted by comets (e.g., Levasseur-Regourd et al. 2007).

Observations of the thermal infrared spectral energy distributions (SEDs) within comae allow the characterization of cometary dust, as observed from Spitzer after Deep Impact hit 9P/Tempel 1 (Lisse et al. 2006, 2007). The interpretation of SEDs strongly relies on laboratory measurements of the thermal emissivity of powdered compounds relevant in terms of analogues for dust in comets, such as minerals, carbonaceous compounds and ices, essential to infer the composition, temperature, size distribution and porosity of the dust (Wooden 2008).

Light scattering observations of comets, and especially their linear polarization, provide information on cometary dust, as already presented in Sect. 1.2. The variations of the polarization with the phase angle and the wavelength are extensively studied in the laboratory on cometary dust analogues to point out comparisons with observations. Figure 8 presents instruments measuring the light scattering properties of randomly-oriented dust particles in suspension. PROGRA[2] corresponds to a series of gonio-polarimeters operating either in the laboratory under an air-draught technique for particles below about 10 μm, or under microgravity conditions during parabolic flight campaigns (Levasseur-Regourd et al. 1996; Levasseur-Regourd and Hadamcik 2003; Hadamcik et al. 2009a; Levasseur-Regourd et al. 2015). CODULAB, adapted from a previous instrument (Hovenier and Muñoz 2009), measures the Stokes parameters of dust particles in levitation under a gas flow (Muñoz et al. 2011, 2012, Muñoz and Hovenier 2015).

The polarimetric phase curves of cometary analogues made of porous aggregates of submicron-sized Mg-silicates, Fe-silicates and carbon black grains mixed with compact Mg-silicates grains measured with these setups were found comparable to those observed on comae of comets (Hadamcik et al. 2007). Submicron-sized grains in aggregates best fit the higher polarization observed in cometary jets and after fragmentation or disruption events, while a mixture of porous aggregates and compact grains was needed to fit whole comae observations (Hadamcik et al. 2006). The influence of the size of dust grains (from nm to μm) and agglomerates (from μm to cm) on the maximum polarization of phase curves has been studied (Hadamcik et al. 2009b), allowing the interpretation of polarimetric images of the coma of 103P/Hartley 2 in term of progressive ice sublimation and sequences of particles fragmentation with increasing distance from the nucleus (Hadamcik et al. 2013). Comparison of observations and laboratory measurements indicate that, at least for some comets, compact particles may dominate the polarization (Muñoz and Hovenier 2015).

Complementary to the interpretation of coma observations, laboratory experiments are used to study the surface and subsurface of comets, where cometary dust may be formed, deposited and eventually ejected (Kochan et al. 1999; Poch et al. 2016; Jost et al. 2017; Brouet et al. 2017).

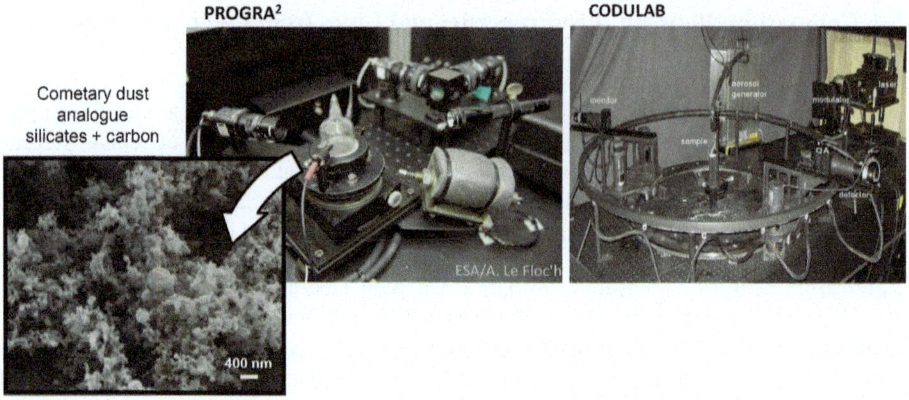

Fig. 8 Instruments PROGRA^2 (Hadamcik et al. 2009a) and CODULAB (Muñoz et al. 2011), used to measure, through experimental studies in the laboratory or under microgravity conditions, the light scattering properties of various analogues of cometary dust in suspension

Numerical Simulations Numerical simulations can complement the laboratory studies by reproducing and validating their results, by increasing the range of experimental conditions available and by confirming the physical interpretations of the properties necessary to reproduce the observations. Whenever intensity and polarization phase curves are obtained on a large enough range of phase angles and at different wavelengths, numerical and experimental simulations may be combined to infer some average properties, such as size and size distribution, ratio between transparent materials and absorbing materials (similar to dark organics), and ratio between fluffy aggregates and compact particles (Levasseur-Regourd et al. 2008; Lasue et al. 2009; Zubko 2012; Hines and Levasseur-Regourd 2016). Compact aggregates usually have a fractal dimension close to 3 and a porosity close to 75% (Bertini et al. 2009), while fluffy aggregates include many voids that translate into a large porosity ($> 90\%$ for $D_f \approx 2$, Bertini et al. 2009), which can be obtained in several ways, either by including voids in particles (Zubko 2012) or by considering specific fractal aggregation processes (see e.g. Levasseur-Regourd et al. 2007; Blum and Wurm 2008).

Numerous authors have tried to constrain the scattering properties of cometary dust. It was first demonstrated that spheres or spheroids cannot, even with various size distributions and compositions, reproduce the observational data (e.g., Kolokolova et al. 2007 and references within). With the development of more versatile T-matrix codes and Discrete Dipole Approximation techniques, models with aggregates of smaller particles were developed and found to provide satisfactory results (e.g. part 7 of Kiselev et al. 2015 review). From combined numerical simulations of polarization data and silicate emission features, (Kolokolova et al. 2007) concluded that the dust in cometary comae presenting a high maximum in polarization consists of highly porous aggregates that may be associated with fresh dust as in new comets. The dust in comets with a lower maximum polarization consists of less porous particles which may be associated with more highly processed dust such as expected for the surfaces of Jupiter Family comets. However, recent simulations strongly suggest that cometary dust is a mixture of aggregates and of compact particles of varying absorption properties (Lasue and Levasseur-Regourd 2006; Lasue et al. 2009).

Comparison with Rosetta Ground-Truth Rosetta data have established, within the coma and nucleus of 67P/C-G, that the relative abundance of organics in dust is large (about 70%

of the total volume, Sect. 3). It has also shown that the dust particles, the size distribution of which may vary, are irregular aggregates, some of them extremely fluffy and other ones extremely compact (Sect. 4). Such observational facts fit very nicely the clues derived from experimental and numerical simulations, once it is noticed that extremely compact aggregates may, from a light scattering point of view, behave like compact particles. The composition, size and structure of dust particles need to be taken into account to interpret the observations (Kolokolova et al. 2017).

While it is quite unlikely that future space missions will rendezvous numerous comets, remote observations of comets, complemented by elaborate experimental and numerical simulations, may thus be a unique approach to study the diversity of dust in comets.

5.3 Comparison Between Properties of Cometary Dust and of Some IDPs and Micrometeorites

Cosmic Dust Link with Cometary Sources Every year, the Earth accretes about 4×10^4 tons of extraterrestrial material less than 1 mm in size on its surface (e.g. Love and Brownlee 1993). These dust particles originate from active comets, from asteroids and may also be coming from interstellar space for the very small particles (for a review, see (Brownlee 1985; Rietmeijer 1998) and references therein).

While the exact origin of interplanetary dust accreted by the Earth is unknown, a significant contribution from cometary dust particles is expected, not only from comparisons between orbits of meteors and comets (e.g., Jenniskens 2006), but also from classical meteoroids models based on dust impacts in the Solar System (Grün et al. 1985; Divine 1993). Recent work on dust particle dynamics and observations also tend to estimate that a major fraction of the small dust particles is dominated by dust from active comets (from 70 to 90% (Lasue et al. 2007; Nesvorny et al. 2010; Rowan-Robinson and May 2013; Poppe 2016)).

Cosmic dust is mostly collected in the Earth's stratosphere by NASA for small particles ranging from 1 to 50 μm (IDPs for "Interplanetary Dust Particles", Brownlee 1985) and in polar caps as "MicroMeteorites" (MMs) in the size ranges from about 20 to > 400 μm (e.g. Maurette et al. 1991; Duprat et al. 2007; Noguchi et al. 2015; Taylor et al. 2016; Rochette et al. 2008). Chondritic porous IDPs (CP-IDPs) collected from the stratosphere (Ishii et al. 2008a) and ultracarbonaceous Antarctic micrometeorites (UCAMMs) (Nakamura et al. 2005; Duprat et al. 2010) are proposed to be of cometary origin. Given that a significant portion of the interplanetary dust particles comes from active comets, it is reasonable to compare their properties with the ones measured at comets, keeping in mind the expected differences due to sampling biases and alteration of the material.

Comparison of Optical Properties and Particles Morphologies The observational properties of the interplanetary dust cloud suggest compositions of silicates and carbon compounds similar to cometary dust particles (Fixsen and Dwek 2002; Reach et al. 2003); the properties from linear and circular polarization of light scattered by interplanetary dust, i.e. zodiacal light, suggest an albedo about 6% at 1 au (Levasseur-Regourd et al. 2001; Lasue et al. 2015 and references therein). The reflectivity of cometary dust particles from 67P/C-G was estimated from COSIMA on Rosetta to be between 3% and 23%, depending on the type of particle (Langevin et al. 2017). CP-IDPs are optically black (Brownlee 1985), with reflectivity < 15% over the visible range (Bradley et al. 1996). The reflectivity of unmelted MMs has not been measured, but they appear as black objects as well.

Optical properties of the zodiacal light indicate that it originates from sunlight scattered by irregular dust particles of a few microns to several tens of microns in size. The fact that

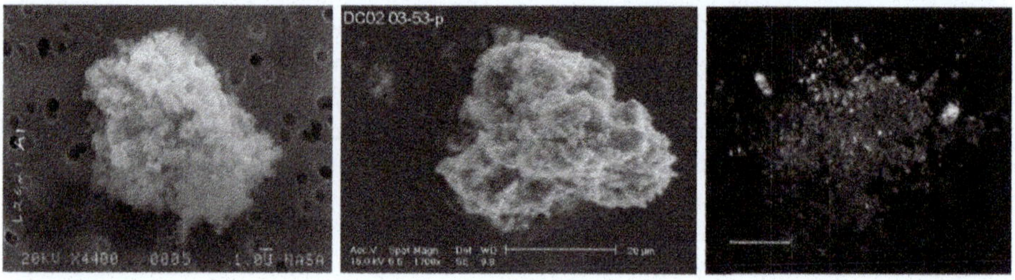

Fig. 9 From left to right, ordered by increasing sizes: Electron micrograph of stratospheric IDP (Particle L2021A1, about 13 × 10 µm, credit NASA); Electron micrograph of Antarctic micrometeorite DC02-03-53 collected near the Concordia station (size about 30 × 40 µm, Duprat et al. 2007); Optical image of the giant cluster IDP U2-20GCA (scale bar 100 µm) Messenger et al. 2015)

cometary dust particles are constituted of irregular fluffy and compact dust particles has been established from the results obtained with the Stardust samples (Hörz et al. 2006; Burchell et al. 2008) and the in-situ studies performed by Rosetta. Cometary dust is therefore inferred to be compact or fluffy and constituted of subcomponents that can be fragmented down to the micrometer scale (Rotundi et al. 2015; Langevin et al. 2016) or even lower (Bentley et al. 2016; Mannel et al. 2016); for details, see Sect. 4.

Analysis of the Stardust aerogel tracks showed the typical size range of dust particles collected from 81P/Wild 2 to be between 5 and 25 µ m. These particles are a mixture of compact and cohesive grains (65%) and friable less cohesive aggregated structures (35%) with constituent grains of a size less than 1 µm and a size distribution consistent with the ones derived in the comae of comets (Hörz et al. 2006). The typology of fragmented dust particles found by Rosetta is similar to some of the structures obtained on CP-IDPs, which have a similar tendency to fragment upon collection, and appear constituted mainly of submicrometer-sized components (Brownlee 1985). Similarly, UCAMMs with a size ranging from 20 to 200 µm exhibit substructure down to the 50 nm scale, with a relatively tight packing of sub-components (Engrand and Maurette 1998; Dobrica et al. 2012). The morphology of the particles collected is illustrated in Fig. 9, the similarity between such particles and the ones collected and those analysed by Stardust and Rosetta is apparent.

Comparison of Elemental Composition and Mineralogy Most cosmic dust particles have undifferentiated (chondritic-like) bulk elemental compositions close to CI and CM meteorites (about 2% of falls), different from the more common ordinary chondrites (>80% of falls) (Kurat et al. 1994; Taylor et al. 2016; Brownlee 2016; Flynn et al. 2016). Some dust particles, like ultracarbonaceous particles, do not even have meteoritic counterparts. These compositions are compatible with that of Stardust samples and of 67P/C-G (see Sect. 3).

The mineralogy of CP-IDPs, MMs and UCAMMs (crystalline silicates dominated by olivine, low-Ca pyroxene, and Fe–Ni sulfides is compatible with that of Stardust particles, although the Fe–Mg range of ferromagnesian silicates in 67P/C-G appears larger than for other comets (Zolensky et al. 2008; Frank et al. 2014). The range of olivine and low-Ca pyroxene compositions indicates a wide range of formation conditions, reflecting a large scale mixing between the inner and outer protoplanetary disk (Zolensky et al. 2006; Ishii et al. 2008b; Frank et al. 2014). The pyroxene to olivine abundance ratio in CP-IDPs and UCAMMs is on average larger than 1, similar to 81P/Wild 2 samples. The presence of glassy phases like GEMS (Bradley 2013) in Stardust samples is difficult to assess, as GEMS-like material was created during capture by melting and intermixing of aerogel with crystalline minerals. Some GEMS inclusions might however be genuine (Gainsforth et al. 2016), but no enstatite whiskers or platelets, like the ones in CP-IDPs, were identified in Wild 2 samples (Ishii et al. 2008b).

Hydrous silicates have not been identified in Stardust samples, which suggests a lack of aqueous processing of 81P/Wild 2 dust (Keller et al. 2006; Zolensky et al. 2008). However, Berger et al. (2011) found several sulfides (e.g. cubanite) in Stardust samples, which were interpreted as evidence for low-temperature hydrothermal processing in 81P/Wild 2, or on a parent body prior to the incorporation of these minerals in 81P/Wild 2. No hydrated minerals have been found in the anhydrous CP-IDPs, nor in UCAMMs. Small Mg-carbonates, similar to the ones identified in CP-IDPs (Flynn et al. 2009b) were identified in Stardust samples.

Comparison of Carbon Content and Isotopic Composition CP-IDPs and UCAMMs are enriched in carbon with regard to the chondritic composition (Schramm et al. 1989; Bradley et al. 1989; Thomas et al. 1993; Dartois et al. 2018). The mean C/Si elemental ratio of CP-IDPs is about 7 times larger than the CI chondrite Orgueil (Thomas et al. 1993). In UCAMMs the C/Si ratio vary from about 50 times CI up to several hundred times the chondritic ratio. Refractory organics in IDPs and UCAMMs can be mixed with rocky components, and constitute a matrix gluing the different minerals (Flynn et al. 2003, 2013; Dobrica et al. 2012; Charon et al. 2017). They can also coat the dust grains (Flynn et al. 2013), a process possibly having taken place in the interstellar medium (Greenberg and Hage 1990) or in the Solar protoplanetary disk (Ciesla and Sandford 2012). The major fraction of carbon consists of polyaromatic organic matter often exhibiting isotopically anomalous H and N (e.g. Floss et al. 2004; Duprat et al. 2010; Bardin et al. 2015). The observed large enrichments in D and ^{15}N can stem from cold chemistry processes at work in the presolar dust cloud which formed our Solar System, or in the cold regions of the early protoplanetary disk (for a review see Sandford et al. 2016).

The organic compounds in Stardust samples are rather scarce, possibly as a result of the harsh collection conditions that destroyed the most thermally unstable phases. Some organic phases are however indigenous to the comet and possess the same kind of polyaromatic composition (de Gregorio et al. 2011). Isotopic anomalies in the H and N compositions are observed in Stardust samples like in the organic matter from CP-IDPs and UCAMMs (Sandford et al. 2006; Matrajt et al. 2008; de Gregorio et al. 2011; Duprat et al. 2014). Results from Rosetta show carbon enrichment in the dust fraction of the comet probably due to high molecular weight organic matter (Fray et al. 2016; Herique et al. 2016; Bardyn et al. 2017). It is however difficult to assess for the time being whether the isotopic composition of organic matter in 67P/C-G is consistent with the variations measured for CP-IDPs and MMs.

Many CP-IDPs also contain sub-μm sized isotopically anomalous minerals, which represent circumstellar dust predating the formation of the Solar System. These pre-solar grains are found on average at the level of ≈500 ppm in CP-IDPs and UCAMMs, with up to concentrations of ≈1% in some IDPs (e.g. Floss and Haenecour 2016; Busemann et al. 2009; Davidson et al. 2012; Kakazu et al. 2014). These values exceed that of the most primitive meteorites, which have pre-solar grain abundances of a few 100 ppm (e.g. Nguyen et al. 2007, 2010). Analysis of the Stardust samples have shown that the concentration of presolar material within the solid fraction of 81P/Wild 2 is about 600 ppm (after correction from possible destruction upon collection), consistent with the average concentration measured in CP-IDPs (Floss et al. 2013).

To summarize, in the least altered IDPs and micrometeorites, the presence of vapor-deposited minerals, abundant organic matter, sub-grains of interstellar materials with strong isotopic anomalies, and both high and low-temperature nebular solids indicate that these particles are pristine, extraterrestrial materials. Their porous fragile structures and their non-homogeneous compositions suggest that they have not been processed in large objects. Their similarity to cometary dust material studied by Stardust and Rosetta indicates a possible link to primitive icy bodies as a source for these dust particles.

5.4 Possible Clues on Origin of Dust in Comets

According to gravitational instability models of proto-planetary discs, the 67P/C-G nucleus is probably a mixture of pristine pebbles (the sources of compact and porous particles), and pre-solar fractals of $D_f < 2$ stored in the voids among pebbles (Mannel et al. 2016; Fulle et al. 2016b; Fulle and Blum 2017; Blum et al. 2017; Barucci and Fulchignoni 2017). Submicron-sized grains were accreted into cm-sized fractal aggregates of $D_f < 2$ in the protosolar nebula at collision speeds $<1\ \mathrm{mm\,s^{-1}}$ (Blum 2000) and later compacted into cm-sized pebbles in the protoplanetary disc at collisions speeds $<1\ \mathrm{m\,s^{-1}}$ (Zsom et al. 2010) well before comets were formed (Fulle and Blum 2017; Blum et al. 2017). Fractal aggregates of $D_f < 2$ are expected to fill about 37% of the nucleus volume (Fulle and Blum 2017), in agreement with the percentage of the type B+C bulbous tracks observed in the Stardust aerogel, 35% of the total tracks (Burchell et al. 2008), and the rate of GIADA showers, about 30% of the GDS detections (Fulle and Blum 2017). Within this scenario, cometary nuclei are statistically homogeneous bodies, so that their erosion provides dust fluxes proportional to the volume abundances of pebbles and fractals.

Pebbles, coming from the compaction by bouncing collisions of pre-solar fractals in the protoplanetary disc, are expected to have a microporosity (Zsom et al. 2010) matching that inferred by GIADA (close to 50%); the dust microporosity measured by GIADA is larger than in the pebbles because many dust voids may be filled by ices in the nucleus pebbles (Fulle et al. 2017). Above the pebble size, the bouncing barrier in the protoplanetary disk may have stopped any further hierarchical structure (Zsom et al. 2010). In this case, the results of dust collection experiments, MIDAS and COSIMA, suggesting a hierarchical structure of dust, cannot be extended to larger chunks observed by the imaging cameras of the Rosetta spacecraft and its Philae lander. If the comet nucleus is composed of cm-sized pebbles (Blum et al. 2017) (Fulle and Blum 2017), then the dust bulk density should show a discontinuity at the pebbles size, because it should depend on the microporosity only at smaller sizes, and on both the micro- and macroporosity, expected to be $\approx$40% (Fulle et al. 2016b; Blum et al. 2017; Fulle and Blum 2017) at larger ones. Regarding the mineral composition, the in depth analysis of the Stardust samples revealed that the dust particles of comet 81P/Wild 2 are an aggregate of material that has been heated at rather high temperature in the inner protosolar nebula and material that has always remained in the outer regions of the Solar System at rather cold temperature. The sub-mm and dense GIADA samples (left-hand data in Fig. 6) suggest as well the presence of aggregates heated in the inner protoplanetary disc. This shows that comets incorporated minerals with various histories and is a signature of a very complex mixing of processed material at the formation of our Solar System.

Regarding the organic composition of the cometary dust particles, putting aside the 81P/Wild 2 particles, for which the carbonaceous content could not be properly quantified, and keeping in mind that the quantification of this organic refractory phase has only been made in two comets (1P/Halley and 67P/C-G), one can derive as a general conclusion that comets are an important reservoir of carbon and organic matter in the Solar System. The refractory organic phase in cometary particles is dominated by a high molecular weight organic component, the actual structure of which bears similarities with insoluble organic matter extracted from carbonaceous chondrites. Such a high molecular weight material has probably been the form under which a large proportion of carbon has been delivered to Earth, as discussed in the next Subsection.

5.5 Possible Clues to Early Delivery of Complex Organics to Telluric Planets

As mentioned in Sect. 5.3, the Earth's accretion rate of interplanetary dust is nowadays of about 40,000 tons per year, and a significant proportion of it is cometary dust. In the Late Heavy Bombardment (LHB) epoch, the spatial density of dust in the inner interplanetary dust cloud was orders of magnitude greater than today, and the brightness of the resulting zodiacal light was possibly more than 10^4 times higher (Nesvorny et al. 2010). Given the composition and morphology of cometary dust particles, which present significant carbon enrichments and porosities (Sect. 3 and 4), their accretion on early telluric planets could have enhanced the delivery of prebiotic material, generating conditions favourable for the emergence of life (Elsila et al. 2009; Altwegg et al. 2016). The morphology of cometary dust particles, which are irregular and consist of low-density aggregates (Sect. 4.1 and 4.2), may have allowed their temperatures to remain low enough to enable the survival of significant amounts of organics in the atmosphere.

The theory of meteoritic ablation during atmospheric entry, including the effects of thermal radiation, heat capacity and deceleration for solid particles, has been described in a number of publications (e.g. for a review Jenniskens 2006). Assuming, as a first approximation, the transfer heat coefficient and the emissivity to be equal to unity (Jones and Kaiser 1966) and a typical particle to be small enough to reach a uniform temperature (Murad 2001), the relationship between the heat transfer from atmospheric molecules to the particle and the light emission and heating of the particle, under thermal equilibrium, leads to a surface temperature of the particle T approximated by $T^4 \approx (V_E \rho_a)/(\sigma_S \xi)$, where V_E represents the entry velocity of the particle, ρ_a the atmospheric density at the considered altitude, σ_S the Stefan constant and ξ the ratio of the total surface of the particle to its projected surface. For spherical particles $\xi = 4$, while it can be 1.7 times higher for spheroidal particles (oblate with a ratio of semi major axes of 2) and even $\approx\pi$ times higher for aggregated fractal particles (Meakin and Donn 1988). Given that the material sublimation temperature could be about 2100 K (typical for rocks and metal, Öpik 1958) and the entry velocity should be about 30 km s^{-1} on average, then the altitude where sublimation starts is about 100, 97 and 93 km respectively for a sphere, a spheroidal particle and an aggregated fractal particle entering the Earth's atmosphere. Furthermore, the deceleration of the particle from collisions with the atmospheric molecules should also be taken into account. Assuming that they stick to the particle and thus transmit all their momentum to it, the maximum value of the temperature and thus the critical radius of the particles that may enter the atmosphere of Earth without being completely ablated can be determined (Jones and Kaiser 1966). The effect of the particles' shape on the equilibrium temperature reached during atmospheric entry can be seen in Fig. 10. While the radius for which spherical particles reach the ground without being ablated is about 3.8 μm, the largest equivalent volume radius of irregularly shaped particles can reach up to 12 μm. All parameters staying the same, fluffy aggregates, which are the most efficiently decelerated particles, may bring up to π^3 times more material in volume without being ablated to the Earth's surface than compact spherical particles (Levasseur-Regourd and Lasue 2011). For particles larger than a few tens of μm, thermal gradients can be observed inside the particles, indicating that thermal equilibrium was not reached during atmospheric entry heating, probably because of the short duration of the heating event (Genge et al. 2008). In that case it is not trivial to model the proportion of material that would survive atmospheric entry, but measurements of the cosmic dust flux on Earth (e.g.

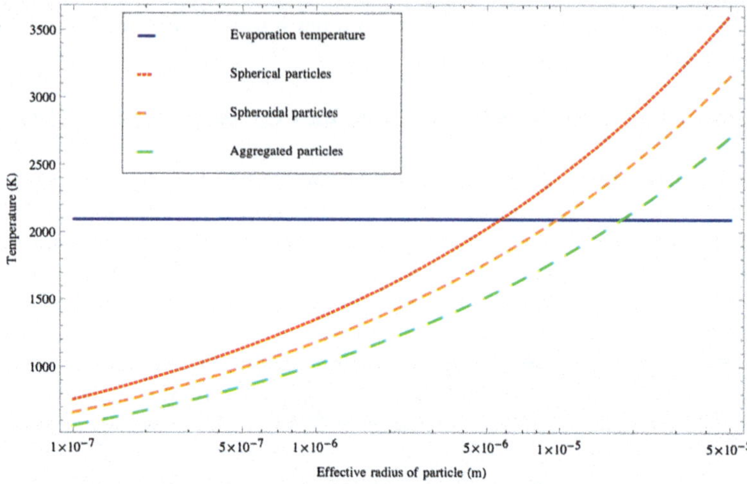

Fig. 10 Maximum equilibrium temperatures for particles entering the Earth atmosphere at 30 $km\,s^{-1}$. The horizontal line corresponds to a typical temperature of sublimation of 2100 K (adapted from Levasseur-Regourd and Lasue 2011)

Taylor et al. 1998; Duprat et al. 2001; Plane 2012) suggest that ≈80% of the mass of the incoming flux is vaporized upon atmospheric entry, the remaining 20% reaching the Earth as unmelted particles for ≈40% of them, the remaining ≈60% being partially melted or fully melted (e.g. Dobrica et al. 2010).

The increased density of interplanetary dust in the LHB epoch may have significantly increased the amount of cometary dust entering the Earth's atmosphere. It cannot be excluded that a significant proportion of the particles would, because of their structure, have brought pristine carbonaceous compounds from comets to the surface of early Earth; it may even be speculated that they could have contributed to the origins of life on our planet.

6 Conclusions and Open Questions

Our understanding of cometary dust, thanks to space missions to comets, to remote observations, to theoretical studies and to laboratory experiments, has remarkably progressed. However, as anticipated by A'Hearn (2017), "As we have dramatically increased our knowledge, we have also opened up many new questions".

It is now recognized that the refractory organic phase in cometary dust is dominated by high molecular weight organic components and that comets are an important reservoir of carbon and organic matter; they could have delivered a significant amount of complex organic components to the Earth, about 4 Ga ago. This hypothesis is strongly reinforced by the evidence that dust particles within many comets are aggregates of grains, with morphologies ranging from extremely porous to extremely compact, and volume filling factors covering many orders of magnitude; their hierarchical structure has been clearly imaged by the Rosetta mission. It may be suggested, from the composition and physical properties of cometary dust particles, that they are aggregates of material from the outer regions of the Solar System mixed with material reprocessed in the inner protosolar nebula and transported to the outer regions of the protoplanetary disk. They are likely to be the most pristine solid matter available from the early stages of its formation. Studies of dust in comets provide constraints to the processes of dust condensation in the protosolar nebula and the protoplanetary disk.

Some exciting and fascinating results, already obtained about the composition and physical properties of dust in comets are opening new questions. First, is the composition of the dust released by the nucleus perfectly representative of the bulk composition of the dust deep inside it, and how could the organic high molecular weight component be better characterized? How are organic and inorganic phases related inside cometary dust? Secondly, what are the smallest grains or monomers building cometary dust aggregates, and what about the composition, size and morphology of these monomers? Did they form within the early Solar System or before? Thirdly, how were the components and the structure of the dust particles processed and altered by the formation and evolution of the nuclei? How did the processes of dust ejection from the nucleus, which remain highly speculative, alter the particles? Has the collection process significantly affected our measurements?

Similarly, links with other cosmic dust particles are already pointed out or need further investigations. While clear similarities appear between cometary dust and CP-IDPs collected in the stratosphere or UCAMMs collected in Antarctica, what about their organics fraction and possible asteroidal sources? To what extent are some outer main-belt primitive asteroids similar to comets, and could future in-situ characterization or sample returns address this issue?

More progress requires sophisticated cross-platform analysis combining the insight of past space missions, elaborate remote observations of comets from Earth and near-Earth based telescopes, interpreted through improved numerical and experimental simulations, and hopefully the development of technological steps allowing sample returns of fragile cometary material in a not too distant future. NASA has preselected in December 2017 two proposals as possible New Frontiers 4 missions, one of them (CAESAR) proposing to return a sample from 67P/C-G; a final selection between the two missions expected in 2019, for a launch sometime in 2025. Meanwhile however, some of the above-mentioned open questions should be solved by further studies from the enormous amount of data that the Rosetta mission has provided.

Acknowledgements We would like to warmly acknowledge the ISSI for having organized a Cosmic dust workshop in November 2016, and ESA for making possible the first rendezvous with a comet through the Rosetta mission. ACLR acknowledges support from Centre National d'Etudes Spatiales (CNES) in the realization of instruments devoted to space exploration of comets and in their scientific analysis. H.C. acknowledges support from CNES. C.E. thanks CNES, PNP, DIM-ACAV, Labex P2IO and ANR for support. G.J.F. is supported by NASA Emerging Worlds grant NNX17AE59G. J.L. acknowledges support from CNES. T.M. acknowledges funding by the Austrian Science Fund FWF P 28100-N36. OP acknowledges a postdoctoral research fellowship from CNES. N.T. is supported through Swiss National Science Foundation grant number 20020–165684.

References

J. Agarwal, M. Müller, W.T. Reach, M.V. Sykes, H. Boehnhardt, E. Grün, The dust trail of comet 67P/Churyumov-Gerasimenko between 2004 and 2006. Icarus **207**, 992–1012 (2010)

J. Agarwal, V. Della Corte, P.D. Feldman, B. Geiger, S. Merouane, I. Bertini, D. Bodewits, S. Fornasier, E. Grün, P. Hasselmann, M. Hilchenbach, S. Höfner, S. Ivanovski, L. Kolokolova, M. Pajola, A. Rotundi, H. Sierks, A.J. Steffl, N. Thomas, M.F. A'Hearn, C. Barbieri, M.A. Barucci, J.-L. Bertaux, S. Boudreault, G. Cremonese, V. Da Deppo, B. Davidsson, S. Debei, M. De Cecco, J.F. Deller, L.M. Feaga, H. Fischer, M. Fulle, A. Gicquel, O. Groussin, C. Güttler, P.J. Gutiérrez, M. Hofmann, K. Hornung, S.F. Hviid, W.-H. Ip, L. Jorda, H.U. Keller, J. Kissel, J. Knollenberg, A. Koch, D. Koschny, J.-R. Kramm, E. Kührt, M. Küppers, P.L. Lamy, Y. Langevin, L.M. Lara, M. Lazzarin, Z.-Y. Lin,

J.J. Lopez Moreno, S.C. Lowry, F. Marzari, S. Mottola, G. Naletto, N. Oklay, J.W. Parker, R. Rodrigo, J. Rynö, X. Shi, O. Stenzel, C. Tubiana, J.-B. Vincent, H.A. Weaver, B. Zaprudin, Evidence of sub-surface energy storage in comet 67P from the outburst of 2016 July 03. Mon. Not. R. Astron. Soc. **469**, S606–S625 (2017)

M.F. A'Hearn, Comets: looking ahead. Phil. Trans. R. Soc A **375**, 20160261 (2017)

M.F. A'Hearn, M.J.S. Belton, W.A. Delamere, J. Kissel, K.P. Klaasen, L.A. McFadden, K.J. Meech, H.J. Melosh, P.H. Schultz, J.M. Sunshine, P.C. Thomas, J. Veverka, D.K. Yeomans, M.W. Baca, I. Busko, C.J. Crockett, S.M. Collins, M. Desnoyer, C.A. Eberhardy, C.M. Ernst, T.L. Farnham, L. Feaga, O. Groussin, D. Hampton, S.I. Ipatov, J.-Y. Li, D. Lindler, C.M. Lisse, N. Mastrodemos, W.M. Owen, J.E. Richardson, D.D. Wellnitz, R.L. White, Deep impact: excavating comet Tempel 1. Science **310**, 258–264 (2005)

C.M.OD. Alexander, R. Bowden, M.L. Fogel, K.T. Howard, C.D.K. Herd, L.R. Nittler, The provenances of asteroids, and their contributions to the volatile inventories of the terrestrial planets. Science **337**, 721 (2012)

K. Altwegg, H. Balsiger, A. Bar-Nun, J.-J. Berthelier, A. Bieler, P. Bochsler, C. Briois, U. Calmonte, M.R. Combi, H. Cottin, J. De Keyser, F. Dhooghe, B. Fiethe, S.A. Fuselier, S. Gasc, T.I. Gombosi, K.C. Hansen, M. Haessig, A. Jackel, E. Kopp, A. Korth, L. Le Roy, U. Mall, B. Marty, O. Mousis, T. Owen, H. Reme, M. Rubin, T. Semon, C.-Y. Tzou, J.H. Waite, P. Wurz, Prebiotic chemicals—amino acid and phosphorus—in the coma of comet 67P/Churyumov-Gerasimenko. Sci. Adv. **2**, e1600285 (2016)

K. Altwegg, H. Balsiger, J.J. Berthelier, A. Bieler, U. Calmonte, S.A. Fuselier, F. Goesmann, S. Gasc, T.I. Gombosi, L. Le Roy, J. de Keyser, A. Morse, M. Rubin, M. Schuhmann, M.G.G.T. Taylor, C.-Y. Tzou, I. Wright, Organics in comet 67P—a first comparative analysis of mass spectra from ROSINA-DFMS, COSAC and Ptolemy. Mon. Not. R. Astron. Soc. **469**, S130–S141 (2017)

R.G. Arendt, DIRBE comet trails. Astron. J. **148**, 135 (2014)

A.-T. Auger, O. Groussin, L. Jorda, S. Bouley, R. Gaskell, P.L. Lamy, C. Capanna, N. Thomas, A. Pommerol, H. Sierks, C. Barbieri, R. Rodrigo, D. Koschny, H. Rickman, H.U. Keller, J. Agarwal, M.F. A'Hearn, M.A. Barucci, J.-L. Bertaux, I. Bertini, G. Cremonese, V. Da Deppo, B. Davidsson, S. Debei, M. De Cecco, M.R. El-Maarry, S. Fornasier, M. Fulle, P.J. Gutiérrez, C. Güttler, S. Hviid, W.-H. Ip, J. Knollenberg, J.-R. Kramm, E. Kührt, M. Küppers, F. La Forgia, L.M. Lara, M. Lazzarin, J.J. Lopez Moreno, S. Marchi, F. Marzari, M. Massironi, H. Michalik, G. Naletto, N. Oklay, M. Pajola, L. Sabau, C. Tubiana, J.-B. Vincent, K.-P. Wenzel, Geomorphology of the Imhotep region on comet 67P/Churyumov-Gerasimenko from OSIRIS observations. Astron. Astrophys. **583**, A35 (2015)

N. Bardin, J. Duprat, C. Engrand, G. Slodzian, D. Baklouti, E. Dartois, R. Brunetto, L. Delauche, M. Godard, T.D. Wu, J.L. Guerquin-Kern, D/H and 15N/14N isotopic ratios in organic matter of ultracarbonaceous Antarctic micrometeorites. Meteorit. Planet. Sci. **50**(Suppl), 5275 (2015)

A. Bardyn, D. Baklouti, H. Cottin, N. Fray, C. Briois, J. Paquette, O. Stenzel, C. Engrand, H. Fischer, K. Hornung, R. Isnard, Y. Langevin, H. Lehto, L. Le Roy, N. Ligier, S. Merouane, P. Modica, F.-R. Orthous-Daunay, J. Rynö, R. Schulz, J. Silén, L. Thirkell, K. Varmuza, B. Zaprudin, J. Kissel, M. Hilchenbach, Carbon-rich dust in comet 67P/Churyumov-Gerasimenko measured by COSIMA/Rosetta. Mon. Not. R. Astron. Soc. **469**, S712–S722 (2017)

M.A. Barucci, M. Fulchignoni, Major achievements of the Rosetta mission in connection with the origin of the solar system. Astron. Astrophys. Rev. **25**, 5 (2017)

M.A. Barucci, E. Dotto, A.C. Levasseur-Regourd, Space missions to small bodies: asteroids and cometary nuclei. Astron. Astrophys. Rev. **19**, 48 (2011)

M.J.S. Belton, P.D. Feldman, M.F. A'Hearn, B. Carcich, Cometary cryo-volcanism: source regions and a model for the UT 2005 June 14 and other mini-outbursts on comet 9P/Tempel 1. Icarus **198**, 189–207 (2008)

M.J.S. Belton, P. Thomas, B. Carcich, A. Quick, J. Veverka, H. Jay Melosh, M.F. A'Hearn, J.-Y. Li, D. Brownlee, P. Schultz, K. Klaasen, G. Sarid, The origin of pits on 9P/Tempel 1 and the geologic signature of outbursts in Stardust-NExT images. Icarus **222**, 477–486 (2013)

M. Benna, P.R. Mahaffy, J.M. Grebowsky, J.M.C. Plane, R.V. Yelle, B.M. Jakosky, Metallic ions in the upper atmosphere of Mars from the passage of comet C/2013 A1 (Siding Spring). Geophys. Res. Lett. **42**, 4670–4675 (2015)

M.S. Bentley, R. Schmied, T. Mannel, K. Torkar, H. Jeszenszky, J. Romstedt, A.C. Levasseur-Regourd, I. Weber, E.K. Jessberger, P. Ehrenfreund, C. Koeberl, O. Havnes, Aggregate dust particles at comet 67P/Churyumov-Gerasimenko. Nature **537**, 73–75 (2016)

E.L. Berger, T.J. Zega, L.P. Keller, D.S. Lauretta, Evidence for aqueous activity on comet 81P/Wild 2 from sulfide mineral assemblages in Stardust samples and CI chondrites. Geochim. Cosmochim. Acta **75**, 3501–3513 (2011)

I. Bertini, N. Thomas, C. Barbieri, Modeling of the light scattering properties of cometary dust using fractal aggregates. Astron. Astrophys. **461**, 351–364 (2007)

I. Bertini, P.J. Gutierrez, W. Sabolo, The influence of the monomer shape in the first stage of dust growth in the protoplanetary disk. Astron. Astrophys. **504**, 625–633 (2009)

I. Bertini, F. La Forgia, C. Tubiana, C. Güttler, M. Fulle, F. Moreno, E. Frattin, G. Kovacs, M. Pajola, H. Sierks, C. Barbieri, P.L. Lamy, R. Rodrigo, D. Koschny, H. Rickman, H.U. Keller, J. Agarwal, M.F. A'Hearn, M.A. Barucci, J.-L. Bertaux, D. Bodewits, G. Cremonese, V. Da Deppo, B. Davidsson, S. Debei, M. De Cecco, E. Drolshagen, S. Ferrari, F. Ferri, S. Fornasier, A. Gicquel, O. Groussin, P.J. Gutierrez, P.H. Hasselmann, S.F. Hviid, W.-H. Ip, L. Jorda, J. Knollenberg, J.R. Kramm, E. Kührt, M. Küppers, L.M. Lara, M. Lazzarin, Z.-Y. Lin, J.J. Moreno, A. Lucchetti, F. Marzari, M. Massironi, S. Mottola, G. Naletto, N. Oklay, T. Ott, L. Penasa, N. Thomas, J.-B. Vincent, The scattering phase function of comet 67P/Churyumov-Gerasimenko coma as seen from the Rosetta/OSIRIS instrument. Mon. Not. R. Astron. Soc. **469**, S404–S415 (2017)

A. Bieler, K. Altwegg, H. Balsiger, J.-J. Berthelier, U. Calmonte, M.R. Combi, J. De Keyser, B. Fiethe, N. Fougere, S. Fuselier, S. Gasc, T.I. Gombosi, K.C. Hansen, M. Hässig, Z. Huang, A. Jäckel, X. Jia, L. Le Roy, U.A. Mall, H. Réme, M. Rubin, V. Tenishev, G. Tóth, C.-Y. Tzou, P. Wurz, Comparison of 3D kinetic and hydrodynamic models to ROSINA-COPS measurements of the neutral coma of 67P/Churyumov-Gerasimenko. Astron. Astrophys. **583**, A7 (2015)

J. Blum, Laboratory experiments on preplanetary dust aggregation. Space Sci. Rev. **92**, 265–278 (2000)

J. Blum, G. Wurm, The growth mechanisms of macroscopic bodies in protoplanetary disks. Annu. Rev. Astron. Astrophys. **46**, 21–56 (2008)

J. Blum, B. Gundlach, M. Krause, M. Fulle, A. Johansen, J. Agarwal, I. von Borstel, X. Shi, X. Hu, M.S. Bentley, F. Capaccioni, L. Colangeli, V. Della Corte, N. Fougere, S.F. Green, S. Ivanovski, T. Mannel, S. Merouane, A. Migliorini, A. Rotundi, R. Schmied, C. Snodgrass, Evidence for the formation of comet 67P/Churyumov-Gerasimenko through gravitational collapse of a bound clump of pebbles. Mon. Not. R. Astron. Soc. **469**, S755–S773 (2017)

D. Bockelée-Morvan, G. Rinaldi, S. Erard, C. Leyrat, F. Capaccioni, P. Drossart, G. Filacchione, A. Migliorini, E. Quirico, S. Mottola, G.P. Tozzi, G. Arnold, N. Biver, M. Combes, J. Crovisier, A. Longobardo, M. Blecka, M.-T. Capria, Comet 67P outbursts and quiescent coma at 1.3 au from the Sun: dust properties from Rosetta/VIRTIS-H observations. Mon. Not. R. Astron. Soc. **469**, S443–S458 (2017)

H. Boehnhardt, H. Fechtig, Electrostatic charging and fragmentation of dust near P/Giacobini-Zinner and P/Halley. Astron. Astrophys. **187**, 824–828 (1987)

J.P. Bradley, How and where did gems form? Geochim. Cosmochim. Acta **107**(0), 336–340 (2013)

J.P. Bradley, M.S. Germani, D.E. Brownlee, Automated thin-film analyses of anhydrous interplanetary dust particles in the analytical electron microscope. Earth Planet. Sci. Lett. **93**(1), 1–13 (1989)

J.P. Bradley, L.P. Keller, D.E. Brownlee, K.L. Thomas, Reflectance spectroscopy of interplanetary dust particles. Meteorit. Planet. Sci. **31**(3), 394–402 (1996)

J.C. Brandt, R.D. Chapman, *Introduction to Comets* (Cambridge University Press, Cambridge, 1981)

G. Briani, M. Gounelle, Y. Marrocchi, S. Mostefaoui, H. Leroux, E. Quirico, A. Meibom, Pristine extraterrestrial material with unprecedented nitrogen isotopic variation. Proc. Natl. Acad. Sci. **106**, 10522–10527 (2009)

J.C. Bridges, H.G. Changela, S. Nayakshin, N.A. Starkey, I.A. Franchi, Chondrule fragments from comet Wild2: evidence for high temperature processing in the outer Solar System. Earth Planet. Sci. Lett. **341**, 186–194 (2012)

Y. Brouet, L. Neves, P. Sabouroux, A.C. Levasseur-Regourd, P. Encrenaz, O. Poch, A. Pommerol, N. Thomas, W. Kofman, Characterization of the permittivity of controlled porous water ice/dust mixtures to support the radar exploration of icy bodies. J. Geophys. Res., Planets **121**, 2426–2443 (2017)

D.E. Brownlee, Cosmic dust: collection and research. Annu. Rev. Earth Planet. Sci. **13**(1), 147–173 (1985)

D.E. Brownlee, The Stardust mission: analyzing samples from the edge of the solar system. Annu. Rev. Earth Planet. Sci. **42**, 179–205 (2014)

D.E. Brownlee, Cosmic dust: building blocks of planets falling from the sky. Elements **12**(3), 165–170 (2016)

D.E. Brownlee, F. Hörz, R.L. Newburn, M. Zolensky, T.C. Duxbury, S. Sandford, Z. Sekanina, P. Tsou, M.S. Hanner, B.C. Clark, S.F. Green, J. Kissel, Surface of young Jupiter family comet 81 P/Wild 2: view from the Stardust spacecraft. Science **304**, 1764–1769 (2004)

D.E. Brownlee, P. Tsou, J. Aléon, C.M.OD. Alexander, T. Araki, S. Bajt, G.A. Baratta, R. Bastien, P. Bland, P. Bleuet, J. Borg, J.P. Bradley, A. Brearley, F. Brenker, S. Brennan, J.C. Bridges, N.D. Browning, J.R. Brucato, E. Bullock, M.J. Burchell, H. Busemann, A. Butterworth, M. Chaussidon, A. Cheuvront, M. Chi, M.J. Cintala, B.C. Clark, S.J. Clemett, G. Cody, L. Colangeli, G. Cooper, P. Cordier, C. Daghlian, Z. Dai, L. D'Hendecourt, Z. Djouadi, G. Dominguez, T. Duxbury, J.P. Dworkin, D.S. Ebel, T.E. Economou, S. Fakra, S.A.J. Fairey, S. Fallon, G. Ferrini, T. Ferroir, H. Fleckenstein, C. Floss, G. Flynn, I.A. Franchi, M. Fries, Z. Gainsforth, J.-P. Gallien, M. Genge, M.K. Gilles, P. Gillet,

J. Gilmour, D.P. Glavin, M. Gounelle, M.M. Grady, G.A. Graham, P.G. Grant, S.F. Green, F. Grossemy, L. Grossman, J.N. Grossman, Y. Guan, K. Hagiya, R. Harvey, P. Heck, G.F. Herzog, P. Hoppe, F. Hörz, J. Huth, I.D. Hutcheon, K. Ignatyev, H. Ishii, M. Ito, D. Jacob, C. Jacobsen, S. Jacobsen, S. Jones, D. Joswiak, A. Jurewicz, A.T. Kearsley, L.P. Keller, H. Khodja, A.L.D. Kilcoyne, J. Kissel, A. Krot, F. Langenhorst, A. Lanzirotti, L. Le, L.A. Leshin, J. Leitner, L. Lemelle, H. Leroux, M.-C. Liu, K. Luening, I. Lyon, G. MacPherson, M.A. Marcus, K. Marhas, B. Marty, G. Matrajt, K. McKeegan, A. Meibom, V. Mennella, K. Messenger, S. Messenger, T. Mikouchi, S. Mostefaoui, T. Nakamura, T. Nakano, M. Newville, L.R. Nittler, I. Ohnishi, K. Ohsumi, K. Okudaira, D.A. Papanastassiou, R. Palma, M.E. Palumbo, R.O. Pepin, D. Perkins, M. Perronnet, P. Pianetta, W. Rao, F.J.M. Rietmeijer, F. Robert, D. Rost, A. Rotundi, R. Ryan, S.A. Sandford, C.S. Schwandt, T.H. See, D. Schlutter, J. Sheffield-Parker, A. Simionovici, S. Simon, I. Sitnitsky, C.J. Snead, M.K. Spencer, F.J. Stadermann, A. Steele, T. Stephan, R. Stroud, J. Susini, S.R. Sutton, Y. Suzuki, M. Taheri, S. Taylor, N. Teslich, K. Tomeoka, N. Tomioka, A. Toppani, J.M. Trigo-Rodríguez, D. Troadec, A. Tsuchiyama, A.J. Tuzzolino, T. Tyliszczak, K. Uesugi, M. Velbel, J. Vellenga, E. Vicenzi, L. Vincze, J. Warren, I. Weber, M. Weisberg, A.J. Westphal, S. Wirick, D. Wooden, B. Wopenka, P. Wozniakiewicz, I. Wright, H. Yabuta, H. Yano, E.D. Young, R.N. Zare, T. Zega, K. Ziegler, L. Zimmerman, E. Zinner, M. Zolensky, Comet 81P/Wild 2 under a microscope. Science **314**, 1711 (2006)

J.L. Burch, T.I. Gombosi, G. Clark, P. Mokashi, R. Goldstein, Observation of charged nanograins at comet 67P/Churyumov-Gerasimenko. Geophys. Res. Lett. **42**, 6575–6581 (2015)

M.J. Burchell, S.A.J. Fairey, P. Wozniakiewicz, D.E. Brownlee, F. Hörz, A.T. Kearsley, T.H. See, P. Tsou, A.J. Westphal, S.F. Green, J.M. Trigo-Rodríguez, G. Domingúez, Characteristics of cometary dust tracks in Stardust aerogel and laboratory calibrations. Meteorit. Planet. Sci. **43**, 23–40 (2008)

H. Busemann, S. Lorenzetti, O. Eugster, Noble gases in D'Orbigny, Sahara 99555 and D'Orbigny glass—evidence for early planetary processing on the angrite parent body. Geochim. Cosmochim. Acta **70**, 5403–5425 (2006)

H. Busemann, A.N. Nguyen, G.D. Cody, P. Hoppe, A.L.D. Kilcoyne, R.M. Stroud, T.J. Zega, L.R. Nittler, Ultra-primitive interplanetary dust particles from the comet 26P/Grigg-Skjellerup dust stream collection. Earth Planet. Sci. Lett. **288**, 44–57 (2009)

A.L. Butterworth, Z. Gainsforth, C.E. Jilly-Rehak, K. Righter, A.J. Westphal, Developing V-XANES oxybarometry for probing materials formed, in *Reducing Environments in the Early Solar Disk. Lunar and Planetary Science Conference* (2017)

S. Chakraborty, Diffusion in solid silicates: a tool to track timescales of processes comes of age. Annu. Rev. Earth Planet. Sci. **36**, 153–190 (2008)

E. Charon, C. Engrand, K. Benzerara, H. Leroux, S. Swaraj, R. Belkhou, J. Duprat, E. Dartois, M. Godard, L. Delauche, Organic phases and mineralogy of ultracarbonaceous Antarctic micrometeorites identified by C- and N-XANES /STXM and TEM. Lunar Planet. Sci. **48**, 2085 (2017)

V. Ciarletti, A. Herique, J. Lasue, A.C. Levasseur-Regourd, D. Plettemeier, F. Lemmonier, C. Guiffaut, P. Pasquero, W. Kofman, CONSERT constrains the internal structure of 67P at a few metres size scale. Mon. Not. R. Astron. Soc. **469**, S805–S817 (2017)

F.J. Ciesla, Residence times of particles in diffusive protoplanetary disk environments. II. radial motions and applications to dust annealing. Astrophys. J. **740**, 9 (2011)

F.J. Ciesla, S.A. Sandford, Organic synthesis via irradiation and warming of ice grains in the solar nebula. Science **336**, 452 (2012)

R.N. Clayton, Oxygen isotopes in meteorites. Annu. Rev. Earth Planet. Sci. **21**, 115–149 (1993)

R.N. Clayton, T.K. Mayeda, Oxygen isotope studies of carbonaceous chondrites. Geochim. Cosmochim. Acta **63**, 2089–2104 (1999)

A.L. Cochran, A.C. Levasseur-Regourd, M. Cordiner, E. Hadamcik, J. Lasue, A. Gicquel, D.G. Schleicher, S.B. Charnley, M.J. Mumma, L. Paganini, D. Bockelée-Morvan, N. Biver, Y.-J. Kuan, The composition of comets. Space Sci. Rev. **197**, 9–46 (2015)

L. Colangeli, J.J. Lopez-Moreno, P. Palumbo, J. Rodriguez, M. Cosi, V. Della Corte, F. Esposito, M. Fulle, M. Herranz, J.M. Jeronimo, A. Lopez-Jimenez, E. Mazzotta Epifani, R. Morales, F. Moreno, E. Palomba, A. Rotundi, The grain impact analyser and dust accumulator (GIADA) experiment for the Rosetta mission: design, performances and first results. Space Sci. Rev. **128**, 803–821 (2007)

A. Coradini, F. Capaccioni, P. Drossart, G. Arnold, E. Ammannito, F. Angrilli, A. Barucci, G. Bellucci, J. Benkhoff, G. Bianchini, J.P. Bibring, M. Blecka, D. Bockelee-Morvan, M.T. Capria, R. Carlson, U. Carsenty, P. Cerroni, L. Colangeli, M. Combes, M. Combi, J. Crovisier, M.C. De Sanctis, E.T. Encrenaz, S. Erard, C. Federico, G. Filacchione, U. Fink, S. Fonti, V. Formisano, W.-H. Ip, R. Jaumann, E. Kührt, Y. Langevin, G. Magni, T. McCord, V. Mennella, S. Mottola, G. Neukum, P. Palumbo, G. Piccioni, H. Rauer, B. Saggin, B. Schmitt, D. Tiphene, G.P. Tozzi, Virtis: an imaging spectrometer for the Rosetta mission. Space Sci. Rev. **128**, 529–559 (2007)

T.K. Croat, C. Floss, B.A. Haas, M.J. Burchell, A.T. Kearsley, Survival of refractory presolar grain analogs during Stardust-like impact into Al foils: implications for Wild 2 presolar grain abundances and study of the cometary fine fraction. Meteorit. Planet. Sci. **50**, 1378–1391 (2015)

J. Crovisier, Infrared observations of volatile molecules in comet Hale-Bopp. Earth Moon Planets **79**, 125–143 (1997)

E. Dartois, C. Engrand, J. Duprat, M. Godard, E. Charon, L. Delauche, C. Sandt, F. Borondics, Dome C ultracarbonaceous antarctic micrometeorites. Infrared and raman fingerprints. Astron. Astrophys. **609**, A65 (2018)

J. Davidson, H. Busemann, I.A. Franchi, A NanoSIMS and Raman spectroscopic comparison of interplanetary dust particles from comet Grigg-Skjellerup and non-Grigg Skjellerup collections. Meteorit. Planet. Sci. **47**, 1748–1771 (2012)

J.K. Davies, S.F. Green, B.C. Stewart, A.J. Meadows, H.H. Aumann, The IRAS fast-moving object search. Nature **309**, 315–319 (1984)

J.K. Davies, M.V. Sykes, W.T. Reach, F. Boulanger, F. Sibille, C.J. Cesarsky, ISOCAM observations of the comet P/Kopff dust trail. Icarus **127**, 251–254 (1997)

B.T. de Gregorio, R.M. Stroud, G.D. Cody, L.R. Nittler, A.L.D. Kilcoyne, S. Wirick, Correlated microanalysis of cometary organic grains returned by Stardust. Meteorit. Planet. Sci. **46**, 1376–1396 (2011)

C. Defouilloy, D. Nakashima, D.J. Joswiak, D.E. Brownlee, T.J. Tenner, N.T. Kita, Origin of crystalline silicates from comet 81P/Wild 2: combined study on their oxygen isotopes and mineral chemistry. Earth Planet. Sci. Lett. **465**, 145–154 (2017)

V. Della Corte, A. Rotundi, M. Accolla, R. Sordini, P. Palumbo, L. Colangeli, J.J. Lopez–Moreno, J. Rodriguez, F.J.M. Rietmeijer, M. Ferrari, F. Lucarelli, E. Mazzotta Epifani, S. Ivanovski, A. Aronica, M. Cosi, E. Bussoletti, J.F. Crifo, F. Esposito, M. Fulle, S.F. Green, E. Grün, M.L. Herranz, J.M. Jeronimo, P.L. Lamy, A. Lopez Jimenez, J.A.M. McDonnell, V. Mennella, A. Molina, R. Morales, F. Moreno, E. Palomba, J.M. Perrin, R. Rodrigo, P. Weissman, V. Zakharov, J.C. Zarnecki, Giada: its status after the Rosetta cruise phase and on-ground activity in support of the encounter with comet 67P/CHURYUMOV-GERASIMENKO. J. Astron. Instrum. **3**, 1350011-110 (2014)

V. Della Corte, A. Rotundi, M. Fulle, E. Grün, P. Weissman, R. Sordini, M. Ferrari, S. Ivanovski, F. Lucarelli, M. Accolla, V. Zakharov, E. Mazzotta Epifani, J.J. Lopez-Moreno, J. Rodriguez, L. Colangeli, P. Palumbo, E. Bussoletti, J.F. Crifo, F. Esposito, S.F. Green, P.L. Lamy, J.A.M. McDonnell, V. Mennella, A. Molina, R. Morales, F. Moreno, J.L. Ortiz, E. Palomba, J.M. Perrin, F.J.M. Rietmeijer, R. Rodrigo, J.C. Zarnecki, M. Cosi, F. Giovane, B. Gustafson, M.L. Herranz, J.M. Jeronimo, M.R. Leese, A.C. Lopez-Jimenez, N. Altobelli, GIADA: shining a light on the monitoring of the comet dust production from the nucleus of 67P/Churyumov-Gerasimenko. Astron. Astrophys. **583**, A13 (2015)

N. Divine, Five populations of interplanetary meteoroids. J. Geophys. Res., Planets **98**(E9), 17029–17048 (1993)

E. Dobrica, C. Engrand, J. Duprat, M. Gounelle, H. Leroux, E. Quirico, J.-N. Rouzaud, Connection between micrometeorites and Wild 2 particles: from Antarctic snow to cometary ices. Meteorit. Planet. Sci. **44**, 1643–1661 (2009)

E. Dobrica, C. Engrand, J. Duprat, M. Gounelle, A statistical overview of concordia antarctic micrometeorites. Meteorit. Planet. Sci. Suppl. **45**, A46 (2010)

E. Dobrica, C. Engrand, H. Leroux, J.N.R.J. Duprat, Transmission electron microscopy of concordia ultracarbonaceous Antarctic micrometeorites (ucamms): mineralogical properties. Geochim. Cosmochim. Acta **76**, 68–82 (2012)

A. Dollfus, P. Bastien, J.-F. Le Borgne, A.C. Levasseur-Regourd, T. Mukai, Optical polarimetry of P/Halley—synthesis of the measurements in the continuum. Astron. Astrophys. **206**, 348–356 (1988)

G. Dominguez, A heterogeneous chemical origin for the ^{16}O-enriched and ^{16}O-depleted reservoirs of the early solar system. Astrophys. J. Lett. **713**, L59–L63 (2010)

G. Dominguez, A.S. McLeod, Z. Gainsforth, P. Kelly, H.A. Bechtel, F. Keilmann, A.J. Westphal, M. Thiemens, D.N. Basov, Nanoscale infrared spectroscopy as a non-destructive probe of extraterrestrial samples. Nat. Commun. **5**, 5445 (2014)

J. Duprat, M. Maurette, C. Engrand, G. Matrajt, G. Immel, C. Hammer, M. Gounelle, G. Kurat, An estimation of the contemporary micrometeorite flux obtained from surface snow samples collected in central antarctica. Meteorit. Planet. Sci. Suppl. **36**, A52 (2001)

J. Duprat, C. Engrand, M. Maurette, G. Kurat, M. Gounelle, C. Hammer, Micrometeorites from central Antarctic snow: the CONCORDIA collection. Adv. Space Res. **39**(4), 605–611 (2007)

J. Duprat, E. Dobrica, C. Engrand, J. Aléon, Y. Marrocchi, S. Mostefaoui, F. Robert, Extreme deuterium excesses in ultracarbonaceous micrometeorites from central Antarctic snow. Science **328**, 742–745 (2010)

J. Duprat, N. Bardin, C. Engrand, D. Baklouti, R. Brunetto, E. Dartois, L. Delauche, M. Godard, J.-L. Guerquin-Kern, G. Slodzian, T.-D. Wu, Isotopic analysis of organic matter in ultra-carbonaceous Antarctic micrometeorites. Meteorit. Planet. Sci. **49**(Suppl), 5341 (2014)

T.E. Economou, S.F. Green, D.E. Brownlee, B.C. Clark, Dust flux monitor instrument measurements during Stardust-NExT flyby of comet 9P/Tempel 1. Icarus **222**, 526–539 (2013)

M.R. El-Maarry, O. Groussin, N. Thomas, M. Pajola, A.-T. Auger, B. Davidsson, X. Hu, S.F. Hviid, J. Knollenberg, C. Güttler, C. Tubiana, S. Fornasier, C. Feller, P. Hasselmann, J.-B. Vincent, H. Sierks, C. Barbieri, P.L. Lamy, R. Rodrigo, D. Koschny, H.U. Keller, H. Rickman, M.F. A'Hearn, M.A. Barucci, J.-L. Bertaux, I. Bertini, S. Besse, D. Bodewits, G. Cremonese, V. Da Deppo, S. Debei, M. De Cecco, J. Deller, J.D.P. Deshapriya, M. Fulle, P.J. Gutierrez, M. Hofmann, W.-H. Ip, L. Jorda, G. Kovacs, J.-R. Kramm, E. Kührt, M. Küppers, L.M. Lara, M. Lazzarin, Z.-Y. Lin, J.J. Lopez Moreno, S. Marchi, F. Marzari, S. Mottola, G. Naletto, N. Oklay, A. Pommerol, F. Preusker, F. Scholten, X. Shi, Surface changes on comet 67P/Churyumov-Gerasimenko suggest a more active past. Science **355**, 1392–1395 (2017)

L.E. Ellerbroek, B. Gundlach, A. Landeck, C. Dominik, J. Blum, S. Merouane, M. Hilchenbach, M.S. Bentley, T. Mannel, H. John, H.A. van Veen, The footprint of cometary dust analogues—I. Laboratory experiments of low-velocity impacts and comparison with Rosetta data. Mon. Not. R. Astron. Soc. **469**, S204–S216 (2017)

J.E. Elsila, D.P. Glavin, J.P. Dworkin, Cometary glycine detected in samples returned by Stardust. Meteorit. Planet. Sci. **44**, 1323–1330 (2009)

C. Engrand, M. Maurette, Carbonaceous micrometeorites from Antarctica. Meteorit. Planet. Sci. **33**(4), 565–580 (1998)

M. Ferrari, V. Della Corte, A. Rotundi, F.J.M. Rietmeijer, Single minerals, carbon- and ice-coated single minerals for calibration of GIADA onboard ROSETTA to comet 67P/Churyumov-Gerasimenko. Planet. Space Sci. **101**, 53–64 (2014)

M.J. Finson, R.F. Probstein, A theory of dust comets. I. Model and equations. Astrophys. J. **154**, 327–352 (1968)

D.J. Fixsen, E. Dwek, The zodiacal emission spectrum as determined by COBE and its implications. Astrophys. J. **578**, 1009 (2002)

C. Floss, P. Haenecour, Presolar silicate grains: abundances, isotopic and elemental compositions, and the effects of secondary processing. Geochem. J. **50**, 3–25 (2016)

C. Floss, F.J. Stadermann, J. Bradley, Z.R. Dai, S. Bajt, G. Graham, Carbon and nitrogen isotopic anomalies in an anhydrous interplanetary dust particle. Science **303**, 1355–1358 (2004)

C. Floss, F.J. Stadermann, A.T. Kearsley, M.J. Burchell, W.J. Ong, The abundance of presolar grains in comet 81P/Wild 2. Astrophys. J. **763**, 140 (2013)

G.J. Flynn, Physical, chemical, and mineralogical properties of comet 81P/Wild 2 particles collected by Stardust. Earth Moon Planets **102**, 447–459 (2008)

G.J. Flynn, L.P. Keller, M. Feser, S. Wirick, C. Jacobsen, The origin of organic matter in the solar system: evidence from the interplanetary dust particles. Geochim. Cosmochim. Acta **67**(24), 4791–4806 (2003)

G.J. Flynn, D.D. Durda, L.E. Sandel, M.M. Strait, Dust production from the hypervelocity impact disruption of hydrated targets. Earth Moon Planets **97**, 213–231 (2005)

G.J. Flynn, P. Bleuet, J. Borg, J.P. Bradley, F.E. Brenker, S. Brennan, J. Bridges, D.E. Brownlee, E.S. Bullock, M. Burghammer, B.C. Clark, Z.R. Dai, C.P. Daghlian, Z. Djouadi, S. Fakra, T. Ferroir, C. Floss, I.A. Franchi, Z. Gainsforth, J.-P. Gallien, P. Gillet, P.G. Grant, G.A. Graham, S.F. Green, F. Grossemy, P.R. Heck, G.F. Herzog, P. Hoppe, F. Hörz, J. Huth, K. Ignatyev, H.A. Ishii, K. Janssens, D. Joswiak, A.T. Kearsley, H. Khodja, A. Lanzirotti, J. Leitner, L. Lemelle, H. Leroux, K. Luening, G.J. MacPherson, K.K. Marhas, M.A. Marcus, G. Matrajt, T. Nakamura, K. Nakamura-Messenger, T. Nakano, M. Newville, D.A. Papanastassiou, P. Pianetta, W. Rao, C. Riekel, F.J.M. Rietmeijer, D. Rost, C.S. Schwandt, T.H. See, J. Sheffield-Parker, A. Simionovici, I. Sitnitsky, C.J. Snead, F.J. Stadermann, T. Stephan, R.M. Stroud, J. Susini, Y. Suzuki, S.R. Sutton, S. Taylor, N. Teslich, D. Troadec, P. Tsou, A. Tsuchiyama, K. Uesugi, B. Vekemans, E.P. Vicenzi, L. Vincze, A.J. Westphal, P. Wozniakiewicz, E. Zinner, M.E. Zolensky, Elemental compositions of comet 81P/Wild 2 samples collected by Stardust. Science **314**, 1731 (2006)

G.J. Flynn, D.D. Durda, L.E. Sandel, J.W. Kreft, M.M. Strait, Dust production from the hypervelocity impact disruption of the Murchison hydrous CM2 meteorite: implications for the disruption of hydrous asteroids and the production of interplanetary dust. Planet. Space Sci. **57**, 119–126 (2009a)

G.J. Flynn, S. Wirick, L.P. Keller, C. Jacobsen, S.A. Sandford, Carbonate grains in anhydrous IDPs. Meteorit. Planet. Sci. Suppl. **72**, 5296 (2009b)

G.J. Flynn, S. Wirick, L.P. Keller, Organic grain coatings in primitive interplanetary dust particles: implications for grain sticking in the solar nebula. Earth Planets Space **65**(10), 13 (2013)

G.J. Flynn, L.R. Nittler, C. Engrand, Composition of cosmic dust: sources and implications for the early solar system. Elements **12**(3), 177–183 (2016)

N.M. Fomenkova, S. Chang, L.M. Mukhin, Carbonaceous components in the comet Halley dust. Geochim. Cosmochim. Acta **58**, 4508–4512 (1994)

N. Fougere, K. Altwegg, J.-J. Berthelier, A. Bieler, D. Bockelée-Morvan, U. Calmonte, F. Capaccioni, M.R. Combi, J. De Keyser, V. Debout, S. Erard, B. Fiethe, G. Filacchione, U. Fink, S.A. Fuselier, T.I. Gombosi, K.C. Hansen, M. Hässig, Z. Huang, L. Le Roy, C. Leyrat, A. Migliorini, G. Piccioni, G. Rinaldi, M. Rubin, Y. Shou, V. Tenishev, G. Toth, C.-Y. Tzou, Direct simulation Monte Carlo modelling of the major species in the coma of comet 67P/Churyumov-Gerasimenko. Mon. Not. R. Astron. Soc. **462**, S156–S169 (2016)

D.R. Frank, M.E. Zolensky, L. Le, Olivine in terminal particles of Stardust aerogel tracks and analogous grains in chondrite matrix. Geochim. Cosmochim. Acta **142**, 240–259 (2014)

E. Frattin, G. Cremonese, E. Simioni, I. Bertini, M. Lazzarin, T. Ott, E. Drolshagen, F. La Forgia, H. Sierks, C. Barbieri, P.L. Lamy, R. Rodrigo, D. Koschny, H. Rickman, H.U. Keller, J. Agarwal, M.F. A'Hearn, M.A. Barucci, J.-L. Bertaux, V. Da Deppo, B. Davidsson, S. Debei, M. De Cecco, J. Deller, S. Ferrari, F. Ferri, S. Fornasier, M. Fulle, A. Gicquel, O. Groussin, P.J. Gutierrez, C. Güttler, M. Hofmann, S.F. Hviid, W.-H. Ip, L. Jorda, J. Knollenberg, J.-R. Kramm, E. Kührt, M. Küppers, L.M. Lara, J.J. Lopez Moreno, A. Lucchetti, F. Marzari, M. Massironi, S. Mottola, G. Naletto, N. Oklay, M. Pajola, L. Penasa, X. Shi, N. Thomas, C. Tubiana, J.-B. Vincent, Post-perihelion photometry of dust grains in the coma of 67P Churyumov-Gerasimenko. Mon. Not. R. Astron. Soc. **469**, S195–S203 (2017)

N. Fray, A. Bardyn, H. Cottin, K. Altwegg, D. Baklouti, C. Briois, L. Colangeli, C. Engrand, H. Fischer, A. Glasmachers, E. Grün, G. Haerendel, H. Henkel, H. Höfner, K. Hornung, E.K. Jessberger, A. Koch, H. Krüger, Y. Langevin, H. Lehto, K. Lehto, L. Le Roy, S. Merouane, P. Modica, F.-R. Orthous-Daunay, J. Paquette, F. Raulin, J. Rynö, R. Schulz, J. Silén, S. Siljeström, W. Steiger, O. Stenzel, T. Stephan, L. Thirkell, R. Thomas, K. Torkar, K. Varmuza, K.-P. Wanczek, B. Zaprudin, J. Kissel, M. Hilchenbach, High-molecular-weight organic matter in the particles of comet 67P/Churyumov-Gerasimenko. Nature **538**, 72–74 (2016)

N. Fray, A. Bardyn, H. Cottin, D. Baklouti, C. Briois, C. Engrand, H. Fischer, K. Hornung, R. Isnard, Y. Langevin, H. Lehto, L. Le Roy, E.M. Mellado, S. Merouane, P. Modica, F.-R. Orthous-Daunay, J. Paquette, J. Rynö, R. Schulz, J. Silén, S. Siljeström, O. Stenzel, L. Thirkell, K. Varmuza, B. Zaprudin, J. Kissel, M. Hilchenbach, Nitrogen-to-carbon atomic ratio measured by COSIMA in the particles of comet 67P/Churyumov-Gerasimenko. Mon. Not. R. Astron. Soc. **469**, S506–S516 (2017)

M. Fulle, Injection of large grains into orbits around comet nuclei. Astron. Astrophys. **325**, 1237–1248 (1997)

M. Fulle, Motion of cometary dust, in *Comets II*, ed. by M.C. Festou, H.U. Keller, H.A. Weaver (University of Arizona Press, Tucson, 2004), pp. 565–575

M. Fulle, The ice content of Kuiper belt objects. Nat. Astron. **1**, 0018 (2017)

M. Fulle, J. Blum, Fractal dust constrains the collisional history of comets. Mon. Not. R. Astron. Soc. **469**, S39–S44 (2017)

M. Fulle, L. Colangeli, V. Mennella, A. Rotundi, E. Bussoletti, The sensitivity of the size distribution to the grain dynamics: simulation of the dust flux measured by GIOTTO at P/Halley. Astron. Astrophys. **304**, 622 (1995)

M. Fulle, A.C. Levasseur-Regourd, N. McBride, E. Hadamcik, In situ dust measurements from within the coma of 1P/Halley: first order approximation with a dust dynamical model. Astron. J. **119**, 1968–1977 (2000)

M. Fulle, C. Barbieri, G. Cremonese, H. Rauer, M. Weiler, G. Milani, R. Ligustri, The dust environment of comet 67P/Churyumov-Gerasimenko. Astron. Astrophys. **422**, 357–368 (2004)

M. Fulle, L. Colangeli, J. Agarwal, A. Aronica, V. Della Corte, F. Esposito, E. Grün, M. Ishiguro, R. Ligustri, J.J. Lopez Moreno, E. Mazzotta Epifani, G. Milani, F. Moreno, P. Palumbo, J. Rodríguez Gómez, A. Rotundi, Comet 67P/Churyumov-Gerasimenko: the GIADA dust environment model of the Rosetta mission target. Astron. Astrophys. **522**, A63 (2010)

M. Fulle, V. Della Corte, A. Rotundi, P. Weissman, A. Juhasz, K. Szego, R. Sordini, M. Ferrari, S. Ivanovski, F. Lucarelli, M. Accolla, S. Merouane, V. Zakharov, E. Mazzotta Epifani, J.J. López-Moreno, J. Rodríguez, L. Colangeli, P. Palumbo, E. Grün, M. Hilchenbach, E. Bussoletti, F. Esposito, S.F. Green, P.L. Lamy, J.A.M. McDonnell, V. Mennella, A. Molina, R. Morales, F. Moreno, J.L. Ortiz, E. Palomba, R. Rodrigo, J.C. Zarnecki, M. Cosi, F. Giovane, B. Gustafson, M.L. Herranz, J.M. Jerónimo, M.R. Leese, A.C. López-Jiménez, N. Altobelli, Density and charge of pristine fluffy particles from comet 67P/Churyumov-Gerasimenko. Astrophys. J. Lett. **802**, L12 (2015a)

M. Fulle, S. Ivanovski, I. Bertini, P. Gutierrez, L.M. Lara, H. Sierks, V. Zakharov, V. Della Corte, A. Rotundi, C. Barbieri, P.L. Lamy, R. Rodrigo, D. Koschny, H. Rickman, H.U. Keller, J. Agarwal, M.F. A'Hearn, M.A. Barucci, J.-L. Bertaux, D. Bodewits, G. Cremonese, V. Da Deppo, B. Davidsson, S. Debei, M. De Cecco, S. Fornasier, O. Groussin, C. Güttler, S.F. Hviid, W.-H. Ip, L. Jorda, J. Knollenberg, R. Kramm, E. Kührt, M. Küppers, M. Lazzarin, J.J. Lopez-Moreno, F. Marzari, H. Michalik, G. Naletto, N. Oklay, L. Sabau, N. Thomas, C. Tubiana, J.-B. Vincent, K.-P. Wenzel, Rotating dust particles in the coma of comet 67P/Churyumov-Gerasimenko. Astron. Astrophys. **583**, A14 (2015b)

M. Fulle, N. Altobelli, B. Buratti, M. Choukroun, M. Fulchignoni, E. Grün, M.G.G.T. Taylor, P. Weissman, Unexpected and significant findings in comet 67P/Churyumov-Gerasimenko: an interdisciplinary view. Mon. Not. R. Astron. Soc. **462**, S2–S8 (2016a)

M. Fulle, V. Della Corte, A. Rotundi, F.J.M. Rietmeijer, S.F. Green, P. Weissman, M. Accolla, L. Colangeli, M. Ferrari, S. Ivanovski, J.J. Lopez-Moreno, E. Mazzotta Epifani, R. Morales, J.L. Ortiz, E. Palomba, P. Palumbo, J. Rodriguez, R. Sordini, V. Zakharov, Comet 67P/Churyumov-Gerasimenko preserved the pebbles that formed planetesimals. Mon. Not. R. Astron. Soc. **462**, S132–S137 (2016b)

M. Fulle, F. Marzari, V. Della Corte, S. Fornasier, H. Sierks, A. Rotundi, C. Barbieri, P.L. Lamy, R. Rodrigo, D. Koschny, H. Rickman, H.U. Keller, J.J. López-Moreno, M. Accolla, J. Agarwal, M.F. A'Hearn, N. Altobelli, M.A. Barucci, J.-L. Bertaux, I. Bertini, D. Bodewits, E. Bussoletti, L. Colangeli, M. Cosi, G. Cremonese, J.F. Crifo, V. Da Deppo, B. Davidsson, S. Debei, M. De Cecco, F. Esposito, M. Ferrari, F. Giovane, B. Gustafson, S.F. Green, O. Groussin, E. Grün, P. Gutierrez, C. Güttler, M.L. Herranz, S.F. Hviid, W.-H. Ip, S. Ivanovski, J.M. Jerónimo, L. Jorda, J. Knollenberg, R. Kramm, E. Kührt, M. Küppers, L.M. Lara, M. Lazzarin, M.R. Leese, A.C. López-Jiménez, F. Lucarelli, E. Mazzotta Epifani, J.A.M. McDonnell, V. Mennella, A. Molina, R. Morales, F. Moreno, S. Mottola, G. Naletto, N. Oklay, J.L. Ortiz, E. Palomba, P. Palumbo, J.-M. Perrin, F.J.M. Rietmeijer, J. Rodríguez, R. Sordini, N. Thomas, C. Tubiana, J.-B. Vincent, P. Weissman, K.-P. Wenzel, V. Zakharov, J.C. Zarnecki, Evolution of the dust size distribution of comet 67P/Churyumov-Gerasimenko from 2.2 au to perihelion. Astrophys. J. **821**, 19 (2016c)

M. Fulle, V. Della Corte, A. Rotundi, S.F. Green, M. Accolla, L. Colangeli, M. Ferrari, S. Ivanovski, R. Sordini, V. Zakharov, The dust-to-ices ratio in comets and Kuiper belt objects. Mon. Not. R. Astron. Soc. **469**, S45–S49 (2017)

M. Fulle, I. Bertini, V. Della Corte, C. Güttler, S. Ivanovski, F. La Forgia, J. Lasue, A.C. Levasseur-Regourd, F. Marzari, F. Moreno, S. Mottola, G. Naletto, A. Rotundi, C. Barbieri, P.L. Lamy, R. Rodrigo, D. Koschny, H. Rickman, M.A. Barucci, J.-L. Bertaux, D. Bodewits, G. Cremonese, V. Da Deppo, B. Davidsson, S. Debei, M. De Cecco, J. Deller, S. Fornasier, O. Groussin, P.J. Gutiérrez, S.H. Hviid, W.-H. Ip, L. Jorda, H.U. Keller, J. Knollenberg, J.R. Kramm, E. Kührt, M. Küppers, L.M. Lara, M. Lazzarin, J.J. López-Moreno, X. Shi, N. Thomas, C. Tubiana, The phase function and density of the dust observed at comet 67P/Churyumov-Gerasimenko. Mon. Not. R. Astron. Soc. **476**, 2835–2839 (2018)

R. Furusho, Y. Ikeda, D. Kinoshita, W.-H. Ip, H. Kawakita, T. Kasuga, Y. Sato, H.-C. Lin, M.-S. Chang, Z.-Y. Lin, J. Watanabe, Imaging polarimetry of comet 9P/Tempel before and after the deep impact. Icarus **190**, 454–458 (2007)

Z. Gainsforth, A.L. Butterworth, J. Stodolna, A.J. Westphal, G.R. Huss, K. Nagashima, R. Ogliore, D.E. Brownlee, D. Joswiak, T. Tyliszczak, A.S. Simionovici, Constraints on the formation environment of two chondrule-like igneous particles from comet 81P/Wild 2. Meteorit. Planet. Sci. **50**, 976–1004 (2015)

Z. Gainsforth, A.L. Butterworth, C.E. Jilly-Rehak, A.J. Westphal, D.E. Brownlee, D. Joswiak, R.C. Ogliore, M.E. Zolensky, H.A. Bechtel, D.S. Ebel, G.R. Huss, S.A. Sandford, A.J. White, Possible gems and ultra-fine grained polyphase units in comet Wild 2. Lunar Planet. Sci. **47**, 2366 (2016)

M.J. Genge, C. Engrand, M. Gounelle, S. Taylor, The classification of micrometeorites. Meteorit. Planet. Sci. **43**(3), 497–515 (2008)

K.-H. Glassmeier, H. Boehnhardt, D. Koschny, E. Kührt, I. Richter, The Rosetta mission: flying towards the origin of the solar system. Space Sci. Rev. **128**, 1–21 (2007)

F. Goesmann, H. Rosenbauer, J.H. Bredehöft, M. Cabane, P. Ehrenfreund, T. Gautier, C. Giri, H. Krüger, L. Le Roy, A.J. MacDermott, S. McKenna-Lawlor, U.J. Meierhenrich, G.M. Muñoz Caro, F. Raulin, R. Roll, A. Steele, H. Steininger, R. Sternberg, C. Szopa, W. Thiemann, S. Ulamec, Organic compounds on comet 67P/Churyumov-Gerasimenko revealed by COSAC mass spectrometry. Science **349**, 6247 (2015)

T.I. Gombosi, M. Horanyi, Time-dependent numerical modeling of dust halo formation at comets. Astrophys. J. **311**, 491–500 (1986)

J.M. Greenberg, J.I. Hage, From interstellar dust to comets-a unification of observational constraints. Astrophys. J. **361**, 260–274 (1990)

E. Grün, H.A. Zook, H. Fechtig, R.H. Giese, Collisional balance of the meteoritic complex. Icarus **62**, 244–272 (1985)

E. Grün, J. Agarwal, N. Altobelli, K. Altwegg, M.S. Bentley, N. Biver, V. Della Corte, N. Edberg, P.D. Feldman, M. Galand, B. Geiger, C. Götz, B. Grieger, C. Güttler, P. Henri, M. Hofstadter, M. Horanyi, E. Jehin, H. Krüger, S. Lee, T. Mannel, E. Morales, O. Mousis, M. Müller, C. Opitom, A. Rotundi, R. Schmied, F. Schmidt, H. Sierks, C. Snodgrass, R.H. Soja, M. Sommer, R. Srama, C.-Y. Tzou, J.-B. Vincent, P. Yanamandra-Fisher, M.F. A'Hearn, A.I. Erikson, C. Barbieri, M.A. Barucci, J.-L. Bertaux, I. Bertini, J. Burch, L. Colangeli, G. Cremonese, V. Da Deppo, B. Davidsson, S. Debei, M. De Cecco, J. Deller, L.M. Feaga, M. Ferrari, S. Fornasier, M. Fulle, A. Gicquel, M. Gillon, S.F. Green,

O. Groussin, P.J. Gutiérrez, M. Hofmann, S.F. Hviid, W.-H. Ip, S. Ivanovski, L. Jorda, H.U. Keller, M. Knight, J. Knollenberg, D. Koschny, J.-R. Kramm, E. Kührt, M. Küppers, P.L. Lamy, L.M. Lara, M. Lazzarin, J.J. Lòpez-Moreno, J. Manfroid, E. Mazzotta Epifani, F. Marzari, G. Naletto, N. Oklay, P. Palumbo, J.Wm. Parker, H. Rickman, R. Rodrigo, J. Rodrìguez, E. Schindhelm, X. Shi, R. Sordini, A.J. Steffl, S.A. Stern, N. Thomas, C. Tubiana, H.A. Weaver, P. Weissman, V. Zakharov, M.G.G.T. Taylor, Feb. 19 outburst of comet 67P/CG: an ESA Rosetta multi-instrument study, Mon. Not. R. Astron. Soc. **462**, S220–S234 (2016)

B. Gundlach, J. Blum, H.U. Keller, V.Y. Skorov, What drives the dust activity of comet 67P/Churyumov-Gerasimenko? Astron. Astrophys. **583**, A12 (2015)

C. Güttler, M. Krause, R.J. Geretshauser, R. Speith, J. Blum, The physics of protoplanetesimal dust agglomerates. IV. Toward a dynamical collision model. Astrophys. J. **701**, 130–141 (2009)

E. Hadamcik, A.C. Levasseur-Regourd, Dust evolution of comet C/1995 O1 (Hale-Bopp) by imaging polarimetric observations. Astron. Astrophys. **403**, 757–768 (2003)

E. Hadamcik, A.C. Levassuer-Regourd, J.B. Renard, Ccd polarimetric imaging of comet Hale-Bopp (C/1995 O1). Earth Moon Planets **78**, 365–371 (1997)

E. Hadamcik, J.-B. Renard, A.C. Levasseur-Regourd, J. Lasue, Light scattering by fluffy particles with the PROGRA2 experiment: mixtures of materials. J. Quant. Spectrosc. Radiat. Transf. **100**, 143–156 (2006)

E. Hadamcik, J.-B. Renard, F.J.M. Rietmeijer, A.C. Levasseur-Regourd, H.G.M. Hill, J.M. Karner, J.A. Nuth, Light scattering by fluffy Mg–Fe–SiO and C mixtures as cometary analogs (PROGRA2 experiment). Icarus **190**, 660–671 (2007)

E. Hadamcik, J.-B. Renard, A.C. Levasseur-Regourd, J.-C. Worms, Laboratory measurements of the light scattered by clouds of solid particles by imaging technique, in *Light Scattering Reviews*, vol. 4 (Springer, Berlin, 2009a), pp. 31–70

E. Hadamcik, J.-B. Renard, A.C. Levasseur-Regourd, J. Lasue, G. Alcouffe, M. Francis, Light scattering by agglomerates: interconnecting size and absorption effects (PROGRA2 experiment). J. Quant. Spectrosc. Radiat. Transf. **110**, 1755–1770 (2009b)

E. Hadamcik, A.K. Sen, A.C. Levasseur-Regourd, R. Gupta, J. Lasue, Polarimetric observations of comet 67P/Churyumov-Gerasimenko during its 2008–2009 apparition. Astron. Astrophys. **517**, A86 (2010)

E. Hadamcik, A.K. Sen, A.C. Levasseur-Regourd, R. Gupta, J. Lasue, R. Botet, Dust in comet 103P/Hartley 2 coma during EPOXI mission. Icarus **222**, 774–785 (2013)

E. Hadamcik, A.C. Levasseur-Regourd, D.C. Hines, A.K. Sen, J. Lasue, J.-B. Renard, Properties of dust particles in comets from photometric and polarimetric observations of 67P. Mon. Not. R. Astron. Soc. **462**, S507–S515 (2016)

M.S. Hanner, R.L. Newburn, Infrared photometry of comet Wilson (1986l) at two epochs. Astron. J. **97**, 254–261 (1989)

J.K. Harmon, M.C. Nolan, S.J. Ostro, D.B. Campbell, Radar studies of comet nuclei and grain comae, in *Comets II*, ed. by M.C. Festou, H.U. Keller, H.A. Weaver (University of Arizona Press, Tucson, 2004), pp. 265–279

C.M. Hartzell, X. Wang, D.J. Scheeres, M. Horányi, Experimental demonstration of the role of cohesion in electrostatic dust lofting. Geophys. Res. Lett. **40**, 1038–1042 (2013)

P.R. Heck, P. Hoppe, J. Huth, Sulfur four isotope NanoSIMS analysis of comet-81P/Wild 2 dust in impact craters on aluminum foil C2037N from NASA's Stardust mission. Meteorit. Planet. Sci. **47**, 649–659 (2012)

P.R. Heck, F.J. Stadermann, D. Isheim, O. Auciello, T.L. Daulton, A.M. Davis, J.W. Elam, C. Floss, J. Hiller, D.J. Larson, J.B. Lewis, A. Mane, M. Pellin, M.R. Savina, D.N. Seidman, T. Stephan, Atom-probe analyses of nanodiamonds from Allende. Meteorit. Planet. Sci. **49**, 453–467 (2014)

A. Herique, W. Kofman, P. Beck, L. Bonal, I. Buttarazzi, E. Heggy, J. Lasue, A.C. Levasseur-Regourd, E. Quirico, S. Zine, Cosmochemical implications of CONSERT permittivity characterization of 67P/CG. Mon. Not. R. Astron. Soc. **462**, S516–S532 (2016)

M. Hilchenbach, J. Kissel, Y. Langevin, C. Briois, H. von Hoerner, A. Koch, R. Schulz, J. Silén, K. Altwegg, L. Colangeli, H. Cottin, C. Engrand, H. Fischer, A. Glasmachers, E. Grün, G. Haerendel, H. Henkel, H. Höfner, K. Hornung, E.K. Jessberger, H. Lehto, K. Lehto, F. Raulin, L. Le Roy, J. Rynö, W. Steiger, T. Stephan, L. Thirkell, R. Thomas, K. Torkar, K. Varmuza, K.-P. Wanczek, N. Altobelli, D. Baklouti, A. Bardyn, N. Fray, H. Krüger, N. Ligier, Z. Lin, P. Martin, S. Merouane, F.R. Orthous-Daunay, J. Paquette, C. Revillet, S. Siljeström, O. Stenzel, B. Zaprudin, Comet 67P/Churyumov-Gerasimenko: close-up on dust particle fragments. Astrophys. J. Lett. **816**, L32 (2016)

M. Hilchenbach, H. Fischer, Y. Langevin, S. Merouane, J. Paquette, J. Rynö, O. Stenzel, C. Briois, J. Kissel, A. Koch, R. Schulz, J. Silen, N. Altobelli, D. Baklouti, A. Bardyn, H. Cottin, C. Engrand, N. Fray, G. Haerendel, H. Henkel, H. Höfner, K. Hornung, H. Lehto, E.M. Mellado, P. Modica, L. Le Roy, S. Siljeström, W. Steiger, L. Thirkell, R. Thomas, K. Torkar, K. Varmuza, B. Zaprudin, Mechanical

and electrostatic experiments with dust particles collected in the inner coma of comet 67P by COSIMA onboard Rosetta. Philos. Trans. R. Soc. A **375, id**, 20160255 (2017)
J.R. Hill, D.A. Mendis, Charged dust in the outer planetary magnetospheres. III—Satellite impact geometries. Earth Moon Planets **25**, 427–436 (1981)
D.C. Hines, A.C. Levasseur-Regourd, Polarimetry observations of comets: status, questions, future pathways. Planet. Space Sci. **123**, 41–50 (2016)
K. Hornung, S. Merouane, M. Hilchenbach, Y. Langevin, E.M. Mellado, V. Della Corte, J. Kissel, C. Engrand, R. Schulz, J. Ryno, J. Silen (COSIMA Team), A first assessment of the strength of cometary particles collected in-situ by the COSIMA instrument onboard ROSETTA. Planet. Space Sci. **133**, 63–75 (2016)
F. Hörz, R. Bastien, J. Borg, J.P. Bradley, J.C. Bridges, D.E. Brownlee, M.J. Burchell, M. Chi, M.J. Cintala, Z.R. Dai, Z. Djouadi, G. Dominguez, T.E. Economou, S.A.J. Fairey, C. Floss, I.A. Franchi, G.A. Graham, S.F. Green, P. Heck, P. Hoppe, J. Huth, H. Ishii, A.T. Kearsley, J. Kissel, J. Leitner, H. Leroux, K. Marhas, K. Messenger, C.S. Schwandt, T.H. See, C. Snead, F.J. Stadermann, T. Stephan, R. Stroud, N. Teslich, J.M. Trigo-Rodríguez, A.J. Tuzzolino, D. Troadec, P. Tsou, J. Warren, A.J. Westphal, P. Wozniakiewicz, I. Wright, E. Zinner, Impact features on Stardust: implications for comet 81P/Wild 2 dust. Science **314**, 1716 (2006)
H.L.F. Houpis, D.A. Mendis, W.-H. Ip, The chemical differentiation of the cometary nucleus—the process and its consequences. Astrophys. J. **295**, 654–667 (1985)
J.W. Hovenier, O. Muñoz, Light scattering in the Solar System: an introductory review. J. Quant. Spectrosc. Radiat. Transf. **110**, 1280–1292 (2009)
D.W. Hughes, Cometary outbursts—a review. Q. J. R. Astron. Soc. **31**, 69–94 (1990)
M. Humayun, J.I. Goldstein, A. Mubarok, A.J. Westphal, Z. Gainsforth, S.R. Sutton, B. Lai, E. Silver, R.M. Stroud, Preliminary analysis of Simeio: a low Ni–Ir kamacite grain of unusual origin from comet Wild2, in *Lunar and Planetary Science Conference* (2015), p. 2298
S.I. Ipatov, M.F. A'Hearn, The outburst triggered by the deep impact collision with comet Tempel 1. Mon. Not. R. Astron. Soc. **414**, 76–107 (2011)
M. Ishiguro, Cometary dust trail associated with Rosetta mission target: 67P/Churyumov Gerasimenko. Icarus **193**, 96–104 (2008)
M. Ishiguro, J. Watanabe, F. Usui, T. Tanigawa, D. Kinoshita, J. Suzuki, R. Nakamura, M. Ueno, T. Mukai, First detection of an optical dust trail along the orbit of 22P/Kopff. Astrophys. J. **572**, L117–L120 (2002)
M. Ishiguro, S.M. Kwon, Y. Sarugaku, S. Hasegawa, F. Usui, S. Nishiura, Y. Nakada, H. Yano, Discovery of the dust trail of the Stardust comet sample return mission target: 81P/Wild 2. Astrophys. J. Lett. **589**, L101–L104 (2003)
M. Ishiguro, Y. Sarugaku, S. Nishihara, Y. Nakada, S. Nishiura, T. Soyano, K. Tarusawa, T. Mukai, S.M. Kwon, S. Hasegawa, F. Usui, M. Ueno, Report on the Kiso cometary dust trail survey. Adv. Space Res. **43**, 875–879 (2009)
H.A. Ishii, J.P. Bradley, Z.R. Dai, M. Chi, A.T. Kearsley, M.J. Burchell, N.D. Browning, F. Molster, Comparison of comet 81P/Wild 2 dust with interplanetary dust from comets. Science **319**, 447 (2008a)
H.A. Ishii, S. Brennan, J.P. Bradley, K. Luening, K. Ignatyev, P. Pianetta, Recovering the elemental composition of comet Wild 2 dust in five Stardust impact tracks and terminal particles in aerogel. Meteorit. Planet. Sci. **43**, 215–231 (2008b)
P.M.M. Jenniskens, *Meteor Showers and Their Parent Comets* (Cambridge University Press, Cambridge, 2006)
E.K. Jessberger, Halley's grains, interplanetary and interstellar dust. Meteoritics **24**, 281 (1989)
E.K. Jessberger, A. Christoforidis, J. Kissel, Aspects of the major element composition of Halley's dust. Nature **332**, 691–695 (1988)
J. Jones, T.R. Kaiser, The effects of thermal radiation, conduction and meteoroid heat capacity on meteoric ablation. Mon. Not. R. Astron. Soc. **133**, 411–420 (1966)
L. Jorda, R. Gaskell, C. Capanna, S.F. Hviid, P.L. Lamy, J. Durech, G. Faury, O. Groussin, P. Gutiérrez, C. Jackman, S.J. Keihm, H.U. Keller, J. Knollenberg, E. Kührt, S. Marchi, S. Mottola, E. Palmer, F.P. Schloerb, H. Sierks, J.-B. Vincent, M.F. A'Hearn, C. Barbieri, R. Rodrigo, D. Koschny, H. Rickman, M.A. Barucci, J.-L. Bertaux, I. Bertini, G. Cremonese, V. Da Deppo, B. Davidsson, S. Debei, M. De Cecco, S. Fornasier, M. Fulle, C. Güttler, W.-H. Ip, J.R. Kramm, M. Küppers, L.M. Lara, M. Lazzarin, J.J. Lopez Moreno, F. Marzari, G. Naletto, N. Oklay, N. Thomas, C. Tubiana, K.-P. Wenzel, The global shape, density and rotation of comet 67P/Churyumov-Gerasimenko from preperihelion Rosetta/OSIRIS observations. Icarus **277**, 257–278 (2016)
B. Jost, A. Pommerol, O. Poch, Z. Yoldi, S. Fornasier, P.H. Hasselmann, C. Feller, N. Carrasco, C. Szopa, N. Thomas, Bidirectional reflectance and VIS-NIR spectroscopy of cometary analogues under simulated space conditions. Planet. Space Sci. **145**, 14–27 (2017)
D.J. Joswiak, D.E. Brownlee, A.N. Nguyen, S. Messenger, Refractory materials in comet samples. Meteorit. Planet. Sci. **52**, 1612–1648 (2017)

Y. Kakazu, C. Engrand, J. Duprat, G. Briani, N. Bardin, S. Mostefaoui, R. Duhamel, L. Remusat, Bulk oxygen isotopic composition of ultracarbonaceous Antarctic micrometeorites with the NanoSIMS. Meteorit. Planet. Sci. **49**(5286), A195 (2014)

L.P. Keller, E.L. Berger, A transmission electron microscope study of Itokawa regolith grains. Earth Planets Space **66**, 71 (2014)

H.U. Keller, J. Knollenberg, W.J. Markiewicz, Collimation of cometary dust jets and filaments. Planet. Space Sci. **42**, 367–382 (1994)

L.P. Keller, S. Bajt, G.A. Baratta, J. Borg, J.P. Bradley, D.E. Brownlee, H. Busemann, J.R. Brucato, M. Burchell, L. Colangeli, L. D'Hendecourt, Z. Djouadi, G. Ferrini, G. Flynn, I.A. Franchi, M. Fries, M.M. Grady, G.A. Graham, F. Grossemy, A. Kearsley, G. Matrajt, K. Nakamura-Messenger, V. Mennella, L. Nittler, M.E. Palumbo, F.J. Stadermann, P. Tsou, A. Rotundi, S.A. Sandford, C. Snead, A. Steele, D. Wooden, M. Zolensky, Infrared spectroscopy of comet 81P/Wild 2 samples returned by Stardust. Science **314**, 1728 (2006)

H.U. Keller, C. Barbieri, P.L. Lamy, H. Rickman, R. Rodrigo, K.-P. Wenzel, H. Sierks, M.F. A'Hearn, F. Angrilli, M. Angulo, M.E. Bailey, P. Barthol, M.A. Barucci, J.-L. Bertaux, G. Bianchini, J.-L. Boit, V. Brown, J.A. Burns, I. Büttner, J.M. Castro, G. Cremonese, W. Curdt, V. da Deppo, S. Debei, M. de Cecco, K. Dohlen, S. Fornasier, M. Fulle, D. Germerott, F. Gliem, G.P. Guizzo, S.F. Hviid, W.-H. Ip, L. Jorda, D. Koschny, J.R. Kramm, E. Kührt, M. Küppers, L.M. Lara, A. Llebaria, A. López, A. López-Jimenez, J.J. López-Moreno, R. Meller, H. Michalik, M.D. Michelena, R. Müller, G. Naletto, A. Origné, G. Parzianello, M. Pertile, C. Quintana, R. Ragazzoni, P. Ramous, K.-U. Reiche, M. Reina, J. Rodríguez, G. Rousset, L. Sabau, A. Sanz, J.-P. Sivan, K. Stöckner, J. Tabero, U. Telljohann, N. Thomas, V. Timon, G. Tomasch, T. Wittrock, M. Zaccariotto, OSIRIS the scientific camera system onboard Rosetta. Space Sci. Rev. **128**, 433–506 (2007)

M.S. Kelley, W.T. Reach, D.J. Lien, The dust trail of comet 67P/Churyumov Gerasimenko. Icarus **193**, 572–587 (2008)

N. Kiselev, K. Jockers, T. Bonev, CCD imaging polarimetry of comet 2P/Encke. Icarus **168**, 385–391 (2004)

N. Kiselev, V. Rosenbush, A.C. Levasseur-Regourd, L. Kolokolova, Comets, in L. Kolokolova, J. Hough, A.C. Levasseur-Regourd *Polarimetry of Stars and Planetary Systems* (Cambridge University Press, Cambridge, 2015), p. 379

J. Kissel, F.R. Krueger, Organic dust in comet Halley. Nature **328**, 117 (1987)

J. Kissel, R.Z. Sagdeev, J.-L. Bertaux, V.N. Angarov, J. Audouze, J.E. Blamont, K. Buchler, E.N. Evlanov, H. Fechtig, M.N. Fomenkova, H. von Hoerner, N.A. Inogamov, V.N. Khromov, W. Knabe, F.R. Krueger, Y. Langevin, B. Leonasv, A.C. Levasseur-Regourd, G.G. Managadze, S.N. Podkolzin, V.D. Shapiro, S.R. Tabaldyev, B.V. Zubkov, Composition of comet Halley dust particles from VEGA observations. Nature **321**, 280–282 (1986)

J. Kissel, F.R. Krueger, J. Silén, B.C. Clark, The cometary and interstellar dust analyzer at comet 81P/Wild 2. Science **304**, 1774–1776 (2004)

J. Kissel, K. Altwegg, B.C. Clark, L. Colangeli, H. Cottin, S. Czempiel, J. Eibl, C. Engrand, H.M. Fehringer, B. Feuerbacher, M. Fomenkova, A. Glasmachers, J.M. Greenberg, E. Grün, G. Haerendel, H. Henkel, M. Hilchenbach, H. von Hoerner, H. Höfner, K. Hornung, E.K. Jessberger, A. Koch, H. Krüger, Y. Langevin, P. Parigger, F. Raulin, F. Rüdenauer, J. Rynö, E.R. Schmid, R. Schulz, J. Silén, W. Steiger, T. Stephan, L. Thirkell, R. Thomas, K. Torkar, N.G. Utterback, K. Varmuza, K.P. Wanczek, W. Werther, H. Zscheeg, Cosima high resolution time-of-flight secondary ion mass spectrometer for the analysis of cometary dust particles onboard Rosetta. Space Sci. Rev. **128**, 823–867 (2007)

N.T. Kita, Q.-Z. Yin, G.J. MacPherson, T. Ushikubo, B. Jacobsen, K. Nagashima, E. Kurahashi, A.N. Krot, S.B. Jacobsen, ^{26}Al–^{26}Mg isotope systematics of the first solids in the early solar system. Meteorit. Planet. Sci. **48**, 1383–1400 (2013)

J. Knollenberg, Z.Y. Lin, S.F. Hviid, N. Oklay, J.-B. Vincent, D. Bodewits, S. Mottola, M. Pajola, H. Sierks, C. Barbieri, P.L. Lamy, R. Rodrigo, D. Koschny, H. Rickman, M.F. A'Hearn, M.A. Barucci, J. L. Bertaux, I. Bertini, G. Cremonese, B. Davidsson, V. Da Deppo, S. Debei, M. De Cecco, S. Fornasier, M. Fulle, O. Groussin, P.J. Gutiérrez, W.-H. Ip, L. Jorda, H.U. Keller, E. Kührt, J.R. Kramm, M. Küppers, L.M. Lara, M. Lazzarin, J.J. Lopez Moreno, F. Marzari, G. Naletto, N. Thomas, C. Güttler, F. Preusker, F. Scholten, C. Tubiana, A mini outburst from the nightside of comet 67P/Churyumov-Gerasimenko observed by the OSIRIS camera on Rosetta. Astron. Astrophys. **596**, A89 (2016)

M.D. Knudson, R.W. Lemke, Shock response of low-density silica aerogel in the multi Mbar regime. J. Appl Phys. **114**, 053510 (2013)

H.W. Kochan, W.F. Huebner, D.W.G. Sears, Simulation experiments with cometary analogous material, in *Laboratory Astrophysics and Space Research*, ed. by P. Ehrenfreund, C. Krafft, H. Kochan, V. Pirronello (Springer, Berlin, 1999), pp. 623–665

W. Kofman, A. Herique, J.-P. Goutail, T. Hagfors, I.P. Williams, E. Nielsen, J.-P. Barriot, Y. Barbin, C. Elachi, P. Edenhofer, A.C. Levasseur-Regourd, D. Plettemeier, G. Picardi, R. Seu, V. Svedhem, The Comet

Nucleus Sounding Experiment by Radiowave Transmission (CONSERT): a short description of the instrument and of the commissioning stages. Space Sci. Rev. **128**, 413–432 (2007)

W. Kofman, A. Herique, Y. Barbin, J.P. Barriot, V. Ciarletti, S. Clifford, P. Edenhoffer, C. Elachi, C. Eyraud, J.P. Goutail, E. Heggy, L. Jorda, J. Lasue, A.C. Levasseur-Regourd, E. Nielsen, P. Pasquero, F. Preusker, P. Puget, D. Plettemeier, Y. Rogez, H. Sierks, C. Statz, H. Svedhem, I. Williams, S. Zine, J. Van Zyl, Properties of 67P/ Churyumov-Gerasimenko interior revealed by CONSERT radar. Science **349**, 6247 aab0639 (2015)

L. Kolokolova, H. Kimura, Comet dust as a mixture of aggregates and solid particles: model consistent with ground-based and space-mission results. Earth Planets Space **62**, 17–21 (2010)

L. Kolokolova, M.S. Hanner, A.C. Levasseur-Regourd, B.A.S. Gustafson, Physical properties of cometary dust from light scattering and thermal emission, in *Comets II*, ed. by M.C. Festou, H.U. Keller, H.A. Weaver (University of Arizona Press, Tucson, 2004), pp. 577–604

L. Kolokolova, H. Kimura, N. Kiselev, V. Rosenbush, Two different evolutionary types of comets proved by polarimetric and infrared properties of their dust. Astron. Astrophys. **463**, 1189–1196 (2007)

L. Kolokolova, L. Nagdimunov, D. Mackowski, Light scattering by hierarchical aggregates. J. Spectrosc. Radiat. Transf. **204**, 138–143 (2017)

L. Kööp, A.M. Davis, D. Nakashima, C. Park, A.N. Krot, K. Nagashima, T.J. Tenner, P.R. Heck, N.T. Kita, A link between oxygen, calcium and titanium isotopes in ^{26}Al-poor hibonite-rich CAIs from Murchison and implications for the heterogeneity of dust reservoirs in the solar nebula. Geochim. Cosmochim. Acta **189**, 70–95 (2016)

D. Koschny et al., IP dust, meteors and meteorites. Space Sci. Rev. (2017, to be submitted for the special issue on Cosmic Dust)

A. Kouchi, S. Sirono, Crystallization heat of impure amorphous H_2O ice. Geophys. Res. Lett. **28**, 827–830 (2001)

L. Kresak, Cometary dust trails and meteor storms. Astron. Astrophys. **279**, 646–660 (1993)

F.R. Krueger, A. Korth, J. Kissel, The organic matter of comet Halley as inferred by joint gas phase and solid phase analyses. Space Sci. Rev. **56**, 167–175 (1991)

E. Kührt, H.U. Keller, On the importance of dust in cometary nuclei. Earth Moon Planets **72**, 79–89 (1996)

G. Kurat, C. Koeberl, T. Presper, F. Brandstätter, M. Maurette, Petrology and geochemistry of Antarctic micrometeorites. Geochim. Cosmochim. Acta **58**, 3879–3904 (1994)

Y. Langevin, M. Hilchenbach, N. Ligier, S. Merouane, K. Hornung, C. Engrand, R. Schulz, J. Kissel, J. Rynö, P. Eng, Typology of dust particles collected by the COSIMA mass spectrometer in the inner coma of 67P/Churyumov Gerasimenko. Icarus **271**, 76–97 (2016)

Y. Langevin, M. Hilchenbach, M. Vincendon, S. Merouane, K. Hornung, N. Ligier, C. Engrand, R. Schulz, J. Kissel, J. Rynö, Optical properties of cometary particles collected by the COSIMA mass spectrometer on-board Rosetta during the rendezvous phase around comet 67P/Churyumov-Gerasimenko. Mon. Not. R. Astron. Soc. **469**, S535–S549 (2017)

A. Lanzirotti, S.R. Sutton, G.J. Flynn, M. Newville, W. Rao, Chemical composition and heterogeneity of Wild 2 cometary particles determined by synchrotron X-ray fluorescence. Meteorit. Planet. Sci. **43**, 187–213 (2008)

J. Lasue, A.C. Levasseur-Regourd, Porous irregular aggregates of sub-micron sized grains to reproduce cometary dust light scattering observations. J. Quant. Spectrosc. Radiat. Transf. **100**, 220–236 (2006)

J. Lasue, A.C. Levasseur-Regourd, N. Fray, H. Cottin, Inferring the interplanetary dust properties-from remote observations and simulations. Astron. Astrophys. **473**, 641–649 (2007)

J. Lasue, A.C. Levasseur-Regourd, E. Hadamcik, G. Alcouffe, Cometary dust properties retrieved from polarization observations: application to C/1995 O1 Hale Bopp and 1P/Halley. Icarus **199**, 129–144 (2009)

J. Lasue, A.C. Levasseur-Regourd, A. Lazarian, Interplanetary dust, in *Polarimetry of Stars and Planetary Systems*, ed. by L. Kolokolova, J. Hough, A.C. Levasseur-Regourd (2015), pp. 419–436

M.E. Lawler, D.E. Brownlee, CHON as a component of dust from comet Halley. Nature **359**, 810–812 (1992)

T. Le Duin, J.F. Crifo, D. Le Queau, F. Crifo, A quantitative interpretation of the in-situ observations of the dust coma of comet P/Grigg-Skjellerup by the OPE photopolarimeter. Astron. Astrophys. **308**, 261–272 (1996)

J. Leitner, P. Hoppe, P.R. Heck, First discovery of presolar material of possible supernova origin in impact residues from comet 81P/Wild 2, in *Lunar and Planetary Science Conference 1607* (2010)

J. Leitner, J. Kodolányi, P. Hoppe, C. Floss, Laboratory analysis of presolar silicate stardust from a nova. Astrophys. J. Lett. **754**, L41 (2012)

H. Leroux, Fine-grained material of 81P/Wild 2 in interaction with the Stardust aerogel. Meteorit. Planet. Sci. **47**, 613–622 (2012)

H. Leroux, D. Jacob, Fine-grained material encased in microtracks of Stardust samples. Meteorit. Planet. Sci. **48**, 1607–1617 (2013)

H. Leroux, F.J.M. Rietmeijer, M.A. Velbel, A.J. Brearley, D. Jacob, F. Langenhorst, J.C. Bridges, T.J. Zega, R.M. Stroud, P. Cordier, R.P. Harvey, M. Lee, M. Gounelle, M.E. Zolensky, A TEM study of thermally modified comet 81P/Wild 2 dust particles by interactions with the aerogel matrix during the stardust capture process. Meteorit. Planet. Sci. **43**, 97–120 (2008)

H. Leroux, M. Roskosz, D. Jacob, Oxidation state of iron and extensive redistribution of sulfur in thermally modified Stardust particles. Geochim. Cosmochim. Acta **73**, 767–777 (2009)

A.C. Levasseur-Regourd, Polarization of light scattered by cometary dust particles: observations and tentative interpretations. Space Sci. Rev. **90**, 163–168 (1999)

A.C. Levasseur-Regourd, E. Hadamcik, Light scattering by irregular dust particles in the solar system: observations and interpretation by laboratory measurements. J. Quant. Spectrosc. Radiat. Transf. **79–80**, 903–910 (2003)

A.C. Levasseur-Regourd, J. Lasue, Inferring sources in the interplanetary dust cloud, from observations and simulations of zodiacal light and thermal emission, in Meteoroids: the smallest Solar System bodies (NASA/CP-2011-216469) (2011), pp. 66–75

A.C. Levasseur-Regourd, B. Goidet, T. Le Duin, C. Malique, J.B. Renard, J.-L. Bertaux, Short communication: optical probing of dust in comet Grigg-Skjellerup from the Giotto spacecraft. Planet. Space Sci. **41**, 167–169 (1993)

A.C. Levasseur-Regourd, E. Hadamcik, J.B. Renard, Evidence for two classes of comets from their polarimetric properties at large phase angles. Astron. Astrophys. **313**, 327–333 (1996)

A.C. Levasseur-Regourd, N. McBride, E. Hadamcik, M. Fulle, Similarities between in situ measurements of local dust light scattering and dust flux impact data within the coma of 1P/Halley. Astron. Astrophys. **348**, 636–641 (1999)

A.C. Levasseur-Regourd, I. Mann, R. Dumont, M.S. Hanner, Optical and thermal properties of interplanetary dust, in *Interplanetary Dust*, ed. by E. Grün, B.A.S. Gustafson, S. Dermott, H. Fechtig (Springer, Berlin, 2001), pp. 57–94

A.C. Levasseur-Regourd, T. Mukai, J. Lasue, Y. Okada, Physical properties of cometary and interplanetary dust. Planet. Space Sci. **55**, 1010–1020 (2007)

A.C. Levasseur-Regourd, M. Zolensky, J. Lasue, Dust in cometary comae: present understanding of the structure and composition of dust particles. Planet. Space Sci. **56**, 1719–1724 (2008)

A.C. Levasseur-Regourd, J.-B. Renard, Y. Shkuratov, E. Hadamcik, Laboratory studies, in *Polarimetry of Stars and Planetary Systems*, ed. by L. Kolokolova, J. Hough, A.C. Levasseur-Regourd (Cambridge University Press, Cambridge, 2015), pp. 62–76

C.M. Lisse, J. VanCleve, A.C. Adams, M.F. A'Hearn, Y.R. Fernández, T.L. Farnham, L. Armus, C.J. Grillmair, J. Ingalls, M.J.S. Belton, O. Groussin, L.A. McFadden, K.J. Meech, P.H. Schultz, B.C. Clark, L.M. Feaga, J.M. Sunshine, Spitzer spectral observations of the deep impact ejecta. Science **313**, 635–640 (2006)

C. Lisse, K.E. Kraemer, J.A. Nuth III., A. Li, D. Joswiak, Comparison of the composition of the tempel 1 ejecta to the dust in comet C/Hale–Bopp 1995 O1 and YSO HD 100546. Icarus **187**, 69–86 (2007)

N. Liu, T. Stephan, P. Boehnke, L.R. Nittler, C.M.OD. Alexander, J. Wang, A.M. Davis, R. Trappitsch, M.J. Pellin, J-type carbon stars: a dominant source of ^{14}N-rich presolar SiC grains of type AB. Astrophys. J. Lett. **844**, L12 (2017)

K. Lodders, Solar system abundances and condensation temperatures of the elements. Astrophys. J. **591**, 1220–1247 (2003)

K. Lodders, Solar system abundances of the elements, in *Principles and Perspectives in Cosmochemistry*. Astrophysics and Space Science Proceedings (Springer, Berlin, 2010), p. 379

S.G. Love, D.E. Brownlee, A direct measurement of the terrestrial mass accretion rate of cosmic dust. Science **262**, 550–553 (1993)

J.R. Lyons, E.D. Young, CO self-shielding as the origin of oxygen isotope anomalies in the early solar nebula. Nature **435**, 317–320 (2005)

I. Mann, H. Kimura, L. Kolokolova, A comprehensive model to describe light scattering properties of cometary dust. J. Quant. Spectrosc. Radiat. Transf. **89**, 291–301 (2004)

T. Mannel, M.S. Bentley, R. Schmied, H. Jeszenszky, A.C. Levasseur-Regourd, J. Romstedt, K. Torkar, Fractal cometary dust - a window into the early solar system. Mon. Not. R. Astron. Soc. **462**, S304–S311 (2016)

R. Marschall, C.C. Su, Y. Liao, N. Thomas, K. Altwegg, H. Sierks, W.-H. Ip, H.U. Keller, J. Knollenberg, E. Kührt, I.L. Lai, M. Rubin, Y. Skorov, J.S. Wu, L. Jorda, F. Preusker, F. Scholten, A. Gracia-Berná, A. Gicquel, G. Naletto, X. Shi, J.-B. Vincent, Modelling observations of the inner gas and dust coma of comet 67P/Churyumov-Gerasimenko using ROSINA/COPS and OSIRIS data: first results. Astron. Astrophys. **589**, A90 (2016)

R. Marschall, S. Mottola, C.C. Su, Y. Liao, M. Rubin, J.S. Wu, N. Thomas, K. Altwegg, H. Sierks, W.-H. Ip, H.U. Keller, J. Knollenberg, E. Kührt, I.L. Lai, Y. Skorov, L. Jorda, F. Preusker, F. Scholten, J.-B. Vincent (Osiris Team Rosina Team), Cliffs versus plains: can ROSINA/COPS and OSIRIS data of comet 67P/Churyumov-Gerasimenko in autumn 2014 constrain inhomogeneous outgassing? Astron. Astrophys. **605**, A112 (2017)

B. Marty, R.L. Palma, R.O. Pepin, L. Zimmermann, D.J. Schlutter, P.G. Burnard, A.J. Westphal, C.J. Snead, S. Bajt, R.H. Becker, J.E. Simones, Helium and neon abundances and compositions in cometary matter. Science **319**, 75 (2008)

G. Matrajt, M. Ito, S. Wirick, S. Messenger, D.E. Brownlee, D. Joswiak, G.J. Flynn, S.A. Sandford, C. Snead, A.J. Westphal, Carbon investigation of two Stardust particles: a TEM, NanoSIMS, and XANES study. Meteorit. Planet. Sci. **43**, 315–334 (2008)

G. Matrajt, S. Messenger, D. Joswiak, D.E. Brownlee, Textures and isotopic compositions of carbonaceous materials in A and B-type Stardust tracks: track 130 (Bidi), track 141 (Coki) and track 80 (Tule). Geochim. Cosmochim. Acta **117**, 65–79 (2013)

J.E.P. Matzel, H.A. Ishii, D. Joswiak, I.D. Hutcheon, J.P. Bradley, D.E. Brownlee, P.K. Weber, N. Teslich, G. Matrajt, K.D. McKeegan, G.J. MacPherson, Constraints on the formation age of cometary material from the NASA Stardust mission. Science **328**, 483 (2010)

M. Maurette, C. Olinger, M.C. Michel-Levy, G. Kurat, M. Pourchet, F. Brandstätter, M. Bourot-Denise, A collection of diverse micrometeorites recovered from 100 tonnes of Antarctic blue ice. Nature **351**, 44–47 (1991)

J.A.M. McDonnell, S.F. Green, E. Grün, J. Kissel, S. Nappo, G.S. Pankiewicz, C.H. Perry, In situ exploration of the dusty coma of comet P/Halley at Giotto's encounter—flux rates and time profiles from 10 to the −19th KG to 10 to the −5th KG. Adv. Space Res. **9**, 277–280 (1989)

K.D. McKeegan, J. Aléon, J. Bradley, D.E. Brownlee, H. Busemann, A. Butterworth, M. Chaussidon, S. Fallon, C. Floss, J. Gilmour, M. Gounelle, G. Graham, Y. Guan, P.R. Heck, P. Hoppe, I.D. Hutcheon, J. Huth, H.A. Ishii, M. Ito, S.B. Jacobsen, A. Kearsley, L.A. Leshin, M.-C. Liu, I. Lyon, K. Marhas, B. Marty, G. Matrajt, A. Meibom, S. Messenger, S. Mostefaoui, S. Mukhopadhyay, K. Nakamura-Messenger, L. Nittler, R. Palma, R.O. Pepin, D.A. Papanastassiou, F. Robert, D. Schlutter, C.J. Snead, F.J. Stadermann, R. Stroud, P. Tsou, A.J. Westphal, E.D. Young, K. Ziegler, L. Zimmermann, E. Zinner, Isotopic compositions of cometary matter returned by Stardust. Science **314**, 1724 (2006)

K.D. McKeegan, A.P.A. Kallio, V.S. Heber, G. Jarzebinski, P.H. Mao, C.D. Coath, T. Kunihiro, R.C. Wiens, J.E. Nordholt, R.W. Moses, D.B. Reisenfeld, A.J.G. Jurewicz, D.S. Burnett, The oxygen isotopic composition of the Sun inferred from captured solar wind. Science **332**, 1528 (2011)

S.M.P. McKenna-Lawlor, V.V. Afonin, A flux enhancement measured in energetic particles ($E \approx 60$–100 keV) by the Epona instrument aboard Giotto close to P/Grigg-Skjellerup, and its interpretation as the signature of a companion comet. Planet. Space Sci. **47**, 557–576 (1999)

P. Meakin, B. Donn, Aerodynamic properties of fractal grains. Astrophys. J. **329**, L39–L41 (1988)

D.A. Mendis, M. Horányi, Plasma processes at comet Churyumov-Gerasimenko: expectations for Rosetta. J. Plasma Phys. **79**, 1067–1070 (2013)

S. Merouane, B. Zaprudin, O. Stenzel, Y. Langevin, N. Altobelli, V. Della Corte, H. Fischer, M. Fulle, K. Hornung, J. Silén, N. Ligier, A. Rotundi, J. Ryno, R. Schulz, M. Hilchenbach, J. Kissel (Cosima Team), Dust particle flux and size distribution in the coma of 67P/Churyumov-Gerasimenko measured in situ by the COSIMA instrument on board Rosetta. Astron. Astrophys. **596**, A87 (2016)

S. Merouane, O. Stenzel, M. Hilchenbach, R. Schulz, N. Altobelli, H. Fischer, K. Hornung, J. Kissel, Y. Langevin, E. Mellado, J. Rynö, B. Zaprudin, Evolution of the physical properties of dust and cometary dust activity from 67P/Churyumov-Gerasimenko measured in situ by Rosetta/COSIMA. Mon. Not. R. Astron. Soc. **469**, S459–S474 (2017)

S. Messenger, D. Joswiak, M. Ito, G. Matrajt, D.E. Brownlee, *Discovery of Presolar SiC from Comet Wild-2. Lunar and Planetary Science Conference* (2009)

S. Messenger, D.E. Brownlee, D.J. Joswiak, A.N. Nguyen, Presolar materials in a giant cluster idp of probable cometary origin. Lunar Planet. Sci. **46**, 2603 (2015)

T. Mikouchi, O. Tachikawa, K. Hagiya, K. Ohsumi, Y. Suzuki, K. Uesugi, A. Takeuchi, M.E. Zolensky, Mineralogy and crystallography of comet 81P/Wild 2 particles, in *Lunar and Planetary Science Conference, 1946* (2007)

M. Montalto, A. Riffeser, U. Hopp, S. Wilke, G. Carraro, The comet 17P/Holmes 2007 outburst: the early motion of the outburst material. Astron. Astrophys. **479**, L45–L49 (2008)

F. Moreno, L.M. Lara, O. Muñoz, J.J. López-Moreno, A. Molina, Dust in comet 67P/Churyumov-Gerasimenko. Astrophys. J. **613**, 1263–1269 (2010)

F. Moreno, C. Snodgrass, O. Hainaut, C. Tubiana, H. Sierks, C. Barbieri, P.L. Lamy, R. Rodrigo, D. Koschny, H. Rickman, H.U. Keller, J. Agarwal, M.F. A'Hearn, M.A. Barucci, J.-L. Bertaux, I. Bertini, S. Besse, D. Bodewits, G. Cremonese, V. Da Deppo, B. Davidsson, S. Debei, M. De Cecco, F. Ferri, S. Fornasier, M. Fulle, O. Groussin, P.J. Gutiérrez, P. Gutiérrez-Marques, C. Güttler, S.F. Hviid, W.-H. Ip, L. Jorda, J. Knollenberg, G. Kovacs, J.-R. Kramm, E. Kührt, M. Küppers, L.M. Lara, M. Lazzarin,

J.J. López-Moreno, F. Marzari, S. Mottola, G. Naletto, N. Oklay, M. Pajola, N. Thomas, J.-B. Vincent, V. Della Corte, A. Fitzsimmons, S. Faggi, E. Jehin, C. Opitom, G.-P. Tozzi, The dust environment of comet 67P/Churyumov-Gerasimenko from Rosetta OSIRIS and VLT observations in the 4.5 to 2.9 AU heliocentric distance range inbound. Astron. Astrophys. **587**, A155 (2016)

F. Moreno, O. Muñoz, P.J. Gutiérrez, L.M. Lara, C. Snodgrass, Z.Y. Lin, V. Della Corte, A. Rotundi, M. Yagi, The dust environment of comet 67P/Churyumov-Gerasimenko: results from Monte Carlo dust tail modelling applied to a large ground-based observation data set. Mon. Not. R. Astron. Soc. **469**, S186–S194 (2017)

O. Mousis, A. Guilbert-Lepoutre, B. Brugger, L. Jorda, J.S. Kargel, A. Bouquet, A.-T. Auger, P.L. Lamy, P. Vernazza, N. Thomas, H. Sierks, Pits formation from volatile outgassing on 67P/Churyumov-Gerasimenko. Astrophys. J. Lett. **814**, L5 (2015)

O. Muñoz, J.W. Hovenier, Experimental scattering matrices of clouds of randomly oriented particles, in *Polarimetry of Stars and Planetary Systems*, ed. by L. Kolokolova, J. Hough, A.C. Levasseur-Regourd (Cambridge University Press, Cambridge, 2015), pp. 130–143

O. Muñoz, F. Moreno, D. Guirado, J.L. Ramos, H. Volten, J.W. Hovenier, The IAA cosmic dust laboratory: experimental scattering matrices of clay particles. Icarus **211**, 894–900 (2011)

O. Muñoz, F. Moreno, D. Guirado, D.D. Dabrowska, H. Volten, J.W. Hovenier, The Amsterdam–Granada light scattering database. J. Quant. Spectrosc. Radiat. Transf. **113**, 565–574 (2012)

E. Murad, Physics and chemistry of meteoroids in the upper atmosphere, in *Proceedings of Meteoroids 2001, ESA SP-495*, Noordwijk (2001), pp. 229–236

T. Nakamura, T. Noguchi, Y. Ozono, T. Osawa, K. Nagao, Mineralogy of ultracarbonaceous large micrometeorites. Meteorit. Planet. Sci. **40**(Suppl), 5046 (2005)

T. Nakamura, T. Noguchi, A. Tsuchiyama, T. Ushikubo, N.T. Kita, J.W. Valley, M.E. Zolensky, Y. Kakazu, K. Sakamoto, E. Mashio, K. Uesugi, T. Nakano Tsukasa, Chondrulelike objects in short-period comet 81P/Wild 2. Science **321**, 1664 (2008)

D. Nakashima, T. Ushikubo, D.J. Joswiak, D.E. Brownlee, G. Matrajt, M.K. Weisberg, M.E. Zolensky, N.T. Kita, Oxygen isotopes in crystalline silicates of comet Wild 2: a comparison of oxygen isotope systematics between Wild 2 particles and chondritic materials. Earth Planet. Sci. Lett. **357**, 355–365 (2012)

D. Nakashima, T. Ushikubo, N.T. Kita, M.K. Weisberg, M.E. Zolensky, D.S. Ebel, Late formation of a comet Wild 2 crystalline silicate particle, Pyxie, inferred from Al–Mg chronology of plagioclase. Earth Planet. Sci. Lett. **410**, 54–61 (2015)

D. Nesvorny, P. Jenniskens, H.F. Levison, W.F. Bottke, D. Vokrouhlicky, M. Gounelle, Cometary origin of the zodiacal cloud and carbonaceous micrometeorites. Implications for hot debris disks. Astrophys. J. **713**, 816–836 (2010)

A.N. Nguyen, F.J. Stadermann, E. Zinner, R.M. Stroud, C.M.D. Alexander, L.R. Nittler, Characterization of presolar silicate and oxide grains in primitive carbonaceous chondrites. Astrophys. J. **656**, 1223 (2007)

A.N. Nguyen, L.R. Nittler, F.J. Stadermann, R.M. Stroud, C.M.D. Alexander, Coordinated analyses of presolar grains in the Allan Hills 77307 and Queen Elizabeth range 99177 meteorites. Astrophys. J. **719**, 166 (2010)

A.N. Nguyen, E.L. Berger, K. Nakamura-Messenger, S. Messenger, Sulfur and oxygen isotopic analysis of a cosmic symplectite from a comet Wild 2 Stardust terminal particle, in *78th Annual Meeting of the Meteoritical Society, No. 5375*, (2015)

H. Nilsson, G. Stenberg Wieser, E. Behar, C.S. Wedlund, H. Gunell, M. Yamauchi, R. Lundin, S. Barabash, M. Wieser, C. Carr, E. Cupido, J.L. Burch, A. Fedorov, J.-A. Sauvaud, H. Koskinen, E. Kallio, J.-P. Lebreton, A. Eriksson, N. Edberg, R. Goldstein, P. Henri, C. Koenders, P. Mokashi, Z. Nemeth, I. Richter, K. Szego, M. Volwerk, C. Vallat, M. Rubin, Birth of a comet magnetosphere: a spring of water ions. Science **347**, aaa0571 (2015)

T. Noguchi, N. Ohashi, S. Tsujimoto, T. Mitsunari, J.P. Bradley, T. Nakamura, S. Toh, T. Stephan, N. Iwata, N. Imae, Cometary dust in Antarctic ice and snow: past and present chondritic porous micrometeorites preserved on the Earth's surface. Earth Planet. Sci. Lett. **410**, 1–11 (2015)

P. Notni, H. Tiersch, Charging of dust particles in comets and in interplanetary space. Astron. Astrophys. **187**, 796 (1987)

R.C. Ogliore, A.L. Butterworth, S.C. Fakra, Z. Gainsforth, M.A. Marcus, A.J. Westphal, Comparison of the oxidation state of Fe in comet 81P/Wild 2 and chondritic-porous interplanetary dust particles. Earth Planet. Sci. Lett. **296**, 278–286 (2010)

R.C. Ogliore, G.R. Huss, K. Nagashima, A.L. Butterworth, Z. Gainsforth, J. Stodolna, A.J. Westphal, D. Joswiak, T. Tyliszczak, Incorporation of a late-forming chondrule into comet Wild 2. Astrophys. J. Lett. **745**, L19 (2012)

R.C. Ogliore, K. Nagashima, G.R. Huss, A.J. Westphal, Z. Gainsforth, A.L. Butterworth, Oxygen isotopic composition of coarse- and fine-grained material from comet 81P/Wild 2. Geochim. Cosmochim. Acta **166**, 74–91 (2015)

E.J. Öpik, *Physics of Meteor Flight in the Atmosphere* (Dover, Phoenix, 1958)

T. Ott, E. Drolshagen, D. Koschny, C. Güttler, C. Tubiana, E. Frattin, J. Agarwal, H. Sierks, I. Bertini, C. Barbieri, P.L. Lamy, R. Rodrigo, H. Rickman, M.F. A'Hearn, M.A. Barucci, J.-L. Bertaux, S. Boudreault, G. Cremonese, V. Da Deppo, B. Davidsson, S. Debei, M. De Cecco, J. Deller, C. Feller, S. Fornasier, M. Fulle, B. Geiger, A. Gicquel, O. Groussin, P.J. Gutiérrez, M. Hofmann, S.F. Hviid, W.-H. Ip, L. Jorda, H.U. Keller, J. Knollenberg, G. Kovacs, J.R. Kramm, E. Kührt, M. Küppers, L.M. Lara, M. Lazzarin, Z.-Y. Lin, J.J. López-Moreno, F. Marzari, S. Mottola, G. Naletto, N. Oklay, M. Pajola, X. Shi, N. Thomas, J.-B. Vincent, B. Poppe, Dust mass distribution around comet 67P/Churyumov-Gerasimenko determined via parallax measurements using Rosetta's OSIRIS cameras. Mon. Not. R. Astron. Soc. **469**, S276–S284 (2017)

M. Pajola, A. Lucchetti, M. Fulle, S. Mottola, M. Hamm, V. Da Deppo, L. Penasa, G. Kovacs, M. Massironi, X. Shi, C. Tubiana, C. Güttler, N. Oklay, J.-B. Vincent, I. Toth, B. Davidsson, G. Naletto, H. Sierks, C. Barbieri, P.L. Lamy, R. Rodrigo, D. Koschny, H. Rickman, H.U. Keller, J. Agarwal, M.F. A'Hearn, M.A. Barucci, J.-L. Bertaux, I. Bertini, G. Cremonese, S. Debei, M. De Cecco, J. Deller, M.R. El Maarry, S. Fornasier, E. Frattin, A. Gicquel, O. Groussin, P.J. Gutierrez, S. Höfner, M. Hofmann, S.F. Hviid, W.-H. Ip, L. Jorda, J. Knollenberg, J.R. Kramm, E. Kührt, M. Küppers, L.M. Lara, M. Lazzarin, J.J. López Moreno, F. Marzari, H. Michalik, F. Preusker, F. Scholten, N. Thomas, The pebbles/boulders size distributions on Sais: Rosetta's final landing site on comet 67P/Churyumov-Gerasimenko. Mon. Not. R. Astron. Soc. **469**, S636–S645 (2017a)

M. Pajola, S. Höfner, J.-B. Vincent, N. Oklay, F. Scholten, F. Preusker, S. Mottola, G. Naletto, S. Fornasier, S. Lowry, C. Feller, P.H. Hasselmann, C. Güttler, C. Tubiana, H. Sierks, C. Barbieri, P.L. Lamy, R. Rodrigo, D. Koschny, H. Rickman, H.U. Keller, J. Agarwal, M.F. A'Hearn, M.A. Barucci, J.-L. Bertaux, I. Bertini, S. Besse, S. Boudreault, G. Cremonese, V. da Deppo, B. Davidsson, S. Debei, M. de Cecco, J. Deller, J.D.P. Deshapriya, M.R. El-Maarry, S. Ferrari, F. Ferri, M. Fulle, O. Groussin, P. Gutierrez, M. Hofmann, S.F. Hviid, W.-H. Ip, L. Jorda, J. Knollenberg, G. Kovacs, J.R. Kramm, E. Kührt, M. Küppers, L.M. Lara, Z.-Y. Lin, M. Lazzarin, A. Lucchetti, J.J. López Moreno, F. Marzari, M. Massironi, H. Michalik, L. Penasa, A. Pommerol, E. Simioni, N. Thomas, I. Toth, E. Baratti, The pristine interior of comet 67P revealed by the combined Aswan outburst and cliff collapse. Nat. Astron. **1**, 0092 (2017b)

R.L. Palma, R.O. Pepin, D. Schlutter, J. Simones, Helium and neon isotopic compositions from stardust aerogel particle tracks, in *Lunar and Planetary Science Conference, 2032* (2007)

R.L. Palma, R.O. Pepin, D.J. Schlutter, J. Stodolna, A.J. Westphal, Z. Gainsforth, Q-gases in an unusual IDP: a noble gas link to carriers in Stardust Track 41, in *Lunar and Planetary Science Conference, 1694* (2013)

R.L. Palma, R.O. Pepin, D.J. Schlutter, D.R. Frank, R. Bastien, M. Rodriguez, Neon and helium in the surface of Stardust Cell C2028, in *Lunar and Planetary Science Conference, 2378* (2015)

J.A. Paquette, K. Hornung, O.J. Stenzel, J. Rynö, J. Silen, J. Kissel, M. Hilchenbach, The $^{34}S/^{32}S$ isotopic ratio measured in the dust of comet 67P/Churyumov-Gerasimenko by Rosetta/COSIMA. Mon. Not. R. Astron. Soc. **469**, S230–S237 (2017)

M. Pätzold, T. Andert, M. Hahn, S.W. Asmar, J.-P. Barriot, M.K. Bird, B. Häusler, K. Peter, S. Tellmann, E. Grün, P.R. Weissman, H. Sierks, L. Jorda, R. Gaskell, F. Preusker, F. Scholten, A homogeneous nucleus for comet 67P/Churyumov-Gerasimenko from its gravity field. Nature **530**, 63–65 (2016)

J.M.C. Plane, Cosmic dust in the Earth's atmosphere. Chem. Soc. Rev. **41**(19), 6507–6518 (2012)

O. Poch, A. Pommerol, B. Jost, N. Carrasco, C. Szopa, N. Thomas, Sublimation of water ice mixed with silicates and tholins: evolution of surface texture and reflectance spectra, with implications for comets. Icarus **267**, 154–173 (2016)

A.R. Poppe, An improved model for interplanetary dust fluxes in the outer solar system. Icarus **264**, 369–386 (2016)

D. Prialnik, A. Bar-Nun, Crystallization of amorphous ice as the cause of comet P/Halley's outburst at 14 AU. Astron. Astrophys. **258**, L9–L12 (1992)

M.C. Price, A.T. Kearsley, M.J. Burchell, F. Hörz, J. Borg, J.C. Bridges, M.J. Cole, C. Floss, G. Graham, S.F. Green, P. Hoppe, H. Leroux, K.K. Marhas, N. Park, R. Stroud, F.J. Stadermann, N. Telisch, P.J. Wozniakiewicz, Comet 81P/Wild 2: the size distribution of finer (sub-10 μm) dust collected by the Stardust spacecraft. Meteorit. Planet. Sci. **45**, 1409–1428 (2010)

E. Quirico, L.V. Moroz, B. Schmitt, G. Arnold, M. Faure, P. Beck, L. Bonal, M. Ciarniello, F. Capaccioni, G. Filacchione, S. Erard, C. Leyrat, D. Bockelée-Morvan, A. Zinzi, E. Palomba, P. Drossart, F. Tosi, M.T. Capria, M.C. De Sanctis, A. Raponi, S. Fonti, F. Mancarella, V. Orofino, M.A. Barucci, M.I. Blecka, R. Carlson, D. Despan, A. Faure, S. Fornasier, M.S. Gudipati, A. Longobardo, K. Markus, V. Mennella, F. Merlin, G. Piccioni, B. Rousseau, F. Taylor, Refractory and semi-volatile organics at the

surface of comet 67P/Churyumov-Gerasimenko: insights from the VIRTIS/Rosetta imaging spectrometer. Icarus **272**, 32–47 (2016)
W.T. Reach, M.V. Sykes, D. Lien, J.K. Davies, The formation of Encke meteoroids and dust trail. Icarus **148**, 80–94 (2000)
W.T. Reach, P. Morris, F. Boulanger, K. Okumura, The mid-infrared spectrum of the zodiacal and exozodiacal light. Icarus **164**, 384–403 (2003)
W.T. Reach, M.S. Kelley, M.V. Sykes, A survey of debris trails from short-period comets. Icarus **191**, 298–322 (2007)
J.-B. Renard, E. Hadamcik, A.C. Levasseur-Regourd, Polarimetric CCD imaging of comet 47P/Ashbrook-Jackson and variability of polarization in the inner coma of comets. Astron. Astrophys. **316**, 263–269 (1996)
W. Riedler, K. Torkar, H. Jeszenszky, J. Romstedt, H.S.C. Alleyne, H. Arends, W. Barth, J.V.D. Biezen, B. Butler, P. Ehrenfreund, M. Fehringer, G. Fremuth, J. Gavira, O. Havnes, E.K. Jessberger, R. Kassing, W. Klöck, C. Koeberl, A.C. Levasseur-Regourd, M. Maurette, F. Rüdenauer, R. Schmidt, G. Stangl, M. Steller, I. Weber, MIDAS the micro-imaging dust analysis system for the Rosetta mission. Space Sci. Rev. **128**, 869–904 (2007)
F.J.M. Rietmeijer, Interplanetary dust particles, in *Planetary Materials*, ed. by J.J. Papike (1998), pp. 1–96
J. Robertson, Diamond-like amorphous carbon. Mater. Sci. Eng., R **37**, 129–281 (2002)
P. Rochette, L. Folco, C. Suavet, M. van Ginneken, J. Gattacceca, N. Perchiazzi, R. Braucher, R.P. Harvey, Micrometeorites from the transantarctic mountains. Proc. Natl. Acad. Sci. **105**(47), 18206–18211 (2008)
V.K. Rosenbush, O.V. Ivanova, N.N. Kiselev, L. Kolokolova, V.L. Afanasiev, Spatial variations of brightness, colour and polarization of dust in comet 67P/Churyumov-Gerasimenko. Mon. Not. R. Astron. Soc. **469**, S475–S491 (2017)
A. Rotundi, H. Sierks, V. Della Corte, M. Fulle, P.J. Gutierrez, L.M. Lara, C. Barbieri, P.L. Lamy, R. Rodrigo, D. Koschny, H. Rickman, H.U. Keller, J.J. López-Moreno, M. Accolla, J. Agarwal, M.F. A'Hearn, N. Altobelli, F. Angrilli, M.A. Barucci, J.-L. Bertaux, I. Bertini, D. Bodewits, E. Bussoletti, L. Colangeli, M. Cosi, G. Cremonese, J.-F. Crifo, V. Da Deppo, B. Davidsson, S. Debei, M. De Cecco, F. Esposito, M. Ferrari, S. Fornasier, F. Giovane, B. Gustafson, S.F. Green, O. Groussin, E. Grün, C. Güttler, M.L. Herranz, S.F. Hviid, W.-H. Ip, S. Ivanovski, J.M. Jerónimo, L. Jorda, J. Knollenberg, R. Kramm, E. Kührt, M. Küppers, M. Lazzarin, M.R. Leese, A.C. López-Jiménez, F. Lucarelli, S.C. Lowry, F. Marzari, E. Mazzotta Epifani, J.A.M. McDonnell, V. Mennella, H. Michalik, A. Molina, R. Morales, F. Moreno, S. Mottola, G. Naletto, N. Oklay, J.L. Ortiz, E. Palomba, P. Palumbo, J.-M. Perrin, J. Rodríguez, L. Sabau, C. Snodgrass, R. Sordini, N. Thomas, C. Tubiana, J.-B. Vincent, P.R. Weissman, K.-P. Wenzel, V. Zakharov, J.C. Zarnecki, Dust measurements in the coma of comet 67P/Churyumov-Gerasimenko inbound to the sun. Science **347**, aaa3905 (2015)
M. Rowan-Robinson, B. May, An improved model for the infrared emission from the zodiacal dust cloud: cometary, asteroidal and interstellar dust. Mon. Not. R. Astron. Soc. **429**, 2894–2902 (2013)
M. Rubin, V.M. Tenishev, M.R. Combi, K.C. Hansen, T.I. Gombosi, K. Altwegg, H. Balsiger, Monte Carlo modeling of neutral gas and dust in the coma of comet 1P/Halley. Icarus **213**, 655–677 (2011)
M. Rubin, K. Altwegg, H. Balsiger, J.-J. Berthelier, A. Bieler, U. Calmonte, M. Combi, J. De Keyser, C. Engrand, B. Fiethe, S.A. Fuselier, S. Gasc, T.I. Gombosi, K.C. Hansen, M. Hässig, L. Le Roy, K. Mezger, C.-Y. Tzou, S.F. Wampfler, P. Wurz, Evidence for depletion of heavy silicon isotopes at comet 67P/Churyumov-Gerasimenko. Astron. Astrophys. **601**, A123 (2017)
N. Sakamoto, Y. Seto, S. Itoh, K. Kuramoto, K. Fujino, K. Nagashima, A.N. Krot, H. Yurimoto, Remnants of the early solar system water enriched in heavy oxygen isotopes. Science **317**, 231 (2007)
S.A. Sandford, J. Aléon, C.M.OD. Alexander, T. Araki, S. Bajt, G.A. Baratta, J. Borg, J.P. Bradley, D.E. Brownlee, J.R. Brucato, M.J. Burchell, H. Busemann, A. Butterworth, S.J. Clemett, G. Cody, L. Colangeli, G. Cooper, L. D'Hendecourt, Z. Djouadi, J.P. Dworkin, G. Ferrini, H. Fleckenstein, G.J. Flynn, I.A. Franchi, M. Fries, M.K. Gilles, D.P. Glavin, M. Gounelle, F. Grossemy, C. Jacobsen, L.P. Keller, A.L.D. Kilcoyne, J. Leitner, G. Matrajt, A. Meibom, V. Mennella, S. Mostefaoui, L.R. Nittler, M.E. Palumbo, D.A. Papanastassiou, F. Robert, A. Rotundi, C.J. Snead, M.K. Spencer, F.J. Stadermann, A. Steele, T. Stephan, P. Tsou, T. Tyliszczak, A.J. Westphal, S. Wirick, B. Wopenka, H. Yabuta, R.N. Zare, M.E. Zolensky, Organics captured from comet 81P/Wild 2 by the Stardust spacecraft. Science **314**, 1720 (2006)
S.A. Sandford, S. Bajt, S.J. Clemett, G.D. Cody, G. Cooper, B.T. Degregorio, V. de Vera, J.P. Dworkin, J.E. Elsila, G.J. Flynn, D.P. Glavin, A. Lanzirotti, T. Limero, M.P. Martin, C.J. Snead, M.K. Spencer, T. Stephan, A.J. Westphal, S. Wirick, R.N. Zare, M.E. Zolensky, Assessment and control of organic and other contaminants associated with the Stardust sample return from comet 81P/Wild 2. Meteorit. Planet. Sci. **45**, 406–433 (2010)
S.A. Sandford, C. Engrand, A. Rotundi, Organic matter in cosmic dust. Elements **12**(3), 185–189 (2016)

Y. Sarugaku, M. Ishiguro, J. Pyo, N. Miura, Y. Nakada, F. Usui, M. Ueno, Detection of a long-extended dust trail associated with short-period comet 4P/Faye in 2006 return. Publ. Astron. Soc. Jpn. **59**, L25–L28 (2007)

Y. Sarugaku, M. Ishiguro, M. Ueno, F. Usui, W.T. Reach, Infrared and optical imagings of the comet 2P/Encke dust cloud in the 2003 return. Astrophys. J. **804**, 127 (2015)

L.S. Schramm, D.E. Brownlee, M.M. Wheelock, Major element composition of stratospheric micrometeorites. Meteorit. Planet. Sci. **24**, 99–112 (1989)

S.B. Simon, D.J. Joswiak, H.A. Ishii, J.P. Bradley, M. Chi, L. Grossman, J. Aléon, D.E. Brownlee, S. Fallon, I.D. Hutcheon, G. Matrajt, K.D. McKeegan, A refractory inclusion returned by Stardust from comet 81P/Wild 2. Meteorit. Planet. Sci. **43**, 1861–1877 (2008)

Y.V. Skorov, J. Blum, Dust release and tensile strength of the non-volatile layer of cometary nuclei. Icarus **221**, 1–11 (2012)

Y.V. Skorov, L. Rezac, P. Hartogh, A.T. Bazilevsky, H.U. Keller, A model of short-lived outbursts on the 67P from fractured terrains. Astron. Astrophys. **593**, A76 (2016)

C. Snodgrass, E. Jehin, J. Manfroid, C. Opitom, A. Fitzsimmons, G.P. Tozzi, S. Faggi, B. Yang, M.M. Knight, B.C. Conn, T. Lister, O. Hainaut, D.M. Bramich, S.C. Lowry, A. Rozek, C. Tubiana, A. Guilbert-Lepoutre, Distant activity of 67P/Churyumov-Gerasimenko in 2014: ground-based results during the Rosetta pre-landing phase. Astron. Astrophys. **588**, A80 (2016a)

C. Snodgrass, C. Opitom, M. de Val-Borro, E. Jehin, J. Manfroid, T. Lister, J. Marchant, G.H. Jones, A. Fitzsimmons, I.A. Steele, R.J. Smith, H. Jermak, T. Granzer, K.J. Meech, P. Rousselot, A.C. Levasseur-Regourd, The perihelion activity of comet 67P/Churyumov-Gerasimenko as seen by robotic telescopes. Mon. Not. R. Astron. Soc. **462**, S138–S145 (2016b)

C. Snodgrass, M.F. A'Hearn, F. Aceituno, V. Afanasiev, S. Bagnulo, J. Bauer, G. Bergond, S. Besse, N. Biver, D. Bodewits, H. Boehnhardt, B.P. Bonev, G. Borisov, B. Carry, V. Casanova, A. Cochran, B.C. Conn, B. Davidsson, J.K. Davies, J. de León, E. de Mooij, M. de Val-Borro, M. Delacruz, M.A. DiSanti, J.E. Drew, R. Duffard, N.J.T. Edberg, S. Faggi, L. Feaga, A. Fitzsimmons, H. Fujiwara, E.L. Gibb, M. Gillon, S.F. Green, A. Guijarro, A. Guilbert-Lepoutre, P.J. Gutiérrez, E. Hadamcik, O. Hainaut, S. Haque, R. Hedrosa, D. Hines, U. Hopp, F. Hoyo, D. Hutsemékers, M. Hyland, O. Ivanova, E. Jehin, G.H. Jones, J.V. Keane, M.S.P. Kelley, N. Kiselev, J. Kleyna, M. Kluge, M.M. Knight, R. Kokotanekova, D. Koschny, E.A. Kramer, J.J. López-Moreno, P. Lacerda, L.M. Lara, J. Lasue, H.J. Lehto, A.C. Levasseur-Regourd, J. Licandro, Z.-Y. Lin, T. Lister, S.C. Lowry, A. Mainzer, J. Manfroid, J. Marchant, A.J. McKay, A. McNeill, K.J. Meech, M. Micheli, I. Mohammed, M. Monguió, F. Moreno, O. Muñoz, M.J. Mumma, P. Nikolov, C. Opitom, J.L. Ortiz, L. Paganini, M. Pajuelo, F.J. Pozuelos, S. Protopapa, T. Pursimo, B. Rajkumar, Y. Ramanjooloo, E. Ramos, C. Ries, A. Riffeser, V. Rosenbush, P. Rousselot, E.L. Ryan, P. Santos-Sanz, D.G. Schleicher, M. Schmidt, R. Schulz, A.K. Sen, A. Somero, A. Sota, A. Stinson, J.M. Sunshine, A. Thompson, G.P. Tozzi, C. Tubiana, G.L. Villanueva, X. Wang, D.H. Wooden, M. Yagi, B. Yang, B. Zaprudin, T.J. Zegmott, The 67P/Churyumov-Gerasimenko observation campaign in support of the Rosetta mission. Philos. Trans. R. Soc. A **375**, 20160249 (2017)

R.H. Soja, M. Sommer, J. Herzog, J. Agarwal, J. Rodmann, R. Srama, J. Vaubaillon, P. Strub, A. Hornig, L. Bausch, E. Grün, Characteristics of the dust trail of 67P/Churyumov-Gerasimenko: an application of the IMEX model. Astron. Astrophys. **583**, A18 (2015)

F.J. Stadermann, P. Hoppe, C. Floss, P.R. Heck, F. Hörz, J. Huth, A.T. Kearsley, J. Leitner, K.K. Marhas, K.D. McKeegan, T. Stephan, Stardust in stardust—the C, N, and O isotopic compositions of Wild 2 cometary matter in Al foil impacts. Meteorit. Planet. Sci. **43**, 299–313 (2008)

O.J. Stenzel, M. Hilchenbach, S. Merouane, J. Paquette, K. Varmuza, C. Engrand, F. Brandstätter, C. Koeberl, L. Ferriére, P. Filzmoser, S. Siljeström, Similarities in element content between comet 67P/Churyumov-Gerasimenko coma dust and selected meteorite samples. Mon. Not. R. Astron. Soc. **469**, S492–S505 (2017)

T. Stephan, G.J. Flynn, S.A. Sandford, M.E. Zolensky, TOF-SIMS analysis of cometary particles extracted from Stardust aerogel. Meteorit. Planet. Sci. **43**, 285–298 (2008)

R. Stevenson, J.M. Bauer, E.A. Kramer, T. Grav, A.K. Mainzer, J.R. Masiero, Lingering grains of truth around comet 17P/Holmes. Astrophys. J. **787**, 116 (2014)

J. Stodolna, D. Jacob, H. Leroux, Mineralogy and petrology of Stardust particles encased in the bulb of track 80: TEM investigation of the Wild 2 fine-grained material. Geochim. Cosmochim. Acta **87**, 35–50 (2012)

J. Stodolna, Z. Gainsforth, A.L. Butterworth, A.J. Westphal, Characterization of preserved primitive fine-grained material from the Jupiter family comet 81P/Wild 2—a new link between comets and CP-IDPs. Earth Planet. Sci. Lett. **388**, 367–373 (2014)

J.M. Sunshine, M.F. A'Hearn, O. Groussin, J.-Y. Li, M.J.S. Belton, W.A. Delamere, J. Kissel, K.P. Klaasen, L.A. McFadden, K.J. Meech, H.J. Melosh, P.H. Schultz, P.C. Thomas, J. Veverka, D.K. Yeomans,

J.C. Busko, M. Desnoyer, T. Farnham, L.M. Feaga, D.L. Hampton, D.J. Lindler, C.M. Lisse, D.D. Wellnitz, Exposed water ice deposits on the surface of comet 9P/Tempel 1. Science **311**, 1453–1455 (2006)
M.V. Sykes, R.G. Walker, Cometary dust trails. I Survey. Icarus **95**, 180–210 (1992)
M.V. Sykes, D.M. Hunten, F.J. Low, Preliminary analysis of cometary dust trails. Adv. Space Res. **6**, 67–78 (1986)
M.V. Sykes, D.J. Lien, R.G. Walker, The Tempel 2 dust trail. Icarus **86**, 236–247 (1990)
S. Taylor, J. Lever, R. Harvey, Accretion rate of cosmic spherules measured at the South Pole. Nature **392**, 899–900 (1998)
S. Taylor, S. Messenger, L. Folco, Cosmic dust: finding a needle in a haystack. Elements **12**(3), 171–176 (2016)
M.H. Thiemens, Mass-independent isotope effects in planetary atmospheres and the early solar system. Science **283**, 341 (1999)
M.H. Thiemens, J.E. Heidenreich, The mass-independent fractionation of oxygen—A novel isotope effect and its possible cosmochemical implications. Science **219**, 1073–1075 (1983)
K.L. Thomas, G.E. Blanford, L.P. Keller, W. Klöck, D. McKay, Carbon abundance and silicate mineralogy of anhydrous interplanetary dust particles. Geochim. Cosmochim. Acta **57**, 1551–1566 (1993)
N. Thomas, B. Davidsson, M.R. El-Maarry, S. Fornasier, L. Giacomini, A.G. Gracia-Berná, S.F. Hviid, W.-H. Ip, L. Jorda, H.U. Keller, J. Knollenberg, E. Kührt, F. La Forgia, I.L. Lai, Y. Liao, R. Marschall, M. Massironi, S. Mottola, M. Pajola, O. Poch, A. Pommerol, F. Preusker, F. Scholten, C.C. Su, J.S. Wu, J.-B. Vincent, H. Sierks, C. Barbieri, P.L. Lamy, R. Rodrigo, D. Koschny, H. Rickman, M.F. A'Hearn, M.A. Barucci, J.-L. Bertaux, I. Bertini, G. Cremonese, V. Da Deppo, S. Debei, M. de Cecco, M. Fulle, O. Groussin, P.J. Gutierrez, J.-R. Kramm, M. Küppers, L.M. Lara, M. Lazzarin, J.J. Lopez Moreno, F. Marzari, H. Michalik, G. Naletto, J. Agarwal, C. Güttler, N. Oklay, C. Tubiana, Redistribution of particles across the nucleus of comet 67P/Churyumov-Gerasimenko. Astron. Astrophys. **583**, A17 (2015a)
N. Thomas, H. Sierks, C. Barbieri, P.L. Lamy, R. Rodrigo, H. Rickman, D. Koschny, H.U. Keller, J. Agarwal, M.F. A'Hearn, F. Angrilli, A.-T. Auger, M.A. Barucci, J.-L. Bertaux, I. Bertini, S. Besse, D. Bodewits, G. Cremonese, V. Da Deppo, B. Davidsson, M. De Cecco, S. Debei, M.R. El-Maarry, F. Ferri, S. Fornasier, M. Fulle, L. Giacomini, O. Groussin, P.J. Gutierrez, C. Güttler, S.F. Hviid, W.-H. Ip, L. Jorda, J. Knollenberg, J.-R. Kramm, E. Kührt, M. Küppers, F. La Forgia, L.M. Lara, M. Lazzarin, J.J. Lopez Moreno, S. Magrin, S. Marchi, F. Marzari, M. Massironi, H. Michalik, R. Moissl, S. Mottola, G. Naletto, N. Oklay, M. Pajola, A. Pommerol, F. Preusker, L. Sabau, F. Scholten, C. Snodgrass, C. Tubiana, J.-B. Vincent, K.-P. Wenzel, The morphological diversity of comet 67P/Churyumov-Gerasimenko. Science **347**, aaa0440 (2015b)
J.M. Trigo-Rodríguez, G. Domínguez, M.J. Burchell, F. Hörz, J. Llorca, Bulbous tracks arising from hypervelocity capture in aerogel. Meteorit. Planet. Sci. **43**, 75–86 (2008)
A.J. Tuzzolino, T.E. Economou, B.C. Clark, P. Tsou, D.E. Brownlee, S.F. Green, J.A.M. McDonnell, N. McBride, M. Colwell, T.S.H. Melusine, Dust measurements in the coma of comet 81P/Wild 2 by the dust flux monitor instrument. Science **304**, 1776–1780 (2004)
J.J. Vaubaillon, W.T. Reach, Spitzer space telescope observations and the particle size distribution of comet 73P/Schwassmann Wachmann 3. Astron. J. **139**, 1491–1498 (2010)
J.-B. Vincent, L.M. Lara, G.P. Tozzi, Z.-Y. Lin, H. Sierks, Spin and activity of comet 67P/Churyumov-Gerasimenko. Astron. Astrophys. **549**, A121 (2013)
J.-B. Vincent, M.F. A'Hearn, Z.-Y. Lin, M.R. El-Maarry, M. Pajola, H. Sierks, C. Barbieri, P.L. Lamy, R. Rodrigo, D. Koschny, H. Rickman, H.U. Keller, J. Agarwal, M.A. Barucci, J.-L. Bertaux, I. Bertini, S. Besse, D. Bodewits, G. Cremonese, V. Da Deppo, B. Davidsson, S. Debei, M. De Cecco, J. Deller, S. Fornasier, M. Fulle, A. Gicquel, O. Groussin, P.J. Gutiérrez, P. Gutiérrez-Marquez, C. Güttler, S. Höfner, M. Hofmann, S.F. Hviid, W.-H. Ip, L. Jorda, J. Knollenberg, G. Kovacs, J.-R. Kramm, E. Kührt, M. Küppers, L.M. Lara, M. Lazzarin, J.J. Lopez Moreno, F. Marzari, M. Massironi, S. Mottola, G. Naletto, N. Oklay, F. Preusker, F. Scholten, X. Shi, N. Thomas, I. Toth, C. Tubiana, Summer fireworks on comet 67P. Mon. Not. R. Astron. Soc. **462**, S184–S194 (2016a)
J.-B. Vincent, N. Oklay, M. Pajola, S. Höfner, H. Sierks, X. Hu, C. Barbieri, P.L. Lamy, R. Rodrigo, D. Koschny, H. Rickman, H.U. Keller, M.F. A'Hearn, M.A. Barucci, J.-L. Bertaux, I. Bertini, S. Besse, D. Bodewits, G. Cremonese, V. Da Deppo, B. Davidsson, S. Debei, M. De Cecco, M.R. El-Maarry, S. Fornasier, M. Fulle, O. Groussin, P.J. Gutiérrez, P. Gutiérrez-Marquez, C. Güttler, M. Hofmann, S.F. Hviid, W.-H. Ip, L. Jorda, J. Knollenberg, G. Kovacs, J.-R. Kramm, E. Kührt, M. Küppers, L.M. Lara, M. Lazzarin, Z.-Y. Lin, J.J. Lopez Moreno, S. Lowry, F. Marzari, M. Massironi, F. Moreno, S. Mottola, G. Naletto, F. Preusker, F. Scholten, X. Shi, N. Thomas, I. Toth, C. Tubiana, Are fractured cliffs the source of cometary dust jets? Insights from OSIRIS/Rosetta at 67P/Churyumov-Gerasimenko. Astron. Astrophys. **587**, A14 (2016b)
A.J. Westphal, S.C. Fakra, Z. Gainsforth, M.A. Marcus, R.C. Ogliore, A.L. Butterworth, Mixing fraction of inner Solar System material in comet 81P/Wild2. Astrophys. J. **694**, 18–28 (2009)

A.J. Westphal, Z. Gainsforth, J.I. Goldstein, M. Humayun, M. Kunz, M.A. Marcus, R.M. Stroud, N.S. Tamura, An ultra Ni-poor metal particle from comet 81P/Wild2. Meteorit. Planet. Sci. Suppl. **75**, 5309 (2012)

A.J. Westphal, G.F. Herzog, G.J. Flynn, Cosmic dust toolbox: microanalytical instruments and methods. Elements **12**, 197–202 (2016)

A.J. Westphal, J.C. Bridges, D.E. Brownlee, A.L. Butterworth, B.T. de Gregorio, G. Dominguez, G.J. Flynn, Z. Gainsforth, H.A. Ishii, D. Joswiak, L.R. Nittler, R.C. Ogliore, R. Palma, R.O. Pepin, T. Stephan, M.E. Zolensky, The future of Stardust science. Meteorit. Planet. Sci. **52**, 1859–1898 (2017)

F.L. Whipple, A comet model: the acceleration of comet Encke. Astrophys. J. Lett. **111**, 375–394 (1950)

S. Wirick, H. Leroux, K. Tomeoka, M.E. Zolensky, G.J. Flynn, T. Tyliszczak, A. Butterworth, N. Tomioka, I. Ohnishi, K. Nakamura-Messenger, S.A. Sandford, L. Keller, C. Jacobsen, Carbonates Found in Stardust Aerogel Tracks, LPSC, 1534 (CD-ROM, 2007)

D.H. Wooden, Cometary refractory grains: interstellar and nebular sources. Space Sci. Rev. **138**, 75–108 (2008)

D.H. Wooden, D.E. Harker, C.E. Woodward, H.M. Butner, C. Koike, F.C. Witteborn, C.W. McMurtry, W. Craig, Silicate mineralogy of the dust in the inner coma of comet C/1995 01 (Hale-Bopp) pre- and postperihelion. Astrophys. J. **517**, 1034–1058 (1999)

D.H. Wooden, H.A. Ishii, M.E. Zolensky, Cometary dust: the diversity of primitive refractory grains. Philos. Trans. R. Soc. A **375, id**, 20160260 (2017)

I.P. Wright, S. Sheridan, S.J. Barber, G.H. Morgan, D.J. Andrews, A.D. Morse, CHO-bearing organic compounds at the surface of 67P/Churyumov-Gerasimenko revealed by Ptolemy. Science **349**, 6247 (2015)

M.E. Zolensky, T.J. Zega, H. Yano, S. Wirick, A.J. Westphal, M.K. Weisberg, I. Weber, J.L. Warren, M.A. Velbel, A. Tsuchiyama, P. Tsou, A. Toppani, N. Tomioka, K. Tomeoka, N. Teslich, M. Taheri, J. Susini, R. Stroud, T. Stephan, F.J. Stadermann, C.J. Snead, S.B. Simon, A. Simionovici, T.H. See, F. Robert, F.J.M. Rietmeijer, W. Rao, M.C. Perronnet, D.A. Papanastassiou, K. Okudaira, K. Ohsumi, I. Ohnishi, K. Nakamura-Messenger, T. Nakamura, S. Mostefaoui, T. Mikouchi, A. Meibom, G. Matrajt, M.A. Marcus, H. Leroux, L. Lemelle, L. Le, A. Lanzirotti, F. Langenhorst, A.N. Krot, L.P. Keller, A.T. Kearsley, D. Joswiak, D. Jacob, H.A. Ishii, R. Harvey, K. Hagiya, L. Grossman, J.N. Grossman, G.A. Graham, M. Gounelle, P. Gillet, M.J. Genge, G.J. Flynn, T. Ferroir, S. Fallon, D.S. Ebel, Z.R. Dai, P. Cordier, B. Clark, M. Chi, A.L. Butterworth, D.E. Brownlee, J.C. Bridges, S. Brennan, A. Brearley, J.P. Bradley, P. Bleuet, P.A. Bland, R. Bastien, Mineralogy and petrology of comet 81P/Wild 2 nucleus samples. Science **314**, 1735 (2006)

M.E. Zolensky, K. Nakamura-Messenger, F.J.M. Rietmeijer, H. Leroux, T. Mikouchi, K. Ohsumi, M. Velbel, Comparing Wild 2 particles to chondrites and IDPs. Meteorit. Planet. Sci. **43**, 261–272 (2008)

A. Zsom, C.W. Ormel, C. Güttler, J. Blum, C.P. Dullemond, The outcome of protoplanetary dust growth: pebbles, boulders, or planetesimals? II. Introducing the bouncing barrier. Astron. Astrophys. **513**, A57 (2010)

E.S. Zubko, Light scattering by irregularly shaped particles with sizes comparable to the wavelength, in *Light Scattering Reviews*, vol. 6 (Springer, Berlin, 2012), pp. 39–74

E. Zubko, K. Muinonen, Y. Shkuratov, E. Hadamcik, A.C. Levasseur-Regourd, G. Videen, Evaluating the carbon depletion found by the Stardust mission in comet 81P/Wild 2. Astron. Astrophys. **544**, L8 (2012)

Space Sci Rev (2018) 214:98
https://doi.org/10.1007/s11214-018-0527-0

Dust Phenomena Relating to Airless Bodies

J.R. Szalay[1] · A.R. Poppe[2] · J. Agarwal[3] · D. Britt[4] · I. Belskaya[5] · M. Horányi[6] · T. Nakamura[7] · M. Sachse[8] · F. Spahn[8]

Received: 8 August 2017 / Accepted: 10 July 2018 / Published online: 7 August 2018

Abstract Airless bodies are directly exposed to ambient plasma and meteoroid fluxes, making them characteristically different from bodies whose dense atmospheres protect their surfaces from such fluxes. Direct exposure to plasma and meteoroids has important consequences for the formation and evolution of planetary surfaces, including altering chemical makeup and optical properties, generating neutral gas and/or dust exospheres, and leading to the generation of circumplanetary and interplanetary dust grain populations. In the past two decades, there have been many advancements in our understanding of airless bodies and their interaction with various dust populations. In this paper, we describe relevant dust phenomena on the surface and in the vicinity of airless bodies over a broad range of scale sizes from $\sim 10^{-3}$ km to $\sim 10^{3}$ km, with a focus on recent developments in this field.

Keywords Dust · Airless bodies · Interplanetary dust

Cosmic Dust from the Laboratory to the Stars
Edited by Rafael Rodrigo, Jürgen Blum, Hsiang-Wen Hsu, Detlef Koschny, Anny-Chantal Levasseur-Regourd, Jesús Martín-Pintado, Veerle Sterken and Andrew Westphal

✉ J.R. Szalay
jszalay@princeton.edu

A.R. Poppe
poppe@ssl.berkeley.edu

1 Department of Astrophysical Sciences, Princeton University, Princeton, NJ 08540, USA

2 Space Sciences Laboratory, U.C. Berkeley, Berkeley, CA, USA

3 Max Planck Institute for Solar System Research, Göttingen, Germany

4 University of Central Florida, Orlando, FL 32816, USA

5 Institute of Astronomy, V.N. Karazin Kharkiv National University, Kharkiv, Ukraine

6 Department of Physics, University of Colorado Boulder, Boulder, CO, USA

7 Division of Earth and Planetary Materials Science, Tohoku University, Miyagi, Japan

8 Institut für Physik und Astronomie, Universität Potsdam, Potsdam, Germany

1 Introduction

The term "airless body" refers to any object in the solar system without a global, collisional atmosphere. These bodies comprise a diverse class of objects with great variation in size, shape, surface properties, composition, and orbital characteristics. The interaction between these bodies and their dust environment is complex and occurs on many spatial and temporal scales. Dust is pervasive throughout the observable universe; it comprises the building blocks of all planetary bodies, which are formed from the accretion of dust particles aggregating into much larger objects (Birnstiel et al. 2012; Levison et al. 2015). While dust particles impacting airless bodies at very low speeds contribute to their mass through accretion, high speed impacts can liberate orders of magnitude more mass than the impacting particle and a fraction of those ejecta particles may escape the target bodies' gravity wells. In this way, all airless bodies can be both sources and sinks of dust grains.

This review focuses on dust phenomena relating to airless bodies, highlighting recent advancements in our understanding of such phenomena in the last two decades. To highlight the diverse nature of these objects and frame the subsequent discussions, we begin the review by describing the orbital distribution of airless bodies in the solar system in Sect. 2. We continue by describing dust evolution on airless surfaces and proceed with subsequent sections ordered phenomenologically outward from the surface. In Sect. 3, we summarize the current understanding of how dust on the surface of an airless body evolves and highlight recent asteroidal regolith sample return analyses. In Sect. 4, we specifically describe processes related to meteoroid bombardment, such as impact gardening, exospheric generation, and surface contamination. A global, direct consequence of meteoroid bombardment is the formation of ejecta clouds, and is discussed in depth in Sect. 5. Escaping, unbound ejecta becomes a source of planetary or interplanetary meteoroids and is the focus of Sect. 6. We briefly address electrostatic effects on surface processes in Sect. 7. Finally, we conclude in Sect. 8 with a summary of current open questions and discuss future dust measurements.

2 Orbital Distribution of Airless Bodies

To highlight the diverse nature of airless bodies, here we describe their orbital distribution in the solar system. Airless bodies can be divided into three groups: airless planets, airless moons, and small bodies (e.g. asteroids and comets). For small airless bodies throughout the solar system, there are three reservoirs that are dynamically stable over the age of the solar system: (1) the main Asteroid Belt between Mars and Jupiter; (2) the Jupiter Trojans clustered in the L4 and L5 Lagrangian points of Jupiter's orbit; and (3) the Trans Neptunian Objects (TNOs) outside the orbit of Neptune. These reservoirs contribute bodies to the transitional classes, which include the Near Earth Objects (NEOs) and Centaurs, which consist of bodies that have been perturbed from the stable reservoirs into short-lived planet crossing orbits. When perturbed even further into orbits crossing the inner solar system, Centaurs can develop into comets.

Small bodies orbiting the Sun in the inner solar system are typically classified as either comets or asteroids via three separate criteria: 1) activity, via the detection of observable atmospheres; 2) ice content; and 3) the Tisserand parameter, $T_J = \frac{a_J}{a} + 2\sqrt{((1-e^2)\frac{a}{a_J})}\cos i$, where a, e, i and the semi-major axis, eccentricity, and inclination of the body and Jupiter, denoted by subscript "J" (Jewitt 2012). Comets typically have high activity levels, are predominantly composed of icy materials, and have $T_J < 3$. Asteroids, on the other hand, typically have lower levels of activity, are more rocky, and have $T_J > 3$. The discrimination between comets and asteroids is not entirely precise, as will be discussed in Sect. 6.2.2 for the case of active asteroids. See review by Jewitt (2012) for a comprehensive description of these criteria.

All of these dynamical reservoirs are essentially survivors of the small bodies in the solar system's original accretion disk that subsequently formed the terrestrial and gas giant planets. These populations have been modified, migrated, and depleted not only by accretion onto the planets and impacts with other small bodies, but also by orbital disruption from gravitational resonances with the giant planets and giant planet orbital migration (Morbidelli et al. 2015). This section provides a brief description of the dynamical classes of airless bodies along with their history. As there is an entire review paper dedicated to comets in this Topical Collection, we omit discussion of cometary-specific processes in this paper.

2.1 Dynamical Classes and Origin of Airless Bodies

2.1.1 Near-Earth Objects (NEOs)

NEO's are a population of relatively small asteroids and comets that have been perturbed by dynamical processes into short-lived orbits that cross the paths of the inner planets (Gladman et al. 2000). The largest NEO is the ~ 32 km diameter (1036) Ganymed and is joined by approximately 1000 objects larger than 1 km diameter. In total, there are probably over 1 million NEOs greater than 40 meters in diameter (DeMeo et al. 2015b). These objects are dominated by chaotic interactions with the inner planets and as a result this is a transitory population with mean lifetimes of approximately 10 million years (Gladman et al. 2000). The population is being constantly replenished by a flux of bodies perturbed from the main asteroid belt, primarily through orbital resonances with Jupiter. Gravitational interactions with Jupiter can rapidly excite the orbit of the small body, increasing its eccentricity until it becomes planet crossing. Early in the study of the asteroid belt it was observed that the dynamical resonance zones called the Kirkwood gaps were remarkably clear of asteroids. Small bodies are moved into the Kirkwood gaps by several processes including collisions with other asteroids and, for objects on the order of a kilometer or smaller, by the Yarkovsky effect (Bottke et al. 2002). As a result, the NEOs effectively sample the zones of the main asteroid belt close to the major resonance zones of the inner asteroid belt.

2.1.2 Main Asteroid Belt

The main asteroid belt lies between 2–4 AU and is probably a strongly depleted and collisionally evolved remnant of what was likely a much more massive population of small bodies (Morbidelli et al. 2015). The current total mass of the asteroid belt is approximately 10^{-4} Earth masses. The presence of Jupiter is the driving force in the structure of the asteroid belt; perturbations from Jupiter early in the solar system formation era limited planetesimal growth and cleared out almost all asteroids within two AU of Jupiter. Orbital mean motion resonances with Jupiter have opened up the Kirkwood gaps and continue to perturb and deplete the asteroid belt, either into the inner solar system or into orbits that interact with Jupiter and are eventually ejected from the solar system. Because of the relatively high eccentricities of asteroidal orbits, collisions are relatively frequent (Morbidelli et al. 2010). Catastrophic collisions between large objects can produce clouds of fragments that are dynamically related. These fragments form the roughly 120 collisional families of the asteroid belt (Masiero et al. 2013; Nesvorný et al. 2015). Differences in thermal inertia observations of main belt asteroids suggest there is a major transition at 100 km diameter in the total dust content of these objects. Smaller asteroids tend to be similar to the NEOs and show rocky surfaces, while larger asteroids tend to have very small thermal inertias indicating that they are even dustier than the Earth's moon (Delbo et al. 2015).

2.1.3 *Jupiter Trojans*

These objects are a large population of asteroids trapped in the L4 and L5 Lagrangian points which lead and trail Jupiter in its orbit by 60 degrees. The orbits slowly librate around these points and are characterized by high inclinations. These zones are affected by a number of weak resonances with Jupiter and Saturn which induce chaotic orbital behavior over the age of the solar system (e.g., Levison et al. 1997; Marzari et al. 2003; de Elía and Brunini 2007). Jupiter Trojans can transiently become temporary satellites of Jupiter or join the population of Jupiter family comets. The populations in both zones are roughly similar with about 60,000 objects larger than 2 km (Yoshida and Nakamura 2005). Jupiter Trojans are thought to have two possible origins: (1) during the runaway, accretionary growth of Jupiter, the rapid increase of Jupiter's gravity cause planetesimals that were in or near the Jovian orbit to become trapped in the Trojan zones, or (2) Jupiter Trojans were captured during the migration of the giant planets 500–600 million years after the solar system formation (Morbidelli et al. 2015). Migration during this period greatly disrupted the population of small bodies around the solar system, probably causing the heavy bombardment seen on the Moon and greatly increasing the flux of scattered planet crossing small bodies.

2.1.4 *Centaurs*

These objects are a transitional group of small bodies that orbit between Jupiter and Neptune. Centaurs have their origins from TNOs that are perturbed onto planet-crossing orbits that are not protected by orbital resonances (Horner et al. 2004). These orbits are chaotic from multiple gravitational encounters with the giant plants and are unstable on a timescale of 10^6–10^7 years. Centaurs evolve into Jupiter-family comets or are ejected from the solar system by encounters with Jupiter (Morbidelli et al. 2015).

2.1.5 *Trans-Neptunian Objects*

TNOs are a general term for any object beyond the orbit of Neptune and includes the Kuiper belt, scattered disk, and Oort cloud. The Kuiper Belt and the scattered disk probably formed as the result of the migration of the gas giants (Levison et al. 2008). The Oort cloud was filled from small bodies ejected by Jupiter into long elliptic or parabolic orbits during the early stages of solar system accretion. Since the Oort cloud extends over 50,000 AU from the Sun, an additional source of material is from an exchange of ejected small bodies between the Sun and its sibling stars in a stellar cluster during the earliest days of solar system formation (Morbidelli et al. 2008). TNO reservoirs contribute objects to the inner solar system from a number of processes. Objects in the Oort Cloud are perturbed by gravitational interactions with passing stars and molecular clouds in Milky Way. This is probably the source of long-period and Halley-family comets as well as at least some of the Centaurs and Jupiter-family comets (Levison et al. 2008). The scattered disk, because of its interactions with Neptune, high orbital eccentricities, and high inclinations is dynamically less stable and the likely origin for most periodic comets. The Kuiper Belt, on the other hand, is dynamically stable and only rarely contributes objects to planet crossing orbits.

2.1.6 Mercury and Moons

The final category of airless bodies includes the airless planet Mercury as well as numerous moons in the solar system. Origins depend upon the accretionary and chaotic processes of early solar system formation. While Mercury formed from the accretionary disk of the solar nebula, the giant planet large moons probably formed from accretionary discs that formed around these large gravitational concentrations. Other large moons formed from major impacts on the primary (Earth and Pluto). The numerous small moons of the giant planets have their origins either from collisional fragments of previous moons or from small bodies captured during giant planet migration (Nesvorný et al. 2014).

3 Evolution of Dust on Surfaces of Airless Bodies

The surfaces of airless bodies are covered by granular materials called regolith. The term 'regolith' comes from the Greek *rhegos* (blanket) + *lithos* (stone) and is defined by the Glossary of Geology as, "A general term for the layer or mantle of fragmental and unconsolidated rock material, whether residual or transported and of highly varied character, that nearly everywhere overlies bedrock. It includes rock debris of all kinds" (Jackson 1997). In planetary studies, the term regolith is usually used in a broader sense meaning a loose unconsolidated material of the upper surface of a planetary body, including icy bodies. The thickness, texture, particle size, distribution, chemistry, and evolution of regolith depends strongly on the geologic context of the body. The creation and depletion of regolith is driven by weathering processes, including from the meteoroid, thermal, and energetic particle environments.

The best studied example is the lunar regolith, which consists of a very loose, fluffy powder about 20 cm deep overlying a fine-grained and more compact layer that can vary between 4–10 meters in depth (Taylor 1982). This loose layer is in equilibrium between processes that disaggregate particles such as thermal shock and meteoroid bombardment, and processes that fuse particles together such as melting. Below the regolith layer is the so-called megaregolith, which consists of a deep shattered layer of large scale impact ejecta (roughly 2 km thick), structurally disturbed and displaced crust (again from impacts, up to 10 km thick), and finally in situ fractured crust down to 25 km depth (Taylor 1982).

3.1 Regolith Evolution due to Space Weathering

All airless surfaces undergo space weathering, defined as the physical, mineralogical, and chemical alterations suffered by solid materials when exposed to the space environment. This includes a broad range of processes that are fundamentally driven by energetic processes such as impacts, radiation, and the solar wind. The effects of space weathering depend on the chemistry and crystal structure of the target material. Here we briefly review the major components of space weathering and their implications for the regoliths of airless bodies, which are graphically summarized in Fig. 1.

Comminution is the physical process of breaking of rocks and minerals into smaller particles; it is the primary process by which dust is generated on airless bodies. In space this can occur in several ways. Impacts at all scales grind down particle size. Major impacts that do not completely disrupt the parent body (e.g., Asphaug et al. 1998; Michel et al. 2003; Goodrich et al. 2004) produce ejecta blocks and fracture the bedrock, while meteoroids grind down gravel and blocks to dust. For objects in interplanetary space (as opposed

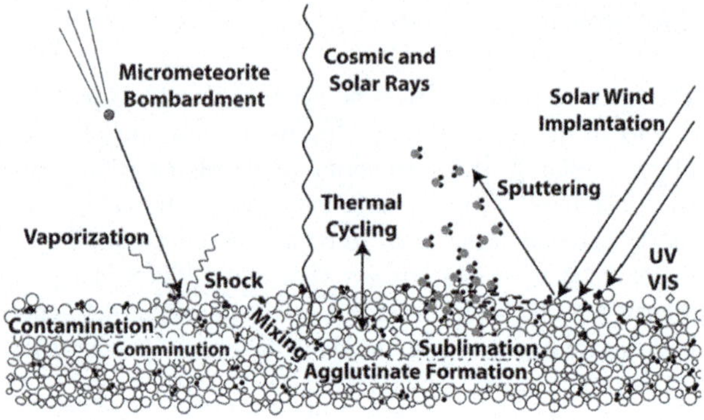

Fig. 1 A summary of the many space weathering processes occurring on airless bodies. *Reproduced from Pieters and Noble (2016) with permission from the Journal of Geophysical Research*

to objects within planetary gravity wells), this effect is a function of heliocentric distance with meteoroid impact velocities dropping off rapidly with distance. Another comminution process is thermal fracturing. In the inner solar system, strong solar heating on the sunward side of rotating asteroids followed by sudden cooling on the dark side, produces substantial thermal shock due to either the different thermal expansion characteristics of the constituent minerals or the thermal gradient within a homogeneous rock which leads to tensile stress between the outer and inner layers. This thermal shock has been shown to propagate fractures in meteorites and, in the inner solar system, can rapidly disaggregate materials (Delbo et al. 2014). This effect is a strong function of heliocentric distance since it depends upon solar heating. These comminution processes push airless surfaces to accumulate fine-grained dust which is seen, for example, in the powdery lunar soil (Hapke 2001).

Dust accumulation, however, is also a strong function of local gravity and heliocentric distance. For small asteroids with low gravity and hence, low escape velocity, a significant portion of impact ejecta will escape the body. Thermal inertia studies show that smaller asteroids are fundamentally coarser grained and rockier than larger asteroids (Delbo et al. 2015). For small near-Earth asteroids, electrostatic interactions may preferentially shed small grains, making the surfaces of these objects even rockier than similar sized bodies in the asteroid belt, since both UV and plasma fluxes decrease with the square of heliocentric distance (Lee 1996). Another hypothesis is that the large surface roughness on small asteroids is caused by the inwards migration of regolith fines due to the Brazil Nut Effect (e.g. Perera et al. 2016).

Acting against comminution is agglutination. Agglutination is the process of melting and welding mineral and rock fragments together by micrometeorite-impact-produced glass. The primary example of agglutination are the high velocity impacts into lunar soil that produce enough heating to melt or vaporize material and weld fragments back together (Taylor 1982). This process is present at the Moon and Mercury, since impacts are energetic enough for the required melting and vaporization; however, it may also be relevant for satellites deep within the gravity wells of the large outer planets. Agglutinates are only rarely found in meteorites because average impact velocities in the asteroid belt are too low to produce extensive melting. Since agglutination welds particles together, it competes with the generation of dust by comminution. Over time, an equilibrium develops where comminution is balanced by agglutination.

In addition to kinetic energy inputs from impacts, energy can also be delivered by a range of processes associated with radiation and the solar wind (Brunetto et al. 2015). Without an atmosphere, solar system materials on airless bodies are directly subjected to the cosmic and solar particle environment. Electrons and ions can carry substantial energy flux and change the chemical environment of surfaces to which they are exposed. Energetic solar and galactic cosmic rays can penetrate minerals and cause crystal damage, producing dislocations and dangling bonds that make crystal structures more disordered and potentially more reactive. Cosmic ray spallation is a result of the collision between an energetic cosmic ray particle and a nucleus, which results in the modification of the atomic makeup of surface elements. This changes the elemental chemistry of the target atom and, because of nuclear interactions with the cloud of ejected nuclear particles, potentially the elemental chemistry of surrounding atoms. Sputtering is a slightly less energetic process, whereby atoms are ejected from the target material due to bombardment by thermal and supra thermal ions (Biersack and Eckstein 1984; Berisch and Eckstein 2007). The sputtered atoms mostly recondense on nearby grain surfaces but the general trend is to increase crystal disorder, disrupt crystal structure, and create amorphous rims of recondensed material (e.g., Christoffersen et al. 1996; Keller and McKay 1997; Carrez et al. 2002; Keller and Berger 2014; Matsumoto et al. 2015; Poppe et al. 2018). Airless bodies are typically effective cold traps for the recondensing vaporized material.

For terrestrial bodies, the net effect of space weathering from charged particle irradiation and meteoroid bombardment is to both darken and redden the surface (e.g., Pieters et al. 1993, 1994; Noble et al. 2007; Pieters and Noble 2016). Many airless bodies are directly exposed to the solar wind, which is composed of approximately 97% protons (H^+) and 3% alpha particles (He^{++}) at typical speeds of 420 km/s (Dmitriev et al. 2011). Solar wind material is implanted in the surfaces and shallow interiors of exposed regolith. The fundamental chemical implication of solar wind implantation is to set up a chemical disequilibrium between the reducing veneer of the solar wind hydrogen and the oxidized silicates that make up airless bodies, adding to the already low oxygen fugacity of the hard space vacuum. This chemical disequilibrium can persist as long as temperatures are low enough to preclude any rapid reactions; however, if the regolith material is heated by, for example, a meteoroid impact, then the implanted hydrogen can drive reduction reactions in the silicates. These reactions convert iron-rich silicates into elemental iron and iron-poor silicates, producing nanophase iron (npFe) particles often found suspended in the agglutinitic glass of lunar regolith (Noble et al. 2007). These npFe particles are a source of one of the characteristics of space weathering, namely, the very strong red slope in lunar reflectance spectra (Pieters and Noble 2016).

While both charged particle irradiation and meteoroid bombardment play a role in space weathering as a whole, the relative importance of these two processes remains uncertain. Furthermore, the rates of irradiation and bombardment may actually change as a function of heliocentric distance, as fluxes and velocity distributions of both interplanetary dust grains and ambient charged particles change. For example, spectra obtained at Mercury are consistent with both an overall greater amount of sub-microscopic iron and a higher ratio of larger (> 50 nm) to smaller (< 50 nm) iron particles than at the Moon, which is attributed to increased production of impact vapor and melt due to meteoroid bombardment at Mercury (Lucey and Riner 2011). This is furthermore consistent with the expected increase in impact velocities of meteoroids at Mercury given its smaller heliocentric distance (Cintala 1992; Hahn et al. 2002). Other optical effects of weathering are more subtle. On asteroids, weathering creates crystal damage in silicates and decreases spectral contrast, which darkens reflectance spectra and shallows absorption bands. Shock can also redistribute the particle size of opaques in asteroid material, resulting in increased darkening (Britt and Pieters 1994).

Space weathering by meteoroid impacts in the outer solar system likely takes on a different character than in the inner solar system due to the predominantly icy surface compositions of outer solar system bodies. For example, many icy surfaces in the outer solar system are in the form of crystalline water ice rather than amorphous solid water (e.g., Grundy et al. 1999; Brown and Calvin 2000; Jewitt and Luu 2004; Merlin et al. 2007; Trujillo et al. 2007; Dalle Ore et al. 2015; Terai et al. 2016). Amorphous solid water would typically be expected given the relatively cold temperatures in the outer solar system and amorphizing processes such as proton and energetic ion irradiation (e.g., Jenniskens and Blake 1996; Leto and Baratta 2003; Mastrapa and Brown 2006). The formation of crystalline water ice requires either fresh deposition of water ice (i.e. via cryovolcanism; e.g., Jewitt and Luu 2004; Cook et al. 2007; Desch et al. 2009; Desch and Neveu 2017) or a source of external heat to transform the amorphous ice into a crystalline state. While one possible source of external heat is from large-scale impacts (e.g., Dalle Ore et al. 2015), meteoroid bombardment may play a role in producing crystalline water ice on planetary surfaces in the outer solar system (Porter et al. 2010). Annealing via meteoroid bombardment has also been suggested as an active process in extrasolar debris disks that display evidence of crystalline water ice signatures (McClure et al. 2015).

3.2 Laboratory Studies of Dust Samples from Airless Bodies

Given the complexity of processes occurring on regolith that covers airless bodies, the ability to examine actual samples of this material in a laboratory setting is extremely valuable. However, returning regolith samples from airless bodies is a challenging endeavor and has only been performed on a small number of airless bodies in the solar system.

The Apollo lunar sample returns revolutionized our understanding of space weathering processes occurring on the lunar surface, and have served as an analogue for weathering processes on many other airless bodies. Our understanding of most of the concepts outlined in the previous section evolved dramatically due to these lunar sample returns and their subsequent analysis via Earth-based techniques. There is large body of literature covering a multitude of lunar regolith topics. Comprehensive reviews on space weathering are given in Hapke (2001) and Pieters and Noble (2016), for example. For detailed discussion of the lunar mare soils, nanophase iron, and particle size distributions, see for example Taylor et al. (2001), Park et al. (2008), Liu et al. (2008), Taylor et al. (2010), Liu and Taylor (2011). In this section, we focus on the more recent Itokawa mission, which returned the first asteroidal regolith sample to Earth, allowing for direct comparisons between the lunar regolith and that of a much smaller airless body at 1 AU.

In September 2005, the Japanese Hayabusa spacecraft visited 25143 Itokawa (Fujiwara et al. 2006), an S(IV)-type, ellipsoidal (500 × 294 × 209 m) asteroid on an Apollo and Mars-crossing orbit. Itokawa was found to be covered with boulders that have a variety of sizes and morphologies. The current shape of Itokawa suggests it is a rubble pile asteroid comprised of reassembled pieces of a much larger parent asteroid (about 20 km in radius), which most likely broke up due to a catastrophic impact (Fujiwara et al. 2006; Nakamura et al. 2011). One smooth area, MUSES-C Regio, was found at the center of the asteroid (Yano et al. 2006). Hayabusa landed twice on the MUSES-C Regio to collect dust particles. In June 2010, Hayabusa returned to the Earth with a sample of < 1 g of Itokawa dust particles in a recovery capsule, allowing for their study and characterization in a terrestrial laboratory setting. Of the recovered particles, the largest are several hundred microns, while most are smaller than 10 μm in diameter (Nakamura et al. 2011). The size distribution and morphology of the particles suggest that meteoroid impacts on the asteroid surface is the primary source of these regolith particles, and that seismic-induced grain motion in the smooth terrain abrades them over time (Tsuchiyama et al. 2011). Itokawa particles retain morphological and isotopic evidence suggestive of a dynamic asteroidal regolith environment.

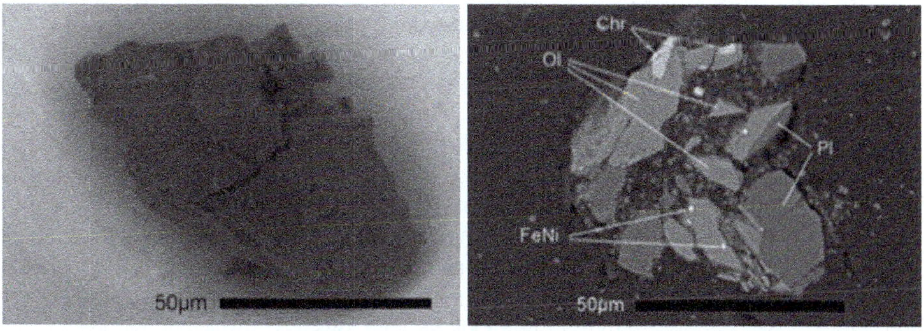

Fig. 2 Regolith particle RA-QD02-0242 recovered from MUSES-C Regio of asteroid 25143 Itokawa. (**a**) The particle shows edged morphology and contains some large cracks, suggesting that it is an impact-fragmented piece of a larger rock. (**b**) Cross section of the particle shows that it consists of well crystalline olivine (Ol), plagioclase (Pl), chromite (Chr), and FeNi metal (FeNi). Large cracks go through the entire particle

A typical particle (Fig. 2) consists mainly of anhydrous silicates such as olivine, low and high Ca pyroxene, and plagioclase with chemical compositions similar to LL ordinary chondrites, indicating S-type asteroids are parent objects of ordinary chondrites, which are the most abundant meteorites recovered on Earth (Nakamura et al. 2011). Most particles showed mineralogical evidence of long-term thermal metamorphism. The peak temperature of the metamorphism was around 800 °C based on pyroxene thermometry (Nakamura et al. 2011). The short-lived radionuclide ^{26}Al was probably a major heat source when Itokawa's parent body formed in the early solar system. At ~ 5 million years after the formation of calcium-aluminum-rich inclusions (CAI) at various depths, the asteroid experienced its maximum temperature and subsequently slowly cooled down. Thermally-metamorphosed particles were initially formed deep inside of the asteroid, but excavated and deposited on the newly accreted body after the putative impact event. Ar-Ar dating of several particles implies that the final disrupting catastrophic impact occurred around 1.3 Gyr ago (Park et al. 2015).

Analysis of the recovered particles showed high concentrations of light solar wind noble gases (Nagao et al. 2011). High resolution electron-microscope observations indicated that a thin layer of Fe-rich nanoparticles developed at the surface of ferromagnesian silicates, which suggests reduction of FeO in silicates by solar wind protons (Noguchi et al. 2011). Some particles showed vapor deposition of sulfur-bearing Fe-rich nanoparticles on the surface, suggestive of sputtering of FeS from nearby minerals (Noguchi et al. 2011). Meteoroid impacts on the surface of regolith produced microcraters and melt splashes on particle surfaces that are similar to those found in lunar soils (Nakamura et al. 2012). The residence time of regolith particles can be estimated from concentrations of spallogenic noble gases that were produced in the particles during exposure to galactic cosmic rays. Low concentrations of noble gases suggest the particles had relatively short residence times of $\lesssim 8$ million years at $\lesssim 1$ m depth (Nagao et al. 2011). This implies loss of small dust particles from Itokawa's surface due to low gravity and therefore suggests continuous surface depletion and refreshment over time.

A wealth of information has already been garnered from the Itokawa sample return, allowing for characterization of asteroidal regolith in a manner only possible with advanced ground-based laboratory techniques. This sample will continue to provide insight into asteroidal evolutionary processes for decades to come. Future asteroidal missions, such as OSIRIS-REx (Lauretta et al. 2015) and Hayabusa-2 (Yoshikawa et al. 2011), aim to return samples from their respective targets. These missions would further diversify our asteroidal sample set and greatly improve our understanding of asteroidal surface and dust grain evolution.

3.3 Remote-Based Estimations of Regolith Depth and Particle Size

While asteroidal samples allow for highly sophisticated laboratory techniques to analyze the surface of an asteroid, acquiring such samples is challenging. Many remote-based techniques have been developed, such as phase-resolved photometry and polarimetry, to characterize regolith properties by more indirect methods. Measurements of phase angle effects in brightness and polarization give a first look into the top-most surface properties. Optical polarization parameters of the lunar surface, asteroids, Centaurs, and TNOs are found to be quite varied (see reviews by Shkuratov et al. 2011; Belskaya and Bagnulo 2015; Belskaya et al. 2015). These polarization differences, as well as observations of thermal inertia (e.g., Gundlach and Blum 2013), have been interpreted to suggest the average size of asteroid regolith particles is greater than that of lunar regolith particles. For example, the Moon and Mercury were found to have small average particle sizes on the order of 10's of μm, while smaller bodies like the Martian moons Phobos and Deimos, as well as other asteroids of similar sizes to those moons were found to have larger average particles sizes on the order of 1–10's mm (Gundlach and Blum 2013). The polarization properties of the surfaces of TNOs and Centaurs also suggests the presence of a thin frost layer of sub-micron particles (Belskaya et al. 2010).

The values of thermal inertia determined from infrared measurements provide insight into the thickness and texture of the regolith layer on the surfaces of airless bodies. It was found that the thermal inertia of asteroids typically increases substantially with decreasing size, suggesting different regolith properties between large asteroids and small objects without the gravity necessary to retain fine regolith fragments generated by impacts (Delbo et al. 2015). Measurements of thermal inertia of TNOs are 2–3 orders of magnitude lower than expected for compact ices and suggest highly porous surfaces (Lellouch et al. 2013).

More detailed information on regolith properties can be obtained from analysis of crater morphology using high-resolution images taken close to a planetary body. Recent space missions to several asteroids have revealed complex dynamics of regolith on their surfaces (see Murdoch et al. 2015 for a review). The depth of the regolith layer was estimated as $\sim$ 600–800 m for the large asteroids 4 Vesta and 21 Lutetia and less than 50 m for smaller size asteroids (433 Eros, 243 Ida).

4 Meteoroid Bombardment

Meteoroid bombardment of planetary surfaces by interplanetary and planetary dust grains is a major driver of various processes that shape and influence the surfaces of airless bodies throughout the solar system. These processes include the vaporization and melting of surface material, the injection of neutral atoms and molecules into exospheres or coronae, the production of micron and sub-micron sized ejecta, and the compositional modification of the surface material via the introduction of exogenous material.

4.1 Meteoroid Fluxes onto Airless Bodies

Fluxes of meteoroids that contribute to the bombardment of airless surfaces originate from three main sources: planetary, interplanetary, and interstellar. Planetary dust grains originate from bodies within the gravity well of a parent planet or object. Examples of this include the massive Phoebe ring at Saturn (Verbiscer et al. 2009; Hamilton et al. 2015), the active plumes of Enceladus (Spahn et al. 2006; Porco et al. 2006), and theoretically, the minor moons of the Pluto-Charon system (Thiessenhusen et al. 2002; Poppe and Horányi 2011; Pires dos Santos et al. 2013). The resulting mass fluxes from any of these various sources can be highly dependent on the dynamical details specific to any system.

Interplanetary dust grains (discussed in detail in the Interplanetary Dust review) are typically the main source of impacting meteoroids for airless bodies throughout the solar system. In the inner solar system, the dust complex is thought to be dominantly formed by cometary dust originating from Jupiter-family comets (e.g., Nesvorný et al. 2010; Nesvorný et al. 2011a) with additional contributions from both asteroids and longer period comets such as Halley type and Oort Cloud comets (Wiegert et al. 2009; Nesvorný et al. 2011b; Pokorný et al. 2014). These fluxes are typically dominated by the sporadic component of the interplanetary background, whose grains experience a significantly different evolution than their parent object such that any dynamical link is erased (Jones and Brown 1993). However, meteoroid showers can be dominant over short periods of time at an airless body (e.g., Killen and Hahn 2015; Colaprete et al. 2016; Szalay and Horányi 2016a).

In the outer solar system, dust grains originate from four main sources: Jupiter-family comets (also the main source for the inner solar system), Halley-type comets, Oort Cloud comets, and Edgeworth-Kuiper Belt objects (e.g. Poppe 2016). Modeling of the equilibrium dust distributions in the outer solar system has suggested varying degrees of contribution from each of these sources, with Jupiter-family cometary dust dominating near Jupiter, Oort Cloud cometary dust dominating between Saturn and Neptune, and Edgeworth-Kuiper Belt grains dominating from Neptune outwards (e.g., Liou and Zook 1999; Vitense et al. 2012; Poppe 2016). At Jupiter, interplanetary dust mass fluxes are on the order of $10^{-13}\ \mathrm{g\,m^{-2}\,s^{-1}}$ (Sremčević et al. 2005), tapering off to approximately $10^{-14}\ \mathrm{g\,m^{-2}\,s^{-1}}$ between 10 and 50 AU (with a slight enhancement in the Edgeworth-Kuiper Belt between 30–45 AU) (Poppe 2016).

Finally, interstellar grains flow through the entire solar system due to the relative motion between our solar system and the local interstellar cloud (e.g., Grün et al. 1994; Mann 2010; Sterken et al. 2012, 2015; Strub et al. 2015; Krüger et al. 2015). The typical mass flux of interstellar dust grains is $\approx 5 \times 10^{-17}\ \mathrm{g\,m^{-2}\,s^{-1}}$, several orders-of-magnitude less than interplanetary dust; however, at far heliocentric distances where the IDP impact speeds will be on the order of a few km/s, the relatively high velocity of interstellar grains (≈ 26 km/s) implies that interstellar grains may still be an important source for processes that scale strongly with impact velocity.

4.2 Gravitational Focusing

Gravitational focusing, the process by which the speed, density, and flux of a stream of particles is increased in the vicinity of a central mass such as a planet, is an important phenomenon when considering the bombardment of planetary surfaces. Particles that enter the Hill sphere of an object are accelerated towards the attracting mass and, as they are pulled towards the same central point, get closer to each other. The amplification of the impactors' kinematic properties is quantified by focusing factors, each defined as the ratio of a variable at a given distance from the body to its value at infinity. These expressions were first derived by Colombo et al. (1966) with later corrections and independent verification by Spahn et al. (2006b). The focusing factors for the incident particle density n_{imp}, velocity v_{imp}, and flux

F_{imp}, relative to their values at infinity ($n_{\mathrm{imp}}^{\infty}$, $v_{\mathrm{imp}}^{\infty}$, and $F_{\mathrm{imp}}^{\infty}$, respectively) are,

$$\frac{v_{\mathrm{imp}}}{v_{\mathrm{imp}}^{\infty}} = \sqrt{1 + \frac{2GM_{\mathrm{p}}}{r(v_{\mathrm{imp}}^{\infty})^2}} \tag{1}$$

$$\frac{n_{\mathrm{imp}}}{n_{\mathrm{imp}}^{\infty}} = 0.5\sqrt{1 + \frac{2GM_{\mathrm{p}}}{r(v_{\mathrm{imp}}^{\infty})^2}} + 0.5\sqrt{1 + \frac{2GM_{\mathrm{p}}}{r(v_{\mathrm{imp}}^{\infty})^2} - \left(\frac{R_{\mathrm{p}}}{r}\right)^2\left[1 + \frac{2GM_{\mathrm{p}}}{R_{\mathrm{p}}(v_{\mathrm{imp}}^{\infty})^2}\right]} \tag{2}$$

$$\frac{F_{\mathrm{imp}}}{F_{\mathrm{imp}}^{\infty}} = \frac{v_{\mathrm{imp}}}{v_{\mathrm{imp}}^{\infty}}\frac{n_{\mathrm{imp}}}{n_{\mathrm{imp}}^{\infty}}, \tag{3}$$

where G is the gravitational constant, r is the radial distance from the center of the body, and M_{p} and R_{p} are the mass and the radius of the central object (e.g., planet), respectively. Since gravitational focusing is inversely proportional to the impact velocity, it will enhance the magnitude and modify the direction of slow-moving dust population fluxes more than faster populations.

Figure 3 (left) shows these enhancement factors for the Galilean moons as a function of radial distance to illustrate how significant the effect can be for a moon orbiting a large planet. The right panel of Fig. 3 shows the flux enhancement factors for a variety airless bodies in the solar system. At Io, deep inside Jupiter's gravity well, gravitational focusing can increase incident dust grain fluxes up to two orders-of-magnitude. Note that gravitational focusing from satellites themselves is typically negligible compared to the focusing from their parent body. Bodies such as Mercury and the Moon have lower gravitational enhancements than the outer planet moons, due to the shallower gravity wells of Mercury itself and the Earth relative to Jupiter and Saturn. At the Moon, Earth's gravitational enhancement is negligible for impact speeds above a few km/s.

Such an effect can be important when calculating relative fluxes between different populations outside and inside of planetary gravity wells. For example, cometary dust populations typically have greater impact speeds (due to their higher eccentricities) than typical, prograde, heliocentric populations (e.g., Edgeworth-Kuiper Belt dust grains in the outer solar system); however, even if cometary sources may dominate the flux in interplanetary space, gravitational focusing can cause heliocentric populations to be the dominant impactor flux deep within a planetary gravity well (e.g., see Fig. 8, Poppe 2016).

4.3 Impact Gardening

A direct consequence of meteoroid bombardment is impact gardening, whereby impacts stimulate lateral and vertical redistribution of regolith on an airless body. Impact gardening rates were estimated for the Moon based on Apollo sample returns and lunar cores, as well as Monte Carlo models that tracked the movement and redistribution of lunar soil subject to meteoroid impacts (e.g., Gault et al. 1974; Arnold 1975; Morris 1978). Morris (1978) derived an empirical relationship between the gardening (or "reworking") depth, defined as the depth at which there exists a 50% probability of having been gardened by an impact, and time given by $D_R = 2.2t^{0.45}$, where D_R is in units of cm and t is in units of Myr. This rate is fairly significant, implying that a full meter of soil has been gardened over the age of the Moon. Much more recently, Szalay and Horányi (2016b) have estimated the accumulation timescale (i.e., the time period over which the surface is covered by a single layer of impact ejecta) and depth accumulation rate of regolith gardening at the Moon near the equator using data from the Lunar Atmosphere and Dust Environment Explorer (LADEE) mission (Elphic

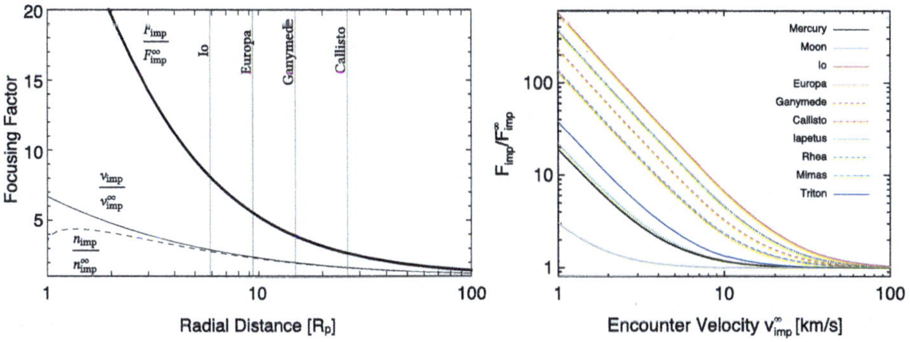

Fig. 3 *Left*. Focusing factors in the Jovian system as a function of distance from the planet for impacts of interplanetary dust particles with $v^\infty_{imp} = 9$ km/s. The thin vertical lines are the positions of the Galilean moons. *Right*. Flux focusing factor for Mercury, the Moon, and a selection of moons of Jupiter (orange), Saturn (green), and Neptune (blue)

et al. 2014) to be approximately 10^4 yrs and 40 μm/Myr, respectively. Analysis of crater production using data from the Lunar Reconnaissance Orbiter (LRO) suggests the top two centimeters are churned every 81,000 yrs (Speyerer et al. 2016). Comparing the LADEE and LRO gardening estimates remains challenging as each estimate addresses different ejecta regimes; LRO images allow for characterization of the redistribution of the low velocity portion of the ejecta, while the LADEE data infers the amount of reblanketing occurring due to particles which are ejected to significantly high altitudes ($\sim$ 1–250 km).

The process of impact gardening will occur on all airless bodies in the solar system in one form or another. At the Moon, which is sufficiently large to retain a large fraction of impact ejecta, meteoroid impacts serve to rework and redistribute the lunar soil to different locations and depths. At smaller bodies, such as the majority of asteroids, impact bombardment will most likely result in a large fraction of ejecta that is unbound and will not return to the surface. In this regime, the impacts directly serve to expose fresh material. The transition between these two regimes is dependent on a variety of factors such as the mass of the body and the speed distribution of impact ejecta. The relative fraction of bound vs. unbound ejecta has been previously estimated for NEAs (Szalay and Horányi 2016c), however, constraints on this ratio for all airless bodies in the solar system have not yet been undertaken.

Impact gardening has important implications for the presence of volatile material known to be sequestered in polar cold traps at Mercury (e.g., Harmon et al. 2011; Neumann et al. 2013), the Moon (Colaprete et al. 2010; Hayne et al. 2015), and Ceres (Platz et al. 2016). Impact vaporization of meteoroids on airless surfaces brings significant amounts of volatile species (e.g., H_2O) (Berezhnoy et al. 2012; Füri et al. 2012) which can be deposited into polar cold traps following migration across the planetary surface. When meteoroid bombardment occurs on the polar cold trapped regions themselves, two processes can occur: (1) ejection and loss or (2) burial and further sequestration of volatile material. Vapor production from meteoroid bombardment is a relatively "hot" process with characteristic temperatures of 2500–5000 K (see Sect. 4.4) and thus, any vaporized material can easily be lost from polar regions. On the other hand, secondary ejecta produced by meteoroid bombardment will be redeposited across the surface and can effectively bury volatile material. Monte Carlo simulations have explored this process at both Mercury and the Moon and found that significant layers of volatile deposits can be buried several meters deep due to ejecta emplacement from bombardment (e.g., Crider and Vondrak 2003; Crider and Killen 2005; Hurley et al. 2012). Reserves of polar hydrogen have been observed at the Moon, which have survived billions of years of impact gardening, adding further evidence to the notion that impact gardening may help preserve volatile content in the polar regions (Siegler et al. 2016).

4.4 Surface Bounded Exospheres

Meteoroid bombardment is an important source in the production of neutral surface-bounded-exospheres and/or coronae at airless bodies throughout the solar system. Upon impact on a planetary surface, meteoroids can be significantly, if not completely, vaporized, along with varying fractions of the surface material as well (Cintala 1992). Any vaporized material can then be injected onto ballistic or escaping trajectories about the body. Laboratory experiments have quantified the degree of vaporization and the characteristic temperatures of the resulting neutral cloud, finding that the vaporization yield scales as $v^{2.4}$, where v is the impact velocity (Collette et al. 2013, 2014). While the experiments observed a vaporization fraction of 5–10% for the slowest impact velocities (2 km/s), extrapolation of the power-law fit suggested that total vaporization yields equal to and greater than the incoming meteoroid mass (indicating that surface material is also being vaporized in appreciable quantities) occurred for impact velocities greater than approximately 20 km/s. These experiments determined the impact vaporization cloud temperature to be between 2500–5000 K, with a power-law dependence on the impact velocity.

In the inner solar system, surface-bounded exospheres have been detected at Mercury and the Moon (see extensive reviews by Stern 1999; Killen and Ip 1999), and predicted for the moons of Mars (Cipriani et al. 2011; Poppe et al. 2016) and various asteroids (Morgan and Killen 1998; Schläppi et al. 2008). As with impact gardening, the extent to which a surface bounded exosphere exists at an asteroid depends on a variety of factors, such as the mass of the body, the relative sources of elements on the surface available to populate the exosphere, and the speed distribution of outgoing neutrals. Meteoroid bombardment plays an important role, along with charged-particle sputtering, in liberating refractory materials from the surface regolith into the exosphere, including species such as O, Na, K, Al, Mg, Ca, Ti, and Fe (e.g., Sarantos et al. 2012; Colaprete et al. 2016). The spatial and temporal variability of the incident meteoroid bombardment flux is often reflected in the vaporized neutral distributions, as observed at Mercury and the Moon (e.g., Hunten et al. 1998; Smith et al. 1999; Kameda et al. 2009; Killen and Hahn 2015; Szalay et al. 2016; Colaprete et al. 2016; Pokorný et al. 2017), including variations from both the sporadic interplanetary dust background and meteoroid streams (such as the Leonid and Geminid streams). Finally, incoming meteoroids can also serve as a source of exogenous species at airless bodies; for example, meteoroids are believed to introduce approximately 5×10^5 kg/yr of water to the lunar surface, potentially the dominant source of such influx (Füri et al. 2012).

4.5 Surface Contamination

Meteoroid bombardment of airless bodies can also alter the composition of planetary surfaces via the introduction of exogenous material. Perhaps the most striking example of this process is Saturn's moon Iapetus, displayed in Fig. 4, which possesses a hemispherical dichotomy in both color and albedo (Cruikshank et al. 1983). Among various theories, the deposition of dust from Saturn's outer, irregular, retrograde moon Phoebe onto the surface

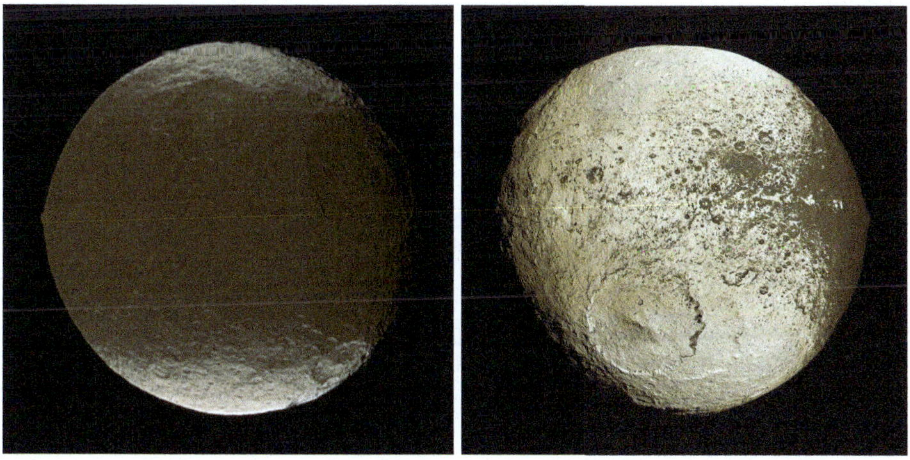

Fig. 4 The leading and trailing hemisphere of the saturnian moon Iapetus, respectively, showing the preferential dark dust deposition on the leading hemisphere. [Courtesy: NASA/JPL/Cassini mission]

of Iapetus stands as the most plausible explanation, especially given the recent discovery of a dust ring around Saturn co-located with Phoebe (Verbiscer et al. 2009). Phoebe's dark and complex surface spectrally matches the dark surface material seen on Iapetus while the bright and icy surface is most likely Iapetus' "true" composition (Clark et al. 2012; Dalle Ore et al. 2012). Dynamical modeling of Phoebe's dust ring and its deposition onto Iapetus strongly supports this hypothesis, with the modeled surface deposition region closely aligning with that observed by Cassini (Tamayo et al. 2011). Surface contamination of satellites is not limited to Iapetus, but seen at most of the mid-sized saturnian icy satellites (Rhea, Dione, Tethys), albeit in far less dramatic fashion (e.g., Clark et al. 2008; Scipioni et al. 2013, 2014). While these satellite surfaces are predominantly (> 99%) water ice, a "dark material" on the order of 1% by composition persists across these moons. Most analyses have suggested that this material is of exogenous origin to the saturnian system, suggesting that material transported into the system from interplanetary dust may be a possible cause (e.g., Mendis and Axford 2008); however, the exact source and method of deposition onto the icy saturnian satellites is not well understood. Dynamical models of dust in the Phoebe ring suggest that while the grains are easily transported to Iapetus (the next major moon inwards of Phoebe), any grains that pass Iapetus are efficiently swept up and/or dispersed from the saturnian system by Titan and thus, unable to reach the inner icy satellites (Tamayo et al. 2011). In contrast, interplanetary dust penetrates throughout the entire saturnian system; however, it is not clear yet that preferential directions in the incoming IDP flux at Saturn (i.e., similar to that suggested at the Galilean satellites (Sremčević et al. 2005)) will yield asymmetric patterns on the inner icy satellites due to the tidally locked nature of these moons. Continued study of dust dynamics and surface contamination of the saturnian satellites, and indeed at all the satellites of the giant planets, is clearly still warranted. Surface material exchange and contamination has also been studied for other multiple satellite systems, including Phobos and Deimos (e.g., Nayak et al. 2016) and the Pluto-Charon system (e.g., Stern 2009; Poppe and Horányi 2011; Porter and Grundy 2015); however, definitive identification of surface ejecta exchange at these multiple satellite systems has not yet been shown.

5 Impact Ejecta Clouds

Ejecta clouds exist in the vicinity of all airless bodies in the solar system due to continual bombardment by fast meteoroids. Comets also experience impact ejecta processes, however, more detailed discussion of additional loss processes at comets is beyond the scope of this review. The impactors are interplanetary dust particles with typical sizes of 1–100 μm (Divine 1993) and typical speeds of a few 10 km/s, particularly after gravitational focusing near a large planet (Colombo et al. 1966; Spahn et al. 2006b). Owing to the impactor's high kinetic energy, the surface of the target is eroded and a cloud of ejecta, typically more numerous and massive than the impactor, is released into space. Depending on the ejection speed and the target mass, the grains either fall back to the surface or they escape the target's gravitational influence and add to the planetary or interplanetary dust environment.

5.1 The Impact Ejection Process

Physically, the impact-ejection process is a combination of an inelastic collision between the impactor and the target, and a fragmentation of the impactor and/or the target. The momentum of the incoming dust particle is conserved and distributed over the ejecta and target, where the target body's velocity is unchanged due to its high mass. However, the impactor's kinetic energy is not conserved and largely (typically > 70%) used to disintegrate and heat the material (Asada 1985; Hartmann 1985). Modeling the impact-ejection process from basic physics is prohibitively complex. Instead, impact-ejection models are based on findings of laboratory experiments (e.g., Koschny and Grün 2001a,b) and on theoretical constraints. Future measurements at high-speed dust accelerators may further elucidate fundamental impact ejection characteristics (Shu et al. 2012).

Given the impactor's mass, speed, and flux at the target, the amount of material ejected by a given impactor can be estimated. Impact experiments show that the volume of the impact crater is a measure for the kinetic energy of the impactor. The relation can be fitted by a power law in the form $V_e = c_1 K_{imp}^{c_2}$, where the parameters c_1 and c_2 depend on the target material. For impacts on the Jovian moons, the surface material is a mixture of ice and silicate, where Koschny and Grün (2001a) find a prefactor of $c_1 = 6.69 \times 10^{-8}(0.015)^{x/100}$, where x is the fraction of silicate in percent (0% = pure ice, 100% = pure silicate), and exponents $c_2 = 1.08$, 1.27, and 1.27 for 5%, 10%, and 20% silicate content, respectively.

The yield is defined as the ratio of the total mass ejected by an impactor to the impactor's mass, $Y = \frac{\sum m_e}{m_{imp}}$. As the mass of the ejecta is proportional to the volume of the impact crater, the yield is therefore a function of the impactor mass, the impact speed, and the composition of the target's surface (in SI units),

$$Y = 6.69 \times 10^{-8}(0.015)^{x/100} \rho_t \, 2^{-b} \, m_{imp}^{b-1} \, v_{imp}^{2b}, \tag{4}$$

where

$$\rho_t = \left(\frac{1 - x/100}{927} + \frac{x/100}{2800} \right)^{-1} \tag{5}$$

is the mean mass density of the material at the target surface assuming a linear mixing model with mass densities $\rho_{ice} = 927\ \mathrm{kg/m^3}$ for ice and $\rho_{sil} = 2800\ \mathrm{kg/m^3}$ for silicate (Koschny and Grün 2001a). The yield for a pure ice target as a function of impactor radius and impact

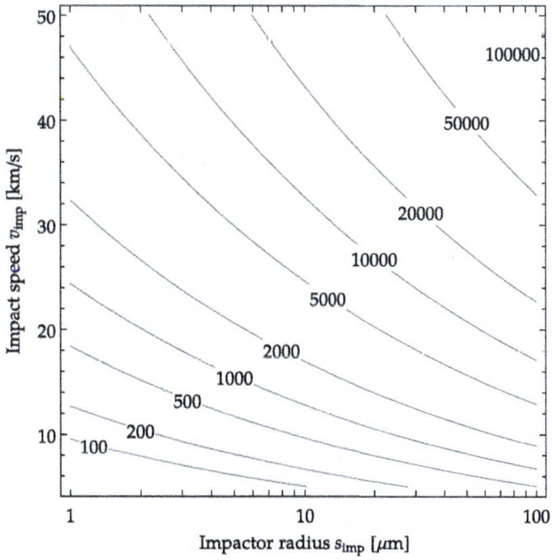

Fig. 5 Yield as a function of impactor radius and impact speed for pure ice, calculated with the empirical relation (4) from Koschny and Grün (2001a)

speed is shown in Fig. 5. For typical sizes and speeds, it varies by about three orders of magnitude. The total mass ejected from the target surface per unit time is,

$$M^{+} = F_{\mathrm{imp}} Y S, \tag{6}$$

where F_{imp} is the impactor mass flux and S is the target surface area.

The cumulative size distribution of the debris is assumed to be a power law with exponent $-q$, so that the number of particles with sizes larger than a_{min} ejected from the target surface per unit time is given by,

$$N(> a_{\mathrm{min}}) = \frac{3-q}{q} \frac{M^{+}}{m_{\mathrm{max}}} \left(\frac{a_{\mathrm{max}}}{a_{\mathrm{min}}}\right)^{q}, \tag{7}$$

where a_{min} (m_{min}) and a_{max} (m_{max}) are the minimal and the maximal ejecta size (mass), respectively. The index q depends on the target material and ranges from 1.5 for loose material to 3 for solid targets (Krivov and Jurewicz 1998). Buhl et al. (2014) suggest that the different impact conditions, rather than the different materials are the main reason for the various indexes. They distinguish between impacts on three different targets:

1. For a semi-infinite target, the strongly affected material is excavated and typically exhibits an index of 2.5. The weakly affected material has a smaller index and remains in the crater.
2. For a large disrupted target, the amount of strongly affected material (index 2.5) is approximately the same, but the weakly affected material (index < 2.5) is also released, which in total results in an index *smaller* than for the semi-infinite target.
3. In the case of a small disrupted target, the target volume is smaller than the volume of the strongly affected material, but the same amount of energy is released into a smaller volume, which leads to smaller fragments and an index *larger* than for the semi-infinite target.

Impact experiments with relatively small and fast projectiles (Takasawa et al. 2011), which better resemble impacts of interplanetary dust particles (micron-sized particles with speeds

Fig. 6 Sketch of the impactor-ejecta process. A fast impactor hits a moon and releases material from its surface within a cone with a half-opening angle $\psi \in [0°, 90°]$. Individual particles are ejected at a distance r_m from the center of the moon with velocity **u** and direction ψ measured from the normal to the moon's surface at the ejection point

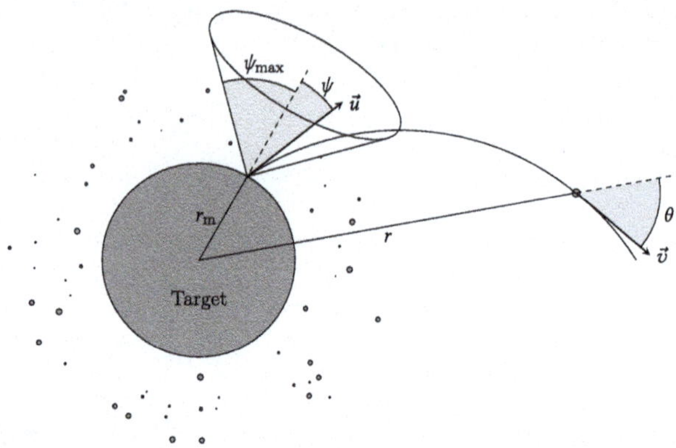

of a few 10 km/s), show a trend to steeper power laws. The *differential* size distribution is given by $n(a) = \frac{dN(>a)}{da}$, and the distribution is normalized to the total mass ejected per unit time as,

$$\frac{4}{3}\pi\rho_t \int_0^{a_{max}} a^3 n(a)\, da = M^+. \tag{8}$$

With estimates for the total mass ejected, a steady state ejecta model can be derived. To model the ejecta distribution, ejecta are started from the target, independently of the launch position, with speed u at an angle ψ measured from the normal to the surface. A typical assumption is that the fragments are ejected uniformly distributed (in terms of solid angle) in a cone of half-opening angle $\psi_{max} \in [0°, 90°]$, with the angular distribution given by (e.g., Horányi et al. 2015),

$$f_\psi(\psi) = \frac{\sin\psi}{1 - \cos\psi_{max}}\, \Theta(\psi_{max} - \psi). \tag{9}$$

The impact-ejection process and relevant variables are depicted in Fig. 6. For initial conditions, it is usually assumed that the particle sizes and speeds at the moment of ejection are uncorrelated. In this case, all ejecta are launched with the same speeds regardless of their size. Impact experiments and scaling laws (Housen and Holsapple 2011) show that the *differential* speed distribution is proportional to a power law with exponent $-(\gamma + 1)$,

$$f_u(u) = \frac{\gamma}{u_{min}^{-\gamma} - u_{max}^{-\gamma}}\, u^{-\gamma-1}\, \Theta(u - u_{min})\Theta(u_{max} - u), \tag{10}$$

where u_{min} and u_{max} and the minimum and maximum velocities respectively, and $\Theta(x)$ is the unit step function, which is unity for $x \geq 0$ and zero otherwise. The index γ depends on properties of the target material and ranges from $\gamma = 1$ for (highly) porous to $\gamma = 2$ for nonporous materials (Krivov et al. 2003).

Conservation of mechanical energy requires the ejecta to have less kinetic energy than the impactor. Assuming separable size and speed distributions, the relation between Y, γ, and u_{min} reads (Krüger et al. 2000),

$$\frac{K_e}{K_{imp}} = Y\, \frac{\gamma}{2-\gamma} \left(\frac{u_{min}}{v_{imp}}\right)^2 \left[\left(\frac{u_{min}}{u_{max}}\right)^{\gamma-2} - 1\right] \quad \text{for } \gamma \neq 2 \tag{11}$$

or

$$\frac{K_e}{K_{imp}} = 2Y\left(\frac{u_{min}}{v_{imp}}\right)^2 \ln\left(\frac{u_{max}}{u_{min}}\right) \quad \text{for } \gamma = 2, \tag{12}$$

where the subscripts '*imp*' and '*e*' refer to impactor and ejecta related quantities, respectively. Hard surfaces (e.g. ice) are generally less dissipative than soft surfaces (e.g. snow, regolith). The minimum ejection speed is found by assuming a plausible value for the kinetic energy ratio of a few tens of percent of the kinetic energy of the impactor (Asada 1985; Hartmann 1985) and solving the above equations for u_{min}. We note the constraint of a minimum velocity is an approximation to simplify the mathematical formalism; the true velocity distribution most likely does not has an exact cutoff velocity.

Integrating Equation (10) for $u > v_{esc}$ gives the fraction of escaping ejecta ("efficiency") as,

$$F(> v_{esc}) \approx \left(\frac{u_{min}}{v_{esc}}\right)^{\gamma}, \tag{13}$$

which is a few percent for the larger moons and (close to) unity for moons smaller than about one hundred kilometers. Ascending and descending ejecta form a dust cloud around the target body. Some ejecta can escape the target's gravity, while others fall back to the surface. In (dynamical) steady state, escaping and re-impacting grains are continuously removed from the cloud and replaced at the same rate by subsequent ejecta.

5.2 Dust Clouds Around the Galilean Satellites

Dust clouds have been observed for the Galilean moons Europa (Krüger et al. 2003), Ganymede (Krüger et al. 1999), and Callisto (Krüger et al. 2003). The spacecraft Galileo arrived at Jupiter in 1995 and studied the Jovian system until 2003. Galileo performed several flybys of all four Galilean moons, with distances at closest approach between a few hundred and a few thousand kilometers. After mid-1999, the flybys were not suitable for the observation of the clouds as the orientation of the spacecraft prevented the detection of dust particles close to the moons. During many of the earlier flybys, the impact rate measured with the Dust Detector System (DDS) (Grün et al. 1992) showed a sharp peak centered around closest approach. This indicates increased concentrations of dust in the vicinity of the Galilean moons, while the symmetry of the measured profile suggests a global source.[1]

During the single suitable Io flyby only 4 dust particles were detected, but 64, 38, and 35 particles with masses larger than 10^{-16} kg (detector threshold) were detected during the 8, 4, and 3 flybys of the moons Europa, Ganymede, and Callisto, respectively. The particle size distribution can be fit by a power law with exponents $q = 1.74 \pm 0.12, 2.46 \pm 0.12$, and 2.58 ± 0.42 (Krüger et al. 2003) for ejecta from the three outer moons assuming the spacecraft velocity relative to each moon as impact speed of the dust particles. Using the speed obtained from the calibration of the dust detector, the exponents are closer together but generally smaller (1.5–1.8). These slopes are in agreement with the typical slopes expected for particles generated by the process of impact-ejection (1.5–3). The average particle radius is between 0.5 and 1 microns, which is consistent with ejecta sizes generated by impactors of 1 μm to 100 μm. This suggests that impact-ejection is the most likely source for the dust clouds around the Galilean satellites.

[1] In comparison to a local source such as cryovolcanism at the south polar terrain of Saturn's moon Enceladus (Spahn et al. 2006).

The number densities in the clouds around the moons (10^{-3} m^{-3} at $r = 0.1\,r_m$) are up to three orders of magnitude larger than the number densities of the dust in the region between them (10^{-6} m^{-3}, see Krivov et al. 2002). The density in the clouds is dominated by bound particles as only a small fraction of the ejecta is launched with speeds larger than the high escape velocity of the Galilean satellites.

A model for a steady state dust cloud around a planetary satellite was first developed by Krüger et al. (2000) and Krivov et al. (2003). For interplanetary meteoroids at Jupiter, we use (Krivov et al. 2003)

$$F_{\text{imp}}^{\infty} = 7.6 \times 10^{-16}\ \text{kg}/\left(\text{m}^2\,\text{s}\right) \tag{14}$$

$$v_{\text{imp}}^{\infty} = 9\ \text{km/s}, \tag{15}$$

and we modify these values for the effect of gravitational focusing as previously described. We note the impactors will arrive with a speed distribution, rather than a single velocity, and utilize this delta function velocity distribution for mathematical simplicity. Because impact-ejecta from the Galilean moons are the most likely source for dust detected between their orbits (Krivov et al. 2002), we assume an upper speed limit larger than the escape velocity of these bodies ($u_{\text{max}} > 3$ km/s). This estimate possibly ignores a very small fraction of high-speed ejecta, but the results depend only weakly on u_{max} owing to the steepness of the speed distribution.

For the case of a moving target—where the spherical symmetry is broken—Sremčević et al. (2003) developed an asymmetric model. In these models, it is assumed (for simplicity and according to the current impact-ejection model) that the distribution of the particle sizes and their speeds at the moment of ejection is uncorrelated. The distribution obeys phase space conservation, with the dynamics in the vicinity of the parent body (to a good approximation) governed by the two-body problem. Krivov et al. (2003) give for the number density of bound ejecta at a distance r (in units of the satellite radius r_m) from the center of the body,

$$n_{\text{bound}}(\tilde{r}) = \frac{N^+}{2\pi r_m^2 v_{\text{esc}}}\,\gamma \tilde{u}_{\text{min}}^{\gamma}\,\tilde{r}^{-2.5} K_{\text{bound}}(\tilde{r}), \tag{16}$$

and for unbound ejecta,

$$n_{\text{unbound}}(\tilde{r}) = \frac{N^+}{4\pi r_m^2 v_{\text{esc}}}\,\gamma \tilde{u}_{\text{min}}^{\gamma} c_0(\gamma)\,\tilde{r}^{-2} K_{\text{unbound}}(\tilde{r}), \tag{17}$$

where the tilde denotes normalized variables, $\tilde{r} = r/r_m$ and $\tilde{u}_{\text{min}} = u_{\text{min}}/v_{\text{esc}}$. $K_{\text{bound}}(\tilde{r})$ and $K_{\text{unbound}}(\tilde{r})$ are form factors only weakly depending on distance and tending to unity far from the moon. $c_0(\gamma) = \frac{\sqrt{\pi}\,\Gamma((\gamma+1)/2)}{\gamma\,\Gamma(\gamma/2)}$ is a monotonically decreasing function with relevant values between $c_0(1) = 1$ and $c_0(2) = \frac{\pi}{4} \approx 0.79$. $\Gamma(x+1) = x \cdot \Gamma(x)$ with $\Gamma(1) = 1$ is the Gamma function. The dust densities decrease with increasing height above the moon's surface, with $r^{-2.5}$ and r^{-2} for large $\tilde{r}$ and particles in bound and unbound orbits, respectively. Figure 7 shows the observed dust densities for the dust cloud around the moon Ganymede and the calculated dust densities predicted by the model from Krivov et al. (2003).

Given estimates for the bound and unbound densities of the ejecta clouds, the total content of the bound exosphere and loss rate can be calculated. For the surface properties of the moons, we adopt numbers from Showman and Malhotra (1999). For bright Europa, we use pure ice ($x = 0\%$); for the volcanically active moon Io, we use pure silicate ($x = 100\%$);

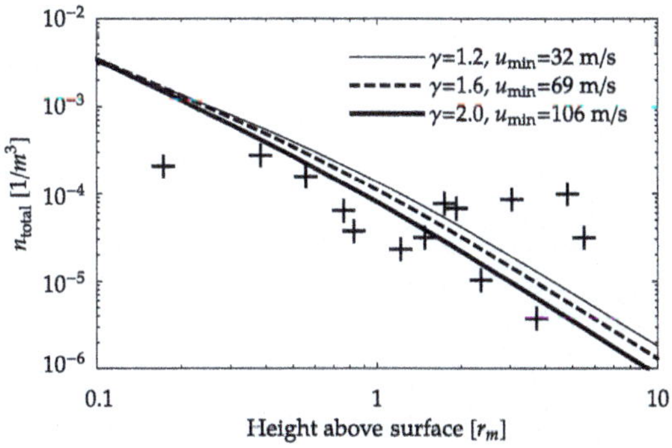

Fig. 7 Number density ejecta at a distance r (in $r_m = 2634$ km) from the surface of Ganymede. The crosses are dust densities obtained by Krüger et al. (2000) from *in situ* measurements of the Galileo DDS during flybys of Ganymede. The solid lines are number densities calculated using the model from Krivov et al. (2003) with $Y = 1000$, $\alpha = 2.5$, $m_{\text{min}} = 10^{-16}$ kg, and $m_{\text{max}} = 10^{-8}$ kg, and for three different sets of parameters for the speed distribution

and for Ganymede and Callisto intermediate values are used, $x = 70\%$ and $x = 30\%$ respectively. With this and assuming an impactor population dominated by 100 μm-sized projectiles (Divine 1993), we find a yield of $Y \approx 1000, 20000, 1000$, and 2000 for the respective moons. As a result, the estimated amount of material ejected from each moon is in the order of 10–100 kg/s. The lifetime of the particles depends on the parameters of the ejecta speed distribution. It is dominated by the large number of slow ejecta but only weakly depends on the maximal ejection speed. For the Galilean satellites, the average lifetime is between several tens of seconds and a few hundred seconds. Then the mass of the dust in a steady state cloud is on the order of 10^3–10^4 kg. Due to the large escape velocity of the moons, the fraction of escaping ejecta is relatively small (on the order of 0.1–1%). As a result, the estimated amount of material ejected from each moon into circumplanetary space is between 0.01 kg/s and 1 kg/s. Escaping ejecta form a tenuous ring across the orbits of the Galilean moons (Krivov et al. 2002).

In the current impact-ejection model, it is assumed that the particle sizes and speeds at the moment of ejection are uncorrelated. In contrast, at least for bigger particles theoretical modeling (Melosh 1984) and impact experiments (Nakamura et al. 1994; Onose and Fujiwara 2004) suggest that larger fragments are on average launched with lower speeds. Sachse et al. (2015) have dropped the assumption that the ejecta sizes and ejection speeds are uncorrelated and applied the new model to dust clouds around planetary satellites. They find that both models can fit *in situ* measurements of the total density equally well, but differences are visible for micron-sized particles or larger. However, the exact form of the correlation is poorly constrained as there is only a small number of studies focusing on this aspect of the impact-ejection process. A further refinement requires additional impact experiments, namely with projectile sizes and impact speeds similar to micron-sized interplanetary impactors.

5.3 The Lunar Dust Cloud

While an ejecta cloud surrounding the Moon was expected from observations of the Galilean satellites, until recently, it had not been explored. The Lunar Atmosphere and Dust Environment Explorer (LADEE) mission orbited the Moon from October 2013 to April 2014, with

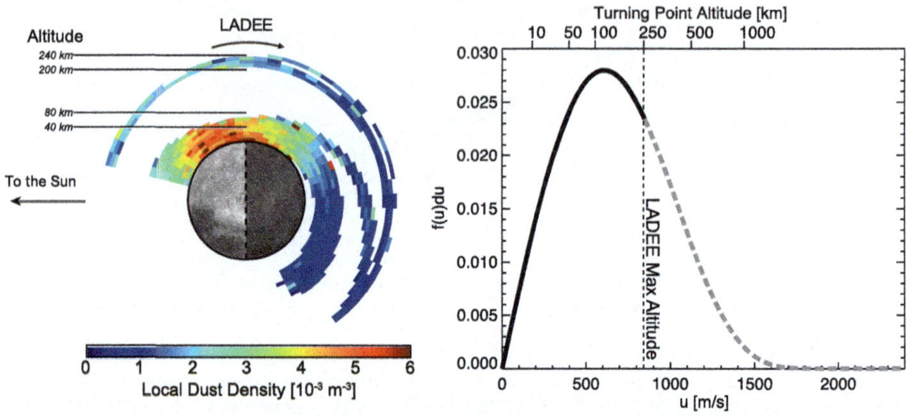

Fig. 8 *Left.* The dust density distribution about the Moon, calculated using $\sim 140{,}000$ impacts averaged over LADEE's 6 months of operations in lunar orbit. White regions indicate locations either LADEE did not visit or that LDEX could not take measurements due to sun pointing constraints. Altitude bands are not to scale. *Right.* The inferred surface ejecta velocity distribution from the altitude density fit. The dotted line indicates the maximum altitude explored by LADEE, above which the distribution function is an extrapolation

a primary science objective to characterize the neutral and dust environment of the Moon (Elphic et al. 2014). Onboard, it carried the Lunar Dust Experiment (LDEX), an impact ionization dust detector capable of individually detecting impacts from grains with radii $a \geq 0.3$ μm (Horányi et al. 2014). LDEX discovered a permanently present, asymmetric lunar dust cloud (Horányi et al. 2015). It detected $\sim 140{,}000$ individual dust impacts during its six months of observations with $\sim 2{,}000$ orbits. The observed dust cloud was found to be most dense on the apex hemisphere (in the direction of orbital motion of the Earth-Moon system about the Sun), with a slight cant towards the Sun, and is generated by the continual bombardment of meteoroids to the lunar surface (Fig. 8).

As described in further detail in the Meteoroids review paper, the distribution of meteoroids impacting the Earth-Moon system is typically divided into two categories, meteoroid showers and the sporadic background. Meteoroid showers are periods where highly collimated beams of particles impact a body from a narrow region in the sky known as the radiant. The sporadic background is larger in overall flux than the total contribution from meteoroid showers and is further broken up into distinct sources. In the ecliptic plane, these sources are the Helion (HE), Apex (AP), and Anti-Helion (AH) (Jones and Brown 1993), with an additional weak Anti-Apex (AA) source (Janches et al. 2000).

The lunar dust density distribution in the equatorial plane was found to be primarily generated by the HE, AP, and AH sporadic sources (Szalay and Horányi 2015a), with a minor contribution potentially coming from the AA source (Szalay and Horányi 2016b). As the Moon orbits the Earth with a relative velocity of 1 km/s, it is continually moving into and away from these sporadic meteoroid sources at different lunar phases. Due to this relative motion, the lunar dust cloud was found to wax and wane on the lunar synodic period (Szalay and Horányi 2015a). Further analyses of the ejecta cloud structure incorporating a dynamical meteoroid model enabled estimations of the relative quantities of short-period and long-period comets impacting the lunar surface (Janches et al. 2018). During several of the known meteoroid showers, such as the Geminids and Quadrantids, LDEX observed temporary enhancements of the lunar dust cloud, localized on the hemisphere exposed to the incident meteoroid shower flux (Szalay and Horányi 2016a; Szalay et al. 2018). Most notably during these periods, LDEX intermittently observed "bursts" of particles, which are interpreted as direct measurements of transits through dense ejecta plumes.

Unlike at the Galilean moons, an exponential fit was found to be in good agreement with the altitude distribution, with the form $n(h) = n_0 e^{-h/h_0}$, where h is the altitude in km above the lunar surface, n_0 is the density at $h = 0$, and $h_0 = 200$ km is the scale height (Szalay and Horányi 2016b). The cumulative size index of lunar impact ejecta was found to be $q = 2.7$ and did not vary significantly with time or altitude (Horányi et al. 2015). Factoring the altitude, size, and local time dependencies, the average lunar dust density in the equatorial was found to be well represented by,

$$n(h, \varphi, a) = e^{-h/h_0} a_\mu^{-q} n_w \sum_s w_s \cos^3(\varphi - \varphi_s) \Theta\big(|\varphi - \varphi_s| - \pi/2\big), \tag{18}$$

where a_μ is the particle radius in μm, s represents each ecliptic sporadic source, Θ is the Heaviside function, φ_s is the characteristic angle for each of the four source radiants (65°, 0°, −65°, 180°) (Campbell-Brown 2008), $n_w = 3.3 \times 10^{-4}$ m^{-3}, and w_s is the relative weight (0.24, 0.49, 0.24, 0.03) for the HE, AP, AH, & AA sources respectively. The relative ratios of AP to HE/AH sources were taken from a combination of Earth-based radar measurements (Campbell-Brown 2008) and lunar dust measurements (Szalay and Horányi 2016b). These ratios may vary throughout the year, following any relative variations in their respective sources; the values quoted here give an average case for the lunar dust cloud.

The dust density altitude distribution can be inverted to determine the initial velocity distribution function for dust grains ejected from the lunar surface,

$$f(\tilde{u}) = \frac{\eta \tilde{u}}{(1 - \tilde{u}^2)^2} e^{-\frac{\beta \tilde{u}^2}{1-\tilde{u}^2}} \tag{19}$$

where $\tilde{u} = u/u_m$ with $u_m = 2.4$ km/s (lunar escape speed), $\beta = 8.69$, and $\eta = 7.2 \times 10^{-3}$ s/m (Szalay and Horányi 2016b), shown in Fig. 8. This distribution was derived assuming LDEX only detected bound grains at their turning points; it therefore represents an approximation of $f(\tilde{u})$ at the Moon.

There is an apparent discrepancy between this velocity distribution and the power law consistent with the ejecta clouds of the Galilean moons. However, Galileo and LADEE visited a different range of altitudes above their respective airless bodies. LADEE visited a range of a few to ∼ 250 km in altitude such that it observed the ejecta cloud up to $\tilde{r} = 0.14$. The inferred velocity distribution monotonically increases for the majority of the altitude range covered by LADEE, with a peak at ∼ 600 m/s (corresponding to a turning point altitude of ∼ 120 km). In the velocity range from 600 m/s to 840 m/s, the lunar velocity distribution monotonically decreases, however, only a small portion of the descending portion of this distribution is sampled. In comparison, the Galileo DDS detected 141 impacts throughout its 16 flybys of the Galilean moons at distances of $\tilde{r}$ in the range of 0.13 to 8 (Krüger et al. 2003), with about half of the flybys above $\tilde{r} = 2$. Therefore, the two missions probed different ejecta cloud regions and we can only infer information about the velocity distribution relative to the applicable initial velocities required to reach these given altitudes. It is possible that the lunar ejecta velocity distribution trends toward a power law for large velocities and that LADEE simply did not visit high enough altitudes to address this issue. Additionally, as previously discussed, the surface ejection properties and subsequent velocity distributions between the icy Galilean moons and the silicate regolith dominated lunar surface are most likely to be characteristically different. Lastly, relaxing the assumption of a single typical impactor/ejecta plume and allowing the speed and mass distributions to be correlated (Sachse et al. 2015) may aid in understanding the impact ejecta measurements discussed here.

5.4 Bound Ejecta Clouds Around Other Airless Bodies

In addition to the observed ejecta clouds at the Galilean satellites and the Moon, impact ejecta should be generated at airless bodies throughout the solar system. Mercury likely possesses a relatively dense impact ejecta cloud from meteoroid bombardment, given its similarity to the Moon and the increased meteoroid flux deeper in the heliosphere (e.g., Hahn et al. 2002; Pokorný et al. 2017). The overall magnitude of the meteoroid flux at Mercury is still fairly uncertain (e.g., Cintala 1992; Borin et al. 2009, 2016); however, future in-situ measurements by the Mercury Dust Monitor on-board the BepiColombo spacecraft (to be launched to Mercury in late-2018) may reveal the presence of an impact ejecta exosphere at Mercury (Nogami et al. 2010).

Similar to the Galilean satellites, one also expects impact ejecta clouds at the major saturnian satellites such as Iapetus, Rhea, Dione, Tethys, Enceladus, and Mimas. To date, however, no such positive detection has occurred, with the exception of the massive Phoebe ring (Verbiscer et al. 2009). Perhaps the biggest obstacle to detection of impact ejecta clouds at the saturnian satellites is the presence of the large and dusty E-ring generated from the Enceladus plumes. Analysis of Cassini dust impact rate observations during fly-bys of Enceladus have fit both plume ejecta and impactor ejecta to the total flux measurements (Spahn et al. 2006a; Kempf 2008); however, the resulting impact ejecta contributions are typically small with respect to the plume measurements. Thus, no clear evidence has been found for clouds around the E ring moons of Saturn,[2] which suggests that the ejecta cloud dust densities are lower than previously predicted by extrapolation of results at the Galilean satellites. Possible reasons for this may be a lower impactor flux (e.g., Poppe 2016), impactor size and/or impact speed, and/or a more dissipative surface, possibly due to a layer of icy E ring particles ('snow') soaking up more of the impactor's kinetic energy.

Smaller bodies in the solar system will also generate ejecta in their near vicinity, however, much of this ejecta will be unbound (e.g., Szalay and Horányi 2016c). The main factor that determines the total ratio of bound to unbound ejecta is the mass of the body, however, the speed distribution also plays an important role. Assuming the speed distribution of ejecta is the same as that of the Moon and a mass density of 3 g/cc for airless bodies, an airless body would need to be larger than 500 km to retain at least 50% of its impact ejecta. Hence, and as discussed in Sect. 6.2.1, most small bodies near 1 AU shed the large majority of their impact ejecta and therefore the bound component of ejecta at smaller bodies is particularly tenuous.

6 Airless Bodies as Sources of Dust

6.1 Phobos and Deimos

Mars' moons Phobos and Deimos have long been suspected to be parents of a pair of tenuous dust tori encircling Mars due to escaping impact ejecta (Soter 1971). The Russian PHOBOS-2 mission observed unusual variations in the interplanetary magnetic field near the location

[2]Rhea and the E ring moons in general are currently debated for having or not having dust clouds that can be measured against the relatively strong E ring background (Spahn et al. 2006a), which is supplied by cryovolcanism on Enceladus (Schmidt et al. 2008). Two *in situ* measurements have been conducted so far by the spacecraft Cassini. The first one is inconclusive as it depends on a single data point, the second one does show a profile but the measured peak is much wider than predicted by the model from Krivov et al. (2003).

of Phobos' orbit, which were at the time interpreted as evidence of solar wind mass loading by either a neutral gas or dust torus (Dubinin et al. 1990, 1991; Sauer et al. 1993; Barabash and Lundin 1994). This suggestion gave rise to a prolific series of investigations into the underlying theoretical physics of the martian dust tori by a number of authors (e.g. Ip and Banaszkiewicz 1990; Horányi et al. 1990, 1991; Tátrallyay et al. 1992; Kholshevnikov et al. 1993; Krivov 1994; Juhász et al. 1993; Juhász and Horányi 1995; Ishimoto 1996; Krivov and Hamilton 1997; Makuch et al. 2005). The underlying prediction from these models is a pair of dust tori around Mars: one from Phobos ejecta that is strongly confined to Phobos' orbital plane and one from Deimos ejecta that is more diffuse and inclined. Despite this wealth of modeling, no observational signatures, either from in-situ or remote sensing observations, have detected the presence of these putative dust rings (Showalter et al. 2006; Krivov et al. 2006; Øieroset et al. 2010). These attempted measurements have placed fairly stringent upper limits to the Phobos and Deimos tori, with maximum normal optical depths of $\approx 3 \times 10^{-8}$ and 10^{-7}, respectively. In-situ measurements by impact ionization dust detectors may be required for a final determination as to the presence and density of these tori.

6.2 Asteroids

In the discussion below, we focus on dust phenomena relating to asteroids (some of which have cometary-like activity), leaving the cometary discussions for the Comets review paper in this Topical Collection.

6.2.1 Impact-Ejecta Production Rates in the Asteroid Belt

Ejecta clouds for Near Earth Asteroids (NEAs) have been predicted based on lunar measurements (Szalay and Horányi 2016c), and the spatial distribution of ejecta dust around Main Belt asteroids is expected to be qualitatively similar to that around NEAs. Asteroidal ejecta clouds are predicted to be predominantly unbound given the relatively low asteroidal gravity. For example, a 10 km radius asteroid near 1 AU is predicted to lose $> 99.9\%$ of the ejecta generated by meteoroid bombardment. A quantitative assessment requires knowledge of the meteoroid flux in the asteroid belt, such as predicted by ESA's IMEM model (Dikarev et al. 2005) and the surface response to such fluxes. The spatial density of meteoroids in the asteroid belt is likely one order of magnitude lower than at 1 AU (Landgraf et al. 2002), suggesting lower unbound ejecta densities around asteroids in this region compared to NEAs. Despite these lower production rates for any individual asteroid, Main Belt asteroids steadily produce ejecta that contribute to supplying the interplanetary dust complex, although at a share of $< 10\%$ (Nesvorný et al. 2010).

Collisions with larger asteroids are less frequent with increasing impactor size but produce large amounts of dust that can be observed with Earth-based telescopes. The typical relative velocity between Main Belt asteroids is $5.3\ \mathrm{km\,s^{-1}}$ (Bottke et al. 1994), and the mass of particles ejected faster than the impacted body's escape speed is comparable to the mass of the projectile (Housen and Holsapple 2011; Jewitt et al. 2011). The most direct observation of an impact on a Main Belt asteroid is that of a decameter-sized projectile onto 57 km-diameter asteroid (596) Scheila (Jewitt et al. 2011; Bodewits et al. 2011; Ishiguro et al. 2011a,b). The impact excavated a dust cloud of 0.1–100 µm-sized grains having a total mass of $\sim 10^8$ kg. The dust was dispersed by radiation pressure within ~ 2 months. Impacts onto smaller asteroids are likely to eject larger debris due to the lower escape speed.

The frequency of collisions between Main Belt asteroids is uncertain. Collisional models suggest a rate of one catastrophic collision per year for 100-m-sized objects, with subcatastrophic impacts being more frequent (Bottke et al. 2005). Denneau et al. (2015) instead

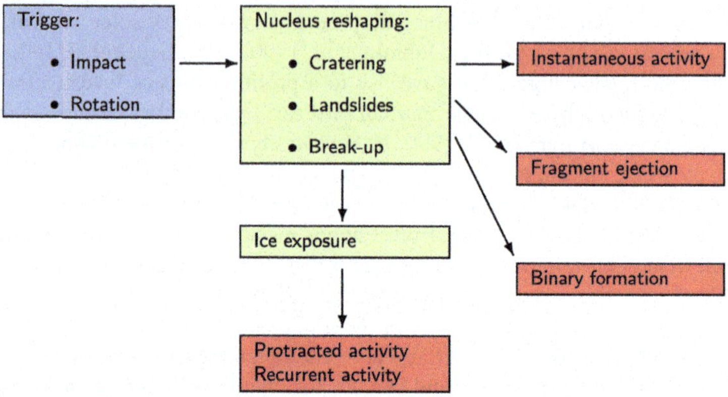

Fig. 9 An overview of the processes involved in the activation of asteroids

find that a rate of one catastrophic collision per year is consistent with Pan-STARRS detection rates only for 7 m-sized objects, while encounters of 100 m-sized objects would be less frequent. Bodewits et al. (2014) interpret a change in Scheila's rotational lightcurve as the signature of surface alteration due to the impact. They find that the < 100 m impactor affected the albedo of an area of 3.5–10 m in radius, likely by covering it in a thin (2 mm–2 cm) ejecta blanket of bright material.

Assuming that at least two out of the 17 active asteroids listed in Table 1 were impact-triggered (176P and Scheila, because their slow rotation rates exclude rotational disruption), we expect that at least 10% of the active asteroids are impact-triggered. Combining this with the estimated total number of active asteroids of 100 (Sheppard and Trujillo 2015) and assuming that the generated dust remains visible for $\sim 10\%$ of an orbital period, we obtain a (telescopically detectable) impact rate of one per year and an impact-generated dust production of order 10^8 kg per year, within an uncertainty of at least one order of magnitude.

6.2.2 Active Asteroids

The emission of dust from asteroids has been detected by telescopic observations in ~ 20 objects (Hsieh and Jewitt 2006; Jewitt 2012; Jewitt et al. 2015d). These active asteroids temporarily display comet-like comae and tails, but have asteroidal orbits with $T_J > 3$ (Kresak 1982; Kosai 1992). Figure 9 summarizes many of the processes occurring on these airless bodies, as described below.

A prime tool to understand the processes effectuating the dust emission is the analysis of the dust tail morphology and temporal evolution through model calculations of the dust dynamics under the influence of initial velocity, solar radiation pressure, and gravity (Finson and Probstein 1968; Agarwal et al. 2010; Fulle et al. 2010; Moreno et al. 2017). The activity shows strong variation between objects, probably reflecting the diversity of underlying processes. In general, asteroid activity seems to be triggered by some violent process of very short duration, such as an impact or re-shaping due to centrifugal forces, where both processes are likely to have been observed in different objects. This trigger process is accompanied by the ejection of dust, debris, and sometimes larger fragments visible to telescopes. In addition, it can lead to the catastrophic breakup of the object or the formation of a binary system. In some active asteroids, such a trigger process has apparently uncovered ice previously hidden in the interior of the asteroid. Exposed to enhanced solar irradiation around perihelion, the ice sublimates and carries along dust, leading to comet-like activity in the classical sense.

At least one of the active asteroids, (596) Scheila, has been impacted by a second, smaller asteroid (Jewitt et al. 2011; Bodewits et al. 2011; Moreno et al. 2011b). The ejected dust formed an impact cone and a downrange plume, and was subsequently driven away from the nucleus by solar radiation pressure (Ishiguro et al. 2011a,b). Due to Scheila's comparatively large size (113 km diameter), only small dust grains could escape from the nucleus' gravity field, which were subsequently dispersed by radiation pressure within approximately one month.

The rotation period of a small asteroid of irregular shape or albedo can be changed by a torque from the asymmetric emission of thermal radiation (so called YORP-effect) (Rubincam 2000) or by localised outgassing. If the centrifugal force compensates gravity at the equatorial surface, the body will shed material from its surface or even breakup entirely (Hirabayashi and Scheeres 2013; Veras et al. 2014; Pravec et al. 2010), depending on its shape and composition (Hirabayashi 2015). The critical rotation rate at which a body will disintegrate depends on its bulk density and cohesion. The lack of observations of asteroids > 100 m in diameter rotating at periods < 2.2 h (Warner et al. 2009) is interpreted as evidence that these bodies are nearly-strengthless rubble piles bound only by gravity, while smaller asteroids seem to have a considerably higher internal strength. The 2.2 h spin barrier seems to reflect rotational disruption as an effective process of asteroid disintegration or re shaping.

Several active asteroids are suspected to have undergone rotational mass shedding or disintegration. The evidence is mainly indirect, based on the exclusion of alternatives such as impact or sublimation (311P: Jewitt et al. 2013, 2015a; Hainaut et al. 2014; Moreno et al. 2014, P/2013 R3: Jewitt et al. 2014a), near-critical rotation periods (331P: Drahus et al. 2015, 133P: Hsieh et al. 2004, (62412): Sheppard and Trujillo 2015), large fragments (P/2010 A2: Jewitt et al. 2010, 331P: Drahus et al. 2015), ejection speeds comparable to the gravitational escape speed (133P: Jewitt et al. 2014b, 288P: Agarwal et al. 2015, P/2010 A2: Jewitt et al. 2010, P/2013 R3: Jewitt et al. 2014a), and in one case planar ejection velocities (P/2010 A2: Agarwal et al. 2013). In some of these objects (133P, 288P), the fast rotation seems to support gas drag in lifting dust, while in others it only triggered instantaneous (P/2010 A2, 331P) or repeated (311P) mass shedding without uncovering ice, or a catastrophic disruption (P/2013 R3). A detailed discussion of individual objects is given in Jewitt et al. (2015d).

Currently nine known asteroids have shown activity for an extended period of several months around perihelion, and for at least five of these the activity recurred in subsequent perihelion passages (133P, 238P, 288P, 313P, 324P), while the remaining four had at the time of writing not been observed during a second perihelion passage (P/2012 T1, P/2015 X6, 259P, P/2016 J1). Protracted and recurrent activity are taken as strong indicators for dust ejection by a flow of sublimating gas, as this is the only known process modulated by the heliocentric distance that is able to lift dust of the observed quantity and size over an extended period of time in the Main Belt (Jewitt 2012). Thermal models of asteroid interiors have shown that the conservation of primordial water ice in the interiors of Main Belt asteroids is possible, while more volatile species are likely depleted (Schorghofer 2008, 2016; Prialnik and Rosenberg 2009; Snodgrass et al. 2017). The gas production rates required to lift the observed amount of dust are below the detection limit of currently available instruments, such that the non-detection of water vapor around active asteroids is not a proof against its presence. Sublimation-driven dust activity is discussed in more detail in the Comets review paper.

Dust can also be accelerated above an asteroid's escape speed by thermal fracturing and dehydration at temperatures exceeding $\sim$ 1000 K (Jewitt 2012). While this process is unlikely to play a significant role in the Main Belt, it is probably responsible for the recurrent ejection of micron-sized dust from asteroid (3200) Phaethon during its 0.14 AU perihelion passage (Jewitt and Li 2010; Li and Jewitt 2013; Jewitt et al. 2013b; Hui and Jewitt 2017). (3200) Phaethon has been suggested to be the parent body of the Geminids stream.

No direct measurements exist concerning the bulk density, porosity, composition, or optical properties of dust from active asteroids. Assuming a size-independent albedo and density, dust ejection models are usually consistent with power-law cumulative size distributions having exponents between −2.0 and −2.7. Jewitt et al. (2015a) found a size distribution steepening towards smaller sizes in 311P. Maximum particle sizes found in sublimation-driven asteroids are below 1 cm, while dust ejected from asteroids suspected to have been impacted or rotationally disrupted reaches up to 1 m in size. P/2010 A2 and 331P displayed a debris trail of decimeter- to meter-sized chunks at least several years after the dust-producing events (Jewitt et al. 2013a; Drahus et al. 2015). In four active asteroids, the dust terminal speeds have been found to be less than the gravitational escape speed from the nucleus. At least two of these are fast rotators, such that it is suspected that in these objects the escape of dust is facilitated by the centrifugal force, the dust leaves the nucleus at a speed comparable to the escape speed and decelerates with increasing nucleus distance. In most other active asteroids, inferred dust speeds range between 1 and 10 m/s. No obvious correlation has been found between the nucleus size and the sizes of ejected dust.

The typical dust mass ejected from an active asteroid per orbit is of order 10^7 kg, with a variation by a factor 100 between individual objects. This is approximately a factor of 1000 less than in comets. Assuming a total number of 100 currently active asteroids (Sheppard and Trujillo 2015) and a typical orbital period of 5 years, this corresponds to a dust production rate of 6 $\mathrm{kg\,s^{-1}}$, only 0.006–0.06% of the rate of 10^4–10^5 kg/s required to sustain the zodiacal cloud (Nesvorný et al. 2011a). It is therefore likely that active asteroids contribute only a minor fraction to the interplanetary dust complex.

As discussed in the Meteoroids review paper, the Geminid meteor stream is associated with the orbit of asteroid (3200) Phaethon (Whipple 1983), but the current level of activity of Phaethon (Jewitt et al. 2013b) is not sufficient to explain the particle sizes in the meteor stream. It is therefore likely that Phaethon lost a large amount of large debris in the past within a relatively short period of time (Ryabova 2017). Several other meteor streams in asteroidal orbits may exist (Jenniskens 2015), but their parent bodies are unknown. The detection of debris trails following active asteroids P/2010 A2 and 311P indicates that processes underlying asteroid activity can lead to the formation of a meteor stream that would be detectable as a meteor shower if the object were on an Earth-crossing orbit.

6.3 Minor Planetary Satellites

While discussed in more detail in the Rings review paper, the minor moons of the outer satellites are also known to be prodigious sources of dust grains. The tenuous, dusty rings of Jupiter are formed from ejecta produced from meteoroid bombardment of four small Jovian moons, Metis, Adrastea, Amalthea, and Thebe, and consist of three parts: the main ring, the halo, and the outer gossamer rings (e.g., Smith et al. 1979a,b). Grains ejected from these bodies undergo orbital decay from Poynting-Robertson drag and perturbations from electromagnetic interactions (Burns et al. 1999; Hamilton and Krüger 2008).

At Saturn, several of the minor moons in the inner system, including Methone, Anthe, and Pallene, have been associated with either complete rings or "arcs" of dust grains that are

Table 1 Properties of ejected dust for 17 active asteroids. Types: c: continuous, r: recurrent, i: instantaneous, e: episodic activity, f: fragments, t: debris trail, m: multiple nucleus; v_{esc}: nucleus escape speed; R_{rot}: nucleus rotation period; β: ratio of radiation pressure and solar gravity, related to dust radius s by $\beta = 5.77 \times 10^{-4} Q_{pr}/\rho/s$, and $Q_{pr} \sim 1$; M_{dust}: dust mass produced per orbit; v_{dust}: dust terminal velocities; q: exponent of power-law cumulative size distribution. References: [a]Moreno et al. (2013); [b]Agarwal and Mommert (2017); [c]Moreno et al. (2016a); [d]Hui et al. (2017); [e]Moreno et al. (2017); [f]Sheppard and Trujillo (2015); [g]Jewitt et al. (2009); [h]MacLennan and Hsieh (2012); [i]Hsieh et al. (2004); [j]Hsieh et al. (2010); [k]Jewitt et al. (2014b); [l]Hsieh et al. (2008); [m]Agarwal et al. (2015); [n]Licandro et al. (2013); [o]Agarwal et al. (2017); [p]Jewitt et al. (2015b); [q]Jewitt et al. (2015c); [r]Hsieh et al. (2015); [s]Pozuelos et al. (2015); [t]Moreno et al. (2011a); [u]Moreno et al. (2012); [v]Drahus et al. (2015); [w]Moreno et al. (2014); [x]Jewitt et al. (2015a); [y]Jewitt et al. (2014a); [z]Jewitt et al. (2010); [aa]Snodgrass et al. (2010); [ab]Hainaut et al. (2012); [ac]Jewitt et al. (2013a); [ad]Agarwal et al. (2013); [ae]Kim et al. (2017b); [af]Kim et al. (2017a); [ag]Moreno et al. (2016b); [ah]Hsieh et al. (2011); [ai]Bodewits et al. (2011); [aj]Jewitt et al. (2011); [ak]Ishiguro et al. (2011b); [al]Ishiguro et al. (2011a); [am]Moreno et al. (2011b)

Object	Type	v_{esc} [m/s]	P_{rot} [h]	β	M_{dust} [kg]	v_{dust} [m/s]	$-q$
358P[a]	c	0.2[b]		10^{-4}–0.2	1.5×10^7	0.3–12.2	−2.5
P/2015 X6[c]	c			6×10^{-5}–0.6	7×10^6	0.3–10	−2.3
P/2016 J1[d,e]	c, m	0.5			1.4×10^7	0.5–0.9	
(62412)[f]		3.9	3.33				
259P[g,h]	c	0.1			1.8×10^4	1–10	
133P[i,j,k]	c, r	1.8	3.47	6×10^{-5}–3×10^{-3}	1.8×10^7	0.02–0.13	−2.5– −2.25
238P[l]	c, r	0.3		10^{-4}–0.1	10^5	0.2–3	
288P[m]	c, r, m[n]	0.7		0.001–0.025	9×10^6	0.06–0.3	−2.1[o]
313P[p,q,r,s]	c, r	0.5		0.01– > 0.03	10^7	0.4–2.5	−2.1– −2.5
324P[t]	c, r	0.4		0.005–0.02	1.3×10^6	0.2–9.1	−2.5
331P[u]	i, f, t	0.7[v]	3.24	–0.03	10^8	0.5–0.7	−2.7
311P[w,x]	e	0.2		2×10^{-5}–0.6	10^7	0.02–0.07	−2.5
P/2013 R3[y]	m	0.2		$< 3 \times 10^{-5}$	2×10^{10}	0.3	
354P[z,aa,ab,ac,ad,ae,af]	i, f, t	0.04	11.36	10^{-6}–2×10^{-3}	5×10^8	0.02–0.3	−2.6– −2.1
P/2016 G1[ag]	i	0.03			1.7×10^7	0.015–0.14	−2.0
176P[ah]	c	1.5	22.22	9×10^{-2}–0.1	7×10^4	11.2–11.8	−2.5
(596) Scheila[ai,aj,ak,al,am]	i	75	15.85	2×10^{-5}–0.2	6×10^8	50–100	−2.5– −2.0

presumably being actively shed via meteoroid bombardment (Hedman et al. 2009; Sun et al. 2017). In the outer saturnian system, the massive Phoebe ring is produced via meteoroid bombardment (Verbiscer et al. 2009; Hamilton et al. 2015) and it is suspected that Saturn's other irregular satellites may also contribute to dust in the saturnian system (Tamayo et al. 2011). Uranus plays host to a complex and potentially unstable set of moon-ring systems (Smith et al. 1986; Showalter et al. 2006; de Pater et al. 2006a,b; Sfair and Winter 2009), such as the μ and ν rings associated with the moons Mab, Puck, Rosalind, and Portia.

6.4 Dust Production from Edgeworth-Kuiper Belt Objects

Edgeworth-Kuiper Belt objects (EKBOs) also may play host to dusty exospheric ejecta clouds or tori. The Pluto-Charon system has been most studied in this regards from both observational and theoretical standpoints. An early simulation by Thiessenhusen et al. (2002) predicted a dust cloud from Pluto and Charon while later investigations focused on dust torus generation by Nix and Hydra (e.g., Poppe and Horányi 2011), two of Pluto's small moons discovered in 2005 (Weaver et al. 2006). Qualitatively similar to the proposed dust torus at Mars from Phobos and Deimos, the Nix and Hydra dust tori at Pluto are believed to be generated from impact ejecta from meteoroid bombardment which are subsequently still bound in orbit around the Pluto-Charon barycenter (Poppe and Horányi 2011; Pires dos Santos et al. 2013). Pluto's two other small moons, Styx and Kerberos, may also contribute significantly to the torus density given their negligible surface gravity. To date, however, neither remote sensing observations (Steffl and Stern 2007; Throop et al. 2015) nor in-situ measurements by the New Horizons spacecraft (Bagenal et al. 2016) have detected any positive evidence of the theorized dust torus. Other EKBOs, especially those in binary or multiple systems (e.g., Noll et al. 2008; Benecchi et al. 2010), may also maintain tenuous dust exospheres and/or tori.

Despite the lack of direct observations of dust production in the Pluto-Charon system, Edgeworth-Kuiper Belt objects are a significant source of interplanetary dust grains in the solar system. Three potential production mechanisms for interplanetary dust from EKBOs have been discussed: (i) mutual collisions and grinding (e.g., Stern 1995, 1996; Stern and Colwell 1997; Davis and Farinella 1997; Durda and Stern 2000), (ii) bombardment of EKBOs from interstellar dust grains (Yamamoto and Mukai 1998), and (iii) bombardment of EKBOs by interplanetary dust grains (Poppe 2015).

Collisional evolution is thought to be an important and on-going process in the EKB, despite the relatively slow velocities compared to the asteroidal main belt. Stern (1996) calculated that collisions produce a cumulative 9.5×10^8 to 3.2×10^{11} $\mathrm{g\,s^{-1}}$ of debris over sizes ranging from submicron-sized dust to 10 km sized blocks. Assuming a collisional power-law size distribution of $a^{-3.5}$, this yields a mass production rate of 8.6×10^4–2.9×10^7 $\mathrm{g\,s^{-1}}$ for grains between 0.1 and 10 µm.

Interstellar grains flow through the entire heliosphere (e.g., Grün et al. 1994; Krüger et al. 2007; Krüger and Grün 2009; Sterken et al. 2012) and may also be a significant source of ejecta from EKBOs, based on calculations by Yamamoto and Mukai (1998), which yielded 0.1–10 µm production rates in the EKB of 3.7×10^5–2.4×10^6 $\mathrm{g\,s^{-1}}$, similar to that estimated for mutual collisions.

Finally, bombardment of EKBOs by interplanetary dust grains may also be a contributor to the total production rate as suggested by Poppe (2015). This process is somewhat reminiscent of the "self-sustained" production models for the saturnian E-ring (e.g., Hamilton and Burns 1994), whereby interplanetary EKB grains ejected from an earlier event have a non-negligible probability of impacting another EKBO. Given that impact ejecta yields can be

Table 2 Theoretical and observationally-constrained 0.1–10 μm dust production rates [$g\,s^{-1}$] from the Edgeworth-Kuiper Belt

Theoretical	
Stern (1996)	8.6×10^4–2.9×10^7
Yamamoto and Mukai (1998)	3.7×10^5–2.4×10^6
Observational	
Landgraf et al. (2002)	5×10^7
Han et al. (2011)	9×10^5
Vitense et al. (2012)	2×10^6
Poppe (2016)	6×10^6–1×10^7

significantly larger than unity even for the relatively low relative impact speeds in the EKB (Koschny and Grün 2001a), interplanetary dust grain bombardment may be an important process. Indeed, Poppe (2015) estimated that interplanetary dust bombardment produces between 3 and 200 times the amount of escaping ejecta from the Pluto-Charon system than does interstellar bombardment; however, it is currently unclear how this process compares to ISD bombardment when extrapolated over the entire EKB population.

Comparison between dynamical dust models (Landgraf et al. 2002; Han et al. 2011; Vitense et al. 2012; Poppe 2016) and in-situ dust flux measurements by several spacecraft, including Pioneer 10 and 11 (Humes 1980), Voyager 1 and 2 (Gurnett et al. 1997), Galileo (Sremčević et al. 2005), and New Horizons (Poppe et al. 2010; Szalay et al. 2013), have placed constraints on the total dust production rate from the EKB between 0.1 and 1.0 μm, listed in Table 2. In general, these observations are remarkably consistent with theoretical predictions.

7 Electrostatically Lofted Dust

In addition to impact processes which eject dust from airless bodies, there has been evidence for electrostatic processes contributing to the evolution and transport of dust on airless bodies. Electrostatic dust mobilization on the lunar surface has remained a controversial topic since the Apollo era. In situ and/or low-altitude observations (Rennilson and Criswell 1974; Severny et al. 1975; Berg 1978; Grün and Horányi 2013) as well as high-altitude remote sensing observations (McCoy and Criswell 1974; McCoy 1976; Glenar et al. 2011) have potentially suggested efficient lofting of charged dust particles near the lunar terminators.

One example of evidence for such a process occurring at low altitudes is from images taken just after sunset by the Surveyors 5, 6, and 7 cameras. In many of these images, a horizon glow was observed due west (Rennilson and Criswell 1974; Colwell et al. 2007a), an example of which is shown in Fig. 10. This glow has been interpreted as forward scattered light from a cloud of dust particles levitating less than 1 meter above the surface with particle radii on the order of 5 μm (Rennilson and Criswell 1974). However, as of yet, no mechanism capable of generating sufficiently large electric fields or dust grain charges to levitate such heavy particles has been identified (Hartzell and Scheeres 2011). The Lunar Ejecta and Meteorites experiment (LEAM) registered a multitude of unexpected hits during lunar sunrise and sunset (Berg et al. 1976), however, recent analysis suggests these signals were not caused by slow moving and highly charged dust grains transported across the lunar surface (Grün and Horányi 2013).

Features on asteroid (433) Eros that morphologically resembled ponds (Fig. 10) may indicate that small-scale electrostatic mobilization is a significant process on asteroidal sur-

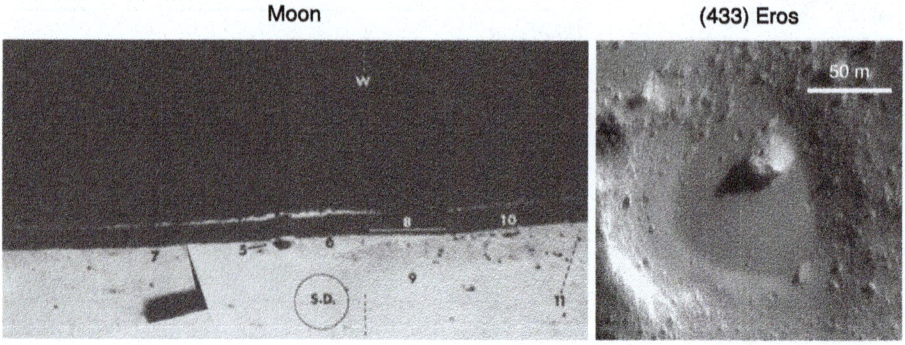

Fig. 10 Examples of possible evidence for electrostatic dust phenomena on airless bodies. *Left.* Surveyor 7 image showing horizon glow just after sunset. The sunlit foreground has been superimposed to show the surface features of the Moon relative to the putative levitated dust cloud. *Reproduced from Rennilson and Criswell (1974) with permission from Earth, Moon, and Planets. Right.* An example of dust ponds on (433) Eros, consisting of smooth, fine grained surface features that are suggested to be formed due to electrostatic mobilization and accumulation in crater depressions. *Reproduced from Richardson et al. (2005) with permission from Icarus*

faces (Veverka et al. 2001; Hughes et al. 2008; Thomas et al. 2015). We note that a competing theory suggests boulder erosion could lead to a similar phenomena (Dombard et al. 2010). Observations of Saturn's small moon Atlas have also suggested similar electrostatic processes occur (Hirata and Miyamoto 2012). For airless bodies with relatively low gravity, electrostatic transport may be part of the process that depletes surface dust (Lee 1996). Recent laboratory studies have shown that dust exposed to UV light and/or incident plasma experiences a high degree of small-scale mobilization and transport (Wang et al. 2009, 2010, 2011, 2016). Dust cohesion may also play an important role in dictating the dynamics of electrostatically lofted dust (Hartzell and Scheeres 2011; Hartzell et al. 2013).

Several decades of modeling efforts have addressed the theoretical possibility of electrostatic dust levitation and/or transport at airless bodies such as asteroids, the Moon, and the rings of Saturn. Early work, motivated by the Apollo observations by LEAM and Surveyor, focused on the possibility that intense electric fields may develop near the terminator regions of airless bodies as sunlit and shadowed portions of the surface lay in close proximity (e.g., De and Criswell 1977; Criswell and De 1977; Pelizzari and Criswell 1978). In theory, photoelectrons emitted from a sunlit patch of surface may be recollected on neighboring shadowed regions, and thus, trapped and unable to return through the highly resistive lunar soil. These simulations suggested that electric fields in the terminator region could reach up to 3×10^5 V/m, many orders-of-magnitude higher than estimated near noon, thereby facilitating the mobilization of lunar dust grains. These early simulations suffered from a number of limitations, most notably the lack of charge neutralization currents from the solar wind electrons. Thus, despite these early predictions, it is unlikely that such strong electric fields develop near the lunar terminator region (see e.g., Poppe et al. 2012; Zimmerman et al. 2014; Piquette and Horányi 2017, and discussion below).

Later analytic models of the electrostatic environment above asteroids, the Moon, and the main rings of Saturn suggested the possibility that dust grains may become electrostatically "trapped" in non-monotonic electrostatic potential layers above a surface (Nitter and Havnes 1992; Nitter et al. 1994, 1998) or preferentially deposited in craters due to differing surface electric fields (Colwell et al. 2005, 2009; Hughes et al. 2008). Farrell et al. (2007, 2008) explored the possibility that regions on the lunar surface just past the terminator may have enhanced electric fields due to the presence of ambipolar separation of solar wind electrons and ions on the edge of the lunar wake; however, in contrast to the supercharging electric fields estimates of $\sim 10^5$, this work derived maximum electric fields near the terminator region of less than 10 V/m.

More recent studies used one- and three-dimensional electrostatic particle-in-cell models to further quantify the lunar plasma sheath both on a flat surface and near the presence of a typical, 10-m sized crater (Poppe and Horányi 2010; Poppe et al. 2012; Piquette and Horányi 2017). Poppe and Horányi (2010) found that non-monotonic potentials can theoretically exist and be energetically preferred for the lunar sheath (in theory allowing for "trapping" of levitated dust grains); however, the electric field strengths proved too weak to stably levitate and/or transport grains larger than ≈ 0.1 μm. For bodies with low gravity, modeling efforts found significant ponding can occur in the shadowed regions of craters, with approximately an order of magnitude more accumulation predicted for small bodies like (433) Eros than at the Moon for similar crater morphologies and plasma conditions (Piquette and Horányi 2017). Grain sizes this small are unable to explain the estimation of 5–10 μm sized levitated dust grains from Surveyor observations (Rennilson and Criswell 1974; Colwell et al. 2007a). Additionally, Hartzell and Scheeres (2013) determined stability regions for electrostatically levitated dust grains. Most recently, Wang et al. (2016) has proposed a "patched surface charge" model that describes how individual micron and submicron sized dust grains may attain charges far in excess of that predicted from a simple application of Gauss' Law (Schwan et al. 2017), thereby allowing significant electrostatic acceleration of micron-sized and larger grains onto ballistic trajectories above a charged surface. Three-dimensional electrostatic simulations performed by Zimmerman et al. (2016) have begun to elucidate this microphysical charging process and may provide the explanation for efficient electrostatic dust transport on airless bodies throughout the solar system.

Lunar and asteroidal observations as well as laboratory experiments have all suggested that small-scale electrostatic mobilization and transport occurs on airless bodies throughout the solar system. However, this phenomenon represents an active field of research and our understanding of this process is continually evolving. While high altitude observations also indicated the existence of lofted dust at tens of km above the surface, the case for such a population is weaker than for low altitudes. The first high altitude, remote sensing optical observations were made during the Apollo 15–17 missions, which took a series of calibrated images to analyze the zodiacal light and the solar corona (McCoy 1976). Some of these images indicated an excess brightness that has been interpreted as forward scattered light from small grains with characteristic radii $a \simeq 0.1$ μm lofted over the terminator regions of the Moon by electrostatic effects. The density of such a dust population was first calculated to be on the order of 10^4 m^{-3} near the surface using Apollo data (McCoy 1976; Glenar et al. 2011). Recent remote sensing surveys by Clementine (Glenar et al. 2014) and LRO/LAMP (Feldman et al. 2014) have significantly lowered the upper limit of the lofted dust density to ~ 1 m^{-3} near the surface. While LDEX could not individually detect dust impacts from 0.1 μm grains, it was able to search for such a population by integrating the cumulative current of impact plasma (Horányi et al. 2014). From data taken throughout LADEE's $\sim 2{,}000$ sunrise terminator transits, LDEX put an in situ upper limit on the dust density distribution of 40–100 m^{-3} (Szalay and Horányi 2015b).

8 Conclusions and Outlook

In this review paper, we have reviewed the current state of knowledge of dust processes occurring in the vicinity of airless bodies. There have been many advancements made in our understanding of these processes in the last two decades, which subsequently opened up new open questions for future studies and measurements. We conclude here with a brief summary of the paper and discuss open issues looking forward.

Our understanding of asteroidal surface properties has been dramatically increased by Itokawa asteroidal samples from the Hayabusa mission. These samples have been studied using advanced ground-based laboratory techniques, the results of which suggest asteroidal surfaces may be continually refreshed over time. Advanced ground-based techniques have been developed, such as phase-resolved photometry and polarimetry, to characterize regolith properties remotely in lieu of having asteroidal samples to study in the lab. There are many upcoming missions, such as OSIRIS-REx and Hayabusa-2, that plan to return additional asteroidal samples to Earth, which will further refine our understanding of these bodies and the surface weathering processes occurring on objects with different compositions and morphologies.

Ejecta clouds around both the Galilean satellites and the Moon have been explored and characterized by in-situ dust detectors. The two datasets have illuminated the nature of dust ejecta clouds and opened up new lines of study related to these clouds. The difference between the inferred ejecta velocity distributions at the Moon and the Galilean satellites remains an open issue that may be resolved by the Surface Dust Analyzer (SUDA) aboard the Europa Clipper mission, set to launch in the early 2020's. This mission will probe Europa's ejecta cloud down to altitudes of 25 km, allowing for direct comparison between the Moon's and Europa's impact ejecta clouds. Additionally, LDEX for the very first time directly transited through dense ejecta plumes comprised of hundreds of impacts. Analysis of these impact plume observations is ongoing and could shed light into the substructure of impact ejecta plumes in a manner unique to dust detector data.

Airless bodies can also be a significant source of interplanetary dust. Due to the large yields for high velocity meteoroid impacts, airless bodies exposed to these impactors can effectively shed their regolith into the IDP population. Quantifying the primary mechanisms and total production rates of IDP production from airless bodies is an important on-going research topic. In the inner solar system, such research is progressing somewhat faster, while in the outer solar system, greater uncertainty persists. Additionally, recent observational campaigns have been unraveling the mystery of active asteroids, which also produce interplanetary dust in characteristically different ways than comets. The relative importance of collisions and rotational disruption for activation of asteroids is a topic of ongoing research.

Electrostatic dust related processes have been a controversial topic since the Apollo era. Such phenomena can be divided into low-altitude and high-altitude populations. As our analysis tools and datasets have evolved, evidence for a high altitude population has continued to look less promising. However, recent laboratory measurements indicate that small-scale electrostatic dust mobilization is likely to occur at all airless bodies in the solar system. Future surface-based instrumentation suites, that would combine dust and plasma measurements, would greatly increase our understanding of these small-scale processes.

Acknowledgements The authors gratefully acknowledge the support of ISSI to produce this review paper. JRS was also supported by NASA's Lunar Data Analysis Program, Grant 80NSSC17K0702. ARP acknowledges NASA Planetary Atmospheres grant #NNX13AG55G and the NASA SSERVI Institute, grant #NNX14AG16A. We thank two anonymous reviewers for their constructive comments evaluating this paper.

References

J. Agarwal, M. Mommert, The nucleus of active asteroid 358P/Pan-STARRS (P/2012 T1). Astron. Astrophys. **549**, 357–359 (2017). https://doi.org/10.1038/nature23892

J. Agarwal, M. Müller, W.T. Reach, M.V. Sykes, H. Boehnhardt, E. Grün, The dust trail of Comet 67P/Churyumov-Gerasimenko between 2004 and 2006. Icarus **207**, 992–1012 (2010). https://doi.org/10.1016/j.icarus.2010.01.003
J. Agarwal, D. Jewitt, H. Weaver, Dynamics of large fragments in the tail of active asteroid P/2010 A2. Astrophys. J. **769**(1), 46 (2013)
J. Agarwal, D. Jewitt, H. Weaver, M. Mutchler, S. Larson, Hubble and Keck Telescope observations of active asteroid 288P/300163 (2006 VW139). Astron. J. **151**(1), 12 (2015)
J. Agarwal, D.C. Jewitt, M. Mutchler, H. Weaver, S. Larson, A binary main-belt comet. Nature **549**(7672), 357–359 (2017)
J.R. Arnold, A Monte Carlo model for the gardening of the lunar regolith. Moon **13**, 159–172 (1975)
N. Asada, Fine fragments in high-velocity impact experiments. J. Geophys. Res. **90**, 12445 (1985). https://doi.org/10.1029/JB090iB14p12445
E. Asphaug, S.J. Ostro, R. Hudson, D.J. Scheeres, W. Benz, Disruption of kilometre-sized asteroids by energetic collisions. Nature **393**(6684), 437 (1998)
F. Bagenal, M. Horányi, D.J. McComas, R.L. McNutt, H.A. Elliot, M.E. Hill, L.E. Brown, P.A. Delamere, P. Kollman, S.M. Krimigis, M. Kusterer, C.M. Lisse, D.G. Mitchell, M. Piquette, A.R. Poppe, D.F. Strobel, J.R. Szalay, P. Valek, J. Vandegriff, S. Weidner, E.J. Zirnstein, S.A. Stern, K. Ennico, C.B. Olkin, H.A. Weaver, L.A. Young, N.H.S. Team, Pluto's interaction with its space environment: Solar wind, energetic particles, and dust. Science **351**(6279), aad9045 (2016)
S. Barabash, R. Lundin, On a possible dust-plasma interaction at Mars. IEEE Trans. Plasma Sci. **22**(2), 173–178 (1994)
I. Belskaya, S. Bagnulo, Transneptunian objects and centaurs, in *Polarimetry of Stars and Planetary Systems*, ed. by L. Kolokolova, J. Hough, A.-C. Levasseur-Regourd (2015), p. 405
I.N. Belskaya, S. Bagnulo, M.A. Barucci, K. Muinonen, G.P. Tozzi, S. Fornasier, L. Kolokolova, Polarimetry of centaurs (2060) Chiron, (5145) Pholus and (10199) Chariklo. Icarus **210**(1), 472–479 (2010). https://doi.org/10.1016/j.icarus.2010.06.005. http://www.sciencedirect.com/science/article/pii/S0019103510002277
I. Belskaya, A. Cellino, R. Gil-Hutton, K. Muinonen, Y. Shkuratov, Asteroid polarimetry, in *Asteroids IV*, ed. by P. Michel, F.E. DeMeo, W.F. Bottke (2015), pp. 151–163
S.D. Benecchi, K.S. Noll, W.M. Grundy, H.F. Levison, (47171) 1999 TC_{36}, a transneptunian triple. Icarus **207**, 978–991 (2010)
A.A. Berezhnoy, E.A. Kozlova, M.P. Sinitsyn, A.A. Shangaraev, V.V. Shevchenko, Origin and stability of lunar polar volatiles. Adv. Space Res. **50**, 1638–1646 (2012)
O.E. Berg, Lunar terminator configuration. Earth Planet. Sci. Lett. **39**(3), 377–381 (1978)
O.E. Berg, H. Wolf, J. Rhee, Lunar soil movement registered by the Apollo 17 cosmic dust experiment, in *Interplanetary Dust and Zodiacal Light*, ed. by H. Elsaesser, H. Fechtig Lecture Notes in Physics, vol. 48 (Springer, Berlin, 1976), pp. 233–237
R. Berisch, W. Eckstein (eds.), *Sputtering by Particle Bombardment: Experiments and Computer Calculations from Threshold to MeV Energies*. Topics in Applied Physics, vol. 110 (Springer, Berlin, 2007)
J.P. Biersack, W. Eckstein, Sputtering studies with the Monte Carlo program TRIM.SP. Appl. Phys. A **34**, 73–94 (1984)
T. Birnstiel, H. Klahr, B. Ercolano, A simple model for the evolution of the dust population in protoplanetary disks. Astron. Astrophys. **539**, 148 (2012)
D. Bodewits, M. Kelley, J.-Y. Li, W. Landsman, S. Besse, M. A'Hearn, Collisional excavation of asteroid (596) Scheila. Astrophys. J. Lett. **733**(1), 3 (2011)
D. Bodewits, J.-B. Vincent, M.S. Kelley, Scheila's scar: direct evidence of impact surface alteration on a primitive asteroid. Icarus **229**, 190–195 (2014)
P. Borin, G. Cremonese, F. Marzari, M. Bruno, S. Marchi, Statistical analysis of micrometeoroids flux on Mercury. Astron. Astrophys. **503**, 259–264 (2009)
P. Borin, G. Cremonese, F. Marzari, Statistical analysis of the flux of micrometeoroids at Mercury from both cometary and asteroidal components. Astron. Astrophys. **585**, A106 (2016)
W.F. Bottke, M.C. Nolan, R. Greenberg, R.A. Kolvoord, Velocity distributions among colliding asteroids. Icarus **107**(2), 255–268 (1994)
W.F. Bottke, D. Vokrouhlicky, D.P. Rubincam, M. Broz, The effect of Yarkovsky thermal forces on the dynamical evolution of asteroids and meteoroids, in *Asteroids III* (2002), p. 395
W.F. Bottke, D.D. Durda, D. Nesvorný, R. Jedicke, A. Morbidelli, D. Vokrouhlický, H.F. Levison, Linking the collisional history of the main asteroid belt to its dynamical excitation and depletion. Icarus **179**(1), 63–94 (2005)
D. Britt, C. Pieters, Darkening in black and gas-rich ordinary chondrites: The spectral effects of opaque morphology and distribution. Geochim. Cosmochim. Acta **58**(18), 3905–3919 (1994)

M.E. Brown, W.M. Calvin, Evidence for crystalline water and ammonia ices on Pluto's satellite Charon. Science **287**, 107–109 (2000)
R. Brunetto, M.J. Loeffler, D. Nesvorný, S. Sasaki, G. Strazulla, Asteroid surface alteration by space weathering processes, in *Asteroids IV* (2015), pp. 597–616
E. Buhl, F. Sommer, M.H. Poelchau, G. Dresen, T. Kenkmann, Ejecta from experimental impact craters: particle size distribution and fragmentation energy. Icarus **237**, 131–142 (2014). https://doi.org/10.1016/j.icarus.2014.04.039
J.A. Burns, M. Showalter, D.P. Hamilton, P.D. Nicholson, I. de Pater, M.E. Ockert-Bell, P.C. Thomas, The formation of Jupiter's faint rings. Science **284**, 1146–1150 (1999)
M.D. Campbell-Brown, High resolution radiant distribution and orbits of sporadic radar meteoroids. Icarus **196**, 144–163 (2008). https://doi.org/10.1016/j.icarus.2008.02.022
P. Carrez, K. Demyk, P. Cordier, L. Gengembre, J. Grimblot, L. D'Hendecourt, A.P. Jones, H. Leroux, Low-energy helium ion irradiation-induced amorphization and chemical changes in olivine: insights for silicate dust evolution in the interstellar medium. Meteorit. Planet. Sci. **37**(11), 1599–1614 (2002)
R. Christoffersen, D.S. McKay, L.P. Keller, Microstructure, chemistry, and origin of grain rims on ilmenite from the lunar soil finest fraction. Meteorit. Planet. Sci. **31**(6), 835–848 (1996)
M.J. Cintala, Impact-induced thermal effects in the Lunar and Mercurian regoliths. J. Geophys. Res. **97**(E1), 947–973 (1992)
F. Cipriani, O. Witasse, F. Leblanc, R. Modolo, R.E. Johnson, A model of interaction of Phobos' surface with the Martian environment. Icarus **212**, 643–648 (2011)
R.N. Clark, J.M. Curchin, R. Jaumann, D.P. Cruikshank, R.H. Brown, T.M. Hoefen, K. Stephan, J.M. Moore, B.J. Buratti, K.H. Baines, P.D. Nicholson, R.M. Nelson, Compositional mapping of Saturn's satellite Dione with Cassini VIMS and implications of dark material in the Saturn system. Icarus **193**, 372–386 (2008)
R.N. Clark, D.P. Cruikshank, R. Jaumann, R.H. Brown, K. Stephan, C.M. Dalle Ore, K.E. Livo, N. Pearson, J.M. Curchin, T.M. Hoefen, B.J. Buratti, G. Filacchione, K.H. Baines, P.D. Nicholson, The surface composition of Iapetus: mapping results from Cassini VIMS. Icarus **218**, 831–860 (2012)
A. Colaprete et al., Detection of water in the LCROSS ejecta plume. Science **330**(6003), 463–468 (2010)
A. Colaprete, M. Sarantos, D.H. Wooden, T.J. Stubbs, A.M. Cook, M. Shirley, How surface composition and meteoroid impacts mediate sodium and potassium in the lunar exosphere. Science **351**(6270), 249–252 (2016)
A. Collette, K. Drake, A. Mocker, Z. Sternovsky, T. Munsat, M. Horányi, Time-resolved temperature measurements in hypervelocity dust impacts. Planet. Space Sci. **89**, 58–62 (2013)
A. Collette, Z. Sternovsky, M. Horányi, Production of neutral gas by micrometeoroid impacts. Icarus **227**, 89–93 (2014)
G. Colombo, D.A. Lautman, I.I. Shapiro, The Earth's Dust Belt: Fact or fiction?: 2. Gravitational focusing and Jacobi capture. J. Geophys. Res. **71**, 5705–5717 (1966). https://doi.org/10.1029/JZ071i023p05705
J.E. Colwell, A.A.S. Gulbis, M. Horányi, S. Robertson, Dust transport in photoelectron layers and the formation of dust ponds on Eros. Icarus **175**, 159–169 (2005)
J.E. Colwell, M. Horányi, S. Robertson, X. Wang, A. Haugsjaa, P. Wheeler, Behavior of charged dust in plasma and photoelectron sheaths, in *Dust in Planetary Systems*. ESA SP, vol. 643 (2007a), pp. 171–175
J.E. Colwell, S. Robertson, M. Horányi, X. Wang, A. Poppe, P. Wheeler, Lunar dust levitation. J. Aerosp. Eng. **1**, 2–9 (2009)
J.C. Cook, S.J. Desch, T.L. Roush, C.A. Trujillo, T.R. Geballe, Near-infrared spectroscopy of Charon: possible evidence for cryovolcanism on Kuiper Belt Objects. Astrophys. J. **663**, 1406–1419 (2007)
D. Crider, R.M. Killen, Burial rate of Mercury's polar volatile deposits. Geophys. Res. Lett. **32**, L12201 (2005)
D.H. Crider, R.R. Vondrak, Space weathering effects on lunar cold trap deposits. J. Geophys. Res. **108**(E7), 5079 (2003)
D.R. Criswell, B.R. De, Intense localized photoelectric charging in the Lunar sunset terminator region: 2. Supercharging at the progression of sunset. J. Geophys. Res. **82**(7), 1005–1007 (1977)
D.P. Cruikshank, J.F. Bell, M.J. Gaffey, R.H. Brown, R. Howell, C. Beerman, M. Rognstad, The dark side of Iapetus. Icarus **53**, 90–104 (1983)
C.M. Dalle Ore, D.P. Cruikshank, R.N. Clark, Infrared spectroscopic characterization of the low-albedo materials on Iapetus. Icarus **221**, 735–743 (2012)
C.M. Dalle Ore, D.P. Cruikshank, R.M.E. Mastrapa, E. Lewis, O.L. White, Impact craters: an icy study on Rhea. Icarus **261**, 80–90 (2015)
D.R. Davis, P. Farinella, Collisional evolution of Edgeworth-Kuiper belt objects. Icarus **125**, 50–60 (1997)
G. de Elía, A. Brunini, Collisional and dynamical evolution of the L_4 Trojan asteroids. Astron. Astrophys. **475**(1), 375–389 (2007)

I. de Pater, S.G. Gibbard, H.B. Hammel, Evolution of the dusty rings of Uranus. Icarus **180**, 186–200 (2006a)
I. de Pater, H.B. Hammel, S.G. Gibbard, M.R. Showalter, New dust belts of Uranus: one ring, two ring, red ring, blue ring. Science **312**, 92–94 (2006b)
B.R. De, D.R. Criswell, Intense localized photoelectric charging in the lunar sunset terminator region: 1. Development of potentials and fields. J. Geophys. Res. **82**(7), 999–1004 (1977)
M. Delbo, G. Libourel, J. Wilkerson, N. Murdoch, P. Michel, K. Ramesh, C. Ganino, C. Verati, S. Marchi, Thermal fatigue as the origin of regolith on small asteroids. Nature **508**(7495), 233–236 (2014)
M. Delbo, M. Mueller, J.P. Emery, B. Rozitis, M.T. Capria, Asteroid thermophysical modeling, in *Asteroids IV* (2015), p. 107
F. DeMeo, C. Alexander, K. Walsh, C. Chapman, R. Binzel, The compositional structure of the asteroid belt, in *Asteroids IV* (2015b), pp. 13–42
L. Denneau, R. Jedicke, A. Fitzsimmons, H. Hsieh, J. Kleyna, M. Granvik, M. Micheli, T. Spahr, P. Vereš, R. Wainscoat, et al., Observational constraints on the catastrophic disruption rate of small main belt asteroids. Icarus **245**, 1–15 (2015)
S.J. Desch, M. Neveu, Differentiation and cryovolcanism on Charon: a view before and after New Horizons. Icarus **287**, 175–186 (2017)
S.J. Desch, J.C. Cook, T.C. Doggett, S.B. Porter, Thermal evolution of Kuiper belt objects, with implications for cryovolcanism. Icarus **202**, 694–714 (2009)
V. Dikarev, E. Grün, J. Baggaley, D. Galligan, M. Landgraf, R. Jehn, The new ESA meteoroid model. Adv. Space Res. **35**(7), 1282–1289 (2005)
N. Divine, Five populations of interplanetary meteoroids. J. Geophys. Res. **98**, 17029–17048 (1993). https://doi.org/10.1029/93JE01203
A.V. Dmitriev, A.V. Suvorova, I.S. Veselovsky, Statistical Characteristics of the Heliospheric Plasma and Magnetic Field at Earth's Orbit during Four Solar Cycles 20-23, in *Handbook on Solar Wind: Effects, Dynamics, and Interactions*, ed. by H.E. Johannson (NOVA Science Publishers, New York, 2011), pp. 81–144. Chap. 2
A.J. Dombard, O.S. Barnouin, L.M. Prockter, P.C. Thomas, Boulders and ponds on the Asteroid 433 Eros. Icarus **210**(2), 713–721 (2010)
M. Drahus, W. Waniak, S. Tendulkar, J. Agarwal, D. Jewitt, S.S. Sheppard, Fast rotation and trailing fragments of the active asteroid P/2012 F5 (Gibbs). Astrophys. J. Lett. **802**(1), 8 (2015)
E. Dubinin, R. Lundin, N.F. Pissarenko, S.V. Barabash, A.V. Zakharov, H. Koskinen, K. Schwingenschuh, Y.G. Yeroshenko, Indirect evidences for a gas/dust torus along the Phobos orbit. Geophys. Res. Lett. **17**(6), 861–864 (1990)
E.M. Dubinin, N.F. Pissarenko, S.V. Barabash, A.V. Zakharov, R. Lundin, R. Pellinen, K. Schwingenschuh, Plasma and magnetic field effects associated with Phobos and Deimos tori. Planet. Space Sci. **39**(1/2), 113–121 (1991)
D.D. Durda, S.A. Stern, Collision rates in the present-day Kuiper belt and Centaur regions: applications to surface activation and modification on comets, Kuiper belt objects, Centaurs, and Pluto-Charon. Icarus **145**, 220–229 (2000)
R.C. Elphic, G.T. Delory, B.P. Hine, P. Mahaffy, M. Horanyi, A. Colaprete, M. Benna, S. Noble, The Lunar Atmosphere and Dust Environment Explorer Mission. Space Sci. Rev. **185**, 3–25 (2014)
W.M. Farrell, T.J. Stubbs, R.R. Vondrak, G.T. Delory, J.S. Halekas, Complex electric fields near the lunar terminator: the near-surface wake and accelerated dust. Geophys. Res. Lett. **34**, L14201 (2007)
W.M. Farrell, T.J. Stubbs, J.S. Halekas, G.T. Delory, M.R. Collier, R.R. Vondrak, R.P. Lin, Loss of solar wind plasma neutrality and affect on surface potentials near the lunar terminator and shadowed polar regions. Geophys. Res. Lett. **35**, L05105 (2008)
P.D. Feldman, D.A. Glenar, T.J. Stubbs, K.D. Retherford, G. Randall Gladstone, P.F. Miles, T.K. Greathouse, D.E. Kaufmann, J.W. Parker, S. Alan Stern, Upper limits for a lunar dust exosphere from far-ultraviolet spectroscopy by LRO/LAMP. Icarus **233**, 106–113 (2014)
M. Finson, R. Probstein, A theory of dust comets. I. Model and equations. Astrophys. J. **154**, 327–352 (1968)
A. Fujiwara, J. Kawaguchi, D.K. Yeomans, M. Abe, T. Mukai, T. Okada, J. Saito, H. Yano, M. Yoshikawa, D.J. Scheeres, O. Barnouin-Jha, A.F. Cheng, H. Demura, R.W. Gaskell, N. Hirata, H. Ikeda, T. Kominato, H. Miyamoto, A.M. Nakamura, R. Nakamura, S. Sasaki, K. Uesugi, The rubble-pile asteroid Itokawa as observed by Hayabusa. Science **312**, 1330–1334 (2006). https://doi.org/10.1126/science.1125841
M. Fulle, L. Colangeli, J. Agarwal, A. Aronica, V. Della Corte, F. Esposito, E. Grün, M. Ishiguro, R. Ligustri, J.J. Lopez Moreno, E. Mazzotta Epifani, G. Milani, F. Moreno, P. Palumbo, J. Rodríguez Gómez, A. Rotundi, Comet 67P/Churyumov-Gerasimenko: the GIADA dust environment model of the Rosetta mission target. Astron. Astrophys. **522**, 63 (2010). https://doi.org/10.1051/0004-6361/201014928
E. Füri, B. Marty, S.S. Assonov, Constraints on the flux of meteoritic and cometary water on the Moon from volatile element (N-Ar) analysis of single lunar soil grains, Luna 24 core. Icarus **218**, 220–229 (2012)

D.E. Gault, F. Hörz, D.E. Brownlee, J.B. Hartung, Mixing of the lunar regolith, in *Proc. 5th Lunar Conf.*, vol. 5 (1974)

B. Gladman, P. Michel, C. Froeschlé, The near-Earth object population. Icarus **146**(1), 176–189 (2000)

D.A. Glenar, T.J. Stubbs, J.E. McCoy, R.R. Vondrak, A reanalysis of the Apollo light scattering observations, and implications for lunar exospheric dust. Planet. Space Sci. **59**, 1695–1707 (2011)

D.A. Glenar, T.J. Stubbs, M. Hahn, Y. Wang, Search for a high altitude lunar dust exosphere using Clementine navigational star tracker measurements. J. Geophys. Res. **119**, 2548–2567 (2014)

C.A. Goodrich, E.R. Scott, A.M. Fioretti, Ureilitic breccias: clues to the petrologic structure and impact disruption of the ureilite parent asteroid. Chem. Erde **64**(4), 283–327 (2004)

E. Grün, M. Horányi, A new look at Apollo 17 LEAM data: nighttime dust activity in 1976. Planet. Space Sci. **89**, 2–14 (2013). https://doi.org/10.1016/j.pss.2013.10.005

E. Grün, H. Fechtig, M.S. Hanner, J. Kissel, B.-A. Lindblad, D. Linkert, D. Maas, G.E. Morfill, H.A. Zook, The Galileo dust detector. Space Sci. Rev. **60**, 317–340 (1992). https://doi.org/10.1007/BF00216860

E. Grün, B. Gustafson, I. Mann, M. Baguhl, G.E. Morfill, P. Staubach, A. Taylor, H.A. Zook, Interstellar dust in the heliosphere. Astron. Astrophys. **286**, 915–924 (1994)

W.M. Grundy, M.W. Buie, J.A. Stansberry, J.R. Spencer, B. Schmitt, Near-infrared spectra of icy outer Solar System surfaces: remote determination of H_2O ice temperatures. Icarus **142**, 536–549 (1999)

B. Gundlach, J. Blum, A new method to determine the grain size of planetary regolith. Icarus **223**(1), 479–492 (2013)

D.A. Gurnett, J.A. Ansher, W.S. Kurth, L.J. Granroth, Micron-sized dust particles detected in the outer solar system by the Voyager 1 and 2 plasma wave instruments. Geophys. Res. Lett. **24**(24), 3125–3128 (1997)

J.M. Hahn, H.A. Zook, B. Cooper, B. Sunkara, Clementine observations of the zodiacal light and the dust content of the inner solar system. Icarus **158**, 360–378 (2002)

O. Hainaut, J. Kleyna, G. Sarid, B. Hermalyn, A. Zenn, K. Meech, P. Schultz, H. Hsieh, G. Trancho, J. Pittichová, et al., P/2010 A2 LINEAR-I. An impact in the asteroid main belt. Astron. Astrophys. **537**, 69 (2012)

O. Hainaut, H. Boehnhardt, C. Snodgrass, K. Meech, J. Deller, M. Gillon, E. Jehin, E. Kuehrt, S. Lowry, J. Manfroid, et al., Continued activity in P/2013 P5 PANSTARRS—unexpected comet, rotational break-up, or rubbing binary asteroid? Astron. Astrophys. **563**, 75 (2014)

D.P. Hamilton, J.A. Burns, Origin of Saturn's E ring: self-sustained, naturally. Science **264**, 550–553 (1994)

D.P. Hamilton, H. Krüger, The sculpting of Jupiter's gossamer rings by its shadow. Nature **453**, 72–75 (2008)

D.P. Hamilton, M.F. Skrutskie, A.J. Verbiscer, F.J. Masci, Small particles dominate Saturn's Phoebe ring to surprisingly large distances. Nature **522**, 185–187 (2015)

D. Han, A.R. Poppe, M. Piquette, E. Grün, M. Horányi, Constraints on dust production in the Edgeworth-Kuiper Belt from Pioneer 10 and New Horizons measurement. Geophys. Res. Lett. **38**, L24102 (2011)

B. Hapke, Space weathering from Mercury to the asteroid belt. J. Geophys. Res. **106**(E), 10039–10074 (2001)

J.K. Harmon, M.A. Slade, M.S. Rice, Radar imagery of Mercury's putative polar ice: 1999–2005 Arecibo results. Icarus **211**, 37–50 (2011)

W.K. Hartmann, Impact experiments. I—Ejecta velocity distributions and related results from regolith targets. Icarus **63**, 69–98 (1985). https://doi.org/10.1016/0019-1035(85)90021-1

C.M. Hartzell, D.J. Scheeres, The role of cohesive forces in particle launching on the moon and asteroids. Planet. Space Sci. **59**(14), 1758–1768 (2011). Lunar Dust, Atmosphere and Plasma: The Next Steps

C.M. Hartzell, D.J. Scheeres, Dynamics of levitating dust particles near asteroids and the moon. J. Geophys. Res., Planets **118**(1), 116–125 (2013)

C.M. Hartzell, X. Wang, D.J. Scheeres, M. Horányi, Experimental demonstration of the role of cohesion in electrostatic dust lofting. Geophys. Res. Lett. **40**(6), 1038–1042 (2013)

P.O. Hayne, A. Hendrix, E. Sefton-Nash, M.A. Siegler, P.G. Lucey, K.D. Retherford, J.-P. Williams, B.T. Greenhagen, D.A. Paige, Evidence for exposed water ice in the Moon's south polar regions from Lunar Reconnaissance Orbiter ultraviolet albedo and temperature measurements. Icarus **255**, 58–69 (2015)

M.M. Hedman, C.D. Murray, N.J. Cooper, M.S. Tiscareno, K. Beurle, M.W. Evans, J.A. Burns, Three tenuous rings/arcs for three tiny moons. Icarus **199**, 378–386 (2009)

M. Hirabayashi, Failure modes and conditions of a cohesive, spherical body due to YORP spin-up. Mon. Not. R. Astron. Soc. **454**(2), 2249–2257 (2015)

M. Hirabayashi, D.J. Scheeres, Analysis of asteroid (216) Kleopatra using dynamical and structural constraints. Astrophys. J. **780**(2), 160 (2013)

N. Hirata, H. Miyamoto, Dust levitation as a major resurfacing process on the surface of a saturnian icy satellite, Atlas. Icarus **220**, 106–113 (2012)

M. Horányi, J.A. Burns, M. Tatrallyay, J.G. Luhmann, Toward understanding the fate of dust lost from the martian satellites. Geophys. Res. Lett. **17**(6), 853–856 (1990)

M. Horányi, M. Tatrallyay, A. Juhász, J.G. Luhmann, The dynamics of submicron-sized dust particles lost from Phobos. J. Geophys. Res. **96**(A7), 11283–11290 (1991)

M. Horányi, Z. Sternovsky, M. Lankton, C. Dumont, S. Gagnard, D. Gathright, E. Grün, D. Hansen, D. James, S. Kempf, B. Lamprecht, J. Szalay, R. Srama, G. Wright, The Lunar Dust Experiment (LDEX) onboard the Lunar Atmosphere and Dust Environment Explorer (LADEE) mission. Space Sci. Rev. **185**, 93–113 (2014)

M. Horányi, J.R. Szalay, S. Kempf, J. Schmidt, E. Grün, R. Srama, Z. Sternovsky, A permanent, asymmetric dust cloud around the Moon. Nature **522**(7556), 324 (2015)

J. Horner, N. Evans, M. Bailey, Simulations of the population of centaurs—I. the bulk statistics. Mon. Not. R. Astron. Soc. **354**(3), 798–810 (2004)

K.R. Housen, K.A. Holsapple, Ejecta from impact craters. Icarus **211**(1), 856–875 (2011)

H.H. Hsieh, D. Jewitt, A population of comets in the main asteroid belt. Science **312**(5773), 561–563 (2006)

H.H. Hsieh, D.C. Jewitt, Y.R. Fernández, The strange case of 133P/Elst-Pizarro: a comet among the asteroids. Astron. J. **127**(5), 2997 (2004)

H.H. Hsieh, D. Jewitt, M. Ishiguro, Physical properties of main-belt comet P/2005 U1 (Read). Astron. J. **137**(1), 157 (2008)

H.H. Hsieh, D. Jewitt, P. Lacerda, S.C. Lowry, C. Snodgrass, The return of activity in main-belt Comet 133P/Elst-Pizarro. Mon. Not. R. Astron. Soc. **403**, 363–377 (2010). https://doi.org/10.1111/j.1365-2966.2009.16120.x

H.H. Hsieh, M. Ishiguro, P. Lacerda, D. Jewitt, Physical properties of main-belt comet 176P/LINEAR. Astron. J. **142**(1), 29 (2011)

H.H. Hsieh, O. Hainaut, B. Novaković, B. Bolin, L. Denneau, A. Fitzsimmons, N. Haghighipour, J. Kleyna, R. Kokotanekova, P. Lacerda, et al., Sublimation-driven activity in main-belt comet 313P/Gibbs. Astrophys. J. Lett. **800**(1), 16 (2015)

A.L.H. Hughes, J.E. Colwell, A.W. Dewolfe, Electrostatic dust transport on Eros: 3-D simulations of pond formation. Icarus **195**, 630–648 (2008)

M.-T. Hui, D. Jewitt, Non gravitational acceleration of the active asteroids. Astron. J. **153**(2), 80 (2017)

M.-T. Hui, D. Jewitt, X. Du, Split active asteroid P/2016 J1 (Panstarrs). Astron. J. **153**(4), 141 (2017). http://stacks.iop.org/1538-3881/153/i=4/a=141

D.H. Humes, Results of Pioneer 10 and 11 meteoroid experiments: interplanetary and near-Saturn. J. Geophys. Res. **85**(A11), 5841–5852 (1980)

D.M. Hunten, G. Cremonese, A.L. Sprague, R.E. Hill, S. Verani, R.W.H. Kozlowski, The Leonid meteor shower and the lunar sodium atmosphere. Icarus **136**, 298–303 (1998)

D.M. Hurley, D.J. Lawrence, D.B.J. Bussey, R.R. Vondrak, R.C. Elphic, G.R. Gladstone, Two-dimensional distribution of volatiles in the lunar regolith from space weathering simulations. Geophys. Res. Lett. **39**, L09203 (2012)

W.-H. Ip, M. Banaszkiewicz, On the dust/gas tori of Phobos and Deimos. Geophys. Res. Lett. **17**(6), 857–860 (1990)

M. Ishiguro, H. Hanayama, S. Hasegawa, Y. Sarugaku, J.-i. Watanabe, H. Fujiwara, H. Terada, H.H. Hsieh, J.J. Vaubaillon, N. Kawai, et al., Interpretation of (596) Scheila's triple dust tails. Astrophys. J. Lett. **741**(1), 24 (2011a)

M. Ishiguro, H. Hanayama, S. Hasegawa, Y. Sarugaku, J.-I. Watanabe, H. Fujiwara, H. Terada, H.H. Hsieh, J.J. Vaubaillon, N. Kawai, et al., Observational evidence for an impact on the main-belt Asteroid (596) Scheila. Astrophys. J. Lett. **740**(1), 11 (2011b)

H. Ishimoto, Formation of Phobos/Deimos dust rings. Icarus **122**, 153–165 (1996)

J. Jackson, Glossary of geology. Technical report. American Geological Institute, Alexandria, VA (1997), 769 pp., ISBN 0-922152-34-9

D. Janches, J.D. Mathews, D.D. Meisel, V.S. Getman, Q.H. Zhou, Doppler studies of near-antapex UHF radar micrometeors. Icarus **143**(2), 347–353 (2000)

D. Janches, P. Pokorný, M. Sarantos, J.R. Szalay, M. Horányi, D. Nesvorný, Constraining the ratio of micrometeoroids from short and long period comets at 1 AU from LADEE observations of the lunar dust cloud. Geophys. Res. Lett. (2018). https://doi.org/10.1002/2017GL076065.

P. Jenniskens, Meteoroid streams and the zodiacal cloud, in *Asteroids IV* (University of Arizona Press, Tucson, 2015), p. 281

P. Jenniskens, D.F. Blake, Crystallization of amorphous water ice in the Solar System. Astrophys. J. **473**, 1104–1113 (1996)

D. Jewitt, The active asteroids. Astron. J. **143**(3), 66 (2012)

D. Jewitt, J. Li, Activity in Geminid Parent (3200) Phaethon. Astron. J. **140**(5), 1519 (2010)

D.C. Jewitt, J.X. Luu, Crystalline water ice on the Kuiper belt object (50000) Quaoar. Nature **432**, 731–733 (2004)

D. Jewitt, B. Yang, N. Haghighipour, Main-belt comet P/2008 R1 (Garradd). Astron. J. **137**(5), 4313 (2009)

D. Jewitt, H. Weaver, J. Agarwal, M. Mutchler, M. Drahus, A recent disruption of the main-belt asteroid P/2010 A2. Nature **467**(7317), 817–819 (2010)

D. Jewitt, H. Weaver, M. Mutchler, S. Larson, J. Agarwal, Hubble Space Telescope observations of main-belt comet (596) Scheila. Astrophys. J. Lett. **733**(1), 4 (2011)
D. Jewitt, J. Agarwal, H. Weaver, M. Mutchler, S. Larson, The extraordinary multi-tailed main-belt Comet P/2013 P5. Astrophys. J. Lett. **778**(1), L21 (2013)
D. Jewitt, M. Ishiguro, J. Agarwal, Large particles in active asteroid P/2010 A2. Astrophys. J. Lett. **764**(1), 5 (2013a)
D. Jewitt, J. Li, J. Agarwal, The dust tail of asteroid (3200) Phaethon. Astrophys. J. Lett. **771**(2), 36 (2013b)
D. Jewitt, J. Agarwal, J. Li, H. Weaver, M. Mutchler, S. Larson, Disintegrating asteroid P/2013 R3. Astrophys. J. Lett. **784**(1), 8 (2014a)
D. Jewitt, M. Ishiguro, H. Weaver, J. Agarwal, M. Mutchler, S. Larson, Hubble Space Telescope investigation of main-belt comet 133P/Elst-Pizarro. Astron. J. **147**(5), 117 (2014b)
D. Jewitt, J. Agarwal, H. Weaver, M. Mutchler, S. Larson, Episodic ejection from active asteroid 311P/PANSTARRS. Astrophys. J. **798**(2), 109 (2015a)
D. Jewitt, J. Agarwal, N. Peixinho, H. Weaver, M. Mutchler, M.-T. Hui, J. Li, S. Larson, New active asteroid 313P/Gibbs. Astron. J. **149**(2), 81 (2015b)
D. Jewitt, J. Li, J. Agarwal, H. Weaver, M. Mutchler, S. Larson, Nucleus and mass loss from active asteroid 313P/Gibbs. Astron. J. **150**(3), 76 (2015c)
D. Jewitt, H. Hsieh, J. Agarwal, et al., The active asteroids, in *Asteroids IV* (University of Arizona, Tucson, 2015d), pp. 221–241
J. Jones, P. Brown, Sporadic meteor radiant distributions—orbital survey results. Mon. Not. R. Astron. Soc. **265**, 524 (1993)
A. Juhász, M. Horányi, Dust torus around Mars. J. Geophys. Res. **100**(E2), 3277–3284 (1995)
A. Juhász, M. Tátrallyay, G. Gévai, M. Horányi, On the density of the dust halo around Mars. J. Geophys. Res. **98**(E1), 1205–1211 (1993)
S. Kameda, I. Yoshikawa, M. Kagitani, S. Okano, Interplanetary dust distribution and temporal variability of Mercury's atmospheric Na. Geophys. Res. Lett. **36**, L15201 (2009)
L.P. Keller, E.L. Berger, A transmission electron microscope study of Itokawa regolith grains. Earth Planets Space **66**(1), 71 (2014)
L.P. Keller, D.S. McKay, The nature and origin of rims on lunar soil grains. Geochim. Cosmochim. Acta **61**(11), 2331–2341 (1997)
S. Kempf, Interpretation of high rate dust measurements with the Cassini dust detector CDA. Planet. Space Sci. **56**, 378–385 (2008). https://doi.org/10.1016/j.pss.2007.11.022
K.V. Kholshevnikov, A.V. Krivov, L.L. Sokolov, V.B. Titov, The dust torus around Phobos orbit. Icarus **105**, 351–362 (1993)
R.M. Killen, J.M. Hahn, Impact vaporization as a possible source of Mercury's calcium exosphere. Icarus **250**, 230–237 (2015)
R.M. Killen, W.-H. Ip, The surface-bounded atmosphere of Mercury and the Moon. Rev. Geophys. **37**(3), 361–406 (1999)
Y. Kim, M. Ishiguro, M.G. Lee, New observational evidence of active asteroid P/2010 A2: slow rotation of the largest fragment. Astrophys. J. Lett. **842**, 23 (2017a). https://doi.org/10.3847/2041-8213/aa7944
Y. Kim, M. Ishiguro, T. Michikami, A.M. Nakamura, Anisotropic ejection from active asteroid P/2010 A2: an implication of impact shattering on an asteroid. Astron. J. **153**, 228 (2017b). https://doi.org/10.3847/1538-3881/aa69bb
H. Kosai, Short-period comets and Apollo-Amor-Aten type asteroids in view of Tisserand invariant, in *Dynamics and Evolution of Minor Bodies with Galactic and Geological Implications* (Springer, Berlin, 1992), pp. 237–240
D. Koschny, E. Grün, Impacts into ice-silicate mixtures: crater morphologies, volumes, depth-to-diameter ratios, and yield. Icarus **154**, 391–401 (2001a)
D. Koschny, E. Grün, Impacts into ice-silicate mixtures: ejecta mass and size distributions. Icarus **154**, 402–411 (2001b). https://doi.org/10.1006/icar.2001.6708
L. Kresak, On the similarity of orbits of associated comets, asteroids and meteoroids. Bull. Astron. Inst. Czechoslov. **33**, 104–110 (1982)
A.V. Krivov, On the dust belts of Mars. Astron. Astrophys. **291**, 657–663 (1994)
A.V. Krivov, D.P. Hamilton, Martian dust belts: waiting for discovery. Icarus **128**, 335–353 (1997)
A. Krivov, A. Jurewicz, The ethereal dust envelopes of the Martian moons. Planet. Space Sci. **47**, 45–56 (1998). https://doi.org/10.1016/S0032-0633(98)00101-9
A.V. Krivov, H. Krüger, E. Grün, K.-U. Thiessenhusen, D.P. Hamilton, A tenuous dust ring of Jupiter formed by escaping ejecta from the Galilean satellites. J. Geophys. Res., Planets **107**, 5002 (2002). https://doi.org/10.1029/2000JE001434
A.V. Krivov, M. Sremčević, F. Spahn, V.V. Dikarev, K.V. Kholshevnikov, Impact-generated dust clouds around planetary satellites: spherically symmetric case. Planet. Space Sci. **51**, 251–269 (2003)

A.V. Krivov, A.G. Feofilov, V.V. Dikarev, Search for the putative dust belts of Mars: The late 2007 opportunity. Planet. Space Sci. **54**, 871–878 (2006)
H. Krüger, E. Grün, Interstellar dust inside and outside the heliosphere. Space Sci. Rev. **143**, 347–356 (2009)
H. Krüger, A.V. Krivov, D.P. Hamilton, E. Grün, Detection of an impact-generated dust cloud around Ganymede. Nature **399**, 558–560 (1999). https://doi.org/10.1038/21136
H. Krüger, A.V. Krivov, E. Grün, A dust cloud of Ganymede maintained by hypervelocity impacts of interplanetary micrometeoroids. Planet. Space Sci. **48**, 1457–1471 (2000). https://doi.org/10.1016/S0032-0633(00)00092-1
H. Krüger, A.V. Krivov, M. Sremčević, E. Grün, Impact-generated dust clouds surrounding the Galilean moons. Icarus **164**(1), 170–187 (2003)
H. Krüger et al., Interstellar dust in the solar system. Space Sci. Rev. **130**(1–4), 401–408 (2007)
H. Krüger, P. Strub, E. Grün, V.J. Sterken, Sixteen years of Ulysses interstellar dust measurements in the Solar System. I. Mass distribution and gas-to-dust mass ratio. Astrophys. J. **812**(2), 1–16 (2015)
M. Landgraf, J.-C. Liou, H.A. Zook, E. Grün, Origins of solar system dust beyond Jupiter. Astron. J. **123**, 2857–2861 (2002)
D.S. Lauretta, A.E. Bartels, M.A. Barucci, E.B. Bierhaus, R.P. Binzel, W.F. Bottke, H. Campins, S.R. Chesley, B.C. Clark, B.E. Clark, E.A. Cloutis, H.C. Connolly, M.K. Crombie, M. Delbó, J.P. Dworkin, J.P. Emery, D.P. Glavin, V.E. Hamilton, C.W. Hergenrother, C.L. Johnson, L.P. Keller, P. Michel, M.C. Nolan, S.A. Sandford, D.J. Scheeres, A.A. Simon, B.M. Sutter, D. Vokrouhlický, K.J. Walsh, The OSIRIS-REx target asteroid (101955) Bennu: constraints on its physical, geological, and dynamical nature from astronomical observations. Meteorit. Planet. Sci. **50**(4), 834–849 (2015). https://doi.org/10.1111/maps.12353
P. Lee, Dust levitation on asteroids. Icarus **124**, 181–194 (1996). https://doi.org/10.1006/icar.1996.0197
E. Lellouch, P. Santos-Sanz, P. Lacerda, M. Mommert, R. Duffard, J. Ortiz, T. Müller, S. Fornasier, J. Stansberry, C. Kiss, et al., "TNOs are Cool": a survey of the trans-Neptunian region-IX. Thermal properties of Kuiper belt objects and Centaurs from combined Herschel and Spitzer observations. Astron. Astrophys. **557**, 60 (2013)
G. Leto, G.A. Baratta, Ly-α photon induced amorphization of Ic water ice at 16 Kelvin. Astron. Astrophys. **397**, 7–13 (2003)
H.F. Levison, E.M. Shoemaker, C.S. Shoemaker, Dynamical evolution of Jupiter's Trojan asteroids. Nature **385**(6611), 42 (1997)
H.F. Levison, A. Morbidelli, C. VanLaerhoven, R. Gomes, K. Tsiganis, Origin of the structure of the Kuiper belt during a dynamical instability in the orbits of Uranus and Neptune. Icarus **196**(1), 258–273 (2008)
H.F. Levison, K.A. Kretke, M.J. Duncan, Growing the gas-giant planets by the gradual accumulation of pebbles. Nature **524**(7565), 322–324 (2015)
J. Li, D. Jewitt, Recurrent perihelion activity in (3200) Phaethon. Astron. J. **145**(6), 154 (2013)
J. Licandro, F. Moreno, J. de León, G. Tozzi, L. Lara, A. Cabrera-Lavers, Exploring the nature of new main-belt comets with the 10.4 m GTC telescope: (300163) 2006 VW139. Astron. Astrophys. **550**, A17 (2013)
J.-C. Liou, H.A. Zook, Signatures of the giant planets imprinted on the Edgeworth-Kuiper Belt dust disk. Astron. J. **118**, 580–590 (1999)
Y. Liu, L.A. Taylor, Characterization of lunar dust and a synopsis of available lunar simulants. Planet. Space Sci. **59**(14), 1769–1783 (2011)
Y. Liu, J. Park, D. Schnare, E. Hill, L.A. Taylor, Characterization of lunar dust for toxicological studies. II: Texture and shape characteristics. J. Aerosp. Eng. **21**(4), 272–279 (2008)
P.G. Lucey, M.A. Riner, The optical effects of small iron particles that darken but do not redden: evidence of intense space weathering on Mercury. Icarus **212**, 451–462 (2011)
E.M. MacLennan, H.H. Hsieh, The nucleus of main-belt comet 259P/Garradd. Astrophys. J. Lett. **758**(1), L3 (2012)
M. Makuch, A.V. Krivov, F. Spahn, Long-term dynamical evolution of dusty ejecta from Deimos. Planet. Space Sci. **53**, 357–369 (2005)
I. Mann, Interstellar dust in the solar system. Annu. Rev. Astron. Astrophys. **48**, 173–203 (2010)
F. Marzari, P. Tricarico, H. Scholl, Stability of Jupiter Trojans investigated using frequency map analysis: the MATROS project. Mon. Not. R. Astron. Soc. **345**(4), 1091–1100 (2003)
J.R. Masiero, A.K. Mainzer, J.M. Bauer, T. Grav, C.R. Nugent, R. Stevenson, Asteroid family identification using the hierarchical clustering method and WISE/NEOWISE physical properties. Astrophys. J. **770**, 7 (2013). https://doi.org/10.1088/0004-637X/770/1/7
R.M.E. Mastrapa, R.H. Brown, Ion irradiation of crystalline H_2O-ice: Effect on the 1.65-µm band. Icarus **183**, 207–214 (2006)

T. Matsumoto, A. Tsuchiyama, A. Miyake, T. Noguchi, M. Nakamura, K. Uesugi, A. Takeuchi, Y. Suzuki, T. Nakano, Surface and internal structures of a space-weathered rim of an Itokawa regolith particle. Icarus **257**, 230–238 (2015)

M.K. McClure, C. Espaillat, N. Calvet, E. Bergin, P. D'Alessio, D.M. Watson, P. Manoj, B. Sargent, L.I. Cleeves, Detections of trans-neptunian ice in protoplanetary disks. Astrophys. J. **799**, 162 (2015)

J.E. McCoy, Photometric studies of light scattering above the lunar terminator from Apollo solar corona photography, in *Lunar and Planetary Science Conference Proceedings* (1976), pp. 1087–1112

J.E. McCoy, D.R. Criswell, Evidence for a high altitude distribution of lunar dust. Proc. Lunar Planet. Sci. Conf. **5**, 2991–3005 (1974)

H.J. Melosh, Impact ejection, spallation, and the origin of meteorites. Icarus **59**, 234–260 (1984). https://doi.org/10.1016/0019-1035(84)90026-5

D.A. Mendis, W.I. Axford, Revisiting Iapetus following recent Cassini observations. J. Geophys. Res. **113**, A11 (2008)

F. Merlin, A. Guilbert, C. Dumas, M.A. Barucci, C. de Bergh, P. Vernazza, Properties of the icy surface of the TNO 136108 (2003 EL_{61}). Astron. Astrophys. **466**, 1185–1188 (2007)

P. Michel, W. Benz, D.C. Richardson, Disruption of fragmented parent bodies as the origin of asteroid families. Nature **421**(6923), 608 (2003)

A. Morbidelli, H.F. Levison, R. Gomes, The dynamical structure of the Kuiper belt and its primordial origin, in *The Solar System Beyond Neptune* (2008), pp. 275–292

A. Morbidelli, R. Brasser, R. Gomes, H.F. Levison, K. Tsiganis, Evidence from the asteroid belt for a violent past evolution of Jupiter's orbit. Astron. J. **140**(5), 1391 (2010)

A. Morbidelli, K.J. Walsh, D.P. O'Brien, D.A. Minton, W.F. Bottke, The dynamical evolution of the Asteroid belt, in *Asteroids IV*, ed. by P. Michel, F.E. DeMeo, W.F. Bottke (2015), pp. 493–508

F. Moreno, L. Lara, J. Licandro, J. Ortiz, J. De Léon, V. Alí-Lagoa, B. Agís-González, A. Molina, The dust environment of main-belt comet P/2010 R2 (La Sagra). Astrophys. J. Lett. **738**(1), 16 (2011a)

F. Moreno, J. Licandro, J.L. Ortiz, L.M. Lara, V. Alí-Lagoa, O. Vaduvescu, N. Morales, A. Molina, Z.-Y. Lin, (596) Scheila in outburst: a probable collision event in the Main Asteroid Belt. Astrophys. J. **738**(2), 130 (2011b)

F. Moreno, J. Licandro, A. Cabrera-Lavers, A short-duration event as the cause of dust ejection from main-belt comet P/2012 F5 (Gibbs). Astrophys. J. Lett. **761**(1), 12 (2012)

F. Moreno, A. Cabrera-Lavers, O. Vaduvescu, J. Licandro, F. Pozuelos, The dust environment of main-belt comet P/2012 T1 (PANSTARRS). Astrophys. J. Lett. **770**(2), 30 (2013)

F. Moreno, J. Licandro, C. Álvarez-Iglesias, A. Cabrera-Lavers, F. Pozuelos, Intermittent dust mass loss from activated asteroid P/2013 P5 (PANSTARRS). Astrophys. J. **781**(2), 118 (2014)

F. Moreno, J. Licandro, A. Cabrera-Lavers, F.J. Pozuelos, Dust loss from activated asteroid P/2015 X6. Astrophys. J. **826**(2), 137 (2016a)

F. Moreno, J. Licandro, A. Cabrera-Lavers, F.J. Pozuelos, Early evolution of disrupted asteroid P/2016 G1 (PANSTARRS). Astrophys. J. Lett. **826**, L22 (2016b)

F. Moreno, F.J. Pozuelos, B. Novaković, J. Licandro, A. Cabrera-Lavers, B. Bolin, R. Jedicke, B.J. Gladman, M.T. Bannister, S.D.J. Gwyn, P. Vereš, K. Chambers, S. Chastel, L. Denneau, H. Flewelling, M. Huber, E. Schunová-Lilly, E. Magnier, R. Wainscoat, C. Waters, R. Weryk, D. Farnocchia, M. Micheli, The splitting of double-component active asteroid P/2016 J1 (PANSTARRS). Astrophys. J. Lett. **837**, 3 (2017). https://doi.org/10.3847/2041-8213/aa6036

T.H. Morgan, R.M. Killen, Production mechanisms for faint by possibly detectable coronae about asteroids. Planet. Space Sci. **46**(8), 843–850 (1998)

R.V. Morris, In situ reworking (gardening) of the lunar surface: evidence from the Apollo cores, in *Proc. 9th Lunar Planet. Sci. Conf.* (1978), pp. 1801–1811

N. Murdoch, P. Sánchez, S.R. Schwartz, H. Miyamoto, Asteroid surface geophysics, in *Asteroids IV*, ed. by P. Michel, F.E. DeMeo, W.F. Bottke (2015), pp. 767–792

K. Nagao, R. Okazaki, T. Nakamura, Y.N. Miura, T. Osawa, K.-i. Bajo, S. Matsuda, M. Ebihara, T.R. Ireland, F. Kitajima, H. Naraoka, T. Noguchi, A. Tsuchiyama, H. Yurimoto, M.E. Zolensky, M. Uesugi, K. Shirai, M. Abe, T. Yada, Y. Ishibashi, A. Fujimura, T. Mukai, M. Ueno, T. Okada, M. Yoshikawa, J. Kawaguchi, Irradiation history of Itokawa regolith material deduced from noble gases in the Hayabusa samples. Science **333**, 1128 (2011). https://doi.org/10.1126/science.1207785

A.M. Nakamura, A. Fujiwara, T. Kadono, Velocity of finer fragments from impact. Planet. Space Sci. **42**, 1043–1052 (1994). https://doi.org/10.1016/0032-0633(94)90005-1

T. Nakamura, T. Noguchi, M. Tanaka, M.E. Zolensky, M. Kimura, A. Tsuchiyama, A. Nakato, T. Ogami, H. Ishida, M. Uesugi, T. Yada, K. Shirai, A. Fujimura, R. Okazaki, S.A. Sandford, Y. Ishibashi, M. Abe, T. Okada, M. Ueno, T. Mukai, M. Yoshikawa, J. Kawaguchi, Itokawa dust particles: a direct link between S-type asteroids and ordinary chondrites. Science **333**, 1113 (2011). https://doi.org/10.1126/science.1207758

E. Nakamura, A. Makishima, T. Moriguti, K. Kobayashi, R. Tanaka, T. Kunihiro, T. Tsujimori, C. Sakaguchi, H. Kitagawa, T. Ota, et al., Space environment of an asteroid preserved on micrograins returned by the Hayabusa spacecraft. Proc. Natl. Acad. Sci. **109**(11), 624–629 (2012)
M. Nayak, F. Nimmo, B. Udrea, Effects of mass transfer between Martian satellites on surface geology. Icarus **267**, 220–231 (2016)
D. Nesvorný, P. Jenniskens, H.F. Levison, W.F. Bottke, D. Vokrouhlický, M. Gounelle, Cometary origin of the zodiacal cloud and carbonaceous micrometeorites. Implications for hot debris disks. Astrophys. J. **713**(2), 816 (2010)
D. Nesvorný, D. Janches, D. Vokrouhlický, P. Pokorný, W.F. Bottke, P. Jenniskens, Dynamical model for the zodiacal cloud and sporadic meteors. Astrophys. J. **743**, 129 (2011a)
D. Nesvorný, D. Vokrouhlický, P. Pokorný, D. Janches, Dynamics of dust particles released from Oort Cloud Comets and their contribution to radar meteors. Astrophys. J. **743**, 37 (2011b)
D. Nesvorný, D. Vokrouhlický, R. Deienno, Capture of irregular satellites at Jupiter. Astrophys. J. **784**(1), 22 (2014)
D. Nesvorný, M. Brož, V. Carruba et al., Identification and dynamical properties of asteroid families, in *Asteroids IV* (2015), pp. 297–321
G.A. Neumann, J.F. Cavanaugh, X. Sun, E.M. Mazarico, D.E. Smith, M.T. Zuber, D. Mao, D.A. Paige, S.C. Solomon, C.M. Ernst, O.S. Barnouin, Bright and dark polar deposits on Mercury: evidence for surface volatiles. Science **339**, 296–300 (2013)
T. Nitter, O. Havnes, Dynamics of dust in a plasma sheath and injection of dust into the plasma sheath above Moon and asteroidal surfaces. Earth Moon Planets **56**, 7–34 (1992)
T. Nitter, T.K. Aslaksen, F. Melandsø, O. Havnes, Levitation and dynamics of a collection of dust particles in a fully ionized plasma sheath. IEEE Trans. Plasma Sci. **22**(2), 159–172 (1994)
T. Nitter, O. Havnes, F. Melandsø, Levitation and dynamics of charged dust in the photoelectron sheath above surfaces in space. J. Geophys. Res. **103**(A4), 6605–6620 (1998)
S.K. Noble, C.M. Pieters, L.P. Keller, An experimental approach to understanding the optical effects of space weathering. Icarus **192**, 629–642 (2007)
K. Nogami, M. Fujii, H. Ohashi, Y. Miyachi, S. Sasaki, S. Hasegawa, H. Yano, H. Shibata, T. Iwai, S. Minami, S. Takechi, E. Grün, R. Srama, Development of the Mercury Dust Monitor (MDM) onboard the BepiColombo mission. Planet. Space Sci. **58**, 108–115 (2010)
T. Noguchi, T. Nakamura, M. Kimura, M.E. Zolensky, M. Tanaka, T. Hashimoto, M. Konno, A. Nakato, T. Ogami, A. Fujimura, M. Abe, T. Yada, T. Mukai, M. Ueno, T. Okada, K. Shirai, Y. Ishibashi, R. Okazaki, Incipient space weathering observed on the surface of Itokawa dust particles. Science **333**, 1121 (2011). https://doi.org/10.1126/science.1207794
K.S. Noll, W.M. Grundy, E.I. Chiang, J.-L. Margot, S.D. Kern, Binaries in the Kuiper Belt, in *The Solar System Beyond Neptune*, ed. by M.A. Barucci, H. Boehnhardt, D.P. Cruikshank, A. Morbidelli (University of Arizona Press, Tucson, 2008)
M. Øieroset, D.A. Brain, E. Simpson, D.L. Mitchell, T.D. Phan, J.S. Halekas, R.P. Lin, M.H. Acuña, Search for Phobos and Deimos gas/dust tori using in situ observations from Mars Global Surveyor MAG/ER. Icarus **206**, 189–198 (2010)
N. Onose, A. Fujiwara, Mass-velocity distributions of fragments in oblique impact cratering on gypsum. Meteorit. Planet. Sci. **39**, 321–331 (2004). https://doi.org/10.1111/j.1945-5100.2004.tb00343.x
J. Park, Y. Liu, K.D. Kihm, L.A. Taylor, Characterization of lunar dust for toxicological studies. I: Particle size distribution. J. Aerosp. Eng. **21**(4), 266–271 (2008)
J. Park, B.D. Turrin, G.F. Herzog, F.N. Lindsay, J.S. Delaney, C.C. Swisher, M. Uesugi, Y. Karouji, T. Yada, M. Abe, et al., ^{40}Ar/^{39}Ar age of material returned from asteroid 25143 Itokawa. Meteorit. Planet. Sci. **50**(12), 2087–2098 (2015)
M.A. Pelizzari, D.R. Criswell, Lunar dust transport by photoelectric charging at sunset, in *Proc. 9th Lunar Sci. Conf.* (1978), pp. 3225–3237
V. Perera, A.P. Jackson, E. Asphaug, R.-L. Ballouz, The spherical Brazil Nut Effect and its significance to asteroids. Icarus **278**(C), 194–203 (2016)
C.M. Pieters, S.K. Noble, Space weathering on airless bodies. J. Geophys. Res., Planets **121**(1), 1865–1884 (2016)
C.M. Pieters, E.M. Fischer, O. Rode, A. Basu, Optical effects of space weathering: the role of the finest fraction. J. Geophys. Res. **98**(E11), 20817–20824 (1993)
C.M. Pieters, O. Rode, E.M. Fischer, A. Basu, Lunar space weathering and the optical properties of lunar regolith. Astron. Vestn. **28**, 109–119 (1994)
M. Piquette, M. Horányi, The effect of asymmetric surface topography on dust dynamics on airless bodies. Icarus **291**, 65–74 (2017)
P.M. Pires dos Santos, S.M.G. Winter, R. Sfair, D.C. Mourão, Small particles in Pluto's environment: effects of the solar radiation pressure. Mon. Not. R. Astron. Soc. **430**(4), 2761–2767 (2013)

T. Platz, A. Nathues, N. Schorghofer, F. Preusker, E. Mazarico, S.E. Schröder, S. Byrne, T. Kneissl, N. Schmedemann, J.-P. Combe, M. Schäfer, G.S. Thangjam, M. Hoffman, P. Guitierrez-Marques, M.E. Landis, W. Dietrich, J. Ripken, K.-D. Matz, C.T. Russell, Surface water-ice deposits in the northern shadowed regions of Ceres. Nat. Astron. **1**, 7 (2016)
P. Pokorný, D. Vokrouhlický, D. Nesvorný, M. Campbell-Brown, P. Brown, Dynamical model for the toroidal sporadic meteors. Astrophys. J. **789**, 25 (2014)
P. Pokorný, M. Sarantos, D. Janches, Reconciling the Dawn-Dusk Asymmetry in Mercury's Exosphere with the Micrometeoroid Impact Directionality. Astrophys. J. Lett. **842**, 17 (2017). https://doi.org/10.3847/2041-8213/aa775d
A.R. Poppe, Interplanetary dust influx to the Pluto-Charon system. Icarus **246**, 352–359 (2015)
A.R. Poppe, An improved model for interplanetary dust fluxes in the outer Solar System. Icarus **264**, 369–386 (2016)
A. Poppe, M. Horányi, Simulations of the photoelectron sheath and dust levitation on the Lunar surface. J. Geophys. Res. **115**, A08106 (2010)
A. Poppe, M. Horányi, The effect of Nix and Hydra on the putative Pluto-Charon dust cloud. Planet. Space Sci. **59**, 1647–1653 (2011)
A. Poppe, D. James, B. Jacobsmeyer, M. Horányi, First results from the Venetia Burney Student Dust Counter on the New Horizons mission. Geophys. Res. Lett. **37**, L11101 (2010)
A.R. Poppe, M. Piquette, A. Likhanskii, M. Horányi, The effect of surface topography on the lunar photoelectron sheath and electrostatic dust transport. Icarus **221**, 135–146 (2012)
A.R. Poppe, S.M. Curry, S. Fatemi, The Phobos neutral and ionized torus. J. Geophys. Res., Planets **121**(5), 770–783 (2016)
A. Poppe, W. Farrell, J. Halekas, Formation timescales of amorphous rims on lunar grains derived from Artemis observations. J. Geophys. Res., Planets **123**, 37–46 (2018)
C.C. Porco, et al., Cassini observes the active south pole of Enceladus. Science **311**, 1393–1401 (2006)
S.B. Porter, W.M. Grundy, Ejecta transfer in the Pluto system. Icarus **246**, 360–368 (2015)
S.B. Porter, S.J. Desch, J.C. Cook, Micrometeorite impact annealing of ice in the outer Solar System. Icarus **208**, 492–498 (2010)
F.J. Pozuelos, A. Cabrera-Lavers, J. Licandro, F. Moreno, On the dust environment of Main-Belt comet 313 P/Gibbs. Astrophys. J. **806**, 102 (2015)
P. Pravec, D. Vokrouhlický, D. Polishook, D.J. Scheeres, A.W. Harris, A. Galád, O. Vaduvescu, F. Pozo, A. Barr, P. Longa, F. Vachier, F. Colas, D.P. Pray, J. Pollock, D. Reichart, K. Ivarsen, J. Haislip, A. LaCluyze, P. Kušnirák, T. Henych, F. Marchis, B. Macomber, S.A. Jacobson, Y.N. Krugly, A.V. Sergeev, A. Leroy, Formation of asteroid pairs by rotational fission. Nature **466**(7), 1085–1088 (2010)
D. Prialnik, E.D. Rosenberg, Can ice survive in main-belt comets? Long-term evolution models of comet 133P/Elst-Pizarro. Mon. Not. R. Astron. Soc. Lett. **399**(1), 79–83 (2009)
J.J. Rennilson, D.R. Criswell, Surveyor observations of lunar horizon-glow. Moon **10**, 121–142 (1974). https://doi.org/10.1007/BF00655715
J.E. Richardson, H.J. Melosh, R.J. Greenberg, D.P. O'Brien, The global effects of impact-induced seismic activity on fractured asteroid surface morphology. Icarus **179**(2), 325–349 (2005)
D.P. Rubincam, Radiative spin-up and spin-down of small asteroids. Icarus **148**(1), 2–11 (2000)
G.O. Ryabova, The mass of the Geminid meteoroid stream. Planet. Space Sci. **143**, 125–131 (2017)
M. Sachse, J. Schmidt, S. Kempf, F. Spahn, Correlation between speed and size for ejecta from hypervelocity impacts. J. Geophys. Res., Planets **120**, 1847–1858 (2015). https://doi.org/10.1002/2015JE004844
M. Sarantos, R.M. Killen, D.A. Glenar, M. Benna, T.J. Stubbs, Metallic species, oxygen and silicon in the lunar exosphere: upper limits and prospects for LADEE measurements. J. Geophys. Res. **117**, A03103 (2012)
K. Sauer, K. Baumgärtel, U. Motschmann, Phobos events as precursors of solar wind-dust interaction. Geophys. Res. Lett. **20**(2), 165–168 (1993)
B. Schläppi, K. Altwegg, P. Wurz, Asteroid exosphere: a simulation for the ROSETTA flyby targets (2867) Steins and (21) Lutetia. Icarus **195**, 674–685 (2008)
J. Schmidt, N. Brilliantov, F. Spahn, S. Kempf, Slow dust in Enceladus' plume from condensation and wall collisions in tiger stripe fractures. Nature **451**, 685–688 (2008). https://doi.org/10.1038/nature06491
N. Schorghofer, The lifetime of ice on main belt asteroids. Astrophys. J. **682**(1), 697 (2008)
N. Schorghofer, Predictions of depth-to-ice on asteroids based on an asynchronous model of temperature, impact stirring, and ice loss. Icarus **276**, 88–95 (2016)
J. Schwan, X. Wang, H.W. Hsu, E. Grün, M. Horanyi, The charge state of electrostatically transported dust on regolith surfaces. Geophys. Res. Lett. **44**(7), 3059–3065 (2017)
F. Scipioni, F. Tosi, K. Stephan, G. Filacchione, M. Ciarniello, F. Capaccioni, P. Cerroni (The VIMS Team), Spectroscopic classification of icy satellites of Saturn I: identification of terrain units on Dione. Icarus **226**, 1331–1349 (2013)

F. Scipioni, F. Tosi, K. Stephan, G. Filacchione, M. Ciarniello, F. Capaccioni, P. Cerroni (The VIMS Team), Spectroscopic classification of icy satellites of Saturn II: identification of terrain units on Rhea. Icarus **234**, 1–16 (2014)
A.B. Severny, E.I. Terez, A.M. Zvereva, The measurements of sky brightness on Lunokhod-2. Earth Moon Planets **14**(1), 123–128 (1975)
R. Sfair, S.M.G. Winter, Orbital evolution of the μ and ν dust ring particles of Uranus. Astron. Astrophys. **505**, 845–852 (2009)
S.S. Sheppard, C. Trujillo, Discovery and characteristics of the rapidly rotating active asteroid (62412) 2000 SY178 in the main belt. Astron. J. **149**(2), 44 (2015)
Y. Shkuratov, V. Kaydash, V. Korokhin, Y. Velikodsky, N. Opanasenko, G. Videen, Optical measurements of the Moon as a tool to study its surface. Planet. Space Sci. **59**(13), 1326–1371 (2011)
M.R. Showalter, D.P. Hamilton, P.D. Nicholson, A deep search for martian dust rings and inner moons using the Hubble Space Telescope. Planet. Space Sci. **54**, 844–854 (2006)
A.P. Showman, R. Malhotra, The Galilean satellites. Science **296**, 77–84 (1999)
A. Shu, A. Collette, K. Drake, E. Grün, M. Horanyi, S. Kempf, A. Mocker, T. Munsat, P. Northway, R. Srama, Z. Sternovsky, E. Thomas, 3 MV hypervelocity dust accelerator at the Colorado Center for Lunar Dust and Atmospheric Studies. Rev. Sci. Instrum. **83**(7), 075108 (2012)
M. Siegler, R.S. Miller, J.T. Keane, M. Laneuville, D.A. Paige, I. Matsuyama, D.J. Lawrence, A. Crotts, M.J. Poston, Lunar true polar wander inferred from polar hydrogen. Nature **531**(7), 480–484 (2016)
B.A. Smith, et al., The Galilean satellites and Jupiter: Voyager 2 imaging science results. Science **206**, 927–950 (1979a)
B.A. Smith, et al., The Jupiter system through the eyes of Voyager 1. Science **204**, 951–972 (1979b)
B.A. Smith, et al., Voyager 2 in the Uranian system: imaging science results. Science **233**(4759), 43–64 (1986)
S.M. Smith, J.K. Wilson, J. Baumgardner, M. Mendillo, Discovery of the distant lunar sodium tail and its enhancement following the Leonid meteor shower of 1998. Geophys. Res. Lett. **26**(12), 1649–1652 (1999)
C. Snodgrass, C. Tubiana, J.-B. Vincent, H. Sierks, S. Hviid, R. Moissl, H. Boehnhardt, C. Barbieri, D. Koschny, P. Lamy, et al., A collision in 2009 as the origin of the debris trail of asteroid P/2010 A2. Nature **467**(7317), 814–816 (2010)
C. Snodgrass, G.H. Jones, H. Boehnhardt, A. Gibbings, M. Homeister, N. Andre, P. Beck, M.S. Bentley, I. Bertini, N. Bowles, M.T. Capria, C. Carr, M. Ceriotti, A.J. Coates, V.D. Corte, K.L.D. Hanna, A. Fitzsimmons, P.J. Gutierrez, O.R. Hainaut, A. Herique, M. Hilchenbach, H.H. Hsieh, E. Jehin, O. Karatekin, W. Kofman, L.M. Lara, K. Laudan, J. Licandro, S.C. Lowry, F. Marzari, A. Masters, K.J. Meech, F. Moreno, A. Morse, R. Orosei, A. Pack, D. Plettemeier, D. Prialnik, A. Rotundi, M. Rubin, J.P. Sanchez, S. Sheridan, M. Trieloff, A. Winterboer, The Castalia mission to main belt comet 133p/Elst-Pizarro. Adv. Space Res. (2017). https://doi.org/10.1016/j.asr.2017.09.011. http://www.sciencedirect.com/science/article/pii/S0273117717306622
S. Soter, The dust belts of Mars. Cornell Center Radiophys. Space Phys. Rept. 472 (1971)
F. Spahn, et al., Cassini dust measurements at Enceladus and implications for the origin of the E ring. Science **311**, 1416–1418 (2006)
F. Spahn, J. Schmidt, N. Albers, M. Hörning, M. Makuch, M. Seiß, S. Kempf, R. Srama, V. Dikarev, S. Helfert, G. Moragas-Klostermeyer, A.V. Krivov, M. Sremčević, A.J. Tuzzolino, T. Economou, E. Grün, Cassini dust measurements at Enceladus and implications for the origin of the E ring. Science **311**, 1416–1418 (2006a). https://doi.org/10.1126/science.1121375
F. Spahn, N. Albers, M. Hörning, S. Kempf, A.V. Krivov, M. Makuch, J. Schmidt, M. Seiß, M. Sremčević, E ring dust sources: implications from Cassini's dust measurements. Planet. Space Sci. **54**, 1024–1032 (2006b). https://doi.org/10.1016/j.pss.2006.05.022
E.J. Speyerer, R.Z. Povilaitis, M.S. Robinson, P.C. Thomas, R.V. Wagner, Quantifying crater production and regolith overturn on the Moon with temporal imaging. Nature Publishing Group **538**(7624), 215–218 (2016)
M. Sremčević, A.V. Krivov, F. Spahn, Impact-generated dust clouds around planetary satellites: asymmetry effects. Planet. Space Sci. **51**, 455–471 (2003)
M. Sremčević, A.V. Krivov, H. Krüger, F. Spahn, Impact-generated dust clouds around planetary satellites: model versus Galileo data. Planet. Space Sci. **53**, 625–641 (2005). https://doi.org/10.1016/j.pss.2004.10.001
A.J. Steffl, S.A. Stern, First constraints on rings in the Pluto system. Astron. J. **133**, 1485–1489 (2007)
V.J. Sterken, N. Altobelli, S. Kempf, G. Schwehm, R. Srama, E. Grün, The flow of interstellar dust into the solar system. Astron. Astrophys. **538**, A102 (2012)

V.J. Sterken, P. Strub, H. Krüger, R. von Steiger, P.C. Frisch, Sixteen years of Ulysses interstellar dust measurements in the Solar System. III. Simulations and data unveil new insights into local interstellar dust. Astrophys. J. **812**(2), 1–24 (2015)

S.A. Stern, Collisional time scales in the Kuiper Disk and their implications. Astron. J. **110**(2), 856 (1995)

S.A. Stern, Signatures of collisions in the Kuiper Disk. Astron. Astrophys. **310**, 999–1010 (1996)

S.A. Stern, The Lunar atmosphere: history, status, current problems, and context. Rev. Geophys. **37**(4), 453–491 (1999)

S.A. Stern, Ejecta exchange and satellite color evolution in the Pluto system, with implications for KBOs and asteroids with satellites. Icarus **199**, 571–573 (2009)

S.A. Stern, J.E. Colwell, Collisional erosion in the primordial Edgeworth-Kuiper Belt and the generation of the 3050 AU Kuiper Gap. Astrophys. J. **490**, 879–882 (1997)

P. Strub, H. Krüger, V.J. Sterken, Sixteen years of Ulysses interstellar dust measurements in the Solar System. II. Fluctuations in the dust flow from the data. Astrophys. J. **812**(2), 140 (2015)

K.-L. Sun, M. Seiß, M.M. Hedman, F. Spahn, Dust in the arcs of Methone and Anthe. Icarus **284**, 206–215 (2017)

J.R. Szalay, M. Horányi, Annual variation and synodic modulation of the sporadic meteoroid flux to the Moon. Geophys. Res. Lett. **42**(24), 10580–10584 (2015a). https://doi.org/10.1002/2015GL066908

J.R. Szalay, M. Horányi, The search for electrostatically lofted grains above the Moon with the Lunar Dust Experiment. Geophys. Res. Lett. **42**(13), 5141–5146 (2015b). https://doi.org/10.1002/2015GL064324

J.R. Szalay, M. Horányi, Detecting meteoroid streams with an in-situ dust detector above an airless body. Icarus **275**, 221–231 (2016a). https://doi.org/10.1016/j.icarus.2016.04.024

J.R. Szalay, M. Horányi, Lunar meteoritic gardening rate derived from in situ LADEE/LDEX measurements. Geophys. Res. Lett. **43**(10), 4893–4898 (2016b). https://doi.org/10.1002/2016GL069148

J.R. Szalay, M. Horányi, The impact ejecta environment of near Earth asteroids. Astrophys. J. Lett. **830**(2), 29 (2016c). https://doi.org/10.3847/2041-8205/830/2/L29

J.R. Szalay, M. Piquette, M. Horányi, The Student Dust Counter: status report at 23 AU. Earth Planets Space **65**, 1145–1149 (2013). https://doi.org/10.5047/eps.2013.02.005

J.R. Szalay, M. Horányi, A. Colaprete, M. Sarantos, Meteoritic influence on sodium and potassium abundance in the lunar exosphere measured by LADEE. Geophys. Res. Lett. (2016). https://doi.org/10.1002/2016GL069541

J.R. Szalay, P. Pokorný, P. Jenniskens, M. Horanyi, Activity of the 2013 Geminid meteoroid stream at the Moon. Mon. Not. R. Astron. Soc. **474**(3), 4225–4231 (2018)

S. Takasawa, A.M. Nakamura, T. Kadono, M. Arakawa, K. Dohi, S. Ohno, Y. Seto, M. Maeda, K. Shigemori, Y. Hironaka, T. Sakaiya, S. Fujioka, T. Sano, K. Otani, T. Watari, K. Sangen, M. Setoh, N. Machii, T. Takeuchi, Silicate dust size distribution from hypervelocity collisions: implications for dust production in debris disks. Astrophys. J. Lett. **733**, 39 (2011). https://doi.org/10.1088/2041-8205/733/2/L39

D. Tamayo, J.A. Burns, D.P. Hamilton, M.M. Hedman, Finding the trigger to Iapetus' odd global albedo patter: dynamics of dust from Saturn's irregular satellites. Icarus **215**, 260–278 (2011)

M. Tátrallyay, M. Horányi, A. Juhász, J.G. Luhmann, Submicron-sized dust grains in the Martian environment. Adv. Space Res. **12**(9), 927–930 (1992)

S.R. Taylor, *Planetary Science: A Lunar Perspective* (Lunar and Planetary Institute, Houston, 1982)

L.A. Taylor, C.M. Pieters, L.P. Keller, R.V. Morris, D.S. McKay, Lunar mare soils: space weathering and the major effects of surface-correlated nanophase Fe. J. Geophys. Res. **106**(E), 27985–28000 (2001)

L.A. Taylor, C. Pieters, A. Patchen, D.-H.S. Taylor, R.V. Morris, L.P. Keller, D.S. McKay, Mineralogical and chemical characterization of lunar highland soils: insights into the space weathering of soils on airless bodies. J. Geophys. Res. **115**(E), 02002 (2010)

T. Terai, Y. Itoh, Y. Oasa, R. Furusho, J. Watanabe, Photometric measurements of H_2O ice crystallinity on trans-neptunian objects. Astrophys. J. **827**, 65 (2016)

K.-U. Thiessenhusen, A.V. Krivov, H. Krüger, E. Grün, A dust cloud around Pluto and Charon. Planet. Space Sci. **50**, 79–87 (2002)

N. Thomas et al., The morphological diversity of comet 67P/Churyumov-Gerasimenko. Science **347**, 6220 (2015)

H.B. Throop, R.G. French, K. Shoemaker, C.B. Olkin, T.R. Ruhland, L.A. Young, Limits on Pluto's ring system from the June 12 2006 stellar occultation and implications for the New Horizons impact hazard. Icarus **246**, 345–351 (2015)

C.A. Trujillo, M.E. Brown, K.M. Barkume, E.L. Schaller, D.L. Rabinowitz, The surface of 2003 EL_{61} in the near infrared. Astrophys. J. **655**, 1172–1178 (2007)

A. Tsuchiyama, M. Uesugi, T. Matsushima, T. Michikami, T. Kadono, T. Nakamura, K. Uesugi, T. Nakano, S.A. Sandford, R. Noguchi, et al., Three-dimensional structure of Hayabusa samples: origin and evolution of Itokawa regolith. Science **333**(6046), 1125–1128 (2011)

D. Veras, S.A. Jacobson, B.T. Gänsicke, Post-main-sequence debris from rotation-induced YORP break up of small bodies. Mon. Not. R. Astron. Soc. **445**(3), 2794–2799 (2014). https://doi.org/10.1093/mnras/stu1926.
A.J. Verbiscer, M.F. Skrutskie, D.P. Hamilton, Saturn's largest ring. Nature **461**, 1098–1100 (2009)
J. Veverka et al., Imaging of small-scale features on 433 Eros from NEAR: evidence for a complex regolith. Science **292**, 484 (2001)
C. Vitense, A.V. Krivov, H. Kobayashi, T. Löhne, An improved model of the Edgeworth-Kuiper debris disk. Astron. Astrophys. **540**, A30 (2012)
X. Wang, M. Horányi, S. Robertson, Experiments on dust transport in plasma to investigate the origin of the lunar horizon glow. J. Geophys. Res. **114**, A05103 (2009)
X. Wang, M. Horányi, S. Robertson, Investigation of dust transport on the lunar surface in a laboratory plasma with an electron beam. J. Geophys. Res. **115**, A11102 (2010)
X. Wang, M. Horányi, S. Robertson, Dust transport near electron beam impact and shadow boundaries. Planet. Space Sci. **59**, 1791–1794 (2011)
X. Wang, J. Schwan, H.-W. Hsu, E. Grün, M. Horányi, Dust charging and transport on airless planetary bodies. Geophys. Res. Lett. **43**, 6103–6110 (2016)
B.D. Warner, A.W. Harris, P. Pravec, The asteroid lightcurve database. Icarus **202**(1), 134–146 (2009)
H.A. Weaver, S.A. Stern, M.J. Mutchler, A.J. Steffl, M.W. Buie, W.J. Merline, J.R. Spencer, E.F. Young, L.A. Young, Discovery of two new satellites of Pluto. Nature **439**, 943–945 (2006)
F. Whipple, 1983 TB and the Geminid meteors. International Astronomical Union Circular, No. 3881 (1983)
P. Wiegert, J. Vaubaillon, M. Campbell-Brown, A dynamical model of the sporadic meteoroid complex. Icarus **201**, 295–310 (2009)
S. Yamamoto, T. Mukai, Dust production by impacts of interstellar dust on Edgeworth-Kuiper Belt objects. Astron. Astrophys. **329**, 785–791 (1998)
H. Yano, T. Kubota, H. Miyamoto, T. Okada, D. Scheeres, Y. Takagi, K. Yoshida, M. Abe, S. Abe, O. Barnouin-Jha, et al., Touchdown of the Hayabusa spacecraft at the Muses sea on Itokawa. Science **312**(5778), 1350–1353 (2006)
F. Yoshida, T. Nakamura, Size distribution of faint Jovian l4 Trojan asteroids. Astron. J. **130**(6), 2900 (2005)
M. Yoshikawa, et al., Outline of the next asteroid sample return mission of Japan-Hayabusa-2, in *28th International Symposium on Space Technology and Science* (2011)
M.I. Zimmerman, W.M. Farrell, A. Poppe, Grid-free 2D plasma simulations of the complex interaction between the solar wind and small, near-Earth asteroids. Icarus **238**, 77–85 (2014)
M.I. Zimmerman, W.M. Farrell, C.M. Hartzell, X. Wang, M. Horányi, D.M. Hurley, K. Hibbits, Grain-scale supercharging and breakdown on airless regoliths. J. Geophys. Res., Planets **121**, 2150–2165 (2016)

Space Sci Rev (2019) 215:34
https://doi.org/10.1007/s11214-019-0597-7

Interplanetary Dust, Meteoroids, Meteors and Meteorites

Detlef Koschny[1,2] · Rachel H. Soja[3] · Cecile Engrand[4] · George J. Flynn[5] · Jérémie Lasue[6] · Anny-Chantal Levasseur-Regourd[7] · David Malaspina[8] · Tomoki Nakamura[9] · Andrew R. Poppe[10] · Veerle J. Sterken[11,12] · Josep M. Trigo-Rodríguez[13]

Received: 23 April 2018 / Accepted: 23 April 2019 / Published online: 21 May 2019

Abstract Interplanetary dust particles and meteoroids mostly originate from comets and asteroids. Understanding their distribution in the Solar system, their dynamical behavior and their properties, sheds light on the current state and the dynamical behavior of the Solar system. Dust particles can endanger Earth-orbiting satellites and deep-space probes, and a good understanding of the spatial density and velocity distribution of dust and meteoroids in the Solar system is important for designing proper spacecraft shielding. The study of inter-

Cosmic Dust from the Laboratory to the Stars
Edited by Rafael Rodrigo, Jürgen Blum, Hsiang-Wen Hsu, Detlef Koschny, Anny-Chantal Levasseur-Regourd, Jesús Martín-Pintado, Veerle Sterken and Andrew Westphal

✉ D. Koschny
detlef.koschny@esa.int

1 SCI-S, European Space Agency, Keplerlaan 1, 2200 AZ Noordwijk ZH, The Netherlands

2 Lehrstuhl für Raumfahrttechnik, Technische Universität München, Boltzmannstr. 15, 85748 Garching, Germany

3 Institut für Raumfahrtsysteme, Universität Stuttgart, Pfaffenwaldring 29, 70049 Stuttgart, Germany

4 CSNSM CNRS/Univ. Paris Sud, Univ. Paris-Saclay, Batiment 104, 91405 Orsay Campus, France

5 Dept of Physics, State University of New York at Plattsburgh, 101 Broad St., Plattsburgh, NY 12901, USA

6 IRAP, Université de Toulouse, CNRS, UPS, 9 avenue du Colonel Roche, 31400, Toulouse, France

7 LATMOS, Sorbonne Univ., CNRS, UVSQ, Campus Pierre et Marie Curie, 4 place Jussieu, 75005 Paris, France

8 Laboratory for Atmospheric and Space Physics, University of Colorado, 1234 Innovation Dr., Boulder, CO 80303, USA

9 Tohoku University, Aoba, Sendai, Miyagi 980-8578, Japan

10 Space Sciences Laboratory, University of California at Berkeley, 7 Gauss Way, Berkeley, CA 94720, USA

11 Institute of Applied Physics, University of Bern, Hochschulstrasse 4, 3012 Bern, Switzerland

12 Astronomical Institutute, University of Bern, Hochschulstrasse 4, 3012 Bern, Switzerland

planetary dust and meteoroids provides clues to the formation of the Solar system. Particles having formed 4.5 billion years ago can survive planetary accretion and those that survived until now did not evolve significantly since then. Meteoroids and interplanetary dust can be observed by measuring the intensity and polarization of the zodiacal light, by observing meteors entering the Earth's atmosphere, by collecting them in the upper atmosphere, polar ices and snow, and by detecting them with in-situ detectors on space probes.

Keywords Interplanetary dust · Meteors · Meteorites · Zodiacal light · Dynamics · Formation · Evolution

1 Introduction

The interplanetary dust and meteoroid cloud is the ensemble of refractory and possibly icy particles in the Solar system in the size range from a few microns to about a meter. The main sources are asteroids, comets, and, in the outer Solar system, in-spiraling particles coming from objects in the Edgeworth-Kuiper Belt. The goal of this paper is to summarize our current knowledge in this field and refer to other relevant papers.

When an asteroid or a comet gets close to the Sun, sublimating gases carry solid particles away from the object at a certain ejection velocity. Solar radiation pressure quickly moves the less massive particles away from the parent, larger particles remain near to the source (see Sect. 2). Their ejection velocity causes their orbits to disperse away from the parent with time. When an object (usually a comet) produces a continuous release of particles over a period of time, these particles may form a visible trail that appears to follow the parent along its orbit (Kresak 1993), see Fig. 1. When they are slightly dispersed mainly by non-gravitational forces, they form meteoroid streams. Meteoroid streams entering the Earth's atmosphere generate so-called meteor showers.

Fig. 1 Spitzer space telescope image of the dust trail of comet Encke. The trail is marked with white lines. The more horizontally distributed material is from recent cometary activity. Image size is about $3' \times 3'$ (from Kelley et al. 2006)

[13] Institute of Space Sciences (CSIC-IEEC), Campus UAB, c/Can Magrans s/n, 08193 Cerdanyola del Vallès (Barcelona), Catalonia, Spain

Fig. 2 A Perseid meteor, recorded with an image-intensified meteor camera. Image size is about 30° in diameter (image: D. Koschny)

Over long time periods, the orbits of the dust particles and meteoroids will change under the effect of both gravitational and non-gravitational forces. In particular the small dust will leave the vicinity of the meteoroid stream and form the almost, but not quite randomly distributed interplanetary background dust cloud. It can be observed from the Earth as the zodiacal light. Astronomical observations of the zodiacal light in visual and infrared from space-based instrumentation provide a comprehensive picture of the distribution of particles and their polarization. In-situ detectors on spacecraft, such as Galileo and Ulysses, probe the dust cloud at the smallest sizes up to a few microns. The larger meteoroids form the sporadic meteoroid complex, which can be observed as sporadic meteors when they enter the Earth's atmosphere. Interplanetary Dust Particles and micrometeorites—typically in the size range from micrometers to 1 mm—can be collected in the Earth's stratosphere, and by extracting them from ocean water or ice and snow in the polar caps. These collection techniques allow to directly obtain extraterrestrial samples without the need of a sample-return spacecraft mission.

Meteorites—rocky or metallic material that reaches the ground—provide examples of generally asteroidal material, in some cases also with a record of its flight through the atmosphere. The synthesis of the observations provides us with an overall picture of the interplanetary dust and meteoroid cloud—its mass flux distribution; chemical, material and structural properties; and the importance of different cometary and asteroidal sources in populating the inner Solar system with dust. This information has implications for the interpretation of the meteoroid deposition in planetary atmospheres (e.g Plane et al. 2014), on the Moon (e.g. Szalay and Horányi 2015), and on other planets and airless bodies (e.g. Christou et al. 2015). In this paper we describe the current state of knowledge of the interplanetary dust and meteoroid cloud. There is also an interstellar contribution, which is described in Sect. 4.4.

In this and the following chapters, we use the terminology as defined by the IAU Commission F1 on Meteors, Meteorites and Interplanetary Dust:

- 'Meteor' is the light and associated physical phenomena (heat, shock, ionization), which result from the high speed entry of a solid object from space into a gaseous atmosphere (see Fig. 2).
- 'Meteoroid' is a solid natural object of a size roughly between 30 μm and 1 meter, moving in, or coming from, interplanetary space.

- 'Interplanetary dust' is finely divided solid matter, with particle sizes in general smaller than meteoroids, moving in, or coming from, interplanetary space.
- 'Meteorite' is any natural solid object that survived the meteor phase in a gaseous atmosphere without being completely vaporized. Meteorites smaller than 1 mm in size are also called micrometeorite. These may be too small to experience ablation in an atmosphere.

For more explanations, see the IAU web site.[1] Note that in particular in the cometary science community, the term 'dust' is used in a wider context and can also be used for larger particles. In that case, the term 'particles' is used without any size qualifier like 'dust' or 'meteoroid'.

2 Formation and Evolution of the Interplanetary Dust Cloud

2.1 Overview

The main sources of interplanetary dust and meteoroids are asteroids and comets. These minor bodies are relatively fragile and generate dust particles due to collisions and the evaporation of volatiles. They accreted from primitive material in cold regions of the protoplanetary disk that included high-temperature components, around the Sun about 4.5 Gyr ago. They formed km-sized rocky planetesimals, that kept growing by mutual collisions. Early-formed asteroids, hundreds of km or larger in diameter, experienced melting and planetary differentiation, thus modifying their rock-forming materials into igneous counterparts. Later-formed smaller asteroids released their radiogenic heat at a rate quick enough to retain primitive features and to avoid differentiation.

Dust particles and meteoroids are refractory and possibly icy particles released by emission, impact or collision processes from these minor bodies. After release and escape from the environment of their parent body (Fig. 1), they are subject to the same set of dynamical forces as their parent. These forces are size-dependent, and they may change if the particle undergoes transformative processes such as collisional shattering or erosion, or sublimation near the Sun (Mukai et al. 2001). Here we examine the evolution of the emitted particles into the interplanetary dust and meteoroid cloud.

2.2 Dynamics

Gravitational, radiative, collisional, and electromagnetic forces alter the orbits of dust and meteoroid particles after their emission from the parent object. In terms of gravitational interaction with larger bodies, they act as massless objects. Radiative forces, on the other hand, are dependent on the particle size and mass. The composition determines the material density and thus the mass-to-surface area ratio of the particle, as well as absorption and scattering properties. The shape and structure of the particle will also play a role (Gustafson 1989; Kimura et al. 2002b). Here we provide an overview of the important effects and for what particle sizes they are relevant.

2.2.1 Gravitational Forces

Gravitational forces dominate the trajectories of larger particles (> 100 μm). Solar gravity dominates for heliocentric orbits, unless the particle passes within the Hill sphere of a

[1] https://www.iau.org/science/scientific_bodies/commissions/F1/.

planet, in which planetary gravity is the dominant effect. Close planetary encounters can shift their orbits to hyperbolic trajectories. This process is an important loss mechanism for in-spiraling particles in the outer Solar system (Liou et al. 1996a). Planetary encounters also disrupt and perturb meteoroid streams (e.g. Vaubaillon and Colas 2005; Vaubaillon et al. 2006). Particles with orbits within the scattering zones of Jupiter or Saturn suffer regular perturbations from these planets. Mean motion and secular resonances with respect to the giant planets are important factors influencing their long term dynamics (e.g. Liou and Zook 1997, 1999). Large particles trapped in resonances may remain there for significant time periods, protected from further close interactions with the planet and temporarily halting their evolution due to Poynting-Robertson drag (e.g. Sekhar and Asher 2014; Beauge and Ferraz-Mello 1994). Chaotic processes may in some cases remove particles from a resonant region. These processes create structures in the outer Solar system dust cloud (e.g. Poppe 2016). Particles trapped in resonant filaments in meteoroid streams may produce outbursts at Earth, due to their enhanced flux density relative to the rest of the stream.

2.2.2 *Radiative Forces*

Radiative forces generally are important for particles < 100 μm. Radiation pressure results from the momentum exchange of the incoming Solar radiation with the particle, producing a small force away from the Sun (Burns et al. 1979). Because Solar gravity and radiation pressure both decrease with the inverse of the heliocentric distance, the relative value of the radiation pressure is often quantified using the β parameter, which is the ratio between the radiation and gravitational forces F_r and F_g:

$$\beta = \frac{F_r}{F_g} = \frac{3L_\odot Q_{pr}}{16\pi G M_\odot c\rho s} = 5.7 \times 10^{-4} \frac{Q_{pr}}{\rho s}$$

where $L_\odot$ is the Solar luminosity in W, G is the gravitational constant in $m^3\ kg^{-1}\ s^{-2}$, c is the speed of light in $m\ s^{-1}$, ρ is the particle bulk density in $kg\ m^{-3}$, and s the particle equivalent radius in m. Q_{pr} is the efficiency factor for radiation pressure, defined by Mie scattering or measurements (Gustafson 1994; Kimura and Mann 1999). It has a maximum, along with β, at a particle equivalent radius of about 0.1 μm, although this depends on the particle's optical properties. For circular orbits, particles with $\beta = 1$ have a radiation force in excess of gravity and therefore escape the Solar system on hyperbolic trajectories. In the more general case, particles released from a parent body immediately assume an orbit different to that of the parent body, described by

$$a' = a \times \frac{1-\beta}{1-2a\beta/r_h}$$

and

$$e' = \left| 1 - \frac{(1-2a\beta/r_h)(1-e^2)}{(1-\beta)^2} \right|$$

where a the semi-major axis and e the eccentricity of the particles' orbit. Because β is size dependent, particles undergo a different evolution, depending on their size after release from their parent body. Larger meteoroids, which are not strongly affected by radiation pressure, stay near the orbit of their parent body. Smaller particles begin with orbits outside the parent body's orbit and form a tail (Fulle 2004). Very small micrometer-sized dust may be quickly blown out of the Solar system. Additionally, radiation pressure affects the trapping of particles into resonances. E.g., Weidenschilling and Jackson (1993) show that resonances under

Poynting-Robertson drag are "metastable" because trapped grains inevitably acquire eccentricities large enough to cause their orbits to cross that of the planet causing the resonance.

2.2.3 Poynting-Robertson Effect

The Poynting-Robertson effect results from the non-isotropic reemission of photons from a moving particle due to relativistic effects (Robertson 1937). Radiation in the direction of motion is blue-shifted, resulting in a higher momentum of the radiation in the direction of motion compared to the opposite direction. The result is a force similar to a drag force, causing the orbit to loose energy and spiral into the Sun. The eccentricity also drops and the orbit is circularized. This process is size and orbit dependent. The variations in the particle semi-major axis and eccentricity as a result of the Poynting-Robertson drag force are given by Wyatt and Whipple (1950):

$$\dot{a}_{PR} = -\frac{\alpha}{a}\frac{2+3e^2}{(1-e^2)^{3/2}}$$

$$\dot{e}_{PR} = -2.5\frac{\alpha}{a^2}\frac{e}{(1-e^2)^{1/2}}$$

where $\alpha = 3.55 \times 10^{-8}/(s\rho)$ au^2 yr^{-1}. Other orbital elements are not affected. For the case of circular orbits, these equations can be solved to provide an approximate analytic formula for the in-spiraling time in years from an initial semi-major axis a_0: $\tau_{PR,circ} \approx 400a_0^2/\beta = 7.0 \times 10^5 a_0^2 \rho s/Q_{pr}$. This timescale defines the importance of the Poynting-Robertson effect for the dynamical evolution of particles of different sizes. For a bulk density of 1000 kg m^{-3}, the in-spiraling lifetimes ranges from $\sim$700 yr for particles of equivalent radius 1 μm and 70000 yr for 100 μm radii, to $\sim$0.7 Myr for radii of 1 mm. Scattering Solar wind particles creates a similar drag force. It is generally taken to have an effect of $\sim$30% of the Poynting-Robertson drag force (Dermott et al. 2001), varying with the 11-year Solar cycle.

2.2.4 Yarkovsky Effect

The Yarkovsky effect results from the small recoil force that results when photons are re-radiated anisotropically due to thermal effects. The diurnal Yarkovsky effect occurs due to anisotropic emission of radiation from a rotating body (Bottke et al. 2002). While it can be the dominant dispersive force for meter-sized asteroids, it is in general not significant for small particles that spin fast relative to their size. However, the seasonal Yarkovsky force, which occurs due to a tilt in the spin axis, can be important for meteoroids (Rubincam 1995). Both result in a force comparable to a drag force, which will shrink the semi-major axis and circularize the orbit.

2.2.5 Lorentz Forces

Lorentz forces—the result of the traverse of a particle through the Solar magnetosphere—are important for dust particles with sizes a few microns or lower. Particles in the electromagnetic field have a charge $q = \epsilon\eta U m^{1/3}$, where $\epsilon = 8.85 \times 10^{-12}$ CV^{-1} m^{-1} is the permittivity, η is a constant describing physical parameters of the particle, and m is the particle mass in kg. The surface potential U is about 5 V for a spherical particle, largely as a result of the photoelectric effect. The Lorentz force is then described by $F_L = \frac{q}{c}|\mathbf{v}_{rel} \times \mathbf{B}|$ where $\mathbf{v}_{rel}$ is the relative velocity between the particle and the Solar wind, and **B** are the components of the interplanetary magnetic field.

Hamilton et al. (1996) showed that the magnitude of the Lorentz force depends on the Solar cycle. They modeled the trajectory of dust starting on circular, un-inclined orbits. They would reach high ecliptic latitudes with the configuration of the magnetic field as in Solar cycle no. 22 or 24, and finally escaping the Solar system. Currently we are in cycle 25, where the orientation of the magnetic field is reversed. This would result in orbits confined closer to the ecliptic, with particles still leaving the Solar system, but less readily.

2.3 Transformative Processes

2.3.1 Collisional Disruption

Larger interplanetary meteoroids are broken up by collisions with smaller, more abundant ones. These collisions result in momentum transfer, altering the orbits of the particles, and producing smaller fragments, which are more significantly influenced by Solar radiation. Steel and Elford (1986) modeled the orbital evolution of 1-mm meteoroids released into the orbits of 28 specific meteoroid streams and found catastrophic collision lifetimes varying from 20 kyr to 400 kyr, shorter than the Poynting-Robertson lifetimes for many of these streams. Leinert et al. (1983a) found that at 1 au the maximum catastrophic collision lifetimes of 100- μm particles, 1-mm particles, and 1-cm particles are 200 kyr, 40 kyr, and 800 kyr, respectively, as a result of collisions with other interplanetary dust particles. These collision lifetimes are shorter than or comparable to the time required for that size particle to spiral into the Sun under Poynting-Robertson drag for the 1-mm and 1-cm particles, indicating that a significant fraction of these particles are fragmented into smaller particles over their lifetimes. Thus consideration of collisional disruption is critical to the understanding of the evolution of the interplanetary dust and meteoroid complex.

Detailed modeling of the collisional lifetimes of interplanetary dust and meteoroids requires a knowledge not only of the size-frequency distribution of the particles, which has been measured at 1 au, but also of the distribution of orbital eccentricities and inclinations, which is not well established. As a result, models of the collisional lifetimes by different investigators differ by an order-of-magnitude or more, depending on the orbital parameters that are assumed in the modeling.

The interplanetary dust and meteoroid mass-frequency distribution, measured at 1 au by impacts onto the Long Duration Exposure Facility during its 5.7 years in low-Earth orbit, peaks at about 250 μm in diameter (Love and Brownlee 1993). Particles near the peak size are disrupted but there are few larger particles to resupply the distribution. Using these measurements, Dermott et al. (2001) conclude that the mean particle diameter s is expected to decrease with decreasing heliocentric distance.

Interstellar grains passing through the Solar system, which are expected to be much smaller than most interplanetary dust, also contribute to the collisional fragmentation of the interplanetary dust. The interstellar grains are generally < 1 μm in diameter, so they only fragment the smallest interplanetary dust particles. Liou et al. (1996a) analyze the contribution of dust grains coming from Edgeworth-Kuiper belt objects towards the inner Solar system. They conclude that for particles up to about 25 μm in size and migrating inwards from the Edgeworth-Kuiper Belt, collisions with interstellar grains are more important than mutual collisions in determining the collisional lifetimes. They find collisional lifetimes of interplanetary dust particles 1, 2, 4, 9 and 23 microns in size due to impacts by ~ 1 μm diameter interstellar grains to be 104, 49, 19, 4.8 and 0.86 Myr, respectively.

Modeling by Wyatt (2005) of the dust disks in other planetary systems indicates that in some massive disks mutual collisions are so frequent that they effectively prevent dust from reaching the inner regions of these systems under Poynting-Robertson drag. Similar conditions could have occurred early in Solar system history if the dust flux density was significantly higher than in the current era.

2.3.2 Disaggregation

Some interplanetary dust and meteoroid particles are composed of ices or are aggregates of mineral grains held together by ices or other relatively weak materials. These particles may easily disaggregate after release from their parent bodies, as a result of Solar heating, photoelectric charging, or other mechanisms. Active dust counters on fast flyby missions through dust clouds of 1P/Halley by the Giotto, VEGA1, and VEGA2 spacecraft; 26P/Grigg-Skjellerup by the Giotto spacecraft; 81P/Wild 2 by the Stardust spacecraft; and 9P/Tempel 1 by the Stardust NexT spacecraft were interpreted as indicating widespread grain disaggregation in the coma (Green et al. 2004; Clark et al. 2004; Tuzzolino et al. 2004; Economou et al. 2013), although the cause of this disaggregation could not be determined.

2.3.3 Sublimation

Grains are lost from the interplanetary dust cloud when they are heated to their sublimation temperatures by Solar radiation. The temperature reached by a grain depends on its distance from the Sun, its composition, which determines its efficiency in absorbing the Solar spectrum, and its size, with very small particles being less efficient at re-radiating energy at any specific grain temperature. Although the most detailed modeling has been performed for water ice, all small grains, even silicates, can sublimate before reaching the Sun.

Ice grains are modeled to sublime between 1 and 4 au (Kobayashi et al. 2011). Even beyond the snow line, there is significant loss of water ice because of Solar wind and Solar ultraviolet-induced sputtering (Grigorieva et al. 2007a).

The two most common minerals in interplanetary dust are olivine and pyroxene. The pyroxene grains sublimate at 4 to 6 Solar radii; olivine grains at 10 to 13 Solar radii (Kimura et al. 2002a). For porous dust particles consisting of a mixture silicates and absorbing material, Mann et al. (1994) have shown that, depending on the amount of absorbing material versus silicate material, they can reach as close as 2 to 3 Solar radii.

2.4 Modeling the Interplanetary Dust Cloud

Various efforts have been made to model the structure and evolution of the interplanetary dust and meteoroid distribution under the above processes, both to understand its formation and major constituents, and to assist in the development of impact hazard models for spacecraft. It should be noted that some of the models are called dust models, other meteoroid models. Nevertheless they normally cover the complete size range from micron-sized particles up to mm- or cm-sized particles.

The early development of space flight technology by NASA was supported by meteoroid penetration experiments on the Explorer XV1 and XXIII and Pegasus satellites (Naumann 1966). These were used to develop initial models of the near-Earth meteoroid risk to early missions (Cour-Palais 1969; Kessler 1970). The mass flux distribution at Earth orbit by Grün et al. (1985) also utilized these data, along with lunar crater counts (Morrison and Clanton 1979), and spacecraft dust measurements: Pioneer 8 and 9 (Berg and Richardson

1969; Berg and Gerloff 1971), Helios (Grün et al. 1980), and HEOS-2 (Hoffmann et al. 1975a,b). Love and Brownlee (1993) derived a mass flux at Earth based on impact craters on the Long Duration Exposure Facility (LDEF) that is higher than the Pegasus-derived flux. It should be noted that different characteristic velocities and densities were assumed in the different studies, which may well be the cause for the higher flux resulting from the LDEF measurements.

The first comprehensive model of the interplanetary dust flux was the Divine model (Divine 1993). The Staubach model provided an extension to small particles (Staubach et al. 1997). These models used the mass flux from Grün et al. (1985) for the meteoroids at the Earth, and used in-situ and zodiacal light data to constrain the flux of particles at other heliocentric distances.

The current meteoroid engineering models for space applications in the inner Solar system are NASA's Meteoroid Engineering Model or MEM (McNamara et al. 2004; Jones 2004; Moorhead et al. 2015) and ESA's Interplanetary Meteoroid Environment Model or IMEM (Dikarev et al. 2005, 2004). The NASA MEM uses a simple dynamical model, along with data from the Canadian Meteor Orbit Radar (CMOR), and zodiacal light observations from Helios to define the radial distribution. It can therefore be used for predictions of the flux, speed, and direction of dust and meteoroids within 0.2–2 au. The ESA IMEM model starts from the distribution of asteroidal and short period comets as sources of meteoroids and propagates them in space and time. However, the model assumes that only particles with $m \leq 10^{-9}$ kg are transported by the Poynting-Robertson effect towards the Sun, while larger particles remain on their initial orbits, creating a sharp discontinuity in the flux profile of particle sizes around this cutoff mass. IMEM uses infrared brightness data and in-situ spacecraft data to calibrate the model (Dikarev et al. 2005). Drolshagen et al. (2008) and Grün et al. (2013) discuss the differences between these existing meteoroid engineering models.

3 Astronomical Observations

3.1 The Zodiacal Light and Thermal Emission

The zodiacal light is a faint glow visible along the ecliptic plane in the night sky. It appears to the naked eye mainly as bright Solar-colored cones above the horizon at dusk and dawn as shown in Fig. 3. While this geometry is the most readily visible, the zodiacal light can be detected in every direction and results from the Solar light scattered by the small solid dust particles in the Solar system. As shown in Fig. 3, it is enhanced by backscattering in the anti-Solar direction, a phenomenon known as the *gegenschein*.

The zodiacal light intensity, I, varies with the position of the observer, the epoch of observation, the wavelength and the viewing direction. This is related to the distribution of the dust in the lenticular interplanetary dust cloud. The dust distribution essentially follows a mirror symmetry with respect to a plane close to the ecliptic, therefore heliocentric ecliptic coordinates are introduced as shown in Fig. 4a. The measurements integrated over the line-of sight are described with the heliocentric ecliptic latitude β and longitude $(\lambda_e - \lambda_{e0})$, the phase angle being α.

For an observer located on Earth's orbit, small amplitude variations of I are observed over a year. Annual oscillations up to $+10\%$ at high and medium latitudes originate from the slight inclination of the zodiacal cloud's symmetry plane in relation to the ecliptic and the eccentricity of the Earth's orbit (Dumont and Levasseur-Regourd 1978). After correcting

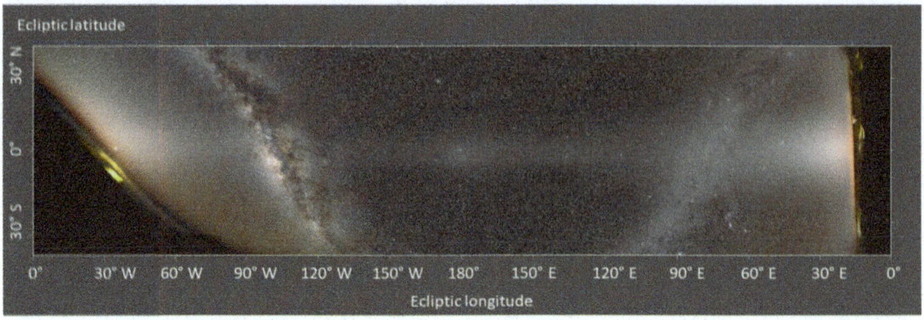

Fig. 3 360 degree panorama illustrating the zodiacal light along the Solar system's ecliptic plane. Mosaic of images taken during a single night at Mauna Kea. The zodiacal light cones are seen at dawn on the left and dusk on the right. The *gegenschein* enhanced backscattering is visible near 180° elongation. Venus and Saturn are marked. Image credit: Druckmüller and Habbal 2012, taken from Horálek et al. (2016)

for such geometrically induced variations, the zodiacal light intensity observed at 1 au remains consistently stable within 1.5% over a whole Solar cycle (Levasseur-Regourd 1996a). Taking into account the fact that the zodiacal light has a Solar spectrum, its intensity I at 550 nm is then found to be close to $2.4 \times 10^{-5}\ \mathrm{W\,m^{-2}\,sr^{-1}\,\mu m^{-1}}$ at 30° Solar elongation in the ecliptic plane (which outshines the brightest part of the Milky Way). It reaches a minimum of $7.6 \times 10^{-7}\ \mathrm{W\,m^{-2}\,sr^{-1}\,\mu m^{-1}}$ in the vicinity of the ecliptic pole. In the ecliptic plane, a minimum of $1.8 \times 10^{-6}\ \mathrm{W\,m^{-2}\,sr^{-1}\,\mu m^{-1}}$ is reached at around 140°, which is still 2.3 times higher than the ecliptic pole value. In the *gegenschein* direction, the intensity increases again to reach a value about 1.3 times larger than the ecliptic minimum, because of a backscattering effect.

The survey from Tenerife by Dumont and Sanchez (1975a,b) remained the most reliable and complete reference until the end of the twentieth century. It was found to be in excellent agreement with spacecraft data obtained at 90° phase angle, without any atmospheric contamination (Levasseur-Regourd and Dumont 1980). Figure 4b presents a compilation of whole sky observations as published in the review by Leinert et al. (1998). The symmetries in the data are used to present just one fourth of the sky. The relative uncertainty in the intensity data remains below 5%, the resolution of the data is between 5° and 15°. The symmetry plane at 1 au has been determined to have an ascending node at $95° \pm 20°$ with an inclination of $1.5° \pm 0.4°$ (Dumont and Levasseur-Regourd 1978).

A re-analysis of Weinberg's observations from Hawaii has recently yielded a new intensity map with a 2° resolution from which the ascending node position gives around 80° while the inclination of the cloud's symmetry plane is around 2° (Kwon et al. 2004), see Fig. 4c. With the continuous photometric observations of the Solar Mass Ejection Imager onboard the Coriolis spacecraft over the whole sky, an updated map of the zodiacal light has been generated with the unprecedented precision of 0.5°, as shown in Fig. 4d (Buffington et al. 2016). Based on the 8.5 years of data collected by the spacecraft, an upper limit to zodiacal light intensity changes of 0.3% was obtained for this time period. Using the same satellite data, Buffington et al. (2009) determined that the *gegenschein* is located at the anti-Solar point, but varies by $\approx 10\%$ of its intensity over time, with a portion of the variation repeating seasonally. Similarly, the latest high-resolution (5' per pixel) survey of the *gegenschein* using a wide-angle camera showed that the maximum scattered intensity is consistently located at the anti-Solar point. The models fitted to the *gegenschein* constrain a very low geometric albedo of 0.06 for those particles (Ishiguro et al. 2013).

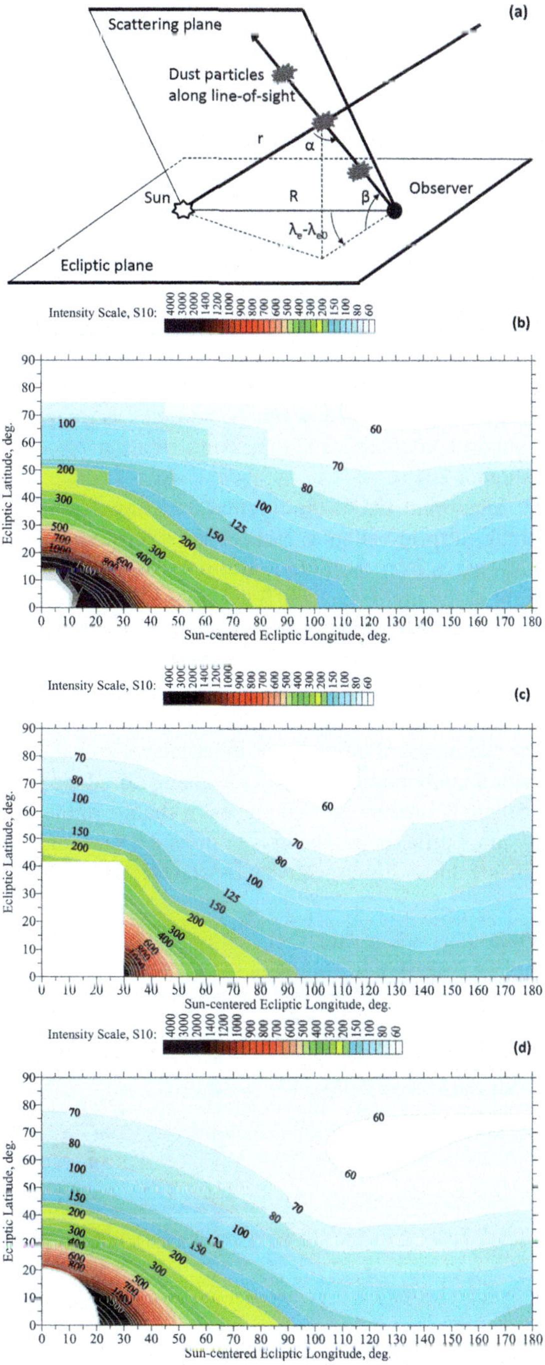

Fig. 4 (**a**) Geometry of observation of the zodiacal light with representation of the different values described in the text (adapted from Levasseur-Regourd et al. 2007). (**b**) Map of the zodiacal light compiled by Leinert et al. (1998), 5° to 15° precision. (**c**) Reanalysis of the Weinberg's observations by Kwon et al. (2004), 2° precision. (**d**) Modeled results of intensity observations from the Solar Mass Ejection Imager averaged over a period of a year in S_{10} units. Measurements limited to > 20° from the Sun with angular bins of 1° × 1°, figure taken from Buffington et al. (2016). Brightnesses are given in the so-called intensity scale, where $1 S_{10}$ unit corresponds to 1.18×10^{-8} W m^{-2} sr^{-1} μm^{-1} at 550 nm

Short-term enhancements of the zodiacal light had tentatively been ascribed to the optical detection of meteoroid streams (e.g. Levasseur and Blamont 1973), as later confirmed by the discovery of faint cometary trails by the Spitzer telescope in the infrared (Reach et al. 2007), as well as of linear dust features in the optical domain (Ishiguro et al. 1999).

Extensive descriptions of the results and methods of zodiacal light surveys can be found in the reviews previously published (Leinert 1975; Weinberg and Sparrow 1978; Leinert and Grün 1990; Leinert et al. 1998; Levasseur-Regourd et al. 2001; Mann et al. 2004; Lasue et al. 2015).

3.2 Spectral Observations

In the visible part of the spectrum, due to scattering, the zodiacal light shows a Solar spectrum. At large elongations, it is possible to determine the spectral gradient of the zodiacal light from space probes. Leinert et al. (1998) compiled a set of data from 8 different missions that indicated a possible reddening of the zodiacal light from 200 nm to 2 μm. Such a spectral gradient for low albedo particles as inferred from their scattering properties would be most consistent for D-type primitive asteroids or comets as indicated in Fig. 5 (Yang and Ishiguro 2015).

Measurements of the thermal emission originating from the zodiacal cloud have been obtained from balloons, rockets and satellites. The zodiacal thermal emission, which gradually takes over the Solar spectrum at 2 μm, is actually the most prominent source in the light of the night sky from 5 to 100 μm, at least away from the galactic plane (Leinert et al. 1998). For wavelengths larger than 100 μm, the sky brightness gets dominated by the interstellar medium, the cosmic diffuse infrared background and finally the cosmic microwave background.

Based on those long-wavelengths measurements, Fixsen and Dwek (2002) have modeled the thermal emission of the interplanetary dust particles using a single blackbody curve with a temperature of 240 K. A change of slope in the spectrum at a wavelength of about 150 μm is interpreted to represent a change in the dust size distribution at a radius of about 30 μm. Furthermore, the fit is improved if one considers a mixture of amorphous carbon and silicate dust with respective temperatures of 280 and 274 K at 1 au. Under those conditions,

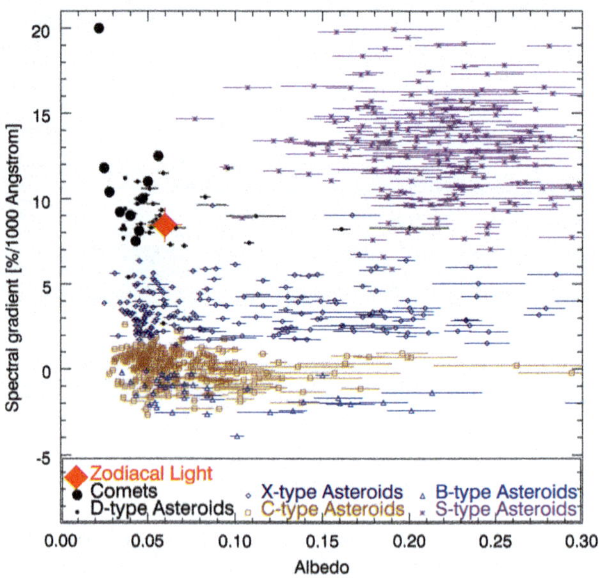

Fig. 5 Comparison of the spectral gradient with the albedos of asteroids, comets, and zodiacal light. The measurements variance of spectral gradients is estimated to be $\approx 0.7\%/1000$ Angstrom (Yang and Ishiguro 2015)

the Pointing-Robertson drag of the particles towards the Sun limits their lifetime to about 10^5 yrs, which requires a replenishment rate of $\approx 10^{11}$ kg yr^{-1} from asteroids and comets.

Further analysis of the zodiacal light spectrum at a wavelength of around 10 μm from the camera on-board the ISO satellite indicate a dust distribution dominated by large (> 10 μm radius), low-albedo (less than 0.08), rapidly-rotating, gray particles at 1 au from the Sun. After continuum subtraction, a 6% excess in the 9 to 11 μm range remains that suggests the presence of small particles, with an equivalent radius of < 1 μm, of amorphous and crystalline silicates (Reach et al. 2003).

3.3 Polarimetric Observations

The zodiacal light is light scattered by an optically thin cloud of small dust particles and therefore shows a partial linear polarization, P_Q. It can reach values as high as 20% along the integrated line of sight as observed from the Earth, see Fig. 4a, Fig. 6 and Fig. 7a. P_Q, by definition within the $[-1, +1]$ interval, is positive if the direction of the electric field vector is perpendicular to the scattering plane, which is the case away from the backscattering region. In the ecliptic plane, P_Q reaches a maximum of 19% at 60° heliocentric ecliptic longitude. The error in the measured polarization is estimated to be $\approx 2\%$. Around the *gegenschein*, at helio-ecliptic longitudes greater than 160°, the curve is smooth, with a slightly negative polarization in the backscattering region and an inversion at $15° \pm 5°$. The slope at inversion is (0.2 ± 0.1) percent per degree (Leinert et al. 1998). Such a polarimetric phase curve is commonly associated with scattering by irregular dust particles and/or fluffy aggregates of submicronic absorbing particles (Levasseur-Regourd et al. 1997; Lumme et al. 1997). This behavior is comparable to that of cometary dust or of C-type asteroids, which are the likely sources of interplanetary dust particles (e.g. Lasue et al. 2015; Lazarian et al. 2015; Kolokolova et al. 2015).

P_Q remains practically constant as a function of the wavelength in the visible range (450 to 800 nm), but infrared observations suggest a weak decreasing trend with increasing wavelength (Berriman et al. 1994). As different spectral polarimetric trends are seen for different types of asteroids surfaces (Belskaya et al. 2017) and comets (Levasseur-Regourd et al.

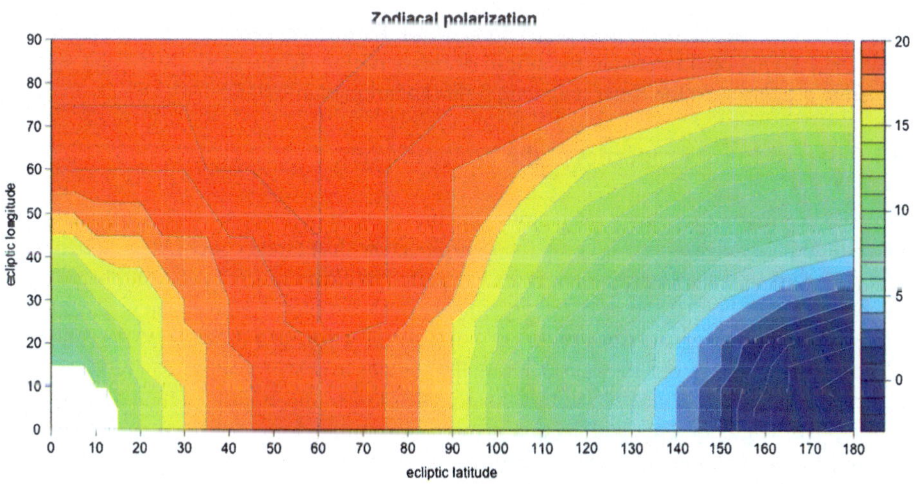

Fig. 6 Polarization map of the zodiacal light (adapted from Dumont and Sanchez 1975a; Leinert et al. 1998)

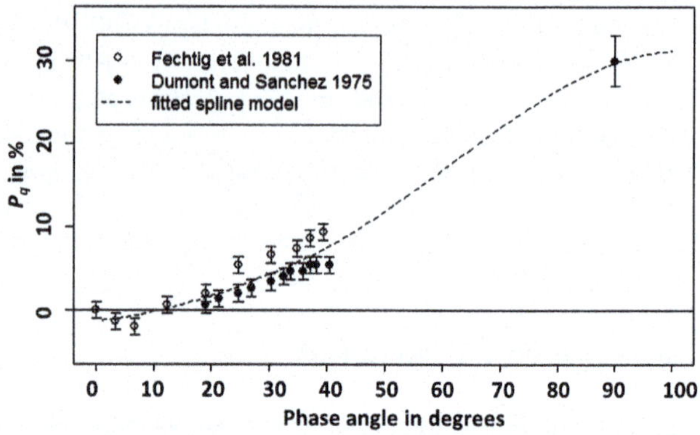

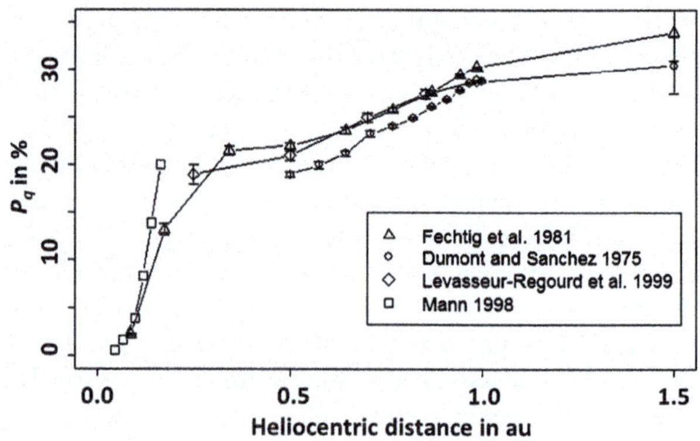

Fig. 7 Top: Local polarization phase curve obtained at the 1.5 au node (adapted from Levasseur-Regourd et al. 2001). Bottom: local polarization as a function of the heliocentric distance between 0.1 and 1.5 au (adapted from Levasseur-Regourd et al. 2001)

1996; Hadamcik and Levasseur-Regourd 2003), further observations may indicate potential genetic links to sources.

Similarly, while the light scattered by the zodiacal dust is mostly linearly polarized, multiple scattering, partial alignment, or asymmetrical particles can introduce a circular polarization component (Mishchenko et al. 2002). Measurements of the circular polarization have been attempted but remain inconclusive as they are inconsistent or not reproducible (Staude and Schmidt 1972; Wolstencroft and Kemp 1972). Further observations may be of importance as a signal could indicate the origin of the particles (Lasue et al. 2015) and specific dynamic properties in the Solar magnetic field (Lazarian et al. 2015).

3.4 Inferred Local Properties of the Zodiacal Light

Because the observations integrated along the line-of-sight correspond to different Solar distances and phase angles and because the interplanetary dust cloud cannot be assumed to be homogeneous, it is necessary to invert the integrated intensity and polarization observations to obtain the local properties of the interplanetary dust particles. An inversion is feasible rigorously along the tangent to the observer's motion direction as a differential measurement (Dumont 1973; Levasseur-Regourd 1996b). The values at other locations can be derived by inversion with good accuracy based on the nodes of lesser uncertainty method as a function of the phase angle, α. This is done at a given distance of 1.5 au, and for $\alpha = 90°$ between 0.1 and 1.5 au (Dumont and Levasseur-Regourd 1985; Renard et al. 1995). A summary of the local properties of the interplanetary dust (intensity, polarization, temperature, albedo, number density) in the symmetry plane as inverted from the data and at $\alpha = 90°$ is presented

Table 1 Variation of the local properties of the dust in the symmetry plane. The properties are described as a function of Solar distance with a power law assumption. The optical properties (linear polarization, albedo, and intensity in $W\,m^{-2}\,sr^{-1}\,\mu m^{-1}\,rad^{-1}$ at 550 nm) are retrieved at $\alpha = 90°$ (Levasseur-Regourd et al. 1996, 2001). The spatial density of the particles is inferred from the inverted values in intensity and polarization

	Value at 1 au	Gradient	Domain in au
Intensity	$\approx 23 \times 10^{-7}$	-1.25 ± 0.02	0.5 to 1.4
Linear polarization	0.3 ± 0.03	$+0.5 \pm 0.1$	0.5 to 1.4
Temperature	$250\ K \pm 10\ K$	-0.36 ± 0.03	1.1 to 1.4
Albedo	0.07 ± 0.03	-0.34 ± 0.05	1.1 to 1.4
Spatial density	$10^{-19}\ kg\,m^{-3}$	-0.93 ± 0.07	1.1 to 1.4

in Table 1. The local values for the geometric albedo at a given phase angle are retrieved by combining the local values of the flux emission, the temperature and the intensity (Hanner et al. 1981).

The local polarization phase curve of the zodiacal light at 1.5 au near the ecliptic can readily be compared to numerical simulations to constrain the physical properties of the scatterers. Early models using Mie scattering and power-law size distributions indicated compositions consistent with pyroxenes (see e.g. Giese 1963, 1973). However, collections of interplanetary dust particles in the stratosphere (Brownlee 1979) have soon established that models relying on sphere and Mie theory are hardly relevant. More realistic light scattering behavior of irregular particles such as aggregates of spheres, were calculated by different groups and confirmed the results of previous studies with respect to the presence of silicates and carbonaceous compounds (Giese et al. 1978; Haudebourg et al. 1999; Nakamura and Okamoto 1999; Kimura 2001). The simulation results show that typical aggregate sizes are of the order of a micron, while constituting monomer sizes appear to be one order of magnitude lower at about 0.1 μm. Typical optical indices have a real part in the range of 1.5–2, and an imaginary part up to 0.6 for strongly absorbing material (see e.g. Kimura et al. 2003, 2006). A link between red polarimetric color variation of light scattered by fractal aggregates and a lower porosity of the aggregate has been demonstrated, indicating the likely presence of compact particles in the interplanetary dust cloud (Kolokolova and Kimura 2010, Lasue et al. 2007).

As shown in Table 1, important variations in local scattered intensity and polarization are measured as a function of the heliocentric distance. Such a variation is probably linked with a change in the local properties of the dust particles, either their composition or their size distribution. The variation in spatial density with heliocentric distance is consistent with dust particles spiraling under Poynting-Robertson drag inside the production region, which ranges from 0.1 to 10 or 20 au (Leinert et al. 1983b; Levasseur-Regourd et al. 1991). However, observations of the Solar F-corona indicates a sharp decrease in polarization below 0.3 au, suggesting a drastic change closer to the Sun (Mann 1996). The radial polarimetric variations may be interpreted as a change in particle composition with more absorbent carbonaceous particles dominant far from the Sun and less absorbing silicates particles more heat-resistant closer to the Sun (Lasue et al. 2007).

Other effects, such as the size distribution of the particles, their morphology and structure, and their optical index variations, may also contribute to changes in polarization with Solar distance.

Assuming a stable interplanetary dust cloud, interplanetary dust and meteoroid particles vanish while spiraling into the Sun and require sources to replenish them. Particle ejection from comets as well as asteroidal collisions could easily replenish the inner zodiacal cloud (Whipple 1955; Dermott et al. 1984). Other minor contributions come from the environment of the giant planets, from the Kuiper belt, and from intruding interstellar dust particles (see e.g. Grün et al. 1993b). We have seen above that models of the light scattering properties of interplanetary dust particles are consistent with a population of dark, absorbent dust particles and/or fluffy aggregates of cometary and asteroidal origin (somehow comparable in volumetric contributions, see Lasue et al. 2007). A dynamical model of dust in the inner Solar system (Nesvorný et al. 2010, 2011a) and an improved model of the zodiacal dust cloud infrared emission (Rowan-Robinson and May 2013) both indicate a large contribution ($> 70\%$) of cometary dust particles to the interplanetary dust cloud (Levasseur-Regourd et al. 2018). These characteristics also appear consistent with the properties of dust particle collections described in Sect. 8.

4 Meteor Observations

4.1 Overview

Meteor phenomena are produced when meteoroids, typically larger than a few tens of microns in size, impact a planetary atmosphere (e.g. Ceplecha et al. 1998; Plane et al. 2018, and references therein). The ablation of the meteoroid body in the Earth's atmosphere produces light and heat that is observable on the Earth visually, as well as being detectable by photographic and video cameras, radar systems, and infrasonic and seismic detectors. Different methods and equipment probe different particle sizes.

Meteor studies utilize the Earth's atmosphere as a large detector of the meteoroid cloud and its structure at 1 au. Various types of information are derived from meteor observations. Optical and radar systems can determine fluxes of meteors, along with their directionality and speeds. Masses can be estimated from the optical brightness or radar electron line density, but the accuracy is limited due to uncertainty in the ionization and luminous efficiencies (Campbell-Brown et al. 2012). They can also be estimated by combining light curve data with deceleration measurements, when available (e.g. Gritsevich and Koschny 2011). Spectra can provide compositional information (Vojáček et al. 2015).

Meteor showers are caused by meteoroid streams, appearing to come from the same direction in the sky, called 'radiant'. Associated particles will have similar incoming velocities. These meteoroids were recently ejected from a comet or asteroid, their 'parent body'. Particles that are not associated with any meteor shower are termed 'sporadic meteors'; they dominate the meteoroid flux at Earth. Here we discuss what meteor observations can tell us about the distribution of interplanetary meteoroids at 1 au, and how the structure of meteor showers and outbursts can provide information on the early history of these particles.

4.2 Meteor Sporadic Background

Sporadic particles dominate the environment at Earth. They represent particles from the interplanetary meteoroid cloud, with its origin in asteroid and comet dust production processes (Levasseur-Regourd et al. 2018, and as described earlier in this paper), along with a small proportion of interstellar particles. Their orbits have been altered by gravitational and radiation forces as well as collisional processes (see Sect. 2) such that it is difficult to link the parent object to the particle orbit. The largest datasets of these particles are detected using optical and radar meteor detection techniques. These methods allow the study of the flux and directional characteristics of interplanetary meteoroids of sizes from a few tens of microns to a few cm.

The proportion of sporadic meteors increases with the limiting magnitude of the meteor detection method. This is a result of different ejection velocities and radiation forces on the small particles, and of collisions affecting the larger particles emitted by comets. Jenniskens et al. (2016c) found that about 36% of all meteors detected by the CAMS system of video cameras (generally between -1 and $+4$ visual magnitude) are in meteor showers. Koschny et al. (2017) find a comparable value of 31% down to $+5$ visual magnitude with their CILBO video camera system on the Canary Islands. The meteor database of the AMOR radar in New Zealand (limiting magnitude $\approx +14$) only contains about 1% shower meteors (Galligan and Baggaley 2005).

The appearance of the interplanetary meteoroid population in the meteor record is affected by the movement of the Earth around the Sun. This results in six sporadic 'sources' at the Earth. While the term 'sources' is widely used, they describe concentrations in radiant space that do not necessarily correspond to physical sources of meteoroids. The helion and anti-helion sources, found in the ecliptic plane at 70° from the apex direction, are usually the strongest sources, although the helion source is only detectable by radar. The north and south apex sources are centered towards the direction of the Earth's motion, with ecliptic latitude $\approx \pm 20°$ (Hawkins 1956; Weiss and Smith 1960; Jones and Brown 1993). The highly-inclined north and south toroidal sources are found 60° north and south of the apex direction (Hawkins 1962, 1963; Jones and Brown 1993). The high-resolution radiant distribution maps in Campbell-Brown (2008), produced from five years of CMOR data, show also a ring at 55° that is likely the result of Kozai oscillations (Kozai 1962) on high-inclination trajectories that are in-spiraling due to the Poynting-Robertson effect (Wiegert et al. 2009). Such a ring is also observed in SonotaCo video meteor data (Jakšová et al. 2015). An asymmetry between the strengths of the anti-helion and helion sources is suspected to be caused by the influence of comet 2P/Encke on the inner interplanetary dust cloud (Wiegert et al. 2009). New results from the SAAMER radar in Argentina provide data on the south toroidal source (Janches et al. 2015), which is not observable by systems in the northern hemisphere.

Different types of orbits are required to populate each source. Meteoroids observed in the helion and anti-helion sources have orbits that suggest an origin in short-period Jupiter-family comets (Jones et al. 2001; Wiegert et al. 2009; Nesvorný et al. 2010). Apex meteoroids have high speed or retrograde orbits that likely stem from Halley-type or Oort cloud comets (Wiegert et al. 2009; Nesvorný et al. 2011b; Pokorný et al. 2014). Recent work demonstrated the ability of evolved Halley-type comets to populate the toroidal sources (Pokorný et al. 2014).

The velocity distribution of meteoroids at the Earth, once corrected for observational biases, shows that most particles impact the Earth at relatively low velocities of typically $\approx 15\ \mathrm{km\,s^{-1}}$ (e.g. McBride et al. 1999; Huang et al. 2015; Moorhead et al. 2017, and references therein). The semi-major axis distribution at sizes below ≈ 1 mm predominantly peaks at 1 au: these particles are therefore likely to be the result of significantly evolved cometary dust populations (Nesvorný et al. 2011b). Even larger cm-sized meteoroids, such as those detected by the CAMS video meteor system, show a only fraction of such evolved orbits (Jenniskens et al. 2016c). This likely demonstrates, as described also by Campbell-Brown (2008), that even mm- to cm-sized meteoroids survive in interplanetary space for long durations and do not suffer collisions with other dust particles in lifetimes of tens of thousands of years, as originally suggested by Grün et al. (1985) or Jenniskens (2015).

Of particular interest is the mass-dependent flux of particles at the Earth. The distribution given by Grün et al. (1985) is somewhat of a standard model for the mass flux, derived from lunar crater counts (Hörz et al. 1975) and meteor data (Hawkins and Upton 1958; Whipple 1967). It gives a size distribution index $s = 2.34$ for sporadic meteors (where the number of meteor N is a function of mass M such that $dN \propto M^{-s} dM$). Newer radar studies of the sporadic meteor population have suggested this mass index may be lower, around $s \sim 2.0$–2.1 (Galligan and Baggaley 2004; Blaauw et al. 2011b; Pokorný and Brown 2016). Meteor showers have generally lower mass indices due to their young age (Blaauw et al. 2011a). The brightness of meteors provides direct information on their mass, in theory allowing similar information to be determined from meteor flux counts. However, large uncertainties in the efficiency in generating photons or electrons—the luminous efficiency for photographic or video meteors and the ionization efficiency for radar meteors—restricts the accuracy of such measurements. In particular, the behavior for higher velocities is not well understood. Weryk and Brown (2013) use simultaneous video and radar measurements to derive the ratio of the luminous and ionization efficiencies. Koschny et al. (2017) use these results to determine the flux of optical double-station meteors, and find a distribution comparable to that given by Grün et al. (1985). Recent laboratory experiments improve our understanding of the ionization efficiency and provide new constraints for it (Thomas et al. 2016).

Information on the physical and chemical properties of the meteoroids is obtained by studying their atmospheric ablation. Both for dynamically cometary and asteroidal meteors, a high proportion of small objects is seen, demonstrating evidence of fragmentation. Furthermore, the types of fragmentation are similar, although this could be indicative of a misunderstanding in the classification of the source population of these objects (Subasinghe et al. 2016). The beginning and end heights of the meteor are related to the impact speed (slower meteors start to ablate lower in the atmosphere), and also to the fragility of the material. Jenniskens et al. (2016c) demonstrated that meteoroids from asteroidal-type showers (Geminids, Quandrantids) have lower beginning heights, which could be consistent with their current or past low perihelion orbits. Kikwaya et al. (2011) used the dustball model of Campbell-Brown and Koschny (2004) and simultaneous optical meteor observations to derive statistics on the bulk density of meteoroids: on average 4200 kg m^{-3} for asteroidal meteoroids and 3100 ± 300 kg m^{-3} for meteoroids with Jupiter-family cometary orbits, and 360 to 1500 kg m^{-3} for Halley-type cometary (HTC) orbits. Moorhead et al. (2017a) developed a two-population spatial density distribution model, describing densities for HTC-like ($T_j < 2$) and non-HTC-like ($T_j > 2$) orbits. More complex models that describe the composition and structure of meteoroids in order to correctly describe their fragmentation and ablation processes are needed (Campbell-Brown et al. 2013; Stokan and Campbell-Brown 2015).

Meteor spectra, with lines from both the ablated meteor atoms and atmospheric constituents, can provide important information about the meteoroid composition and parent body (Trigo-Rodriguez et al. 2003; Jenniskens 2007). Borovička et al. (2005) and Vojáček et al. (2015) use a classification system based on the strength of the Na, Fe and Mg lines given in a ternary diagram, which can be used to study the differences between Jupiter-family, HTC-like, and asteroidal meteoroids. Sodium depletion is found to occur in cometary meteoroids on orbits with low perihelia or potentially due to longer exposure times in the Oort cloud, or asteroidal meteoroids with high iron content. Jupiter-family comet meteoroids tend to have a higher Na content. Variations in the Na content of Geminid meteors might indicate variations in their emission times from their parent. Large-scale video spectra campaigns provide spectra for meteors and fireballs, in order to improve our understanding of the chemical composition and origin of sporadic and shower objects (Jenniskens et al. 2014; Rudawska et al. 2016).

4.3 Meteor Showers, Outbursts and Storms

If a cluster of meteors is observed to appear from a similar radiant in the sky and with a similar velocity, it is presumed that their orbits are related and that they have a common parent object. The directional and velocity information can be propagated backwards to the Hill sphere of the Earth in order to determine heliocentric trajectories, which can then be compared to known parent bodies - comets or near-Earth asteroids. Note that in older work, often an analytical correction called zenithal attraction was used instead of a backward orbit propagation (for a more detailed discussion, see e.g. Gural 2001). In some cases there are clear associations between meteor showers and parent objects, such as for the Perseid meteor shower and the Halley-type comet 109P/Swift-Tuttle. In other cases no apparent source object has been observed. Such a link might be established by a simple comparison of orbital elements, with the aid of a dissimilarity criterion, e.g. the D-criterion. It describes the dissimilarity between two orbits by using either a combination of orbital elements (e.g. Southworth and Hawkins 1963; Drummond 1981) or geometric parameters such as velocity or radiant (Valsecchi et al. 1999; Jopek et al. 2008). The D-criterion has to be used with a cutoff value, which has been tested recently by Moorhead (2016) and Neslušan and Hajduková (2017). They attempted to characterize the false-positive rate for D-parameter-based shower association. Various other authors have defined dissimilarity criteria (Southworth and Hawkins 1963; Drummond 1981; Jopek et al. 2008). Rudawska et al. (2015) use a combination of a dissimilarity criterion and other techniques to improve meteor shower searches in the EDMOND database. Wavelet transform techniques have been used to extract shower structure within large radar datasets (Galligan and Baggaley 2002; Brown et al. 2008; Pokorný et al. 2017).

The result of such searches is a database of meteor showers at the Earth, and of the structure in the interplanetary meteoroid cloud, as observed at the Earth. In the last ≈ 15 years large datasets of video and meteor data from systems such as CMOR, SonataCo, CAMS, and SAAMER have expanded the existing database of known stream structure, with meteor radars such as CMOR and SAAMER probing the 100- μm to mm-size range (Brown et al. 2008; Pokorný et al. 2017), while video camera systems such as SonataCo, EDMOND and CAMS detect streams at cm sizes (Jenniskens et al. 2016a,b,c; Koschny et al. 2017). In particular, the southern hemisphere location of SAAMER and some CAMS stations have facilitated detection of a larger number of showers in a previously poorly surveyed region of the sky. The Meteor Data Center, administered by commission F1 of the IAU, maintains an online database of 'established', 'working' and 'pro tempore' showers (Jopek and Jenniskens 2011; Jopek and Kaňuchová 2014). As of 1 Jan 2017, 701 showers are listed: 112 established, 507 working, and 35 pro tempore (with 23 to be removed). There are also 24 (4 established, 1 pro tempore) shower groups. Less than half of these established showers have a known parent body, demonstrating the limits of knowledge on the origin of the structure in the interplanetary meteoroid cloud. Resolution of this issue requires a comprehensive study of the dynamical and physical (dust emission) history of potential parent bodies.

Modeling of the meteor showers has improved significantly over the last years (e.g., Vaubaillon et al. 2005) and benefits from enhanced computer resources, which now allow integration of the individual orbits of millions of meteoroids. However, our ability to fully model their dynamics and evolution is hampered by the lack of knowledge about the parent comets and their emission processes. Necessary parameters such as the ejection velocity are still poorly constrained (Ryabova 2013). Nonetheless, studies of specific meteor showers, coupled with modeling, provides important information on the parent bodies (comets and asteroids) and their dust emission processes, and on the dynamical and collisional processes that govern the behavior of these dust particles after emission. Observations of the

Camelopardalids in 2015 demonstrated that the mass distribution of particles in the stream leans toward untypically small particles, that are more easily observable by radar (Campbell-Brown et al. 2016). Ye et al. (2016) use the combination of meteor observations of the Camelopardalids, comet observations of the parent 209P/LINEAR, and dynamical studies of the dust emission to demonstrate that the comet may be near the end of its active lifetime. Studies of meteor shower groups such as the Taurid complex can illuminate the dynamical history of these streams, which may result from the dispersion of a parent body into a cascade of smaller objects and particles (Asher and Steel 1995; Dubietis and Arlt 2007; Jenniskens et al. 2016a). A full review of meteor shower studies beyond these individual examples is found in Jenniskens (2017).

4.4 Interstellar Meteors and Meteoroids

The presence of interstellar dust in the Solar system was confirmed by in-situ spacecraft measurements from Ulysses in 1992 (Grün et al. 1993a), and later Galileo (Baguhl et al. 1996; Altobelli et al. 2005), Helios (Altobelli et al. 2006) and Cassini (Altobelli et al. 2003, 2016).

Also, three likely contemporary interstellar dust particles of micron-size were brought back to the Earth by the Stardust sample return mission in 2006 (Westphal et al. 2014) and infrared space telescope observations indicated the presence of interstellar dust in the Solar system (Rowan-Robinson and May 2013). The measured and captured interstellar dust particles in the Solar system are typically smaller than a few microns. Their apparent speed relative to the Solar system is about 26 $km\,s^{-1}$ (Grün et al. 1994) and they generally come from a direction of 259° ecliptic longitude and −8° ecliptic latitude (Landgraf 1998; Frisch et al. 1999; Strub et al. 2015), due to the motion of the Solar system relative to the local interstellar cloud.

For the past two decades, a debate has existed about the presence of interstellar meteoroids in the meteor data from various radar, video and optical systems. This debate is important, because the detection of interstellar dust larger than a couple of microns may have large implications for the mass distribution and thus for the dust-to-gas mass ratio in the diffuse nearby interstellar medium. Reliable detection of big interstellar dust would provide more insight in the mixing of dust populations and about processes in the interstellar medium. Grün and Landgraf (2000) elaborate on the implications of big interstellar meteoroids in the interstellar medium.

Sub-micron-sized interstellar dust particles are coupled to the interplanetary magnetic field on scales of a few tens to a few hundreds of au when they travel through the Solar system. The dynamics of interstellar meteoroids in the Solar system differs from sub-micron particles in that they are influenced by mainly the Solar gravity and to a smaller extent Solar radiation pressure force, instead of by the Lorentz forces. Particles larger than a few tens of microns are thus considered uncoupled from the (local) interstellar and interplanetary magnetic fields because of their low charge-to-mass ratio (Grün and Landgraf 2000; Landgraf 2000). As a consequence, these could come from different directions than the stream of interstellar dust currently known from the in-situ data (depending on their ejection velocity) and these may directly represent their source direction. Meteoroids with velocities above about 73 $km\ s^{-1}$ are on unbound orbits from the Sun. This is the sum of the escape speed from the Solar system (42.1 $km\ s^{-1}$), the Earth orbital velocity in a head-on collision (30.3 $km\ s^{-1}$), and the effect of the terrestrial gravitational field (Taylor et al. 1996). Meteoroids with smaller velocities but above about 11 $km\ s^{-1}$ may still be interstellar, but cannot be classified as such by examining their velocity alone.

4.4.1 *Radar Observations of Interstellar Particles*

When cosmic dust particles larger than a couple of microns enter the atmosphere at high velocities, they ablate and leave a streak of ionized air that reflects the radar signals. Even though the radar limit of a minimum particle size of about 5 μm (Baggaley et al. 2007) is a disadvantage to detect interstellar dust because it typically is micron or sub-micron in size, the large detection surface of the Earth and the ability to detect large particles of several tens of microns and bigger, is an advantage and complementary to in-situ instruments and astronomical observations. A major challenge is the reliable estimation of the meteoroid velocity which determines whether a meteoroid is hyperbolic and hence, whether it is interstellar.

Taylor et al. (1996) announced the discovery of the first interstellar dust detections by radar in the Advanced Meteor Orbit Radar (AMOR) meteor data. These were coming from two distinct directions other than the known radiant of the (smaller) interstellar dust from the local interstellar cloud determined by in-situ measurements. Baggaley (2000) reported on the existence of a flux of particles coming from southern ecliptic latitudes, and on a discrete stream of interstellar dust particles coming from the direction of β-Pictoris: a main-sequence star with a debris disk. These particles are much bigger (tens of microns, about 40 μm) than the particles detected by in-situ instruments. However, the results—like from other radar-based studies—are highly debated (Hajduková 2016; Hajduková et al. 2018).

The Canadian Meteor Orbit Radar (CMOR) data have a larger minimum detectable radius of 100 μm (assuming particle bulk densities of $3000\ \mathrm{kg\,m^{-3}}$ for masses of $1 \cdot 10^{-8}$ kg) (Weryk and Brown 2004). Two and a half years of data have been analyzed and a small fraction of 0.0008% was determined to be hyperbolic and thus not bound to the Solar system. The total statistics from this dataset were too low to distinguish a collimated stream of particles (Weryk and Brown 2004). The flux of particles derived from the observations is in line with the extrapolation of the size distribution of the data from Ulysses and Galileo (Weryk and Brown 2004), although they are likely from different populations. For all of these radar-based studies, meteoroids were only selected as "interstellar dust" when their heliocentric speed was larger than 3σ above the hyperbolic threshold, thus the numbers presented in these studies are lower boundaries. However, Moorhead et al. (2017a) re-analyzed data from CMOR with updated ionization efficiencies. Especially high-velocity meteoroids have lower speeds as a result. This was confirmed by Moorhead (2018), who shows that after applying a sharpening algorithm to the velocity distribution, no meteoroids with velocities larger than escape velocity are left.

Meisel et al. (2002) reports on interstellar meteoroids detected by the Arecibo radar, and uses different techniques for orbit and velocity determination especially as these particles are generally smaller than with the AMOR or CMOR radars.

4.4.2 *Video and Optical Observations of Interstellar Particles*

Video and optical methods have a higher accuracy for the velocity and radiant determination than radar (Baggaley 1999) and thus these are valuable complementary datasets for verification. Hajduková (1994) and Hajduková (2008) analyzed photographic and radar measurements from the IAU Meteor Data Center catalog and concluded that the majority of interstellar meteors were from velocity determination errors. Hawkes and Woodworth (1997) report on the detection of two interstellar meteoroids with video detectors. The meteoroids have estimated velocities of 4σ and 8σ above the hyperbolic escape speed of the Solar system. Hill et al. (2005) discusses meteor ablation for high velocities and concludes that these, with mass larger than about 10^{-8} kg, should be observable with optical methods. Musci et al.

(2012) reports on the search for interstellar particles in one year of data from a two station automated electro-optical system (the *Canadian Automated Meteor Observatory*), but found no convincing evidence for a collinear stream of interstellar dust, perhaps because of low statistics. Hajduková (2011) and Hajduková et al. (2014a,b) analyzed the SonotaCo and EDMOND video meteor data and also concluded from this study that the majority of the hyperbolic meteoroids were due to velocity measurement errors. For instance, about 50% of the identified hyperbolic meteors belonged to meteor showers. For SonotaCo and EDMOND, the limiting magnitudes are +2 and +4 respectively, corresponding to limiting masses of roughly 10^5 kg, but this largely depends on meteor velocity (Hajdukova 2018, pers. comm.). If at all, then planetary perturbations have little influence on the hyperbolicity of the dust, hence most hyperbolic meteors were due to velocity determination errors. Wiegert (2014) confirmed that hyperbolic meteoroids can be generated by planetary perturbations of Solar system dust, but that the velocity of such locally generated hyperbolic meteoroids in the Solar system is only marginally larger than the heliocentric escape velocity at the Earth, and that the rate of locally generated hyperbolic meteors is relatively low (about one in every 10^4 optical meteors).

Musci et al. (2012) summarize a few of the observations discussed above, together with in-situ data from spacecraft, in one plot. Although in many observations, a percentage of hyperbolic meteors are found, it seems from many studies, that the majority of these observations resulted from measurement errors. A comprehensive review of the radar observations of hyperbolic meteoroids is given in Hajduková (2016) and Hajduková et al. (2018), who show in their Fig. 2 the estimated interstellar dust flux from various observation methods.

5 In-Situ Measurements

The previous sections have described remote methods of studying the interplanetary meteoroid cloud. Here we discuss recent developments in the measurement of the meteoroid distribution in-situ. A history of in-situ detection up until the year 2000 is given in Grün et al. (2001).

5.1 Antenna Detections Using the STEREO and Wind Spacecraft

Cosmic dust striking a spacecraft surface at high velocity (more than a few $\mathrm{km\,s^{-1}}$) vaporizes and ionizes upon impact, creating a transient plasma cloud near the spacecraft. The generation of this cloud perturbs the spacecraft floating potential and can be detected by instruments designed to measure electric fields in space.

Detections of cosmic dust using electric field antennas have been reported in a variety of dusty environments, including near the gas giant planets: Jupiter, (Tsintikidis et al. 1996), Saturn (Gurnett et al. 1983, 2004; Kurth et al. 2006; Meyer-Vernet et al. 2009a), Uranus (Meyer-Vernet et al. 1986; Gurnett et al. 1987), Neptune (Gurnett et al. 1991; Pedersen et al. 1991), and near comets (Giacobini-Zinner (Gurnett et al. 1986), Halley (Laakso et al. 1989; Neubauer et al. 1990), and P/Borrelly (Tsurutani et al. 2003).

More recently, dust has been observed using electric field instruments in the open Solar wind at 1 au, far from any known dust sources, both on the STEREO and Wind spacecraft (Meyer-Vernet et al. 2009b; St Cyr et al. 2009; Zaslavsky et al. 2012; Malaspina et al. 2014; Kellogg et al. 2016).

STEREO observes two populations of interplanetary meteoroids: beta-meteoroids, micron-sized dust coming from the direction of the Sun due to Solar radiation pressure (Zaslavsky et al. 2012; Malaspina et al. 2015), and nanometer dust (Meyer-Vernet et al. 2009b;

Zaslavsky et al. 2012). The Wind spacecraft observed micron-sized interplanetary dust, primarily grains one to two orders larger than the beta-meteoroids detected on STEREO (Malaspina et al. 2015; Kellogg et al. 2016). Wind does not observe nanometer dust (Kellogg et al. 2016).

The nanometer dust detected by STEREO has important implications for the dynamics of heliospheric dust and Solar wind plasma. Nanometer dust has a charge to mass ratio large enough that it can act as a massive pickup ion, getting accelerated to Solar wind velocities (about 400 $\mathrm{km\,s^{-1}}$) away from the Sun through electromagnetic interactions with the interplanetary magnetic field (Czechowski and Mann 2010; Juhász and Horányi 2013). Because nano-dust can efficiently exchange momentum with the Solar wind, it can slow the Solar wind near its injection point. This effect has been observed in relation to destructive collisions between large bodies in the Solar wind (Lai et al. 2014), and could potentially play an important role in the evolution of the near-Sun Solar wind. Variations in the flux of nanometer dust observed at 1 au may also contain information about the rate of collisional grinding occurring in the near-Sun environment.

Finally, it is important to mention that a publicly accessible database of all dust impacts recorded by the Wind spacecraft over its 22-year mission ($>$ 107 000 impacts as of 2015) was recently created (Malaspina and Wilson 2016). This database provides an unprecedented opportunity to study the flux and variation of interplanetary dust near 1 au over two full Solar cycles. The Wind dust database was recently made available (early 2017) through NASA's Space Science Data Facility Coordinated Data Analysis Web.[2]

5.2 Impact Dust Detector on New Horizons

The New Horizons mission was launched in January 2006 on an initial reconnaissance of the Pluto-Charon system (Stern 2008). After launch, New Horizons underwent a gravitational assist from Jupiter in February 2007 followed by nearly eight-and-a-half years of interplanetary cruise from Jupiter to Pluto. The Pluto-Charon fly-by was successfully executed in July 2015 and the spacecraft re-entered a period of interplanetary cruise through the Edgeworth-Kuiper Belt. An additional fly-by of Kuiper Belt object 2014 $\mathrm{MU_{69}}$ is now scheduled for 01 January 2019.

While not part of the original payload, the Venetia Burney Student Dust Counter (SDC) was added to the manifest as a student-led project (Horányi et al. 2008). SDC is a polyvinylidene fluoride (PVDF) based impact dust detector, with heritage from laboratory designs (e.g. Simpson and Tuzzolino 1985; Simpson et al. 1989; Simpson and Tuzzolino 1989; Tuzzolino 1991, 1992; James et al. 2010) and previous instruments such as Space Dust (SPADUS, Tuzzolino et al. 2001a,b), the Dust Flux Monitor Instrument on Stardust (DFMI, Tuzzolino et al. 2003, 2004), the Cassini High Rate Detector (HRD, Srama et al. 2004), and the Cosmic Dust Experiment on-board the Aeronomy of Ice in the Mesosphere mission (Poppe et al. 2011). SDC operates by detecting individual dust impact events on twelve PVDF detectors mounted on the ram side of the New Horizons spacecraft. Two additional, identical detectors operate under the main deck of SDC, protected from any dust impacts in order to serve as references for the background noise rate from spacecraft vibrations (which can register as hits on PVDF detectors) and/or cosmic ray hits in the instrument electronics. When dust grains impact one of the twelve exposed PVDF films, they create micron-sized craters in the material inducing a change in electrostatic surface charge density on the PVDF detectors (Poppe et al. 2010a; Shu et al. 2013). This change in surface charge density is recorded by

[2] https://cdaweb.sci.gsfc.nasa.gov/index.html/.

Fig. 8 The spatial density of interplanetary dust grains with radii greater than 0.5 μm as measured by the New Horizons Venetia Burney Student Dust Counter as of 2013. The blue continuous line is a model prediction; the black bars indicate the measurements including their errors (reproduced from Szalay et al. 2013)

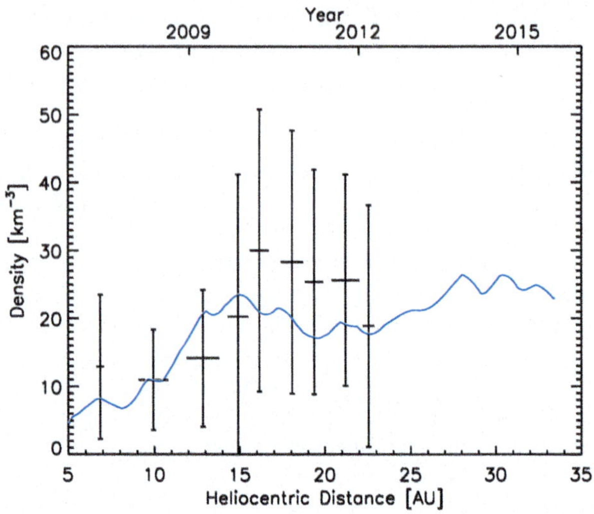

associated electronics and empirically calibrated to the dust grain mass assuming an impact speed. Impact speeds are calculated from the assumption that most grains can be approximated as being on circular, keplerian orbits. The keplerian velocity is then vector added to the New Horizons velocity to determine the impact speed. Based on ground calibrations, SDC is sensitive to dust grains with radii larger than approximately 0.5 μm. Between 0.5 and 5 μm, SDC can resolve the mass of the impacting dust grain, while for grains larger than ~ 5 μm, a hit is registered, however, internal electronics saturate and the size is only constrained to be greater than 5 μm.

SDC has operated successfully throughout the New Horizons mission, including through interplanetary cruise before the Jupiter encounter, through interplanetary cruise before and after the Pluto-Charon encounter, and for limited periods of time during the main Pluto encounter itself. First results from SDC were reported in 2010, approximately half-way through the interplanetary cruise to Pluto at 15 au (Poppe et al. 2010b) with later interplanetary results presented before the Pluto encounter at 23 au (Szalay et al. 2013). SDC measured fluxes of interplanetary dust grains greater than 0.5 μm between (2–5) $\times 10^{-4}$ m^{-2} s^{-1}. Assuming heliocentric Kepler velocities for the dust grains and appropriately computing the relative impact velocity of these grains onto SDC, this equates to cumulative IDP densities > 0.5 μm of ≈ 10 km^{-3} between 5 and 15 au with an increase to $\approx$ 20–30 km^{-3} between 15 and 23 au. Figure 8 shows the interplanetary dust spatial density measured by SDC as a function of heliocentric distance (Szalay et al. 2013). These invaluable data have been used to constrain models of interplanetary dust grain production in the outer Solar system (Han et al. 2011; Vitense et al. 2014; Poppe 2016), see Sect. 6 for further discussion. Within the Pluto-Charon system, SDC detected a single valid dust grain impact greater than ≈ 1.4 μm in radius. Note that the instrument minimum detection threshold was raised for the Pluto encounter in order to filter out anticipated background noise (Bagenal et al. 2016). This is consistent with densities for grains > 1.4 μm in radius measured between 20 and 30 au before the Pluto encounter. SDC will continue to make measurements of the interplanetary dust flux through the Edgeworth-Kuiper Belt and beyond.

5.3 Impact Ionization Detectors—Recent Reanalyses

Impact ionization detectors have been flown on Ulysses, Galileo and Cassini. These detectors are able to provide orbital and mass information for detected grains, and in the case of the Cassini instrument, also compositional information. Those on Galileo and Cassini were designed to study the circumplanetary environments of Jupiter and Saturn. All three instruments, however, also sampled the interplanetary dust cloud, providing information on very small particle sizes up to a few microns. In addition, all three instruments also observed nano-dust particle streams originating from Io's volcanoes in interplanetary space near Jupiter (Grün et al. 1993c, 1996; Graps et al. 2000).

Ulysses flew a highly inclined orbit in order to explore the Solar polar regions. This allowed it to sample meteoroids outside the ecliptic plane, and in particular to observe interstellar dust (see Sect. 4.4). Outside of the ecliptic plane, accelerated beta-meteoroids were a dominant source (Wehry and Mann 1999; Wehry et al. 2004). The Staubach model of the interplanetary dust cloud (Grün et al. 1997; Staubach et al. 1997) is partially based on five years of Ulysses data, and is supported by further Ulysses dust data (Krüger et al. 2006, 2010).

Galileo and Cassini remained in the ecliptic plane during their cruise phases. Galileo's Dust Detection System (DDS) detected interplanetary particles during the cruise phase (Krüger et al. 1998). Cassini's Cosmic Dust Analyzer (CDA) detected both interplanetary particles in the inner planet region and between Jupiter and Saturn (Altobelli et al. 2007). The Chemical Analyzer of the CDA detected two iron-rich interplanetary dust particles, which have most probable orbits similar to Aten and Apollo asteroids, but the orbit uncertainty means they could also have Jupiter-family comet-like orbits (Hillier et al. 2007).

Normally particles are detected when they hit the back wall of the detector, then the direction of the particle can be constrained by the relative opening angle as seen from the back wall through the detector aperture. However, impacts may also generate a charge—and thus be detected—when hitting the side walls of the detector. This has only recently been taken into account (Altobelli et al. 2004). Neglect of this effect could lead to an overestimation of the rates of interplanetary particles (Willis et al. 2005).

6 Outer Interplanetary Dust and Meteoroid Cloud

In the outer Solar system, typically defined as outside Jupiter's orbit at 5.2 au, meteoroids originate from several sources, including Edgeworth-Kuiper Belt objects, Halley-type comets, Jupiter-family comets, and Oort Cloud comets (e.g. Liou et al. 1996a; Landgraf et al. 2002). Cometary bodies are known to release significant amounts of dust and gas through heating, sublimation, and disruption. Such activity is typically strongest in the inner Solar system, although cometary outgassing and disruption has been observed in the outer Solar system as well (e.g. Fulle 1992; Sekanina 1996; Kelley et al. 2013; Epifani et al. 2016). Dust production from Edgeworth Kuiper Belt objects has also been long hypothesized, with mutual collisions (Stern 1995, 1996), bombardment by interstellar dust grains (Yamamoto and Mukai 1998), and bombardment by interplanetary dust grains (Poppe 2015) all theorized as possible production mechanisms. The relative balance between these hypothesized production mechanisms at EKBs is currently unclear and it is possible that all three contribute substantially. Dust grains can also be produced in the outer Solar system through grain-grain collisions, discussed in further detail below. Thus, the equilibrium distribution of dust in the outer Solar system is comprised of contributions from various sources, each with their own individual spatial and velocity distributions.

As dust grains are shed from their parent bodies, several forces and processes affect both their dynamical and compositional evolution. Just as in the inner Solar system, meteoroids in the outer Solar system are subject to Solar and planetary gravity, Solar radiation pressure, Poynting-Robertson and Solar wind drag, and the electromagnetic Lorentz force as a result of grains becoming electro-statically charged (see review by Gustafson 1994). In the outer Solar system, gravitational forces from the Sun and the giant planets dominate. In the absence of planetary gravity, meteoroids will slowly spiral in towards the Sun due to Poynting-Robertson and Solar wind drag on the particle, with lifetimes in years as described in Sect. 2. For typical grains from the Kuiper Belt, the Poynting-Robertson drag lifetime assuming no planetary perturbations is about one million years per micron of radius (i.e., a 1-μm Kuiper Belt dust grain would spiral into the Sun in about 1 million years).

The role of the giant planets in sculpting the equilibrium spatial density distribution of interplanetary dust grains in the outer Solar system has long been recognized (Liou et al. 1996a; Liou and Zook 1997, 1999). As dust grains spiral inwards from their birth due to Poynting-Robertson and Solar wind drag, they encounter mean motion resonances (MMR) with planets. For Kuiper Belt grains, typically born outside the orbit of Neptune, trapping occurs mainly in neptunian MMRs. Such trapping can extend the lifetime of the dust grains by temporarily halting their inward progression due to Poynting-Robertson drag (e.g., Liou and Zook 1997, 1999; Moro-Martín and Malhotra 2002, 2003), potentially forming a circumstellar ring of dust grains just outside the orbit of Neptune.

In addition to dynamical perturbations, dust grains are subject to various mass loss processes that, over time, can significantly reduce the size of the grain relative to its size at birth. These mechanisms primarily include charged particle sputtering (e.g., Mukai and Schwehm 1981), photo-sputtering (Westley et al. 1995; Grigorieva et al. 2007b; Öberg et al. 2009), and grain-grain collisions (e.g., Borkowski and Dwek 1995; Stark and Kuchner 2009). Sputtering is the process of energetic particles or photons hitting the surface of a grain, releasing individual atoms, ions, or molecules from this grain.

Sublimation, a mass loss process that is highly effective in the inner Solar system, does not operate efficiently in the outer Solar system due to the far colder equilibrium grain temperatures (e.g., see Mukai and Schwehm 1981). It has been hypothesized by Kobayashi et al. (2010), however, that sublimation of icy grains from Edgeworth-Kuiper Belt objects may be responsible for the fairly flat distribution of interplanetary dust measurements by Pioneer 10 (Humes 1980) and Voyager (Gurnett et al. 1997). Such calculations would appear to be at odds with simulations by Grigorieva et al. (2007b), which showed that photodesorption (atomic or molecular species leaving a surface) is a highly efficient process even in the outer Solar system. Thus, icy grains would rapidly become β-meteoroids before being able to migrate inwards from their birth place in the Edgeworth-Kuiper Belt. The ultimate fate of icy grains generated from the Edgeworth-Kuiper Belt is still unclear and warrants further study.

Grain-grain collisions are also a critical process for the evolution of interplanetary dust grains, yet have only recently been successfully included in dynamical models (e.g., Stark and Kuchner 2009; Vitense et al. 2012). Stark and Kuchner (2009) have described a novel algorithm for incorporating the effects of grain-grain collisions on the calculation of equilibrium dust grain distributions in N-body simulations. This so-called "collisional grooming" technique operates by first producing a non-collisional spatial density distribution and then repeatedly recalculating the collisional degradation of the dust disk by computing the collisional history of each individual dust grain through the previous equilibrium spatial density distribution. Kuchner and Stark (2010) used this technique to model the Kuiper Belt, finding

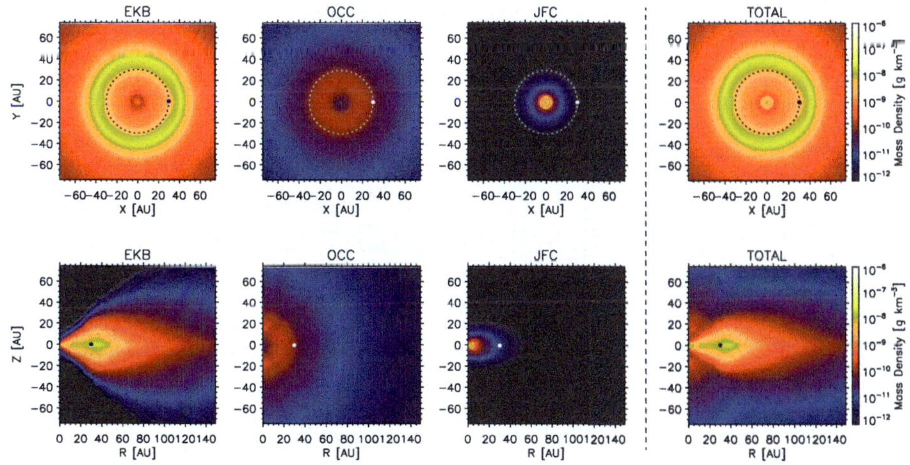

Fig. 9 (top row) The modeled mass density of interplanetary dust grains from Edgeworth-Kuiper Belt objects (EKB), Oort Cloud comets (OCC), Jupiter-family comets (JFC), and total in the ecliptic plane. The dot and dotted line denotes the position and orbit of Neptune, respectively. (bottom row) Same as top row but in the meridional plane. From the model of Poppe (2016)

that collisions are an important process even in today's EKB dust disk. Later models of the Kuiper Belt debris disks have incorporated various forms of this technique for collisions, yielding similar results (e.g., Poppe 2016).

Several dynamical models have explored various aspects of the spatial density and velocity distributions of interplanetary dust in the outer Solar system including those by Liou and Zook (1999), Landgraf et al. (2002), Moro-Martín and Malhotra (2002, 2003), Kuchner and Stark (2010), Vitense et al. (2012, 2014), Han et al. (2011), and Poppe (2015, 2016). These models have progressively increased in both theoretical and computational sophistication over time. As an example, Fig. 9 shows the ecliptic and meridional mass density distributions of 0.5–100 μm dust grains for three separate families (Edgeworth-Kuiper Belt objects, or EKBs, Oort Cloud comets, or OCCs, and Jupiter-family comets, or JFCs) and the total, respectively, from the Poppe (2016) dynamical model.

EKB grains form a broad, ≈ 10 au thick ring centered at 40 au, just outside the orbit of Neptune, while extending both into the inner Solar system and out past 150 au. The peak in mass density of $\approx 10^{-6}$ g km^{-3} near 40 au indicates the primary birth location centered on the main Edgeworth-Kuiper Belt.

Reflecting the nearly isotropic inclination distribution of parent body comets (e.g., Dones et al. 2004), the OCC dust grains form a three-dimensional halo (or shell) in the outer Solar system centered between the orbits of Jupiter and Neptune ($\approx$ 5–30 au). Mass densities of OCC grains are estimated to be lower than that of Kuiper Belt grains, based on fits to Pioneer 10 and New Horizons SDC measurements (Poppe 2016).

JFC dust grains are highly concentrated within the orbit of Jupiter and do not extend beyond the orbit of Neptune in appreciable numbers. Nevertheless, JFC grains form the dominant dust species in the inner Solar system as shown by Nesvorný et al. (2010, 2011a).

Finally, the right-most column of Fig. 9 displays the total mass density in the outer Solar system. Kuiper Belt grains dominate the mass density in the ecliptic plane in the outer Solar system and beyond, with OCC grains dominating over Kuiper Belt grains only at high ecliptic latitudes.

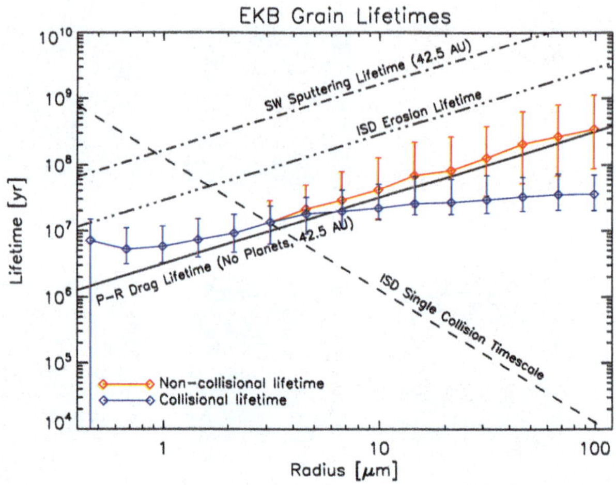

Fig. 10 A comparison of relevant time scales for Edgeworth-Kuiper Belt grains, more explanations in the text. The red and blue lifetime curves are taken from the model of Poppe (2016), combining all the different processes

Figure 10 shows a comparison of various theoretical and dynamically modeled lifetimes for EKB grains as a function of grain radius in the outer Solar system from the model of Poppe (2016). Theoretically derived lifetimes include the Poynting-Robertson drag lifetime (solid black line) assuming zero initial eccentricity and an initial semi-major axis of 42.5 au (Wyatt and Whipple 1950; Burns et al. 1979), the lifetime due to erosion by interstellar dust grain impacts (dash-triple dot line), and the lifetime due to charged particle sputtering from the Solar wind assuming values at 42.5 au (dash-dot line). The time-scale for a single stochastic collision with an interstellar dust grain is shown as the dashed line. The ISD erosion lifetime is calculated from a characteristic ISD flux of $\Gamma_{ISD} \approx 10^{-4}\ \mathrm{m}^{-2}\,\mathrm{s}^{-1}$ and grain size of 0.5 μm (Grün et al. 1997), and an impact cratering yield from Borkowski and Dwek (1995). The Solar wind sputtering lifetime is calculated assuming a Solar wind composition of 97% protons, 3% alphas (He^{++}), a flux at 1 au of $\Gamma_{sw} = 10^{8}\ \mathrm{cm}^{-2}\,\mathrm{s}^{-1}$ with radial dependence as r^{-2}, and proton and alpha sputtering yields of 10^{-2} and 10^{-1}, respectively (Biersack and Eckstein 1984; Behrisch and Eckstein 2007). The dynamical lifetimes of dust grains subjected to the full panoply of forces as described in Poppe (2016) is shown in color for both non-collisional (red) and collisional (blue) cases. For all sizes, the non-collisional lifetimes are slightly higher than the theoretical P-R drag lifetime. The relatively broad error bars (representing the 10% and 90% levels) are due to the broad range of initial semi-major axes and eccentricities in the dynamical model. As the grain size increases, the non-collisional lifetime scales as the radius, in agreement with theory. In the collisional case, dust grain lifetimes are identical to those in the non-collisional case for radii of ≈ 3 μm or less. Grains with radii larger than ≈ 3 μm have lifetimes that asymptote to a maximum of approximately 3×10^{7} years, significantly shorter than either the dynamical or theoretical lifetimes. While subject to some uncertainty given model assumptions, the transition between transport-dominated dust grains and collisionally-dominated dust grains in the Kuiper Belt occurs near 3 μm. Note also that Solar wind sputtering and ISD collisional erosion are much longer processes than Poynting-Robertson drag transport and interplanetary grain-grain collisions. The grain radius at which the collisional lifetime (blue) is equal to the approximate Poynting-Robertson drag lifetime in the absence of planets (solid line), termed the "critical radius", occurs near ≈ 7 μm, similar to that found by Kuchner and Stark (2010) in their dynamical, collisionally groomed model of the EKB dust disk. Strictly speaking, Kuchner and Stark (2010) found a critical radius for the EKB near 15 μm. However, given various model assumptions, the agreement between Kuchner and Stark (2010) and Poppe (2016) is good.

7 Meteorites

7.1 Overview

Meteorites are the surviving pieces of a meteoroid or asteroid passing through the atmosphere. Most of them come from chondritic bodies, primitive asteroids formed in the early Solar system which never reached the melting limit temperature in their interiors. Because of this, they have preserved chemical clues of the early Solar system (see e.g. Trieloff et al. 2003; Scott and Krot 2003) which can be directly measured in ground-based laboratories. Figure 11 shows a graphical representation of the classification of these objects, whose relevant properties will be addressed in detail in this section. Besides chondrites, stony achondrites, stony-iron, and iron meteorites exist. However, these comprise only about 7% of our meteorite collection, and are therefore not discussed here.

When linking meteorite properties to those of meteoroids, selection effects need to be taken into account. Only meteoroids larger than typically tens of centimeters in size will survive the passage through the atmosphere; smaller objects will completely disintegrate and not reach the ground. The material properties will also play a role - very fragile material will break apart more easily in the atmosphere than compact material. Only under rare circumstances, e.g. very shallow entry angles, can fragile material be recovered on the ground, as has happened for the Tagish Lake meteorite fall (Brown et al. 2000).

'Chondritic' means that they contain chondrules, about 1 mm-sized objects of glass and crystalline silicates formed by melting followed by rapid cooling. They also contain Ca- and Al-rich inclusions (CAIs) formed by condensation of high-temperature nebular gas. CAIs are composed mainly by refractory oxides and silicates. Other components in chondrites

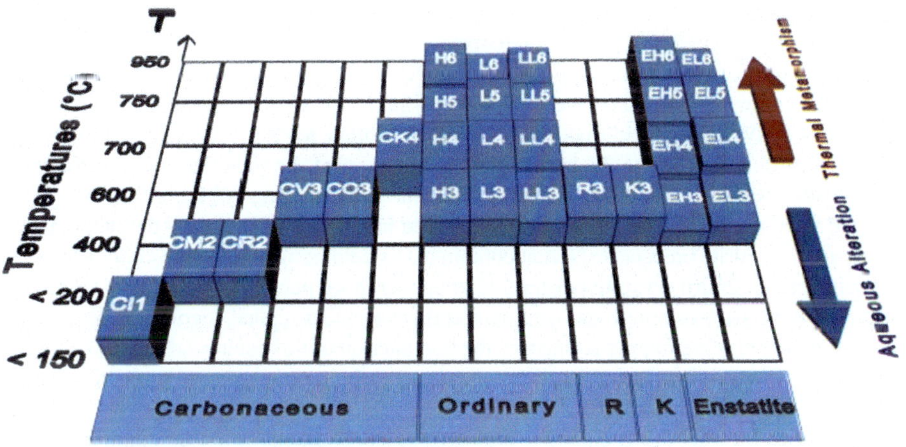

Fig. 11 Classification of chondritic meteorites. This basic diagram is useful for a general understanding of the key processes modifying asteroidal compositions over time. They are grouped in classes (carbonaceus, ordinary, R, K, and enstatite). The letters indicate so-called groups. The number indicates the thermal metamorphic grade (image E. Dotto)

are fine-grained matrix material (Brearley and Jones 1998) and a very small amount of tiny particles formed in other stars (Huss 1997; Messenger et al. 2003; Zinner 2003).

The classification is done based on mineralogical and petrologic characteristics, as well as chemical and isotopic compositions, see Fig. 11. The first letter in the figure denotes the class, the second a subclass. For carbonaceous meteorites, the second letter is based on the name of the first meteorite found in that subclass. L, H, and LL stand for low iron, medium iron, and low iron and low metal, respectively. 15 subclasses are known (Weisberg et al. 2006), 13 of which are shown in the figure.

7.2 Petrologic Types

Within the chondritic group, six different so-called thermal metamorphic grades or petrologic types have been defined, depending on their thermal or aqueous processing. These are marked by numbers in Fig. 11. Types 1 and 2 denote meteorites affected by aqueous alteration at low temperatures (Trigo-Rodríguez 2015). These groups come from water-bearing early-formed small asteroids, or larger asteroids formed later that never suffered significant radioactive heating since most of the nuclides responsible for heating were already gone. Type 3 is represented by chondrites that experienced little aqueous alteration and metamorphism, so they can be considered the most pristine chondrites. Thermal metamorphism increasingly affected petrologic types 3 to 6, type 6 being the most affected by in-situ parent body heating and shock-induced melting. A comprehensive review on the thermal metamorphism experienced by all chondrite groups has been published by Huss et al. (2006). In summary, the degree of thermal metamorphism and aqueous alteration is highly variable among the different chondrite groups as a consequence of different degrees of processing on the parent body.

7.3 Evolution of Meteorites

After formation of the chondritic bodies through accretion of planetesimals in the early Solar system, most of them soon became exposed to impacts with other bodies (Beitz et al. 2016). After being bombarded over 4.5 Gyr, these bodies exhibit impact-excavated structures, and have suffered significant erosion of their surfaces.

Progressive collisional gardening excavates craters and produces dust and meteoroids with enough energy to escape the gravitational field of their parent bodies. These erosive processes deliver chondritic material to Earth and other terrestrial planets. The observed chemical differences in the 15 known subclasses support the idea that each of them was formed in a different reservoir, probably from the separate rings observed in disks found around young stars (Zhang et al. 2015). This idea has been reinforced by the fact that relatively few chondrite breccias exist that contain clasts belonging to different chondrite groups (Bischoff et al. 2006). Primitive asteroids are covered by pebbles produced by continuous impacts that have excavated and fragmented their surfaces, called regolith. Consequently, chondrite breccias are mainly formed by the compaction of this asteroidal regolith under the action of impacts, whereas at the typical encounter velocities the original projectile material would be preferentially vaporized.

Meteorite breccias have complex and diverse constituents, often containing different lithologies, as measured e.g. in the Almahatta Sitta meteorite (Bischoff et al. 2010). This provides evidence for the substantial mixing of material in the early Solar system, as predicted by dynamical models as a consequence of the migration of the giant planets (see e.g. DeMeo and Carry 2014).

7.4 Ordinary Chondrites

It is estimated that about 500 meteorites reach the surface of our planet every year, most as small rocks with a mass of few grams (Halliday et al. 1996). Nowadays, meter-sized ordinary chondritic meteoroids (hereafter OCs) dominate the current terrestrial flux, but their parent asteroids (S spectral type, Gaffey and McCord 1978) represent only about the 17% of the asteroids in the main belt. The reason for OCs being the most common meteorites reaching our planet is because the composition is dominant for asteroids located between the inner belt and the near-Earth region (Gradie and Tedesco 1982). In addition, most asteroids are covered by regolith grains that have been exposed to space weathering over the eons, significantly modifying their reflectance properties and mineralogy. This was first measured by the Hayabusa sample return mission from asteroid 25143 Itokawa, (e.g. Nakamura et al. 2011b).

From the study of almost fifty thousand OCs available in meteorite collections,[3] we know that they have experienced different degrees of thermal metamorphism. The reason is that they come from moderately large chondritic asteroids that were fractured by large impacts, ejecting chondritic materials that were buried at different depths in their interiors. It is thus not easy to find pristine chondrites among the different subclasses of ordinary chondrites (H, L, and LL). Only a few examples exist, e.g. LL3.0 Semarkona, and LL3.1 Bishunpur (Mostefaoui et al. 2004). These few well-preserved rocks contain highly unequilibrated material, whose study reveals that most chemically homogenized OCs have suffered significant mineralogical and chemical modifications. This is associated with heating induced by radiogenic decay, and shock-induced annealing from impacts. The heat associated with these processes was cementing OCs together, i.e. they chemically reacted to produce asteroids with higher strength, less bulk porosity and higher bulk density (Trigo-Rodriguez and Blum 2009; Moyano-Cambero et al. 2017). The collisional products of such asteroids have been proved to be different of those coming from fragile aggregates (Blum et al. 2006).

Chondrite materials that have experienced thermal processing over the eons become progressively "cooked" and compacted (Huss et al. 2003; Trigo-Rodríguez and Blum 2008). These properties are important because the collisional products of these bodies depend on the processes having occurred on asteroids (Flynn and Durda 2004). In fact, laboratory experiments to produce aggregates as proxies of chondrites have demonstrated that the parent bodies of chondrites accreted as highly porous bodies (Blum et al. 2006). Subsequent large impacts collapsed pore spaces and produced shock heating, resulting in fragmented and compacted objects (Beitz et al. 2016). This thermal processing homogenized the chemical composition of the chondritic components, resulting in different levels of asteroid compaction depending on the amount of volatiles which acted as a buffer for the impact-delivered heat (Trigo-Rodriguez and Blum 2009; Beitz et al. 2016).

7.5 Carbonaceous Chondrites

Carbonaceous chondrites (CCs) are associated with bodies formed around or outwards the protoplanetary disk snow line (Brearley and Jones 1998). Their main parent asteroids are called C-type, constituting about 80% of the main belt asteroids. Their spectra indicate a relatively high amount of carbon compounds, that make them extremely dark objects (Fornasier et al. 1999). These asteroids are located in the outer half of the asteroid belt, and their surfaces have very low visible albedos of 0.03–0.09 (Jewitt et al. 2007; Trigo-Rodríguez

[3]Meteoritical Bull. Database: https://www.lpi.usra.edu/meteor/metbull.php.

et al. 2014). The asteroids have low densities, similar to the CCs they produce, ranging from about 2 $g\,cm^{-3}$ for the ones with a significant content of water and porosity, to 3.5 $g\,cm^{-3}$ for the driest and Fe-rich ones (Britt and Consolmagno 2000). Among these CCs, the CI and CM groups are associated with transitional objects with significant abundance of water, mostly as hydrated minerals that could have exhibited cometary activity at some point of their evolution (Trigo-Rodríguez 2015). They experienced extensive aqueous alteration, and many of them escaped thermal metamorphism.

They contain phyllosilicates as major phases, serpentine in CM and sponite-serpentine mixed layers for CI chondrites. Some CM and CI chondrites were heated and dehydrates after aqueous alteration, as summarized in Nakamura et al. (2005). This indicates that a dehydration process is common in the hydrated C-type asteroids, as also suggested by reflectance spectra (Hiroi et al. 1993).

CO, CV, and CR chondrites suffered much less severe aqueous alteration, but some CVs and CRs are certainly moderately aqueously altered as well. Thermal metamorphic grades for these groups are ranging from low (3.0) to nearly type 4 (Scott and Jones 1990; Russell et al. 1998), except for the CV and CK groups that can have members with higher values. In general, carbonaceous chondrites can be considered "pristine" as they retain a bulk chemistry close to Solar abundance. But only a few meteorites retain highly unequilibrated mineral assemblages relatively unaffected by thermal and aqueous modifications. The most primitive CCs identified so far are the CO chondrite ALHA77307, the CM2 chondrite Yamato791198, and the CM-like un-grouped chondrite Acfer094 (Rubin et al. 2007).

The degree of metamorphism in chondrites reflects thermal effects associated with burial due to self-gravity, or with the propagation of heat in shock waves generated in collisions (Scott and Krot 2003). Chondritic materials have then experienced such thermal processing over the eons, and become progressively "cooked" and compacted (Huss et al. 2003; Trigo-Rodriguez and Blum 2009). These properties are important as the meteorites are the collisional products of asteroids on which these processes have occurred (Flynn and Durda 2004), thus letting us constrain the asteroid's evolution. Laboratory experiments to produce aggregates as proxies of chondrites have demonstrated that the parent bodies of chondrites accreted as highly porous bodies, see Blum et al. (2006). Large impacts contributed to collapse pore spaces and produced shock heating, so most asteroids are fragmented and compacted objects (Beitz et al. 2016). This thermal processing homogenized the chemical composition of the chondritic components forming asteroids. They were compacted to different extents, depending on the amount of volatile material acting as a buffer of the impact-delivered heat (Trigo-Rodriguez and Blum 2009; Beitz et al. 2016).

7.6 Other Classes

Other classes are less relevant here, but mentioned briefly for completeness: Rumuruti (R) chondrites are a new chondrite group formed mostly by brecciated meteorites with chemical patterns that differ from carbonaceous, ordinary, and enstatite chondrites. The Kakangari (K) chondrite group have a similar set of petrologic and oxygen-isotopic characteristics that distinguishes them from other chondrite groups. Enstatite chondrites are chemically reduced meteorites, with most of their iron taking the form of metal or sulfide rather than an oxide, and being mostly formed by the mineral enstatite (MgSiO3), from which they derive their name.

7.7 Summary

Summarizing, as a direct result of the collisional processing of chondritic bodies which dominates the population of asteroids and possibly comets, most of the interplanetary space contains rocks and particles of chondritic nature. Fragile, water-bearing bodies such as C- and D-type asteroids and comets are the main contributors to the interplanetary dust and meteoroid cloud, see e.g. Engrand and Maurette (1998). As a direct consequence, most meteoroids ablate in the terrestrial atmosphere producing a luminous phase that usually exhibits chondritic bulk elemental chemistry (Trieloff et al. 2003; Trigo-Rodríguez et al. 2004). Only recently, researchers started to directly study the ablation of chondrites in the laboratory. This allows obtaining new clues about the delivery of chondritic materials to Earth by simulating the factors regulating the ablation of meteoroids in the Earth's atmosphere (see e.g. Gómez Martín JC et al. 2017). Chondritic material is also important from an astrobiological perspective. Its rock-forming minerals, under aqueous alteration and in presence of Nitrogen-bearing species, are able to catalyze extremely diverse organic species (Rotelli et al. 2016). These findings are relevant as they show a natural way of increasing the chemical complexity as needed to explain the ubiquity of life in planetary bodies (Trigo-Rodríguez et al. 2017).

8 Dust Particle Collection

8.1 Introduction

The interplanetary dust environment in Earth orbit was investigated by the deployment of dust collectors on the Gemini 10 mission, the Long Duration Exposure Facility (LDEF), the MIR Space Station, and the Space Shuttle, as well as by the examination of space-exposed surfaces retrieved during the repair of the Solar Maximum satellite and the Hubble Space Telescope. Elemental analyses of the residue left in impact craters from the LDEF experiment identified three major classes of extraterrestrial materials, particles with roughly chondritic abundances of Fe, Mg, and Si suggestive of fine-grained aggregates, mono-mineralic silicates, and Ni-Fe-sulfides (Horz and Bernhard 1992), as well as a population of man-made particles.

The LDEF experiment yielded a measure of the meteoroid mass flux at the top of the Earth's atmosphere. This flux is dominated by particles around 250 μm diameter, and accounts for $30\,000 \pm 20\,000$ tons/year (from Love and Brownlee 1993). This is ~ 1000 times the contemporary flux of meteorites (Bland et al. 1996).

As discussed previously, interplanetary dust particles can enter the terrestrial atmosphere with velocities varying between $\sim 11\ \mathrm{km\,s^{-1}}$ to $\sim 73\ \mathrm{km\,s^{-1}}$, depending on the origin of the particles. As early as 1937, Opik recognized that small particles entering the Earth's atmosphere could radiate away heat as rapidly as it was being regenerated by friction in the atmosphere, allowing them to survive entry. Detailed modeling by Whipple (1950) indicated that particles around 10 μm in size could survive atmospheric entry without severe heating, particularly those particles having the lowest bulk densities and entry speeds. Although small particles at the surface of the Earth are dominated by local dust, the anticipated survival of interplanetary dust motivated high-altitude collection efforts by rockets, balloons, and eventually aircraft. For larger particles ($\gtrsim 25$ μm), the high speed impact with the atmosphere causes the evaporation of a large fraction of the incoming dust, but $\sim 20\%$ of them reach the terrestrial surface (Taylor et al. 1998; Duprat et al. 2006; Plane 2012) and can be collected on Earth, either in the stratosphere (Brownlee 1985), in the polar caps from ice/snow (Maurette et al. 1987, 1991; Yada and Kojima 2000; Nakamura et al. 2001; Duprat et al. 2007) or by being trapped in ice in the transantarctic mountains (Rochette et al. 2008). Dust particles reaching the Earth from space deposit on the surface to form sediments, at a rate as slow as 1 particle m^{-2} day^{-1} (Brownlee et al. 2006). The effects of the vaporized material on the atmosphere are discussed in Plane et al. (2018).

For historical reasons, particles collected in the stratosphere are called "Interplanetary Dust Particles" (IDPs). Their sizes range typically from a few μm to $\sim$ 50 μm. Particles collected in polar caps are named "micrometeorites" (MMs), and have larger sizes, from $\sim$ 25 μm to $\sim$ 1 mm. The difference in the size ranges between IDPs and MMs is due to the collection mode, and these two collection techniques are complementary.

The stratospheric collection of IDPs is inherently biased. Particles smaller than a micron or so in diameter are generally trapped in the air stream, and flow past the collector, while the abundance of particles > 50 μm in diameter is so low that few are ever collected. After flight, particles are removed individually from the collectors and examined optically and by electron beam instruments, allowing them to be characterized as "cosmic", "possibly cosmic", "natural terrestrial", or "man-made terrestrial" material. Even at the collection altitude, a majority of the particles are terrestrial material, either naturally occurring dust or man-made particles such as rocket exhaust. Due to their very low bulk densities, the more porous particles settle more slowly, so they are preferentially collected at these altitudes. Further, all Earth collections are biased by gravitational focusing, which significantly enhances the flux of particles with the lowest geocentric velocities.

IDPs and MMs collected on Earth originates both from asteroids and comets. Recognizing their origin requires comparing their characteristics to those of meteorites and comets. So far two families of cosmic dust of probable cometary origin have been proposed both in the IDP and micrometeorite collections. The abundance of icy and carbon-rich asteroids in the asteroid belt has recently been revisited (DeMeo and Carry 2014) and they are underrepresented in the meteorite collections, possibly due to preferential destruction of carbonaceous chondrites on Earth, during atmospheric entry or by weathering on Earth. The cosmic dust collected on Earth could be more representative of the interplanetary matter than meteorites.

8.2 Interplanetary Dust Particles (IDPs) Collected in the stratosphere

NASA has been collecting IDPs from the Earth's stratosphere since 1981. U2, ER2 and WB-57 aircrafts collect IDPs at altitudes between 17–20 km, with flight durations from a few hours to 65 hours (Brownlee 1985). Collector plates covered with a layer of high viscosity silicon oil, which needs later to be rinsed from the particles, are deployed to collect particles from the stratosphere by impact.

Due to their small sizes, IDPs, like the ones shown in Fig. 12, usually do not suffer from extreme atmospheric entry heating effects. Ceplecha et al. (1998) give a size limit of roughly 10 μm; the precise values depend on the velocity. IDPs are classified according to their composition into two main categories: chondritic IDPs, with a bulk chemical composition compatible with the average chondrite composition; and non-chondritic IDPs, e.g. large single crystals of silicates, Fe-sulphide minerals, or refractory minerals (e.g. McKeegan 1987a; Zolensky 1987). Chondritic IDPs are in turn classified according to their mineralogy, i.e. olivine-, pyroxene-, or layer-lattice-silicate-dominated, or by their textures: (i) chondritic porous, abbreviated CP; (ii) chondritic filled, less porous than CP IDPs; and (iii) chondritic smooth IDPs, abbreviated CS (e.g. Bradley 2005; Flynn et al. 2016).

CP IDPs are unequilibrated aggregates of mostly sub-micron grains of diverse compositions, sampling a wide variety of formation conditions. Most CP IDPs have never experienced any significant hydrous or thermal parent body processing, gravitational compaction, or impact shock, which have affected most meteorites, and many were minimally heated during atmospheric deceleration. They have a chondritic composition for the major rock-forming elements (Fig. 13). These CP IDPs consist mainly of anhydrous crystalline phases,

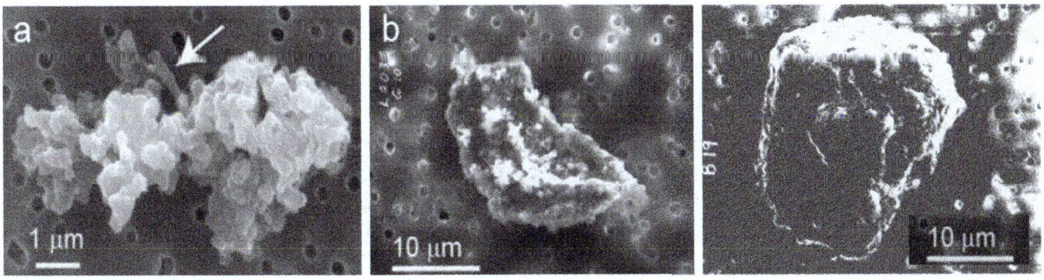

Fig. 12 Typical IDPs include: (left) anhydrous aggregate CP-IDPs, (center) single mineral IDPs, and (right) compact, hydrous IDPs. The arrow in the frame points to a pyroxene whisker

Fig. 13 Si- and CI-normalized elemental abundance patterns in representative samples of IDPs, porous (CP) IDPs, smooth IDPs, micrometeorites, and Larger Cluster IDPs. The IDPs are collected in silicone oil, which is never completely removed by hexane washing, so the IDP data is typically Fe-normalized. The Si-normalized values shown were generated assuming a Si/Fe ratio as in CI meteorites (adapted from Flynn et al. 2016)

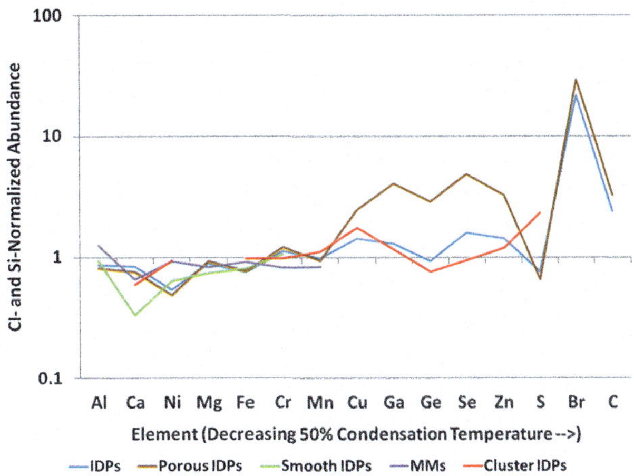

mostly Mg-rich olivines, pyroxenes and low-Ni Fe sulphides (Bradley 2005), as well as primitive glassy components called Glass Embedded with Metals and Sulphides (GEMS) that may be of pre-Solar interstellar origin (Bradley 1994; Klöck and Stadermann 1994; Bradley et al. 1999). These highly unequilibrated IDPs, containing individual olivine and pyroxene grains, span a wide range of Fe contents, with Mg/(Mg + Fe) varying from nearly 100 to ~ 50 mol%. The mineralogy of CP IDPs is compatible with that of comets Halley and Hale-Bopp (e.g. Hanner and Zolensky 2010, and references therein).

Detailed studies of the CP IDPs indicate that they are the most primitive materials of the Solar system currently available for laboratory study (Ishii et al. 2008), providing the best available samples of the materials present in the Solar protoplanetary disk. Low-Fe, Mn-enriched olivines, commonly observed in CP IDPs and sometimes found in unequilibrated ordinary chondrite meteorites, as well as exotic whisker (Fig. 12) and platelet morphologies of the crystalline silicate enstatite found in CP IDPs have been proposed to form directly as a high temperature condensates from the disk (Ishii et al. 2008, and references therein). A comparative study of the sizes of the crystalline Mg-rich silicate and the Fe-sulfides, the two most abundant phases in CP IDPs, shows that the mean constituent grain size varies from one CP IDP to another. They have a size-density relationship that can be explained with an aerodynamic-type sorting mechanism operating in the disk prior to grain aggregation (Wozniakiewicz 2012). In many CP IDPs the individual mineral grains are coated with a thin (~ 100 nm) layer of organic carbon (Flynn et al. 2013), consistent with modeling by Ciesla and Sandford (2012) showing that ultraviolet and thermal processing of ice coated grains in the cold, outer Solar nebula can produce complex organic molecules. These organic rims likely aided in the grain aggregation in the Solar nebula (Flynn et al. 2013), since bare crystalline grains can stick together only in very slow collisions.

The composition of hydrated IDPs is broadly chondritic, with a depletion in Ca (Fig. 12), likely related to an aqueous alteration event on their parent body (Schramm et al. 1989). The mineralogy of hydrated IDPs is dominated by phyllosilicates and carbonates, and also contains Ni-rich Fe sulphides like pentlandite. The phyllosilicates are usually smectite, possibly linking some of them to CI chondrites (Keller et al. 1992), although most hydrated IDPs still contain anhydrous parts, with is not the case for CI chondrites that have been fully aqueously altered.

Both the CP IDPs and the CS IDPs are significantly enriched in C over the most carbon-rich meteorites, with C contents ranging from ~ 5 to ~ 45 wt% (Keller et al. 1994).

Large isotopic anomalies interpreted as pre-Solar interstellar signatures have been found in IDPs, both in the organic matter and in the mineral components, including GEMS and silicates (Messenger 2000; Messenger et al. 2003; Floss and Stadermann 2004). Oxygen, which is the most abundant element in chondritic materials, varies with heliocentric distance within the Solar system. Isotopic imaging of IDPs demonstrates that some sub-micron grains have O isotopic ratios outside the range seen in Solar system materials, suggesting these grains formed in outflows or explosions of previous generations of stars and survived the incorporation into the Solar system without equilibration (Messenger et al. 2003). Some of these IDPs also exhibit elevated D/H ratios (Zinner et al. 1983; McKeegan et al. 1985; McKeegan 1987b; Messenger 2000).

It has been proposed that a large proportion (if not all) of CP IDPs could be of cometary origin (e.g. Bradley and Brownlee 1986; Ishii et al. 2008, and references therein). Vernazza et al. (2015) proposed that anhydrous IDPs could originate from icy asteroids.

8.3 Micrometeorites Collected in the Polar Caps

Micrometeorites ranging from ~ 25 μm to ~ 500 μm have been most efficiently collected from ice or snow in the polar caps: in Greenland (Maurette et al. 1986, 1987) and in Antarctica (Maurette et al. 1991; Taylor et al. 1998, 2000; Yada and Kojima 2000; Nakamura et al. 2001; Duprat et al. 2007). Larger micrometeorites up to ~ 1 mm with older terrestrial ages in the order of Myr have been collected at the surface of the transantarctic mountains (Rochette et al. 2008).

Since MMs are larger than IDPs, they suffer more extensively from atmospheric entry effects. They are first classified according to the textures resulting from atmospheric entry heating (Genge et al. 2008):

- unmelted particles comprising fine-grained particles (Noguchi et al. 2017);
- partially melted scoriaceous particles;
- and fully melted cosmic spherules (CS), see Fig. 15 and Duprat et al. (2007).

In terms of composition, two main families are found in the MM collections: the chondritic particles ($\sim 99\%$ of the particles, both anhydrous and hydrated), and the ultracarbonaceous particles that have been identified in the last decade (Nakamura et al. 2005; Duprat et al. 2010), see Fig. 14.

Chondritic micrometeorites $< \sim 500$ μm are mostly related to the carbonaceous chondrites (Kurat et al. 1994; Engrand and Maurette 1998; Herzog et al. 1999; Engrand et al. 2005; Genge 2008; Genge et al. 2008; Dobrică et al. 2009), whereas larger micrometeorites show more affinity with ordinary chondrites, which dominate the meteorite collections (van

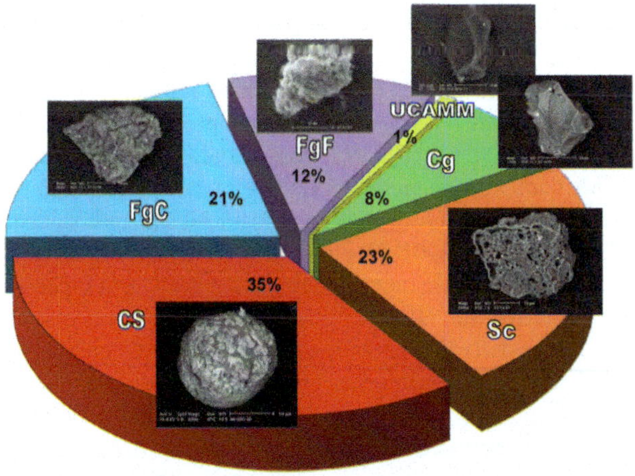

Fig. 14 Proportion and types of MMs from the Concordia collection (updated from Dobrică et al. 2010)

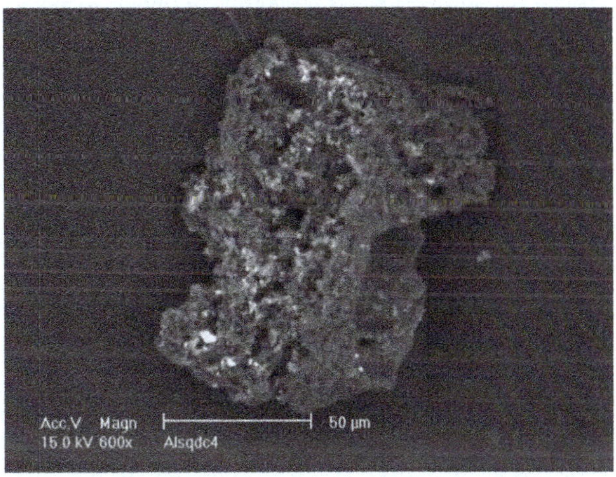

Fig. 15 Backscattered electron micrograph of Concordia Ultracarbonaceous Antarctic Micrometeorite (UCAMM) DC02-11-19. All dark gray patches are constituted of organic matter. Brighter flakes are silicates or Fe-Ni sulphides

Ginneken et al. 2017). The chondritic MMs exhibit a composition for major and minor elements that is comparable within a factor of 2 to the bulk CI composition (Dobrică et al. 2009). The mineralogy of anhydrous MMs is dominated by crystalline silicates, mainly olivine and low-Ca pyroxene, followed by Fe–Ni sulfides in abundance. The olivine compositions range from almost pure forsterite (Mg-olivine) to relatively high Fe-olivine, up to Fe/Mg about 1. The low-Ca pyroxenes show a similar range of Fe contents (van Ginneken et al. 2017) The mineralogical compositions, in particular the composition of the olivines, show that the constituents of cosmic dust have sampled a wide range of conditions in the protoplanetary disk.

The number ratio of pyroxene versus olivine in micrometeorites is about 1, while it is on the order of 0.2 in CM chondrites, which are the closest relatives to MMs in the meteorite collections (Kurat et al. 1994; Engrand and Maurette 1998). The CR chondrites are the only carbonaceous chondrites with a high abundance of pyroxenes relative to olivines. Since Fe-Ni sulfides are ubiquitous, they can be used for comparison between families of extraterrestrial objects. In MMs, most Fe-Ni sulfides show a composition close to that of troilite, containing less than 3 wt% of Ni. Pentlandite is only rarely detected, mainly in scoriaceous MMs (Engrand et al. 2007). Refractory minerals (spinel, anorthite, melilite, hibonite, Ca-Al-rich pyroxene, and perovskite) or refractory inclusions have been reported in a few dozen of micrometeorites (e.g. Hoppe et al. 1995; Engrand et al. 1999c; Genge et al. 2008). The relative abundance of MMs containing refractory minerals or inclusions is less than 1%. Chondrules are also rare in MMs, but present in the coarse-grained population of micrometeorites (Walter et al. 1995; Kurat et al. 1996; Genge et al. 2005; Genge 2008). Hydrated MMs contain saponite or serpentine in association with organic matter, that show similarities but also differences with phyllosilicates in carbonaceous chondrites (Genge et al. 1997; Nakamura et al. 2001; Noguchi et al. 2002; Battandier et al. 2016; Noguchi et al. 2017).

The hydrogen isotopic composition of hydrous phases in micrometeorites is centered on the terrestrial value (Engrand et al. 1999a), suggesting that MMs could have played a role in the delivery of water to Earth (Maurette et al. 2000). Fossil micrometeorites trapped as clasts in HED meteorites (HED = howardite, eucrite, diogenite, thought to come from asteroid (4) Vesta) also bear the same H isotopes (Gounelle et al. 2005a). The oxygen isotopic composition of anhydrous minerals in MMs < 500 μm show similarities with that of carbonaceous chondrites (Engrand et al. 1999b). For larger micrometeorites—melted into cosmic spherules—a larger proportion of ordinary chondrite-related particles is found (van Ginneken et al. 2017).

The carbon content of MMs (except for UCAMMs) has a range between 0.2 to 2.8 wt% (Maurette et al. 2000; Matrajt et al. 2003). The MMs exhibit Raman signatures of the first-order carbon bands G and D, suggesting the presence of polyaromatic carbonaceous matter. The width and position of these bands, as well as high-resolution Transmission Electron Microscope imaging, show a very disordered organic matter that did not experience thermal metamorphism (Dobrică et al. 2011). MMs also contain nanoglobules comparable to those observed in primitive meteorites and Wild 2 samples (Maurette et al. 1995; Nakamura-Messenger et al. 2006; de Gregorio BT et al. 2011). MMs also contain complex organics in the form of PAHs (Clemett et al. 1998) and amino acids (Brinton et al. 1998; Matrajt et al. 2004).

The origin of chondritic micrometeorites—asteroidal or cometary—is still debated (e.g. Maurette 2006; Genge 2008; Genge et al. 2008; Dobrică et al. 2009, and references therein). Models predict that a large fraction of dust on Earth could originate from Jupiter-family comets (e.g. Nesvorný et al. 2010; Poppe 2016).

UCAMMs are most probably of cometary origin. They contain up to 85% of organic matter in volume, thus showing carbon contents up to 10 times that of the most C-rich carbonaceous chondrites (Kerridge 1985; Grady and Pillinger 1986; Pizzarello et al. 2001). Such concentrations of carbonaceous matter are comparable with that of the most C-rich IDPs (Thomas et al. 1993; Keller et al. 2004), and bear similarities with CHON grains detected in comet Halley by the Giotto and Vega spatial missions in 1986 showing the association of dominant organic matter with a minor rocky component (Kissel and Krueger 1987; Lawler and Brownlee 1992), and with cometary grains from 67P/Churyumov-Gerasimenko (Fray et al. 2016; Bardyn et al. 2017; Fray et al. 2017).

The hydrogen isotopic composition of the UCAMM organic matter shows extreme deuterium enrichments (up to 30 times the terrestrial value—i.e. 300 times the proto-Solar value, Duprat et al. 2010). Their organic matter is a disordered polyaromatic carbonaceous phase, and contains at least two organic phases with different nitrogen contents. The bulk N concentration of their organic matter reaches 20 wt%, that is about 5 times that of insoluble organic matter (IOM) extracted from meteorites and the O/C ratio of UCAMM organic matter is lower than that of meteoritic IOM (Dartois et al. 2013; Engrand et al. 2015; Charon et al. 2017; Dartois et al. 2018). UCAMMs contain both low and high temperature components, and their micro-structure is intriguing: unequilibrated mineral assemblages with individual submicrometer sizes are enclosed in the organic matter (Dobrică et al. 2012). The formation of the UCAMMs nitrogen-rich organic matter may result from galactic cosmic ray irradiation of N2- and CH4-rich surfaces of icy bodies in the outer regions of the Solar system (Dartois et al. 2013; Augé et al. 2016).

The minerals in UCAMMs cover a size range between 50 nm and several hundreds of nanometers (Dobrică et al. 2012). The minerals are usually associated to the organic matter that does not show very high N contents (Engrand et al. 2015; Charon et al. 2017). Their mineralogy mainly consist of Mg-rich olivines and pyroxenes, Fe-Ni sulphides, and GEMS. The pyroxene to olivine abundance ratio in UCAMMs is on average larger than 1. In addition, UCAMMs contain a significant abundance of silica-rich glass. Accessory mineral phases such as Mg-Al spinel, sphalerite, niningerite, perryite and Cr-rich pyroxenes are also observed (Dobrică et al. 2012). So far, no carbonate nor phyllosilicate minerals have been found in UCAMMs. The analysis of the hydrogen and nitrogen isotopic composition of organic matter in UCAMMs from the Concordia collection does not show clear spatial correlation between the D-rich and 15N-rich phases (Duprat et al. 2014; Bardin et al. 2015).

The unusually large concentration of organics in UCAMMs is of utmost interest for the early Earth prebiotic matter input.

8.4 Interplanetary Dust Versus Meteorites as Representatives of Interplanetary Matter

One of the most striking results of micrometeorite research consists in their relationship to carbonaceous chondrites that represent only 4% of the meteorites (e.g. Kurat et al. 1994; Engrand and Maurette 1998; Engrand et al. 1999c). In other words, 96% of all meteorite types are not represented in the incoming cosmic dust flux. Based on the following two points, it is suggested that collected dust particles are better representatives of cosmic dust and meteoroids than meteorites:

- The dynamical evolution of grains resulting in IDPs and MMs is less biased than that of meteorites. It is dominated by non-gravitational forces like Poynting-Robertson drag (Burns et al. 1979), whereas meteorites are delivered to Earth as a result of interactions with resonances, into which they are brought by the Yarkowsky effect (Vokrouhlický and Farinella 2000).
- The high proportion of carbonaceous chondrite-like objects (CC) in the collected IDPs better agrees with the proportions of C-type versus S-type asteroids than those of meteorite collections: More than 80% of the IDPs are carbonaceous chondrites (e.g. Kurat et al. 1994; Engrand and Maurette 1998; Engrand et al. 1999c), but only 5% of the meteorite collection is of this type, whereas > 55% of the asteroids are C-type, < 20% are S-type. The meteorite collections could in fact under-represent the CCs with regard to ordinary chondrites linked to S-type asteroids (Fujiwara et al. 2006; Buczkowski et al. 2008; Lim and Nittler 2009; Nakamura et al. 2011a), as these fragile CC meteorites may be preferentially destroyed during atmospheric entry and by weathering at the Earth's surface before recovery (e.g. Meibom and Clark 1999).

Analyzing the composition of micrometeorites in old ("fossile") rocks gives evidence that the composition did not change significantly over time (Gounelle et al. 2003; Engrand et al. 2005; Gounelle et al. 2005b) or toward smaller sizes (Gounelle et al. 2005b; Aléon et al. 2009).

This view was recently challenged by new dynamical models that favor a preferential formation of large amounts of dust from Jupiter-family comets that could explain the high abundance of carbonaceous chondrite-like objects in the micrometeorite collections (Bottke et al. 2008; Nesvorný et al. 2010). This work would thus imply that most—if not all—micrometeorites with sizes < 500 μm are of cometary origin, and thus that the mineral components of cometary material are close to that of carbonaceous chondrites (Dobrică et al. 2009). This is a result also suggested by analyses of Stardust samples (e.g. Brownlee et al. 2006).

Larger micrometeorites (> 500 μm) collected in the transantarctic mountains (Rochette et al. 2008) seem to form a population that is intermediary between micrometeorites with sizes < 500 μm and meteorites, as large amounts of ordinary chondritic material ($\sim 30\%$, Suavet et al. 2010) and objects from differentiated bodies (Cordier et al. 2011) are present in this collection. In the following we present the limits of the classical distinction between inner and outer regions of the protoplanetary disk.

Comets contain Mg-rich silicates, olivines and pyroxenes, identified by infrared spectroscopy (e.g. Wooden et al. 2000). Stardust samples showed that Wild 2 cometary dust also contain very high-temperature phases like CAIs and chondrules (Brownlee et al. 2006). More than 50% of Wild 2 material recovered by Stardust was processed in the inner regions of the Solar system (Ogliore et al. 2009; Westphal et al. 2009). This corresponds mainly to the fraction > 2 μm, as the finer matrix was mostly destroyed during the impact with the collection device. The presence of high-temperature phases (formed or severely processed close to the protosun) in the comet-forming regions requires large-scale radial mixing mechanism(s) in the early Solar system.

Since both CAIs and chondrules have been found in Stardust samples, this mechanism(s) had to operate for at least a few million years, as chondrules formed a few million years after CAIs (Brownlee et al. 2006; Zolensky et al. 2006; Nakamura et al. 2008). Several models have been proposed to explain this radial transport, by viscous diffusion by turbulence (Bockelée-Morvan et al. 2002; Ciesla 2007), by jets (Shu et al. 1997), or radiation pressure processes (Vinković 2009). Other evidences for transport in the protoplanetary disk are given by the oxygen isotopic composition of high temperature phases (e.g. Simon et al. 2011), as well as by the hydrogen isotopic composition of low temperature phases (Remusat et al. 2010). Comet Wild 2 present similarities with primitive meteorite matter (carbonaceous chondrites), strongly supporting the hypothesis of a continuum between primitive asteroids and cometary matter. In fact, the outer asteroid belt contains objects that show cometary activity (Bertini 2011), reinforcing the concept of this continuum. Cosmic dust particles, in the form of micrometeorites (Duprat et al. 2010) and primitive IDPs (Ishii et al. 2008) are the preferred objects to study the primitive interplanetary matter, and this continuum between primitive asteroids and comets.

9 Summary and Outlook

This paper summarizes our knowledge of the interplanetary meteoroid and dust cloud. It describes how their properties are constrained by ground- and space-based observations of the zodiacal light, by observing meteors entering the Earth's atmosphere, by in-situ collection in space, in the upper atmosphere, and on the ground by analyzing meteorites.

Interplanetary meteoroids (larger than ~ 30 μm) and dust particles (smaller than ~ 30 μm) are continuously produced by the outgassing of comets and asteroids, and by collisional processes. Depending on the particle size, different things happen: smaller objects decouple from their parent object quickly and are dispersed by Solar radiation pressure.

Larger meteoroids stay close to the orbit of their parent body, forming so-called meteoroid streams. When entering the Earth's atmosphere, they can be observed as meteor showers.

In particular non-gravitational forces (radiation pressure, Poynting-Robertson drag, Lorentz forces) move the orbits of the particles away from their parent, until they form a background interplanetary cloud. The size distribution of the initially escaped particles is modified by collisions, disaggregation, and sublimation.

Detailed observations of the zodiacal light, including polarimetric and spectroscopic measurements, show that most dust particles in the zodiacal cloud present similarities with cometary dust and asteroid regoliths. Particles consist of a mix of amorphous carbon and silicate dust. Polarization measurements indicate that a significant fraction of the particles are irregular and fluffy aggregates.

A few critical open points remain: (a) the exact origin and physical properties of the dust particles contributing to the zodiacal cloud; the possible time evolution of the zodiacal cloud over decades; (b) the circular polarization that can probe the interplanetary magnetic fields structure; the dust evolution in the circum-Solar region; (c) how can we link our knowledge of the interplanetary dust, observed via zodiacal light, to that of exo-zodiacal dust clouds? Their structure could be quite different, taking into account the possible presence of "hot Jupiters" and variety in the orbital planes of exoplanets.

When meteoroids enter the Earth's atmosphere, they generate a streak of light, called a meteor. The emitted light intensity can be used to estimate the mass of the object; spectral observations constrain the chemical composition. The percentage of the kinetic energy converted to light as a function of wavelength is not yet well known, resulting in large uncertainties in these mass estimates. Still, the determined mass flux densities confirm the distributions obtained via other means, such as in-situ measurements or lunar secondary crater counts on returned lunar samples.

The directional distribution of meteors constrains the dynamical behavior of the meteoroid population. It can be seen that the sporadic background is not isotropic, even when taking observational biases into account. Some of these inhomogeneities can be attributed to comet 2P/Encke as a source, others to the Kozai effect. Many meteor showers, re-occurring on a yearly basis, are a result of meteoroids close to the orbit of their parent body. However, for about half of the meteor showers the identification of the related parent bodies has not yet been successful. Attempts to predict meteor showers have become better over the last few decades. While predicting the times of peak activity is possible with good accuracy, predicting the flux density is not. This would need better quantitative models of the ejection of meteoroids from their parent object.

While the observation of meteors with optical means and radar systems is a cost-effective way of studying meteoroids, a big open point is the percentage of kinetic energy converted into light, or radar reflection capability (luminous efficiency and ionization efficiency, respectively). These could be better constrained by combining light curve data with deceleration data, and by combining optical with radar observations. Also, laboratory experiments, properly simulating the typical velocities of meteoroids, have to be performed.

When larger objects penetrate our atmosphere, they can reach the ground as meteorites. However, they may be processed by their interaction with the atmosphere. Small dust can stay fairly unmodified, as it can re-radiate the imparted energy fully. Interplanetary dust can be collected via airplanes, balloons, or sounding rockets in the upper atmosphere. It can also be found in ice at the polar caps and in dust traps in the transantarctic mountains.

Most meteorites contain chondrules, mm-sized spheres of glass and crystalline silicates, formed by melting in the early Solar system. They also contain Ca-Al-rich inclusions, called CAIs, all held together by a fine-grained silicate matrix. 15 chondritic types have been identified, which presumably are linked to different source regions in the early Solar system.

Meteorite types can be linked to spectral classes of asteroids. The collected dust samples can also be classified into chondritic and non-chondritic groups. It is interesting to note that most collected dust grains are related to carbonaceous chondrites - which constitute only $\sim 5\%$ of all meteorites. This indicates that the collected dust in the upper atmosphere or arctic ice and snow could be more representative of interplanetary meteoroids than meteorites.

The flux density of the dust and meteoroid cloud can be described by dynamical models, like NASA's Meteoroid Environment Model (MEM) or ESA's Interplanetary Meteoroid Model (IMEM). However it is still difficult to explain all observational data consistently with one model. More work is needed in this area; possibly adding the meanwhile abundant optical data can help.

Space-based sensors could be used to both increase our observational data of the zodiacal light, or to look back to the Earth's atmosphere and allow continuous meteor and fireball observations without any constraints by clouds or by the Earth's atmosphere that absorbs certain wavelength ranges. So-called 'US government sensors' are providing data in particular on larger (> tens of cm) objects entering the Earth's atmosphere (https://cneos.jpl.nasa.gov/fireballs/). However, no public information is available on the properties of these sensors, making the data difficult to interpret. A dedicated sensor for scientific work could greatly enhance our dataset.

A recently emerging research field to better constrain the flux of meteoroids, in particular in the size range larger than a few cm, is the observation of Lunar impact flashes. NASA has been operating an observing system since many years (e.g. Suggs et al. 2014), and recently new systems went on-line in other countries (e.g. Xilouris et al. 2018). These observations would particularly benefit from space-based observations. A system close to the Moon would increase the sensitivity, and avoid being constrained by observing geometry or clouds (Topputo et al. 2018).

Meteorites could be a good proxy of the properties of interplanetary meteoroids. However, during their passage through the Earth's atmosphere they are modified. To better understand these modifications, ablation experiments in the laboratory—e.g. in wind tunnels or shock tubes, are necessary. Sample return missions from asteroids like OSIRIS-REx (Lauretta 2012), or Hayabusa-2 (Tsuda et al. 2013), or from comets, would allow to return unaltered material into ground-based laboratories.

Interplanetary dust may even be detectable by spacecraft not designed to detect them, and it may be possible to detect and characterize them by analyzing spacecraft attitude or guidance data, e.g. as proposed for the Lisa Pathfinder mission (Thorpe et al. 2016). The authors show that changes in the attitude of the spacecraft can be linked to micro-meteoroid impacts, thus allowing us to study the population in-situ without dedicated dust detectors.

While we already know a lot about interplanetary dust and meteoroids, we still need to know more details to obtain a better understanding of our Solar system and to improve our capabilities of shielding our spacecraft around the Earth or going to another planet.

Acknowledgements This paper was made possible by the International Space Science Institute, Bern, who brought the authors together for a whole week in 2016 and supported the meeting location and stay of the participants. All of the authors acknowledge their respective institutes and funding agencies for their support. In particular, JMTR thanks Spanish Ministry of Science and Innovation under research project AYA2015-67175-P. We acknowledge the hard work done by two referees whose comments were extremely valuable to improve this review.

References

J. Aléon, C. Engrand, L.A. Leshin, K.D. McKeegan, Oxygen isotopic composition of chondritic interplanetary dust particles: a genetic link between carbonaceous chondrites and comets. Geochim. Cosmochim. Acta **73**, 4558–4575 (2009). https://doi.org/10.1016/j.gca.2009.04.034

N. Altobelli, S. Kempf, M. Landgraf, R. Srama, V. Dikarev, H. Krüger, G. Moragas-Klostermeyer, E. Grün, Cassini between Venus and Earth: detection of interstellar dust. J. Geophys. Res. Space Phys. **108**, 8032 (2003). https://doi.org/10.1029/2003JA009874

N. Altobelli, H. Krüger, R. Moissl, M. Landgraf, E. Grün, Influence of wall impacts on the Ulysses dust detector on understanding the interstellar dust flux. Planet. Space Sci. **52**, 1287–1295 (2004). https://doi.org/10.1016/j.pss.2004.07.022

N. Altobelli, S. Kempf, H. Krüger, M. Landgraf, M. Roy, E. Grün, Interstellar dust flux measurements by the Galileo dust instrument between the orbits of Venus and Mars. J. Geophys. Res. Space Phys. **110**(A9), 7102 (2005). https://doi.org/10.1029/2004JA010772

N. Altobelli, E. Grün, M. Landgraf, A new look into the Helios dust experiment data: presence of interstellar dust inside the Earth's orbit. Astron. Astrophys. **448**, 243–252 (2006). https://doi.org/10.1051/0004-6361:20053909

N. Altobelli, V. Dikarev, S. Kempf, R. Srama, S. Helfert, G. Moragas-Klostermeyer, M. Roy, E. Grün, Cassini/Cosmic Dust Analyzer in situ dust measurements between Jupiter and Saturn. J. Geophys. Res. Space Phys. **112**, A07105 (2007). https://doi.org/10.1029/2006JA011978

N. Altobelli, F. Postberg, K. Fiege, M. Trieloff, H. Kimura, V.J. Sterken, H.W. Hsu, J. Hillier, N. Khawaja, G. Moragas-Klostermeyer, J. Blum, M. Burton, R. Srama, S. Kempf, E. Grün, Flux and composition of interstellar dust at Saturn from Cassini's Cosmic Dust Analyzer. Science **352**, 312–318 (2016). https://doi.org/10.1126/science.aac6397

D.J. Asher, D.I. Steel, Theoretical meteor radiants for macroscopic Taurid complex objects. Earth Moon Planets **68**, 155–164 (1995). https://doi.org/10.1007/BF00671504

B. Augé, E. Dartois, C. Engrand, J. Duprat, M. Godard, L. Delauche, N. Bardin, C. Mejía, R. Martinez, G. Muniz, A. Domaracka, P. Boduch, H. Rothard, Irradiation of nitrogen-rich ices by swift heavy ions. Clues for the formation of ultracarbonaceous micrometeorites. Astron. Astrophys. **592**, A99 (2016). https://doi.org/10.1051/0004-6361/201527650

F. Bagenal, M. Horányi, D.J. McComas, R.L. McNutt, H.A. Elliot, M.E. Hill, L.E. Brown, P.A. Delamere, P. Kollman, S.M. Krimigis, M. Kusterer, C.M. Lisse, D.G. Mitchell, M. Piquette, A.R. Poppe, D.F. Strobel, J.R. Szalay, P. Valek, J. Vandegriff, S. Weidner, E.J. Zirnstein, S.A. Stern, K. Ennico, C.B. Olkin, H.A. Weaver, L.A. Young (NHS Team), Pluto's interaction with its space environment: solar wind, energetic particles, and dust. Science **351**, 6279 (2016)

W.J. Baggaley, The interstellar particle component measured by AMOR, in *Meteroids 1998*, ed. by W.J. Baggaley, V. Porubcan (1999), p. 265

W.J. Baggaley, Advanced Meteor Orbit Radar observations of interstellar meteoroids. J. Geophys. Res. **105**, 10353–10362 (2000). https://doi.org/10.1029/1999JA900383

W.J. Baggaley, S.H. Marsh, S. Close, Interstellar meteors. Dust Planet. Syst. **643**, 27–32 (2007)

M. Baguhl, E. Grün, M. Landgraf, In situ measurements of interstellar dust with the ULYSSES and Galileo spaceprobes. Space Sci. Rev. **78**(1/2), 165–172 78:165–172 (1996)

N. Bardin, J. Duprat, C. Engrand, G. Slodzian, D. Baklouti, E. Dartois, R. Brunetto, L. Delauche, M. Godard, T.D. Wu, J.L. Guerquin-Kern, D/H and 15N/14N isotopic ratios in organic matter of ultracarbonaceous Antarctic micrometeorites, in *78th Annual Meeting of the Meteoritical Society*. LPI Contributions, vol. 1856 (2015), p. 5275

A. Bardyn, D. Baklouti, H. Cottin, N. Fray, C. Briois, J. Paquette, O. Stenzel, C. Engrand, H. Fischer, K. Hornung, R. Isnard, Y. Langevin, H. Lehto, L. Le Roy, N. Ligier, S. Merouane, P. Modica, F.R. Orthous-Daunay, J. Rynö, R. Schulz, J. Silén, L. Thirkell, K. Varmuza, B. Zaprudin, J. Kissel, M. Hilchenbach, Carbon-rich dust in comet 67P/Churyumov-Gerasimenko measured by COSIMA/Rosetta. Mon. Not. R. Astron. Soc. **469**, S712–S722 (2017). https://doi.org/10.1093/mnras/stx2640

M. Battandier, L. Bonal, E. Quirico, P. Beck, C. Engrand, J. Duprat, Characterization of the organic matter and hydration state of a series of Antarctic micrometeorites, in *Lunar and Planetary Science Conference*, vol. 47 (2016), p. 1475

C. Beaugé, S. Ferraz-Mello, Capture in exterior mean-motion resonances due to Poynting-Robertson drag. Icarus **110**, 239–260 (1994). https://doi.org/10.1006/icar.1994.1119

R. Behrisch, W. Eckstein, *Sputtering by particle bombardment: experiments and computer calculations from threshold to MeV energies*. Topics in Applied Physics, vol. 110 (Springer, Berlin, 2007)

E. Beitz, J. Blum, M.G. Parisi, J. Trigo-Rodriguez, The collisional evolution of undifferentiated asteroids and the formation of chondritic meteoroids. Astrophys. J. **824**, 12 (2016). https://doi.org/10.3847/0004-637X/824/1/12. arXiv:1604.02340

I. Belskaya, S. Fornasier, G. Tozzi, R. Gil-Hutton, A. Cellino, K. Antonyuk, Y.N. Krugly, A. Dovgopol, S. Faggi, Refining the asteroid taxonomy by polarimetric observations. Icarus **284**, 30–42 (2017)
O.E. Berg, U. Gerloff, More than two years of micrometeorite data from two Pioneer satellites, in *Space Research XI*, ed. by K.Y. Kondratyev, M.J. Rycroft, C. Sagan (1971), pp. 225–235
O.E. Berg, F.F. Richardson, The Pioneer 8 cosmic dust experiment. Rev. Sci. Instrum. **40**, 1333–1337 (1969). https://doi.org/10.1063/1.1683778
G. Berriman, N. Boggess, M. Hauser, T. Kelsall, C. Lisse, S. Moseley, W. Reach, R. Silverberg, Cobe dirbe near-infrared polarimetry of the zodiacal light: initial results. Astrophys. J. **431**, L63–L66 (1994)
I. Bertini, Main Belt comets: a new class of small bodies in the solar system. Planet. Space Sci. **59**, 365–377 (2011). https://doi.org/10.1016/j.pss.2011.01.014
J.P. Biersack, W. Eckstein, Sputtering studies with the Monte Carlo program TRIM.SP. Appl. Phys. A **34**, 73–94 (1984)
A. Bischoff, E.R.D. Scott, K. Metzler, C.A. Goodrich, *Nature and Origins of Meteoritic Breccias* (2006), pp. 679–712
A. Bischoff, M. Horstmann, M. Laubenstein, S. Haberer, Asteroid 2008 TC3—Almahata Sitta: not only a ureilitic meteorite, but a breccia containing many different achondritic and chondritic lithologies, in *Lunar and Planetary Science Conference*, vol. 41 (2010), p. 1763
R.C. Blaauw, M.D. Campbell-Brown, R.J. Weryk, A meteoroid stream survey using the Canadian Meteor Orbit Radar, III: mass distribution indices of six major meteor showers. Mon. Not. R. Astron. Soc. **414**, 3322–3329 (2011a). https://doi.org/10.1111/j.1365-2966.2011.18633.x
R.C. Blaauw, M.D. Campbell-Brown, R.J. Weryk, Mass distribution indices of sporadic meteors using radar data. Mon. Not. R. Astron. Soc. **412**, 2033–2039 (2011b). https://doi.org/10.1111/j.1365-2966.2010.18038.x
P.A. Bland, T.B. Smith, A.J.T. Jull, F.J. Berry, A.W.R. Bevan, S. Cloudt, C.T. Pillinger, The flux of meteorites to the Earth over the last 50,000 years. Mon. Not. R. Astron. Soc. **283**, 551 (1996). https://doi.org/10.1093/mnras/283.2.551
J. Blum, R. Schräpler, B.J.R. Davidsson, J.M. Trigo-Rodríguez, The physics of protoplanetesimal dust agglomerates, I: mechanical properties and relations to primitive bodies in the solar system. Astrophys. J. **652**, 1768–1781 (2006). https://doi.org/10.1086/508017
D. Bockelée-Morvan, D. Gautier, F. Hersant, J.M. Huré, F. Robert, Turbulent radial mixing in the solar nebula as the source of crystalline silicates in comets. Astron. Astrophys. **384**, 1107–1118 (2002). https://doi.org/10.1051/0004-6361:20020086
K.J. Borkowski, E. Dwek, The fragmentation and vaporization of dust in grain-grain collisions. Astrophys. J. **454**, 254–276 (1995)
J. Borovička, P. Koten, P. Spurný, J. Boček, R. Štork, A survey of meteor spectra and orbits: evidence for three populations of Na-free meteoroids. Icarus **174**, 15–30 (2005). https://doi.org/10.1016/j.icarus.2004.09.011
W.F. Bottke, H.F. Levison, A. Morbidelli, K. Tsiganis, The collisional evolution of objects captured in the Outer Asteroid Belt during the late heavy bombardment, in *Lunar and Planetary Science Conference*, vol. 39 (2008), p. 1447
W.F. Bottke Jr, D. Vokrouhlický, D.P. Rubincam, M. Broz, *The Effect of Yarkovsky Thermal Forces on the Dynamical Evolution of Asteroids and Meteoroids* (2002), pp. 395–408
J.P. Bradley, Chemically anomalous, preaccretionally irradiated grains in interplanetary dust from comets. Science **265**, 925–929 (1994). https://doi.org/10.1126/science.265.5174.925
J.P. Bradley, *Interplanetary Dust Particles* (Elsevier, Amsterdam, 2005), p. 689
J.P. Bradley, D.E. Brownlee, Cometary particles—thin sectioning and electron beam analysis. Science **231**, 1542–1544 (1986). https://doi.org/10.1126/science.231.4745.1542
J.P. Bradley, L.P. Keller, T.P. Snow, M.S. Hanner, G.J. Flynn, J.C. Gezo, S.J. Clemett, D.E. Brownlee, J.E. Bowey, An infrared spectral match between GEMS and interstellar grains. Science **285**, 1716–1718 (1999). https://doi.org/10.1126/science.285.5434.1716
A.J. Brearley, R.H. Jones, Chondritic meteorites, in *Planetary Materials*, ed. by J.J. Papike (Mineralogical Society of America, Washington, 1998), pp. 1–398
K.L.F. Brinton, C. Engrand, D.P. Glavin, J.L. Bada, M. Maurette, A search for extraterrestrial amino acids in carbonaceous Antarctic micrometeorites. Orig. Life Evol. Biosph. **28**, 413–424 (1998)
D.T. Britt, G.J. Consolmagno, The porosity of dark meteorites and the structure of low-albedo asteroids. Icarus **146**, 213–219 (2000). https://doi.org/10.1006/icar.2000.6374
P. Brown, R.J. Weryk, D.K. Wong, J. Jones, A meteoroid stream survey using the Canadian Meteor Orbit Radar, I: methodology and radiant catalogue. Icarus **195**, 317–339 (2008). https://doi.org/10.1016/j.icarus.2007.12.002
P.G. Brown, A.R. Hildebrand, M.E. Zolensky, M. Grady, R.N. Clayton, T.K. Mayeda, E. Tagliaferri, R. Spalding, N.D. MacRae, E.L. Hoffman, D.W. Mittlefehldt, J.F. Wacker, J.A. Bird, M.D. Campbell, R.

Carpenter, H. Gingerich, M. Glatiotis, E. Greiner, M.J. Mazur, P.J. McCausland, H. Plotkin, T. Rubak Mazur, The fall, recovery, orbit, and composition of the Tagish lake meteorite: a new type of carbonaceous chondrite. Science **290**, 320–325 (2000). https://doi.org/10.1126/science.290.5490.320

D. Brownlee, Interplanetary dust. Rev. Geophys. **17**(7), 1735–1743 (1979)

D. Brownlee, P. Tsou, J. Aléon, C.M.O. Alexander, T. Araki, S. Bajt, G.A. Baratta, R. Bastien, P. Bland, P. Bleuet, J. Borg, J.P. Bradley, A. Brearley, F. Brenker, S. Brennan, J.C. Bridges, N.D. Browning, J.R. Brucato, E. Bullock, M.J. Burchell, H. Busemann, A. Butterworth, M. Chaussidon, A. Cheuvront, M. Chi, M.J. Cintala, B.C. Clark, S.J. Clemett, G. Cody, L. Colangeli, G. Cooper, P. Cordier, C. Daghlian, Z. Dai, L. D'Hendecourt, Z. Djouadi, G. Dominguez, T. Duxbury, J.P. Dworkin, D.S. Ebel, T.E. Economou, S. Fakra, S.A.J. Fairey, S. Fallon, G. Ferrini, T. Ferroir, H. Fleckenstein, C. Floss, G. Flynn, I.A. Franchi, M. Fries, Z. Gainsforth, J.P. Gallien, M. Genge, M.K. Gilles, P. Gillet, J. Gilmour, D.P. Glavin, M. Gounelle, M.M. Grady, G.A. Graham, P.G. Grant, S.F. Green, F. Grossemy, L. Grossman, J.N. Grossman, Y. Guan, K. Hagiya, R. Harvey, P. Heck, G.F. Herzog, P. Hoppe, F. Hörz, J. Huth, I.D. Hutcheon, K. Ignatyev, H. Ishii, M. Ito, D. Jacob, C. Jacobsen, S. Jacobsen, S. Jones, D. Joswiak, A. Jurewicz, A.T. Kearsley, L.P. Keller, H. Khodja, A.L.D. Kilcoyne, J. Kissel, A. Krot, F. Langenhorst, A. Lanzirotti, L. Le, L.A. Leshin, J. Leitner, L. Lemelle, H. Leroux, M.C. Liu, K. Luening, I. Lyon, G. MacPherson, M.A. Marcus, K. Marhas, B. Marty, G. Matrajt, K. McKeegan, A. Meibom, V. Mennella, K. Messenger, S. Messenger, T. Mikouchi, S. Mostefaoui, T. Nakamura, T. Nakano, M. Newville, L.R. Nittler, I. Ohnishi, K. Ohsumi, K. Okudaira, D.A. Papanastassiou, R. Palma, M.E. Palumbo, R.O. Pepin, D. Perkins, M. Perronnet, P. Pianetta, W. Rao, F.J.M. Rietmeijer, F. Robert, D. Rost, A. Rotundi, R. Ryan, S.A. Sandford, C.S. Schwandt, T.H. See, D. Schlutter, J. Sheffield-Parker, A. Simionovici, S. Simon, I. Sitnitsky, C.J. Snead, M.K. Spencer, F.J. Stadermann, A. Steele, T. Stephan, R. Stroud, J. Susini, S.R. Sutton, Y. Suzuki, M. Taheri, S. Taylor, N. Teslich, K. Tomeoka, N. Tomioka, A. Toppani, J.M. Trigo-Rodríguez, D. Troadec, A. Tsuchiyama, A.J. Tuzzolino, T. Tyliszczak, K. Uesugi, M. Velbel, J. Vellenga, E. Vicenzi, L. Vincze, J. Warren, I. Weber, M. Weisberg, A.J. Westphal, S. Wirick, D. Wooden, B. Wopenka, P. Wozniakiewicz, I. Wright, H. Yabuta, H. Yano, E.D. Young, R.N. Zare, T. Zega, K. Ziegler, L. Zimmerman, E. Zinner, M. Zolensky, Comet 81P/Wild 2 under a microscope. Science **314**, 1711 (2006). https://doi.org/10.1126/science.1135840

D.E. Brownlee, Cosmic dust—collection and research. Annu. Rev. Earth Planet. Sci. **13**, 147–173 (1985). https://doi.org/10.1146/annurev.ea.13.050185.001051

D.L. Buczkowski, O.S. Barnouin-Jha, L.M. Prockter, 433 Eros lineaments: global mapping and analysis. Icarus **193**, 39–52 (2008). https://doi.org/10.1016/j.icarus.2007.06.028

A. Buffington, M.M. Bisi, J.M. Clover, P.P. Hick, B.V. Jackson, T.A. Kuchar, S.D. Price, Measurements of the gegenschein brightness from the solar mass ejection imager (smei). Icarus **203**(1), 124–133 (2009)

A. Buffington, M.M. Bisi, J.M. Clover, P.P. Hick, B.V. Jackson, T.A. Kuchar, S.D. Price, Measurements and an empirical model of the zodiacal brightness as observed by the solar mass ejection imager (smei). Icarus **272**, 88–101 (2016)

J.A. Burns, P.L. Lamy, S. Soter, Radiation forces on small particles in the solar system. Icarus **40**, 1–48 (1979). https://doi.org/10.1016/0019-1035(79)90050-2

M.D. Campbell Brown, High resolution radiant distribution and orbits of sporadic radar meteoroids. Icarus **196**, 144–163 (2008). https://doi.org/10.1016/j.icarus.2008.02.022

M.D. Campbell-Brown, D. Koschny, Model of the ablation of faint meteors. Astron. Astrophys. **418**, 751–758 (2004). https://doi.org/10.1051/0004-6361:20041001-1

M.D. Campbell-Brown, J. Kero, C. Szasz, A. Pellinen-Wannberg, R.J. Weryk, Photometric and ionization masses of meteors with simultaneous EISCAT UHF radar and intensified video observations. J. Geophys. Res. Space Phys. **117**, A09323 (2012). https://doi.org/10.1029/2012JA017800

M.D. Campbell-Brown, J. Borovička, P.G. Brown, E. Stokan, High-resolution modelling of meteoroid ablation. Astron. Astrophys. **557**, A41 (2013). https://doi.org/10.1051/0004-6361/201322005

M.D. Campbell-Brown, R. Blaauw, A. Kingery, Optical fluxes and meteor properties of the camelopardalid meteor shower. Icarus **277**, 141–153 (2016). https://doi.org/10.1016/j.icarus.2016.05.001

Z. Ceplecha, J. Borovička, W.G. Elford, D.O. Revelle, R.L. Hawkes, V. Porubčan, M. Šimek, Meteor phenomena and bodies. Space Sci. Rev. **84**, 327–471 (1998). https://doi.org/10.1023/A:1005069928850

E. Charon, C. Engrand, K. Benzerara, H. Leroux, S. Swaraj, R. Belkhou, J. Duprat, E. Dartois, M. Godard, L. Delauche, A C-, N-, O-XANES/STXM and TEM study of organic matter and minerals in ultracarbonaceous Antarctic micrometeorites (UCAMMs), in *Lunar and Planetary Science Conference*, vol. 48 (2017), p. 2085

A.A. Christou, R.M. Killen, M.H. Burger, The meteoroid stream of comet Encke at Mercury: implications for Mercury surface, space environment, geochemistry, and ranging observations of the exosphere. Geophys. Res. Lett. **42**, 7311–7318 (2015). https://doi.org/10.1002/2015GL065361

F.J. Ciesla, Outward transport of high-temperature materials around the midplane of the solar nebula. Science **318**, 613 (2007). https://doi.org/10.1126/science.1147273

F.J. Ciesla, S.A. Sandford, Organic synthesis via irradiation and warming of ice grains in the solar nebula. Science **336**, 452 (2012). https://doi.org/10.1126/science.1217291

B.C. Clark, S.F. Green, T.E. Economou, S.A. Sandford, M.E. Zolensky, N. McBride, D.E. Brownlee, Release and fragmentation of aggregates to produce heterogeneous, lumpy coma streams. J. Geophys. Res., Planets **109**(E18), E12S03 (2004). https://doi.org/10.1029/2004JE002319

S.J. Clemett, X.D.F. Chillier, S. Gillette, R.N. Zare, M. Maurette, C. Engrand, G. Kurat, Observation of indigenous polycyclic aromatic hydrocarbons in 'giant' carbonaceous Antarctic micrometeorites. Orig. Life Evol. Biosph. **28**, 425–448 (1998)

C. Cordier, L. Folco, C. Suavet, C. Sonzogni, P. Rochette, Major, trace element and oxygen isotope study of glass cosmic spherules of chondritic composition: the record of their source material and atmospheric entry heating. Geochim. Cosmochim. Acta **75**, 5203–5218 (2011). https://doi.org/10.1016/j.gca.2011.06.014

B.G. Cour-Palais, Meteoroid environment model-1969 [Near Earth to Lunar Surface]. NASA SP-8013, NASA (1969)

A. Czechowski, I. Mann, Formation and acceleration of nano dust in the inner heliosphere. Astrophys. J. **714**, 89 (2010). https://doi.org/10.1088/0004-637X/714/1/89

E. Dartois, C. Engrand, R. Brunetto, J. Duprat, T. Pino, E. Quirico, L. Remusat, N. Bardin, G. Briani, S. Mostefaoui, G. Morinaud, B. Crane, N. Szwec, L. Delauche, F. Jamme, C. Sandt, P. Dumas, Ultracarbonaceous Antarctic micrometeorites, probing the solar system beyond the nitrogen snow-line. Icarus **224**, 243–252 (2013). https://doi.org/10.1016/j.icarus.2013.03.002

E. Dartois, C. Engrand, J. Duprat, M. Godard, E. Charon, L. Delauche, C. Sandt, F. Borondics, Dome C ultracarbonaceous Antarctic micrometeorites. Infrared and Raman fingerprints. Astron. Astrophys. **609**, A65 (2018). https://doi.org/10.1051/0004-6361/201731322. arXiv:1711.00647

B.T. de Gregorio, R.M. Stroud, G.D. Cody, L.R. Nittler, A.L. David Kilcoyne, S. Wirick, Correlated microanalysis of cometary organic grains returned by stardust. Meteorit. Planet. Sci. **46**, 1376–1396 (2011). https://doi.org/10.1111/j.1945-5100.2011.01237.x

F.E. DeMeo, B. Carry, Solar system evolution from compositional mapping of the asteroid belt. Nature **505**, 629 (2014)

S. Dermott, P. Nicholson, J. Burns, J. Houck, Origin of the solar system dust bands discovered by iras. Nature **312**(5994), 505–509 (1984)

S.F. Dermott, K. Grogan, D.D. Durda, S. Jayaraman, T.J.J. Kehoe, S.J. Kortenkamp, M.C. Wyatt, *Orbital Evolution of Interplanetary Dust* (2001), p. 569

V. Dikarev, E. Grün, J. Baggaley, D. Galligan, M. Landgraf, R. Jehn, Modeling the sporadic meteoroid background cloud. Earth Moon Planets **95**, 109–122 (2004). https://doi.org/10.1007/s11038-005-9017-y

V. Dikarev, E. Grün, J. Baggaley, D. Galligan, M. Landgraf, R. Jehn, The new ESA meteoroid model. Adv. Space Res. **35**, 1282–1289 (2005). https://doi.org/10.1016/j.asr.2005.05.014

N. Divine, Five populations of interplanetary meteoroids. J. Geophys. Res. **98**(17), 029 (1993). https://doi.org/10.1029/93JE01203. 048

E. Dobrică, C. Engrand, J. Duprat, M. Gounelle, H. Leroux, E. Quirico, J.N. Rouzaud, Connection between micrometeorites and Wild 2 particles: from Antarctic snow to cometary ices. Meteorit. Planet. Sci. **44**, 1643–1661 (2009). https://doi.org/10.1111/j.1945-5100.2009.tb01196.x

E. Dobrică, C. Engrand, J. Duprat, M. Gounelle, A statistical overview of CONCORDIA Antarctic micrometeorites. Meteorit. Planet. Sci. Suppl. **73**, 5213 (2010)

E. Dobrică, C. Engrand, H. Leroux, J.N. Rouzaud, J. Duprat, Transmission electron microscopy of CONCORDIA ultracarbonaceous Antarctic micrometeorites (UCAMMs): mineralogical properties. Geochim. Cosmochim. Acta **76**, 68–82 (2012). https://doi.org/10.1016/j.gca.2011.10.025

E. Dobrică, C. Engrand, E. Quirico, G. Montagnac, J. Duprat, Raman characterization of carbonaceous matter in CONCORDIA Antarctic micrometeorites. Meteorit. Planet. Sci. **46**, 1363–1375 (2011). https://doi.org/10.1111/j.1945-5100.2011.01235.x

L. Dones, P.R. Weissman, H.F. Levison, M.J. Duncan, Oort cloud formation and dynamics, in *Comets II*, ed. by M.C. Festou, H.U. Keller, H.A. Weaver (University of Arizona Press, Tucson, 2004)

G. Drolshagen, V. Dikarev, M. Landgraf, H. Krag, W. Kuiper, Comparison of meteoroid flux models for near earth space. Earth Moon Planets **102**, 191–197 (2008). https://doi.org/10.1007/s11038-007-9199-6

J.D. Drummond, A test of comet and meteor shower associations. Icarus **45**, 545–553 (1981). https://doi.org/10.1016/0019-1035(81)90020-8

A. Dubietis, R. Arlt, Taurid resonant-swarm encounters from two decades of visual observations. Mon. Not. R. Astron. Soc. **376**, 890–894 (2007). https://doi.org/10.1111/j.1365-2966.2007.11488.x

R. Dumont, Phase function and polarization curve of interplanetary scatterers from zodiacal light photopolarimetry. Planet. Space Sci. **21**(12), 2149–2155 (1973)

R. Dumont, A.C. Levasseur-Regourd, Zodiacal light photopolarimetry, IV: annual variations of brightness and the symmetry plane of the zodiacal cloud: absence of solar-cycle variations. Astron. Astrophys. **64**, 9–16 (1978)
R. Dumont, A.L. Levasseur-Regourd, Zodiacal light gathered along the line of sight: retrieval of the local scattering coefficient from photometric surveys of the ecliptic plane. Planet. Space Sci. **33**(1), 1–9 (1985)
R. Dumont, F. Sanchez, Zodiacal light photopolarimetry, I: observations, reductions, disturbing phenomena, accuracy; II: gradients along the ecliptic and the phase functions of interplanetary matter. Astron. Astrophys. **38**, 397–403 (1975a)
R. Dumont, F. Sanchez, Zodiacal light photopolarimetry, II: gradients along the ecliptic and the phase functions of interplanetary matter. Astron. Astrophys. **38**, 405 (1975b)
J. Duprat, C. Engrand, M. Maurette, F. Naulin, G. Kurat, M. Gounelle, The micrometeorite mass flux as recorded in Dome C central Antarctic surface snow. Meteorit. Planet. Sci. Suppl. **41**, 5239 (2006)
J. Duprat, C. Engrand, M. Maurette, G. Kurat, M. Gounelle, C. Hammer, Micrometeorites from central Antarctic snow: the CONCORDIA collection. Adv. Space Res. **39**, 605–611 (2007). https://doi.org/10.1016/j.asr.2006.05.029
J. Duprat, E. Dobrică, C. Engrand, J. Aléon, Y. Marrocchi, S. Mostefaoui, A. Meibom, H. Leroux, J.N. Rouzaud, M. Gounelle, F. Robert, Extreme deuterium excesses in ultracarbonaceous micrometeorites from central Antarctic snow. Science **328**, 742 (2010). https://doi.org/10.1126/science.1184832
J. Duprat, N. Bardin, C. Engrand, D. Baklouti, R. Brunetto, E. Dartois, L. Delauche, M. Godard, J.L. Guerquin-Kern, G. Slodzian, T.D. Wu, Isotopic analysis of organic matter in ultra-carbonaceous antarctic micrometeorites, in *77th Annual Meeting of the Meteoritical Society*. LPI Contributions, vol. 1800 (2014), p. 5341
T.E. Economou, S.F. Green, D.E. Brownlee, B.C. Clark, Dust flux monitor instrument measurements during Stardust-NExT Flyby of Comet 9P/Tempel 1. Icarus **222**, 526–539 (2013). https://doi.org/10.1016/j.icarus.2012.09.019
C. Engrand, M. Maurette, Carbonaceous micrometeorites from Antarctica. Meteorit. Planet. Sci. **33**, 565–580 (1998). https://doi.org/10.1111/j.1945-5100.1998.tb01665.x
C. Engrand, E. Deloule, F. Robert, M. Maurette, G. Kurat, Extraterrestrial water in micrometeorites and cosmic spherules from Antarctica: an ion microprobe study. Meteorit. Planet. Sci. **34**, 773–786 (1999a). https://doi.org/10.1111/j.1945-5100.1999.tb01390.x
C. Engrand, K.D. McKeegan, L.A. Leshin, Oxygen isotopic compositions of individual minerals in Antarctic micrometeorites: further links to carbonaceous chondrites. Geochim. Cosmochim. Acta **63**, 2623–2636 (1999b). https://doi.org/10.1016/S0016-7037(99)00160-X
C. Engrand, K.D. McKeegan, L.A. Leshin, J.P. Bradley, D.E. Brownlee, Oxygen isotopic compositions of interplanetary dust particles: 16O-Excess in a GEMS-rich IDP, in *Lunar and Planetary Science Conference*, vol. 30 (1999c)
C. Engrand, K.D. McKeegan, L.A. Leshin, G.F. Herzog, C. Schnabel, L.E. Nyquist, D.E. Brownlee, Isotopic compositions of oxygen, iron, chromium, and nickel in cosmic spherules: toward a better comprehension of atmospheric entry heating effects. Geochim. Cosmochim. Acta **69**, 5365–5385 (2005) https://doi.org/10.1016/j.gca.2005.07.002
C. Engrand, J. Duprat, M. Maurette, M. Gounelle, Fe-Ni sulfides in Concordia Antarctic micrometeorites, in *Lunar and Planetary Science Conference*, vol. 38 (2007), p. 1668
C. Engrand, K. Benzerara, H. Leroux, J. Duprat, E. Dartois, N. Bardin, L. Delauche, Carbonaceous phases and mineralogy of ultracarbonaceous Antarctic micrometeorites identified by C- and N-XANES/STXM and TEM, in *Lunar and Planetary Science Conference*, vol. 46 (2015), p. 1902
E.M. Epifani, C. Snodgrass, D. Perna, M. Dall'Ora, P. Palumbo, V. Della Corte, A. Alvarez-Candal, M. Melita, A. Rotundi, Photometry of the Oort Cloud comet C/2009 P1 (Garradd): pre-perihelion observations at 5.7 and 2.5 AU. Planet. Space Sci. **132**, 23–31 (2016)
D. Fixsen, E. Dwek, The zodiacal emission spectrum as determined by cobe and its implications. Astrophys. J. **578**(2), 1009 (2002)
C. Floss, F.J. Stadermann, Isotopically primitive interplanetary dust particles of cometary origin: evidence from nitrogen isotopic compositions, in *Lunar and Planetary Science Conference*, ed. by S. Mackwell, E. Stansbery (2004), p. 1281
G.J. Flynn, D.D. Durda, Chemical and mineralogical size segregation in the impact disruption of inhomogeneous, anhydrous meteorites. Planet. Space Sci. **52**, 1129–1140 (2004). https://doi.org/10.1016/j.pss.2004.07.010
G.J. Flynn, S. Wirick, L.P. Keller, Organic grain coatings in primitive interplanetary dust particles: implications for grain sticking in the Solar Nebula. Earth Planets Space **65**, 1159–1166 (2013). https://doi.org/10.5047/eps.2013.05.007

G.J. Flynn, L.R. Nittler, C. Engrand, Composition of cosmic dust: sources and implications for the early solar system. Elements **12**(3), 177 (2016). https://doi.org/10.2113/gselements.12.3.177

S. Fornasier, M. Lazzarin, C. Barbieri, M.A. Barucci, Spectroscopic comparison of aqueous altered asteroids with CM2 carbonaceous chondrite meteorites. Astron. Astrophys. Suppl. Ser. **135**, 65–73 (1999). https://doi.org/10.1051/aas:1999161

N. Fray, A. Bardyn, H. Cottin, K. Altwegg, D. Baklouti, C. Briois, L. Colangeli, C. Engrand, H. Fischer, A. Glasmachers, E. Grün, G. Haerendel, H. Henkel, H. Höfner, K. Hornung, E.K. Jessberger, A. Koch, H. Krüger, Y. Langevin, H. Lehto, K. Lehto, L. Le Roy, S. Merouane, P. Modica, F.R. Orthous-Daunay, J. Paquette, F. Raulin, J. Rynö, R. Schulz, J. Silén, S. Siljeström, W. Steiger, O. Stenzel, T. Stephan, L. Thirkell, R. Thomas, K. Torkar, K. Varmuza, K.P. Wanczek, B. Zaprudin, J. Kissel, M. Hilchenbach, High-molecular-weight organic matter in the particles of comet 67P/Churyumov-Gerasimenko. Nature **538**, 72–74 (2016). https://doi.org/10.1038/nature19320

N. Fray, A. Bardyn, H. Cottin, D. Baklouti, C. Briois, C. Engrand, H. Fischer, K. Hornung, R. Isnard, Y. Langevin, H. Lehto, L. Le Roy, E.M. Mellado, S. Merouane, P. Modica, F.R. Orthous-Daunay, J. Paquette, J. Rynö, R. Schulz, J. Silén, S. Siljeström, O. Stenzel, L. Thirkell, K. Varmuza, B. Zaprudin, J. Kissel, M. Hilchenbach, Nitrogen-to-carbon atomic ratio measured by COSIMA in the particles of comet 67P/Churyumov-Gerasimenko. Mon. Not. R. Astron. Soc. **469**, S506–S516 (2017). https://doi.org/10.1093/mnras/stx2002

P.C. Frisch, J.M. Dorschner, J. Geiss, J.M. Greenberg, E. Grün, M. Landgraf, P. Hoppe, A.P. Jones, W. Krätschmer, T.J. Linde, G.E. Morfill, W. Reach, J.D. Slavin, J. Svestka, A.N. Witt, G.P. Zank, Dust in the local interstellar wind. Astrophys. J. **525**, 492–516 (1999)

A. Fujiwara, J. Kawaguchi, D.K. Yeomans, M. Abe, T. Mukai, T. Okada, J. Saito, H. Yano, M. Yoshikawa, D.J. Scheeres, O. Barnouin-Jha, A.F. Cheng, H. Demura, R.W. Gaskell, N. Hirata, H. Ikeda, T. Kominato, H. Miyamoto, A.M. Nakamura, R. Nakamura, S. Sasaki, K. Uesugi, The rubble-pile asteroid Itokawa as observed by Hayabusa. Science **312**, 1330–1334 (2006). https://doi.org/10.1126/science.1125841

M. Fulle, Dust from short-period comet P/Schwassmann-Wachmann 1 and replenishment of the interplanetary dust cloud. Nature **359**, 42–44 (1992)

M. Fulle, *Motion of Cometary Dust* (2004), pp. 565–575

M.J. Gaffey, T.B. McCord, Asteroid surface materials—mineralogical characterizations from reflectance spectra. Space Sci. Rev. **21**, 555–628 (1978). https://doi.org/10.1007/BF00240908

D.P. Galligan, W.J. Baggaley, Wavelet enhancement for detecting shower structure in radar meteoroid data, I: methodology, in *IAU Colloq. 181: Dust in the Solar System and Other Planetary Systems*, vol. 15, ed. by S.F. Green, I.P. Williams, J.A.M. McDonnell, N. McBride (2002), p. 42

D.P. Galligan, W.J. Baggaley, The orbital distribution of radar-detected meteoroids of the Solar system dust cloud. Mon. Not. R. Astron. Soc. **353**, 422–446 (2004). https://doi.org/10.1111/j.1365-2966.2004.08078.x

D.P. Galligan, W.J. Baggaley, The radiant distribution of AMOR radar meteors. Mon. Not. R. Astron. Soc. **359**, 551–560 (2005). https://doi.org/10.1111/j.1365-2966.2005.08918.x

M.J. Genge, Koronis asteroid dust within Antarctic ice. Geology **36**(9), 687–690 (2008). https://doi.org/10.1130/G24493A.1

M.J. Genge, M.M. Grady, R. Hutchison, The textures and compositions of fine-grained Antarctic micrometeorites—implications for comparisons with meteorites. Geochim. Cosmochim. Acta **61**, 5149 (1997). https://doi.org/10.1016/S0016-7037(97)00308-6

M.J. Genge, A. Gileski, M.M. Grady, Chondrules in Antarctic micrometeorites. Meteorit. Planet. Sci. **40**, 225 (2005). https://doi.org/10.1111/j.1945-5100.2005.tb00377.x

M.J. Genge, C. Engrand, M. Gounelle, S. Taylor, The classification of micrometeorites. Meteorit. Planet. Sci. **43**, 497–515 (2008). https://doi.org/10.1111/j.1945-5100.2008.tb00668.x

R. Giese, Light scattering by small particles and models of interplanetary matter derived from the zodiacal light. Space Sci. Rev. **1**(3), 589–611 (1963)

R. Giese, Optical properties of single-component zodiacal light models. Planet. Space Sci. **21**(3), 513–521 (1973)

R. Giese, K. Weiss, R. Zerull, T. Ono, Large fluffy particles-a possible explanation of the optical properties of interplanetary dust. Astron. Astrophys. **65**, 265–272 (1978)

J.C. Gómez Martín, D.L. Bones, J.D. Carrillo-Sánchez, A.D. James, J.M. Trigo-Rodríguez, B. Fegley Jr., J.M.C. Plane, Novel experimental simulations of the atmospheric injection of meteoric metals. Astrophys. J. **836**, 212 (2017). https://doi.org/10.3847/1538-4357/aa5c8f

M. Gounelle, M.E. Zolensky, J.C. Liou, P.A. Bland, O. Alard, Mineralogy of carbonaceous chondritic microclasts in howardites: identification of C2 fossil micrometeorites. Geochim. Cosmochim. Acta **67**, 507–527 (2003). https://doi.org/10.1016/S0016-7037(02)00985-7

M. Gounelle, C. Engrand, O. Alard, P.A. Bland, M.E. Zolensky, S.S. Russell, J. Duprat, Hydrogen isotopic composition of water from fossil micrometeorites in howardites. Geochim. Cosmochim. Acta **69**, 3431–3443 (2005a). https://doi.org/10.1016/j.gca.2004.12.021

M. Gounelle, C. Engrand, M. Maurette, G. Kurat, K.D. McKeegan, F. Brandsttter, Small Antarctic micrometeorites: a mineralogical and in situ oxygen isotope study. Meteorit. Planet. Sci. **40**, 917 (2005b). https://doi.org/10.1111/j.1945-5100.2005.tb00163.x

J. Gradie, E. Tedesco, Compositional structure of the asteroid belt. Science **216**, 1405–1407 (1982). https://doi.org/10.1126/science.216.4553.1405

M.M. Grady, C.T. Pillinger, Carbon isotope relationships in winonaites and forsterite chondrites. Geochim. Cosmochim. Acta **50**, 255–263 (1986). https://doi.org/10.1016/0016-7037(86)90174-2

A.L. Graps, E. Grün, H. Svedhem, H. Krüger, M. Horányi, A. Heck, S. Lammers, Io as a source of the jovian dust streams. Nature **405**, 48–50 (2000)

S.F. Green, J.A.M. McDonnell, N. McBride, M.T.S.H. Colwell, A.J. Tuzzolino, T.E. Economou, P. Tsou, B.C. Clark, D.E. Brownlee, The dust mass distribution of comet 81P/Wild 2. J. Geophys. Res., Planets **109**(E18), E12S04 (2004). https://doi.org/10.1029/2004JE002318

A. Grigorieva, P. Thébault, P. Artymowicz, A. Brandeker, Survival of icy grains in debris discs. The role of photosputtering. Astron. Astrophys. **475**, 755–764 (2007a). https://doi.org/10.1051/0004-6361:20077686. arXiv:0709.0811

A. Grigorieva, P. Thébault, P. Artymowicz, A. Brandeker, Survival of icy grains in debris disks: the role of photosputtering. Astron. Astrophys. **475**, 755–764 (2007b)

M. Gritsevich, D. Koschny, Constraining the luminous efficiency of meteors. Icarus **212**, 877–884 (2011). https://doi.org/10.1016/j.icarus.2011.01.033

E. Grün, M. Landgraf, Collisional consequences of big interstellar grains. J. Geophys. Res. **105**, 10,291–10,298 (2000)

E. Grün, N. Pailer, H. Fechtig, J. Kissel, Orbital and physical characteristics of micrometeoroids in the inner solar system as observed by HELIOS 1. Planet. Space Sci. **28**, 333–349 (1980). https://doi.org/10.1016/0032-0633(80)90022-7

E. Grün, H.A. Zook, H. Fechtig, R.H. Giese, Collisional balance of the meteoritic complex. Icarus **62**, 244–272 (1985). https://doi.org/10.1016/0019-1035(85)90121-6

E. Grün, H. Zook, M. Baguhl, A. Balogh, S. Bame, H. Fechtig, R. Forsyth, M. Hanner, M. Horanyi, J. Kissel, B.A. Lindblad, D. Linkert, G. Linkert, I. Mann, J. McDonnel, G. Morfill, J. Phillips, C. Polanskey, G. Schwehm, N. Siddique, P. Staubach, J. Svestka, A. Taylor, Discovery of Jovian dust streams and interstellar grains by the Ulysses spacecraft. Nature **362**, 428–430 (1993a)

E. Grün, H.A. Zook, M. Baguhl, A. Balogh, S. Bame, H. Fechtig, R. Forsyth, M. Manner, M. Horanyi, J. Kissel et al., Discovery of jovian dust streams and interstellar grains by the Ulysses spacecraft. Nature **362**(6419), 428–430 (1993b)

E. Grün, H.A. Zook, M. Baguhl, A. Balogh, S.J. Bame, H. Fechtig, R. Forsyth, M.S. Hanner, M. Horanyi, J. Kissel, B.A. Lindblad, D. Linkert, G. Linkert, I. Mann, J.A.M. McDonnell, G.E. Morfill, J.L. Phillips, C. Polanskey, G. Schwehm, N. Siddique, P. Staubach, J. Svestka, A. Taylor, Discovery of Jovian dust streams and interstellar grains by the ULYSSES spacecraft. Nature **362**, 428–430 (1993c). https://doi.org/10.1038/362428a0

E. Grün, B. Gustafson, I. Mann, M. Baguhl, G.E. Morfill, P. Staubach, A. Taylor, H.A. Zook, Interstellar dust in the heliosphere. Astron. Astrophys. **286**, 915–924 (1994)

E. Grün, M. Baguhl, D.P. Hamilton, R. Riemann, H.A. Zook, S. Dermott, H. Fechtig, B.A. Gustafson, M.S. Hanner, M. Horányi, K.K. Khurana, J. Kissel, M. Kivelson, B.A. Lindblad, D. Linkert, G. Linkert, I. Mann, J.A.M. McDonnell, G.E. Morfill, C. Polanskey, G. Schwehm, R. Srama, Constraints from Galileo observations on the origin of jovian dust streams. Nature **381**, 395–398 (1996). https://doi.org/10.1038/381395a0

E. Grün, P. Staubach, M. Baguhl, D.P. Hamilton, H.A. Zook, S. Dermott, B.A. Gustafson, H. Fechtig, J. Kissel, D. Linkert, G. Linkert, R. Srama, M.S. Hanner, C. Polanskey, M. Horányi, B.A. Lindblad, I. Mann, J.A.M. McDonnell, G.E. Morfill, G. Schwehm, South-north and radial traverses through the interplanetary dust cloud. Icarus **129**, 270–288 (1997). https://doi.org/10.1006/icar.1997.5789

E. Grün, M. Baguhl, H. Svedhem, H.A. Zook, *In situ Measurements of Cosmic Dust* (2001), p. 295

E. Grün, R. Srama, M. Horányi, H. Krüger, R. Soja, V. Sterken, Z. Sternovsky, P. Strub, Comparative analysis of the ESA and NASA interplanetary meteoroid enviroment models, in *6th European Conference on Space Debris*. ESA Special Publication, vol. 723 (2013), p. 36

P.S. Gural, Fully correcting for the spread in meteor radiant positions due to gravitational attraction. J. Int.l Meteor Org. **29**, 134–138 (2001)

D.A. Gurnett, E. Grun, D. Gallagher, W.S. Kurth, F.L. Scarf, Micron-sized particles detected near Saturn by the Voyager plasma wave instrument. Icarus **53**, 236–254 (1983). https://doi.org/10.1016/0019-1035(83)90145-8

D.A. Gurnett, T.F. Averkamp, F.L. Scarf, E. Grun, Dust particles detected near Giacobini-Zinner by the ICE plasma wave instrument. Geophys. Res. Lett. **13**, 291–294 (1986). https://doi.org/10.1029/GL013i003p00291

D.A. Gurnett, W.S. Kurth, K.L. Scarf, J.A. Burns, J.N. Cuzzi, Micron-sized particle impacts detected near Uranus by the Voyager 2 plasma wave instrument. J. Geophys. Res. Space Phys. **92**, 14,959–14,968 (1987). https://doi.org/10.1029/JA092iA13p14959

D.A. Gurnett, W.S. Kurth, L.J. Granroth, S.C. Allendorf, R.L. Poynter, Micron-sized particles detected near Neptune by the Voyager 2 plasma wave instrument. J. Geophys. Res. Space Phys. **96**, 19 (1991). https://doi.org/10.1029/91JA01270

D.A. Gurnett, J.A. Ansher, W.S. Kurth, L.J. Granroth, Micron-sized dust particles detected in the outer solar system by the Voyager 1 and 2 plasma wave instruments. Geophys. Res. Lett. **24**(24), 3125–3128 (1997)

D.A. Gurnett, W.S. Kurth, D.L. Kirchner, G.B. Hospodarsky, T.F. Averkamp, P. Zarka, A. Lecacheux, R. Manning, A. Roux, P. Canu, N. Cornilleau-Wehrlin, P. Galopeau, A. Meyer, R. Boström, G. Gustafsson, J.E. Wahlund, L. Åhlen, H.O. Rucker, H.P. Ladreiter, W. Macher, L.J.C. Woolliscroft, H. Alleyne, M.L. Kaiser, M.D. Desch, W.M. Farrell, C.C. Harvey, P. Louarn, P.J. Kellogg, K. Goetz, A. Pedersen, The Cassini radio and plasma wave investigation. Space Sci. Rev. **114**, 395–463 (2004). https://doi.org/10.1007/s11214-004-1434-0

B.A.S. Gustafson, Comet ejection and dynamics of nonspherical dust particles and meteoroids. Astrophys. J. **337**, 945–949 (1989). https://doi.org/10.1086/167164

B.A.S. Gustafson, Physics of zodiacal dust. Annu. Rev. Earth Planet. Sci. **22**, 553–595 (1994). https://doi.org/10.1146/annurev.ea.22.050194.003005

E. Hadamcik, A.C. Levasseur-Regourd, Imaging polarimetry of cometary dust: different comets and phase angles. J. Quant. Spectrosc. Radiat. Transf. **79**, 661–678 (2003)

M. Hajduková Jr, On the frequency of interstellar meteoroids. Astron. Astrophys. **288**, 330–334 (1994)

M. Hajduková, Meteors in the IAU meteor data center on hyperbolic orbits. Earth Moon Planets **102**, 67–71 (2008). https://doi.org/10.1007/s11038-007-9171-5

M. Hajduková Jr, Interstellar meteoroids in the Japanese tv catalogue. Publ. Astron. Soc. Jpn. **63**, 481–487 (2011). https://doi.org/10.1093/pasj/63.3.481

M. Hajduková Jr, The occurrence of interstellar particles in the vicinity of the Sun an overview—25 years of research, in *International Meteor Conference Egmond*, ed. by A. Roggemans, P. Roggemans, the Netherlands, 2–5 June 2016 (2016), pp. 105–110

M. Hajduková, L. Kornoš, J. Tóth, Frequency of hyperbolic and interstellar meteoroids. Meteorit. Planet. Sci. **49**, 63–68 (2014a). https://doi.org/10.1111/maps.12119

M. Hajduková Jr, L. Kornoš, J. Tóth, Hyperbolic orbits in the EDMOND. Meteoroids **2013**, 289–295 (2014b)

M. Hajduková, V.J. Sterken, P. Wiegert, *Interstellar Meteoroids* (2018)

I. Halliday, A.A. Griffin, A.T. Blackwell, Detailed data for 259 fireballs from the Canadian camera network and inferences concerning the influx of large meteoroids. Meteorit. Planet. Sci. **31**, 185–217 (1996). https://doi.org/10.1111/j.1945-5100.1996.tb02014.x

D.P. Hamilton, E. Grün, M. Baguhl, Electromagnetic escape of dust from the solar system. Int. Astron. Union Colloq. **150**, 31 (1996). https://doi.org/10.1017/S0252921100501225

D. Han, A.R. Poppe, M. Piquette, E. Grün, M. Horányi, Constraints on dust production in the Edgeworth-Kuiper Belt from Pioneer 10 and New Horizons measurement. Geophys. Res. Lett. **38**, L24102 (2011)

M. Hanner, R. Giese, K. Weiss, R. Zerull, On the definition of albedo and application to irregular particles. Astron. Astrophys. **104**, 42–46 (1981)

M.S. Hanner, M.E. Zolensky, The mineralogy of cometary dust, in *Lecture Notes in Physics*, vol. 815, ed. by T. Henning (Springer, Berlin, 2010), pp. 203–232. https://doi.org/10.1007/978-3-642-13259-9_4

V. Haudebourg, M. Cabane, A.C. Levasseur-Regourd, Theoretical polarimetric responses of fractal aggregates, in relation with experimental studies of dust in the solar system. Phys. Chem. Earth, Part C, Sol.-Terr. Planet. Sci. **24**(5), 603–608 (1999)

R.L. Hawkes, S.C. Woodworth, Optical detection of two meteoroids from interstellar space. J. R. Astron. Soc. Can. **91**, 218 (1997)

G.S. Hawkins, Variation in the occurrence rate of meteors. Astron. J. **61**, 386 (1956). https://doi.org/10.1086/107367

G.S. Hawkins, Symposium: small meteoric particles in the earth's neighborhood: radar determination of meteor orbits. Astron. J. **67**, 241 (1962). https://doi.org/10.1086/108702

G.S. Hawkins, The Harvard radio meteor project. Smithson. Contrib. Astrophys. **7**, 53 (1963)

G.S. Hawkins, E.K.L. Upton, The influx rate of meteors in the Earth's atmosphere. Astrophys. J. **128**, 727 (1958). https://doi.org/10.1086/146585

G.F.H. Herzog, S. Xue, G.S. Hall, L.E. Nyquist, C. Shih, H. Wiesmann, D.E. Brownlee, Isotopic and elemental composition of iron, nickel, and chromium in type I deep-sea spherules: implications for origin and composition of the parent micrometeoroids. Geochim. Cosmochim. Acta **63**, 1443–1457 (1999). https://doi.org/10.1016/S0016-7037(99)00011-3

K.A. Hill, L.A. Rogers, R.L. Hawkes, High geocentric velocity meteor ablation. Astron. Astrophys. **444**, 615–624 (2005). https://doi.org/10.1051/0004-6361:20053053

J.K. Hillier, S.F. Green, N. McBride, N. Altobelli, F. Postberg, S. Kempf, J. Schwanethal, R. Srama, J.A.M. McDonnell, E. Grün, Interplanetary dust detected by the Cassini CDA Chemical Analyser. Icarus **190**, 643–654 (2007). https://doi.org/10.1016/j.icarus.2007.03.024

T. Hiroi, C.M. Pieters, M.E. Zolensky, M.E. Lipschutz, Evidence of thermal metamorphism on the C, G, B, and F asteroids. Science **261**, 1016–1018 (1993). https://doi.org/10.1126/science.261.5124.1016

H.J. Hoffmann, H. Fechtig, E. Grün, J. Kissel, First results of the micrometeoroid experiment S 215 on the HEOS 2 satellite. Planet. Space Sci. **23**, 215–224 (1975a). https://doi.org/10.1016/0032-0633(75)90080-X

H.J. Hoffmann, H. Fechtig, E. Grün, J. Kissel, Temporal fluctuations and anisotropy of the micrometeoroid flux in the earth-moon system measured by HEOS 2. Planet. Space Sci. **23**, 985–991 (1975b). https://doi.org/10.1016/0032-0633(75)90186-5

P. Hoppe, G. Kurat, J. Walter, M. Maurette, Trace elements and oxygen isotopes in a CAI-bearing micrometeorite from Antarctica, in *Lunar and Planetary Science Conference*, vol. 26 (1995)

P. Horálek, L.L. Christensen, D. Nesvornỳ, R. Davies, Light phenomena over the eso observatories, III: zodiacal light. Messenger **164**, 45–47 (2016)

M. Horányi et al., The student dust counter on the New Horizons mission. Space Sci. Rev. **140**, 387–402 (2008)

F. Horz, R.P. Bernhard, Compositional analysis and classification of projectile residues in LDEF impact craters. NASA STI/Recon technical report No. 93 (1992)

F. Hörz, E. Schneider, D.E. Gault, J.B. Hartung, D.E. Brownlee, Catastrophic rupture of lunar rocks—a Monte Carlo simulation. Moon **13**, 235–258 (1975). https://doi.org/10.1007/BF00567517

W. Huang, X. Chu, C.S. Gardner, J.D. Carrillo-Sánchez, W. Feng, J.M.C. Plane, D. Nesvorný, Measurements of the vertical fluxes of atomic Fe and Na at the mesopause: Implications for the velocity of cosmic dust entering the atmosphere. Geophys. Res. Lett. **42**, 169–175 (2015). https://doi.org/10.1002/2014GL062390

D.H. Humes, Results of Pioneer 10 and 11 meteoroid experiments: interplanetary and near-Saturn. J. Geophys. Res. **85**(A11), 5841–5852 (1980)

G.R. Huss, The survival of presolar grains in solar system bodies, in *American Institute of Physics Conference Series*, vol. 402, ed. by T.J. Bernatowicz, E. Zinner (1997), pp. 721–748. https://doi.org/10.1063/1.53338

G.R. Huss, A.P. Meshik, J.B. Smith, C.M. Hohenberg, Presolar diamond, silicon carbide, and graphite in carbonaceous chondrites: implications for thermal processing in the solar nebula. Geochim. Cosmochim. Acta **67**, 4823–4848 (2003). https://doi.org/10.1016/j.gca.2003.07.019

G.R. Huss, A.E. Rubin, J.N. Grossman, *Thermal Metamorphism in Chondrites* (2006), pp. 567–586

M. Ishiguro, R. Nakamura, Y. Fujii, K. Morishige, H. Yano, H. Yasuda, S. Yokogawa, T. Mukai, First detection of visible zodiacal dust bands from ground-based observations. Astrophys. J. **511**(1), 432 (1999)

M. Ishiguro, H. Yang, F. Usui, J. Pyo, M. Ueno, T. Ootsubo, S.M. Kwon, T. Mukai, High-resolution imaging of the gegenschein and the geometric albedo of interplanetary dust. Astrophys. J. **767**(1), 75 (2013)

H.A. Ishii, J.P. Bradley, Z.R. Dai, M. Chi, A.T. Kearsley, M.J. Burchell, N.D. Browning, F. Molster, Comparison of comet 81P/Wild 2 dust with interplanetary dust from comets. Science **319**, 447 (2008). https://doi.org/10.1126/science.1150683

I. Jakšová, V. Porubčan, J. Klačka, Structure and sources of the sporadic meteor background from video observations. Publ. Astron. Soc. Jpn. **67**, 99 (2015). https://doi.org/10.1093/pasj/psv068

D. James, M. Horányi, V. Hoxie, Polyvinylidene fluoride detector response to particle impacts. Rev. Sci. Instrum. **81**(3), 034501 (2010). https://doi.org/10.1063/1.3340880

D. Janches, S. Close, J.L. Hormaechea, N. Swarnalingam, A. Murphy, D. O'Connor, B. Vandepeer, B. Fuller, D.C. Fritts, C. Brunini, The Southern Argentina Agile MEteor Radar Orbital System (SAAMER-OS): an initial sporadic meteoroid orbital survey in the southern sky. Astrophys. J. **809**, 36 (2015). https://doi.org/10.1088/0004-637X/809/1/36

P. Jenniskens, Quantitative meteor spectroscopy: elemental abundances. Adv. Space Res. **39**, 491–512 (2007). https://doi.org/10.1016/j.asr.2007.03.040

P. Jenniskens, *Meteoroid Streams and the Zodiacal Cloud* (2015), pp. 281–295. https://doi.org/10.2458/azu_uapress_9780816532131-ch015

P. Jenniskens, Meteor showers in review. Planet. Space Sci. **143**, 116–124 (2017). https://doi.org/10.1016/j.pss.2017.01.008

P. Jenniskens, P. Gural, A. Berdeu, CAMSS: a spectroscopic survey of meteoroid elemental abundances. Meteoroids **2013**, 117–124 (2014)

P. Jenniskens, Q. Nénon, J. Albers, P.S. Gural, B. Haberman, D. Holman, R. Morales, B.J. Grigsby, D. Samuels, C. Johannink, The established meteor showers as observed by CAMS. Icarus **266**, 331–354 (2016a). https://doi.org/10.1016/j.icarus.2015.09.013

P. Jenniskens, Q. Nénon, P.S. Gural, J. Albers, B. Haberman, B. Johnson, D. Holman, R. Morales, B.J. Grigsby, D. Samuels, C. Johannink, CAMS confirmation of previously reported meteor showers. Icarus **266**, 355–370 (2016b). https://doi.org/10.1016/j.icarus.2015.08.014

P. Jenniskens, Q. Nénon, P.S. Gural, J. Albers, B. Haberman, B. Johnson, R. Morales, B.J. Grigsby, D. Samuels, C. Johannink, CAMS newly detected meteor showers and the sporadic background. Icarus **266**, 384–409 (2016c). https://doi.org/10.1016/j.icarus.2015.11.009

D. Jewitt, L. Chizmadia, R. Grimm, D. Prialnik, Water in the small bodies of the solar system, in *Protostars and Planets V* (2007), pp. 863–878

J. Jones, Meteoroid engineering model—final report. Tech. rep., NASA/MSFC internal report SEE/CR-2004-400 (2004)

J. Jones, P. Brown, Sporadic meteor radiant distributions - Orbital survey results. Mon. Not. R. Astron. Soc. **265**, 524 (1993). https://doi.org/10.1093/mnras/265.3.524

J. Jones, M. Campbell, S. Nikolova, Modelling of the sporadic meteoroid sources, in *Meteoroids 2001 Conference*, ed. by B. Warmbein. ESA Special Publication, vol. 495 (2001), pp. 575–580

T.J. Jopek, P.M. Jenniskens, The working group on meteor showers nomenclature: a history, current status and a call for contributions, in *Meteoroids: the Smallest Solar System Bodies*, ed. by W.J. Cooke, D.E. Moser, B.F. Hardin, D. Janches (2011), pp. 7–13

T.J. Jopek, Z. Kaňuchová, Current status of the IAU MDC Meteor Showers Database. Meteoroids **2013**, 353–364 (2014)

T.J. Jopek, R. Rudawska, P. Bartczak, Meteoroid stream searching: the use of the vectorial elements. Earth Moon Planets **102**, 73–78 (2008). https://doi.org/10.1007/s11038-007-9197-8

A. Juhász, M. Horányi, Dynamics and distribution of nano-dust particles in the inner solar system. Geophys. Res. Lett. **40**, 2500–2504 (2013). https://doi.org/10.1002/grl.50535

L.P. Keller, K.L. Thomas, D.S. McKay, An interplanetary dust particle with links to CI chondrites. Geochim. Cosmochim. Acta **56**, 1409–1412 (1992). https://doi.org/10.1016/0016-7037(92)90072-Q

L.P. Keller, K.L. Thomas, D.S. McKay, Carbon in primitive interplanetary dust particles, in *Analysis of Interplanetary Dust Particles*, ed. by E. Zolensky, T.L. Wilson, F.J.M. Rietmeijer, G.J. Flynn. American Institute of Physics Conference Series, vol. 310, (1994), p. 159. https://doi.org/10.1063/1.46531

L.P. Keller, S. Messenger, G.J. Flynn, S. Clemett, S. Wirick, C. Jacobsen, The nature of molecular cloud material in interplanetary dust. Geochim. Cosmochim. Acta **68**, 2577–2589 (2004). https://doi.org/10.1016/j.gca.2003.10.044

M.S. Kelley, C.E. Woodward, D.E. Harker, D.H. Wooden, R.D. Gehrz, H. Campins, M.S. Hanner, S.M. Lederer, D.J. Osip, J. Pittichová, E. Polomski, A Spitzer study of comets 2P/Encke, 67P/Churyumov-Gerasimenko, and C/2001 HT50 (LINEAR-NEAT). Astrophys. J. **651**, 1256–1271 (2006). https://doi.org/10.1086/507701. astro-ph/0607416

M.S. Kelley, Y.R. Fernández, J. Licandro, C.M. Lisse, W.T. Reach, M.F. A'Hearn, J. Bauer, H. Campins, A. Fitzsimmons, O. Groussin, P.L. Lamy, S.C. Lowry, K.J. Meech, J. Pittichová, C. Snodgrass, I. Toth, H.A. Weaver, The persistent activity of Jupiter-family comets at 3–7 AU. Icarus **225**, 475–494 (2013)

P.J. Kellogg, K. Goetz, S.J. Monson, Dust impact signals on the wind spacecraft. J. Geophys. Res. Space Phys. **121**, 966–991 (2016). https://doi.org/10.1002/2015JA021124

J.F. Kerridge, Carbon, hydrogen and nitrogen in carbonaceous chondrites abundances and isotopic compositions in bulk samples. Geochim. Cosmochim. Acta **49**, 1707–1714 (1985). https://doi.org/10.1016/0016-7037(85)90141-3

D.J. Kessler, Meteoroid environment model-1970 [Interplanetary and planetary]. NASA SP-8038, NASA (1970)

J.B. Kikwaya, M. Campbell-Brown, P.G. Brown, Bulk density of small meteoroids. Astron. Astrophys. **530**, A113 (2011). https://doi.org/10.1051/0004-6361/201116431

H. Kimura, Light-scattering properties of fractal aggregates: numerical calculations by a superposition technique and the discrete-dipole approximation. J. Quant. Spectrosc. Radiat. Transf. **70**(4), 581–594 (2001)

H. Kimura, I. Mann, Radiation pressure on porous micrometeoroids, in *Meteroids*, ed. by W.J. Baggaley, V. Porubcan (1999), p. 283. 1998

H. Kimura, I. Mann, D.A. Biesecker, E.K. Jessberger, Dust grains in the comae and tails of sungrazing comets: modeling of their mineralogical and morphological properties. Icarus **159**, 529–541 (2002a). https://doi.org/10.1006/icar.2002.6940

H. Kimura, H. Okamoto, T. Mukai, Radiation pressure and the Poynting-Robertson effect for fluffy dust particles. Icarus **157**, 349–361 (2002b). https://doi.org/10.1006/icar.2002.6849

H. Kimura, L. Kolokolova, I. Mann, Optical properties of cometary dust-constraints from numerical studies on light scattering by aggregate particles. Astron. Astrophys. **407**(1), L5–L8 (2003)

H. Kimura, L. Kolokolova, I. Mann, Light scattering by cometary dust numerically simulated with aggregate particles consisting of identical spheres. Astron. Astrophys. **449**(3), 1243–1254 (2006)

J. Kissel, F.R. Krueger, The organic component in dust from comet Halley as measured by the PUMA mass spectrometer on board Vega 1. Nature **326**, 755–760 (1987). https://doi.org/10.1038/326755a0

W. Klöck, F.J. Stadermann, Mineralogical and chemical relationships of interplanetary dust particles micrometeorites and meteorites, in *Analysis of Interplanetary Dust Particles*, ed. by E. Zolensky, T.L. Wilson, F.J.M. Rietmeijer, G.J. Flynn. American Institute of Physics Conference Series, vol. 310 (1994), p. 51. https://doi.org/10.1063/1.46523

H. Kobayashi, J. Kimura, S. Yamamoto, S. Watanabe, T. Yamamoto, Ice sublimation of dust particles and their detection in the outer solar system. Earth Planets Space **62**, 57–61 (2010)

H. Kobayashi, H. Kimura, S.Watanabe, T. Yamamoto, S. Müller, Sublimation temperature of circumstellar dust particles and its importance for dust ring formation. Earth Planets Space **63**, 1067 (2011)

L. Kolokolova, H. Kimura, Effects of electromagnetic interaction in the polarization of light scattered by cometary and other types of cosmic dust. Astron. Astrophys. **513**, A40 (2010)

L. Kolokolova, J. Hough, A.C. Levasseur-Regourd, *Polarimetry of stars and planetary systems* (Cambridge University Press, Cambridge, 2015)

D. Koschny, E. Drolshagen, S. Drolshagen, J. Kretschmer, T. Ott, G. Drolshagen, B. Poppe, Flux densities of meteoroids derived from optical double-station observations. Planet. Space Sci. **143**, 230–237 (2017). https://doi.org/10.1016/j.pss.2016.12.007

Y. Kozai, Secular perturbations of asteroids with high inclination and eccentricity. Astron. J. **67**, 591 (1962). https://doi.org/10.1086/108790

L. Kresak, Cometary dust trails and meteor storms. Astron. Astrophys. **279**, 646–660 (1993)

H. Kruger, E. Grün, D.P. Hamilton, M. Baguhl, S. Dermott, H. Fechtig, B.A. Gustafson, M.S. Hanner, M. Horányi, J. Kissel, B.A. Lindblad, D. Linkert, G. Linkert, I. Mann, J.A.M. McDonnell, G.E. Morfill, C. Polanskey, R. Riemann, G. Schwehm, R. Srama, H.A. Zook, Three years of Galileo dust data, II: 1993–1995. Planet. Space Sci. **47**, 85–106 (1998). https://doi.org/10.1016/S0032-0633(98)00097-X. astro-ph/9809318

H. Krüger, N. Altobelli, B. Anweiler, S.F. Dermott, V. Dikarev, A.L. Graps, E. Grün, B.A. Gustafson, D.P. Hamilton, M.S. Hanner, M. Horányi, J. Kissel, M. Landgraf, B.A. Lindblad, D. Linkert, G. Linkert, I. Mann, J.A.M. McDonnell, G.E. Morfill, C. Polanskey, G. Schwehm, R. Srama, H.A. Zook, Five years of Ulysses dust data: 2000. Planet. Space Sci. **54**, 932–956 (2006). https://doi.org/10.1016/j.pss.2006.04.015. 2004

H. Krüger, V. Dikarev, B. Anweiler, S.F. Dermott, A.L. Graps, E. Grün, B.A. Gustafson, D.P. Hamilton, M.S. Hanner, M. Horányi, J. Kissel, D. Linkert, G. Linkert, I. Mann, J.A.M. McDonnell, G.E. Morfill, C. Polanskey, G. Schwehm, R. Srama, Three years of Ulysses dust data: 2005. Planet. Space Sci. **58**, 951–964 (2010). https://doi.org/10.1016/j.pss.2009.11.002. to 2007. 0908.1279

M.J. Kuchner, C.C. Stark, Collisional grooming models of the Kuiper belt dust cloud. Astron. J. **140**, 1007–1019 (2010)

G. Kurat, C. Koeberl, T. Presper, F. Brandstätter, M. Maurette, Petrology and geochemistry of Antarctic micrometeorites. Geochim. Cosmochim. Acta **58**, 3879–3904 (1994). https://doi.org/10.1016/0016-7037(94)90369-7

G. Kurat, P. Hoppe, C. Engrand, A chondrule micrometeorite from antarctica with vapor-fractionated trace-element abundances. Meteorit. Planet. Sci. Suppl. **31**, A75–A76 (1996)

W.S. Kurth, T.F. Averkamp, D.A. Gurnett, Z. Wang, Cassini RPWS observations of dust in Saturn's E ring. Planet. Space Sci. **54**, 988–998 (2006). https://doi.org/10.1016/j.pss.2006.05.011

S. Kwon, S. Hong, J. Weinberg, An observational model of the zodiacal light brightness distribution. New Astron. **10**(2), 91–107 (2004)

H. Laakso, R. Grard, A. Pedersen, G. Schwehm, Impacts of large dust particles on the VEGA spacecraft. Adv. Space Res. **9**, 269–272 (1989). https://doi.org/10.1016/0273-1177(89)90273-1

H. Lai, C.T. Russell, H. Wei, T. Zhang, The evolution of co-orbiting material in the orbit of 2201 Oljato from 1980 to 2012 as deduced from Pioneer Venus Orbiter and Venus Express magnetic records. Meteorit. Planet. Sci. **49**, 28–35 (2014). https://doi.org/10.1111/maps.12102

M. Landgraf, Modellierung der dynamik und interpretation der in-situ-messung interstellaren staubs in der lokalen umgebung des sonnensystems. PhD thesis, Ruprecht-Karls-Universität Heidelberg (1998)

M. Landgraf, Modeling the motion and distribution of interstellar dust inside the heliosphere. J. Geophys. Res. **105**, 10,303–10,316 (2000)

M. Landgraf, J.C. Liou, H.A. Zook, E. Grün, Origins of solar system dust beyond Jupiter. Astron. J. **123**, 2857–2861 (2002)

J. Lasue, A.C. Levasseur-Regourd, N. Fray, H. Cottin, Inferring the interplanetary dust properties-from remote observations and simulations. Astron. Astrophys. **473**(2), 641–649 (2007)

J. Lasue, A. Levasseur-Regourd, A. Lazarian, Interplanetary dust, in *Polarimetry of Stars and Planetary Systems*, ed. by Kolokolova et al. (2015)
D.S. Lauretta (OSIRIS-Rex Team), An overview of the OSIRIS-REx asteroid sample return mission, in *Lunar and Planetary Science Conference*. Lunar and Planetary Inst. Technical Report, vol. 43 (2012), p. 2491
M.E. Lawler, D.E. Brownlee, CHON as a component of dust from comet Halley. Nature **359**, 810–812 (1992). https://doi.org/10.1038/359810a0
A. Lazarian, B. Andersson, T. Hoang, Grain alignment: role of radiative torques and paramagnetic relaxation, in *Polarimetry of Stars and Planetary Systems*, ed. by Kolokolova et al. (2015)
C. Leinert, Zodiacal light—a measure of the interplanetary environment. Space Sci. Rev. **18**(3), 281–339 (1975)
C. Leinert, E. Grün, Interplanetary dust, in *Physics of the Inner Heliosphere I*, ed. by R. Schwenn, E. Marsch (Springer, Berlin, 1990), pp. 207–275
C. Leinert, S. Roser, J. Buitrago, How to maintain the spatial distribution of interplanetary dust. Astron. Astrophys. **118**, 345–357 (1983a)
C. Leinert, S. Roser, J. Buitrago, How to maintain the spatial distribution of interplanetary dust. Astron. Astrophys. **118**, 345–357 (1983b)
C. Leinert, S. Bowyer, L. Haikala, M. Hanner, M. Hauser, A.C. Levasseur-Regourd, I. Mann, K. Mattila, W. Reach, W. Schlosser et al., The 1997 reference of diffuse night sky brightness. Astron. Astrophys. Suppl. Ser. **127**(1), 1–99 (1998)
A.C. Levasseur, J. Blamont, Satellite observations of intensity variations of the zodiacal light. Nature **246**(5427), 26–28 (1973)
A. Levasseur-Regourd, J. Renard, R. Dumont, Dust optical properties: a comparison between cometary and interplanetary grains. Adv. Space Res. **11**(12), 175–182 (1991)
A. Levasseur-Regourd, M. Cabane, J. Worms, V. Haudebourg, Physical properties of dust in the solar system: relevance of a computational approach and of measurements under microgravity conditions. Adv. Space Res. **20**(8), 1585–1594 (1997)
A.C. Levasseur-Regourd, Optical and thermal properties of zodiacal dust, in *IAU Colloq. 150: Physics, Chemistry, and Dynamics of Interplanetary Dust*, vol. 104, ed. by B.A.S. Gustafson, M.S. Hanner (Astron. Soc. of the Pacific Press, San Francisco, 1996a), p. 301
A.C. Levasseur-Regourd, Physical properties of dust grains deduced by optical probing techniques. Adv. Space Res. **17**(12), 117–122 (1996b)
A.C. Levasseur-Regourd, R. Dumont, Absolute photometry of zodiacal light. Astron. Astrophys. **84**, 277–279 (1980)
A.C. Levasseur-Regourd, E. Hadamcik, J. Renard, Evidence for two classes of comets from their polarimetric properties at large phase angles. Astron. Astrophys. **313**, 327–333 (1996)
A.C. Levasseur-Regourd, I. Mann, R. Dumont, M.S. Hanner, Optical and thermal properties of interplanetary dust, in *Interplanetary Dust*, ed. by E. Grün, B.A.S. Gustafson, S.F. Dermott, H. Fechtig (Springer, Berlin, 2001), pp. 57–94
A.C. Levasseur-Regourd, T. Mukai, J. Lasue, Y. Okada, Physical properties of cometary and interplanetary dust. Planet. Space Sci. **55**(9), 1010–1020 (2007)
A.C. Levasseur-Regourd, J. Agarwal, H. Cottin, C. Engrand, G. Flynn, M. Fulle, T. Gombosi, Y. Langevin, J. Lasue, T. Mannel, S. Merounane, O. Poch, N. Thomas, A. Westphal, Cometary dust. Space Sci. Rev. **214**, 64 (2018). https://doi.org/10.1007/s11214-018-0496-3
L.F. Lim, L.R. Nittler, Elemental composition of 433 eros: new calibration of the NEAR-Shoemaker XRS data. Icarus **200**, 129–146 (2009). https://doi.org/10.1016/j.icarus.2008.09.018
J.C. Liou, H.A. Zook, Evolution of interplanetary dust particles in mean motion resonances with planets. Icarus **128**(2), 354–367 (1997)
J.C. Liou, H.A. Zook, Signatures of the giant planets imprinted on the Edgeworth-Kuiper belt dust disk. Astron. J. **118**, 580–590 (1999)
J.C. Liou, H.A. Zook, S.F. Dermott, Kuiper belt dust grains as a source of interplanetary dust particles. Icarus **124**, 429–440 (1996a). https://doi.org/10.1006/icar.1996.0220
S.G. Love, D.E. Brownlee, A direct measurement of the terrestrial mass accretion rate of cosmic dust. Science **262**, 550–553 (1993). https://doi.org/10.1126/science.262.5133.550
K. Lumme, J. Rahola, J. Hovenier, Light scattering by dense clusters of spheres. Icarus **126**(2), 455–469 (1997)
D.M. Malaspina, L.B. Wilson, A database of interplanetary and interstellar dust detected by the Wind spacecraft. J. Geophys. Res. Space Phys. **121**, 9369–9377 (2016). https://doi.org/10.1002/2016JA023209
D.M. Malaspina, M. Horányi, A. Zaslavsky, K. Goetz, L.B. Wilson, K. Kersten, Interplanetary and interstellar dust observed by the Wind/WAVES electric field instrument. Geophys. Res. Lett. **41**, 266–272 (2014). https://doi.org/10.1002/2013GL058786

D.M. Malaspina, L.E. O'Brien, F. Thayer, Z. Sternovsky, A. Collette, Revisiting STEREO interplanetary and interstellar dust flux and mass estimates. J. Geophys. Res. Space Phys. **120**, 6085–6100 (2015). https://doi.org/10.1002/2015JA021352

I. Mann, Interstellar grains in the solar system: requirements for an analysis. Space Sci. Rev. **78**(1), 259–264 (1996)

I. Mann, H. Okamoto, T. Mukai, H. Kimura, Y. Kitada, Fractal aggregate analogues for near solar dust properties. Astron. Astrophys. **291**, 1011–1018 (1994)

I. Mann, H. Kimura, D.A. Biesecker, B.T. Tsurutani, E. Grün, R.B. McKibben, J.C. Liou, R.M. MacQueen, T. Mukai, M. Guhathakurta et al., Dust near the sun. Space Sci. Rev. **110**(3), 269–305 (2004)

G. Matrajt, S. Taylor, G. Flynn, D. Brownlee, D. Joswiak, A nuclear microprobe study of the distribution and concentration of carbon and nitrogen in Murchison and Tagish lake meteorites, Antarctic micrometeorites, and IDPs: implications for astrobiology. Meteorit. Planet. Sci. **38**, 1585–1600 (2003). https://doi.org/10.1111/j.1945-5100.2003.tb00003.x

G. Matrajt, S. Pizzarello, S. Taylor, D. Brownlee, Concentration and variability of the AIB amino acid in polar micrometeorites: implications for the exogenous delivery of amino acids to the primitive Earth. Meteorit. Planet. Sci. **39**, 1849–1858 (2004). https://doi.org/10.1111/j.1945-5100.2004.tb00080.x

M. Maurette, Cometary micrometeorites in planetology, exobiology, and early climatology, in *Comets and the Origin and Evolution of Life*, ed. by P.J. Thomas, R.D. Hicks, C.F. Chyba, C.P. McKay (2006), p. 69. https://doi.org/10.1007/3-540-33088-7_3

M. Maurette, C. Hammer, N. Reeh, D.E. Brownlee, H.H. Thomsen, Placers of cosmic dust in the blue ice lakes of Greenland. Science **233**, 869–872 (1986). https://doi.org/10.1126/science.233.4766.869

M. Maurette, C. Jehanno, E. Robin, C. Hammer, Characteristics and mass distribution of extraterrestrial dust from the Greenland ice CAP. Nature **328**, 699–702 (1987). https://doi.org/10.1038/328699a0

M. Maurette, C. Olinger, M.C. Michel-Levy, G. Kurat, M. Pourchet, F. Brandstatter, M. Bourot-Denise, A collection of diverse micrometeorites recovered from 100 tonnes of Antarctic blue ice. Nature **351**, 44–47 (1991). https://doi.org/10.1038/351044a0

M. Maurette, C. Engrand, A. Brack, G. Kurat, S. Leach, M. Perreau, Carbonaceous phases in Antarctic micrometeorites and their mineralogical environment. Their contribution to the possible role of micrometeorites as "chondritic chemical reactors" in atmospheres, waters and/or ices, in *Lunar and Planetary Science Conference*, vol. 26 (1995)

M. Maurette, J. Duprat, C. Engrand, M. Gounelle, G. Kurat, G. Matrajt, A. Toppani, Accretion of neon, organics, CO_2, nitrogen and water from large interplanetary dust particles on the early Earth. Planet. Space Sci. **48**, 1117–1137 (2000). https://doi.org/10.1016/S0032-0633(00)00086-6

N. McBride, S.F. Green, J.A.M. McDonnell, Meteoroids and small sized debris in low earth orbit and at 1 Au: results of recent modelling. Adv. Space Res. **23**, 73–82 (1999). https://doi.org/10.1016/S0273-1177(98)00232-4

K.D. McKeegan, Ion microprobe measurements of H, C, O, Mg, and SI isotopic abundances in individual interplanetary dust particles. PhD thesis, Washington Univ., Seattle (1987a)

K.D. McKeegan, Oxygen isotopes in refractory stratospheric dust particles—proof of extraterrestial origin. Science **237**, 1468–1471 (1987b). https://doi.org/10.1126/science.237.4821.1468

K.D. McKeegan, R.M. Walker, E. Zinner, Ion microprobe isotopic measurements of individual interplanetary dust particles. Geochim. Cosmochim. Acta **49**, 1971–1987 (1985). https://doi.org/10.1016/0016-7037(85)90091-2

H. McNamara, J. Jones, B. Kauffman, R. Suggs, W. Cooke, S. Smith, Meteoroid engineering model (MEM): a meteoroid model for the inner solar system. Earth Moon Planets **95**, 123–139 (2004). https://doi.org/10.1007/s11038-005-9044-8

A. Meibom, B.E. Clark, Invited review: evidence for the insignificance of ordinary chondritic material in the asteroid belt. Meteorit. Planet. Sci. **34**, 7–24 (1999)

D.D. Meisel, D. Janches, J.D. Mathews, Extrasolar micrometeors radiating from the vicinity of the local interstellar bubble. Astrophys. J. **567**, 323–341 (2002). https://doi.org/10.1086/322317

S. Messenger, Identification of molecular-cloud material in interplanetary dust particles. Nature **404**, 968–971 (2000). https://doi.org/10.1038/35010053

S. Messenger, L.P. Keller, F.J. Stadermann, R.M. Walker, E. Zinner, Samples of stars beyond the solar system: silicate grains in interplanetary dust. Science **300**, 105–108 (2003). https://doi.org/10.1126/science.1080576

N. Meyer-Vernet, P. Couturier, S. Hoang, C. Perche, J.L. Steinberg, J. Fainberg, C. Meetre, Plasma diagnosis from thermal noise and limits on dust flux or mass in comet Giacobini-Zinner. Science **232**, 370–374 (1986). https://doi.org/10.1126/science.232.4748.370

N. Meyer-Vernet, A. Lecacheux, M.L. Kaiser, D.A. Gurnett, Detecting nanoparticles at radio frequencies: jovian dust stream impacts on Cassini/RPWS. Geophys. Res. Lett. **36**, L03103 (2009a). https://doi.org/10.1029/2008GL036752

N. Meyer-Vernet, M. Maksimovic, A. Czechowski, I. Mann, I. Zouganelis, K. Goetz, M.L. Kaiser, O.C. St Cyr, J.L. Bougeret, S.D. Bale, Dust detection by the wave instrument on STEREO: nanoparticles picked up by the solar wind? Sol. Phys. **256**, 463–474 (2009b). https://doi.org/10.1007/s11207-009-9349-2. arXiv:0903.4141

M.I. Mishchenko, L.D. Travis, A.A. Lacis, *Scattering, Absorption, and Emission of Light by Small Particles* (Cambridge University Press, Cambridge, 2002)

A.V. Moorhead, Performance of D-criteria in isolating meteor showers from the sporadic background in an optical data set. Mon. Not. R. Astron. Soc. **455**, 4329–4338 (2016). https://doi.org/10.1093/mnras/stv2610

A.V. Moorhead, Deconvoluting measurement uncertainty from the meteor speed distribution. Meteorit. Planet. Sci. **1**, 1–7 (2018). https://doi.org/10.1111/maps.13066

A.V. Moorhead, H.M. Koehler, W.J. Cooke, NASA meteoroid engineering model release 2.0. NASA/TM-2015-218214 (2015)

A.V. Moorhead, R.C. Blaauw, D.E. Moser, M.D. Campbell-Brown, P.G. Brown, W.J. Cooke, A two-population sporadic meteoroid bulk density distribution and its implications for environment models. Mon. Not. R. Astron. Soc. **472**, 3833–3841 (2017a). https://doi.org/10.1093/mnras/stx2175

A.V. Moorhead, P.G. Brown, M.D. Campbell-Brown, D. Heynen, W.J. Cooke, Fully correcting the meteor speed distribution for radar observing biases. Planet. Space Sci. **143**, 209–217 (2017). https://doi.org/10.1016/j.pss.2017.02.002

A. Moro-Martín, R. Malhotra, A study of the dynamics of dust from the Kuiper belt: spatial distribution and spectral energy distribution. Astron. J. **124**, 2305–2321 (2002)

A. Moro-Martín, R. Malhotra, Dynamical models of Kuiper belt dust in the inner and outer solar system. Astrophys. J. **125**, 2255–2265 (2003)

D.A. Morrison, U.S. Clanton, Properties of microcraters and cosmic dust of less than 1000 A dimensions, in *Lunar and Planetary Science Conference Proceedings*, vol. 10, ed. by N.W. Hinners (1979), pp. 1649–1663

S. Mostefaoui, G.W. Lugmair, P. Hoppe, A. El Goresy, Evidence for live ^{60}Fe in meteorites. New Astron. Rev. **48**, 155–159 (2004). https://doi.org/10.1016/j.newar.2003.11.022

C.E. Moyano-Cambero, E. Pellicer, J.M. Trigo-Rodríguez, I.P. Williams, J. Blum, P. Michel, M. Küppers, M. Martínez-Jiménez, I. Lloro, J. Sort, Nanoindenting the Chelyabinsk meteorite to learn about impact deflection effects in asteroids. Astrophys. J. **835**, 157 (2017). https://doi.org/10.3847/1538-4357/835/2/157. 1612.07131

T. Mukai, G. Schwehm, Interaction of grains with the solar energetic particles. Astron. Astrophys. **95**, 373–382 (1981)

T. Mukai, A.M. Nakamura, J. Blum, R.E. Johnson, O. Havnes, *Physical Processes on Interplanetary Dust* (2001), p. 445

R. Musci, R.J. Weryk, P. Brown, M.D. Campbell-Brown, P.A. Wiegert, An optical survey for millimeter-sized interstellar meteoroids. Astrophys. J. **745**, 161 (2012). https://doi.org/10.1088/0004-637X/745/2/161. 1110.5882

R. Nakamura, H. Okamoto, Optical properties of fluffy aggregates as analogue of interplanetary dust particles. Adv. Space Res. **23**(7), 1209–1212 (1999)

T. Nakamura, T. Noguchi, T. Yada, Y. Nakamuta, N. Takaoka, Bulk mineralogy of individual micrometeorites determined by X-ray diffraction analysis and transmission electron microscopy. Geochim. Cosmochim. Acta **65**, 4385–4397 (2001). https://doi.org/10.1016/S0016-7037(01)00722-0

T. Nakamura, T. Noguchi, Y. Ozono, T. Osawa, K. Nagao, Mineralogy of ultracarbonaceous large micrometeorites. Meteorit. Planet. Sci. Suppl. **40**, 5046 (2005)

T. Nakamura, T. Noguchi, A. Tsuchiyama, T. Ushikubo, N.T. Kita, J.W. Valley, M.E. Zolensky, Y. Kakazu, K. Sakamoto, E. Mashio, K. Uesugi, T. Nakano, Chondrulelike objects in short-period comet 81P/Wild 2. Science **321**, 1664 (2008). https://doi.org/10.1126/science.1160995

T. Nakamura, T. Noguchi, M. Tanaka, M.E. Zolensky, M. Kimura, A. Nakato, T. Ogami, H. Ishida, A. Tsuchiyama, T. Yada, K. Shirai, R. Okazaki, A. Fujimura, Y. Ishibashi, M. Abe, T. Okada, M. Ueno, T. Mukai, Mineralogy and major element abundance of the dust particles recovered from Muses-C Regio on the asteroid Itokawa, in *Lunar and Planetary Science Conference*, vol. 42 (2011a), p. 1766

T. Nakamura, T. Noguchi, M. Tanaka, M.E. Zolensky, M. Kimura, A. Tsuchiyama, A. Nakato, T. Ogami, H. Ishida, M. Uesugi, T. Yada, K. Shirai, A. Fujimura, R. Okazaki, S.A. Sandford, Y. Ishibashi, M. Abe, T. Okada, M. Ueno, T. Mukai, M. Yoshikawa, J. Kawaguchi, Itokawa dust particles: a direct link between S-type asteroids and ordinary chondrites. Science **333**, 1113 (2011b). https://doi.org/10.1126/science.1207758

K. Nakamura-Messenger, S. Messenger, L.P. Keller, S.J. Clemett, M.E. Zolensky, Organic globules in the Tagish lake meteorite: remnants of the protosolar disk. Science **314**, 1439–1442 (2006). https://doi.org/10.1126/science.1132175

R.J. Naumann, The near earth meteoroid environment. NASA technical note NASA TN D-3717 (1966)
L. Neslušan, M. Hajduková, Separation and confirmation of showers. Astron. Astrophys. **598**, A40 (2017). https://doi.org/10.1051/0004-6361/201629659
D. Nesvorný, P. Jenniskens, H.F. Levison, W.F. Bottke, D. Vokrouhlický, M. Gounelle, Cometary origin of the zodiacal cloud and carbonaceous micrometeorites. Implications for hot debris disks. Astrophys. J. **713**, 816–836 (2010). https://doi.org/10.1088/0004-637X/713/2/816. arXiv:0909.4322
D. Nesvorný, D. Janches, D. Vokrouhlický, P. Pokorný, W.F. Bottke, P. Jenniskens, Dynamical model for the zodiacal cloud and sporadic meteors. Astrophys. J. **743**(2), 129 (2011a)
D. Nesvorný, D. Vokrouhlický, P. Pokorný, D. Janches, Dynamics of dust particles released from Oort cloud comets and their contribution to radar meteors. Astrophys. J. **743**, 37 (2011b). https://doi.org/10.1088/0004-637X/743/1/37. arXiv:1109.2981
F.M. Neubauer, K.H. Glassmeier, A.J. Coates, R. Goldstein, M.H. Acuna, Hypervelocity dust particle impacts observed by the Giotto magnetometer and plasma experiments. Geophys. Res. Lett. **17**, 1809–1812 (1990). https://doi.org/10.1029/GL017i011p01809
T. Noguchi, T. Nakamura, W. Nozaki, Mineralogy of phyllosilicate-rich micrometeorites and comparison with Tagish lake and Sayama meteorites. Earth Planet. Sci. Lett. **202**, 229–246 (2002). https://doi.org/10.1016/S0012-821X(02)00777-X
T. Noguchi, H. Yabuta, S. Itoh, N. Sakamoto, T. Mitsunari, A. Okubo, R. Okazaki, T. Nakamura, S. Tachibana, K. Terada, M. Ebihara, N. Imae, M. Kimura, H. Nagahara, Variation of mineralogy and organic material during the early stages of aqueous activity recorded in Antarctic micrometeorites. Geochim. Cosmochim. Acta **208**, 119–144 (2017). https://doi.org/10.1016/j.gca.2017.03.034
K.I. Öberg, H. Linnartz, R. Visser, E.F. van Dishoeck, Photodesorption of ices, II: H_2O and D_2O. Astrophys. J. **693**, 1209–1218 (2009)
R.C. Ogliore, A.J. Westphal, Z. Gainsforth, A.L. Butterworth, S.C. Fakra, M.A. Marcus, Nebular mixing constrained by the stardust samples. Meteorit. Planet. Sci. **44**, 1675–1681 (2009). https://doi.org/10.1111/j.1945-5100.2009.tb01198.x
B.M. Pedersen, N. Meyer-Vernet, M.G. Aubier, P. Zarka, Dust distribution around Neptune—grain impacts near the ring plane measured by the Voyager planetary radio astronomy experiment. J. Geophys. Res. Space Phys. **96**, 19 (1991). https://doi.org/10.1029/91JA01601
S. Pizzarello, Y. Huang, L. Becker, R.J. Poreda, R.A. Nieman, G. Cooper, M. Williams, The organic content of the Tagish lake meteorite. Science **293**, 2236–2239 (2001). https://doi.org/10.1126/science.1062614
J.M.C. Plane, Cosmic dust in the earth's atmosphere. Chem. Soc. Rev. **41**, 6507–6518 (2012). https://doi.org/10.1039/c2cs35132c
J.M.C. Plane, W. Feng, E. Dawkins, M.P. Chipperfield, J. Höffner, D. Janches, D.R. Marsh, Resolving the strange behavior of extraterrestrial potassium in the upper atmosphere. Geophys. Res. Lett. **41**, 4753–4760 (2014). https://doi.org/10.1002/2014GL060334
J.M.C. Plane, G.J. Flynn, A. Maattanen, J.E. Moores, A.R. Poppe, J.D. Carrillo-Sanchez, C. Listowski, Impacts of cosmic dust on planetary atmospheres and surfaces. Space. Sci. Rev. **214**, 23 (2018)
P. Pokorný, P.G. Brown, A reproducible method to determine the meteoroid mass index. Astron. Astrophys. **592**, A150 (2016). https://doi.org/10.1051/0004-6361/201628134. arXiv:1605.04437
P. Pokorný, D. Vokrouhlický, D. Nesvorný, M. Campbell-Brown, P. Brown, Dynamical model for the toroidal sporadic meteors. Astrophys. J. **789**, 25 (2014). https://doi.org/10.1088/0004-637X/789/1/25
P. Pokorný, D. Janches, P.G. Brown, J.L. Hormaechea, An orbital meteoroid stream survey using the Southern Argentina Agile MEteor Radar (SAAMER) based on a wavelet approach. Icarus **290**, 162–182 (2017). https://doi.org/10.1016/j.icarus.2017.02.025
A.R. Poppe, Interplanetary dust influx to the Pluto-Charon system. Icarus **246**, 352–359 (2015)
A.R. Poppe, An improved model for interplanetary dust fluxes in the outer solar system. Icarus **264**, 369–386 (2016). https://doi.org/10.1016/j.icarus.2015.10.001
A. Poppe, B. Jacobsmeyer, D. James, M. Horányi, Simulation of polyvinylidene fluoride detector response to hypervelocity particle impact. Nucl. Instrum. Methods **622**(3), 583–587 (2010a)
A. Poppe, D. James, B. Jacobsmeyer, M. Horányi, First results from the Venetia Burney Student Dust Counter on the New Horizons mission. Geophys. Res. Lett. **37**, L11101 (2010b)
A. Poppe, D. James, M. Horányi, Measurements of the terrestrial dust influx variability by the cosmic dust experiment. Planet. Space Sci. **59**, 319–326 (2011)
W.T. Reach, P. Morris, F. Boulanger, K. Okumura, The mid-infrared spectrum of the zodiacal and exozodiacal light. Icarus **164**(2), 384–403 (2003)
W.T. Reach, M.S. Kelley, M.V. Sykes, A survey of debris trails from short-period comets. Icarus **191**(1), 298–322 (2007)
L. Remusat, Y. Guan, Y. Wang, J.M. Eiler, Accretion and preservation of D-rich organic particles in carbonaceous chondrites: evidence for important transport in the early solar system nebula. Astron. Astrophys. **713**, 1048–1058 (2010). https://doi.org/10.1088/0004-637X/713/2/1048

J. Renard, A. Levasseur-Regourd, R. Dumont, Properties of interplanetary dust from infrared and optical observations, II: brightness, polarization, temperature, albedo and their dependence on the elevation above the ecliptic. Astron. Astrophys. **304**, 602 (1995)

H.P. Robertson, Dynamical effects of radiation in the solar system. Mon. Not. R. Astron. Soc. **97**, 423 (1937). https://doi.org/10.1093/mnras/97.6.423

P. Rochette, L. Folco, C. Suavet, M. van Ginneken, J. Gattacceca, N. Perchiazzi, R. Braucher, R.P. Harvey, Micrometeorites from the transantarctic mountains. Proc. Natl. Acad. Sci. **105**, 18,206–18,211 (2008). https://doi.org/10.1073/pnas.0806049105

L. Rotelli, J.M. Trigo-Rodríguez, C.E. Moyano-Cambero, E. Carota, L. Botta, E. di Mauro, R. Saladino, The key role of meteorites in the formation of relevant prebiotic molecules in a formamide/water environment. Nat. Sci. Rep. **6**, 38888 (2016). https://doi.org/10.1038/srep38888

M. Rowan-Robinson, B. May, An improved model for the infrared emission from the zodiacal dust cloud: cometary, asteroidal and interstellar dust. Mon. Not. R. Astron. Soc. **429**, 2894 (2013)

A.E. Rubin, J.M. Trigo-Rodríguez, H. Huber, J.T. Wasson, Progressive aqueous alteration of CM carbonaceous chondrites. Geochim. Cosmochim. Acta **71**, 2361–2382 (2007). https://doi.org/10.1016/j.gca.2007.02.008

D.P. Rubincam, Asteroid orbit evolution due to thermal drag. J. Geophys. Res. **100**, 1585–1594 (1995). https://doi.org/10.1029/94JE02411

R. Rudawska, P. Matlovič, J. Tóth, L. Kornoš, Independent identification of meteor showers in EDMOND database. Planet. Space Sci. **118**, 38–47 (2015). https://doi.org/10.1016/j.pss.2015.07.011. arXiv:1406.6598

R. Rudawska, J. Tóth, D. Kalmančok, P. Zigo, P. Matlovič, Meteor spectra from AMOS video system. Planet. Space Sci. **123**, 25–32 (2016). https://doi.org/10.1016/j.pss.2015.11.018

S.S. Russell, G.R. Huss, A.J. Fahey, R.C. Greenwood, R. Hutchison, G.J. Wasserburg, An isotopic and petrologic study of calcium-aluminum-rich inclusions from CO3 meteorites. Geochim. Cosmochim. Acta **62**, 689–714 (1998). https://doi.org/10.1016/S0016-7037(97)00374-8

G.O. Ryabova, Modeling of meteoroid streams: the velocity of ejection of meteoroids from comets (a review). Sol. Syst. Res. **47**, 219–238 (2013). https://doi.org/10.1134/S0038094613030052

L.S. Schramm, D.E. Brownlee, M.M. Wheelock, Major element composition of stratospheric micrometeorites. Meteoritics **24**, 99–112 (1989)

E.R.D. Scott, R.H. Jones, Disentangling nebular and asteroidal features of CO3 carbonaceous chondrite meteorites. Geochim. Cosmochim. Acta **54**, 2485–2502 (1990). https://doi.org/10.1016/0016-7037(90)90235-D

E.R.D. Scott, A.N. Krot, Chondrites and their components. Treatise Geochem. **1**, 711 (2003). https://doi.org/10.1016/B0-08-043751-6/01145-2

Z. Sekanina, Activity of comet Hale-Bopp (1995 O1) beyond 6 AU from the Sun. Astron. Astrophys. **314**, 957–965 (1996)

A. Sekhar, D.J. Asher, Resonant behavior of comet Halley and the Orionid stream. Meteorit. Planet. Sci. **49**, 52–62 (2014). https://doi.org/10.1111/maps.12117. arXiv:1303.2928

A. Shu, S. Bugiel, E. Grün, J. Hillier, M. Horányi, T. Munsat, R. Srama, Cratering studies in polyvinylidene fluoride (PVDF) thin films. Planet. Space Sci. **89**, 29–35 (2013)

F.H. Shu, H. Shang, A.E. Glassgold, T. Lee, X-rays and fluctuating X-winds from protostars. Science **277**, 1475–1479 (1997). https://doi.org/10.1126/science.277.5331.1475

J.I. Simon, I.D. Hutcheon, S.B. Simon, J.E.P. Matzel, E.C. Ramon, P.K. Weber, L. Grossman, D.J. DePaolo, Oxygen isotope variations at the margin of a CAI records circulation within the solar nebula. Science **331**, 1175 (2011). https://doi.org/10.1126/science.1197970

J.A. Simpson, A.J. Tuzzolino, Polarized polymer films as electronic pulse detectors of cosmic dust particles. Nucl. Instrum. Methods Phys. Res., Sect. A **236**, 187–202 (1985)

J.A. Simpson, A.J. Tuzzolino, Cosmic dust investigations, II: instruments for measurement of particle trajectory, velocity and mass. Nucl. Instrum. Methods Phys. Res., Sect. A **279**, 625–639 (1989)

J.A. Simpson, D. Rabinowitz, A.J. Tuzzolino, Cosmic dust investigations, I: PVDF detector signal dependence on mass and velocity for penetrating particles. Nucl. Instrum. Methods Phys. Res., Sect. A **279**, 611–624 (1989)

R.B. Southworth, G.S. Hawkins, Statistics of meteor streams. Smithson. Contrib. Astrophys. **7**, 261 (1963)

R. Srama et al., The Cassini cosmic dust analyzer. Space Sci. Rev. **114**, 465–518 (2004)

O.C. St Cyr, M.L. Kaiser, N. Meyer-Vernet, R.A. Howard, R.A. Harrison, S.D. Bale, W.T. Thompson, K. Goetz, M. Maksimovic, J. Bougeret, D. Wang, S. Crothers, STEREO SECCHI and S/WAVES observations of spacecraft debris caused by micron-size interplanetary dust impacts. Sol. Phys. **256**, 475–488 (2009). https://doi.org/10.1007/s11207-009-9362-5

C.C. Stark, M.J. Kuchner, A new algorithm for self-consistent three-dimensional modeling of collisions in dust debris disks. Astrophys. J. **707**, 543–553 (2009)

P. Staubach, E. Grün, R. John, The meteoroid environment near earth. Adv. Space Res. **19**, 301–308 (1997). https://doi.org/10.1016/S0273-1177(97)00017-3

J. Staude, T. Schmidt, Circular polarisation measurements of the zodiacal light. Astron. Astrophys. **20**, 163 (1972)

D.I. Steel, W.G. Elford, Collisions in the solar system, III: meteoroid survival times. Mon. Not. R. Astron. Soc. **218**, 185–199 (1986). https://doi.org/10.1093/mnras/218.2.185

S.A. Stern, Collisional time scales in the Kuiper disk and their implications. Astron. J. **110**, 2 (1995)

S.A. Stern, Signatures of collisions in the Kuiper disk. Astron. Astrophys. **310**, 999–1010 (1996)

S.A. Stern, The New Horizons Pluto Kuiper belt mission: an overview with historical context. Space Sci. Rev. **140**, 3–21 (2008)

E. Stokan, M.D. Campbell-Brown, A particle-based model for ablation and wake formation in faint meteors. Mon. Not. R. Astron. Soc. **447**, 1580–1597 (2015). https://doi.org/10.1093/mnras/stu2552

P. Strub, H. Krüger, V.J. Sterken, Sixteen years of Ulysses interstellar dust measurements in the solar system, II: fluctuations in the dust flow from the data. Astrophys. J. **812**, 140 (2015). https://doi.org/10.1088/0004-637X/812/2/140. arXiv:1508.03242

C. Suavet, A. Alexandre, I.A. Franchi, J. Gattacceca, C. Sonzogni, R.C. Greenwood, L. Folco, P. Rochette, Identification of the parent bodies of micrometeorites with high-precision oxygen isotope ratios. Earth Planet. Sci. Lett. **293**, 313–320 (2010). https://doi.org/10.1016/j.epsl.2010.02.046

D. Subasinghe, M.D. Campbell-Brown, E. Stokan, Physical characteristics of faint meteors by light curve and high-resolution observations, and the implications for parent bodies. Mon. Not. R. Astron. Soc. **457**, 1289–1298 (2016). https://doi.org/10.1093/mnras/stw019

R.M. Suggs, D.E. Moser, W.J. Cooke, R.J. Suggs, The flux of kilogram-sized meteoroids from lunar impact monitoring. Icarus **238**, 23 (2014)

J.R. Szalay, M. Horányi, Annual variation and synodic modulation of the sporadic meteoroid flux to the Moon. Geophys. Res. Lett. **42**, 10 (2015). https://doi.org/10.1002/2015GL066908

J.R. Szalay, M. Piquette, M. Horányi, The student dust counter: status report at 23 AU. Earth Planets Space **65**, 1145–1149 (2013)

A.D. Taylor, W.J. Baggaley, D.I. Steel, Discovery of interstellar dust entering the Earth's atmosphere. Nature **380**, 323–325 (1996). https://doi.org/10.1038/380323a0

S. Taylor, J.H. Lever, R.P. Harvey, Accretion rate of cosmic spherules measured at the South pole. Nature **392**, 899–903 (1998). https://doi.org/10.1038/31894

S. Taylor, J.H. Lever, R.P. Harvey, Numbers, types, and compositions of an unbiased collection of cosmic spherules. Meteorit. Planet. Sci. **35**, 651–666 (2000). https://doi.org/10.1111/j.1945-5100.2000.tb01450.x

E. Thomas, M. Horányi, D. Janches, T. Munsat, J. Simolka, Z. Sternovsky, Measurements of the ionization coefficient of simulated iron micrometeoroids. Geophys. Res. Lett. **43**, 3645–3652 (2016). https://doi.org/10.1002/2016GL068854

K.L. Thomas, G.E. Blanford, L.P. Keller, W. Klock, D.S. McKay, Carbon abundance and silicate mineralogy of anhydrous interplanetary dust particles. Geochim. Cosmochim. Acta **57**, 1551–1566 (1993). https://doi.org/10.1016/0016-7037(93)90012-I

J.I. Thorpe, C. Parvini, J.M. Trigo-Rodríguez, Detection and measurement of micrometeoroids with LISA Pathfinder. Astron. Astrophys. **586**, A107 (2016). https://doi.org/10.1051/0004-6361/201527658

F. Topputo, D.A. Dei Tos, A. Cipriano, Orbit design of LUMIO, a lunar meteoroid impact observer, in *42nd COSPAR Scientific Assembly, COSPAR Meeting*, vol. 42 (2018), pp. 1–28

M. Trieloff, E.K. Jessberger, I. Herrwerth, J. Hopp, C. Fiéni, M. Ghélis, M. Bourot-Denise, P. Pellas, Structure and thermal history of the H-chondrite parent asteroid revealed by thermochronometry. Nature **422**, 502–506 (2003). https://doi.org/10.1038/nature01499

J.M. Trigo Rodríguez, Aqueous alteration in chondritic asteroids and comets from the study of carbonaceous chondrites, in *European Mineralogical Union Notes in Mineralogy*, vol. 15 (2015), pp. 67–87

J.M. Trigo-Rodríguez, J. Blum, The role of collisional compaction in primitive asteroids and comets, in *European Planetary Science Congress* (2008), p. 29

J.M. Trigo-Rodriguez, J. Blum, Tensile strength as an indicator of the degree of primitiveness of undifferentiated bodies. Planet. Space Sci. **57**, 243–249 (2009). https://doi.org/10.1016/j.pss.2008.02.011

J.M. Trigo-Rodriguez, J. Llorca, J. Borovicka, J. Fabregat, Chemical abundances determined from meteor spectra, I: ratios of the main chemical elements. Meteorit. Planet. Sci. **38**, 1283–1294 (2003). https://doi.org/10.1111/j.1945-5100.2003.tb00313.x

J.M. Trigo-Rodríguez, J. Llorca, J. Fabregat, Chemical abundances determined from meteor spectra, II: evidence for enlarged sodium abundances in meteoroids. Mon. Not. R. Astron. Soc. **348**, 802–810 (2004). https://doi.org/10.1111/j.1365-2966.2004.07389.x

J.M. Trigo-Rodríguez, C.E. Moyano-Cambero, J. Llorca, S. Fornasier, M.A. Barucci, I. Belskaya, Z. Martins, A.S. Rivkin, E. Dotto, J.M. Madiedo, A.A. Jacinto, UV to far-IR reflectance spectra of carbonaceous chondrites, I: implications for remote characterization of dark primitive asteroids targeted by sample-return missions. Mon. Not. R. Astron. Soc. **437**, 227–240 (2014). https://doi.org/10.1093/mnras/stt1873. arXiv:1310.1742

J.M. Trigo-Rodríguez, R. Saladino, E. Di Mauro, L. Rotelli, C.E. Moyano-Cambero, E. Carota, L. Botta, The catalytic role of chondritic meteorites in the prebiotic enrichment of earth and other planetary-rich surfaces, under high meteoritic flux, in *Lunar and Planetary Science Conference*. Lunar and Planetary Inst. Technical Report, vol. 48 (2017), p. 1161

D. Tsintikidis, D.A. Gurnett, W.S. Kurth, L.J. Granroth, Micron-sized particles detected in the vicinity of Jupiter by the Voyager plasma wave instruments. Geophys. Res. Lett. **23**, 997–1000 (1996). https://doi.org/10.1029/96GL00961

Y. Tsuda, M. Yoshikawa, M. Abe, H. Minamino, S. Nakazawa, System design of the Hayabusa 2 asteroid sample return mission to 1999 JU3. Acta Astronaut. **91**, 356–362 (2013). https://doi.org/10.1016/j.actaastro.2013.06.028

B.T. Tsurutani, D.R. Clay, L.D. Zhang, B. Dasgupta, D. Brinza, M. Henry, A. Mendis, S. Moses, K.H. Glassmeier, G. Musmann, I. Richter, Dust impacts at comet P/Borrelly. Geophys. Res. Lett. **30**(22), 2134 (2003). https://doi.org/10.1029/2003GL017580

A.J. Tuzzolino, Two-dimensional position-sensing PVDF dust detectors for measurements of dust particle trajectory, velocity and mass. Nucl. Instrum. Methods Phys. Res., Sect. A **301** (1991)

A.J. Tuzzolino, PVDF copolymer dust detectors: particle response and penetration characteristics. Nucl. Instrum. Methods Phys. Res., Sect. A **316**, 223–237 (1992)

A.J. Tuzzolino, T.E. Economou, R.B. McKibben, J.A. Simpson, J.A.M. McDonnell, M.J. Burchell, B.A.M. Vaughan, P. Tsou, M.S. Hanner, B.C. Clark, D.E. Brownlee, Dust flux monitor instrument for the stardust mission to comet Wild 2. J. Geophys. Res. **108**, E10 (2003)

A.J. Tuzzolino, T.E. Economou, B.C. Clark, P. Tsou, D.E. Brownlee, S.F. Green, J.A.M. McDonnell, N. McBride, M.T.S.H. Colwell, Dust measurements in the coma of comet 81P/Wild 2 by the dust flux monitor instrument. Science **304**, 1776–1780 (2004). https://doi.org/10.1126/science.1098759

A.J. Tuzzolino et al., The Space Dust (SPADUS) instrument aboard the Earth-orbiting ARGOS spacecraft, I: instrument description. Planet. Space Sci. **49**, 689–703 (2001a)

A.J. Tuzzolino et al., The Space Dust (SPADUS) instrument aboard the Earth-orbiting ARGOS spacecraft, II: results form the first 16 months of flight. Planet. Space Sci. **49**, 705–729 (2001b)

A.J. Tuzzolino et al., Dust measurements in the coma of comet 81P/Wild 2 by the dust flux monitor instrument. Science **304**, 1776–1780 (2004)

G.B. Valsecchi, T.J. Jopek, C. Froeschle, Meteoroid stream identification: a new approach, I: theory. Mon. Not. R. Astron. Soc. **304**, 743–750 (1999). https://doi.org/10.1046/j.1365-8711.1999.02264.x

M. van Ginneken, J. Gattacceca, P. Rochette, C. Sonzogni, A. Alexandre, V. Vidal, M.J. Genge, The parent body controls on cosmic spherule texture: evidence from the oxygen isotopic compositions of large micrometeorites. Geochim. Cosmochim. Acta **212**, 196–210 (2017). https://doi.org/10.1016/j.gca.2017.05.008

J. Vaubaillon, F. Colas, Demonstration of gaps due to Jupiter in meteoroid streams. What happened with the 2003 Pi-Puppids? Astron. Astrophys. **431**, 1139–1144 (2005). https://doi.org/10.1051/0004-6361:20041391

J. Vaubaillon, F. Colas, L. Jorda, A new method to predict meteor showers, I: description of the model. Astron. Astrophys. **439**, 751–760 (2005). https://doi.org/10.1051/0004-6361:20041544

J. Vaubaillon, P. Lamy, L. Jorda, On the mechanisms leading to orphan meteoroid streams. Mon. Not. R. Astron. Soc. **370**, 1841–1848 (2006). https://doi.org/10.1111/j.1365-2966.2006.10606.x

P. Vernazza, M. Marsset, P. Beck, R.P. Binzel, M. Birlan, R. Brunetto, F.E. Demeo, Z. Djouadi, C. Dumas, S. Merouane, O. Mousis, B. Zanda, Interplanetary dust particles as samples of icy asteroids. Astrophys. J. **806**, 204 (2015). https://doi.org/10.1088/0004-637X/806/2/204

D. Vinković, Radiation-pressure mixing of large dust grains in protoplanetary disks. Nature **459**, 227 (2009)

C. Vitense, A.V. Krivov, H. Kobayashi, T. Löhne, An improved model of the Edgeworth-Kuiper debris disk. Astron. Astrophys. **540**, A30 (2012). https://doi.org/10.1051/0004-6361/201118551. arXiv:1202.2257

C. Vitense, A.V. Krivov, T. Löhne, Will New Horizons see dust clumps in the Edgeworth-Kuiper belt? Astron. J. **147** (2014)

V. Vojáček, J. Borovička, P. Koten, P. Spurný, R. Štork, Catalogue of representative meteor spectra. Astron. Astrophys. **580**, A67 (2015). https://doi.org/10.1051/0004-6361/201425047

D. Vokrouhlický, P. Farinella, Efficient delivery of meteorites to the Earth from a wide range of asteroid parent bodies. Nature **407**, 606–608 (2000). https://doi.org/10.1038/35036528

J. Walter, F. Brandstaetter, G. Kurat, C. Koeberl, M. Maurette, Cosmic spherules, micrometeorites, and chondrules, in *Lunar and Planetary Science Conference*, vol. 26 (1995)

A. Wehry, I. Mann, Identification of beta -meteoroids from measurements of the dust detector onboard the ULYSSES spacecraft. Astron. Astrophys. **341**, 296–303 (1999)

A. Wehry, H. Krüger, E. Grün, Analysis of Ulysses data: Radiation pressure effects on dust particles. Astron. Astrophys. **419**, 1169–1174 (2004). https://doi.org/10.1051/0004-6361:20035613

S.J. Weidenschilling, A.A. Jackson, Orbital resonances and Poynting-Robertson drag. Icarus **104**, 244–254 (1993). https://doi.org/10.1006/icar.1993.1099

J. Weinberg, J. Sparrow, *Zodiacal Light as an Indicator of Interplanetary Dust* (Wiley, Chichester, 1978), pp. 75–122

M.K. Weisberg, T.J. McCoy, A.N. Krot, *Systematics and Evaluation of Meteorite Classification* (2006), pp. 19–52

A.A. Weiss, J.W. Smith, A southern hemisphere survey of the radiants of sporadic meteors. Mon. Not. R. Astron. Soc. **121**, 5 (1960). https://doi.org/10.1093/mnras/121.1.5

R.J. Weryk, P. Brown, A search for interstellar meteoroids using the Canadian Meteor Orbit Radar (CMOR). Earth Moon Planets **95**, 221–227 (2004). https://doi.org/10.1007/s11038-005-9034-x

R.J. Weryk, P.G. Brown, Simultaneous radar and video meteors, II: photometry and ionisation. Planet. Space Sci. **81**, 32–47 (2013). https://doi.org/10.1016/j.pss.2013.03.012

M.S. Westley, R.A. Baragiola, R.E. Johnson, G.A. Baratta, Photodesorption from low-temperature water ice in interstellar and circumsolar grains. Nature **373**, 405–407 (1995)

A.J. Westphal, S.C. Fakra, Z. Gainsforth, M.A. Marcus, R.C. Ogliore, A.L. Butterworth, Mixing fraction of inner solar system material in comet 81P/Wild2. Astrophys. J. **694**, 18–28 (2009). https://doi.org/10.1088/0004-637X/694/1/18

A.J. Westphal, R.M. Stroud, H.A. Bechtel, F.E. Brenker, A.L. Butterworth, G.J. Flynn, D.R. Frank, Z. Gainsforth, J.K. Hillier, F. Postberg, A.S. Simionovici, V.J. Sterken, L.R. Nittler, C. Allen, D. Anderson, A. Ansari, S. Bajt, R.K. Bastien, N. Bassim, J. Bridges, D.E. Brownlee, M. Burchell, M. Burghammer, H. Changela, P. Cloetens, A.M. Davis, R. Doll, C. Floss, E. Grün, P.R. Heck, P. Hoppe, B. Hudson, J. Huth, A. Kearsley, A.J. King, B. Lai, J. Leitner, L. Lemelle, A. Leonard, H. Leroux, R. Lettieri, W. Marchant, R. Ogliore, W.J. Ong, M.C. Price, S.A. Sandford, J.A.S. Tresseras, S. Schmitz, T. Schoonjans, K. Schreiber, Evidence for interstellar origin of seven dust particles collected by the stardust spacecraft. Science **345**(6198), 786–791 (2014). https://doi.org/10.1126/science.1252496.

F.L. Whipple, The theory of micro-meteorites, part I: in an isothermal atmosphere. Proc. Natl. Acad. Sci. **36**, 687–695 (1950). https://doi.org/10.1073/pnas.36.12.687

F.L. Whipple, A comet model, III: the zodiacal light. Astrophys. J. **121**, 750 (1955)

F.L. Whipple, On maintaining the meteoritic complex. SAO Spec. Rep. **239**, 1 (1967)

P. Wiegert, J. Vaubaillon, M. Campbell-Brown, A dynamical model of the sporadic meteoroid complex. Icarus **201**, 295–310 (2009). https://doi.org/10.1016/j.icarus.2008.12.030

P.A. Wiegert, Hyperbolic meteors: interstellar or generated locally via the gravitational slingshot effect? Icarus **242**, 112–121 (2014). https://doi.org/10.1016/j.icarus.2014.06.031. 1404.2159

M.J. Willis, M.J. Burchell, T.J. Ahrens, H. Krüger, E. Grün, Decreased values of cosmic dust number density estimates in the solar system. Icarus **176**, 440–452 (2005). https://doi.org/10.1016/j.icarus.2005.02.018

R.D. Wolstencroft, J.C. Kemp, Circular polarization of the nightsky radiation. Astrophys. J. **177**, L137 (1972)

D.H. Wooden, H.M. Butner, D.E. Harker, C.E. Woodward, Mg-rich silicate crystals in comet Hale-Bopp: ISM relics or solar nebula condensates? Icarus **143**, 126–137 (2000). https://doi.org/10.1006/icar.1999.6240

P.J. Wozniakiewicz, Grain sorting in cometary dust from the outer solar nebula. Astrophys. J. Lett. **760**, L23 (2012). https://doi.org/10.1088/2041-8205/760/2/L23

M.C. Wyatt, The insignificance of P-R drag in detectable extrasolar planetesimal belts. Astron. Astrophys. **433**, 1007–1012 (2005). https://doi.org/10.1051/0004-6361:20042073. astro-ph/0501038

S.P. Wyatt, F.L. Whipple, The Poynting-Robertson effect on meteor orbits. Astrophys. J. **111**, 134–141 (1950). https://doi.org/10.1086/145244

E.M. Xilouris, A.Z. Bonanos, I. Bellas-Velidis, P. Boumis, A. Dapergolas, A. Maroussis, A. Liakos, I. Alikakos, V. Charmandaris, G. Dimou, A. Fytsilis, M. Kelley, D. Koschny, V. Navarro, K. Tsiganis, K. Tsinganos, NELIOTA: the wide-field, high-cadence, lunar monitoring system at the prime focus of the Kryoneri telescope. Astron. Astrophys. **619**, A141 (2018). https://doi.org/10.1051/0004-6361/201833499. arXiv:1809.00495

T. Yada, H. Kojima, The collection of micrometeorites in the Yamato meteorite ice field of Antarctica in 1998. Antarct. Meteor. Res. **13**, 9 (2000)

S. Yamamoto, T. Mukai, Dust production by impacts of interstellar dust on Edgeworth-Kuiper belt objects. Astron. Astrophys. **329**, 785–791 (1998)

H. Yang, M. Ishiguro, Origin of interplanetary dust through optical properties of zodiacal light. Astrophys. J. **813**(2), 87 (2015)

Q.Z. Ye, M.T. Hui, P.G. Brown, M.D. Campbell-Brown, P. Pokorný, P.A. Wiegert, X. Gao, When comets get old: a synthesis of comet and meteor observations of the low activity comet 209P/LINEAR. Icarus **264**, 48–61 (2016). https://doi.org/10.1016/j.icarus.2015.09.003. arXiv:1509.00560

A. Zaslavsky, N. Meyer-Vernet, I. Mann, A. Czechowski, K. Issautier, G. Le Chat, F. Pantellini, K. Goetz, M. Maksimovic, S.D. Bale, J.C. Kasper, Interplanetary dust detection by radio antennas: mass calibration and fluxes measured by STEREO/WAVES. J. Geophys. Res. Space Phys. **117**(A16), A05102 (2012). https://doi.org/10.1029/2011JA017480

K. Zhang, G.A. Blake, E.A. Bergin, Evidence of fast pebble growth near condensation fronts in the HL Tau protoplanetary disk. Astrophys. J. Lett. **806**, L7 (2015). https://doi.org/10.1088/2041-8205/806/1/L7. 1505.00882

E. Zinner, K.D. McKeegan, R.M. Walker, Laboratory measurements of D/H ratios in interplanetary dust. Nature **305**, 119–121 (1983). https://doi.org/10.1038/305119a0

E.K. Zinner, Presolar grains. Treatise Geochem. **1**, 711 (2003). https://doi.org/10.1016/B0-08-043751-6/01144-0

M.E. Zolensky, Refractory interplanetary dust particles. Science **237**, 1466–1468 (1987). https://doi.org/10.1126/science.237.4821.1466

M.E. Zolensky, T.J. Zega, H. Yano, S. Wirick, A.J. Westphal, M.K. Weisberg, I. Weber, J.L. Warren, M.A. Velbel, A. Tsuchiyama, P. Tsou, A. Toppani, N. Tomioka, K. Tomeoka, N. Teslich, M. Taheri, J. Susini, R. Stroud, T. Stephan, F.J. Stadermann, C.J. Snead, S.B. Simon, A. Simionovici, T.H. See, F. Robert, F.J.M. Rietmeijer, W. Rao, M.C. Perronnet, D.A. Papanastassiou, K. Okudaira, K. Ohsumi, I. Ohnishi, K. Nakamura-Messenger, T. Nakamura, S. Mostefaoui, T. Mikouchi, A. Meibom, G. Matrajt, M.A. Marcus, H. Leroux, L. Lemelle, L. Le, A. Lanzirotti, F. Langenhorst, A.N. Krot, L.P. Keller, A.T. Kearsley, D. Joswiak, D. Jacob, H. Ishii, R. Harvey, K. Hagiya, L. Grossman, J.N. Grossman, G.A. Graham, M. Gounelle, P. Gillet, M.J. Genge, G. Flynn, T. Ferroir, S. Fallon, D.S. Ebel, Z.R. Dai, P. Cordier, B. Clark, M. Chi, A.L. Butterworth, D.E. Brownlee, J.C. Bridges, S. Brennan, A. Brearley, J.P. Bradley, P. Bleuet, P.A. Bland, R. Bastien, Mineralogy and petrology of comet 81P/Wild 2 nucleus samples. Science **314**, 1735 (2006). https://doi.org/10.1126/science.1135842

Space Sci Rev (2018) 214:23
https://doi.org/10.1007/s11214-017-0458-1

Impacts of Cosmic Dust on Planetary Atmospheres and Surfaces

John M.C. Plane[1] · George J. Flynn[2] · Anni Määttänen[3] · John E. Moores[4] · Andrew R. Poppe[5] · Juan Diego Carrillo-Sanchez[1] · Constantino Listowski[3]

Received: 18 May 2017 / Accepted: 11 December 2017 / Published online: 21 December 2017

Abstract Recent advances in interplanetary dust modelling provide much improved estimates of the fluxes of cosmic dust particles into planetary (and lunar) atmospheres throughout the solar system. Combining the dust particle size and velocity distributions with new chemical ablation models enables the injection rates of individual elements to be predicted as a function of location and time. This information is essential for understanding a variety of atmospheric impacts, including: the formation of layers of metal atoms and ions; meteoric smoke particles and ice cloud nucleation; perturbations to atmospheric gas-phase chemistry; and the effects of the surface deposition of micrometeorites and cosmic spherules. There is discussion of impacts on all the planets, as well as on Pluto, Triton and Titan.

Keywords Cosmic dust flux · Meteoric ablation · Planetary atmospheres · Meteoric smoke particles · Metallic layers · Noctilucent clouds · Surface deposition

1 Introduction

Critical to our understanding of how interplanetary dust affects the atmospheres of solar system bodies is an accurate understanding of the mass and velocity distributions of interplanetary dust grains throughout the solar system. These distributions are necessary not only

Cosmic Dust from the Laboratory to the Stars
Edited by Rafael Rodrigo, Jürgen Blum, Hsiang-Wen Hsu, Detlef Koschny, Anny-Chantal Levasseur-Regourd, Jesús Martín-Pintado, Veerle Sterken and Andrew Westphal

✉ J.M.C. Plane
j.m.c.plane@leeds.ac.uk

1 School of Chemistry, University of Leeds, Woodhouse Lane, Leeds LS2 9JT, UK

2 SUNY Plattsburgh, 101 Broad St., Plattsburgh, NY, 12901, USA

3 LATMOS/IPSL, UVSQ Université Paris-Saclay, UPMC Univ. Paris 06, CNRS, Quartier des Garennes, 11 boulevard d'Alembert, 78280 Guyancourt, France

4 Department of Earth and Space Science and Engineering, York University, 4700 Keele Street, Toronto, ON M3J 1P3, Canada

5 Space Sciences Laboratory, University of California at Berkeley, Berkeley, CA 94720, USA

for determining the overall mass influx of exogenous material to planetary atmospheres, but also for accurately quantifying the fraction of the incoming mass that ablates (since not all grains that enter a planetary atmosphere will fully ablate), the altitudes at which ablation occurs, the chemical evolution of the ablated products after injection into an atmosphere, and the degree of modification of the material deposited at the surface. Continual improvements in observations (both *in situ* and remote) and dynamical modelling have slowly but surely defined these quantities throughout the solar system. In parallel, significant advances in understanding the nature of the ablation process—though a combination of experimental simulation and modelling—means that the injection of individual meteoric constituents into a planetary atmosphere, and the consequent impacts, can be determined with much greater confidence.

In this review, the advances in interplanetary dust modelling are described in Sect. 2, followed by a discussion of meteoric ablation in Sect. 3. The next four sections describe a variety of atmospheric impacts, some observed and others predicted by models: layers of metal atoms and ions (Sect. 4); meteoric smoke particles and cloud nucleation (Sect. 5); perturbations to atmospheric gas-phase chemistry (Sect. 6); and the effects of the surface deposition of meteoric debris (Sect. 7). While this is not an exhaustive review, there is discussion of impacts on all planets with atmospheres, as well as on Pluto, Triton and Titan. Note that this review does not cover impacts on airless bodies, apart from a brief mention of the Moon and Mercury.

2 Astronomical Dust Models

In the inner solar system, observations and models over many decades have striven to define the interplanetary dust distribution, including both the velocity distribution and the overall flux. Early models were often phenomenological or empirical in nature and sometimes created *ad hoc* interplanetary dust distributions (i.e., not necessarily associated with a definite source) in order to explain discrete sets of observations (e.g., Divine 1993). Later models have used the orbital parameters of potential parent body populations, such as asteroids (ASTs), Jupiter Family Comets (JFCs), Halley Type Comets (HTCs), and Oort Cloud Comets (OCCs), as conditions to fit observations such as the Clementine star tracker zodiacal dust cloud images (Hahn et al. 2002); however, such an approach does not accurately capture the time evolution of dust grains away from their parents bodies. A more detailed understanding of the zodiacal dust cloud at 1 AU, including both its ultimate origin(s) and its contribution to the cosmic dust flux to the Earth, has come from a recent series of dynamical models (e.g., Wiegert et al. 2009; Nesvorný et al. 2010, 2011a,b; Pokorny et al. 2014). By comparison with remote observations from the Infrared Astronomical Satellite (IRAS), the Cosmic Background Explorer (COBE), and the Spitzer Space Telescope (Hauser et al. 1984; Low et al. 1984; Kelsall et al. 1998), and observations from terrestrial-based meteor surveys including the Canadian Meteor Orbit Radar (CMOR) (Campbell-Brown 2008), the Advanced Meteor Orbit Radar (AMOR) (Galligan and Baggaley 2004, 2005), and High Performance Large Aperture (HPLA) radars (e.g., Janches and Chau 2005; Janches and ReVelle 2005; Janches et al. 2006, 2008, 2014), the Nesvorný et al. (2010, 2011a) dynamical models have concluded that a majority (∼85–95%) of the zodiacal dust cloud near 1 AU originates from JFCs (e.g., Levison and Duncan 1997). AST, HTC, and OCC dust sources contribute the remaining 5–15%. Note also that while meteor showers can produce incident fluxes larger than the sporadic background flux for isolated time periods, the net flux at 1 AU from meteor streams contributes at most ∼10% relative to the sporadic background (e.g., Jones and Brown 1993). Additional in situ measurements continue to be made near 1 AU, such as by the IKAROS dust detector (Hirai et al. 2014, 2017) and the Lunar Dust Experiment (LDEX) (Horányi et al. 2015) onboard NASA's Lunar Atmospheric and Dust Environment Explorer (LADEE) mission.

Recently, Carrillo-Sánchez et al. (2016) used the cosmic spherule accretion rate at the bottom of an ice chamber at the Amundsen-Scott base at South Pole (Taylor et al. 1998), together with recent measurements of the vertical fluxes of Na and Fe atoms above 87 km in the atmosphere (Gardner et al. 2014; Huang et al. 2015; Gardner et al. 2016), to determine the absolute contributions of each of these dust sources to the global input of cosmic dust. This study showed that JFCs contribute $(80 \pm 17)\%$ of the total input mass of $43 \pm 14\ \mathrm{t\,d^{-1}}$, in good accord with COBE and Planck observations of the zodiacal cloud (Nesvorný et al. 2010, 2011a,b; Rowan-Robinson and May 2013; Yang and Ishiguro 2015).

Our understanding of dust distributions in the outer solar system has been slower to coalesce, mainly due to the infrequent opportunities for in situ observations. To date, dust measurements in the outer solar system come from the Pioneer 10 and 11 meteoroid detectors (Humes 1980; Dikarev and Grun 2002), the Galileo and Ulysses DDS experiments (e.g., Grün et al. 1995a, 1995b, Grün et al. 1997; Kruger et al. 1999, 2006; Kuchner and Stark 2010), the Cassini Cosmic Dust Analyzer (Altobelli et al. 2007; Hillier et al. 2007), and the New Horizons Student Dust Counter (SDC) (Horanyi et al. 2008; Poppe et al. 2010; Szalay et al. 2013). Plasma wave instrumentation on-board the Voyager 1 and 2 spacecraft has also detected signals best interpreted as interplanetary dust impacts on the spacecraft body (Gurnett et al. 1997). While each of these datasets are valuable in their own right, significant differences between the sensitivity, location, and operational profiles of each instrument present challenges when attempting to construct and/or constrain an overall picture of the dust influx to planetary atmospheres.

One method for estimating the dust fluxes to planetary atmospheres in the outer solar system is to extrapolate the Grün et al. (1985) interplanetary dust flux model, or its subsequent updates (Dikarev et al. 2004, 2005), from 1 AU to the desired heliocentric distance. In the absence of sufficient in situ measurements in the outer solar system, this methodology is perhaps the only recourse; however, there is no *a priori* reason to justify the assumption that dust dynamics at 1 AU hold throughout the outer solar system, especially given the strongly perturbing presence of the giant planets (e.g., Liou and Zook 1999). Fortunately, on-going measurements and modelling have begun to quantify the outer solar system dust environment in more detail.

The last two decades have seen a progression of increasingly detailed dynamical models for the outer solar system distribution, including those by Liou and Zook (1999), Landgraf et al. (2002), Kuchner and Stark (2010), Vitense et al. (2010, 2012, 2014), and Poppe (2016). The most recent model for the interplanetary dust distribution in the outer solar system comes from Poppe (2016). This model uses a dynamical approach to trace $>10^5$ individual dust grains from their origin under the influence of forces such as solar and planetary gravity, solar radiation pressure, Poynting-Robertson and solar wind drag, and the electromagnetic Lorentz force. Dust grains are launched with initial conditions representative of four families of parent bodies: JFCs, HTCs, OCCs, and Edgeworth-Kuiper Belt (EKB) objects. After applying a collisional grooming algorithm as developed by Stark and Kuchner (2009), the model is constrained by in situ measurements from Pioneer 10 (Humes 1980), Galileo (Sremcevic et al. 2005), and the New Horizons Student Dust Counter (Poppe et al. 2010; Szalay et al. 2013).

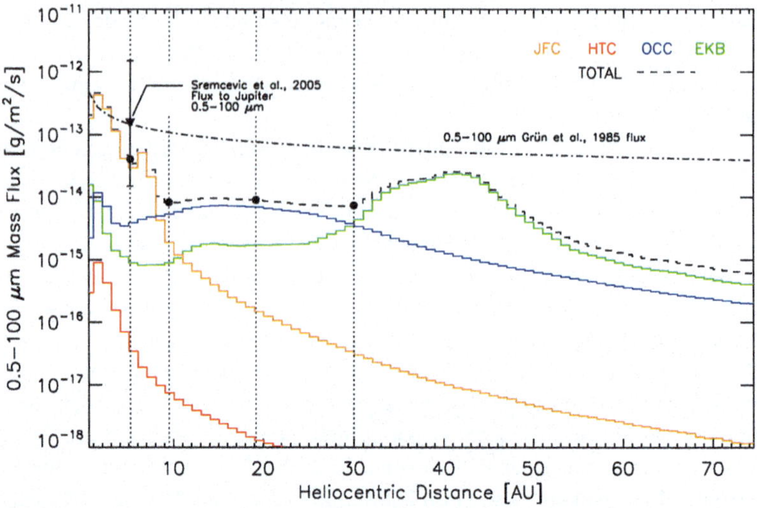

Fig. 1 The *unfocused* interplanetary dust grain mass flux as a function of heliocentric distance. The *black filled circles* refer to the four giant planets' average heliocentric distance. *Coloured lines* are contributions from the JFC, HTC, OCC and EKB parent sources, while the total flux is the *black dashed line*. The *dot-dash line* is the Grün et al. (1985) flux at 1 AU extrapolated outwards for comparison (adapted from Poppe 2016). While not shown, the mass flux of interstellar dust entering the solar system is approximately 5×10^{-17} g m^{-2} s^{-1} (Grün et al. 1994), and thus only a minor contribution relative to interplanetary dust mass fluxes

Figure 1 shows the unfocused interplanetary dust mass flux as a function of heliocentric distance for the four individual populations considered (coloured lines) and the total flux (dashed line). For comparison, Fig. 1 also displays the extrapolated Grün et al. (1985) interplanetary dust flux (dash-dot line). The relative contributions of various dust sources are strongly dependent on heliocentric distance. Within 10 AU, JFC dust dominates by more than an order-of-magnitude. Since a significant fraction (~30–40%) of JFC particles are hydrated, this represents a significant source of hydrous particles into the planetary atmospheres within 10 AU. Between 10 and 30 AU, OCC dust is predicted to be dominant, with some additional contributions (~10–20%) from both JFC and EKB dust. Finally, at distances greater than 30 AU the EKB dust dominates, reflecting the near-by location of the EKB parent bodies themselves (e.g., Petit et al. 2011). One can also see the very large discrepancy between the extrapolated Grün et al. (1985) model prediction and the total flux found in Poppe (2016), with the latter typically an order-of-magnitude less than the former. This model also provides the mass and velocity distributions for each dust grain family with gravitational focusing taken into account (e.g., Spahn et al. (2006)). A remaining open question is whether dust grains from the distant EKB can reach the inner solar system and contribute to cosmic dust accretion at the terrestrial planets (e.g., Flynn 1994; Liou et al. 1996; Moro-Martin and Malhotra 2003; Ipatov and Mather 2006). Lastly, the mass flux of interstellar dust entering the solar system is approximately 5×10^{-17} g m^{-2} s^{-1} (Grün et al. 1994). This is between 2 and 3 orders-of-magnitude lower than the interplanetary dust mass fluxes at the giant planets (Fig. 1), and thus only a minor contribution. Although interstellar dust grains have relatively high velocities compared with interplanetary dust, any differences in ablation behaviour will most likely be obscured by the much higher flux of interplanetary dust grains.

3 Meteoric Ablation

In order to assess the impacts of cosmic dust in a planetary atmosphere, it is necessary to model the rate of injection of ablation products (i.e. metal atoms and ions) as a function of height, location and time. This requires combining an ablation model with an astronomical model which predicts the distributions of particle mass, velocity and entry angle from the different dust sources (Sect. 2). Ablation models typically use the "classical" meteor physics equations of momentum and energy conservation as a meteoroid enters the atmosphere, underwritten by several assumptions: the meteoroid is treated as a homogenous spherical particle; radial heat transfer is assumed fast enough so the particle is isothermal along its whole path through the atmosphere; and the interaction with the atmosphere occurs in the free molecular flow regime in which the molecular collision mean free path is larger than the dimension of the meteoroid and thus no shock structure can develop.

A current example of an ablation model which also predicts the injection rates of individual elements is the Chemical Ablation MODel (CABMOD) (Vondrak et al. 2008). This model estimates the elemental ablation rates as a function of height from a meteoroid of specified mass, entry velocity and zenith angle. CABMOD treats mass loss both by physical sputtering, which is the mass loss mechanism before particle melting when individual atoms are displaced from the surface of the meteoroid by collisions with air molecules, and thermal ablation after melting. Sputtering can be modelled using laboratory ion-sputtering data in the absence of appropriate data on high speed neutral-surface collisions (Vondrak et al. 2008), but in any case turns out to be relatively minor except for small ($r < 1$ μm) particles entering at high velocity (>40 km s^{-1}).

To model thermal ablation, CABMOD contains the MAGMA chemical equilibrium model (Fegley and Cameron 1987; Schaefer and Fegley 2004, 2005) which calculates the equilibrium vapour pressures of the various melt constituents which are treated as eight metal oxides (SiO_2, MgO, FeO, Al_2O_3, TiO_2, CaO, Na_2O and K_2O). The chemical equilibria in the melt are calculated using their Raoultian activities. The mass loss rate through evaporation is estimated by using Langmuir evaporation through the Hertz-Knudsen equation (Markova et al. 1986; Love and Brownlee 1991; McNeil et al. 1998; Vondrak et al. 2008), which assumes that the rate of evaporation into a vacuum is equal to the rate of evaporation needed to balance the rate of uptake of a species in a closed system. The calculated evaporation rate is formally an upper limit, which is probably correct for pure metals (Safarian and Engh 2013), but may be lower in silicate melts because diffusion from the bulk into the surface film may become rate-limiting (Alexander et al. 2002). Vondrak et al. (2008) assumed that meteoroids are mineralogically CI chondrites with an olivine composition (the elemental atomic ratio Fe:Mg is ~0.84), with an onset of melting at 1750 K.

Recently, a Meteoric Ablation Simulator (MASI) has been developed to test the predictions of ablation models like CABMOD (Bones et al. 2016; Gómez Martín et al. 2017). The MASI heats a meteoritic particle ($r = 20$–200 μm) over a temperature ramp (up to 2800 K) that is programmed to match atmospheric entry for a specified velocity, and measures the absolute rates of evaporation of pairs of metal atoms (e.g. Na and Fe) using high repetition rate time resolved laser induced fluorescence. MASI measurements confirm the model prediction of differential ablation i.e. the evaporation of relatively volatile elements such as Na and K before the main elements Fe, Mg and Si, and finally the refractory elements Ca, Ti and Al. However, the ablation profiles of individual species tend to be broader than predicted by a model such as CABMOD because meteorites do not consist of single mineral phases (Gómez Martín et al. 2017).

An important use of differential ablation models is to predict the rate of electron production along the path of a meteor; this occurs as individual elements ablate and then ionize through collisions with air molecules (Janches et al. 2014). The electron production rate is needed to determine the meteor detectability of a particular radar, and hence to correct for the inherent observational bias to fast meteors (which produce electrons much more efficiently) when performing meteor astronomy (Close et al. 2007; Janches et al. 2014). An electrostatic dust accelerator has recently been used to generate metallic particles with velocities of 10–70 $km\,s^{-1}$, which are then introduced into a pressurized chamber where the particle partially or completely ablates over a short distance. The deceleration is used to determine the rate of mass loss, and an array of biased electrodes above and below the ablation path collect the ions and electrons produced along the ablation path, from which the ionization efficiency can be determined (Thomas et al. 2016, 2017).

A model like CABMOD can then be used to calculate the height profile of the injection rates of the meteoritic constituents into a planetary atmosphere. This is achieved by sampling individual particles from the mass and velocity distributions of the cosmic dust flux entering the atmosphere, computing each particle's elemental injection profiles, and performing a weighted sum over the mass/velocity distribution (Carrillo-Sánchez et al. 2015). The left-hand panels of Fig. 2 illustrate the cosmic dust mass flux as a function of entry velocity into the atmospheres of Earth, Mars and Venus, for JFC, AST, and HTC particles (OCC particles are also shown in the case of the Earth). These distributions are derived from the ZoDy model (Nesvorný et al. 2010, 2011a,b; Carrillo-Sánchez et al. 2016). The range of entry velocities are 11.5–71.5 $km\,s^{-1}$ for Earth, 5.5–59.5 $km\,s^{-1}$ for Mars, and 10.5–85.5 $km\,s^{-1}$ for Venus. Most of the JFC and AST mass enters at low velocities ($<20\ km\,s^{-1}$) compared with the HTC and particularly OCC particles, which therefore experience a much higher degree of ablation (Carrillo-Sánchez et al. 2016). The right-hand panels show the ablation rate profiles of various metals as a function of height; these are produced by integrating the ablation profiles calculated by CABMOD for individual silicate particles over the mass, velocity and entry angle distributions of the particle sources (Carrillo-Sánchez et al. 2016). Note that peak mass loss occurs for all three planets when the pressure is around 10^{-6} bar, which is ~80 km on Mars, 92 km on Earth, and 115 km on Venus, in good agreement with the earlier work of Molina-Cuberos et al. (2008). The volatile alkali metals (Na and K) ablate several km higher, and the refractory elements (Ca, Ti, Al) several km lower. The height over which ablation occurs is narrower on Mars because of the smaller range of entry velocities, and the ablated mass is much smaller than on Earth and Venus.

There have been a number of studies of ablation in the atmospheres of the giant planets, including at Jupiter (Kim et al. 2001; Pesnell and Grebowsky 2001), Saturn (Moses and Bass 2000; Moses et al. 2000), and Neptune (Moses 1992). See Molina-Cuberos et al. (2008) for a general review. As discussed in Sect. 2, there is a relative lack of measurements in the outer solar system which has necessitated the use of simplifying assumptions regarding the particle mass/velocity distributions; however, the dust model of Poppe (2016) now provides these distributions for ablation modelling. This was recently applied to Saturn's moon Titan (Frankland et al. 2016). Due to the large masses and hence deep gravitational wells of the giant planets, the minimum meteoroid entry velocities are 59.5, 35.5, 21.3, and 23.5 $km\,s^{-1}$ for Jupiter, Saturn, Uranus, and Neptune, respectively. Any variations in the velocity distributions of meteoroids in interplanetary space are compressed during gravitational focusing as all grains are accelerated at minimum to the planetary escape velocity. In the outer solar system, the three primary families of dust grains are JFC, OCC and EKB dust, each with distinct mass and velocity distributions. These distributions are a function of not only the grain mass and velocity, but also heliocentric distance (Fig. 1). The initial mass and velocity distributions of dust entering the atmospheres of each of the giant planets are shown in Fig. 3, calculated from the interplanetary dust distributions of Poppe (2016).

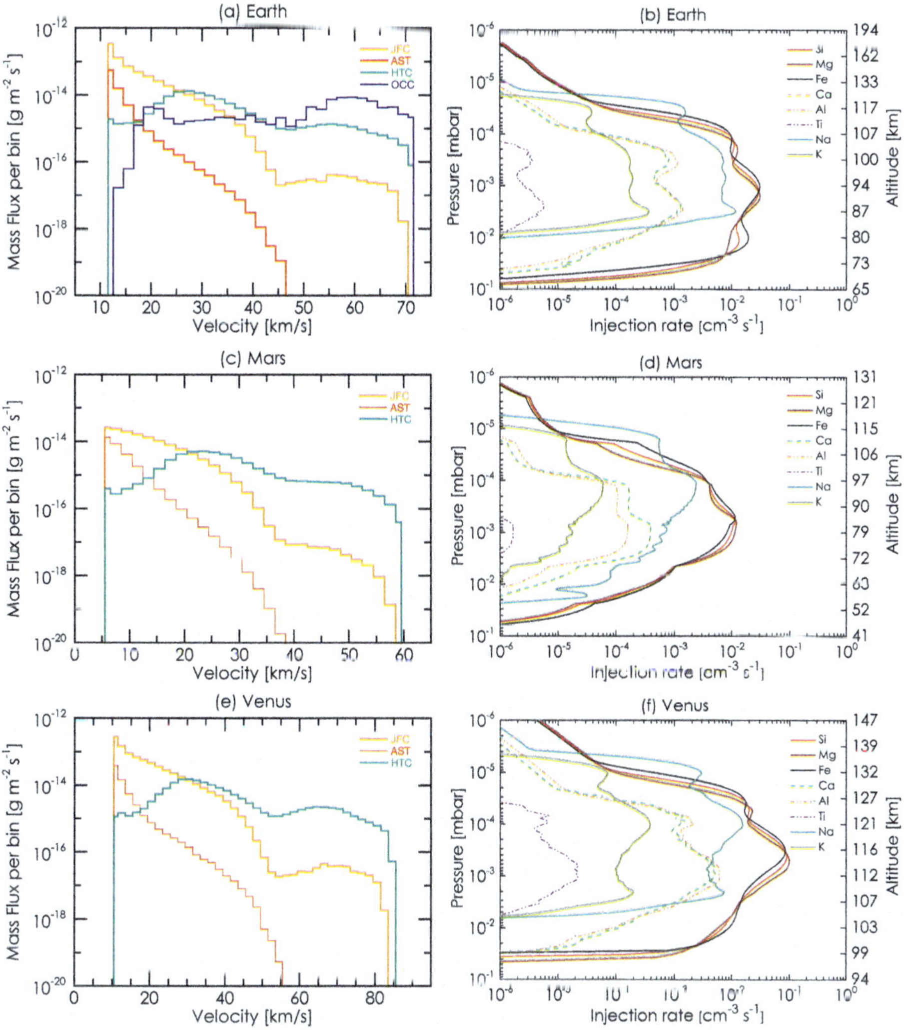

Fig. 2 Panels **a**, **c**, and **e**: mass flux as a function of entry velocity for interplanetary dust from Jupiter-Family comets (JFC, in *yellow*), Asteroid belt (AST, in *red*), Halley-Type comets (HTC, in *cyan*), and Oort-Cloud comets (OCC, in *blue*) for Earth, Mars and Venus. Panels **b**, **d**, and **e**: overall ablation rate profiles of individual elements, integrated and weighted for the JFC, AST and HTC particle populations (Carrillo-Sánchez et al. 2016)

Figure 4 shows the gas injection rate profiles arising from meteoric ablation in the atmospheres of Jupiter, Saturn, Uranus, and Neptune for three dust grain compositions: silicate, organic (using benzoapyrene as a representative species), and water ice (Moses and Poppe 2017). These model results use the interplanetary dust mass and velocity distributions in Fig. 3. Based on the approximate relative mass fractions in cometary nuclei and dust (Greenberg and Li 1999), the bulk dust influx is assumed to comprise 26% silicate, 32% refractory organic, and 42% ice grains. The dust grain parameters—density, mean molecular mass, vaporization temperature, latent heat of vaporization, emissivity etc.—control the differences in the ablation profiles (Moses and Poppe 2017). Several general trends are notable: (1) larger grains penetrate deeper into any atmosphere for any other combination of param-

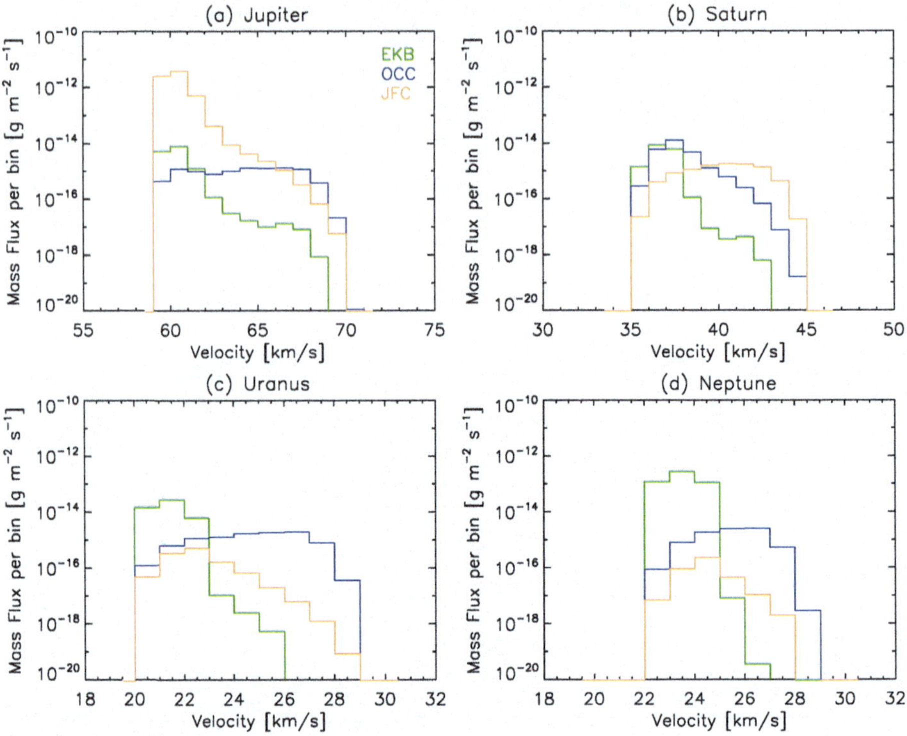

Fig. 3 The mass flux as a function of entry velocity for interplanetary dust grains from three populations (Edgeworth-Kuiper Belt, *green*; Oort Cloud comets, *blue*; Jupiter-family comets, *orange*) for each of the giant planets (adapted from Moses and Poppe 2017)

eters; (2) larger grains also tend to ablate more fully, since their greater cross section leads to more impacts with ambient molecules and thus, greater heating; (3) faster particles tend to ablate at higher altitudes and more fully than their slower counterparts; (4) the model predicts that grains of all compositions ablate on Jupiter and Saturn due to higher minimum entry velocities while silicate grains only partially ablate on Uranus and Neptune (organic and water ice grains completely or nearly completely ablate on all the giant planets); (5) the vaporization temperature typically controls the average altitude that a given composition grain ablates, with water ice beginning to ablate at the highest altitude (i.e., lower vaporization temperature), organics at an intermediate altitude, and silicate grains (with higher vaporization temperatures) ablating at typically lower altitudes; and (6) the latent heat of vaporization influences the range of altitudes over which ablation occurs, with low latent heats associated with narrower ablation profiles and high latent heats associated with broad ablation profiles (i.e., low latent heat implies more inefficient cooling from evaporation which in turn leads to higher grain temperatures and relatively rapid ablation and vaporization).

4 Metallic Layers in Planetary Atmospheres

4.1 Meteoric Metal Layers on Earth

Layers of metallic atoms and ions occur between about 75 and 200 km in the terrestrial atmosphere, as a result of meteoric ablation. These layers have been the subject of two

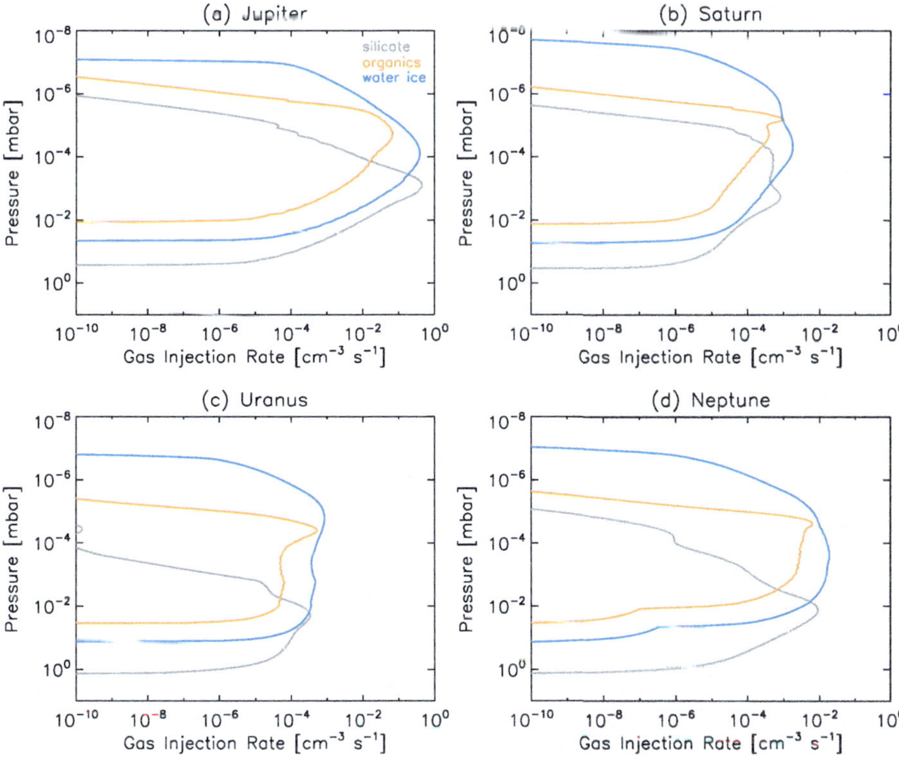

Fig. 4 The gas injection rate due to micrometeoroid ablation in the atmospheres of the giant planets (adapted from Moses and Poppe 2017). Three different dust grain compositions were considered: silicate (*gray*), organic (*orange*), and water ice (*blue*). All panels are scaled identically

major reviews which cover the history of their detection, laboratory studies to unravel the chemistry which creates the layers, and the development of models including recent whole atmosphere chemistry-climate models (Plane 2003; Plane et al. 2015). Hence, the discussion here will be comparatively brief.

Metal ions (Fe^+, Mg^+, Na^+, Si^+ etc.) have been observed since the 1960s mainly by rocket-borne mass spectrometry (Kopp 1997; Grebowsky and Aikin 2002), although Mg^+ can also be observed by satellite optical spectroscopy (Langowski et al. 2015) and Ca^+ by lidar (Gerding et al. 2000). Following the invention of the laser in the 1970s, the layers of neutral Na, K, Fe and Ca have been observed by ground-based resonance lidars because these metals have spectroscopic transitions at wavelengths longer than 330 nm, and hence the laser light and resonance fluorescence are not absorbed by the stratospheric ozone layer (unlike Mg). The metal layers peak between 85 and 95 km, and are only 5–10 km wide with sharp top- and bottom-sides (Plane 2003). Although most lidar measurements have focused on these metal layers below 105 km, recent advances in lidar technology have enabled measurements of Fe atoms up to ~190 km (Chu et al. 2011). Simultaneous lidar measurements of the vertical wind and Na and Fe densities have been used to determine the vertical fluxes of these metals around 87 km (Huang et al. 2015), and hence the absolute inputs into the atmosphere of dust from the different sources discussed in Sect. 2 (Carrillo-Sánchez et al. 2016). Since 2007, satellite optical spectrometers using solar-pumped resonance fluorescence—SCIAMACHY on Envisat (Langowski et al. 2015) and OSIRIS on

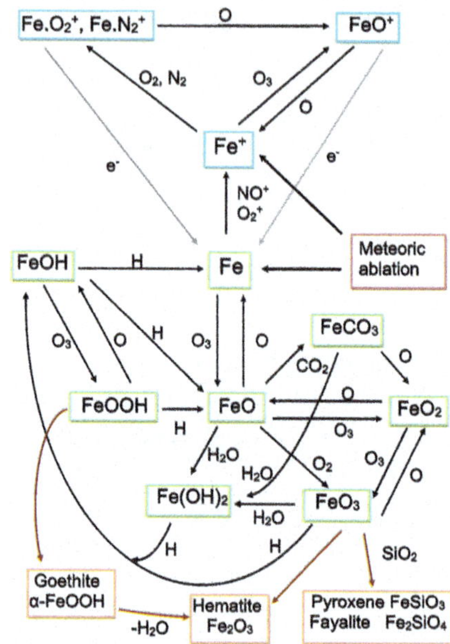

Fig. 5 Schematic diagram of the chemistry of Fe in the terrestrial mesosphere and lower thermosphere (adapted from Plane et al. 2015). Ionized species are shown in *blue boxes*, neutral gas-phase species in *green boxes*, and potential precursors of meteoric smoke particles in *orange boxes*

Odin (Fan et al. 2007; Dawkins et al. 2014), or stellar occultation—GOMOS on Envisat (Fussen et al. 2010)—have provided near-global coverage of the Na, K and Mg layers. We therefore know a great deal about the latitudinal, seasonal and diurnal variations of the neutral metal layers.

Figure 5 is a schematic diagram of the chemistry of Fe in the terrestrial upper atmosphere. Many of the individual reactions in this scheme have been studied in the laboratory under appropriate conditions of temperature, and a sufficient range of pressure to permit confident extrapolation to the low pressures of the upper mesosphere ($<10^{-5}$ bar) where reaction rate coefficients cannot be measured using the conventional pulsed laser photolysis and fast flow tube techniques. Similarly detailed schemes exist for Na, K, Mg and Ca (Plane et al. 2015), as well as Si (Plane et al. 2016). The ionic species in Fig. 5 are shown in blue boxes. These species tend to dominate above 100 km in the lower *E* region. Metal ions are produced directly during meteoric ablation: the metals atoms which evaporate are initially travelling with a speed similar to that of the parent meteoroid, and so undergo hyperthermal collisions with air molecules which can lead to ionization. Metallic ions are also produced by charge transfer with the ambient NO^+ and O_2^+ ions, and photo-ionization. Neutralization occurs through forming molecular ions (for Fe^+ mostly by reaction with O_3), followed by dissociative recombination with electrons; however, this neutralization pathway is interrupted by atomic oxygen, which reduces metal-containing ions back to the atomic metal ions so that the lifetimes of metal ions can be extended to several days. Above ~105 km, where there is very little O_3, neutralization occurs via the slow process of radiative recombination with electrons (Bones et al. 2015).

The neutral species in Fig. 5 are identified in green boxes. All the metal atoms studied to date react rapidly with O_3 to form the metal monoxides, which can then form more stable reservoir species (e.g. hydroxides and carbonates) by further reactions involving O_3, O_2, H_2O and CO_2 (Plane et al. 2015). However, as shown in Fig. 5 these reservoir species are reduced back to neutral metal atoms by atomic O and H. Below 82 km, the concentrations of atomic O and H decrease rapidly, and the reservoir species now survive long enough to polymerize into small particles (orange boxes in Fig. 5) known as Meteoric Smoke Particles (MSPs), which cause permanent removal of the metals from the gas phase (Sect. 5.1). The conversion to MSPs below 82 km, and to ions above 100 km, explains why the metal atoms occur in thin layers that peak between 85 and 95 km in the terrestrial atmosphere. The chemistry for six meteoric metals (Na, K, Fe, Mg, Ca and Si—described by ∼140 reactions) has now been incorporated into a whole atmosphere chemistry-climate model (Plane et al. 2015, 2016).

4.2 Meteoric Metal Layers on Venus and Mars

Until very recently, there was only indirect evidence for metallic layers in other planetary atmospheres. This evidence was obtained from radio occultation measurements, where the attenuation of radio waves transmitted from a spacecraft through a planet's atmosphere and received at Earth can be used to determine the electron density profile in the atmosphere. Pioneer Venus measurements show that the main ion layer in the Venus night-side ionosphere peaks around 142 km, with a second, intermittent peak around 120 km (Kliore et al. 1979). This is close to the peak of meteoric ablation (Fig. 2), and so was tentatively attributed to a layer of metallic ions and electrons, although it was recognized that direct ionization by energetic electron or proton precipitation could also be responsible (Kliore et al. 1979; Molina-Cuberos et al. 2008). The main ionospheric peak in Titan was located at 1180 ± 150 km during the Voyager I fly-by (Bird et al. 1997). More recently, Cassini radio occultation measurements show that there is a secondary intermittent layer between 500 and 600 km, which coincides with the region where ablation is predicted to occur (Frankland et al. 2016).

The atmosphere of Mars has been sounded in more detail. Fjeldbo et al. (1966) analyzed Mariner IV data and found that the daytime ionosphere is characterized by a main layer produced by solar photo-ionization at an altitude of 140 km, and a secondary layer around 100 km with an electron density approximately one order of magnitude lower. Savich et al. (1976) reported a secondary layer at ∼80 km during nighttime using the Soviet spacecraft Mars 4 and Mars 5. Some years later, these layers were confirmed by Mars Express observations of a layer between 65 and 110 km which appeared sporadically in about 10% of measured electron density profiles (Pätzold et al. 2005). A more extensive set of electron density profiles obtained using Mars Global Surveyor exhibited secondary layers between 70 and 105 km, in ∼4% of cases (Withers et al. 2008; Pandya and Haider 2012). Their sporadic occurrence, width of a few km, and appearance over the height range where ablation occurs (Fig. 2), are in many respects similar to sporadic *E* layers in the terrestrial ionososphere which are known to consist of metallic ions (Grebowsky and Aikin 2002). Therefore, these studies of the Mars low-lying layers concluded that they were most likely composed of metallic ions deposited by meteoric ablation, although the mechanism for concentrating metallic ions into a narrow layer in the absence of a permanent magnetic field was unclear.

More recently, metallic ions (Mg^+, Fe^+ and Na^+) have been measured in situ above 120 km in the Mars ionosphere by the Neutral Gas Ion Mass Spectrometer (NGIMS) aboard the Mars Atmosphere and Volatile Evolution (MAVEN) spacecraft (Grebowsky et al. 2017) Another instrument on MAVEN, the Imaging Ultraviolet Spectrograph (IUVS), was able to detect Mg^+ down to around 70 km by observing solar-pumped resonance fluorescence at 280 nm (Crismani et al. 2017b). Several unexpected features were observed. First, the NGIMS observations show that all three metallic ions have the same scale height, in spite of being in a very low pressure region where gravitational separation should occur. Second, the IUVS measurements of the Mg^+ layer peaking around 95 km are reproduced well by a detailed chemistry model with meteoric ablation (Whalley and Plane 2010); however, the model predicts an ever larger neutral Mg layer peaking around 80 km, which is *not* observed. Third, no instances of sporadic Mg^+ layers were observed which would correspond to the sporadic electron density layers recorded by radio occultation, so it appears that they have been wrongly attributed to metallic ion layers in the case of Mars (Crismani et al. 2017b). It is also worth noting that although the IUVS instrument observed Mg, Fe^+ and Fe in addition to Mg^+ after the meteor storm following the close encounter between Comet Siding Spring (C/2013 A1) and Mars in October 2014 (Schneider et al. 2015), only Mg^+ has been observed under background conditions.

5 Meteoric Smoke Particles and Clouds

5.1 Formation of Meteoric Smoke Particles

MSPs form via the polymerization of metal-containing molecules (e.g. FeOH, $Mg(OH)_2$, $MgCO_3$, $NaHCO_3$ and $Si(OH)_2$), which are the relatively long-lived reservoir species on the underside of the metallic layers (Plane et al. 2015, 2016). In the CO_2-rich atmospheres of Mars and Venus, there is probably a higher fraction of carbonates compared with hydroxides (Whalley and Plane 2010). Laboratory studies have shown that these molecules polymerize rapidly because they have large electric dipole moments and, if they contain Fe, their collisions are also governed by long-range magnetic dipole forces (Saunders and Plane 2006, 2010). However, polymerization occurs over several days because the concentrations of these species are low. In the terrestrial atmosphere, most of the particles around 80 km appear to be just large molecular clusters ($\sim$1 nm in effective radius) (Rapp et al. 2007), which have proved very challenging to capture for compositional analysis (Hedin et al. 2014). This has been frustrating given the potentially important roles that MSPs play in the middle atmospheres of planets: acting as ice nuclei for polar mesospheric clouds (Sect. 5.2); providing a significant component of stratospheric sulphate aerosols and polar stratospheric clouds (Sect. 5.3); and providing a reactive surface which can alter the gas-phase composition of a planetary atmosphere through heterogeneous chemistry (Sect. 6).

Because MSPs form in the *D* region of the terrestrial ionosphere, a small fraction ($\sim$6%) are charged by uptake of electrons and this enables them to be detected using Faraday cup detectors on sounding rocket payloads (Gelinas et al. 2005; Rapp et al. 2012; Plane et al. 2014). Faraday detectors collect charged particles by using the rocket payload ram velocity to drive the particles through a series of electrically biased screens, which reject thermal electrons and positive ions and allow heavy charged particles to reach the detector. By modelling the MSP size range—typically between 0.5 and 2 nm radius—that will be collected efficiently according to the aerodynamic flow around the payload, and the fraction that are charged, the total concentration of MSPs can be estimated (Rapp et al. 2007).

MSPs have also been detected between 80 and 95 km using high performance large aperture radars such as the 430 MHz dual-beam Arecibo incoherent scatter radar in Puerto Rico (18°N). The distinctive line shapes of the incoherent scatter radar spectra result from the different diffusion modes in the *D* region plasma caused by the presence of positive ions and relatively heavy charged MSPs (Strelnikova et al. 2007). MSP number densities and size can be retrieved, although a monodisperse MSP population has to be assumed. MSPs have been detected optically between 40 and 75 km by the limb-scanning Solar Occultation For Ice Experiment (SOFIE) spectrometer on the Aeronomy of Ice in the Mesosphere (AIM) satellite (Hervig et al. 2009). The extremely small extinctions due to MSPs ($<1 \times 10^{-8}$ km^{-1}) are measured by solar occultation over a pathlength of $\sim$300 km through the atmosphere.

MSPs are deposited at the Earth's surface 4–5 years after formation in the MLT (Dhomse et al. 2013). Several studies have reported measurements of the MSP deposition flux in polar ice cores. Gabrielli et al. (2004) measured the concentrations of Ir and Pt in the Greenland Ice Core Project (GRIP) ice core from Summit, central Greenland using inductively-coupled plasma sector field mass spectrometry (ICP-MS). The Ir and Pt signals were normalized to Al measured in the ice, which is an element that is a good indicator of crustal dust. This showed that the contribution from crust dust to the fluxes of Ir and Pt during the Holocene was negligible, and that the ratio of these elements was in the expected cosmic abundance ($Ir/Pt = 0.49$), confirming that they were of extra-terrestrial origin. Another method for detecting MSPs in polar ice cores is laboratory-induced remanent magnetization, which measures the magnetization carried by ferromagnetic dust particles in the ice (Lanci et al. 2012). This non-destructive technique provides a way of separating MSPs from larger crustal dust particles, because particles with radii between 3.5 and 10 nm become super-paramagnetic as the ice sample is warmed from 77 K to 255 K. In the case of central Greenland during the Holocene, the concentration of super-paramagnetic Fe was almost identical to that calculated from the Ir and Pt measurements using the relative Fe/Ir and Fe/Pt cosmic abundances. Lanci et al. (2012) demonstrated that wet deposition of the MSPs is more important than dry deposition: the deposition flux is about an order of magnitude higher in central Greenland than the eastern highlands of Antarctica, consistent with the relative snowfall at the two locations. It should be noted, however, that a recent study using a global circulation model to predict the deposition of MSPs over the Earth's surface has found a much lower deposition flux in these polar locations, by more than an order of magnitude, when using a cosmic dust input to the atmosphere of $43\ \mathrm{t\,d^{-1}}$ (Brooke et al. 2017).

5.2 Mesospheric Clouds on Earth and Mars

Polar mesospheric clouds (PMCs) on Earth have received a great deal of attention as sensitive indicators of climate change (Thomas and Olivero 2001). The clouds were first reported during June 1885 over middle and northern Europe. When viewed from the ground after sunset they are often referred to as noctilucent clouds (NLCs). PMCs are H_2O-ice clouds which form at the very cold temperatures (<145 K) found at altitudes between 82 and 86 km during summer at high latitudes ($>55°$) (Rapp and Thomas 2006). One explanation for the increasingly bright clouds is the growing emission of methane since the Industrial Revolution, since this relatively inert gas reaches the stratosphere where it is oxidized to H_2O. In fact, H_2O in the stratosphere and mesosphere has been increasing at a rate of $\sim 1\%\ \mathrm{year^{-1}}$ since the 1950s (Russell et al. 2014). Temperatures in the mesosphere are also decreasing by around -0.5 to $-2.0\ \mathrm{K\,decade^{-1}}$ because of depletion of the stratospheric ozone layer, and increasing greenhouse gases (particularly CO_2) which act as refrigerants in the middle atmosphere (Lübken et al. 2013). The combined trends of cooling temperatures and increasing H_2O concentrations have almost certainly caused the statistically significant increase in both PMC brightness and occurrence frequency which has been observed using Solar Backscatter Ultraviolet (SBUV) instruments on a sequence of nadir viewing satellites (DeLand et al. 2007; Shettle et al. 2009). PMCs can also be detected by radar: Polar Mesospheric Summer Echoes (PMSEs) are intense radar backscatter echoes that are caused mainly by small ice particles ($r < 10$ nm). These particles are negatively charged by the attachment of electrons and the radar is then scattered by the resulting plasma inhomogeneities (Lübken et al. 1998). There has also been an upward trend in PMSE since 1994 (Latteck and Bremer 2013).

One remaining question is what provides the ice nuclei for PMC formation, since the clouds are observed to form at temperatures above 120 K, which is the upper limit for homogeneous nucleation in the very dry mesosphere (Murray and Jensen 2010). Proposed ice nuclei include *D* region proton hydrates (Balsiger et al. 1996) and MSPs (Kalashnikova et al. 2000), although nucleation on proton hydrates is challenging because of competition with dissociative electron recombination destroying the nascent ice particle before it grows large enough. Nucleation presumably occurs at the mesopause (~87 km) and the ice particles then settle gravitationally as they grow to visible clouds with $r = 50$–80 nm, eventually sublimating in the warmer mesosphere below 82 km. Laboratory studies have not yet demonstrated that MSPs of 1–2 nm radius can overcome the Kelvin barrier to nucleation. However, as discussed in Sect. 5.1, around 6% of MSPs should be negatively-charged in above 80 km region, and the presence of even a single charge on a nm-sized particle can make it an effective ice nucleus by reducing the classical free energy barrier associated with the Kelvin effect (Gumbel and Megner 2009). Alternatively, electronic structure theory calculations indicate that metal silicate molecules such as $FeSiO_3$ and $MgSiO_3$ should form readily in the upper atmosphere (Plane 2011). These molecules, which are essentially the smallest MSP unit, have enormous electric dipole moments of 9.5 and 12.2 Debye, respectively. H_2O molecules therefore bind very effectively to them, and model calculations show that they should nucleate ice particles efficiently under polar mesospheric conditions at temperatures around 140 K (Plane 2011). Optical extinction measurements by the SOFIE instrument (Sect. 5.1) show that between 0.01 and 3% of the ice particle mass is meteoric material (Hervig et al. 2012).

Martian Mesospheric Clouds (MMCs) are the highest clouds in the Martian atmosphere; they are the counterparts to terrestrial PMCs (Määttänen et al. 2013). Their composition has been identified as mainly CO_2 ice crystals, with a small component of H_2O ice. The former feature is unique for a telluric planet as it results from the condensation of the main atmospheric constituent. The detection of CO_2 cloud ice crystals was first claimed by Herr and Pimentel (1970) using Mariner 6 and 7 infrared observations. The low altitude of the feature (<30 km) pointed rather to CO_2 fluorescence (Lellouch et al. 2000; Lopez-Valverde et al. 2005). Given the CO_2 supersaturations measured around 80 km during the Mars Pathfinder mission's atmospheric entry (Schofield et al. 1997; Clancy and Sandor 1998) used Earth-based submillimeter observations to infer that the predawn bluish cloud observed by the Imager for Mars Pathfinder (IMP) (Smith et al. 1997) was a CO_2 ice cloud. Four detections of nighttime aerosol layers (90–100 km), at southern mid-latitudes and $L_s = 135°$ were made in stellar occultation mode (Montmessin et al. 2006) with the SPICAM spectrometer onboard Mars Express (MEx). They were identified as CO_2 ice clouds due to the simultaneous presence of supersaturations. The first identification of CO_2 ice spectroscopic features in an infrared atmospheric spectrum was presented by Formisano et al. (2006) in their study of non-LTE emission using the Planetary Fourier Spectrometer (PFS) onboard MEx. However, the first unambiguous observations of CO_2 mesospheric clouds (Montmessin et al. 2007) were made using the spectral imager OMEGA (MEx). Very recently, Aoki et al. (2018) performed unambiguous CO_2 MMC detections using the PFS. They achieved the highest resolution cloud spectra to date, which will provide detailed information about crystal shape and composition.

More generally, the first systematic detections of MMCs (65–75 km) over the Martian years MY = 24–26—without constraints on their composition—were presented by Clancy and co-workers (Clancy et al. 2003, 2007) using limb observations with the Thermal Emission Spectrometer (TES) and the Mars Orbiter Camera (MOC), onboard Mars Global Surveyor (MGS). MMCs were mainly located around the equator (15°S–15°N), and formed after northern spring equinox ($L_s = 0$–$55°$) and during northern summer ($L_s = 105$–$180°$), confined between 240°E and 30°E. Later systematic detection of MMCs largely followed this spatial coverage and seasonality. Määttänen et al. (2010) focused on the systematic detection of CO_2 ice clouds with OMEGA over the years MY = 27–29, characterising the effective crystal size ($r_{eff} = 1$–3 μm) and opacity at 1 μm (<0.5). The nighttime cloud layers detected by Montmessin et al. (2006a) had crystal sizes of $r_{eff} = 80$–110 μm and low opacities (0.01 at 200 nm). The HRSC stereoscopic imager (onboard MEx) allowed for the derivation of the OMEGA daytime CO_2 clouds altitudes (55–85 km), and unique measurements of mesospheric wind speeds (5–40 m s^{-1}) (Määttänen et al. 2010; Scholten et al. 2010). Vincendon et al. (2011) used the imaging spectrometer CRISM onboard the Mars Reconnaissance Orbiter (MRO) to discriminate between CO_2 MMCs, and more infrequent water ice MMCs with similar spatial and seasonal distribution. Water ice was detected as high as 80 km. Systematic detections of aerosol layers have been made using THEMIS onboard Mars Odyssey (McConnochie et al. 2010), and differentiating dust layers from MMCs through radiative transfer analysis led to results consistent with the previous unambiguous MMCs detections. Moreover, numerous detections of MMCs by THEMIS occurred in the northern winter mid-latitudes (McConnochie et al. 2010) where OMEGA had detected very few CO_2 MMCs. Sefton-Nash et al. (2013) used the Mars Climate Sounder (MCS) onboard MRO to detect mesospheric aerosol layers using different wavelength channels and similarities with previous MMCs detections were highlighted, although with less constraints on the cloud composition.

Clancy and Sandor (1998) had suggested that the formation mechanism for CO_2 MMCs required low temperatures created by gravity waves in the temperature minima of the larger scale tidal waves. The correlations between the LMD-GCM-simulated temperature minima in the mesosphere with observations of CO_2 MMCs (Gonzalez-Galindo et al. 2011), and the coincidences between areas where atmospheric conditions are favorable to the upward propagation of gravity waves and those of CO_2 MMCs observations (Spiga et al. 2012), strengthen the case for this proposed formation pathway. In addition, Listowski et al. (2014) validated this scenario with one-dimensional detailed microphysics modelling, arguing for an exogenous source of cloud nuclei—most likely MSPs—to explain the observed MMC opacities. Recently, Nachbar et al. (2016) measured in the laboratory the nucleation and growth of CO_2 ice on small ($r < 4$ nm) iron oxide and silica particles, representing MSPs at conditions close to the mesosphere of Mars. They showed that significant supersaturation is required: the characteristic temperatures for the onset of CO_2 ice nucleation are 8–18 K below the CO_2 frost point temperature, depending on MSP particle size.

5.3 Stratospheric Clouds on Earth and Venus

An aerosol layer of H_2SO_4–H_2O droplets (commonly referred to as the Junge layer) occurs in the Earth's lower stratosphere between ~20 and 28 km (Kremser et al. 2016). Single particle analyses of these aerosols at mid-latitudes show that roughly half the particles contain 0.5–1.0 wt% meteoric iron (Cziczo et al. 2001). Inside the Arctic polar vortex up to 75% of aerosol particles can contain refractory material i.e. thermally stable residuals which are almost certainly MSPs, transported downwards by the prevailing meridional circulation during winter (Curtius et al. 2005; Weigel et al. 2014). These large concentrations of refractory aerosol are a regular feature, with the accumulation starting during December and reaching its highest level during March in the Arctic. Balloon-borne measurements above 30 km over Antarctica show that MSPs most likely nucleate the H_2SO_4 droplets above the main Junge layer (Campbell and Deshler 2014).

Polar stratospheric clouds (PSCs) form in the winter polar stratosphere when the temperature falls below 197 K; they are responsible for activating chlorine during the polar night, leading to severe ozone depletion the following spring (Kremser et al. 2016). PSCs consist of frozen nitric acid trihydrate (NAT) particles, and recent modelling of PSC formation shows that heterogeneous nucleation is necessary to produce good agreement with observations (Engel et al. 2013; Hoyle et al. 2013). However, an unsolved problem is the actual nature of the ice nuclei. Saunders et al. (2012) showed that the Fe and Mg in MSP analogue particles dissolves in concentrated H_2SO_4 at low temperatures, leaving insoluble silicate cores. These cores, which have been observed in single particle mass spectrometry measurements of stratospheric aerosols (Murphy et al. 2014), may provide the PSC nuclei. A recent laboratory study has shown that analogues of silica cores and unablated meteoric material both trigger the nucleation of NAT (James et al. 2017a).

The sulfuric acid clouds of Venus are characterized by their high visible albedo (>0.8) and optical thickness (>30), and the nearly total coverage of the planet by their 20 km thick layer. A salient feature of the clouds is the markings or contrasts observed at ultraviolet wavelengths, although the absorber causing these markings has still not been identified. The clouds are essentially a product of atmospheric chemistry, with the photochemical production of sulfuric acid vapor near the cloud tops functioning as the source for formation of the acid droplets. The ensuing microphysics takes care of the cloud droplet growth through condensation and coagulation, and the droplets are redistributed vertically through sedimentation and vertical atmospheric motions, which are particularly vigorous in the lowest, turbulent cloud layers. The three cloud layers are found between 48 and 70 km altitude in the atmosphere, embedded in a haze that surrounds them both above and below.

Most observations show that the clouds are formed mainly of concentrated sulfuric acid solution droplets. One curious phenomenon is the blue absorption (400–500 nm) in the upper clouds between 55 and 70 km, which is responsible for the yellowish appearance of the planet. Krasnopolsky (2006) concluded that the most plausible candidate is a $\sim$1% solution of ferric chloride ($FeCl_3$) in H_2SO_4 droplets, which matches well the blue absorption. Support for this comes from observations of Fe and Cl by X-ray Fluorescence Spectroscopy, made by the *VENERA 14* and *VEGA* landers.

The Pioneer Venus probe performed the first and only in situ measurements of cloud particle properties within the cloud layers (Knollenberg and Hunten 1980). This showed that the Venus cloud particle size distributions exhibit two modes of sulfuric acid droplets (mode 1: $r \sim 0.2$ μm; mode 2: $r \sim 1$ μm) throughout the clouds (Knollenberg and Hunten 1980). A mode 3 was deduced from the Pioneer Venus measurements as well, but this remains controversial since the measurements could be explained by a large ($r \sim 3$–4 μm), potentially crystalline particle mode, or by a misinterpretation of the data due to instrument calibration (Toon et al. 1984). The presence of crystalline particles and exotic trace substances such as P_4O_{10} and Fe_2Cl_6 has also been suggested (Andreychikov et al. 1987; Krasnopolsky 1989, 2017) to explain these controversial mode 3 particles.

Although the photochemical production of H_2SO_4 molecules provides a source of condensable vapour, the pathways for cloud droplet formation are uncertain. Although direct formation of droplets through homogeneous nucleation is possible, heterogeneous or ion-induced nucleation should be energetically favoured. Recent theoretical and experimental studies of H_2SO_4–H_2O particle formation have developed state-of-the-art descriptions of the homogeneous and ion-induced particle formation processes (Duplissy et al. 2016; Merikanto et al. 2016), enabling their application to Venus. Modelling of ionization in Venus' atmosphere indicates a significant level of ionization at the cloud formation altitudes around 70 km (Michael et al. 2009; Plainaki et al. 2016). Many models have considered heterogeneous nucleation on soluble or insoluble particles (James et al. 1997). The latter could be provided either by: polysulfur (Young 1983), formed through atmospheric chemistry; small particles of $FeCl_3$, whose source is Fe_2Cl_6 vapour produced by the action of HCl on the hot ferric surface planetary surface (Krasnopolsky 2006); or an exogenous source of nuclei provided by MSPs (Gao et al. 2014).

6 Impacts on Atmospheric Chemistry

6.1 Earth

Apart from their role in cloud formation (Sect. 5), MSPs have the potential to influence gas-phase chemistry in the Earth's middle atmosphere. Starting in the upper mesosphere, rocket-borne measurements show that between 75 and 95 km the electron density is significantly depleted compared to the positive ion density. While this had long been attributed to the formation of negative ions, recent observations show that the electrons are actually attached to MSPs (e.g. Friedrich et al. 2011), because negative ion formation is shut down in the presence of atomic oxygen (Plane et al. 2014).

During winter, MSPs are rapidly transported down to the stratosphere within the polar vortex, on a timescale of a month or so (Bardeen et al. 2008). It has been proposed that metal-rich MSPs can remove trace acidic vapours such as H_2SO_4 and HNO_3 in the middle atmosphere. Balloon-borne mass spectrometry measurements show unexpectedly low H_2SO_4 concentrations above 40 km (Arijs et al. 1985), which can be explained if the uptake coefficient for H_2SO_4 on MSPs is greater than 0.01 (Saunders et al. 2012). Frankland et al. (2015) used a laboratory measurement of the uptake coefficient of HNO_3 on MSP analogue particles to show that heterogeneous removal on MSPs in the winter polar vortex between 30 and 60 km should provide an important sink for HNO_3. Similarly, heterogeneous uptake of the HO_2 radical on MSPs, which is sensitive to the fraction of Fe in the particles, should significantly alter the general radical chemistry of the nighttime polar vortex (James et al. 2017b).

6.2 Mars

Observations of methane (CH_4) in the Martian atmosphere have received a great deal of attention because of the possibility of a subsurface biological source, although the serpentinization reaction between water and olivinic rocks is a potential abiotic source. In addition, the organic matter deposited by exogenous sources may be a significant source of CH_4, which has been produced in the laboratory by irradiating simple organics such as glycine mixed with Martian analogue surface materials (e.g. Stoker and Bullock 1997), as well as carbonaceous chondrites (e.g. Murchison CM2) (Keppler et al. 2012; Schuerger et al. 2012). Later work quantified the amount of CH_4 produced, indicating that UV irradiation of accreted cosmic dust could explain a portion of the globally averaged CH_4 abundance (Moores and Schuerger 2012). Since CH_4 has a relatively short lifetime on Mars ($\leq$330 yrs) (Atreya et al. 2007), to be observed in the Martian atmosphere it must be resupplied on a geologically continuous timescale and persist long enough to accumulate to detectable concentrations. On a pure mass balance basis, the flux of organic carbon from interplanetary dust (e.g. Flynn 1996) is probably sufficient to produce up to 11 ppbv of CH_4 in the Martian atmosphere, if it does not ablate on entry. The total amount of organic carbon observed mixed with surface materials, measured in the Viking, Phoenix and MSL missions, is consistent with organic delivery via IDPs (Moores and Schuerger 2012).

However, this source does not easily explain the seasonal, temporal, diurnal and plume fluctuations of CH_4 that have been reported (Schuerger et al. 2012). A large number of measurements of Martian CH_4 have been published over the past 20 years, including telescopic observations from Earth, remote sensing from orbiting spacecraft, and in situ surface measurements (Webster et al. 2015). The values reported have ranged from 0.7 ppbv to over 100 ppbv. In some cases the releases that are observed are isolated events (Mumma et al. 2009;

Webster et al. 2015), while in other cases a repeating seasonal dependence is reported with CH_4 levels declining to near zero in between spikes. In all telescopic cases, large amounts of CH_4 must be invoked—for instance, the ~45 ppbv peak plume observed by Mumma et al. (2009) would have required 19,000 tons of CH_4. The smallest (sub-ppb) levels were measured by tunable diode laser spectroscopy on the Mars Science Laboratory at Gale crater (Webster et al. 2015).

While none of these observations are contemporaneous and therefore no single observation explicitly contradicts another, questions have been raised about the reliability of any single result or combination of results (e.g. Zahnle et al. 2011). Additionally, the long lifetime of CH_4 in the Martian atmosphere under normal conditions means that any large release of CH_4 reported should be rapidly mixed throughout the Martian atmosphere and persist for long periods, unless an unknown rapid CH_4 destruction process exists (Lefevre and Forget 2009). This brings into question the high variability reported in the literature: it is not the larger spikes of CH_4 that are most difficult to reconcile with our understanding of Martian CH_4 chemistry, but rather the incidence of low CH_4 observations that are most challenging to explain. This is especially problematic for exogenous sources. In order for the low-CH_4 baseline of 0.7 ppbv (Webster et al. 2015) to be explicable, either: (1) conversion rates of exogenous organic carbon to CH_4 must be surprisingly low ($<6\%$); (2) the rate of organic infall from IDPs is over-estimated by a factor > 15, or much of the incoming carbon ablates and is converted to CO_2; (3) the carbon content of IDPs is overestimated by a similar factor; (4) somehow the in-falling IDPs are rapidly buried or otherwise prevented from being irradiated by UV; (5) the destruction pathway for organic carbon on Mars does not lead to CH_4; or (6) some combination of these effects.

Recently, Fries et al. (2016) have posited that there is a way to reconcile all of the measurements that have been made to date, including the low values of CH_4. They note an apparent strong correlation between the times of high CH_4 observed concentrations and the timing of cosmic dust delivery via meteor streams. They then hypothesize that these meteor stream deliver most of the carbonaceous material to Mars, which is subsequently photolyzed to CH_4 by the mechanism of Schuerger et al. (2012). In this way, seasonally repeating observations are the result of the inherent seasonality of the meteor streams, which recur at the same point in Mars' orbit. The isolated plumes represent Mars passing through unusually dusty segments of the meteor streams, yielding a higher than usual rate of delivery of organic carbon. In order to prevent all of these events from interfering with one another, Fries et al. (2016) propose that the carbonaceous material is actually converted to CH_4 while still high in the Martian atmosphere where CH_4 photolysis lifetimes are potentially less than a year. However, there have been recent objections to this hypothesis. First, Roos-Serote et al. (2016) have questioned the timing correlation of Fries et al. (2016) by applying statistical methods to a larger catalogue of meteor streams. Second, observations made by the MAVEN spacecraft when Comet Siding Spring passed close to Mars in October 2014 were used to show that between 2700 kg and 16,000 kg of fresh carbon-rich dust entered the Martian atmosphere (Schneider et al. 2015), in only ~5400 s (Tricarico et al. 2014). This global accretion rate of between 0.5 $kg\,s^{-1}$ and 3.0 $kg\,s^{-1}$ is at least an order of magnitude higher than the mass delivered by a typical intense meteor stream, and is still too small by ~3 orders of magnitude to result in a visible CH_4 plume (Crismani et al. 2017a).

The solar wind implants significant concentrations of noble gases (e.g. He and Ne) in cosmic dust particles while they are in interplanetary space (Flynn 1997). During the atmospheric entry of a dust particle, some or all of these implanted noble gases will be released directly into the atmosphere; and, if the particle does not completely ablate and reaches the surface, the remaining noble gases may be released if the particle decomposes, or by episodic surface heating. Because Mars has a much lower atmospheric mass compared with Venus and Earth, the atmosphere of Mars is more significantly influenced by this exogenous source of noble gases: over the past 3.6 billion yr, interplanetary dust particles are estimated to have contributed quantities of ^{3}He, ^{4}He, ^{20}Ne and ^{22}Ne that are comparable to the current total atmospheric inventories of these isotopes; furthermore, the He and Ne isotope ratios are distinctly different than assumed to outgas from the planetary interior (Flynn 1997). See further discussion of this topic in Sect. 7.1.

6.3 The Giant Planets; Titan, Triton and Pluto

The photochemical implications of an influx of external material into the atmospheres of the giant planets are significant. As reported by Feuchtgruber et al. (1997, 1999), all of the giant planets and Titan contain oxygen-bearing species (e.g., H_2O, CO, CO_2) at levels typically too large to be explained by upwelling from the deep interior. Additionally, many other observations of external oxygen-bearing species at the giant planets have been reported, including at Jupiter (from both the Shoemaker-Levy 9 impact and background cosmic dust) (e.g., Prather et al. 1978; Bergin et al. 2000; Bezard et al. 2002; Lellouch et al. 2002, 2006; Cavalie et al. 2008a, 2012, 2013), Saturn (deGraauw et al. 1997; Bergin et al. 2000; Moses and Bass 2000; Moses et al. 2000; Prange et al. 2006; Cavalie et al. 2009; Abbas et al. 2013), Uranus (Marten et al. 1993; Encrenaz et al. 2004; Cavalie et al. 2008b; Teanby and Irwin 2013; Cavalie et al. 2014; Orton et al. 2014), and Neptune (Marten et al. 1993; Naylor et al. 1994; Lellouch et al. 2005; Hesman et al. 2007; Fletcher et al. 2010; Luszcz-Cook and de Pater 2013; Irwin et al. 2014). While each of the giant planets presents unique details with respect to the possible source(s) of the external oxygen and its subsequent chemical processing, observations across all four planets clearly demonstrate both internal and external sources. A more complicated question to answer is the relative contribution amongst these sources, including deep internal oxygen that is upwelled, compared with a variety of exogenous oxygen sources including cometary impacts (e.g., the Shoemaker-Levy 9 impact at Jupiter Zahnle and MacLow 1994), ring sources (e.g., Saturn's E-ring), and the ablation of interplanetary dust.

To assess the relative importance of the interplanetary dust influx to the giant planets one can either, to first order, compare observational-based calculations of the effective external oxygen influx at each planet with predictions from a dynamical dust model; or, for a more detailed approach, one can use ablation profiles as input to photochemical models. Poppe (2016) performed a first-order comparison of oxygen influx to the giant planets from interplanetary dust modeling while Moses and Poppe (2017) recently conducted a photochemical model study using the interplanetary dust ablation profiles in each of the giant planet atmospheres (Fig. 4). In Fig. 6, the effective oxygen influxes from interplanetary dust grains are shown for each planet in black and observational estimates and/or constraints are shown in colour (Feuchtgruber et al. 1997; Moses and Bass 2000; Moses et al. 2000; Bezard et al. 2002; Lellouch et al. 2002, 2005; Cavalie et al. 2010, 2014; Orton et al. 2014).

Starting at Neptune, the IDP model influx of oxygen is consistent with observations of H_2O and CO_2 (Feuchtgruber et al. 1997); however, measurements by Lellouch et al. (2005), Hesman et al. (2007), and more recently by Luszcz-Cook and de Pater (2013) have shown that CO is highly enriched at in Neptune's stratosphere, with an estimated influx of $\sim 5 \times 10^7$ cm^{-2} s^{-1} (green triangle, Fig. 6)—far more than can be attributed to the interplanetary dust flux of $\sim 7 \times 10^5$ cm^{-2} s^{-1}. This discrepancy and the facts that CO is enriched in Neptune's stratosphere relative to its troposphere (e.g., Fletcher et al. 2010; Irwin et al. 2014) and that CO is thermochemically favored at higher temperatures argues strongly for

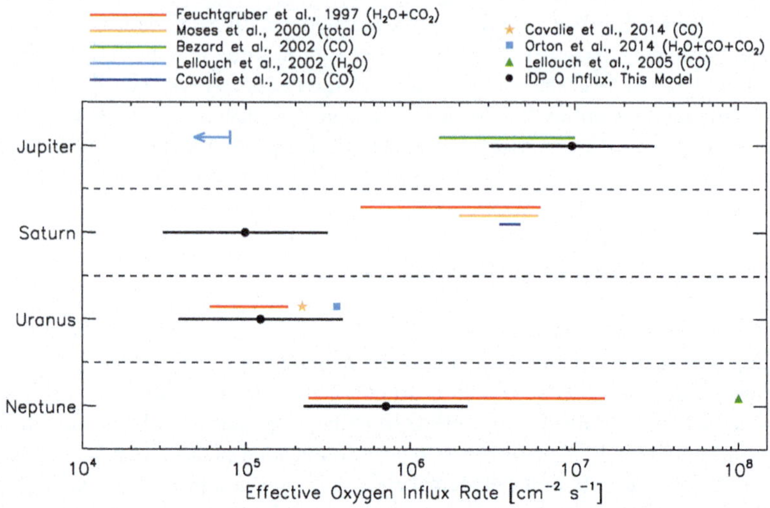

Fig. 6 A comparison of effective molecular oxygen influx rates to each of the giant planets from interplanetary dust grains (*black*) to various observational constraints (*colour lines* and *points*) (adapted from Poppe 2016)

the deposition of oxygen by cometary impacts and formation of CO in the associated high-temperature shock chemistry. At the present moment, interplanetary dust plays a minor role in oxygen-bearing photochemistry at Neptune.

At Uranus, the estimated interplanetary dust influx is consistent with observations by Feuchtgruber et al. (1997), Cavalie et al. (2014), and Orton et al. (2014) for the total amount of exogenous oxygen influx, approximately 10^5 O atom $cm^{-2}\,s^{-1}$. This suggests that the dust influx can deliver the observed stratospheric oxygen; indeed, recent modeling by Moses and Poppe (2017) of oxygen photochemistry from ablated dust grains in the Uranian atmosphere provides a satisfactory fit to the relative observed abundances of H_2O, CO, and CO_2 by freely fitting the relative form in which the ablated oxygen is deposited into the atmosphere (i.e., as H_2O, CO, and/or CO_2). The best-fit ratio is 30% H_2O, 69% CO, and 0.8% CO_2. As the influx of interplanetary dust can thus explain not only the overall oxygen influx but also the species-specific observations, one possible conclusion is that dust is the primary source; however, given the direct observations or evidence of cometary impacts at Jupiter, Saturn, and Neptune (Lellouch 1996; Lellouch et al. 2005; Cavalie et al. 2012) and estimates from Poppe (2016) as to the timescales of cometary CO dissipation relative to estimated impact rates from Levison and Duncan (1997) (i.e., a km-sized comet should impact Uranus every ~700 years whereas eddy diffusion on Uranus would take ~16,000 years to remove the cometary CO), cometary impacts most likely contribute an oxygen influx at Uranus that is comparable to that of interplanetary dust. Note also that dust of planetary origin (potentially from the rings and/or moons of Uranus) can also affect the Uranian atmosphere (e.g., Rizk and Hunten 1990).

At Saturn, the oxygen influx to the atmosphere is potentially a mix of several sources, including the prodigious amount of water from Enceladus, both in the form of neutral water vapor and E-ring grains (Cassidy and Johnson 2010; Fleshman et al. 2013), the main ring system (i.e., "ring rain") (O'Donoghue et al. 2013; Moore et al. 2015), and cometary impacts (Cavalie et al. 2010). Thus, as shown in Fig. 6, it is unsurprising that the effective oxygen influx from interplanetary dust estimated by Poppe (2016) falls more than an order-of-magnitude short in explaining the inferred atmospheric oxygen influx based on remote observations (Feuchtgruber et al. 1997; Moses and Bass 2000; Moses et al. 2000; Cavalié et al. 2010). As further concluded by Moses and Poppe (2017), interplanetary dust plays a minor role in the delivery of oxygen and other exogenous species to Saturn's atmosphere.

Frankland et al. (2016) investigated the role of meteoric material in converting acetylene (C_2H_2) into C_6H_6 in Titan's atmosphere. Because of the relatively small mass of the moon, dust particles enter at low speeds (<29 $km\,s^{-1}$) and encounter an atmosphere with a large scale height (~40 km, so the pressure increases gradually compared to the Earth or Venus). Much of the incoming mass consists of dust particles between $r = 0.4$ and 10 μm from Kuiper belt and Oort Cloud comets; less than 1% of mass loss occurs through sputtering, and few particles reach a temperature above 1750 K where melting and significant evaporative mass loss would occur (Sect. 3). Hence, a much smaller fraction of the particles ablate compared with the terrestrial planets, and unablated meteoroids—rather than MSPs—provide most of the surface area for heterogeneous chemistry. The kinetics of C_2H_2 uptake and cyclo-trimerization into C_6H_6 were measured at low temperatures in the laboratory and input into a 1D model, which shows that the heterogeneous synthesis of C_6H_6 on cosmic dust is likely to be competitive with the gas-phase production of C_6H_6 between 80 and 120 km (Frankland et al. 2016).

Finally, at Jupiter, a more puzzling situation exists. Since the atmospheric state of Jupiter is strongly perturbed by the Shoemaker-Levy 9 impact in the southern hemisphere, constraints have been derived for non-SL9 input mainly from northern hemispheric observations (e.g., Lellouch et al. 2002). The interplanetary dust model of Poppe (2016) predicts an effective oxygen influx rate of $\sim 10^7$ $cm^{-2}\,s^{-1}$; however, Lellouch et al. (2002) have placed a rather stringent limit on the H_2O influx to Jupiter of $<10^5$ $cm^{-2}\,s^{-1}$. As the Jupiter dust mass influx estimates are perhaps the most certain of all four of the giant planets due to observations by the Galileo Dust Detection System (e.g., Krivov et al. 2003; Sremcevic et al. 2003), there seems little reason to doubt the Poppe (2016) values. Thus, the incoming oxygen from interplanetary dust must be sequestered in other forms such as CO or CO_2. Bezard et al. (2002) have estimated a CO influx rate of $(1.5–10) \times 10^6$ $cm^{-2}\,s^{-1}$ (which is consistent with the interplanetary dust influx), but attributed this influx to smaller, km-sized and sub-km sized comets, as the shock chemistry from cometary impacts will thermo-kinetically favor the production of CO over H_2O. While this process could potentially provide the measured CO, Moses and Poppe (2017) provide an alternative solution where interplanetary dust provides the appropriate balance of H_2O, CO, and CO_2 by stipulating relative species deposition fractions from ablated dust grains of 98% CO, 1.4% CO_2, and 0.6% H_2O. The speciation of ablated interplanetary dust grains currently remains an unknown, so this solution provided by Moses and Poppe (2017) awaits confirmation.

Two other objects with atmospheres deserve brief mention here: Triton and Pluto, which both have cold, tenuous, N_2-dominated atmospheres driven in large part by the vapour sublimation of ices on their surfaces (Broadfoot et al. 1989; Elliot et al. 1989; Tyler et al. 1989; Yelle et al. 1991; Krasnopolsky et al. 1993). Both atmospheres also contain other minor species, such as CH_4, CO, and various hydrocarbons (Lellouch et al. 2009, 2010; Gladstone et al. 2016), and both atmospheres are observed to have clouds and/or haze layers (Yelle et al. 1991, 1995). At Triton, observations of ion layers below the main ionospheric peak could be evidence of the presence of metallic ions introduced from meteoric ablation (cf. Sect. 4.2). MSPs could also play a role in the formation of the observed haze layers (Yelle et al. 1995; Pesnell et al. 2004). Pesnell et al. (2004) have to date performed the only study of meteoroid ablation in the atmosphere of Triton. By considering both stony and icy meteoroids at characteristic velocities of 10 and 15 $km\,s^{-1}$, they found that stony meteoroids

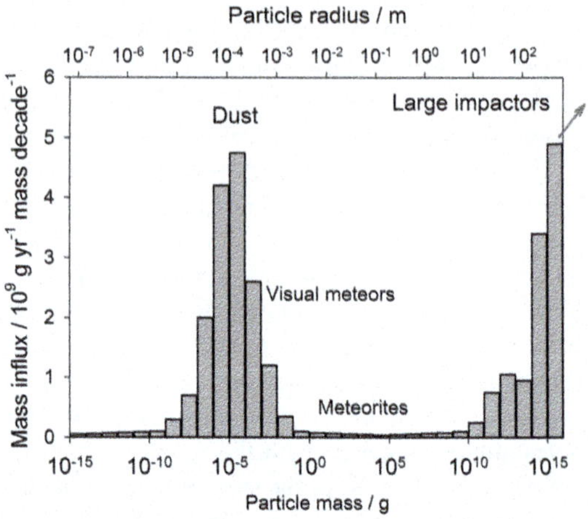

Fig. 7 The estimated mass accretion rates of extraterrestrial objects at the top of the Earth's atmosphere are dominated by two peaks. The peak at small masses is caused by the continuous accretion of cosmic dust, while the peak at large masses results from the infrequent impacts of large bodies (adapted from Kyte and Wasson 1986)

ablate 10–75% of their mass before impacting Triton's surface, while icy meteoroids ablate at least 70% of their mass before surface impact. Combining an estimate of the mainly EKB dust influx (Poppe 2016) which is assumed to be 42% water ice (Moses and Poppe 2017), with the ablation calculations of Pesnell et al. (2004), yields an equivalent O influx to Triton's atmosphere of $\sim 10^5$ cm^{-2} s^{-1}. Krasnopolsky (2012) included an oxygen influx of 2×10^6 cm^{-2} s^{-1} (a factor of $\sim$20 higher) in a photochemical model of Triton's atmosphere and found that meteoric H_2O is efficiently converted to O and ultimately CO via interactions with molecules such as CH_x, CN, and CNN. This production method was found to be higher than loss to space, and so Krasnopolsky (2012) concluded that such CO would condense on the surface of Triton. It is unclear if such a conclusion still holds in the case of the relatively lower meteoric O influx based on the revised dust influx (Poppe 2016).

At Pluto, a similar situation is expected to be present, albeit with less meteoric influx compared to Triton due to the lack of gravitational focusing by a giant parent planet (Triton's meteoric influx is enhanced by a factor of $\sim$10 relative to interplanetary space due to its position deep inside Neptune's gravity well). A preliminary study of meteoric ablation in Pluto's atmosphere, using EKB dust grain mass and velocity distributions from Poppe (2015), indicates that water ice grains will completely ablate before reaching Pluto's surface, while silicate grains only partially ablate, similar to the Pesnell et al. (2004) results for Triton. The Krasnopolsky (2012) model predicts that water introduced from meteoric ablation to Pluto's atmosphere will react (photochemically) with the abundant CO to produce CO_2, which then condenses on the surface. More recently, Wong et al. (2017) have examined a detailed photochemical model of Pluto's atmosphere constrained by New Horizons observations, including a presumed meteoric dust/water flux from Poppe (2015). This influx leads to predicted values for oxygen-bearing species in Pluto's atmosphere such as CO (the third-most abundant gas in the atmosphere), H_2O, H_2CO, and CO_2. While the latter three species have yet to be detected, their eventual identification would provide strong constraints on the dust/oxygen influx predicted by Poppe (2015, 2016).

7 Surface Accretion of Dust and Meteorites

The contribution by dust and meteorites to planetary surfaces was demonstrated by the Apollo samples. Elemental analyses showed that the Lunar regolith and regolith breccias had elevated levels of Ir, Au, Zn, Cd, Ag, Br, Bi, and Tl compared to the ordinary Lunar rocks in a pattern that indicated the addition of 1.5% to 2.0% carbonaceous chondrite-like material to the regolith (Keays et al. 1970; Anders et al. 1973). In the case of the Earth, the mass flux at the top of the atmosphere has been estimated by combining results from satellite impact measurements for small particles, radar meteors for intermediate size objects, and the cratering record for large objects, as discussed by Peucker-Ehrenbrink et al. (2016). As shown in Fig. 7, the mass-frequency distribution is bimodal, with peaks corresponding to the continuous, planet-wide input of dust and the infrequent impact of large bodies, with a minimal contribution from objects in the intermediate size range. Measurements of impacts onto the Long Duration Exposure Facility, which was in low-Earth orbit for about 69 months, indicate that the accretion rate of cosmic dust into the Earth's atmosphere is $110 \pm 55\ \mathrm{t\,d^{-1}}$ in the current era (Love and Brownlee 1993). This is at least 100 times larger than the annual influx of meteorites (Bland et al. 1996), with particles in the narrow mass range from 10^{-8} to 10^{-3} g ($r = \sim 10$–460 μm) contributing more than 80% of the total mass flux of meteoritic material in the 10^{-13} to 10^{6} g mass range incident on the Earth (Hughes 1978; Carrillo-Sánchez et al. 2016). Modeling by Carrillo-Sánchez et al. (2016) of the dust up to 500 μm in diameter indicates that the total mass input is $43 \pm 14\ \mathrm{t\,d^{-1}}$, with $35.4\ \mathrm{t\,d^{-1}}$ surviving as either unmelted particles or melted spherules, and the remaining $7.9\ \mathrm{t\,d^{-1}}$ being deposited in the upper atmosphere as ablated atoms. The dust accretion rate was likely much greater during the first 0.6 billion years of Solar System history, when asteroids and comets were more abundant in the inner Solar System as evidenced by the higher impact rate of large objects on the Moon during the Late Heavy Bombardment (Hartmann et al. 2000).

In the case of planets with atmospheres, some of this extraterrestrial material vaporizes during atmospheric deceleration (Sect. 3). With the exception of the volatile gases released into the atmosphere, all of this material, whether as surviving objects or recondensed MSPs, eventually accretes onto the planet's surface. Unlike the Lunar case, where the impact of extra-Lunar material is the major regolith-generating mechanism, terrestrial soil is generated by a variety of more rapid weathering mechanisms, which results in a greater dilution of the extraterrestrial component, thus making it more difficult to detect. The most dramatic effects on the surface composition of the terrestrial planets are for the siderophile elements, which are depleted in the surface material and concentrated in the metallic core during planetary differentiation but are abundant in undifferentiated extraterrestrial materials, and a few isotopes that are significantly enriched in extraterrestrial materials, generally by cosmic ray interactions in space. The elemental effects, particularly for Ir, Os, and Pt, and isotopic effects, particularly for He, Re–Os, and Cr, of the meteoritic contributions to the composition of terrestrial sediments have been reviewed in detail by Kyte (2002), Peucker-Ehrenbrink (1996), and Peucker-Ehrenbrink et al. (2016).

7.1 Siderophile and Isotope Contributions to the Earth's Surface

Pettersson and Rotschi (1950) found that Ni, the most abundant of the siderophile elements, was present in Central Pacific Ocean sediment cores in significantly higher concentrations than in continental rocks and sediments, and they suggested this Ni might be partly derived from cosmic dust settling over the Earth's surface. Bonner and Lourenqo (1965) attempted to use the Ni abundance in Pacific Ocean cores to measure the accretion rate of extraterrestrial material onto the Earth. However, their value was more than an order of magnitude larger than the currently accepted value, most likely because of Ni mobilization in the oceans.

Iridium has been used most commonly as a tracer of the extraterrestrial component in surface materials because of its high concentration in chondritic material, $\sim 10^{4}$ times the concentration found in crustal rocks (Palme et al. 1978), and its easy detection by neutron activation analysis. Barker and Anders (1968) used the Ir and Os concentrations in dated deep-sea sediments to infer a value for the infall rate of extraterrestrial material that is in good agreement with the current estimates. The most significant elemental perturbations in both deep-sea sediments and ice cores occur for Ir, Os, and Re (Lal and Jull 2003). The available evidence suggests the input rate has been relatively constant over the past 70 million years (Kyte 2002), allowing the Ir concentration to be used to determine the deposition rates of sediments as well as polar ice and snow layers. As discussed in Sect. 5.1, Ir has also been used to determine the surface deposition rate of MSPs.

One major deviation from this long term average was identified in the clay layer at the Cretaceous-Tertiary boundary. The Ir excesses in deep-sea limestones exposed in Italy, Denmark, and New Zealand, which are 30, 160, and 20 times their respective background levels at the time of the Cretaceous-Tertiary extinctions 65 Myr, were linked to the global deposition of debris from the impact of a $\sim$10 kilometer diameter asteroid (Alvarez et al. 1980). Clear confirmation of the meteoritic contribution to these boundary clays was presented by Shukolyukov and Lugmair (1998), who found a ^{53}Cr/^{52}Cr ratio consistent with the ratio in carbonaceous meteorites but inconsistent with the terrestrial value. Localized enhancements of Ir and Au, comparable with those at the Cretaceous-Tertiary boundary, were identified in a sediment from the Antarctic Ocean (Kyte et al. 1981), indicating the impact of an asteroid >1 km in diameter $\sim$2.15 Myr ago (Gersonde et al. 1997).

Because of their high surface area-to-mass ratio compared to larger bodies, interplanetary dust particles have very high concentrations of solar-wind implanted ions, particularly the most abundant solar wind elements, H and He. Since the solar wind ^{3}He/^{4}He ratio is much higher than that in any terrestrial material (Nier and Schlutter 1990), interplanetary dust that survives atmospheric entry without outgassing perturbs the He isotopic ratio in deep-sea sediments. Ozima et al. (1984) found an approximately inverse relationship between ^{3}He concentrations and the accumulation rate of Pacific surface sediments, which they attributed to the influx of interplanetary dust. Ozima et al. (1984) calculated an influx rate of only 5 $t\,d^{-1}$, nearly an order of magnitude lower than the influx estimates derived from most other techniques (Barker and Anders 1968; Kyte and Wasson 1986; Love and Brownlee 1993; Peucker-Ehrenbrink 1996; Plane 2012; Carrillo-Sánchez et al. 2016). This is because the ^{3}He is most concentrated in the fine fraction of the interplanetary dust, $\sim$2 to 5 μm radius (Lal and Jull 2005), and is degassed from most larger particles during atmospheric entry, never making it to the sediments. Thus, the ^{3}He content in sediments is not a measure of the total influx of extraterrestrial matter, rather it is a tracer of the fine fraction of the interplanetary dust complex (Fig. 7).

While Ir excesses serve to identify large impacts onto the Earth, the ^{3}He contribution has been used to identify increases in the flux from major, dust-producing events in the past. Farley (1998) measured a factor of 5 increase in the ^{3}He flux over a 2 Myr interval $\sim$35 Myr ago, and attributed this the accretion of dust from a comet shower. Later work by Mukhopadhyay et al. (2001), covering the interval from 70 to 40 Myr ago, found a factor of 2–4 increase in the ^{3}He accretion rate about 55 Myr ago, but no evidence for variations of more than a factor of 2 over the remainder of that interval. Farley et al. (2006) also identified a factor of 4 increase in the ^{3}He concentration 8.3 Myr ago, which they attributed to the dust produced by the collisional disruption of the 150 km diameter asteroid that created the Veritas family.

Lal and Jull (2003) modeled the contributions of ^{3}He, ^{10}Be, ^{14}C, and ^{26}Al to terrestrial sediments by both interplanetary dust and fragments produced by the atmospheric disintegration of larger meteors. They concluded that the rate of production of ^{10}Be and ^{14}C in the Earth's atmosphere by galactic cosmic rays significantly exceeds the contribution from cosmic dust. In contrast, their modelling indicated that ^{26}Al is mostly meteoric, and they suggested that it should be possible to observe high ^{26}Al concentration pulses in sediments during any periods of enhanced meteoroid fluxes in the past (Lal and Jull 2003).

Lastly, Dhomse et al. (2013) and Brooke et al. (2017) have modelled the mass deposition flux of MSPs over the earth's surface, using two different global circulation models. The strongest deposition occurs over northern and southern mid-latitudes, reflecting the geographical distribution of deep stratosphere-troposphere exchange which is driven both by large mountain ranges and storm tracks over the North Atlantic, North Pacific and Southern Oceans. The Southern Ocean between 50 and 60°S is where the supply of bio-available iron to phytoplankton is limited (Johnson 2001). The modelled input of cosmic Fe contained in MSPs is ~2–5% of the Aeolian dust input into the Southern Ocean. However, unlike continental mineral dust which has a low solubility (estimates vary from <1 to 10%), the MSP Fe should be in the form of highly soluble ferrous/ferric sulfate after processing in the stratospheric sulphate layer (Sect. 5.3) Thus, the input of bio-available Fe from cosmic dust should be 50–400% of the soluble Aeolian dust input. This would have significant climate implications because increased primary production will draw down CO_2 which is then exported to the deep ocean (Dhomse et al. 2013).

7.2 Organics and Water from Dust and Meteoroids

The high temperatures reached during differentiation likely destroyed any organic matter that accreted when the Earth formed from the protoplanetary disk. The presence of solar wind He in $r \sim 5$ μm interplanetary dust particles captured in the stratosphere (Rajan et al. 1977) demonstrates that small, slow particles survive atmospheric deceleration with minimal heating (Whipple 1950; Fraundorf 1980). The gentle accretion of interplanetary dust to the surface of the primitive Earth could have contributed a layer rich in organic matter, including biologically important compounds that did not form by abiotic synthesis on Earth (Anders 1989). While the organic matter in meteorites has been well characterized (Pizzarello et al. 2006), the small size of the interplanetary dust has precluded full characterization. However, these interplanetary dust particles contain a high abundance of carbon (Thomas et al. 1993), from a few to over 90% by volume, including percent-levels of both carbonyl (C=O) and aliphatic hydrocarbons (C–H_3 and C–H_2) (Flynn et al. 2003). Using the Fraundorf (1980) entry heating model and the C concentration from Thomas et al. (1993), Flynn et al. (2004) estimated that, in the current era, interplanetary dust contributes ~0.04 $\mathrm{t\,d^{-1}}$ of unpyrolized organic matter to the surface of the Earth, and this contribution is likely to have been much greater during the Late Heavy Bombardment.

Bradley et al. (2014) demonstrated that solar wind irradiation of the surfaces of interplanetary dust particles deposits H^+ ions, which then react with oxygen in the minerals to produce water. This result indicates that slow, small interplanetary dust particles can simultaneously deliver organic carbon and water to the surface of the Earth.

7.3 Dust and Meteoritic Contributions to Mars

The accretion rate of cosmic dust in the Martian atmosphere has recently been estimated to be 2–3 tonnes $\mathrm{sol^{-1}}$, based on modelling observations of the Mg^+ layer around 90 km (Crismani et al. 2017b). However, the MSP and meteorite contribution to the surface of Mars is less well constrained than for Earth. The first in situ elemental analysis of Mars soils, by the Viking landers, indicated that the Mars soil composition was inconsistent with typical basaltic rock fragments but could be fit by a mixture of basaltic rock fragments and meteoritic material (Clark and Baird 1979).

Flynn and McKay (1990) extrapolated from the measured dust accretion rate at Earth, finding that 2–29% of the Martian regolith was contributed by cosmic dust. Morris et al. (2000) found that the compositions of the soils analyzed by the Sojourner Rover were consistent with up to 22% meteoritic material. The Alpha Particle X-ray Spectrometers on the Mars Exploration Rovers measured 16 elements (Na, Mg, Al, Si, P, S, Cl, K, Ca, Ti, Cr, Mn, Fe, Ni, Zn, and Br) in the soils and rocks near the two landing sites. Yen et al. (2006) used the Ni content to indicate that the Mars soil samples and some of the sedimentary rocks contain, on average, 1–3% contamination from chondritic material. As is the case for the Earth, ^{3}He in the regolith should serve as a tracer of the fine interplanetary dust flux, while siderophiles such as Ir are a tracer of the total dust and meteoritic contribution. In the case of planets with thin atmospheres, like Mars, modelling by Lal and Jull (2003) indicates the contribution from meteoroids could be significant for ^{10}Be, ^{14}C, and ^{26}Al.

The heating pulse experienced during atmospheric deceleration of dust entering at Mars is less severe than on Earth because of the lower surface gravity ($g = 3.7\ \mathrm{m\,s^{-2}}$, compared with $9.8\ \mathrm{m\,s^{-2}}$ on Earth). This results in both a lower entry speed (Sect. 3) and a somewhat larger scale height in the ablation region (7.2 km on Mars, compared with 5.9 km on Earth) allowing dust to decelerate over a greater distance so that larger dust grains survive atmospheric entry with their organic matter intact (Flynn and McKay 1990). Flynn (1996) modelled the entry heating and found that interplanetary dust delivers an order-of-magnitude higher surface concentration of unpyrolized carbon compounds onto Mars than onto Earth, suggesting interplanetary dust may be an important source of organic carbon on Mars. In the current era, modelling indicates that this organic matter is destroyed relatively rapidly by the ultraviolet radiation that reaches the surface (Stoker and Bullock 1997), explaining the low organic concentration in the Mars regolith samples analyzed by the Viking landers and a potential source of atmospheric CH_4 (Sect. 6.2)

The lower impact speed as well as the relatively gentle accretion of material ejected into the dust torous orbiting with the Martian moons Phobos and Diemos likely leads to a higher concentration of chondritic material in their regoliths than was found on the Moon (Flynn 1991). If this concentration is sufficiently high, the surface reflectance spectra used to infer a chondritic composition may not reflect the actual composition of the underlying material.

7.4 Other Terrestrial Planets

The contributions of dust and meteorites to the surfaces of the other terrestrial planets, Venus and Mercury, have not been studied in any detail. However, the presence of a dense atmosphere on Venus should result in similar conditions for deceleration and accretion to those on Earth (Fig. 2). The global influx of cosmic dust in Venus' atmosphere was recently estimated to be 32 tonnes per Earth day (Frankland et al. 2017). Impact craters have been identified on Venus, but little or no modelling has been done on the meteoritic contribution to the surface of the planet. Thus far, the soil and rock composition data from the Venera landers (e.g. Surkov et al. 1988) has not included any of the siderophile elements critical for assessing the meteoritic component.

Mercury, lacking a substantial atmosphere, is more similar to the Moon, however the elements detected by the orbiting MESSENGER spacecraft (Mg, Al, Si, S, Ca, Ti and Fe) did not provide a direct measurement of the meteoritic component. Hence, inferences of the meteoritic contribution are based on modelling, which suggests that 50 times as many carbon-rich micrometeorites per unit surface area are delivered to Mercury, compared with the Moon, resulting in approximately 3–6 wt% carbon at Mercury's surface (Bruck Syal et al. 2015). MESSENGER measurements limit the northern hemisphere C abundance to the range 0–4.1 wt% at the three-sigma significance level (Peplowski et al. 2015). Bruck Syal et al. (2015) suggested the C contributed by micrometeorites could explain Mercury's globally low reflectance and might also contribute to the darkening of other planetary surfaces. Modelling by Bruno et al. (2007) indicates that the neutral Na "atmosphere" detected on Mercury and the Moon could be produced by the impact of meteoroids and that the mass of meteoroids that have impacted the whole surfaces of the Moon and Mercury in the last 3.8 billion years are 8.9×10^{18} g and 2.7×10^{19} g, respectively. Killen and Hahn (2015) estimated the dust deposition rate on Mercury to be around 440 t per Earth day, though this should probably be reduced to $\sim 170\ \mathrm{t\,d^{-1}}$ based on the recent estimate of the dust input to the terrestrial atmosphere (Carrillo-Sánchez et al. 2016). Lastly, Christou et al. (2015) demonstrated that the annual, repeatable Ca emission excess in Mercury's exosphere could result from the impact of millimeter-sized grains that were ejected from Comet 2P/Encke 10,000–20,000 years ago.

7.5 Summary

This review has described the very diverse impacts that interplanetary dust can have throughout a planet's atmosphere and at its surface. Significant progress has been made recently in quantifying the dust flux throughout the solar system. Models such as those by Nesvorný et al. (2010, 2011a, 2011b) and Poppe (2016) are necessary to connect *in situ* measurements of dust densities and fluxes at selected positions in the solar system with an overall picture of dust distributions throughout the solar system. In the inner solar system, continued comparison of dynamical models to available datasets will further refine the available models; although, a dedicated orbital mission around Earth for quantifying the terrestrial meteoric input function and resolving existing discrepancies between Earth- and space-based observations (Plane 2012) is crucial for constraining the overall magnitude and variability of the meteoric influx to Earth. Analysis of existing datasets, especially from the Cassini Cosmic Dust Analyzer and the New Horizons Student Dust Counter, will continue to yield information and constraints on the outer solar system distributions. NASA's forthcoming Europa Clipper mission contains within its payload the Surface Dust Mass Analyzer (SUDA) (Kempf et al. 2015), a powerful dust instrument that, while primarily focused on dust ejected from Europa and the other Galilean moons, will hopefully yield ancillary information regarding the interplanetary dust distribution from its vantage point at Jupiter.

The new laboratory experimental systems for simulating meteoric ablation (Bones et al. 2016; Thomas et al. 2017) are important for testing and improving ablation models, leading to more secure predictions of the fraction of incoming dust that ablates in a planetary atmosphere, the altitudes where different constituents are injected, and the fraction of unablated material that reaches the surface in the form of micrometeorites and cosmic spherules (Carrillo-Sánchez et al. 2016). Future developments in this area should include ablation models that treat multiple mineral phases in a meteoroid, as well as fragmentation during atmospheric entry.

In terms of atmospheric impacts, although the metal atom and ion layers in the terrestrial atmosphere have been studied for decades (Plane 2003), the first direct measurements in another planetary atmosphere have only very recently been obtained for Mars by the MAVEN spacecraft (Crismani et al. 2017b; Grebowsky et al. 2017). There has been a great deal of work in the past decade on meteoric smoke particles. This has included laboratory investigations into MSP formation (Saunders and Plane 2006) and ice-nucleating ability (Nachbar et al. 2016); detection in the terrestrial atmosphere by rocket payloads (Plane et al. 2014), radar (Strelnikova et al. 2007) and optical extinction (Hervig et al. 2009); and modelling impacts through providing a surface for heterogeneous chemistry (Frankland et al. 2016; James et al. 2017b) and ice nuclei for the formation of mesospheric clouds on Earth (Rapp and Thomas 2006; Hervig et al. 2012) and Mars (Määttänen et al. 2013; Listowski et al. 2014). The role of MSPs in the formation of sulphuric acid particles in Venus, and polar stratospheric clouds in the terrestrial atmosphere, still needs to be better understood.

A credible explanation does not yet exist for the low magnitude yet variable nature of CH_4 in the Martian atmosphere (Webster et al. 2015). Following the experience with comet Siding Spring, it is unclear how the photochemical production of CH_4 from cosmic dust settling in the atmosphere (Fries et al. 2016) can be reconciled with the relatively constant delivery of small amounts of dust (Crismani et al. 2017a). With the recent arrival in late 2016 of the ESA Trace Gas Orbiter mission, which will map CH_4 concentrations in the Martian atmosphere with a sensitivity of 20 pptv, the link between Martian CH_4 and the delivery of carbon-rich cosmic dust to the surface is likely to be better understood.

All of the giant planets and Titan are subject to significant fluxes of exogenous material that has profound impacts on their respective chemistry and structure (Moses and Poppe 2017). At Saturn, the main source of this material is the E-ring and Enceladus water vapor torus. Neptune's atmosphere shows strong evidence of a recent cometary impact that currently swamps any interplanetary dust input. Jupiter, despite the dramatic atmospheric changes wrought by the Shoemaker-Levy 9 impact, still shows evidence for additional exogenous input, whether from km-sized comets or interplanetary dust (or both). Meanwhile, Uranus' atmospheric composition can currently be explained solely on the basis of the interplanetary dust influx; however, given the evidence of cometary impacts at the other giant planets, delivery of external material from comets remains a possibility in the Uranian system as well (Moses and Poppe 2017). Despite these recent advances in understanding, there are significant areas for further investigation: reconciling the rates of interplanetary dust delivery of oxygen to Jupiter with the measurements of Lellouch et al. (2002) and Bezard et al. (2002); determining whether all the observed O-bearing species in Uranus' atmosphere originate from interplanetary dust, or these is also evidence of cometary impact; and determining the speciated chemical composition of the ablated vapour in each giant planet's atmosphere, and the role of metallic species in their atmospheres and ionospheres.

Acknowledgements We thank David Nesvorný (Southwest Research Institute) and Petr Pokorný (The Catholic University of America) for supplying output from the ZoDy astronomical dust model that was used in Fig. 2. J.M.C.P. and J.D.C.S. are supported by the European Research Council (project number 291332—CODITA). G.J.F. is supported by NASA Emerging Worlds grant #NNX17AE59G. A.R.P. is supported by the NASA Planetary Atmospheres Program, grant #NNX13AG55G. A.M. and C.L. are supported by CNES, the Programme National de Planétologie, and the ESA mission Mars Express. C.L. also acknowledges CNES for a postdoctoral fellowship.

References

M.M. Abbas, A. LeClair, E. Woodard, M. Young, M. Stanbro, F.M. Flasar, V.G. Kunde, R.K. Achterberg, G. Bjoraker, J. Brasunas, D.E. Jennings, C.T. Cassini, Distribution of CO_2 in Saturn's atmosphere from Cassini/CIRS infrared observations. Astrophys. J. **776**(2), 8 (2013). https://doi.org/10.1088/0004-637x/776/2/73

C.M.O. Alexander, S. Taylor, J.S. Delaney, M. Pixue, G.F. Herzog, Mass-dependent fractionation of Mg, Si, and Fe isotopes in five stony cosmic spherules. Geochim. Cosmochim. Acta **66**(1), 173–183 (2002). https://doi.org/10.1016/S0016-7037(01)00764-5

N. Altobelli, V. Dikarev, S. Kempf, R. Srama, S. Helfert, G. Moragas-Klostermeyer, M. Roy, E. Grun, Cassini/Cosmic Dust Analyzer in situ dust measurements between Jupiter and Saturn. J. Geophys. Res. **112**(A7), 15 (2007). https://doi.org/10.1029/2006ja011978

L.W. Alvarez, W. Alvarez, F. Asaro, H.V. Michel, Extraterrestrial cause for the Cretaceous-Tertiary extinction. Science **208**, 1095–1108 (1980). https://doi.org/10.1126/science.208.4448.1095

E. Anders, Pre-biotic organic matter from comets and asteroids. Nature **342**, 255–257 (1989). https://doi.org/10.1038/342255a0

E. Anders, R. Ganapathy, U. Kähenbühl, J.W. Morgan, Meteoritic material on the Moon. Moon **8**, 3–24 (1973). https://doi.org/10.1007/BF00562747

B.M. Andreychikov, I.K. Akhmetshin, B.N. Korchuganov, L.M. Mukhin, B.I. Ogorodnikov, I.V. Petryanov, V.I. Skitovich, X-ray radiometric analysis of the cloud aerosol of Venus by the Vega 1 and 2 probes. Cosm. Res. **25**, 721–736 (1987)

S. Aoki, Y. Sato, M. Giuranna, P. Wolkenberg, T.M. Sato, H. Nakagawa, Y. Kasaba, Mesospheric CO_2 ice clouds on Mars observed by Planetary Fourier Spectrometer onboard Mars Express. Icarus **302**, 175–190 (2018). https://doi.org/10.1016/j.icarus.2017.10.047

E. Arijs, D. Nevejans, J. Ingels, P. Frederick, Recent stratospheric negative-ion composition measurements between 22 km and 45 km altitude. J. Geophys. Res. **90**(d4), 5891–5896 (1985). https://doi.org/10.1029/JD090iD04p05891

S.K. Atreya, P.R. Mahaffy, A.S. Wong, Methane and related species on Mars: origin, loss, implications for life and habitability. Planet. Space Sci. **55**, 358–369 (2007). https://doi.org/10.1016/j.pss.2006.02.005

F. Balsiger, E. Kopp, M. Friedrich, K.M. Torkar, U. Walchli, G. Witt, Positive ion depletion in a noctilucent cloud. Geophys. Res. Lett. **23**(1), 93–96 (1996). https://doi.org/10.1029/95gl03608

C.G. Bardeen, O.B. Toon, E.J. Jensen, D.R. Marsh, V.L. Harvey, Numerical simulations of the three-dimensional distribution of meteoric dust in the mesosphere and upper stratosphere. J. Geophys. Res. **113**(D17), D17202 (2008). https://doi.org/10.1029/2007JD009515

J.L. Barker Jr., E. Anders, Accretion rate of cosmic matter from iridium and osmium contents of deep-sea sediments. Geochim. Cosmochim. Acta **32**, 627–645 (1968). https://doi.org/10.1016/0016-7037(68)90053-7

E.A. Bergin, E. Lellouch, M. Harwit, M.A. Gurwell, G.J. Melnick, M.L.N. Ashby, G. Chin, N.R. Erickson, P.F. Goldsmith, J.E. Howe, S.C. Kleiner, D.G. Koch, D.A. Neufeld, B.M. Patten, R. Plume, R. Schieder, R.L. Snell, J.R. Stauffer, V. Tolls, Z. Wang, G. Winnewisser, Y.F. Zhang, Submillimeter Wave Astronomy satellite observations of jupiter and Saturn: detection of 557 GHz water emission from the upper atmosphere. Astrophys. J. **539**(2), L147–L150 (2000). https://doi.org/10.1086/312846

B. Bezard, E. Lellouch, D. Strobel, J.P. Maillard, P. Drossart, Carbon monoxide on Jupiter: evidence for both internal and external sources. Icarus **159**(1), 95–111 (2002). https://doi.org/10.1006/icar.2002.6917

M.K. Bird, R. Dutta Roy, S.W. Asmar, T.A. Rebold, Detection of Titan's ionosphere from Voyager 1 radio occultation observations. Icarus **130**(2), 426–436 (1997). https://doi.org/10.1006/icar.1997.5831

P.A. Bland, T.B. Smith, A.J.T. Jull, F.J. Berry, A.W.R. Bevan, S. Cloudt, C.T. Pillinger, The flux of meteorites to the Earth over the last 50000 years. Mon. Not. R. Astron. Soc. **283**, 551–565 (1996). https://doi.org/10.1093/mnras/283.2.551

D. Bones, J.M.C. Plane, W. Feng, Dissociative recombination of FeO^+ with electrons: implications for plasma layers in the ionosphere. J. Phys. Chem. A **120**, 1369–1376 (2015). https://doi.org/10.1021/acs.jpca.5b04947

D.L. Bones, J.C.G. Martín, C.J. Empson, J.D.C. Sánchez, A.D. James, T.P. Conroy, J.M.C. Plane, A novel instrument to measure differential ablation of meteorite samples and proxies: the Meteoric Ablation Simulator (MASI). Rev. Sci. Instrum. **87**, 094504 (2016). https://doi.org/10.1063/1.4962751

F.T. Bonner, A.S. Lourenqo, Nickel content of Pacific Ocean cores. Nature **207**, 933–935 (1965). https://doi.org/10.1038/207933a0

J. Bradley, H. Ishii, J. Gillis-Davis, J. Ciston, M. Nielsen, H. Bechtel, M. Martin, Detection of solar wind produced water in irradiated rims on silicate minerals. Proc. Natl. Acad. Sci. **111**, 1732–1735 (2014). https://doi.org/10.1073/pnas.1320115111

A.L. Broadfoot, S.K. Atreya, J.L. Bertaux, J.E. Blamont, A.J. Dessler, T.M. Donahue, W.T. Forrester, D.T. Hall, F. Herbert, J.B. Holberg, D.M. Hunten, V.A. Krasnopolsky, S. Linick, J.I. Lunine, J.C. McConnell, H.W. Moos, B.R. Sandel, N.M. Schneider, D.E. Shemansky, G.R. Smith, D.F. Strobel, R.V. Yelle, Ultravolet spectrometer observations of Neptune and Triton. Science **246**(4936), 1459–1466 (1989). https://doi.org/10.1126/science.246.4936.1459

J.S.A. Brooke, W. Feng, J.D. Carrillo-Sánchez, G.W. Mann, A.D. James, C.G. Bardeen, J.M.C. Plane, Meteoric smoke deposition in the polar regions: a comparison of measurements with global atmospheric models. J. Geophys. Res., Atmos. **122**, 11,112–111,130 (2017). https://doi.org/10.1002/2017JD027143
M. Bruck Syal, P.H. Schultz, M.A. Riner, Darkening of Mercury's surface by cometary carbon. Nat. Geosci. **8**, 352–356 (2015). https://doi.org/10.1038/ngeo2397
M. Bruno, G. Cremonese, S. Marchi, Neutral sodium atoms release from the surfaces of the Moon and Mercury induced by meteoroid impacts. Planet. Space Sci. **55**, 1494–1501 (2007). https://doi.org/10.1016/j.pss.2006.10.006
P. Campbell, T. Deshler, Condensation nuclei measurements in the midlatitude (1982–2012) and Antarctic (1986–2010) stratosphere between 20 and 35 km. J. Geophys. Res. **119**, 137–152 (2014). https://doi.org/10.1002/2013JD019710
M.D. Campbell-Brown, High resolution radiant distribution and orbits of sporadic radar meteoroids. Icarus **196**(1), 144–163 (2008). https://doi.org/10.1016/j.icarus.2008.02.022
J.D. Carrillo-Sánchez, J.M.C. Plane, W. Feng, D. Nesvorný, D. Janches, On the size and velocity distribution of cosmic dust particles entering the atmosphere. Geophys. Res. Lett. **42**(15), 6518–6525 (2015). https://doi.org/10.1002/2015gl065149
J.D. Carrillo-Sánchez, D. Nesvorný, P. Pokorný, D. Janches, J.M.C. Plane, Sources of cosmic dust in the Earth's atmosphere. Geophys. Res. Lett. **43**, 11,979–911,986 (2016). https://doi.org/10.1002/2016GL071697
T.A. Cassidy, R.E. Johnson, Collisional spreading of Enceladus' neutral cloud. Icarus **209**(2), 696–703 (2010). https://doi.org/10.1016/j.icarus.2010.04.010
T. Cavalie, F. Billebaud, N. Biver, M. Dobrijevic, E. Lellouch, J. Brillet, A. Lecacheux, A. Hjalmarson, A. Sandqvist, U. Frisk, M. Olberg, E.A. Bergin, T. Odin, Observation of water vapor in the stratosphere of Jupiter with the Odin space telescope. Planet. Space Sci. **56**(12), 1573–1584 (2008a). https://doi.org/10.1016/j.pss.2008.04.013
T. Cavalie, F. Billebaud, T. Fouchet, E. Lellouch, J. Brillet, M. Dobrijevic, Observations of CO on Saturn and Uranus at millimeter wavelengths: new upper limit determinations. Astron. Astrophys. **484**(2), 555–561 (2008b). https://doi.org/10.1051/0004-6361:20079170
T. Cavalie, F. Billebaud, M. Dobrijevic, T. Fouchet, E. Lellouch, T. Encrenaz, J. Brillet, G.H. Moriarty-Schieven, J.G.A. Wouterloot, P. Hartogh, First observation of CO at 345 GHz in the atmosphere of Saturn with the JCMT: new constraints on its origin. Icarus **203**(2), 531–540 (2009). https://doi.org/10.1016/j.icarus.2009.05.024
T. Cavalie, P. Hartogh, F. Billebaud, M. Dobrijevic, T. Fouchet, E. Lellouch, T. Encrenaz, J. Brillet, G.H. Moriarty-Schieven, A cometary origin for CO in the stratosphere of Saturn? Astron. Astrophys. **510**, 7 (2010). https://doi.org/10.1051/0004-6361/200912909
T. Cavalie, N. Biver, P. Hartogh, M. Dobrijevic, F. Billebaud, E. Lellouch, A. Sandqvist, J. Brillet, A. Lecacheux, A. Hjalmarson, U. Frisk, M. Olberg, T. Odin, Odin space telescope monitoring of water vapor in the stratosphere of Jupiter. Planet. Space Sci. **61**(1), 3–14 (2012). https://doi.org/10.1016/j.pss.2011.04.001
T. Cavalie, H. Feuchtgruber, E. Lellouch, M. de Val-Borro, C. Jarchow, R. Moreno, P. Hartogh, G. Orton, T.K. Greathouse, F. Billebaud, M. Dobrijevic, L.M. Lara, A. Gonzalez, H. Sagawa, Spatial distribution of water in the stratosphere of Jupiter from Herschel HIFI and PACS observations. Astron. Astrophys. **553**, 16 (2013). https://doi.org/10.1051/0004-6361/201220797
T. Cavalie, R. Moreno, E. Lellouch, P. Hartogh, O. Venot, G.S. Orton, C. Jarchow, T. Encrenaz, F. Selsis, F. Hersant, L.N. Fletcher, The first submillimeter observation of CO in the stratosphere of Uranus. Astron. Astrophys. **562**, 6 (2014). https://doi.org/10.1051/0004-6361/201322297
A.A. Christou, R.M. Killen, M.H. Burger, The meteoroid stream of comet Encke at Mercury: implications for Mercury surface, space environment, geochemistry, and ranging observations of the exosphere. Geophys. Res. Lett. **42**, 7311–7318 (2015). https://doi.org/10.1002/2015GL065361
X. Chu, Z. Yu, C.S. Gardner, C. Chen, W. Fong, Lidar observations of neutral Fe layers and fast gravity waves in the thermosphere (110–155 km) at McMurdo (77.8° S, 166.7° E). Antarctica. Geophys. Res. Lett. **38**, L23807 (2011). https://doi.org/10.1029/2011gl050016
R.T. Clancy, B.J. Sandor, CO_2 ice clouds in the upper atmosphere of Mars. Geophys. Res. Lett. **25**(4), 489–492 (1998). https://doi.org/10.1029/98gl00114
R.T. Clancy, M.J. Wolff, P.R. Christensen, Mars aerosol studies with the MGS TES emission phase function observations: optical depths, particle sizes, and ice cloud types versus latitude and solar longitude. J. Geophys. Res. **108**(E9), 5098 (2003). https://doi.org/10.1029/2003je002058
R.T. Clancy, M.J. Wolff, B.A. Whitney, B.A. Cantor, M.D. Smith, Mars equatorial mesospheric clouds: global occurrence and physical properties from Mars Global Surveyor Thermal Emission Spectrometer and Mars Orbiter Camera limb observations. J. Geophys. Res. **112**(E4), E04004 (2007). https://doi.org/10.1029/2006je002805

B.C. Clark, A.K. Baird, Volatiles in the Martian regolith. Geophys. Res. Lett. **6**, 811–814 (1979). https://doi.org/10.1029/GL006i010p00811
S. Close, P. Brown, M. Campbell-Brown, M. Oppenheim, P. Colestock, Meteor head echo radar data: mass-velocity selection effects. Icarus **186**(2), 547–556 (2007). https://doi.org/10.1016/j.icarus.2006.09.007
M.M.J. Crismani, N.M. Schneider, J.M.C. Plane, Comment on a cometary origin for atmospheric martian methane by Fries et al. (2016). Geochem. Perspect. Lett. **3** (2017a). https://doi.org/10.7185/geochemlet.1715
M.M.J. Crismani, N.M. Schneider, J.M.C. Plane, J.S. Evans, S.K. Jain, M.S. Chaffin, J.D. Carrillo-Sanchez, J.I. Deighan, R.V. Yelle, A.I.F. Stewart, W. McClintock, J. Clarke, G.M. Holsclaw, A. Stiepen, F. Montmessin, B.M. Jakosky, Detection of a persistent meteoric metal layer in the Martian atmosphere. Nat. Geosci. **10**, 401–404 (2017b). https://doi.org/10.1038/ngeo2958
J. Curtius, R. Weigel, H.J. Vossing, H. Wernli, A. Werner, C.M. Volk, P. Konopka, M. Krebsbach, C. Schiller, A. Roiger, H. Schlager, V. Dreiling, S. Borrmann, Observations of meteoric material and implications for aerosol nucleation in the winter Arctic lower stratosphere derived from in situ particle measurements. Atmos. Chem. Phys. **5**, 3053–3069 (2005). https://doi.org/10.5194/acp-5-3053-2005
D.J. Cziczo, D.S. Thomson, D.M. Murphy, Ablation, flux, and atmospheric implications of meteors inferred from stratospheric aerosol. Science **291**(5509), 1772–1775 (2001). https://doi.org/10.1126/science.1057737
E.C.M. Dawkins, J.M.C. Plane, M.P. Chipperfield, W. Feng, J. Gumbel, J. Hedin, J. Höffner, J. Friedman, First global observations of the mesospheric potassium layer. Geophys. Res. Lett. **41**, 5653–5661 (2014). https://doi.org/10.1002/2014GL060801
T. deGraauw, H. Feuchtgruber, B. Bezard, P. Drossart, T. Encrenaz, D.A. Beintema, M. Griffin, A. Heras, M. Kessler, K. Leech, E. Lellouch, P. Morris, P.R. Roelfsema, M. RoosSerote, A. Salama, B. Vandenbussche, E.A. Valentijn, G.R. Davis, D.A. Naylor, First results of ISO-SWS observations of Saturn: detection of CO_2, CH_3C_2H, C_4H_2 and tropospheric H_2O. Astron. Astrophys. **321**(2), L13–L16 (1997). https://doi.org/10.1016/S0032-0633(03)00047-3
M.T. DeLand, E.P. Shettle, G.E. Thomas, J.J. Olivero, Latitude-dependent long-term variations in polar mesospheric clouds from SBUV version 3 PMC data. J. Geophys. Res. **112**(D10), D10315 (2007). https://doi.org/10.1029/2006jd007857
S.S. Dhomse, R.W. Saunders, W. Tian, M.P. Chipperfield, J.M.C. Plane, Plutonium-238 observations as a test of modeled transport and surface deposition of meteoric smoke particles. Geophys. Res. Lett. **40**(16), 4454–4458 (2013). https://doi.org/10.1002/grl.50840
V. Dikarev, E. Grun, New information recovered from the Pioneer 11 meteoroid experiment data. Astron. Astrophys. **383**(1), 302–308 (2002). https://doi.org/10.1051/0004-6361:20011686
V. Dikarev, E. Grün, J. Baggaley, D. Galligan, M. Landgraf, R. Jehn, Modeling the sporadic meteoroid background cloud. Earth Moon Planets **95**, 109–122 (2004). https://doi.org/10.1007/s11038-005-9017-y
V. Dikarev, E. Grün, J. Baggaley, D. Galligan, M. Landgraf, R. Jehn, The new ESA meteoroid model. Adv. Space Res. **35**, 1282–1289 (2005). https://doi.org/10.1016/j.asr.2005.05.014
N. Divine, 5 populations of interplanetary meteoroids. J. Geophys. Res. **98**(E9), 17029–17048 (1993). https://doi.org/10.1029/93je01203
J. Duplissy, J. Merikanto, A. Franchin, G. Tsagkogeorgas, J. Kangasluoma, D. Wimmer, H. Vuollekoski, S. Schobesberger, K. Lehtipalo, R.C. Flagan, D. Brus, N.M. Donahue, H. Vehkamäki, J. Almeida, A. Amorim, P. Barmet, F. Bianchi, M. Breitenlechner, E.M. Dunne, R. Guida, H. Henschel, H. Junninen, J. Kirkby, A. Kürten, A. Kupc, A. Määttänen, V. Makhmutov, S. Mathot, T. Nieminen, A. Onnela, A.P. Praplan, F. Riccobono, L. Rondo, G. Steiner, A. Tome, H. Walther, U. Baltensperger, K.S. Carslaw, J. Dommen, A. Hansel, T. Petäjä, M. Sipilä, F. Stratmann, A. Vrtala, P.E. Wagner, D.R. Worsnop, J. Curtius, M. Kulmala, Effect of ions on sulfuric acid-water binary particle formation: 2. experimental data and comparison with QC-normalized classical nucleation theory. J. Geophys. Res. **121**(4), 1752–1775 (2016). https://doi.org/10.1002/2015JD023539
J.L. Elliot, E.W. Dunham, A.S. Bosh, S.M. Slivan, L.A. Young, L.H. Wasserman, R.L. Millis, Pluto's atmosphere. Icarus **77**(1), 148–170 (1989). https://doi.org/10.1016/0019-1035(89)90014-6
T. Encrenaz, E. Lellouch, P. Drossart, H. Feuchtgruber, G.S. Orton, S.K. Atreya, First detection of CO in Uranus. Astron. Astrophys. **413**(2), L5–L9 (2004). https://doi.org/10.1051/0004-6361:20034637
I. Engel, B.P. Luo, M.C. Pitts, L.R. Poole, C.R. Hoyle, J.U. Grooß, A. Dörnbrack, T. Peter, Heterogeneous formation of polar stratospheric clouds—Part 2: nucleation of ice on synoptic scales. Atmos. Chem. Phys. **13**(21), 10769–10785 (2013). https://doi.org/10.5194/acp-13-10769-2013
Z.Y. Fan, J.M.C. Plane, J. Gumbel, J. Stegman, E.J. Llewellyn, Satellite measurements of the global mesospheric sodium layer. Atmos. Chem. Phys. **7**, 4107–4115 (2007). https://doi.org/10.5194/acpd-7-5413-2007
K.A. Farley, A. Montanari, E.M. Shoemaker, C.S. Shoemaker, Geochemical evidence for a comet shower in the late Eocene. Science **280**, 1250–1253 (1998). https://doi.org/10.1126/science.280.5367.1250

K.A. Farley, D. Vokrouhlický, W.F. Bottke, D. Nesvorný, A late Miocene dust shower from the break-up of an asteroid in the main belt. Nature **439**, 295–297 (2006). https://doi.org/10.1038/nature04391

B. Fegley, A.G.W. Cameron, A vaporization model for iron silicate fractionation in the Mercury protoplanet. Earth Planet. Sci. Lett. **82**(3–4), 207–222 (1987). https://doi.org/10.1016/0012-821x(87)90196-8

H. Feuchtgruber, E. Lellouch, T. deGraauw, B. Bezard, T. Encrenaz, M. Griffin, External supply of oxygen to the atmospheres of the giant planets. Nature **389**(6647), 159–162 (1997). https://doi.org/10.1038/38236

H. Feuchtgruber, E. Lellouch, T. Encrenaz, B. Bezard, A. Coustenis, P. Drossart, A. Salama, T. de Graauw, G.R. Davis, Oxygen in the stratospheres of the giant planets and Titan, in *Universe as Seen by Iso, Vols. I and II*, ed. by P. Cox, M.F. Kessler. Esa Special Publications vol. 427 (European Space Agency, Paris, 1999), pp. 133–136

G. Fjeldbo, W.C. Fjeldbo, V.R. Eshleman, Atmosphere of Mars—Mariner IV models compared. Science **153**(3743), 1518–1523 (1966). https://doi.org/10.1126/science.153.3743.1518

B.L. Fleshman, P.A. Delamere, F. Bagenal, T. Cassidy, A 1-D model of physical chemistry in Saturn's inner magnetosphere. J. Geophys. Res. **118**(8), 1567–1581 (2013). https://doi.org/10.1002/jgre.20106

L.N. Fletcher, P. Drossart, M. Burgdorf, G.S. Orton, T. Encrenaz, Neptune's atmospheric composition from AKARI infrared spectroscopy. Astron. Astrophys. **514**, A17 (2010). https://doi.org/10.1051/0004-6361/200913358

G.J. Flynn, Accretion of meteoritic material onto Mars—implications for the surface, atmosphere, and moons, in *2nd COSPAR Colloquium*, Sopron, Hungary (Pergamon, Oxford, 1991), pp. 121–124

G.J. Flynn, Does the Kuiper belt contribute significantly to the zodiacal cloud and the stratospheric interplanetary dust? in *Abstracts of the 25th Lunar and Planetary Science Conference*, Houston, TX, vol. 25 (1994), p. 379

G.J. Flynn, The delivery of organic matter from asteroids and comets to the early surface of Mars. Earth Moon Planets **72**, 469–474 (1996). https://doi.org/10.1007/BF00117551

G.J. Flynn, The contribution by interplanetary dust to noble gases in the atmosphere of Mars. J. Geophys. Res. **102**, 9175–9182 (1997). https://doi.org/10.1029/96JE03883

G.J. Flynn, D.S. McKay, An assessment of the meteoritic contribution to the Martian soil. J. Geophys. Res. **96**, 14,497–414,509 (1990). https://doi.org/10.1029/JB095iB09p14497

G.J. Flynn, L.P. Keller, M. Feser, S. Wirick, C. Jacobsen, The origin of organic matter in the solar system: evidence from the interplanetary dust particles. Geochim. Cosmochim. Acta **67**, 4791–4806 (2003). https://doi.org/10.1016/j.gca.2003.09.001

G.J. Flynn, L.P. Keller, C. Jacobsen, S. Wirick, An assessment of the amount and types of organic matter contributed to the Earth by interplanetary dust. Adv. Space Res. **33**, 57–66 (2004). https://doi.org/10.1016/j.asr.2003.09.036

V. Formisano, A. Maturilli, M. Giuranna, E. D'Aversa, M.A. Lopez-Valverde, Observations of non-LTE emission at 4–5 microns with the planetary Fourier spectrometer abord the Mars Express mission. Icarus **182**(1), 51–67 (2006). https://doi.org/10.1016/j.icarus.2005.12.022

V.L. Frankland, A.D. James, W.H. Feng, J.M.C. Plane, The uptake of HNO_3 on meteoric smoke analogues. J. Atmos. Sol.-Terr. Phys. **127**, 150–160 (2015). https://doi.org/10.1016/j.jastp.2015.01.010

V.L. Frankland, A.D. James, J.D.C. Sanchez, T.P. Mangan, K. Willacy, A.R. Poppe, J.M.C. Plane, Uptake of acetylene on cosmic dust and production of benzene in Titan's atmosphere. Icarus **278**, 88–99 (2016). https://doi.org/10.1016/j.icarus.2016.06.007

V.L. Frankland, A.D. James, J.D. Carrillo-Sánchez, D. Nesvorný, P. Pokorný, J.M.C. Plane, CO oxidation and O_2 removal on meteoric material in Venus' atmosphere. Icarus **296**, 150–162 (2017). https://doi.org/10.1016/j.icarus.2017.06.005

P. Fraundorf, The distribution of temperature maxima for micrometeorites decelerated in the Earth's atmosphere without melting. Geophys. Res. Lett. **7**, 765–768 (1980). https://doi.org/10.1029/GL007i010p00765

M. Friedrich, M. Rapp, J.M.C. Plane, K.M. Torkar, Bite-outs and other depletions of mesospheric electrons. J. Atmos. Sol.-Terr. Phys. **73**(14–15), 2201–2211 (2011). https://doi.org/10.1016/j.jastp.2010.10.018

M. Fries, A. Christou, D. Archer, P. Conrad, W. Cooke, J. Eigenbrode, I.L. ten Kate, M. Matney, P. Niles, M. Sykes, A. Steele, A. Treiman, A cometary origin for martian atmospheric methane. Geochem. Perspect. Lett. **2**(1), 10–22 (2016). https://doi.org/10.7185/geochemlet.1602

D. Fussen, F. Vanhellemont, C. Tetard, N. Mateshvili, E. Dekemper, N. Loodts, C. Bingen, E. Kyrola, J. Tamminen, V. Sofieva, A. Hauchecorne, F. Dalaudier, J.L. Bertaux, G. Barrot, L. Blanot, O.F. d'Andon, T. Fehr, L. Saavedra, T. Yuan, C.Y. She, A global climatology of the mesospheric sodium layer from GOMOS data during the 2002–2008 period. Atmos. Chem. Phys. **10**(19), 9225–9236 (2010). https://doi.org/10.5194/acp-10-9225-2010

P. Gabrielli, C. Barbante, J.M.C. Plane, A. Varga, S. Hong, G. Cozzi, V. Gaspari, F.A.M. Planchon, W. Cairns, C. Ferrari, P. Crutzen, P. Cescon, C.F. Boutron, Meteoric smoke fallout over the Holocene epoch revealed by iridium and platinum in Greenland ice. Nature **432**(7020), 1011–1014 (2004). https://doi.org/10.1038/nature03137

D.P. Galligan, W.J. Baggaley, The orbital distribution of radar-detected meteoroids of the solar system dust cloud. Mon. Not. R. Astron. Soc. **353**(2), 422–446 (2004). https://doi.org/10.1111/j.1365-2966.2004.08078.x

D.P. Galligan, W.J. Baggaley, The radiant distribution of AMOR radar meteors. Mon. Not. R. Astron. Soc. **359**(2), 551–560 (2005). https://doi.org/10.1111/j.1365-2966.2005.08918.x

P. Gao, X. Zhang, D. Crisp, C.G. Bardeen, Y.L. Yung, Bimodal distribution of sulfuric acid aerosols in the upper haze of Venus. Icarus **231**, 83–98 (2014). https://doi.org/10.1016/j.icarus.2013.10.013

C.S. Gardner, A.Z. Liu, D.R. Marsh, W.H. Feng, J.M.C. Plane, Inferring the global cosmic dust influx to the Earth's atmosphere from lidar observations of the vertical flux of mesospheric Na. J. Geophys. Res. **119**(9), 7870–7879 (2014). https://doi.org/10.1002/2014ja020383

C.S. Gardner, A.Z. Liu, Y. Guo, Vertical and horizontal transport of mesospheric Na: implications for the mass influx of cosmic dust. J. Atmos. Sol.-Terr. Phys. (2016). https://doi.org/10.1016/j.jastp.2016.07.013

L.J. Gelinas, K.A. Lynch, M.C. Kelley, R.L. Collins, M. Widholm, E. MacDonald, J. Ulwick, P. Mace, Mesospheric charged dust layer: implications for neutral chemistry. J. Geophys. Res. **110**(A1), A01310 (2005). https://doi.org/10.1029/2004ja010503

M. Gerding, M. Alpers, U. von Zahn, R.J. Rollason, J.M.C. Plane, Atmospheric Ca and Ca^+ layers: midlatitude observations and modeling. J. Geophys. Res. **105**(A12), 27131–27146 (2000). https://doi.org/10.1029/2000ja900088

R. Gersonde, F.T. Kyte, U. Bleil, B. Diekmann, J.A. Flores, K. Gohl, G. Grahl, R. Hagen, G. Kuhn, F.J. Sierro, D. Völker, A. Abelmann, J.A. Bostwick, Geological record and reconstruction of the late Pliocene impact of the Eltanin asteroid in the Southern Ocean. Nature **390**, 357–363 (1997). https://doi.org/10.1038/37044

G.R. Gladstone, S.A. Stern, K. Ennico, C.B. Olkin, H.A. Weaver, L.A. Young, M.E. Summers, D.F. Strobel, D.P. Hinson, J.A. Kammer, A.H. Parker, A.J. Steffl, I.R. Linscott, J.W. Parker, A.F. Cheng, D.C. Slater, M.H. Versteeg, T.K. Greathouse, K.D. Retherford, H. Throop, N.J. Cunningham, W.W. Woods, K.N. Singer, C.C.C. Tsang, E. Schindhelm, C.M. Lisse, M.L. Wong, Y.L. Yung, X. Zhu, W. Curdt, P. Lavvas, E.F. Young, G.L. Tyler, T. New Horizons Sci, The atmosphere of Pluto as observed by New Horizons. Science **351**(6279), 1280–1285 (2016). https://doi.org/10.1126/science.aad8866

J.C. Gómez Martín, D.L. Bones, J.D. Carrillo-Sánchez, A.D. James, J.M. Trigo-Rodríguez, B. Fegley Jr., J.M.C. Plane, Novel experimental simulations of the atmospheric injection of meteoric metals. Astrophys. J. **836**, 212 (2017). https://doi.org/10.3847/1538-4357/aa5c8f

F. Gonzalez-Galindo, A. Määttänen, F. Forget, A. Spiga, The martian mesosphere as revealed by CO_2 cloud observations and general circulation modeling. Icarus **216**(1), 10–22 (2011). https://doi.org/10.1016/j.icarus.2011.08.006

J.M. Grebowsky, A.C. Aikin, In situ measurements of meteoric ions, in *Meteors in the Earth's Atmosphere*, ed. by E. Murad, I.P. Williams (Cambridge University Press, Cambridge, 2002), pp. 189–214

J.M. Grebowsky, M. Benna, J.M.C. Plane, G.A. Collinson, P.R. Mahaffy, B.M. Jakosky, Unique, non-Earthlike, meteoritic ion behavior in upper atmosphere of Mars. Geophys. Res. Lett. **44**(7), 3066–3072 (2017). https://doi.org/10.1002/2017GL072635

J.M. Greenberg, A. Li, Morphological structure and chemical composition of cometary nuclei and dust. Space Sci. Rev. **90**(1–2), 149–161 (1999). https://doi.org/10.1023/a:1005298014670

E. Grün, H.A. Zook, H. Fechtig, R.H. Giese, Collisional balance of the meteoric complex. Icarus **62**(2), 244–272 (1985). https://doi.org/10.1016/0019-1035(85)90121-6

E. Grün, B. Gustafson, I. Mann, M. Baguhl, G.E. Morfill, P. Staubach, A. Taylor, H.A. Zook, Interstellar dust in the heliosphere. Astron. Astrophys. **286**, 915–924 (1994)

E. Grün, M. Baguhl, N. Divine, H. Fechtig, D.P. Hamilton, M.S. Hanner, J. Kissel, B.A. Lindblad, D. Linkert, G. Linkert, I. Mann, J.A.M. McDonnell, G.E. Morfill, C. Polanskey, R. Riemann, G. Schwehm, N. Siddique, P. Staubach, H.A. Zook, 3 years of Galileo dust data. Planet. Space Sci. **43**(8), 953–969 (1995a). https://doi.org/10.1016/0032-0633(94)00234-i

E. Grün, M. Baguhl, N. Divine, H. Fechtig, D.P. Hamilton, M.S. Hanner, J. Kissel, B.A. Lindblad, D. Linkert, G. Linkert, I. Mann, J.A.M. McDonnell, G.E. Morfill, C. Polanskey, R. Riemann, G. Schwehm, N. Siddique, P. Staubach, H.A. Zook, Two years of Ulysses dust data. Planet. Space Sci. **43**(8), 971–999 (1995b). https://doi.org/10.1016/0032-0633(94)00233-H

E. Grün, P. Staubach, M. Baguhl, D.P. Hamilton, H.A. Zook, S. Dermott, B.A. Gustafson, H. Fechtig, J. Kissel, D. Linkert, G. Linkert, R. Srama, M.S. Hanner, C. Polanskey, M. Horanyi, B.A. Lindblad, I. Mann, J.A.M. McDonnell, G.E. Morfill, G. Schwehm, South–North and radial traverses through the interplanetary dust cloud. Icarus **129**(2), 270–288 (1997). https://doi.org/10.1006/icar.1997.5789

J. Gumbel, L. Megner, Charged meteoric smoke as ice nuclei in the mesosphere: Part 1-A review of basic concepts. J. Atmos. Sol.-Terr. Phys. **71**, 1225–1235 (2009). https://doi.org/10.1016/j.jastp.2009.04.012

D.A. Gurnett, J.A. Ansher, W.S. Kurth, L.J. Granroth, Micron-sized dust particles detected in the outer solar system by the Voyager 1 and 2 plasma wave instruments. Geophys. Res. Lett. **24**(24), 3125–3128 (1997). https://doi.org/10.1029/97gl03228

J.M. Hahn, H.A. Zook, B. Cooper, B. Sunkara, Clementine observations of the zodiacal light and the dust content of the inner solar system. Icarus **158**(2), 360–378 (2002). https://doi.org/10.1006/icar.2002.6881

W.K. Hartmann, G. Ryder, L. Dones, D. Grinspoon, The time-dependent intense bombardment of the primordial Earth/Moon system, in *Origin of the Earth and Moon*, ed. by R.M. Canup, K. Righter (University of Arizona Press, Tucson, 2000), pp. 493–512

M.G. Hauser, F.C. Gillett, F.J. Low, T.N. Gautier, C.A. Beichman, H.H. Aumann, G. Neugebauer, B. Baud, N. Boggess, J.P. Emerson, IRAS observations of the diffuse infrared background. Astrophys. J. **278**, 115–118 (1984). https://doi.org/10.1086/184212 (Letters to the editor)

J. Hedin, F. Giovane, T. Waldemarsson, J. Gumbel, J. Blum, R.M. Stroud, L. Marlin, J. Moser, D.E. Siskind, K. Jansson, R.W. Saunders, M.E. Summers, P. Reissaus, J. Stegman, J.M.C. Plane, M. Horányi, The MAGIC meteoric smoke particle sampler. J. Atmos. Sol.-Terr. Phys. **118**, 127–144 (2014). https://doi.org/10.1016/j.jastp.2014.03.003

K.C. Herr, G.C. Pimentel, Evidence for solid carbon dioxide in upper atmosphere of Mars. Science **167**(3914), 47–49 (1970). https://doi.org/10.1126/science.167.3914.47

M.E. Hervig, L.L. Gordley, L.E. Deaver, D.E. Siskind, M.H. Stevens, J.M. Russell III., S.M. Bailey, L. Megner, C.G. Bardeen, First satellite observations of meteoric smoke in the middle atmosphere. Geophys. Res. Lett. **36**, L18805 (2009). https://doi.org/10.1029/2009gl039737

M.E. Hervig, L.E. Deaver, C.G. Bardeen, J.M. Russell III., S.M. Bailey, L.L. Gordley, The content and composition of meteoric smoke in mesospheric ice particles from SOFIE observations. J. Atmos. Sol.-Terr. Phys. **84–85**, 1–6 (2012). https://doi.org/10.1016/j.jastp.2012.04.005

B.E. Hesman, G.R. Davis, H.E. Matthews, G.S. Orton, The abundance profile of CO in Neptune's atmosphere. Icarus **186**(2), 342–353 (2007). https://doi.org/10.1016/j.icarus.2006.08.025

J.K. Hillier, S.F. Green, N. McBride, N. Altobelli, F. Postberg, S. Kempf, J. Schwanethal, R. Srama, J.A. McDonnell, E. Gruen, Interplanetary dust detected by the cassini CDA chemical analyser. Icarus **190**(2), 643–654 (2007). https://doi.org/10.1016/j.icarus.2007.03.024

T. Hirai, M.J. Cole, M. Fujii, S. Hasegawa, T. Iwai, M. Kobayashi, R. Srama, H. Yano, Microparticle impact calibration of the Arrayed Large-Area Dust Detectors in INterplanetary space (ALADDIN) onboard the solar power sail demonstrator IKAROS. Planet. Space Sci. **100**(Suppl. C), 87–97 (2014). https://doi.org/10.1016/j.pss.2014.05.009

T. Hirai, H. Yano, M. Fujii, S. Hasegawa, N. Moriyama, C. Okamoto, M. Tanaka, Data screening and reduction in interplanetary dust measurement by IKAROS-ALADDIN. Adv. Space Res. **59**(6), 1450–1459 (2017). https://doi.org/10.1016/j.asr.2016.12.023

M. Horanyi, V. Hoxie, D. James, A. Poppe, C. Bryant, B. Grogan, B. Lamprecht, J. Mack, F. Bagenal, S. Batiste, N. Bunch, T. Chanthawanich, F. Christensen, M. Colgan, T. Dunn, G. Drake, A. Fernandez, T. Finley, G. Holland, A. Jenkins, C. Krauss, E. Krauss, O. Krauss, M. Lankton, C. Mitchell, M. Neeland, T. Reese, K. Rash, G. Tate, C. Vaudrin, J. Westfall, The Student Dust Counter on the New Horizons mission. Space Sci. Rev. **140**(1–4), 387–402 (2008). https://doi.org/10.1007/s11214-007-9250-y

M. Horányi, J.R. Szalay, S. Kempf, J. Schmidt, E. Grün, R. Srama, Z. Sternovsky, A permanent, asymmetric dust cloud around the Moon. Nature **522**, 324–326 (2015). https://doi.org/10.1038/nature14479

C.R. Hoyle, I. Engel, B.P. Luo, M.C. Pitts, L.R. Poole, J.U. Grooß, T. Peter, Heterogeneous formation of polar stratospheric clouds—Part 1: nucleation of nitric acid trihydrate (NAT). Atmos. Chem. Phys. **13**(18), 9577–9595 (2013). https://doi.org/10.5194/acp-13-9577-2013

W. Huang, X. Chu, C.S. Gardner, J.D. Carrillo-Sánchez, W. Feng, J.M.C. Plane, D. Nesvorný, Measurements of the vertical fluxes of atomic Fe and Na at the mesopause: implications for the velocity of cosmic dust entering the atmosphere. Geophys. Res. Lett. **42**, 169–175 (2015). https://doi.org/10.1002/2014GL062390

D.E. Hughes, Meteors, in *Cosmic Dust*, ed. by J.A.M. McDonnell (Wiley-Interscience, New York, 1978), pp. 123–185

D.H. Humes, Results of Pioneer 10 and 11 meteoroid experiments: interplanetary and near-Saturn. J. Geophys. Res. **85**(A11), 5841–5852 (1980). https://doi.org/10.1029/JA085iA11p05841

S.I. Ipatov, J.C. Mather, Migration of small bodies and dust to near-Earth space, in *Moon and Near-Earth Objects, Vol. 37*, ed. by P. Ehrenfreund, B. Foing, A. Cellino. Advances in Space Research, vol. 1 (Elsevier, Amsterdam, 2006), pp. 126–137. 1

P.G.J. Irwin, E. Lellouch, C. de Bergh, R. Courtin, B. Bezard, L.N. Fletcher, G.S. Orton, N.A. Teanby, S.B. Calcutt, D. Tice, J. Hurley, G.R. Davis, Line-by-line analysis of Neptune's near-IR spectrum observed with Gemini/NIFS and VLT/CRIRES. Icarus **227**, 37–48 (2014). https://doi.org/10.1016/j.icarus.2013.09.003

E.P. James, O.B. Toon, G. Schubert, A numerical microphysical model of the condensational Venus cloud. Icarus **129**(1), 147–171 (1997). https://doi.org/10.1006/icar.1997.5763

A.D. James, J.S.A. Brooke, T.P. Mangan, T.F. Whale, J.M.C. Plane, B.J. Murray, Nucleation of nitric acid hydrates in Polar stratospheric clouds by meteoric material. Atmos. Chem. Phys. Discuss. **3**, 1–2 (2017a). https://doi.org/10.5194/acp-2017-816

A.D. James, D.R. Moon, W.H. Feng, P.S.J. Lakey, V.L. Frankland, D.E. Heard, J.M.C. Plane, The uptake of HO_2 on meteoric smoke analogues. J. Geophys. Res. **122**(1), 554–565 (2017b). https://doi.org/10.1002/2016jd025882

D. Janches, J.L. Chau, Observed diurnal and seasonal behavior of the micrometeor flux using the Arecibo and Jicamarca radars. J. Atmos. Sol.-Terr. Phys. **67**(13), 1196–1210 (2005). https://doi.org/10.1016/j.jastp.2005.06.011

D. Janches, D.O. ReVelle, Initial altitude of the micrometeor phenomenon: comparison between Arecibo radar observations and theory. J. Geophys. Res. **110**(A8), A08307 (2005). https://doi.org/10.1029/2005JA011022

D. Janches, C.J. Heinselman, J.L. Chau, A. Chandran, R. Woodman, Modeling the global micrometeor input function in the upper atmosphere observed by high power and large aperture radars. J. Geophys. Res. **111**(A7), A07317 (2006). https://doi.org/10.1029/2006JA011628

D. Janches, S. Close, J.T. Fentzke, A comparison of detection sensitivity between ALTAIR and Arecibo meteor observations: can high power and large aperture radars detect low velocity meteor head-echoes. Icarus **193**(1), 105–111 (2008). https://doi.org/10.1016/j.icarus.2007.08.022

D. Janches, J.M.C. Plane, D. Nesvorný, W. Feng, D. Vokrouhlický, Radar detectability studies of slow and small zodiacal dust cloud particles: I. The case of arecibo 430 MHz meteor head echo observations. Astrophys. J. **796**, 41 (2014)

K.S. Johnson, Iron supply and demand in the upper ocean: is extraterrestrial dust a significant source of bioavailable iron? Glob. Biogeochem. Cycles **15**, 61–63 (2001). https://doi.org/10.1029/2000GB001295

J. Jones, P. Brown, Sporadic meteor radiant distributions—orbital survey results. Mon. Not. R. Astron. Soc. **265**(3), 524–532 (1993). https://doi.org/10.1093/mnras/265.3.524

O. Kalashnikova, M. Horanyi, G.E. Thomas, O.B. Toon, Meteoric smoke production in the atmosphere. Geophys. Res. Lett. **27**(20), 3293–3296 (2000). https://doi.org/10.1029/1999gl011338

R.R. Keays, R. Ganapathy, J.C. Laul, E. Anders, G.F. Herzog, P.M. Jeffery, Trace elements and radioactivity in lunar rocks: implications for meteorite infall, solar-wind flux, and formation conditions of Moon. Science **167**, 490–493 (1970). https://doi.org/10.1126/science.167.3918.490

T. Kelsall, J.L. Weiland, B.A. Franz, W.T. Reach, R.G. Arendt, E. Dwek, H.T. Freudenreich, M.G. Hauser, S.H. Moseley, N.P. Odegard, R.F. Silverberg, E.L. Wright, The COBE diffuse infrared background experiment search for the cosmic infrared background. II. Model of the interplanetary dust cloud. Astrophys. J. **508**(1), 44–73 (1998). https://doi.org/10.1086/306380

S. Kempf, Z. Sternovsky, M. Horanyi, K.P. Hand, R. Srama, F. Postberg, N. Altobelli, E. Grün, M.S. Gudipati, J. Schmidt, M.Y. Zolotov, S. Tucker, V.C. Hoxie, R. Kohnert, *Compositional Mapping of Europa's Surface with SUDA. American Geophysical Union, Fall Meeting 2015, abstract #P13E-07* (2015)

F. Keppler, I. Vigano, A. McLeod, U. Ott, M. Früchtl, T. Röckmann, Ultraviolet-radiation-induced methane emissions from meteorites and the Martian atmosphere. Nature **486**, 93–96 (2012). https://doi.org/10.1038/nature11203

R.M. Killen, J.M. Hahn, Impact vaporization as a possible source of Mercury's calcium exosphere. Icarus **250**(Suppl. C), 230–237 (2015). https://doi.org/10.1016/j.icarus.2014.11.035

Y.H. Kim, W.D. Pesnell, J.M. Grebowsky, J.L. Fox, Meteoric ions in the ionosphere of Jupiter. Icarus **150**(2), 261–278 (2001). https://doi.org/10.1006/icar.2001.6590

A.J. Kliore, I.R. Patel, A.F. Nagy, T.E. Cravens, T.I. Gombosi, Initial obervations of the nightside ionosphere of Venus from Pioneer Venus orbiter radio occultations. Science **205**(4401), 99–102 (1979). https://doi.org/10.1126/science.205.4401.99

R.G. Knollenberg, D.M. Hunten, The microphysics of the clouds of Venus: results of the Pioneer Venus Particle Size Spectrometer experiment. J. Geophys. Res. **85**(A13), 8039–8058 (1980). https://doi.org/10.1029/JA085iA13p08039

E. Kopp, On the abundance of metal ions in the lower ionosphere. J. Geophys. Res. **102**(A5), 9667–9674 (1997). https://doi.org/10.1029/97ja00384

V.A. Krasnopolsky, VEGA Mission results and chemical composition of Venusian clouds. Icarus **80**(1), 202–210 (1989). https://doi.org/10.1016/0019-1035(89)90168-1

V.A. Krasnopolsky, Chemical composition of Venus atmosphere and clouds: some unsolved problems. Planet. Space Sci. **54**, 1352–1359 (2006)

V.A. Krasnopolsky, Titan's photochemical model: further update, oxygen species, and comparison with Triton and Pluto. Planet. Space Sci. **73**(1), 318–326 (2012). https://doi.org/10.1016/j.pss.2012.08.013

V.A. Krasnopolsky, On the iron chloride aerosol in the clouds of Venus. Icarus **286**, 134–137 (2017). https://doi.org/10.1016/j.icarus.2016.10.003

V.A. Krasnopolsky, B.R. Sandel, F. Herbert, R.J. Vervack, Temperture, N2, and density profiles of Triton's atmosphere—observations and model. J. Geophys. Res. **98**(E2), 3065–3078 (1993). https://doi.org/10.1029/92je02680

S. Kremser, L.W. Thomason, M. von Hobe, M. Hermann, T. Deshler, C. Timmreck, M. Toohey, A. Stenke, J.P. Schwarz, R. Weigel, S. Fueglistaler, F.J. Prata, J.P. Vernier, H. Schlager, J.E. Barnes, J.C. Antuna-Marrero, D. Fairlie, M. Palm, E. Mahieu, J. Notholt, M. Rex, C. Bingen, F. Vanhellemont, A. Bourassa, J.M.C. Plane, D. Klocke, S.A. Carn, L. Clarisse, T. Trickl, R. Neely, A.D. James, L. Rieger, J.C. Wilson, B. Meland, Stratospheric aerosol-observations, processes, and impact on climate. Rev. Geophys. **54**(2), 278–335 (2016). https://doi.org/10.1002/2015rg000511

A.V. Krivov, M. Sremcevic, F. Spahn, V.V. Dikarev, K.V. Kholshevnikov, Impact-generated dust clouds around planetary satellites: spherically symmetric case. Planet. Space Sci. **51**(3), 251–269 (2003). https://doi.org/10.1016/S0032-0633(02)00147-2

H. Kruger, E. Grun, D.P. Hamilton, M. Baguhl, S. Dermott, H. Fechtig, B.A. Gustafson, M.S. Hanner, M. Horanyi, J. Kissel, B.A. Lindblad, D. Linkert, G. Linkert, I. Mann, J.A.M. McDonnell, G.E. Morfill, C. Polanskey, R. Riemann, G. Schwehm, R. Srama, H.A. Zook, Three years of Galileo dust data: II. 1993–1995. Planet. Space Sci. **47**(1–2), 85–106 (1999). https://doi.org/10.1016/S0032-0633(98)00097-X

H. Kruger, N. Altobelli, B. Anweiler, S.F. Dermott, V. Dikarev, A.L. Graps, E. Grun, B.A. Gustafson, D.P. Hamilton, M.S. Hanner, M. Horanyi, J. Kissel, M. Landgraf, B.A. Lindblad, D. Linkert, G. Linkert, I. Mann, J.A.M. McDonnell, G.E. Morfill, C. Polanskey, G. Schwehm, R. Srama, H.A. Zook, Five years of Ulysses dust data: 2000–2004. Planet. Space Sci. **54**(9–10), 932–956 (2006). https://doi.org/10.1016/j.pss.2006.04.015

M.J. Kuchner, C.C. Stark, Collisional grooming models of the Kuiper belt dust cloud. Astron. J. **140**(4), 1007–1019 (2010). https://doi.org/10.1088/0004-6256/140/4/1007

F. Kyte, Tracers of the extraterrestrial component in sediments and inferences for Earth's accretion history, in *Catastrophic Events and Mass Extinctions: Impacts and Beyond*, vol. Special Paper 356, ed. by C. Koeberl, K.G. MacLeod (Geological Society of America, Boulder, 2002), pp. 21–38

F.T. Kyte, J.T. Wasson, Accretion rate of extraterrestrial matter: iridium deposited 33 to 67 million years ago. Science **232**, 1225–1229 (1986). https://doi.org/10.1126/science.232.4755.1225

F.T. Kyte, Z. Zhou, J.T. Wasson, High noble metal concentrations in a late Pliocene sediment. Nature **292**, 417–420 (1981). https://doi.org/10.1038/292417a0

D. Lal, A.J.T. Jull, Extra-terrestrial influx rates of cosmogenic isotopes and platinum group elements: realizable geochemical effects. Geochim. Cosmochim. Acta **67**, 4925–4933 (2003). https://doi.org/10.1016/S0016-7037(03)00502-7

D. Lal, A.J.T. Jull, On the fluxes and fates of ^{3}He accreted by the Earth with extraterrestrial particles. Earth Planet. Sci. Lett. **235**, 375–390 (2005). https://doi.org/10.1016/j.espl.2005.04.011

L. Lanci, B. Delmonte, D.V. Kent, V. Maggi, P.E. Biscaye, J.R. Petit, Magnetization of polar ice: a measurement of terrestrial dust and extraterrestrial fallout. Quat. Sci. Rev. **33**, 20–31 (2012). https://doi.org/10.1016/j.quascirev.2011.11.023

M. Landgraf, J.C. Liou, H.A. Zook, E. Grun, Origins of solar system dust beyond Jupiter. Astron. J. **123**(5), 2857–2861 (2002). https://doi.org/10.1086/339704

M. Langowski, C.V. Savigny, J.P. Burrows, W. Feng, J.M.C. Plane, D.R. Marsh, D. Janches, M. Sinnhuber, A.C. Aikin, Global investigation of the Mg atom and ion layers using SCIAMACHY/Envisat observations between 70 km and 150 km altitude and WACCM-Mg model results. Atmos. Chem. Phys. **15**, 273–295 (2015). https://doi.org/10.5194/acp-15-273-2015

R. Latteck, J. Bremer, Long-term changes of polar mesosphere summer echoes at 69°N. J. Geophys. Res. **118**(18), 10,441–410,448 (2013). https://doi.org/10.1002/jgrd.50787

F. Lefevre, F. Forget, Observed variations of methane on Mars unexplained by known atmospheric chemistry and physics. Nature **460**(7256), 720–723 (2009). https://doi.org/10.1038/nature08228

E. Lellouch, Chemistry induced by the impacts: observations, in *The Collision of Comet Shoemaker-Levy 9 and Jupiter*, ed. by K.S. Noll, H.A. Weaver, P.D. Feldman. Space Telescope Science Institute Symposium Series (No. 9) (Cambridge University Press, Cambridge, 1996), pp. 213–242

E. Lellouch, T. Encrenaz, T. de Graauw, S. Erard, P. Morris, J. Crovisier, H. Feuchtgruber, T. Girard, M. Burgdorf, The 2.4–45 μm spectrum of Mars observed with the Infrared Space Observatory. Planet. Space Sci. **48**(12–14), 1393–1405 (2000). https://doi.org/10.1016/s0032-0633(00)00118-5

E. Lellouch, B. Bezard, J.I. Moses, G.R. Davis, P. Drossart, H. Feuchtgruber, E.A. Bergin, R. Moreno, T. Encrenaz, The origin of water vapor and carbon dioxide in Jupiter's stratosphere. Icarus **159**(1), 112–131 (2002). https://doi.org/10.1006/icar.2002.6929

E. Lellouch, R. Moreno, G. Paubert, A dual origin for Neptune's carbon monoxide? Astron. Astrophys. **430**(2), L37–L40 (2005). https://doi.org/10.1051/0004-6361:200400127

E. Lellouch, B. Bezard, D.F. Strobel, G.L. Bjoraker, F.M. Flasar, P.N. Romani, On the HCN and CO_2 abundance and distribution in Jupiter's stratosphere. Icarus **184**(2), 478–497 (2006). https://doi.org/10.1016/j.icarus.2006.05.018

E. Lellouch, B. Sicardy, C. de Bergh, H.U. Kaufl, S. Kassi, A. Campargue, Pluto's lower atmosphere structure and methane abundance from high-resolution spectroscopy and stellar occultations. Astron. Astrophys. **495**(3), L17–L21 (2009). https://doi.org/10.1051/0004-6361/200911633

E. Lellouch, C. de Bergh, B. Sicardy, S. Ferron, H.U. Kaufl, Detection of CO in Triton's atmosphere and the nature of surface-atmosphere interactions. Astron. Astrophys. **512**, 6 (2010). https://doi.org/10.1051/0004-6361/201014339

H.F. Levison, M.J. Duncan, From the Kuiper belt to Jupiter-family comets: the spatial distribution of ecliptic comets. Icarus **127**(1), 13–32 (1997). https://doi.org/10.1006/icar.1996.5637

J.C. Liou, H.A. Zook, Signatures of the giant planets imprinted on the Edgeworth-Kuiper belt dust disk. Astron. J. **118**(1), 580–590 (1999). https://doi.org/10.1086/300938

J.C. Liou, H.A. Zook, S.F. Dermott, Kuiper belt dust grains as a source of interplanetary dust particles. Icarus **124**(2), 429–440 (1996). https://doi.org/10.1006/icar.1996.0220

C. Listowski, A. Määttänen, F. Montmessin, A. Spiga, F. Lefevre, Modeling the microphysics of CO_2 ice clouds within wave-induced cold pockets in the martian mesosphere. Icarus **237**, 239–261 (2014). https://doi.org/10.1016/j.icarus.2014.04.022

M.A. Lopez-Valverde, M. Lopez-Puertas, J.J. Lopez-Moreno, V. Formisano, D. Grassi, A. Maturilli, E. Lellouch, P. Drossart, Analysis of CO_2 non-LTE emissions at 4.3 μm in the Martian atmosphere as observed by PFS/Mars Express and SWS/ISO. Planet. Space Sci. **53**(10), 1079–1087 (2005). https://doi.org/10.1016/j.pss.2005.03.007

S.G. Love, D.E. Brownlee, Heating and thermal transformation of micrometeoroids entering the Earth's atmosphere. Icarus **89**(1), 26–43 (1991). https://doi.org/10.1016/0019-1035(91)90085-8

S.G. Love, D.E. Brownlee, A direct measurement of the terrestrial mass accretion rate of cosmic dust. Science **262**, 550–553 (1993). https://doi.org/10.1126/science.262.5133.550

F.J. Low, D.A. Beintema, T.N. Gautier, F.C. Gillett, C.A. Beichman, G. Neugebauer, E. Young, H.H. Aumann, N. Boggess, J.P. Emerson, H.J. Habing, M.G. Hauser, J.R. Houck, M. Rowanrobinson, B.T. Soifer, R.G. Walker, P.R. Wesselius, Infrared Cirrus—new components of the extended infrared-emission. Astrophys. J. **278**(1), L19–L22 (1984). https://doi.org/10.1086/184213

F.J. Lübken, M. Rapp, T. Blix, E. Thrane, Microphysical and turbulent measurements of the Schmidt number in the vicinity of polar mesosphere summer echoes. Geophys. Res. Lett. **25**(6), 893–896 (1998). https://doi.org/10.1029/98GL50479

F.J. Lübken, U. Berger, G. Baumgarten, Temperature trends in the midlatitude summer mesosphere. J. Geophys. Res. **118**(24), 13,347–313,360 (2013). https://doi.org/10.1002/2013jd020576

S.H. Luszcz-Cook, I. de Pater, Constraining the origins of Neptune's carbon monoxide abundance with CARMA millimeter-wave observations. Icarus **222**(1), 379–400 (2013). https://doi.org/10.1016/j.icarus.2012.11.002

A. Määttänen, F. Montmessin, B. Gondet, F. Scholten, H. Hoffmann, F. Gonzalez-Galindo, A. Spiga, F. Forget, E. Hauber, G. Neukum, J.P. Bibring, J.L. Bertaux, Mapping the mesospheric CO_2 clouds on Mars: MEx/OMEGA and MEx/HRSC observations and challenges for atmospheric models. Icarus **209**(2), 452–469 (2010). https://doi.org/10.1016/j.icarus.2010.05.017

A. Määttänen, K. Pérot, F. Montmessin, A. Hauchecorne, Mesospheric clouds on Mars and on Earth, in *Comparative Climatology of Terrestrial Planets*, ed. by S.J. Mackwell, M.A. Bullock, J.W. Harder, A.A. Simon-Miller (University of Arizona Press, Tucson, 2013), pp. 393–413

O.M. Markova, O.L. Yakovlev, G.L. Semenov, A.N. Belov, Some general results on natural melt evaporation in the Knudsen cell. Geokhimiya **11**, 1559–1569 (1986)

A. Marten, D. Gautier, T. Owen, D.B. Sanders, H.E. Matthews, S.K. Atreya, R.P.J. Tilanus, J.R. Deane, First observations of CO and HCN on Neptune and Uranus at millimeter wavelengths and the implications for atmospheric chemistry. Astrophys. J. **406**(1), 285–297 (1993). https://doi.org/10.1086/172440

T.H. McConnochie, J.F. Bell, D. Savransky, M.J. Wolff, A.D. Toigo, H. Wang, M.I. Richardson, P.R. Christensen, THEMIS-VIS observations of clouds in the martian mesosphere: altitudes, wind speeds, and decameter-scale morphology. Icarus **210**(2), 545–565 (2010). https://doi.org/10.1016/j.icarus.2010.07.021

W.J. McNeil, S.T. Lai, E. Murad, Differential ablation of cosmic dust and implications for the relative abundances of atmospheric metals. J. Geophys. Res. **103**(D9), 10899–10911 (1998). https://doi.org/10.1029/98JD00282

J. Merikanto, J. Duplissy, A. Määttänen, H. Henschel, N.M. Donahue, D. Brus, S. Schobesberger, M. Kulmala, H. Vehkamäki, Effect of ions on sulfuric acid-water binary particle formation: 1. Theory for kinetic- and nucleation-type particle formation and atmospheric implications. J. Geophys. Res. **121**(4), 1736–1751 (2016). https://doi.org/10.1002/2015JD023538

M. Michael, S.N. Tripathi, W.J. Borucki, R.C. Whitten, Highly charged cloud particles in the atmosphere of Venus. J. Geophys. Res. **114**, E04008 (2009). https://doi.org/10.1029/2008je003258

G.J. Molina-Cuberos, J.J. Lopez-Moreno, F. Arnold, Meteoric layers in planetary atmospheres. Space Sci. Rev. **137**(1–4), 175–191 (2008). https://doi.org/10.1007/s11214-008-9340-5

F. Montmessin, J.L. Bertaux, E. Quemerais, O. Korablev, P. Rannou, F. Forge, S. Perrier, D. Fussen, S. Lebonnois, A. Reberac, E. Dimarellis, Subvisible CO_2 ice clouds detected in the mesosphere of Mars. Icarus **183**(2), 403–410 (2006). https://doi.org/10.1016/j.icarus.2006.03.015

F. Montmessin, B. Gondet, J.P. Bibring, Y. Langevin, P. Drossart, F. Forget, T. Fouchet, Hyperspectral imaging of convective CO_2 ice clouds in the equatorial mesosphere of Mars. J. Geophys. Res. **112**(E11), E11S90 (2007). https://doi.org/10.1029/2007je002944

L. Moore, J. O'Donoghue, I. Muller-Wodarg, M. Galand, M. Mendillo, Saturn ring rain: model estimates of water influx into Saturn's atmosphere. Icarus **245**, 355–366 (2015). https://doi.org/10.1016/j.icarus.2014.08.041

J.E. Moores, A.C. Schuerger, UV degradation of accreted organics on Mars: IDP longevity, surface reservoir of organics, and relevance to the detection of methane in the atmosphere. J. Geophys. Res. **117**, E08008 (2012). https://doi.org/10.1029/2012JE004060

A. Moro-Martin, R. Malhotra, Dynamical models of Kuiper belt dust in the inner and outer solar system. Astron. J. **125**(4), 2255–2265 (2003). https://doi.org/10.1086/368237

R.V. Morris, D.C. Golden, J.F. Bell, T.D. Shelfer, A.C. Scheinost, N.W. Hinman, G. Furniss, S.A. Mertzman, J.L. Bishop, D.W. Ming, C.C. Allen, D.T. Britt, Mineralogy, composition, and alteration of Mars Pathfinder rocks and soils: evidence from multispectral, elemental, and magnetic data on terrestrial analogue, SNC meteorite, and Pathfinder samples. J. Geophys. Res. **105**, 1757–1817 (2000). https://doi.org/10.1029/1999JE001059

J.I. Moses, Meteoroid ablation in Neptune's atmosphere. Icarus **99**(2), 368–383 (1992). https://doi.org/10.1016/0019-1035(92)90153-X

J.I. Moses, S.F. Bass, The effects of external material on the chemistry and structure of Saturn's ionosphere. J. Geophys. Res. **105**(E3), 7013–7052 (2000). https://doi.org/10.1029/1999je001172

J.I. Moses, A. Poppe, Dust ablation on the giant planets: consequences for stratospheric photochemistry. Icarus **297**, 33–58 (2017). https://doi.org/10.1016/j.icarus.2017.06.002

J.I. Moses, E. Lellouch, B. Bézard, G.R. Gladstone, H. Feuchtgruber, M. Allen, Photochemistry of Saturn's atmosphere. Icarus **145**(1), 166–202 (2000). https://doi.org/10.1006/icar.1999.6320

S. Mukhopadhyay, K.A. Farley, A. Montanari, A 35 Myr record of helium in pelagic limestones from Italy: implications for interplanetary dust accretion from the early Maastrichtian to the middle Eocene. Geochim. Cosmochim. Acta **65**, 653–669 (2001). https://doi.org/10.1016/S0016-7037(00)00555-X

M.J. Mumma, G.L. Villanueva, R.E. Novak, T. Hewagama, B.P. Bonev, M.A. DiSanti, A.M. Mandell, M.D. Smith, Strong release of methane on Mars in Northern Summer 2003. Science **323**(5917), 1041–1045 (2009). https://doi.org/10.1126/science.1165243

D.M. Murphy, K.D. Froyd, J.P. Schwarz, J.C. Wilson, Observations of the chemical composition of stratospheric aerosol particles. Q. J. R. Meteorol. Soc. **140**(681), 1269–1278 (2014). https://doi.org/10.1002/qj.2213

B.J. Murray, E.J. Jensen, Homogeneous nucleation of amorphous solid water particles in the upper mesosphere. J. Atmos. Sol.-Terr. Phys. **72**(1), 51–61 (2010). https://doi.org/10.1016/j.jastp.2009.10.007

M. Nachbar, D. Duft, T.P. Mangan, J.C.G. Martin, J.M.C. Plane, T. Leisner, Laboratory measurements of heterogeneous CO_2 ice nucleation on nanoparticles under conditions relevant to the Martian mesosphere. J. Geophys. Res. **121**(5), 753–769 (2016). https://doi.org/10.1002/2015je004978

D.A. Naylor, G.R. Davis, M.J. Griffin, T.A. Clark, D. Gautier, A. Marten, Broad-band spectroscopic detection of the CO $J = 3$–2 tropospheric absorption in the atmosphere of Neptune. Astron. Astrophys. **291**(3), L51–L53 (1994)

D. Nesvorný, P. Jenniskens, H.F. Levison, W.F. Bottke, D. Vokrouhlický, M. Gounelle, Cometary origin of the zodiacal cloud and carbonaceous micrometeorites. Implications for hot debris disks. Astrophys. J. **713**(2), 816–836 (2010). https://doi.org/10.1088/0004-637X/713/2/816

D. Nesvorný, D. Janches, D. Vokrouhlický, P. Pokorný, W.F. Bottke, P. Jenniskens, Dynamical model for the zodiacal cloud and sporadic meteors. Astrophys. J. **743**(2), 129–144 (2011a). https://doi.org/10.1088/0004-637X/789/1/25

D. Nesvorný, D. Vokrouhlicky, P. Pokorny, D. Janches, Dynamics of dust particles released from Oort Cloud comets and their contribution to radar meteors. Astrophys. J. **743**(1), 37 (2011b). https://doi.org/10.1088/0004-637x/743/1/37

A.O. Nier, D.J. Schlutter, Helium and neon isotopes in stratospheric particles. Meteoritics **25**, 263–267 (1990). https://doi.org/10.1111/j.1945-5100.1990.tb00710.x

J. O'Donoghue, T.S. Stallard, H. Melin, G.H. Jones, S.W.H. Cowley, S. Miller, K.H. Baines, J.S.D. Blake, The domination of Saturn's low-latitude ionosphere by ring 'rain'. Nature **496**(7444), 193–195 (2013). https://doi.org/10.1038/nature12049

G.S. Orton, J.I. Moses, L.N. Fletcher, A.K. Mainzer, D. Hines, H.B. Hammel, J. Martin-Torres, M. Burgdorf, C. Merlet, M.R. Line, Mid-infrared spectroscopy of Uranus from the Spitzer infrared spectrometer: 2. Determination of the mean composition of the upper troposphere and stratosphere. Icarus **243**, 471–493 (2014). https://doi.org/10.1016/j.icarus.2014.07.012

M. Ozima, M. Takayanagi, S. Zashu, S. Amari, High $^3He/^4He$ ratio in ocean sediments. Nature **311**, 448–450 (1984). https://doi.org/10.1038/311448a0

H. Palme, M.J. Janssens, H. Takahashi, E. Anders, J. Hertogen, Meteoritic material at five large impact craters. Geochim. Cosmochim. Acta **42**, 313–323 (1978). https://doi.org/10.1016/0016-7037(78)90184-9

B.M. Pandya, S.A. Haider, Meteor impact perturbation in the lower ionosphere of Mars: MGS observations. Planet. Space Sci. **63–64**, 105–109 (2012). https://doi.org/10.1016/j.pss.2011.09.013

M. Pätzold, S. Tellman, B. Hausler, D. Hinson, R. Schaa, G.L. Tyler, A sporadic third layer in the ionosphere of Mars. Science **310**(5749), 837–839 (2005). https://doi.org/10.1126/science.1117755

P.N. Peplowski, D.J. Lawrence, L.G. Evans, R.L. Klima, D.T. Blewett, J.O. Goldsten, S.L. Murchie, T.J. McCoy, L.R. Nittler, S.C. Solomon, R.D. Starr, S.Z. Weider, Constraints on the abundance of carbon in near-surface materials on Mercury: results from the MESSENGER Gamma-Ray Spectrometer. Planet. Space Sci. **108**, 98–107 (2015). https://doi.org/10.1016/j.pss.2015.01.008

W.D. Pesnell, J.M. Grebowsky, Meteoric ions in planetary ionospheres, Vol. 27, in *Planetary Ionospheres*, ed. by T.E. Cravens, A.F. Nagy. Advances in Space Research, vol. 11 (Elsevier, Amsterdam, 2001), pp. 1807–1814

W.D. Pesnell, J.M. Grebowsky, A.L. Weisman, Watching meteors on Triton. Icarus **169**(2), 482–491 (2004). https://doi.org/10.1016/j.icarus.2004.01.011

J.M. Petit, J.J. Kavelaars, B.J. Gladman, R.L. Jones, J.W. Parker, C. Van Laerhoven, P. Nicholson, G. Mars, P. Rousselot, O. Mousis, B. Marsden, A. Bieryla, M. Taylor, M.L.N. Ashby, P. Benavidez, A.C. Bagatin, G. Bernabeu, The Canada–France Plane Survey-full data release: the orbital strucuture of the Kuiper belt. Astron. J. **142**(4), 24 (2011). https://doi.org/10.1088/0004-6256/142/4/131

H. Pettersson, H. Rotschi, Nickel content of deep-sea deposits. Nature **166**, 308–310 (1950). https://doi.org/10.1016/0016-7037(52)90001-X

B. Peucker-Ehrenbrink, Accretion of extraterrestrial matter during the last 80 million years and its effect on the marine osmium isotope record. Geochim. Cosmochim. Acta **60**, 3187–3196 (1996). https://doi.org/10.1016/0016-7037(96)00161-5

B. Peucker-Ehrenbrink, G. Ravizza, G. Winckler, Geochemical tracers of extraterrestrial matter in sediments. Elements **12**, 191–196 (2016). https://doi.org/10.2113/gselements.12.3.191

S. Pizzarello, G.W. Cooper, G.J. Flynn, The nature and distribution of organic matter in carbonaceous chondrites and interplaetary dust particles, in *Meteorites and the Early Solar Syatem II*, ed. by D.S. Lauretta, H.Y. McSween Jr. (University of Arizona Press, Tucson, 2006), pp. 625–652

C. Plainaki, P. Paschalis, D. Grassi, H. Mavromichalaki, M. Andriopoulou, Solar energetic particle interactions with the Venusian atmosphere. Ann. Geophys. **114**(E4), 595–608 (2016). https://doi.org/10.5194/angeo-34-595-2016

J.M.C. Plane, Atmospheric chemistry of meteoric metals. Chem. Rev. **103**(12), 4963–4984 (2003). https://doi.org/10.1021/cr0205309

J.M.C. Plane, On the role of metal silicate molecules as ice nuclei. J. Atmos. Sol.-Terr. Phys. **73**(14–15), 2192–2200 (2011). https://doi.org/10.1016/j.jastp.2010.07.008

J.M.C. Plane, Cosmic dust in the Earth's atmosphere. Chem. Soc. Rev. **41**(19), 6507–6518 (2012). https://doi.org/10.1039/C2CS35132C

J.M.C. Plane, R.W. Saunders, J. Hedin, J. Stegman, M. Khaplanov, J. Gumbel, K.A. Lynch, P.J. Bracikowski, L.J. Gelinas, M. Friedrich, S. Blindheim, M. Gausa, B.P. Williams, A combined rocket-borne and ground-based study of the sodium layer and charged dust in the upper mesosphere. J. Atmos. Sol.-Terr. Phys. **118**, 151–160 (2014). https://doi.org/10.1016/j.jastp.2013.11.008

J.M.C. Plane, W. Feng, E.C. Dawkins, The mesosphere and metals: chemistry and changes. Chem. Rev. **115**(10), 4497–4541 (2015). https://doi.org/10.1021/cr500501m

J.M.C. Plane, J.C. Gomez-Martin, W. Feng, D. Janches, Silicon chemistry in the mesosphere and lower thermosphere. J. Geophys. Res. **121**, 3718–3728 (2016). https://doi.org/10.1002/2015JD024691

P. Pokorny, D. Vokrouhlicky, D. Nesvorny, M. Campbell-Brown, P. Brown, Dynamical model for the toroidal sporadic meteors. Astrophys. J. **789**(1), 25 (2014). https://doi.org/10.1088/0004-637x/789/1/25

A.R. Poppe, Interplanetary dust influx to the Pluto–Charon system. Icarus **246**, 352–359 (2015). https://doi.org/10.1016/j.icarus.2013.12.029

A.R. Poppe, An improved model for interplanetary dust fluxes in the outer solar system. Icarus **264**, 369–386 (2016). https://doi.org/10.1016/j.icarus.2015.10.001

A.R. Poppe, D. James, B. Jacobsmeyer, M. Horanyi, First results from the Venetia Burney Student Dust Counter on the New Horizons mission. Geophys. Res. Lett. **37**, 5 (2010). https://doi.org/10.1029/2010gl043300

R. Prange, T. Fouchet, R. Courtin, J.E.P. Connerney, J.C. McConnell, Latitudinal variation of Saturn photochemistry deduced from spatially-resolved ultraviolet spectra. Icarus **180**(2), 379–392 (2006). https://doi.org/10.1016/j.icarus.2005.11.005

M.J. Prather, J.A. Logan, M.B. McElroy, Carbon monoxide in Jupiter's upper atmosphere—an extraplanetary source. Astrophys. J. **223**(1), 1072–1081 (1978). https://doi.org/10.1086/156340

R. Rajan, D.E. Brownlee, D. Tomandl, P. Hodge, H. Farrar, R. Britten, Detection of ^{4}He in stratospheric dust particles gives evidence of extraterrestrial origin. Nature **267**, 133–134 (1977). https://doi.org/10.1038/267133a0

M. Rapp, G.E. Thomas, Modeling the microphysics of mesospheric ice particles: assessment of current capabilities and basic sensitivities. J. Atmos. Sol.-Terr. Phys. **68**, 715–744 (2006). https://doi.org/10.1016/j.jastp.2005.10.015

M. Rapp, I. Strelnikova, J. Gumbel, Meteoric smoke particles: evidence from rocket and radar techniques. Adv. Space Res. **40**(6), 809–817 (2007). https://doi.org/10.1016/j.asr.2006.11.021

M. Rapp, J.M.C. Plane, B. Strelnikov, G. Stober, S. Ernst, J. Hedin, M. Friedrich, U.P. Hoppe, In situ observations of meteor smoke particles (MSP) during the Geminids 2010: constraints on MSP size, work function and composition. Ann. Geophys. **30**(12), 1661–1673 (2012). https://doi.org/10.5194/angeo-30-1661-2012

B. Rizk, D.M. Hunten, Solar heating of the uranian mesopause by dust of ring origin. Icarus **88**(2), 429–447 (1990). https://doi.org/10.1016/0019-1035(90)90093-o

M. Roos-Serote, S.K. Atreya, C.R. Webster, P.R. Mahaffy, Cometary origin of atmospheric methane variations on Mars unlikely. J. Geophys. Res. **121**(10), 2108–2119 (2016). https://doi.org/10.1002/2016je005076

M. Rowan-Robinson, B. May, An improved model for the infrared emission from the zodiacal dust cloud: cometary, asteroidal and interstellar dust. Mon. Not. R. Astron. Soc. **429**(4), 2894–2902 (2013). https://doi.org/10.1093/mnras/sts471

J.M. Russell, P. Rong, M.E. Hervig, D.E. Siskind, M.H. Stevens, S.M. Bailey, J. Gumbel, Analysis of northern midlatitude noctilucent cloud occurrences using satellite data and modeling. J. Geophys. Res. **119**(6), 3238–3250 (2014). https://doi.org/10.1002/2013jd021017

J. Safarian, T.A. Engh, Vacuum evaporation of pure metals. Metall. Mater. Trans. A, Phys. Metall. Mater. Sci. **44A**(2), 747–753 (2013). https://doi.org/10.1007/s11661-012-1464-2

R.W. Saunders, J.M.C. Plane, A laboratory study of meteor smoke analogues: composition, optical properties and growth kinetics. J. Atmos. Sol.-Terr. Phys. **68**(18), 2182–2202 (2006). https://doi.org/10.1016/j.jastp.2006.09.006

R.W. Saunders, J.M.C. Plane, The formation and growth of Fe_2O_3 nanoparticles from the photo-oxidation of iron pentacarbonyl. J. Aerosol Sci. **41**(5), 475–489 (2010). https://doi.org/10.1016/j.jaerosci.2010.02.009

R.W. Saunders, S. Dhomse, W.S. Tian, M.P. Chipperfield, J.M.C. Plane, Interactions of meteoric smoke particles with sulphuric acid in the Earth's stratosphere. Atmos. Chem. Phys. **12**(10), 4387–4398 (2012). https://doi.org/10.5194/acp-12-4387-2012

N.A. Savich, V.A. Samovol, M.B. Vasilyev, A.S. Vyshlov, L.N. Samoznaev, A.I. Sidorenko, D.Y. Shtern, The nighttime ionosphere of Mars from Mars-4 and Mars-5 radio occultation dual-frequency measurements, in *Solar-Wind Interaction with the Planets Mercury, Venus, and Mars* (NASA, Goddard Space Flight Center, Greenbelt, 1976), pp. 41–46

L. Schaefer, B. Fegley, A thermodynamic model of high temperature lava vaporization on Io. Icarus **169**(1), 216–241 (2004). https://doi.org/10.1016/j.icarus.2003.08.023

L. Schaefer, B. Fegley, Application of an equilibrium vaporization model to the ablation of chondritic and achondritic meteoroids. Earth Moon Planets **95**(1–4), 413–423 (2005). https://doi.org/10.1007/s11038-005-9030-1

N.M. Schneider, J.I. Deighan, A.I.F. Stewart, W.E. McClintock, S.K. Jain, M.S. Chaffin, A. Stiepen, M. Crismani, J.M.C. Plane, J.D. Carrillo-Sanchez, J.S. Evans, M.H. Stevens, R.V. Yelle, MAVEN IUVS observations of the aftermath of the Comet Siding Spring meteor shower on Mars. Geophys. Res. Lett. **42**(12), 4755–4761 (2015). https://doi.org/10.1002/2015gl063863

J.T. Schofield, J.R. Barnes, D. Crisp, R.M. Haberle, S. Larsen, J.A. Magalhaes, J.R. Murphy, A. Seiff, G. Wilson, The Mars Pathfinder atmospheric structure investigation meteorology (ASI/MET) experiment. Science **278**(5344), 1752–1758 (1997). https://doi.org/10.1126/science.278.5344.1752

F. Scholten, H. Hoffmann, A. Määttänen, F. Montmessin, B. Gondet, E. Hauber, Concatenation of HRSC colour and OMEGA data for the determination and 3D-parameterization of high-altitude CO_2 clouds in the Martian atmosphere. Planet. Space Sci. **58**(10), 1207–1214 (2010). https://doi.org/10.1016/j.pss.2010.04.015

A.C. Schuerger, J.E. Moores, C.A. Clausen, N.G. Barlow, D.T. Britt, Methane from UV-irradiated carbonaceous chondrites under simulated Martian conditions. J. Geophys. Res. **117**, E08007 (2012). https://doi.org/10.1029/2011JE004023

E. Sefton-Nash, N.A. Teanby, L. Montabone, P.G.J. Irwin, J. Hurley, S.B. Calcutt, Climatology and first-order composition estimates of mesospheric clouds from Mars Climate Sounder limb spectra. Icarus **222**(1), 342–356 (2013). https://doi.org/10.1016/j.icarus.2012.11.012

E.P. Shettle, M.T. DeLand, G.E. Thomas, J.J. Olivero, Long term variations in the frequency of polar mesospheric clouds in the Northern Hemisphere from SBUV. Geophys. Res. Lett. **36**(2), L02803 (2009). https://doi.org/10.1029/2008gl036048

A. Shukolyukov, G.W. Lugmair, Isotopic evidence for the Cretaceous-Tertiary impactor and its type. Science **282**, 927–929 (1998). https://doi.org/10.1126/science.282.5390.927

P.H. Smith, J.F. Bell, N.T. Bridges, D.T. Britt, L. Gaddis, R. Greeley, H.U. Keller, K.E. Herkenhoff, R. Jaumann, J.R. Johnson, R.L. Kirk, M. Lemmon, J.N. Maki, M.C. Malin, S.L. Murchie, J. Oberst, T.J. Parker, R.J. Reid, R. Sablotny, L.A. Soderblom, C. Stoker, R. Sullivan, N. Thomas, M.G. Tomasko, W. Ward, E. Wegryn, Results from the Mars Pathfinder camera. Science **278**(5344), 1758–1765 (1997). https://doi.org/10.1126/science.278.5344.1758

F. Spahn, N. Albers, M. Horning, S. Kempf, A.V. Krivov, M. Makuch, J. Schmidt, M. Seiss, M. Sremcevic, E ring dust sources: implications from Cassini's dust measurements. Planet. Space Sci. **54**(9–10), 1024–1032 (2006). https://doi.org/10.1016/j.pss.2006.05.022

A. Spiga, F. Gonzalez-Galindo, M.A. Lopez-Valverde, F. Forget, Gravity waves, cold pockets and CO_2 clouds in the Martian mesosphere. Geophys. Res. Lett. **39**, L02201 (2012). https://doi.org/10.1029/2011gl050343

M. Sremcevic, A.V. Krivov, F. Spahn, Impact-generated dust clouds around planetary satellites: asymmetry effects. Planet. Space Sci. **51**(7–8), 455–471 (2003). https://doi.org/10.1016/s0032-0633(03)00050-3

M. Sremcevic, A.V. Krivov, H. Kruger, F. Spahn, Impact-generated dust clouds around planetary satellites: model versus Galileo. Planet. Space Sci. **53**(6), 625–641 (2005). https://doi.org/10.1016/j.pss.2004.10.001

C.C. Stark, M.J. Kuchner, A new algorithm for self-consistent three-dimensional modeling of collisions in dusty debris disks. Astrophys. J. **707**(1), 543–553 (2009). https://doi.org/10.1088/0004-637x/707/1/543

C.R. Stoker, M.A. Bullock, Organic degradation under simulated Martian conditions. J. Geophys. Res. **102**, 10,881–810,888 (1997). https://doi.org/10.1029/97JE00667

I. Strelnikova, M. Rapp, S. Raizada, M. Sulzer, Meteor smoke particle properties derived from Arecibo incoherent scatter radar observations. Geophys. Res. Lett. **34**(15), L15815 (2007). https://doi.org/10.1029/2007gl030635

Y.A. Surkov, L.P. Moskalyova, V.P. Kharyukova, A.D. Dudin, G.G. Smirnov, S.Y. Zaitseva, Venus rock composition at the VEGA 2 landing site, Proceedings of the Seventeenth Lunar and Planetary Science Conference, Part 1. J. Geophys. Res. **91**, E215–E218 (1988). https://doi.org/10.1029/JB091iB13p0E215

J.R. Szalay, M. Piquette, M. Horanyi, The Student Dust Counter: status report at 23 AU. Earth Planets Space **65**(10), 1145–1149 (2013). https://doi.org/10.5047/eps.2013.02.005

S. Taylor, J.H. Lever, R.P. Harvey, Accretion rate of cosmic spherules measured at the South Pole. Nature **392**(6679), 899–903 (1998). https://doi.org/10.1038/31894

N.A. Teanby, P.G.J. Irwin, An external origin for carbon monoxide on Uranus from Herschel/Spire? Astrophys. J. Lett. **775**(2), L49 (2013). https://doi.org/10.1088/2041-8205/775/2/l49

G.E. Thomas, J.J. Olivero, Noctilucent clouds as possible indicators of global change in the mesosphere. Adv. Space Res. **28**(7), 937–946 (2001). https://doi.org/10.1016/S0273-1177(01)80021-1

K.L. Thomas, G.E. Blanford, L.P. Keller, W. Klock, D.S. McKay, Carbon abundance and silicate mineralogy of anhydrous interplanetary dust particles. Geochim. Cosmochim. Acta **57**, 1551–1566 (1993). https://doi.org/10.1016/0016-7037(93)90012-L

E. Thomas, M. Horanyi, D. Janches, T. Munsat, J. Simolka, Z. Sternovsky, Measurements of the ionization coefficient of simulated iron micrometeoroids. Geophys. Res. Lett. **43**, 3645–3649 (2016). https://doi.org/10.1002/2016GL068854

E. Thomas, J. Simolka, M. DeLuca, M. Horanyi, D. Janches, R.A. Marshall, T. Munsat, J.M.C. Plane, Z. Sternovsky, Experimental setup for the laboratory investigation of micrometeoroid ablation using a dust accelerator. Rev. Sci. Instrum. (2017). https://doi.org/10.1063/1.4977832

O.B. Toon, B. Ragent, D. Colburn, J. Blamont, C. Cot, Large Solid particles in the clouds of Venus: do they exist? Icarus **57**(2), 143–160 (1984). https://doi.org/10.1016/0019-1035(84)90063-0

P. Tricarico, N.H. Samarasinha, M.V. Sykes, J.Y. Li, T.L. Farnham, M.S.P. Kelley, D. Farnocchia, R. Stevenson, J.M. Bauer, R.E. Lock, Delivery of dust grains from comet C/2013 A1 (siding spring) to Mars. Astrophys. J. Lett. **787**, L35 (2014). https://doi.org/10.1088/2041-8205/787/2/l35

G.L. Tyler, D.N. Sweetnam, J.D. Anderson, S.E. Borutzki, J.K. Campbell, V.R. Eshleman, D.L. Gresh, E.M. Gurrola, D.P. Hinson, N. Kawashima, E.R. Kursinski, G.S. Levy, G.F. Lindal, J.R. Lyons, E.A. Marouf, P.A. Rosen, R.A. Simpson, G.E. Wood, Voyager radio science observations of Neptune and Triton. Science **246**(4936), 1466–1473 (1989). https://doi.org/10.1126/science.246.4936.1466

M. Vincendon, C. Pilorget, B. Gondet, S. Murchie, J.P. Bibring, New near-IR observations of mesospheric CO_2 and H_2O clouds on Mars. J. Geophys. Res. **116**, E00J02 (2011). https://doi.org/10.1029/2011je003827

C. Vitense, A.V. Krivov, T. Lohne, The Edgeworth-Kuiper debris disk. Astron. Astrophys. **520**, 18 (2010). https://doi.org/10.1051/0004-6361/201014208

C. Vitense, A.V. Krivov, H. Kobayashi, T. Lohne, An improved model of the Edgeworth–Kuiper debris disk. Astron. Astrophys. **540**, 10 (2012). https://doi.org/10.1051/0004-6361/201118551

C. Vitense, A.V. Krivov, T. Lohne, Will New Horizons see dust clumps in the Edgeworth-Kuiper belt. Astron. J. **147**(6), 8 (2014). https://doi.org/10.1088/0004-6256/147/6/154

T. Vondrak, J.M.C. Plane, S. Broadley, D. Janches, A chemical model of meteoric ablation. Atmos. Chem. Phys. **8**, 7015–7031 (2008). https://doi.org/10.5194/acp-8-7015-2008

C.R. Webster, P.R. Mahaffy, S.K. Atreya, G.J. Flesch, M.A. Mischna, P.Y. Meslin, K.A. Farley, P.G. Conrad, L.E. Christensen, A.A. Pavlov, J. Martin-Torres, M.P. Zorzano, T.H. McConnochie, T. Owen, J.L. Eigenbrode, D.P. Glavin, A. Steele, C.A. Malespin, P.D. Archer, B. Sutter, P. Coll, C. Freissinet, C.P. McKay, J.E. Moores, S.P. Schwenzer, J.C. Bridges, R. Navarro-Gonzalez, R. Gellert, M.T. Lemmon, M.S.L.S. Team, Mars methane detection and variability at Gale crater. Science **347**(6220), 415–417 (2015). https://doi.org/10.1126/science.1261713

R. Weigel, C.M. Volk, K. Kandler, E. Hösen, G. Günther, B. Vogel, J.-U. Grooß, S. Khaykin, G.V. Belyaev, S. Borrmann, Enhancements of the refractory submicron aerosol fraction in the Arctic polar vortex: feature or exception? Atmos. Chem. Phys. **14**, 12,319–312,342 (2014). https://doi.org/10.5194/acp-14-12319-2014

C.L. Whalley, J.M.C. Plane, Meteoric ion layers in the Martian atmosphere. Faraday Discuss. **147**, 349–368 (2010). https://doi.org/10.1039/c003726e

F.L. Whipple, The theory of micro-meteorites. Part I. In an isothermal atmosphere. Proc. Natl. Acad. Sci. **36**, 687–695 (1950). https://doi.org/10.1073/pnas.36.12.687

P. Wiegert, J. Vaubaillon, M. Campbell-Brown, A dynamical model of the sporadic meteoroid complex. Icarus **201**(1), 295–310 (2009). https://doi.org/10.1016/j.icarus.2008.12.030

P. Withers, M. Mendillo, D.P. Hinson, K. Cahoy, Physical characteristics and occurrence rates of meteoric plasma players detected in the Martin ionosphere by the Mars Global Surveyor radio science experiment. J. Geophys. Res. **113**, A12314 (2008). https://doi.org/10.1029/2008JA013636

M.L. Wong, S. Fan, P. Gao, M.-C. Liang, R.-L. Shia, Y.L. Yung, J.A. Kammer, M.E. Summers, G.R. Gladstone, L.A. Young, C.B. Olkin, K. Ennico, H.A. Weaver, S.A. Stern, The photochemistry of Pluto's atmosphere as illuminated by New Horizons. Icarus **287**, 110–115 (2017). https://doi.org/10.1016/j.icarus.2016.09.028

H.G. Yang, M. Ishiguro, Origin of interplanetary dust through optical properties of the zodiacal light. Astrophys. J. **813**(2), 9 (2015). https://doi.org/10.1088/0004-637x/813/2/87

R.V. Yelle, J.I. Lunine, D.M. Hunten, Energy balance and plume dynamics in Triton's lower atmosphere. Icarus **89**(2), 347–358 (1991). https://doi.org/10.1016/0019-1035(91)90182-S

R.V. Yelle, J.I. Lunine, J.B. Pollack, R.H. Brown, Lower atmospheric structure and surface-atmosphere interactions on Triton, in *Neptune and Triton*, ed. by D.P. Cruikshank, M.S. Mathews, A.M. Schumann. Neptune and Triton (University of Arizona Press, Tucson, 1995), pp. 1031–1105

A.S. Yen, D.W. Mittlefehldt, S.M. McLennan, R. Gellert, H.Y. McSween, D.W. Ming, T.J. McCoy, R.V. Morris, M. Golombek, T. Economou, M.B. Madsen, T. Wdowiak, B.C. Clark, B.L. Jolliff, C. Schröder, J. Brückner, J. Zipfel, S.W. Squyres, Nickel on Mars: constraints on meteoritic material at the surface. J. Geophys. Res. **111**, E12S11 (2006). https://doi.org/10.1029/2006JE002797

A.T. Young, Venus cloud microphysics. Icarus **56**, 568–577 (1983). https://doi.org/10.1016/0019-1035(83)90174-4

K. Zahnle, M.M. MacLow, The collision of Jupiter and comet Shoemaker-Levy 9. Icarus **108**(1), 1–17 (1994). https://doi.org/10.1006/icar.1994.1038

K. Zahnle, R.S. Freedman, D.C. Catling, Is there methane on Mars? Icarus **212**, 493–503 (2011). https://doi.org/10.1016/j.icarus.2010.11.027

Space Sci Rev (2018) 214:131
https://doi.org/10.1007/s11214-018-0539-9

Dust Emission by Active Moons

J.K. Hillier[1,2] · J. Schmidt[3] · H.-W. Hsu[4] · F. Postberg[5]

Received: 13 April 2018 / Accepted: 16 August 2018 / Published online: 4 December 2018

Abstract In recent decades, volcanic and cryovolcanic activity on moons within the Solar System has been recognised as an important source of cosmic dust. Two moons, Jupiter's satellite Io and Saturn's satellite Enceladus, are known to be actively emitting dust into circumplanetary and interplanetary space. A third moon, Europa, shows tantalising hints of activity. Here we review current observations and theories concerning the generation, emission and evolution of cosmic dust arising from these objects.

Keywords Cosmic dust · Tidal heating · Planetary satellites · Volcanism

---

Cosmic Dust from the Laboratory to the Stars
Edited by Rafael Rodrigo, Jürgen Blum, Hsiang-Wen Hsu, Detlef Koschny, Anny-Chantal Levasseur-Regourd, Jesús Martín-Pintado, Veerle Sterken and Andrew Westphal

✉ J.K. Hillier
j.hillier@kent.ac.uk

J. Schmidt
jurgen.a.schmidt@oulo.fi

H. W. Hsu
sean.hsu@lasp.colorado.edu

F. Postberg
frank.postberg@fu-berlin.de

1 School of Physical Sciences, University of Kent, Canterbury, CT2 7NH, Kent, UK

2 Present address: Institute of Geological Sciences, Freie Universität Berlin, 12249 Berlin, Germany

3 University of Oulu, 90014 Oulu, Finland

4 LASP, University of Colorado at Boulder, Boulder, CO 80303-7814, USA

5 Institute of Geological Sciences, Freie Universität Berlin, 12249 Berlin, Germany

1 Introduction

In the past few decades, the "traditional" sources of cosmic dust encountered in the Solar System: asteroids, comets, Edgeworth-Kuiper Belt objects and the interstellar environment, have been joined by a fourth source, *active moons*. These objects are now recognised as originators of planetary rings, circumplanetary and interplanetary dust.[1]

In this work we define an *active moon* as a satellite of a primary, currently experiencing volcanism or cryovolcanism maintained by endogenic[2] heat sources. This definition therefore excludes activity driven by insolation or impacts. To be a *cosmic dust emitting* active moon, the satellite must be capable of producing dust which is transported beyond the Hill sphere of the satellite.

Two moons achieve this: Jupiter's moon Io and Saturn's moon Enceladus. There are tentative signs that a second moon of Jupiter, Europa, may also be active, but we discount Neptune's moon Triton as, although geyser-like activity has been observed (Smith et al. 1989; Soderblom et al. 1990), the driving force has not yet been unambiguously identified, with theories favouring insolation (Smith et al. 1989; Brown et al. 1990), or basal heating (Duxbury and Brown 1997) and a deeper source of the emitted gases (Laufer et al. 2013). In particular there is no evidence the small ($\sim$ 8 km high, Smith et al. 1989; Soderblom et al. 1990) geysers are capable of lofting dust beyond Triton's Hill radius.

There are also indications on the surfaces of many other icy moons (e.g. Miranda, Ariel, Titan, Ganymede, Charon) and (dwarf) planets (e.g. Pluto) that cryovolcanism may have occurred there in the past (e.g. Lopes et al. 2013; Kargel 1994; Showman et al. 2004 and Moore et al. 2016), and as such they may once have been active as defined above. However, in the absence of any evidence of extant activity, we will not consider them here.

In this work we will instead review the current state of knowledge regarding the generation, emission and evolution of cosmic dust produced by the currently active moons. Each of the moons considered here has been the subject of decades of research, which could not possibly be fully summarised here. For in depth, non-dust specific, summaries of Io, Europa and Enceladus, we recommend volumes such as those by Lopes and Spencer (2007), Pappalardo et al. (2009) and the upcoming Schenk et al. (2018) book. However, in topics directly related to the cosmic dust itself, we endeavour to provide more detailed references.

Rather than consider Io and Enceladus separately, in order to highlight similarities and differences in the processes producing and affecting the dust we instead consider each body with respect to the lifecycle of the dust produced—namely the energy sources (Sect. 2), dust sources (Sect. 3), dust generation and emission (Sect. 4), and circumplanetary (Sect. 5) and interplanetary (Sect. 6) behaviour. We follow these with a brief review of Europa's suspected/potential activity (Sect. 7) and a short conclusions section (Sect. 8). We begin with a summary of the discovery of dust emission by Io and Enceladus.

For each moon, there were at least two phases of discovery, corresponding to the identification of dust grains emitted by each moon and the discovery of the generating activity itself. In the case of Enceladus, the emitted dust grains form Saturn's E ring, which was discovered telescopically (Feibelman 1967). Although there were early suggestions that liquid water on Enceladus was responsible (Pang et al. 1984; Showalter et al. 1991) for the generation of these grains, it was nearly forty years before the E ring's link to Enceladus' activity

[1] Given the high speed of dust in the Jovian and Saturnian "streams" (although squalls may be a better name—see Sect. 6), it is also likely that active moons inject dust into the local interstellar environment, making them also a source of interstellar dust.

[2] Strictly speaking the primary energy sources for the moons considered here are originally exogenic, being the motion of planet and satellite.

was confirmed (Spencer et al. 2006; Tokar 2006; Dougherty et al. 2006; Spahn et al. 2006b). With Io the situation was reversed—volcanic activity was predicted (Peale et al. 1979) and discovered (La Morabito et al. 1979) twelve years before the first detection of emitted grains (Grün et al. 1993) and 21 years before their source was confirmed (Graps et al. 2000). The major events and their dates[3] are outlined below:

- 27th October 1966: Discovery of Saturn's E ring (Feibelman 1967).
- 8th March 1979: Discovery of Io's volcanic activity (La Morabito et al. 1979).
- 12th December 1991: Discovery of Jovian stream particles (Grün et al. 1993).
- 4th May 2000: Confirmation of Io as the source of Jovian stream particles (Graps et al. 2000).
- January 2004: Discovery of Saturnian stream particles (Kempf et al. 2005a).
- 14th July 2005: Discovery of Enceladus' cryovolcanic activity (Spencer et al. 2006; Tokar 2006).
- 14th July 2005: Discovery of dust emission from Enceladus' south polar region (Dougherty et al. 2006; Porco et al. 2006; Spahn et al. 2006b).

Current values of physical and orbital parameters of Io and Enceladus are shown in Table 1. As can be seen, the two moons are vastly different in size. Despite this, "volcanism" is clearly occurring on them both, maintained by a range of heat sources.

2 Heating

For volcanism or cryovolcanism to occur on the moons, and result in dust emission, phase changes (solid to liquid at least, but solid to liquid to vapour/gas in general) must be present. Particularly in the case of Enceladus, given its small size, current sources of heat are believed to be necessary to maintain these phase changes. Although internal heating may be global or large-scale in nature, cryovolcanism or volcanism, and hence dust emission, is likely to be extremely localised. In the context of the moons considered here, three extant sources of heating are possible:

1. Tidal heating.
2. Radionuclide heating.
3. Chemical heating.

Other potential sources of heat, such as accretion, differentiation, impact or crystallisation processes (e.g. Hussmann et al. 2010), are unlikely to be occurring now, or will provide very low, or extremely limited, episodic, contributions. For all the active (or potentially active) moons discussed here, the dominant source of heating is tidal flexure. Radionuclide heating from long-lived isotopes also contributes to the internal heating of each moon, whilst chemical heating from serpentinisation is only confirmed for Enceladus, but highly likely to occur, or have occurred, within Europa.

2.1 Tidal Heating

Tidal heating occurs when a body dissipates part of the energy, generated by internal flexing, due to internal friction. The periodic flexing is provided by the motion of the body through a varying gravitational potential and as such can be driven by regular variations in satellite/primary distance or orientation, due to:

[3] First observation where known, publication where not.

- Orbital eccentricity
- Orbital inclination (depending on primary shape)
- Obliquity
- Spin and libration

For the active moons considered here, the dominant sources of tidal dissipative heating are due to the moons' orbital eccentricities and changes in satellite-primary orientations. Heating from tidal dissipation due to the regular changing of the gravitational potential across a satellite arising from changes in orbital distance is the conceptually simplest process to visualise. The tidal power generated within a synchronously-rotating satellite is given by (Peale 1999):

$$\frac{dE}{dt} = \frac{42\pi}{19}\frac{\rho_S^2 R_S^2 n^5}{\mu_S Q_S} e^2. \tag{1}$$

Here n is the satellite mean motion, e the satellite eccentricity, R_S is satellite radius, ρ_S is satellite density, μ_S is the shear modulus and Q_S is the tidal quality factor for the satellite. Equation (1) includes energy dissipation from libration, arising from the change in orbital speed of a body on an elliptical orbit. In this situation, dissipation within the satellite introduces a lag between the orientation of the tidal deformation of the satellite and the direction of the primary.

In the specific cases where the eccentricity of an orbit is an equilibrium value (e.g. in a resonance such as that between Dione and Enceladus), it is possible to derive the maximum rate at which tidal energy is dissipated (and hence tidal heating) within the satellite in terms of the satellite's orbital parameters, and the tidal Love number (second order) k_{2P} and tidal quality factor Q_P of the primary (Peale 1999). In many situations, the ratio of k_{2P} and Q_P parameters can be found from astrometric measurements of orbital changes of the primary from a number of satellites (e.g. Lainey et al. 2009, 2012, 2017).

The tidal quality factor, Q, relates the energy stored (E) over one flexing cycle and the energy dissipated (ΔE) (Goldreich and Soter 1966):

$$Q = \frac{2\pi E}{\Delta E}, \tag{2}$$

Q is generally a function of the perturbing frequency. Energy for tidal heating is transferred from the orbital and rotational energy of the system. In the case of the satellite, this means an initial reduction in spin rate Ω_S (despinning, due to spin-induced tidal heating) over a timescale given by Eq. (3) (e.g. Hussmann et al. 2010), under the assumption of a homogeneous satellite:

$$\tau \approx \frac{76}{45}\frac{\Omega_S \mu_S Q_S}{\rho_S R_S^2}\frac{1}{n^4}\,\mathrm{sign}(\Omega_S - n). \tag{3}$$

For initial satellite spin periods less than their orbital period, the spin period will increase, such that the periods enter a resonance, 1:1 spin-orbit synchronicity. All the active moons are in this state.

Tidal heating will also act to circularise the satellite orbit and reduce the orbital semi-major axis at rates given by Eqs. (4) and (5) (Goldreich and Soter 1966; Hussmann et al. 2010):

$$\frac{da}{dt} = -21\left(\frac{k_2}{Q}\right)_S \frac{M_P}{M_S}\left(\frac{R_S}{a_S}\right)^5 n a_S e^2, \tag{4}$$

$$\frac{de}{dt} = -\frac{21}{2}\left(\frac{k_2}{Q}\right)_S \frac{M_P}{M_S}\left(\frac{R_S}{a_S}\right)^5 ne. \tag{5}$$

Here the subscripts refer to the primary (P) and satellite (S). The k_2 term is the tidal Love number of the second order (Love 1944), Q is the tidal quality factor, and M, R, n, a and e correspond to mass, radius, mean-motion, semi-major axis and eccentricity respectively.

The Love numbers k_2 and h_2 describe the tidal potential and radial displacement of a satellite. For the simplest case of a small, elastic, homogeneous body (e.g. Varga et al. 2009) the values are approximately:

$$h_2 = \frac{5f}{2f+1}; \qquad k_2 = \frac{3f}{2f+1} \quad \text{where } f = \frac{g\rho_S R_S}{19\mu_S}. \tag{6}$$

Where ρ_S, R_S, μ_S and g are the mean density, mean radius, shear modulus and surface gravity of the body.

For the active moons to still be active, with tidal heating as their primary energy sources, their orbital eccentricities and semi-major axes must be maintained against this decay (energy transfer) process. In the case of these moons, gravitational resonances with at least one other moon and the corresponding forcing of orbital eccentricity ensures this.

The types of resonance responsible for maintaining tidal dissipation in active moons are mean-motion resonances, which relate the mean-motions (and hence orbital periods) of the moons involved.

The most famous example of a mean-motion resonance is that which maintains tidal dissipation in Io and Europa, the Laplace resonance. This relates the mean motions of Io, Europa and Ganymede such that:

$$n_{\mathrm{Io}} - 3n_{\mathrm{Europa}} + 2n_{\mathrm{Ganymede}} = 0 \tag{7}$$

This resonance excites the orbital eccentricities of both Io and Europa and prevents the circularisation and semi-major axis reduction of their orbits, as would be expected from Eqs. (4) and (5).

A mean-motion resonance also keeps Enceladus' orbit from circularising, although in this case interaction with only one other moon, Dione, is involved.

$$n_{\mathrm{Enceladus}} - 2n_{\mathrm{Dione}} = 0 \tag{8}$$

These resonances are often also expressed in terms of the mean longitudes (λ) and longitudes of pericentre (ω) of the orbits involved, as well as the resonant angle or variable (ϕ) (Dermott 1968). In these terms, the Laplace resonance for Io, Europa and Ganymede is $\phi = 180° = \lambda_{\mathrm{Io}} - 3\lambda_{\mathrm{Europa}} + 2\lambda_{\mathrm{Ganymede}}$, and the resonance for Dione and Enceladus is $\phi = 2\lambda_{\mathrm{Dione}} - \lambda_{\mathrm{Enceladus}} - \omega_{\mathrm{Enceladus}} = 0°$. This 2:1 mean-motion resonance with Dione maintains Enceladus' orbital eccentricity.

The general dependence (Eq. (1)) of the actual (rather than maximum) magnitude of tidal dissipation, and hence tidal heating, on the density, tidal quality factor, second order tidal Love number and shear modulus of the satellite undergoing flexure, indicates that its composition and structure are important in instigating and maintaining heating. Melting and differentiation, in particular, contribute to the degree in which a body is centrally condensed (i.e. the radial density and permeability distributions).

Some degree of melting and differentiation is likely to occur early in the history of the active moons, regardless of initial composition, based purely upon accretional heating and delivery of energy through large scale impacts. However, further heating, and that likely to drive the majority of early-stage melting and differentiation, will have occurred from radionuclide decay.

2.2 Radionuclide Heating

Short-term or early heating can occur via the decay of radionuclides such as ^{26}Al (e.g. Castillo-Rogez et al. 2009) and ^{60}Fe (e.g. Mostefaoui et al. 2005). With a half-life of 7.17×10^5 years, and a heat production value (i.e. excluding energy partitioned into emitted neutrinos, which are highly unlikely to interact further with their immediate surroundings) of 3.12 MeV per decay (Castillo-Rogez et al. 2009), the decay of ^{26}Al is considered the dominant heat source for melting and differentiation in the early solar system. ^{26}Al decays to ^{26}Mg via either positron emission or electron capture, resulting in Mg in one of two excited states. Subsequent transition of the ^{26}Mg, and/or annihilation of the positron, releases further γ photons (Castillo-Rogez et al. 2009).

^{60}Fe, with a half-life of 2.6–2.7 $\times\ 10^6$ years (Ostdiek et al. 2017), decays to ^{60}Ni, with heat production of 2.712 MeV per decay of ^{60}Fe (e.g. Castillo-Rogez et al. 2009).

Radionuclide heating which may still affect moons today, occurs predominantly via the decays of ^{40}K, ^{235}U, ^{238}U and ^{232}Th. ^{40}K decays to ^{40}Ca via β^- emission, or to ^{40}Ar via electron capture and subsequent gamma emission, with a half-life of 1.277×10^9 years (Van Schmus 1995). The heat production value per decay is 0.690 MeV (Van Schmus 1995).

The ^{235}U, ^{232}Th and ^{238}U series are well described in the literature (e.g. Van Schmus 1995). With a half-life of 703.81×10^6 years (Castillo-Rogez and McCord 2010), ^{235}U decays to ^{207}Pb. The heat production value per complete decay to stability is 44.4 MeV (Van Schmus 1995).

Similarly ^{238}U decays to ^{206}Pb. The half-life for this process is 4.468×10^9 years (e.g. Castillo-Rogez and McCord 2010), with a heat production value per decay of 47.5 MeV (Van Schmus 1995).

The decay series with the longest total half-life, 14.01–14.05 $\times\ 10^9$ years (e.g. Castillo-Rogez and McCord 2010), is that of ^{232}Th to ^{208}Pb. In this process, the heat production value is 40.4 MeV per decay.

Typically when dealing with current planetary heating, generated power is quoted for "present-day Chondritic heating", namely that due to current radioactivity levels present in a bulk chondritic composition material. This value is approximately $(4.5 \pm 0.5) \times 10^{-12}\ \mathrm{W\,kg^{-1}}$ (ranging between 3–5 $\times\ 10^{-12}\ \mathrm{W\,kg^{-1}}$ for different compositions) (Spohn and Schubert 2003).

Estimates for present day radionuclide heating for Io, Enceladus and Europa are given in Table 1. Radiogenic heating plays only a minor role in Io, and probably so for Enceladus and Europa[4] in the overall energy budgets and volcanic activities associated with present day dust emission.

2.3 Chemical Heating

Serpentinisation is the conversion of Fe-Mg silicates to forms of serpentine, hydroxides and magnetite, and potentially also clays (saponite) and carbonates, via interaction with water, generating heat, lowering the bulk density and increasing the volume. The process has been considered with respect to Enceladus and Europa (e.g. Vance et al. 2007; Hussmann et al. 2010; Sohl et al. 2010; Hsu et al. 2015; Sekine et al. 2015; Waite et al. 2017). For

[4]Estimates of the exact contribution are strongly model dependent.

serpentinisation to occur, liquid water has to be in contact with a rock surface, so highly porous and permeable rocky cores or mantles will generate greater heat (Vance et al. 2007). Depths to which significant porosity may be maintained have been estimated for Europa and Enceladus as 20–30 km of the upper mantle and 100 km of the core respectively (Sohl et al. 2010). In the case of Enceladus, the porous depth estimate was made prior to interpretations of Cassini gravity data which indicate a fluffy core (e.g. Iess et al. 2014; Roberts 2015), with probably tidal-driven liquid water circulation within it (e.g. Choblet et al. 2017). Laboratory treatment of forsterite indicates that hydration can progress through even large (50 mm) grains, at low (20 °C) temperature and pressure (0.5 kBar), in geologically-short timescales (1 Ma) (e.g. Wegner and Ernst 1983). To maintain serpentinisation in the present epoch, much larger grain sizes, crack propagation through reaction or tidal stressing (e.g. Vance et al. 2007; Malvoisin et al. 2017), periodic disturbances of equilibrium states, such as by the replacement of salt-rich brine with lower salt brine (e.g. Lamadrid et al. 2017) or changes in the concentration of aqueous silica (Tutolo et al. 2018) may be required.

Choblet et al. (2017) (Suppl. Mat.) provide examples of typical serpentinisation processes acting upon olivine and pyroxene:

$$(\mathrm{Mg}_x\mathrm{Fe}_{1-x})_2\mathrm{SiO}_4 + \frac{1}{2}(\mathrm{Mg}_x\mathrm{Fe}_{1-x})_2\mathrm{Si}_2\mathrm{O}_6 + 2\mathrm{H}_2\mathrm{O} \rightarrow (\mathrm{Mg}_x\mathrm{Fe}_{1-x})_3\mathrm{Si}_2\mathrm{O}_5(\mathrm{OH})_4. \tag{9}$$

Here (Eq. (9)) olivine and pyroxene are hydrated, becoming serpentine. Mg and Fe brucite are also formed by a similar reaction (Eq. (10)), as is magnetite (Eq. (11)). Although not shown here, the hydration of ferrosilite and fayalite will also produce aqueous silica (Vance et al. 2007; Sekine et al. 2015).

$$2(\mathrm{Mg}_x\mathrm{Fe}_{1-x})_2\mathrm{SiO}_4 + 3\mathrm{H}_2\mathrm{O} \rightarrow (\mathrm{Mg}_x\mathrm{Fe}_{1-x})_3\mathrm{Si}_2\mathrm{O}_5(\mathrm{OH})_4 + (\mathrm{Mg}_x\mathrm{Fe}_{1-x})(\mathrm{OH})_2, \tag{10}$$

$$6(\mathrm{Mg}_{0.75}\mathrm{Fe}_{0.25})_2\mathrm{SiO}_4 + 7\mathrm{H}_2\mathrm{O} \rightarrow 3\mathrm{Mg}_3\mathrm{Si}_2\mathrm{O}_5(\mathrm{OH})_4 + \mathrm{Fe}_3\mathrm{O}_4 + \mathrm{H}_2. \tag{11}$$

Further reactions, forming talc, saponite and carbonates (e.g. magnesite and dolomite) are possible in the presence of e.g. dissolved CO_2,[5] or under differing pH conditions, with Sekine et al. (2015) reporting that serpentine and brucite tend to form when the Mg/Si ratio is > 1.5, whereas serpentine and saponite form at Mg/Si ratios < 1.5.

The degree to which these processes are capable of providing heat, especially in the current epoch, depends upon both the composition and structure of the moons. The abundance of suitable mineralogy, together with amenable environmental conditions (liquid water, CO_2, pH, temperature) will enable and rate limit the serpentinisation reactions (Vance et al. 2007). Heat production from the serpentinisation of olivine and clinopyroxene at relatively low temperatures (300 K) and pressure ranges of 10–200 MPa is reasonably high, at ~ 83 kJ per mole of serpentine produced, based upon a starting composition of 70% olivine, with $\mathrm{Fe/Mg} = 0.1$ (Vance et al. 2007).

The exact contribution of tidal dissipation, radionuclide decay and serpentinisation to the current heat budgets of the active moons depends on their structures, compositions and other physical characteristics, such as orbital parameters.

3 Heating and Dust Sources

The internal structures and states, and to some degree compositions, of all three moons considered in this review are relatively poorly constrained, as might perhaps be expected for

[5] Methane may also form during the production of magnetite if CO_2 is present.

Fig. 1 A Galileo Solid State Imaging (SSI) image of Io, with North to the top of the image and the bright orange sulfur deposit of Pele clearly visible to the lower left. Image credit: NASA/JPL/University of Arizona (PIA01667)

objects other than the Earth. The majority of the internal structural information has been gleaned from spacecraft flybys (most recently Galileo for Io and Europa, Cassini for Enceladus) (e.g. Schubert et al. 2004; Iess et al. 2014; McKinnon 2015), with further constraints derived by remote observations (e.g. Thomas et al. 2016; Martens et al. 2015; Pappalardo et al. 1998, 1999; Greenberg et al. 1998), modelling—including variable compositions (e.g. Zolotov and Fegley 1999; Fegley and Zolotov 2000; Neveu et al. 2017) and, particularly in the case of Enceladus, in situ observations of cosmic dust (e.g. Postberg et al. 2006, 2008, 2009b, 2011b; Spahn et al. 2006b; Hsu et al. 2015) and gases (Waite et al. 2006, 2009, 2017).

In this section we shall try to summarise the latest theories regarding the structures, compositions and heat sources within Io and Enceladus, with Europa briefly independently discussed in Sect. 7. Pertinent physical characteristics for all three moons are given in Table 1.

3.1 Io

The innermost of Jupiter's Galilean moons, Io (Fig. 1), is larger than Europa, but smaller than Ganymede and Callisto. To match the observed density, shape and moment of inertia factor (Table 1), Anderson et al. (2001) constructed plausible three layer models of Io, considering: a thick ($\leq$ 50 km) low density ($\leq 2600\ \mathrm{kg\,m^{-3}}$) crust over a $3400\ \mathrm{kg\,m^{-3}}$ solid mantle, leading to a core of Fe-FeS, or a thinner low density crust overlying a thicker (100–200 km), melt-rich ($\sim 20\%$ melt fraction, 3000–$3200\ \mathrm{kg^{-3}}$) asthenosphere and mantle ($3600\ \mathrm{kg\,m^{-3}}$), over a similarly-sized core. The core sizes were strongly compositionally dependent, with a low density[6] ($5150\ \mathrm{kg\,m^{-3}}$) Fe-FeS (up to 37% Fe-S) eutectic core requiring a radius of 500–950 km, and a higher density ($\sim 8000\ \mathrm{kg\,m^{-3}}$) pure Fe core only needing to be 350–650 km in radius.

[6] Values given for a core-appropriate pressure of 8 GPa.

Table 1 Physical parameters of known and suspected active moons

Parameter	Io	Enceladus	Europa
Mass (kg)	$(8.8911 \pm 0.042) \times 10^{22}$[a]	$(10.822 \pm 0.01) \times 10^{19}$[b]	$(4.789 \pm 0.047) \times 10^{22}$[a]
Mean radius (km)	1821.6 ± 0.5[c]	252.1 ± 0.2[d]	1562.09[e]
Polar radius (km)	1815.6 ± 0.5[f]	248.3 ± 0.2[d]	1560.76[c]
Equatorial radii (planet-facing/orbit-facing, km)	$1830/1819.2 \pm 0.5$[f]	$256.6 \pm 0.3/251.4 \pm 0.2$[d]	1563.96/1561.23[e]
Mean density ($kg\,m^{-3}$)	3527 ± 2.9[f]	1609 ± 5[d]	2990 ± 5[a]
GM ($km^3\,s^{-2}$)	5959.916 ± 0.012[g]	7.2096 ± 0.0067[b]	3202.739 ± 0.009[g]
Moment of inertia factor	0.37824 ± 0.00022[c]	0.335[h]	0.346 ± 0.005[a]
Orbital semi-major axis (km)	421800[g]	238042[g]	671000[a]
Orbital inclination (°)	0.036[g]	0.00819[i]	0.466[g]
Orbital/spin period (days)	1.769[g]	1.370218[g]	3.551[g]
Orbital eccentricity	0.0041[g]	0.0485 ± 0.00012[j]	0.0101[a]
Obliquity (°)	0.00054–0.0021[k]/0.04[l]	$\leq 4 \times 10^{-4}$[i]	0.033–0.044[k]
Primary mass (kg)	1.8982×10^{27}[g]	$5.6833 \pm 0.0009 \times 10^{26}$[b]	As for Io
Primary spin period (days)	0.41354[m]	0.4393 ± 0.0002[n]	As for Io
Tidal heating power (GW)	80000[p]	10–50[o]	≤ 10000[p]
Radionuclide heating power (GW)	308–514[q]	0.231–0.335[q]	152–254[q]
Chemical heating power (GW)	≈ 0	0.0132–0.022[r]	0.076–0.127[r]

[a]Greenberg (2010). [b]Jacobson et al. (2006). [c]Schubert et al. (2004). [d]Thomas (2010). [e]Schubert et al. (2009). [f]Anderson et al. (2001). [g]https://ssd.jpl.nasa.gov. [h]Iess et al. (2014). [i]Baland et al. (2016). [j]Dourneau (1993). [k]Baland et al. (2012). [l]Davies (2007). [m]Seidelmann et al. (2007). [n]Anderson and Schubert (2007). [o]Choblet et al. (2017). [p]Schubert et al. (2010). [q]Hussmann et al. (2010). [r]Based on Vance et al. (2007) estimates of fractions of radionuclide heating, as given in Hussmann et al. (2010).

The core composition and hence size has not yet been further constrained (e.g. Moore et al. 2007), but progress has been made in understanding the composition and structure of the overlying silicate layer. To support the observed topology, a cold, low-density lithosphere with a thickness of at least 12–15 km (Keszthelyi and McEwen 1997; Schenk et al. 2001; Keszthelyi et al. 2004) is suggested. The depth and melt-state of an upper-mantle asthenosphere below this is still uncertain, with a 50 km thick, 20% melt-fraction, global, conductive asthenosphere, underlying a 30 km crust suggested to explain Galileo magnetometer findings (Khurana et al. 2011), although this result is contested by Roth et al. (2017), who find that a conductive core layer with an outer radius of 1050 km is a better fit to auroral spot data.

A partially molten ($\leq 25\%$ melt fraction) asthenosphere is also suggested by measurements of Ionian maximum eruption temperatures (Sect. 3) by Keszthelyi et al. (2007) and tidal modelling (e.g. Tyler et al. 2015; Bierson and Nimmo 2016), with asthenosphere thicknesses of up to 100 km predicted. It is important to note that the melt-fractions described do not necessarily imply a thin, globally liquid (magma ocean) layer, but instead may more realistically represent a layer in which an interconnected liquid can flow through a deformable, solid, matrix (Tyler et al. 2015). Io's silicate mantle and lithosphere is believed to be predominantly forsteritic olivine (76–85 wt.%) (Sohl et al. 2002). Based on the bulk density of Io, a broadly chondritic composition is expected (e.g. Keszthelyi et al. 2007).

Io was the first planetary moon to be predicted to have active volcanism driven by tidal heating (Peale et al. 1979) maintained by the Laplace resonance with Europa and Ganymede (discussed above), with a total tidal heating power found to be 2.4×10^{13} W. Later estimates of the tidal dissipated power within Io tend to be higher (e.g. in Ross and Schubert 1985; Segatz et al. 1988; Makarov and Efroimsky 2014), up to $9.33 \pm 1.87 \times 10^{13}$ W (Lainey et al. 2009). This latter value indicates that Io's heat generation approximately balances its emitted heat (Veeder et al. 2004), although it is likely that this situation is not constant, with phases of heating and cooling occurring with evolution of the Laplace resonance, over typical timescales of e.g. 10–30 Ma active and 60–200 Ma quiescent (e.g. Ojakangas and Stevenson 1986; Schubert et al. 2010).

Tidal heating in Io is likely to consist of contributions from dissipation in both solid and fluid regions (e.g. Tyler et al. 2015; Bierson and Nimmo 2016). Modelling considering a uniform mantle tends to produce maxima of tidal heating at the poles and corresponding minima at the sub-Jovian and anti-Jovian points (e.g. Tackley et al. 2001; Bierson and Nimmo 2016), with the majority of the heating occurring in the lower mantle (e.g. Tobie et al. 2005; Bierson and Nimmo 2016). This does not match the observed distribution of volcanic hot spots (e.g. Kirchoff et al. 2011; Hamilton et al. 2013). To match the observed pattern, with, for example, increased volcanic activity in equatorial regions, requires a bicomponent mantle (in depth) (Bierson and Nimmo 2016), consistent with a non-zero melt-fraction.

Radionuclide heating rates depend on the original composition of the moon, with estimates for the contribution to Io's present day radionuclide heating power in the range $3.08\text{–}5.14 \times 10^{11}$ W (Hussmann et al. 2010) assuming a 100% rock fraction, approximately 1% of the present day tidal heating rate.

No trace of water has been found on Io, as would be expected from the high temperatures within the moon (Consolmagno 1981), and as such no chemical heating is likely to be occurring at the present time.

The predictions of volcanism on Io were borne out by the discovery of enormous gas/dust plumes rising above the moon's surface (Fig. 2, La Morabito et al. (1979)). These Pele-type plumes are capable of lofting dust to the (≈ 400 km) heights necessary for the electrical charging required for emission from Io's Hill sphere (Flandes 2004a,b), and are the active

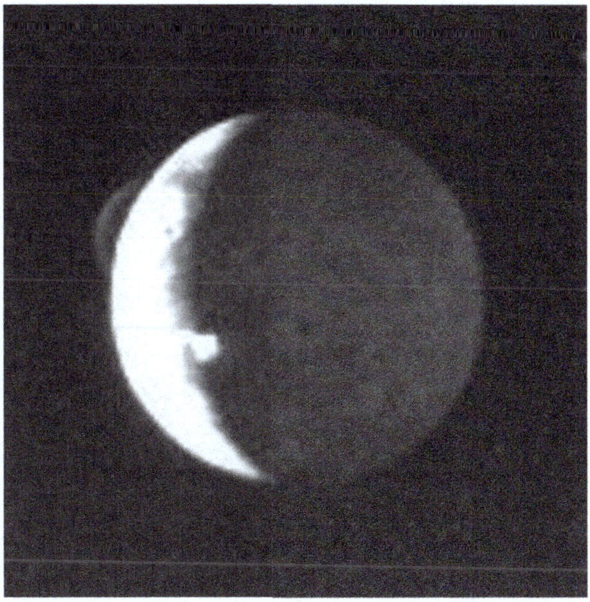

Fig. 2 The Voyager 1 image from which La Morabito et al. (1979) discovered the first plumes emanating from the surface of Io. Image credit: NASA/JPL (PIA00379)

sources of Io's cosmic dust, with emission of Jovian stream particles found to correlate with periods of intense volcanic activity (Krüger et al. 2003a).

Pele, the archetypal cosmic dust emitter on Io, is believed to be an overturning lava lake (Davies et al. 2001), covering approximately 6.5–7.5 km^2 (de Pater et al. 2016), with extensive lava fountaining (e.g. Howell and Lopes 2011) driven by exsolvation of sulfur-rich gases (Smith et al. 1979). Lava temperatures of 795–1450 K (de Pater et al. 2016; Keszthelyi et al. 2007) have been observed. In Galileo imaging the giant plumes, like Pele, were only marginally detected, while they appear clearly in Cassini and Voyager images (Geissler and McMillan 2008). Because the Galileo SSI camera is less sensitive to UV wavelengths than the Cassini and Voyager cameras, Geissler and McMillan (2008) surmise that typical grain sizes in the Pele-type plumes are on average much less than the 100 nm grains observed in Prometheus-type plumes. Estimates of mean grain sizes in the Pele plume range between 50–100 nm (time variable) based on Hubble observations (Jessup and Spencer 2012), although it is important to note that the majority of grains within the Pele plume are believed to be sulfurous in nature. Surface deposition around Pele is typically dark (close to the observed hotspots) or reddish (in a 1200 km diameter ring), believed to represent silicates and sulfur compounds respectively (McEwen and Soderblom 1983; Geissler et al. 1999).

3.2 Enceladus

Spacecraft flybys and remote observations, as with Io, have been used to constrain the interior structure of Saturn's small moon Enceladus. Enceladus' moment of inertia factor (Table 1) is indicative of a differentiated body, with a low density ($\sim 2400\ kg\,m^{-3}$ was assumed) rocky core (Iess et al. 2014), overlain by a lower density water/ice mantle, although no definitive constraints on the state of the mantle could be placed, given the similarity in density between liquid and solid water. That liquid water is present under the surface of Enceladus is no longer in doubt, in part because of the measurements of the composition of the emitted dust and gas (e.g. Waite et al. 2006, 2009, 2017; Postberg et al. 2008, 2009b,

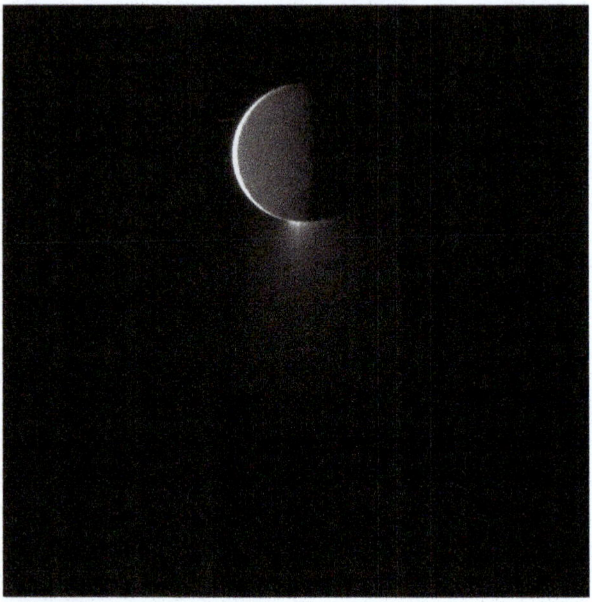

Fig. 3 A high phase angle image of Enceladus, captured by Cassini's Imaging Science Subsystem (ISS). Enceladus is partially illuminated by Saturn shine. North is up, with this image approximately covering the trailing hemisphere of Enceladus. The plume of vapour and ice particles can be clearly seen emanating from the South Polar region. Image credit: NASA/JPL/Space Science Institute (PIA12713)

2011b, 2018; Schneider et al. 2009; Glein et al. 2015; Hsu et al. 2015), although the precise area over which the water may extend has been more difficult to determine.

The salinity (approx. 5–20 $g\,L^{-1}$) reported by Postberg et al. (2009b) has been used to imply a large, non-regional underground ocean (Glein and Shock 2010). Further strands of evidence, such as signs of past fractures in the southern polar region (e.g. Patthoff and Kattenhorn 2011) and measurements of Enceladus' libration (Thomas et al. 2016; Van Hoolst et al. 2016) indicate that such an ocean is global, completely dislocating the rocky core from the ice shell.

To maintain a global ocean, as well as explain the prodigious heat flow in the southern polar region (e.g. Spencer et al. 2006; Howett et al. 2011; Spencer and Howett 2013), mechanisms to increase the tidal heating of Enceladus' interior, as well as provide routes for heat transfer out of the core, have been postulated. These include a fluffy, unconsolidated rocky core in which water percolates (e.g. McKinnon 2015; Roberts 2015; Choblet et al. 2017) as well as revised estimates of frequency-dependent tidal dissipation within Saturn (Lainey et al. 2012, 2017; Shoji and Hussmann 2017; Fuller et al. 2016; Rhoden et al. 2017).

The picture which is forming concerning the interior structure and state of Enceladus is as follows: an unconsolidated "fluffy" rocky core, in which water can percolate, is surrounded by a global "briny" ocean, overlain by an icy crust of varying thickness. As will be seen with Europa (Sect. 7), there are two possible structural models considered, a thin-shell/thick-ocean model, in agreement with e.g. the observed diurnal libration (e.g. Van Hoolst et al. 2016) and a thick-shell/thin-ocean model, in agreement with the gravity data (e.g. Iess et al. 2014; McKinnon 2015). In both models the core radius is $\sim$ 185–190 km (e.g. Čadek et al. 2016). In the thick shell model the ice crust is approximately 50 km deep (McKinnon 2015), overlying a 10 km thick ocean, whereas in the thin-shell model the crust is typically 14–26 km deep overlying a 21–67 km thick ocean (Van Hoolst et al. 2016). In the former case, thinning of the crust to 30 km (McKinnon 2015) in the south polar region is postulated, with equatorial thicknesses of up to 60 km, whereas in the latter the south polar crust is 2.5–17 km thick, increasing to 26 km at the equator (Van Hoolst et al. 2016). Crust densities are typically assumed to be 925 $kg\,m^{-3}$, with the ocean density increased by salinity to approximately 1030 $kg\,m^{-3}$ (Čadek et al. 2016).

Tidal heating within Enceladus is now believed to be concentrated in the porous core, with models such as those by Čadek et al. (2016), Choblet et al. (2017) showing that prolonged maintenance of the liquid ocean are possible, with heat transferred by long-term hot water circulation in the core (Travis and Schubert 2015). As with Io, tidal heating is maintained within Enceladus by an orbital resonance, in this case the 2:1 resonance with Dione, which generates a forced non-zero eccentricity (Table 1). Tidal power dissipated within Enceladus' unconsolidated core can be as high as $10\text{–}30 \times 10^9$ W depending on the shear modulus and dissipation parameter of the core, with additional tidal heating in the thin, fractured south polar region perhaps contributing further. Together it is possible to maintain a global subsurface ocean within Enceladus for $> 20 \times 10^6$ years (Choblet et al. 2017), with larger (50 GW) tidal power produced for 10^9 years under differing orbital evolution and Saturnian tidal dissipation (e.g. Saturn's tidal quality factor) scenarios.

Assuming that Enceladus is 53% rock by weight, Hussmann et al. (2010) estimate present day radiogenic heating to be $2.01\text{–}3.35 \times 10^8$ W. Estimates of the present day heating power provided by serpentinisation are given in Vance et al. (2007) as approximately 6.56% of the radiogenic power, corresponding to $13.2\text{–}22 \times 10^6$ W.

Active dust emission on Enceladus occurs when cracks in the ice shell reach the subsurface ocean, which occurs where the ice crust is thinnest, in the south polar region. Asymmetric crustal thinning at the poles (with the south polar region thinning more than the north polar region) is a prerequisite for any model of Enceladus' inner structure, explaining the depressed topography of the south polar region (Collins and Goodman 2007; Thomas et al. 2007; Thomas 2010), the gravity results (Iess et al. 2014; McKinnon 2015), the heat flux through the crust (e.g. Spencer et al. 2006; Abramov and Spencer 2009; Goguen et al. 2013; Le Gall et al. 2017) and the plume and jet activity (e.g. Porco et al. 2006, 2014; Spitale and Porco 2007; Spitale et al. 2015). Multiple processes may contribute towards the thinning, with hydrothermal flow producing downwelling at equatorial latitudes, thickening the crust, and upwelling at the poles thinning the crust (Travis and Schubert 2015). The insulating properties of plume dust "snowfall" may also aid the retention of subsurface heat nearer the southern polar region, perhaps contributing to the difference in tectonic activity and crustal thickness between the poles (Travis and Schubert 2015; Bland et al. 2015). Increased tidal dissipation in the thinned fractured crust at the south polar region also plays a role in maintaining activity in that region (e.g. Čadek et al. 2016; Souček et al. 2016; Běhounková et al. 2017). It is also possible that the heating or thinning was initiated at a different position (Nimmo and Pappalardo 2006; Collins and Goodman 2007; Matsuyama and Nimmo 2008; Tajeddine et al. 2017; Martin 2016) or that the core is asymmetric (e.g. Iess et al. 2014), but it is not currently clear how any asymmetry arose.

A large plume of vapour and particles (Fig. 3) is emitted from four sulci (the "Tiger Stripes") within the ~ 160 km radius (Crow-Willard and Pappalardo 2015) South Polar Terrain (SPT) centred on the South Pole of Enceladus (e.g. Dougherty et al. 2006; Porco et al. 2006; Spencer et al. 2006; Waite et al. 2006; Spahn et al. 2006b). The plume composition is dominated by water vapour ($\sim$ 96–99%, Waite et al. 2017), with trace amounts of CH_4, CO_2, CO, H_2, NH_3, and other light organics such as acetylene, possibly methanol and propane (Waite et al. 2006, 2009, 2017; Hodyss et al. 2009). Entrained within this plume (Spahn et al. 2006a,b) are dust grains, varying in radius from approximately 2 nm (Jones et al. 2009; Hill et al. 2012; Dong et al. 2015), to greater than[7] (Hedman et al. 2009; Kempf

[7]The large sizes inferred from remote observation (Jaumann et al. 2008) probably result from thermal annealing of grains after deposition.

et al. 2010) 5 μm, which experience stratification within the plume (Hedman et al. 2009; Postberg et al. 2011b) due to a size-dependent ejection speed (Schmidt et al. 2008).

The total length of the four "Tiger Stripe" sulci[8] (Alexandria Sulcus, Baghdad Sulcus, Cairo Sulcus and Damascus Sulcus, in alphabetical order) is approximately 500–590 km (Ingersoll and Pankine 2010; Porco et al. 2006). In the SPT, sulci are either double ridges or subdued double ridges (Nahm and Kattenhorn 2015). Double ridges are typically up to 2 km wide (Porco et al. 2006), with a 500 m central depression flanked by $\sim$ 100 m high ridges. Flank slopes range in gradient from 3°–23°, with an average of 9° (Nahm and Kattenhorn 2015; Patthoff and Kattenhorn 2011). Subdued double ridges are typically narrower (10–100s m wide rather than km) with less pronounced flanks. Although features attributed to tectonic modification are found at other locations (approximately centred on the orbital leading and trailing hemispheres, Bland et al. 2015 for example), the SPT is the only location in which double-ridged "Tiger Stripe" sulci are found (Nahm and Kattenhorn 2015).

The sulci lie within a depressed basin (approximately south of 65°S), which is 344–800 m lower than would be expected for a reference ellipsoid matching Enceladus' axial dimensions (e.g. Thomas et al. 2007; Thomas 2010).

Emission of vapour and dust appears to be often concentrated in jets (e.g. Fig. 4, which shows Baghdad Sulcus and the locations and jet azimuths of geysers observed emitting into the plume) (Hansen et al. 2008, 2006), spatially coincident with small-scale thermal hot spots (e.g. Spitale and Porco 2007; Goguen et al. 2013), although the exact contribution to plume emission from the jets, versus from the SPT in general, is still unclear (Porco et al. 2014; Hedman et al. 2018). Varying estimates of the solid:vapour ratio in the plume have been made, with typical values found to range from low (0.01–0.05 (Schmidt et al. 2008; Ingersoll and Ewald 2011)), to high (0.3–1 (Porco et al. 2014; Hedman et al. 2018)). However, the value important for dust formation models is more likely to be that estimated for the major fast dust emission regions, the jets. Here the solid:vapour ratio may be as high as 6 (Porco et al. 2014).

Plume and jet gas expansion speeds are typically 300 $\mathrm{m\,s^{-1}}$ and $\gg$ 500 $\mathrm{m\,s^{-1}}$ respectively (Tian et al. 2007; Hansen et al. 2008; Schmidt et al. 2008; Dong et al. 2011). Some observations of the plume and "Tiger Stripes" are compatible with "curtain" emission along the fractures (Spitale et al. 2015), with large-scale changing activity levels, although this interpretation is contested (Porco et al. 2015) for gas production, and the partitioning of solids emission between jet and curtain-like features is still poorly understood (Porco et al. 2015; Hedman et al. 2018). The opening angle of the individual jets is approximately 20° half angle (e.g. Teolis et al. 2010), with the combined emission forming a plume more structurally similar to a Mach 2 emission cone (Dong et al. 2011).

The latest estimates, combining both Cassini Ion and Neutral Mass Spectrometer (INMS) and Ultraviolet Imaging Spectrograph (UVIS) datasets, indicate that the plume H_2O production rate is of the order of several hundred $\mathrm{kg\,s^{-1}}$ (Yeoh et al. 2017; Hansen et al. 2017), greater than the 200 $\mathrm{kg\,s^{-1}}$ estimated from just UVIS or INMS data alone (e.g. Hansen et al. 2011), and varies with Enceladus' orbital position (Hurford et al. 2012; Nimmo et al. 2014; Hedman et al. 2013; Ingersoll and Ewald 2017). Normal, rather than shear flexing, appears to modulate the plume activity as a whole (Hedman et al. 2013; Ingersoll and Ewald 2017) with maximum emission near apocentre and minimum near pericentre, although individual jet activity seems to be triggered or halted by a different process (Porco et al. 2014). The direction of emission by jets varies considerably (Fig. 4) (Porco et al. 2014) and appears to

[8] The "Tiger Stripe" sulci appear to have dark flanks in the IR (Porco et al. 2006), which may have given rise to the nickname.

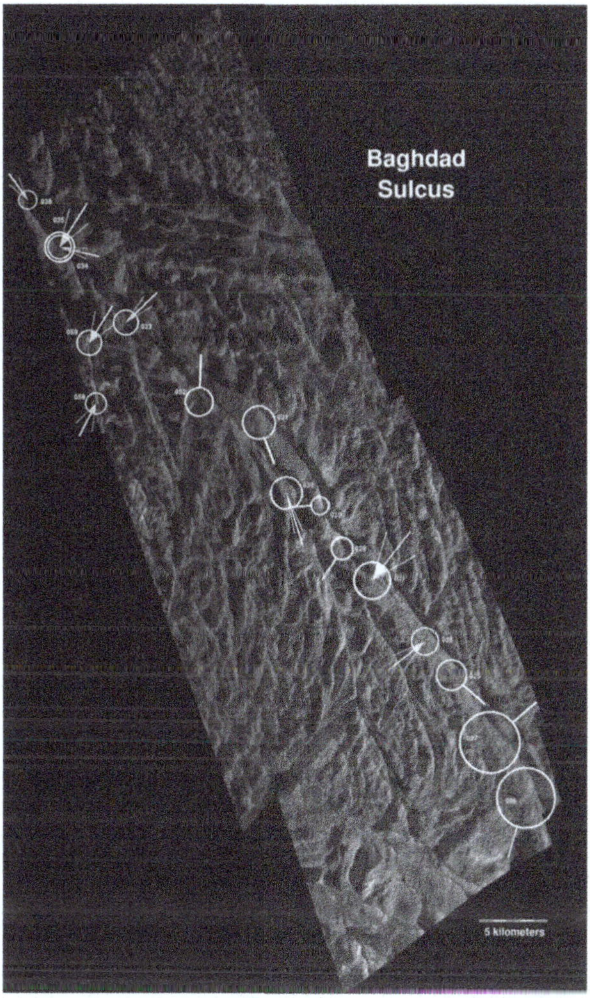

Fig. 4 The locations and emission azimuths of jets located along Baghdad sulcus. P. Helfenstein, C.C. Porco, Enceladus' geysers: relation to geological features. Astron. J. 150 (3) 96 (2015). © AAS Reproduced with Permission

be controlled by local topographical features (Helfenstein and Porco 2015), which may also account for the ~ 5 hour lag between tidal flexing and plume activity changes (Helfenstein and Porco 2015), although turbulent dissipation within the fractures may also contribute to this, by slowing the rate at which fractures refill with liquid during flexing (Kite and Rubin 2016).

The internal structure of the fractures is unknown, although vent widths at the surface have been estimated to be between 1–10 m wide, maintained by a balance of ice sublimation and vapour/ice grain deposition (Ingersoll and Pankine 2010; Goguen et al. 2013; Nakajima and Ingersoll 2016; Ingersoll and Nakajima 2016), narrowing to > 0.075–1 m in the subsurface (Schmidt et al. 2008; Kite and Rubin 2016; Nakajima and Ingersoll 2016; Ingersoll and Nakajima 2016) and then widening again to several metres[9] at the ocean boundary (Postberg et al. 2009b, 2011b, Nakajima and Ingersoll 2016; Ingersoll and Nakajima 2016)

[9]To maintain an ocean surface area of approximately 10 km^2 (Postberg et al. 2009b, 2011b) and assuming subsurface fracture lengths similar to those (500 km) measured at the surface, results in ocean-boundary

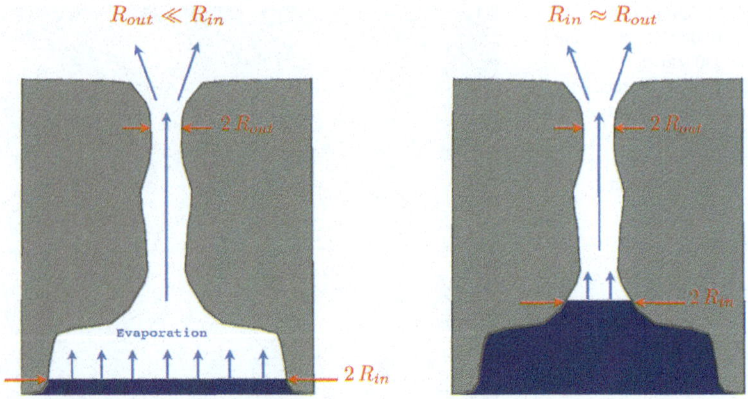

Fig. 5 Schematics of possible fracture models, showing the approximate relative dimensions of "chamber" (left) and "filled-fracture" (right) models. From Postberg et al. (2009b) supplementary materials and Ingersoll and Nakajima (2016). Reprinted with permission of Springer Nature

(Fig. 5). The depths of the fractures are unknown, although limits based upon the density ratios of the ocean brine and overlying ice indicate that the brine cannot reach closer to the surface than approximately 10% $(1 - \rho_{\mathrm{ice}}/\rho_{\mathrm{brine}})$ of the ice thickness without mechanisms such as bubble formation (Matson et al. 2012) or by flexure of brine-filled fractures above a constant-pressure ocean (Kite and Rubin 2016). Modelling of thermal emission from the SPT indicates that a liquid ocean is likely to be at a depth of greater than 1 km below the surface (Abramov and Spencer 2009), with fluid-filled fractures closer to the surface only possible if masked by an insulating layer. Large-scale (i.e. oceanic) liquid water at depths as small as 2 km is also suggested by microwave radiometry (Le Gall et al. 2017). Thin shell models of the SPT (e.g. Van Hoolst et al. 2016) result in minimum surface to ocean-boundary fracture depths of 0.25–1.7 km, whereas thick-shell models (e.g. McKinnon 2015) bring water to within 3 km of the surface. For at least some of the shallower fracture depths alterations in local surface thermal conductivities are required to provide agreement with thermal observations (e.g. Abramov and Spencer 2009).

4 Dust Generation and Emission

Although the process may contribute to particle formation, direct sublimation from solid to gas/vapour, followed by condensation/nucleation and freezing, is not a significant source of cosmic dust from Io or Enceladus. Instead the presence of a liquid phase is critical for dust production.

4.1 Io

The composition of the Jovian stream particles observed by Cassini's CDA as it approached, and passed Jupiter (Postberg et al. 2006), provides insights to their formation process. The particles were found to be dominated by Na, K, S and Cl, probably in the form of NaCl, with

fracture widths of 20 m, although smaller 10 m widths are possible if the fractures taper significantly towards the surface (Nakajima and Ingersoll 2016; Ingersoll and Nakajima 2016).

Fig. 6 (**a**) A coadded (from 30 spectra) Jovian stream dust spectrum, (**b**) a Saturnian stream particle spectrum, from Hsu et al. (2012). Reprinted with permission of Springer Nature

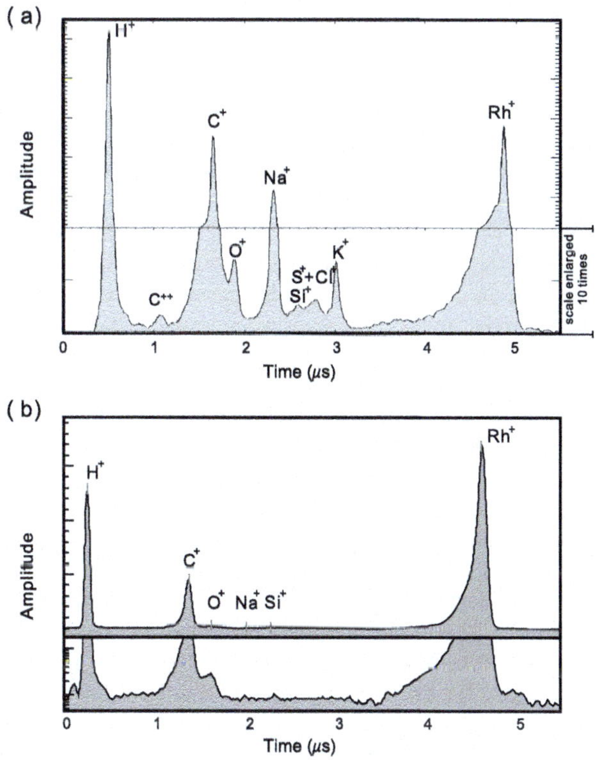

contributions of Na_2SO_4 and K_2SO_4 occurring at mass percentages of approximately 5–10% of the NaCl present in the grains. Minor silicate inclusions could not be ruled out. Average grain sizes were found to be 6 nm radius.[10] A typical Jovian stream particle mass spectrum is shown in Fig. 6. This composition is in excellent agreement with predictions made by Fegley and Zolotov (2000), Lellouch et al. (2003) and Schaefer and Fegley (2005), for gas production and condensation from a Pele-type[11] plume, deriving from "Pillan" lava temperatures now considered likely for silicate volcanism on Io ($\geq$ 1150–1335 K, potentially up to 1450 K (e.g. Spencer et al. 2007; Keszthelyi et al. 2007)). Considering pressures of 0.02 bar, formation of mineral condensates was calculated to occur for Na_2SO_4 ($\sim$ 1350 K), NaF ($\sim$ 1115 K), NaCl ($\sim$ 1050 K), KCl (950–620 K) amongst others (Schaefer and Fegley 2005).

In these models, SO_2 and S_2 gases dominate the plume, exsolved during the lava ascent (e.g. Smith et al. 1979; Cataldo et al. 2002; Leone et al. 2011), followed by SO (Zolotov and Fegley 2000) and then the halogen sulphate and chloride species. The SO_2, S_2 and SO gases condense at much lower temperatures (e.g. 110 K for SO_2 (Kieffer et al. 2000)). Experimental investigation (De Maria et al. 1971) of the vaporisation of lunar samples indicates that Fe and O_2 may also appear in the vapour phase at 1450 K, at lower ($\sim 50\times$) partial pressures than Na. SiO is not expected to appear until at least 1600 K, 150 K higher than the revised Io maximum lava temperatures (Keszthelyi et al. 2007). It is therefore possible that

[10] Corrected from the original 12 nm radius published in error by Postberg et al. (2006).

[11] Although some Prometheus-type plumes can reach similar heights, so it is possible that both the major volcano types on Io may produce cosmic dust.

the tentative traces of Si periodically found in the Jovian stream particles (Postberg et al. 2006) are from physical disruption of the lava (e.g. bubble bursting, fire fountaining) rather than condensation from the vapour phase.

The exact region in the plumes at which suitable temperatures are reached to begin the condensation cascade is uncertain, although the initial condensation is likely to occur close to the lava lake/fire fountains. Models of the plume indicate that the vapour velocity is 850–900 $m\,s^{-1}$ (Zhang et al. 2003, 2004; McDoniel et al. 2015) by a height of 750 m above the lava, with gas temperatures of $\gtrsim 500$ K. Particulates smaller than about 250 nm radius travel at the gas speed, with New Horizons observations of the Tvashtar plume showing grain speeds of 400–700 $m\,s^{-1}$ (Spencer et al. 2007). The entrained mineral particulates will therefore rise and cool, perhaps acting as nucleation cores for more volatile species, such as SO_2 in cooler regions of the plume.

The mean size of the entrained particles across the entire Pele plume is estimated to be 50–110 nm diameter (Jessup and Spencer 2012), although this was found to be variable.

Plumes rise until they stall, at which point a shock region is formed, as lofting gas and grains meet downfalling material (Strom et al. 1981). In this region a small amount of heating is likely, raising the temperature of the gas and entrained grains to > 200 K again, potentially removing overlayers of volatiles (such as SO_2) on the grains (Postberg et al. 2006; McDoniel et al. 2017) and reducing their sizes.

The best fit size distribution for grains which matched the Pele plume appearance and deposition pattern was found by McDoniel et al. (2015) to be a bicomponent sum of two log-normal distributions, leading them to suggest that two different mechanisms may be responsible for particle formation. Plausible formation mechanisms are likely to result in particles with differing compositions, with larger grains directly produced by fire fountaining and smaller grains (the vast majority (Collins 1981)) produced via condensation and growth mechanisms within the plumes.

Perhaps counterintuitively, the very smallest (~ 10 nm radius, based on a density of 800 $kg\,m^{-3}$ in McDoniel et al. (2015), equivalent in mass to a 7.5 nm radius NaCl grain) grains are not those which obtain the highest altitudes, as they remain entrained in the plume gas as it forms the shock front and then returns to the surface. Slightly larger grains (60–160 nm radius for example) decouple and continue ballistically through the slowed gas, slightly overshooting the shock front (McDoniel et al. 2015). However, it is not known whether this significantly affects the ability of the grains to subsequently enter circumjovian space. The largest (micron size upwards) grains reach progressively lower altitudes with increasing mass, leading to a grain size stratification within the plume (Zhang et al. 2003, 2004; McDoniel et al. 2015), which may also represent compositional sorting.

The Ionian ionosphere reaches down to altitudes of ~ 400 km above the surface of Io (Flandes 2004a,b), such that the tops of the giant Pele-type plumes are exposed to a moving plasma environment. Galileo measurements indicate that a 10 eV plasma, with an ion number density of 18,000 cm^{-1}, dominated by S^+, S^{2+}, O^+, O^{2+} and SO_2^+ ions is moving at approximately 10 $km\,s^{-1}$ with respect to a (azimuthally) stationary grain at the top of a plume (Frank et al. 1996). In this plasma the most efficient charging mechanism is by electron capture as other mechanisms are too slow (Flandes 2004a).

Particle escape can occur when the electrostatic force on a grain balances and then exceeds the gravitational force. For the plasma conditions in Io's ionosphere, Flandes (2004a) estimates an equilibrium potential of -34 V. A lower size limit can be obtained by considering the coulomb destruction of the grains (e.g. Hsu et al. 2011b). The destruction radius is given by $r_d(\mu m) = 9.4|\phi|F^{-\frac{1}{2}}$, where ϕ is the potential (V) and F is the tensile strength in $dyne\,cm^{-2}$. Reasonable estimates for the tensile strength of NaCl range between

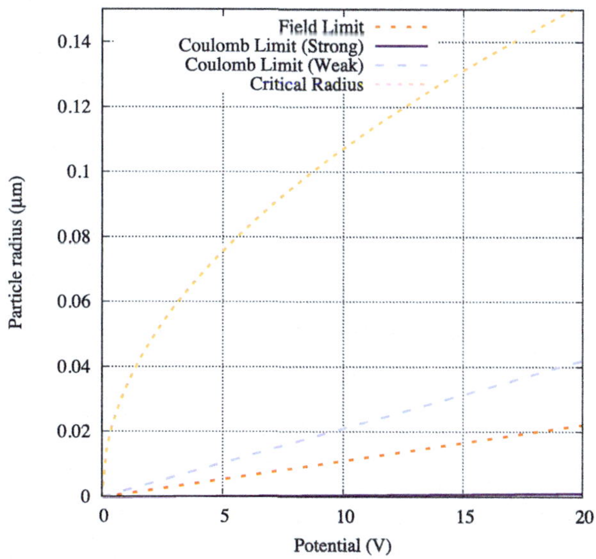

Fig. 7 Maximum radius as a function of grain potential, for particles to be extracted from the top of a plume. Also shown are minimum size limits for charge loss or particle destruction processes. Particles smaller than the critical radius (yellow dotted line) can be accelerated by the planet's corotational field. Particles smaller than the field limit (orange dashed line) lose charges by field emission. Coulomb disruption occurs for grains smaller than the size indicated by the light purple dashed line (internal strength: 2×10^7 dyne cm^{-2}, 2 MPa) or the dark purple line (internal strength: 3×10^{10} dyne cm^{-2}, 3 GPa)

$2 \times 10^7 - 2\text{–}3 \times 10^{10}$ dyne cm^{-2} (2 MPa–3 GPa). For the -34 V equilibrium potential, the minimum particle radius is 1.8–7.1 nm. Field electron emission provides another charge limit such that (Mendis and Axford 1974): ϕ (V) < -900 (a/μm). Thus for 0.01–0.1 μm grains, there is a limiting maximum potential range of -9– -34 V.

Considering the balance of forces, Flandes (2004a) derive an expression (Eq. (12)) for the maximum radius of a particle which could be removed from the plumes by Jupiter's corotational electric field:

$$r_c = 2.246 \times 10^{-5} \sqrt{\frac{(R_{\text{Io}} + H)^2 |\phi|}{\rho}}\ \mu\text{m}. \tag{12}$$

Setting Io's radius R_{Io} to 1821 km, plume height H to 400 km, and density ρ to 2.16 g cm^{-3}, this reduces to $r_c = 3.394 \times 10^{-2} \sqrt{\phi}$. For a reasonable range of potentials ($\leq |20|$ V, with $|5|$ V estimated by e.g. Horányi et al. 1997), the accelerable size ranges are shown in Fig. 7. For a given potential, particles must be below the critical radius and above the coulomb (assuming either strong or weak particles) and field emission limits. The range of values span the particle radii calculated in Postberg et al. (2006) of 3–13 nm. It is therefore clear that particles with a range of sizes, depending on their charging behaviour and timescales[12] can be picked up and removed from Io's influence by interaction with the Jovian magnetosphere. The typical velocity gained by the grains during this interaction process is the difference between Io's orbital velocity and the speed of the corotating magnetic field at Io's orbital radius, ~ 57 km s^{-1} (e.g. Horányi et al. 1993b; Johnson et al. 1980, Morfill et al. 1980).

4.2 Enceladus

As with Io, it is the composition of the dust emitted by Enceladus which tells us most about its formation. Three main compositional types of dust have been observed by Cassini in the

[12] Which in turn depend on the depth of the charging region—for example, at 400 km, a grain may only spend $\approx 10^2$ s in a 10 km thick "charging region" at the top of its trajectory, neglecting gas interaction.

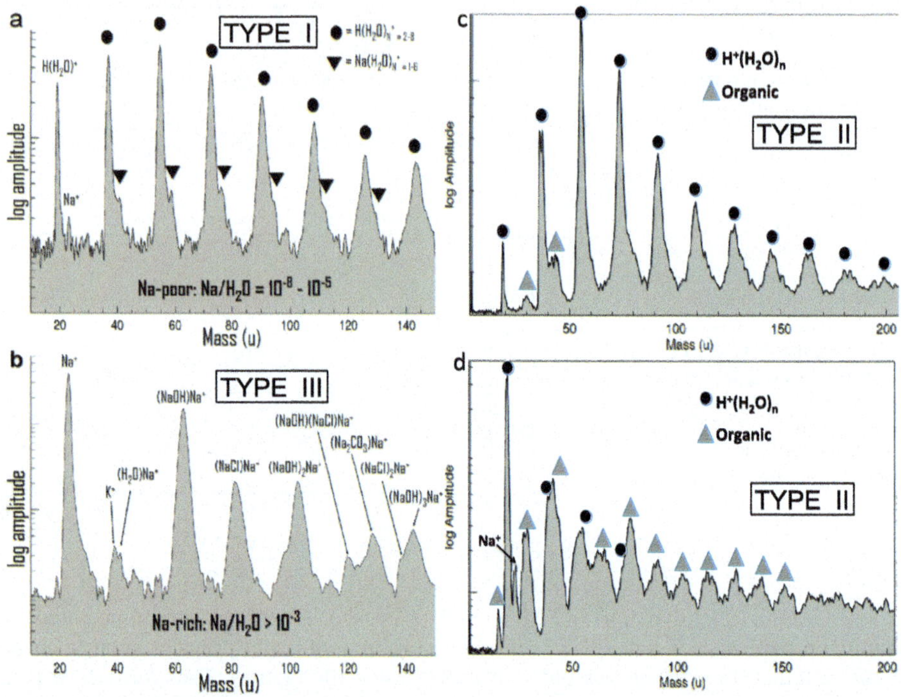

Fig. 8 Typical Type I (panel **a**), Type III (panel **b**) and Type II organic-poor/organic-rich (panels **c** and **d** respectively) particle CDA impact ionisation mass spectra. From *Enceladus and the Icy Moons of Saturn* edited by Paul M. Schenk, Roger N. Clark, Carly J. Howett, Anne J. Verbiscer and J. Hunter Waite. © 2018 The Arizona Board of Regents. Reprinted by permission of the University of Arizona

E ring and Enceladus plume (Postberg et al. 2008, 2009b, 2011b; Hsu et al. 2015; Postberg et al. 2018):

Type I Low salt (NaCl) particles, dominated by water ice and with Na:H_2O concentrations $< 10^{-5}$.

Type II Water ice with organic[13] constituents with drastically varying concentrations from ppm up to of the order of a few % by mass. A small subset of these grains, recognisable by their high mass organic cation (HMOC) spectra (Postberg et al. 2018) have organic components in excess of 1%.

Type III Salt-rich grains, again dominated by water ice, with NaCl, $NaHCO_3$, Na_2CO_3 (and low contributions (approximately 100–200× less[14] than from sodium (Postberg et al. 2011b)) from their potassium equivalents) at Na:H_2O concentrations of $\approx 3 \times 10^{-3}$–10^{-5} and potential silica inclusions (Kempf et al. 2005b; McBride et al. 2007; Hillier et al. 2007; Postberg et al. 2009b; Hsu et al. 2015). Laboratory simulations indicate the NaCl:$NaHCO_3$ ratio is 2–5:1, with 0.05–0.2 $mol\,kg^{-1}$ NaCl and 0.02–0.1 $mol\,kg^{-1}$ Na-carbonates (Postberg et al. 2009b).

[13] We here correct the definition given in Postberg et al. (2008, 2011b), which proposed minor silicate inclusions as an alternative to the organics.

[14] This number reflects the current stage of knowledge. It is under investigation and may change in the future.

Typical grain spectra are shown in Fig. 8. The Type I grains form by homogeneous nucleation and freezing of the water vapour expanding away from the liquid water surface and moving through the ice vents to the surface (Brilliantov et al. 2007; Schmidt et al. 2008). These grains are numerically the largest population in the Enceladus plume (Jones et al. 2009; Hill et al. 2012). Although in situ analysis has not been possible, the composition of Enceladus' plume nanograin population (Jones et al. 2009; Hill et al. 2012) is therefore likely to be similar to that of Type I grains, given the similar expected formation mechanism.

The presence of high concentrations of salts in Type III dust grains places limitations on their formation process. The solvation energy of NaCl makes it unlikely that significant quantities of NaCl will appear in particles formed by the condensation and freezing of a vapour (Schmidt et al. 2008; Postberg et al. 2009b). It is therefore likely that Type III grains formed directly from liquid water droplets. This interpretation is reinforced by the presence of silica nanoparticles and organics (Hillier et al. 2007; Postberg et al. 2008, 2009b, 2018; Hsu et al. 2015) in some of the detected Type II and III grains. The generation mechanism for such droplets requires some degree of turbulence at the water surface, through, for example, gas exsolvation (Postberg et al. 2009b; Matson et al. 2012; Ingersoll and Nakajima 2016; Nakajima and Ingersoll 2016). Whether this process is a result of gases (e.g. CO_2) exsolving due to a reduction in pressure (Postberg et al. 2009b; Matson et al. 2012), or vacuum-driven boiling tempered by overlying vapour pressure (Ingersoll and Nakajima 2016; Nakajima and Ingersoll 2016) is currently not constrained and it may be that both processes contribute.

For the CO_2 gas exsolvation case, a large ($\sim 10\ \mathrm{km}^2$, Postberg et al. 2009a, 2011b), liquid surface area across which bubbles burst, generating spray in a vapour-filled chamber which is then drawn out through fractures to the surface was originally suggested (Postberg et al. 2009b, 2011a), although the process is also compatible with fluid-filled fractures leading directly to the surface, with suitably wide (e.g. 20 m) fractures at the water surface. Gently boiling water sitting at the base of fractures, exsolving water vapour and possibly other dissolved gases, and similarly generating an aerosolised spray, without the need for subsurface chambers has also been postulated by Ingersoll and Nakajima (2016) and Nakajima and Ingersoll (2016), with a far lower water surface area ($\sim 1\ \mathrm{km}^2$) required.

Salts enter the ocean via leaching interactions between water and the porous rocky core and may subsequently enhance the dissolution of some minerals (e.g. Newton and Manning 2006). The water-rock interaction also results in serpentinisation, generating not only gases (e.g. H_2, CO_2, Sect. 2) but also aqueous silica (Sect. 2, Hsu et al. 2015; Sekine et al. 2015). Silica nanograins are formed at the ocean-core interface, where the aqueous SiO_2 cools and precipitates, forming tiny (2–8 nm) nanograins (Hsu et al. 2015; Sekine et al. 2015). The presence of these grains, as well as the appearance of NaOH in Type III spectra (Postberg et al. 2009b), places constraints on the pH of Enceladus' subsurface ocean, indicating that it is basic. Derived values for the alkalinity of the ocean are in general agreement, with Postberg et al. (2009b) inferring a pH of 8.5–9, Hsu et al. (2015) finding a range of 8.5–10.5 and modelling indicating possible ranges of 8–12 (Zolotov 2007; Glein et al. 2015). It may therefore be that there is an alkalinity gradient within the ocean.

Constraints on temperature are also found, with core-ocean interface temperatures found to be > 363 K and ocean surface temperatures expected to be at, or near, 273 K (Hsu et al. 2015; Matson et al. 2012; Ingersoll and Nakajima 2016). Transport timescales from hydrothermal upwelling (at least out of the regions in which further SiO_2 growth via Ostwald ripening is possible) imply that silica nanoparticles are ejected in Type III grains within months to a year after formation. This implies that convection within the subsurface produces currents that move at velocities of a few (~ 3) $\mathrm{cm\,s^{-1}}$ (Hsu et al. 2015; Ingersoll and Nakajima 2016; Matson et al. 2012; Choblet et al. 2017). These currents are likely to be driven by several processes, including heat (rising of water, less dense fluid, sinking of colder), salinity (sinking of denser saltier liquid following preferential vaporisation of water at the surface (Ingersoll and Nakajima 2016)) and gas exsolvation (bubble formation in parcels of rising water reducing density (Matson et al. 2012)).

An ocean salinity of 5–20 $\mathrm{g\,L^{-1}}$ of NaCl estimated from the Type III grain compositions (Postberg et al. 2009b, 2011b) is not only in good agreement with predictions made from first-principles (leaching of a chondritic composition core, Zolotov 2007), but also with the lower limit required to prevent freeze-out of the ocean surface in fractures due to evaporative cooling (Nakajima and Ingersoll 2016; Ingersoll and Nakajima 2016) of 16.2 $\mathrm{g\,L^{-1}}$.

Type II organic-rich[15] grains are believed to form via a similar process to the salt-rich Type III grains, as bubbles of exsolving gas or water vapour burst. In this case, however, the bubbles burst through a thin layer of potentially high molecular weight organics sitting on the brine ocean surface (Postberg et al. 2018). Organic-rich grains then form by aerosolisation, with further condensation and nucleation as they are entrained in the (predominantly water) vapour and emitted from the surface fractures. The composition of the organics is believed to include benzene-like, aromatic subgroups, together with short-chain aliphatic structures and indications of hydroxyl, ethoxy and/or carbonyl functional groups (Postberg et al. 2018, accepted).

The measured (CDA High Rate Detector, Spahn et al. 2006a) size distribution of all grains, assuming formation purely by condensation and nucleation above a source of (still) liquid water, during transfer through a variable width channel (which also acts to reduce the dust speed relative to the gas speed), was successfully reproduced by Schmidt et al. (2008), extending the work of Brilliantov et al. (2007). The resulting grain velocity distribution is (Eq. (13)):

$$P(u_{\text{grain}}) = \frac{R}{R_c}\left[1+\frac{R}{R_c}\right]\frac{u_{\text{grain}}}{u_{\text{gas}}^2}\left[1-\frac{u_{\text{grain}}}{u_{\text{gas}}}\right]^{\frac{R}{R_c}-1}. \tag{13}$$

Here u is the velocity of gas or grain as indicated, and the critical radius, R_c is (Eq. (14)):

$$R_c \equiv \frac{\rho_{\text{gas}}}{\rho_{\text{grain}}}\sqrt{\frac{8k_B T_{\text{gas}}}{\pi m_0}}\left[1+\frac{\pi}{8}(1-\beta)\right]\frac{L_{\text{coll}}}{u_{\text{gas}}}. \tag{14}$$

Where ρ is the density of gas or grain, k_B is Boltzmann's constant, m_0 is the mass of a water molecule, L_{coll} is the mean collision length, T_{gas} is the gas temperature and β is the condensation coefficient of the grains, proportional to their growth rate. R_c corresponds to the transition size below which particle motion is gas dominated and above which it is collision dominated (Schmidt et al. 2008). Values found to match the (Spahn et al. 2006a,b) Cassini High Rate Detector (HRD) data, as well as high phase plume observations, are: $L_{\text{coll}} = 0.1$ m, $\beta = 0.5$, $T_{\text{gas}} = 240$–260 K (the implied source temperature is therefore higher than this, implying a liquid source for the vapour), $\rho_{\text{grain}} = 920\ \mathrm{kg\,m^{-3}}$, $u_{\text{gas}} = 500\ \mathrm{m\,s^{-1}}$. Gas density is that near the triple point saturation pressure, at 0.00485 $\mathrm{kg\,m^{-3}}$ (Hansen et al. 2008). For the above values the critical radius is ≈ 0.7 µm.

However, the Schmidt et al. (2008) model underestimates the flux of grains larger than a few micrometres in radius, compared to Cassini Visible and Infrared Mapping Spectrometer (VIMS) observations (Teolis et al. 2010; Hedman et al. 2009, 2013; Kempf et al. 2010;

[15] These are a small subset (known by their High Mass Organic Cation, HMOC spectra) of the overall Type II grain population, which exhibit a wide range of organic concentrations.

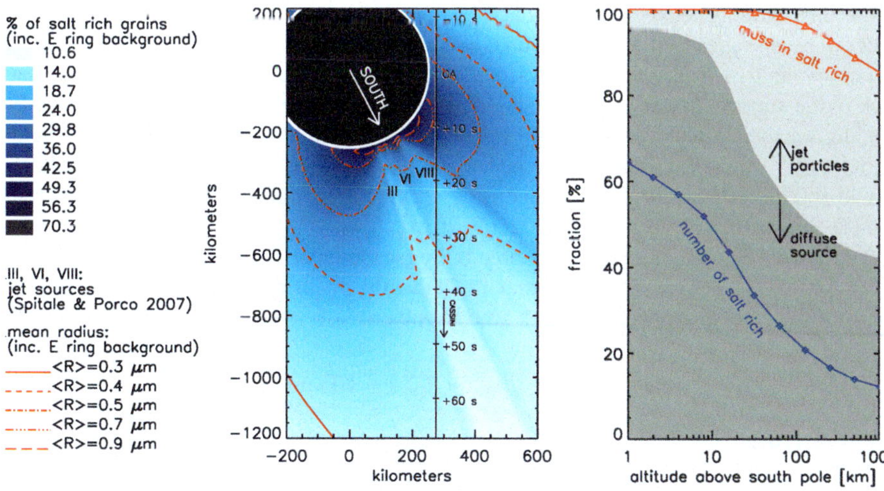

Fig. 9 A modelled compositional and size profile for recently emitted plume grains detected during Cassini's E5 plume traversal (trajectory indicated). The left panel shows the proportion of salt-rich Type III grains within the plume (blue colour), overlaid with contours of constant particle size (red lines). The contribution of the E ring is included for both quantities. The right-hand panel shows the modelled contribution from fast jets versus slow diffuse sources in the plume, as a function of altitude above the South Pole. Modified from Postberg et al. (2011b)

Jaumann et al. 2008) as no direct dust observations were available for such large grains and the primary mechanism for producing larger grains, bubble-bursting, had not then been considered. It is likely that the condensation process also produces far more nanograins (Jones et al. 2009; Meier et al. 2014; Dong et al. 2015; Engelhardt et al. 2015) than expected by the Schmidt et al. (2008) model, for similar reasons, with more than 99% (by number) of the grains present in the plume smaller than 10 nm in radius (Meier et al. 2014), peaking at 2 nm.

Other models for vapour-based dust formation include agglomeration processes (Gao et al. 2016) although these, as with the condensation model of Schmidt et al. (2008) tend to produce solid:vapour ratios (e.g. 0.07 ± 0.01) that, whilst applicable to the averaged plume, underestimate the solid mass in the jets. Gao et al. (2016) suggest that a low value for the solid:vapour ratio indicates a "gentle" source, such as a relatively quiescent vapour, rather than a vigorously boiling liquid directly exposed to vacuum. The high jet values suggested by Porco et al. (2014) and Hedman et al. (2018) thus indicate that a different mechanism is responsible for the generation of at least a fraction of the particles emitted here—namely the bubble-bursting process postulated by Postberg et al. (2009b, 2011b, 2018), Matson et al. (2012), Ingersoll and Nakajima (2016), Nakajima and Ingersoll (2016). It is, however, likely that condensation and agglomeration also occur in these sources, as well as for more distributed sources, contributing to the total plume production.

Although the spectral properties of the emitted dust appear fairly uniform across the SPT, there are minor spectroscopic variations attributed to localised differences in the dust size distribution (Dhingra et al. 2017) and evidence for compositional and size stratification within the plume (e.g. as shown in Fig. 9) (Hansen et al. 2011; Postberg et al. 2011b; Kempf et al. 2010). Although this may indicate that there are subsurface differences in fracture tortuosity/width/length or even regional differences in ocean composition, such differences in size distributions between jets have also been predicted as a consequence of variations in effective escape speeds at different source locations (Kempf et al. 2010).

Grains emitted from the "Tiger Stripes" will follow either a ballistic bound trajectory back to the surface, or escape Enceladus' Hill radius and enter circumplanetary space, depending on their kinetic energy. VIMS observations of plume particles indicate that for particles of 1 μm radius, there is an equal chance they will be ejected at velocities between 80 and 160 $\mathrm{m\,s^{-1}}$. Particles larger than this (2–3 μm radius) have a steeper velocity distribution (Teolis et al. 2010; Hedman et al. 2009, 2013).

These low velocities are below the three-body escape speed (v_{esc}) for Enceladus, as given in e.g. Kempf et al. (2010) (Eq. (15)) and imply that the majority of grains greater than 1 μm radius follow ballistic trajectories back onto the moon.

$$v_{esc} = \sqrt{2\mu_E\left(R_E^{-1} - R_{Hill}^{-1}\right)} = 205 \text{ m s}^{-1}. \tag{15}$$

With (Eq. (16)):

$$R_{Hill} = \langle r_E \rangle \left(\frac{1}{3} \frac{\mu_E}{\mu_E + \mu_S} \right)^{\frac{1}{3}} = 948 \text{ km}. \tag{16}$$

μ_E and μ_S are the standard gravitational parameters of Enceladus and Saturn, 7.2047 $\mathrm{km^3\,s^{-2}}$ and 3.7931×10^7 $\mathrm{km^3\,s^{-2}}$ respectively, $\langle r_E \rangle$ is the mean orbital radius ($3.94 R_S$) and R_E is Enceladus' radius, 252.1 km. The radius in Eq. (15) has been updated with the latest values and differs slightly from that in Kempf et al. (2010).

However, the injection of particles into circumplanetary space occurs from the SPT of Enceladus, roughly orthogonal to the ring plane. This results in slightly inclined orbits with respect to Enceladus, decreasing the chance of impact with the moon's surface (Kempf et al. 2010). Particles with radii of > 0.7 μm can therefore only escape if their emission velocities significantly exceed v_{esc}, whereas smaller particles ($r \lesssim 200$ nm, Meier et al. (2014)), for which the Lorentz force is effective in altering orbits, are swept away by Saturn's corotating magnetic field. The mass fraction of grains escaping into the E ring is estimated to be 7.9–10.4% for individual jets, although the number fraction is much higher (95.3–96.9%, Kempf et al. (2010), 99%, Meier et al. (2014)).

Although triboelectric charging of nanograins during emission has been postulated (Jones et al. 2009), early charging appears to conflict with Cassini CAPS data (Meier et al. 2014), for which initially uncharged grains produce the best explanation. Triboelectric charging to explain positively charged nanograins (Jones et al. 2009; Hill et al. 2012) has also been discounted by Meyer-Vernet (2013), who find that impacts of ambient plasma particles are adequate to explain the population of both positive and negatively charged nanograins. Charging time by exposure to the local plasma environment, for which an equilibrium potential of −3 V (Kempf et al. 2006) is expected, is inversely proportional to grain size (e.g. Kempf et al. 2010). Typical timescales to reach an equilibrium charge when outside of the Enceladus plume are 120, 15 and 2 s for grains of 0.1, 1.0 and 10 μm radius respectively. For the nanograin population charging times are greater still, with 2 nm particles taking several hundred seconds to reach an equilibrium charge, at which point they are significantly distant from Enceladus (0.3–$0.6 r_E$, Meier et al. 2014). Grains much smaller than this (below ~ 0.7 nm radius[16]) are likely to be disrupted by coulomb explosion (Meyer-Vernet 2013).

[16]If water-ice particles this small can even form, as minimum formation sizes are estimated to be 1.3 nm, Meyer-Vernet (2013).

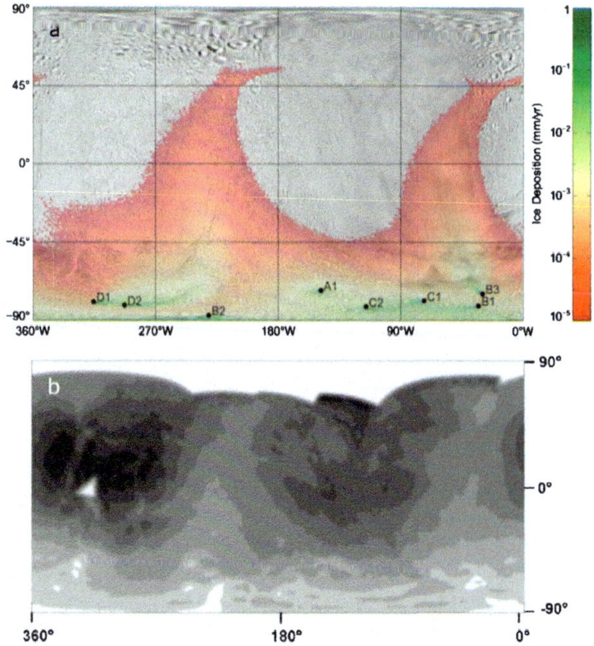

Fig. 10 (**a**) Mass deposition rate of jet particles between 0.5 μm and 5 μm (Kempf et al. 2010). (**b**) UV/IR colour ratio map (Schenk et al. 2011). Vertical axes show Enceladian latitude, horizontal axes show Enceladian longitude. Modified from Kempf et al. (2010) and Schenk et al. (2011)

The fate of larger nanograins is poorly constrained, although confinement within the Saturnian system and eventual sputtering (timescales for the loss of 0.1 μm particles in the Saturnian system are estimated at a few years, Jurac et al. 2001) or collisional destruction is most likely, as postulated for the smallest grains within the Jovian system (e.g. Grün et al. 1998).

The much larger and/or slower grains are preferentially deposited onto the surface of Enceladus, forming a deposition pattern (Fig. 10) of grains primarily landing on old, heavily-cratered surfaces rather than geologically young surfaces (Kempf et al. 2010), which well matches a UV/IR colour ratio plot from surface imagery of Enceladus (Schenk et al. 2011) (Fig. 10). The majority of jet particles, regardless of size (bar nanograins), impact close to the SPT and vents, however, with deposition rates as high as 0.5 mm per year[17] at current activity levels possible (Kempf et al. 2010). It may be that these grains provide the insulation required to disguise thermal emission from the fractures, as required by e.g. Abramov and Spencer (2009) to mask water-filled fractures close to the surface.

The remaining grain population, roughly 0.1–3 μm in size[18] (e.g. Nicholson et al. 1996; Hillier et al. 2007; Spahn et al. 2006b; Kempf et al. 2008) enters the E ring.

5 Circumplanetary Behaviour

The circumplanetary behaviour of the Iogenic and Enceladogenic grains differs significantly, with the majority of the larger E ring grains undergoing a slower evolution than the much smaller, rapidly Lorentz force dominated Ionian dust.

[17] Assuming a compact ice layer—for fluffy snow-like deposition the layer thickness will be higher.

[18] With particles up to 10 μm also possible, Ye et al. (2016).

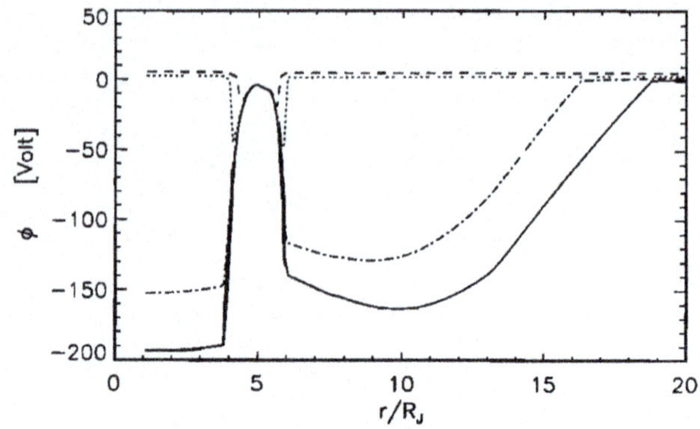

Fig. 11 The equilibrium potential reached by a grain in a circular, equatorial, Keplerian orbit, at a distance r from Jupiter. The secondary electron yield (δ_M) was varied between 0 and 2.0 (0: solid line, 0.5: dot-dashed line, 1.0: dotted line and 2.0: dashed line). The optimum energy parameter (E_M) was set to 500 eV. Modified from Horányi et al. (1993b)

5.1 Ionian Dust

The orbits of the grains following removal from Io's ionosphere are strongly dependent on the charge to mass ratio of the grains (Horányi et al. 1993a) and the local electromagnetic environment to which they are coupled (Grün et al. 1998). Grains are initially in Io's cold plasma torus, in which secondary electron emission is low (e.g. Horányi et al. 1997; Krüger et al. 2003b; Flandes 2004a) and the particles, as previously discussed, become negatively charged.

The plasma torus is not azimuthally symmetric, exhibiting significant dawn-dust asymmetry (Barbosa et al. 1983; Ip and Goertz 1983) which results in higher plasma temperatures on the dusk side (Krüger et al. 2003b). Particles extracted from Io therefore develop different charges depending on where on Io's orbit they enter the torus.

In hotter plasma regions, such as this dusk side, and other regions at different Jovian distances (Fig. 11), secondary electron emission increases, leading to the small dust grains developing a positive charge (Horányi et al. 1997). Particles entering the dusk side of the plasma torus are therefore likely to obtain a positive charge more rapidly than those entering the torus on the dawn side, which may remain captured by the torus for longer periods (Horányi et al. 1997).

Typical equilibrium potentials for tiny (radii between 5–15 nm, density 1000 $kg\,m^{-3}$) grains outside of the plasma torus (~ -5 V) are +5 V (Horányi et al. 1997). Approximately 80% of particles of this radius range manage to escape the torus rapidly, with 20%, dominated by those extracted on the dawn side, staying confined for one day (Horányi et al. 1997). With the generation of a positive potential, grains with radii of $\approx$9–180 nm (assuming a density of 1000 $kg\,m^{-3}$) (Grün et al. 1998) begin to be accelerated radially outwards from Jupiter by the co-rotating electric field (Horányi et al. 1993a,b), until they leave the magnetosphere. Smaller grains remain trapped in ion-like behaviour, gyrating around magnetic field lines, whilst for larger grains the Lorentz force is not strong enough to overcome Jupiter's gravity (Grün et al. 1998).

5.2 Enceladian Dust

Dust grains emitted from Enceladus, which do not immediately couple to the rotating Saturnian magnetosphere, or rapidly impact Enceladus, form Saturn's E ring. Extending from approximately $3R_S$ to over $20R_S$ (Srama et al. 2006; Kempf et al. 2006; Kempf 2008), with a number density maximum just outside the orbit of Enceladus (Kempf et al. 2008), the E ring is the second largest planetary ring in the Solar System, after Saturn's Phoebe ring

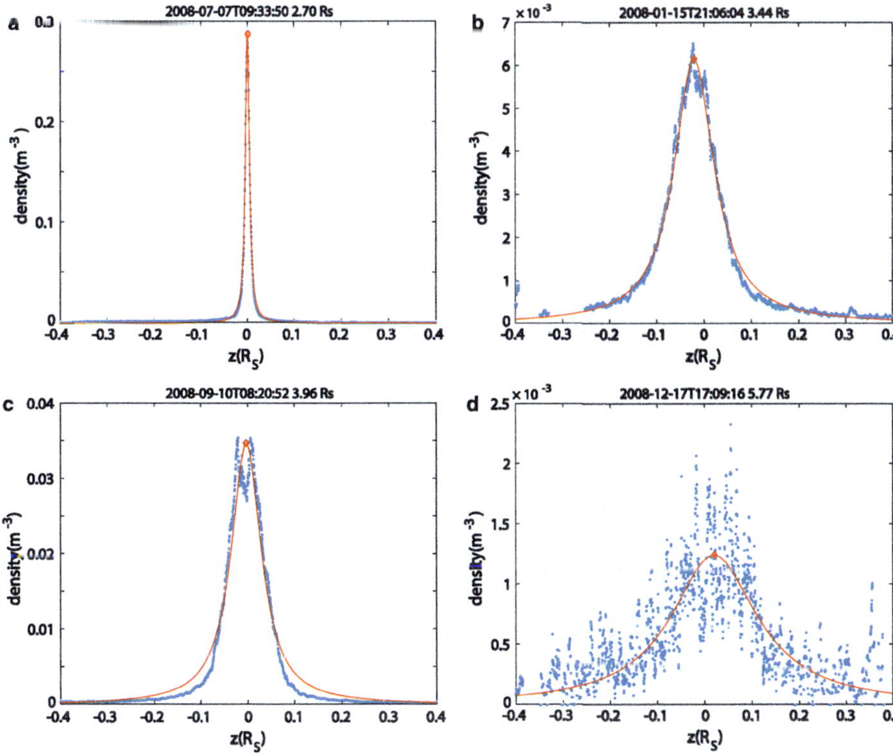

Fig. 12 RPWS-derived number density (m^{-3}) profiles (in R_S) interior to Saturn's G ring (**a**) and within Saturn's E ring (**b**, interior to Enceladus' orbit; **c**, E ring core, showing double-peaked structure and **d**, Exterior to Enceladus' orbit). Red diamonds show the density peaks and red line shows Lorentzian fits to the dust number density profiles. Profiles were calculated based on impacts detected within a 25 s wide moving window. Reprinted from Icarus, 279, Ye S.-Y. et al. pp. 51–61, 2016 with permission from Elsevier

(Verbiscer et al. 2009). Within the E ring, dust grains follow a differential power law size distribution, with a slope of -4– -5.4 (Kempf et al. 2008; Ye et al. 2014, 2016).

In the vicinity of Enceladus (the "core" E ring), the vertical structure of the ring differs from that further within and without, exhibiting two peaks in number density, above and below the ring plane (Fig. 12c), with the decrease in number density at the ring plane identified as a result of the initial emission velocities of grains from Enceladus (e.g. Juhász et al. 2007; Horanyi et al. 2008; Kempf et al. 2010; Graps et al. 2008), essentially protecting a population of grains against impact with the moon. Outside of this region the vertical profile of the ring is approximately Lorentzian (Fig. 12b, d) (e.g. Hedman et al. 2012; Ye et al. 2016), and roughly Gaussian within the orbit of Enceladus for larger grains (Kempf et al. 2008). The ring height is expected to depend on the size of particles considered (e.g. Horanyi et al. 2008), although at sizes of at least 0.9 μm this effect has not been seen in HRD data (Kempf 2008). The ring height (half width half maximum) as measured by RPWS on Cassini is $0.033 R_S$ at Enceladus' orbit, increasing linearly to $0.1 R_S$ at $3 R_S$ and $0.12 R_S$ at $6 R_S$ from Enceladus (Ye et al. 2016). CDA has, however, measured icy grains at heights of $> 1 R_S$ from the ring plane (Srama et al. 2006), in agreement with modelling (Horanyi et al. 2008). The ring height also increases interior to Enceladus' orbit (Kempf et al. 2008).

Owing to the influence of radiation pressure and Saturn's magnetic field, the maximum vertical ring density is in general offset from the ring plane, forming a warped disc. The offsets reported in the literature vary from $-0.03R_S$ at $3R_S$ to $+0.03R_S$ at $6R_S$ (Kempf et al. 2008; Hedman et al. 2012; Ye et al. 2016). Longitudinal variations in ring height with respect to Enceladus are not observed, although longitudinal ring thickness variations occur with respect to the Sun, being thicker interior to Enceladus' orbit on the anti-solar side of Saturn and exterior to Enceladus on the sub-solar side (Hedman et al. 2012).

Evolution of the ring grains to greater orbital distances is believed to occur via the influence of plasma drag (Dikarev 1999) and plasma sputtering (e.g. Horanyi et al. 2008; Jurac et al. 2001).

The Saturnian plasma environment plays a major role in the evolution of E ring grains and the resulting structure of the E ring. Both Enceladus and the inner ($\lessapprox 7.5R_S$) E ring lie within Saturn's cold plasma environment, where electron and ion temperatures are low ($T_e \approx 0.7$–8 eV, $T_i < 100$ eV, Schippers et al. 2008; Wahlund et al. 2009). The plasma components are dominated by electrons and ions originating from the injection of water vapour by Enceladus, namely H_3O^+, H_2O^+, OH^+ and O^+ (e.g. Wilson et al. 2008; Bu et al. 2016). In this cold region, CDA measurements of the potentials gained by grains within the E ring show that at Saturnian distances less than $\sim 6.5R_S$ grains become negatively charged, to approximately -1.6 V (Kempf et al. 2006), by the collection of electrons. Outside of $7.5R_S$ the detected grains were all positively charged, at up to ~ 7 V (Kempf et al. 2006), as the hotter ($T_e = 80$–200 eV, Schippers et al. 2008; Graps et al. 2008) electron population produces secondary electron emission, and UV photoemission increases with decreasing plasma density (e.g. Horanyi et al. 2008; Hsu et al. 2013).

Horanyi et al. (2008) modelled the large scale structure of the E ring, considering plasma drag, sputtering, Lorentz, gravitational and solar radiation pressure forces. Radiation pressure effects affect positively charged grains more than negatively charged grains, resulting in increases in grain orbital inclinations and thickening the ring (e.g. Graps et al. 2008) outside ≈ 7–$8R_S$. Plasma drag increases the semi-major axes of dust grains and is inversely proportional to the size of the dust grains, with smaller grains migrating outwards faster than larger grains (e.g. Horanyi et al. 2008; Graps et al. 2008).

Particle sputtering also plays a crucial role, with lifetimes of 1 μm ice particles emitted from Enceladus estimated to be 50 years at approximately $4R_S$ (Jurac et al. 2001). The timescale required for plasma drag to transport an E ring grain to Titan's orbit ($\sim 20R_S$) is estimated to be 200 hundred years, with particles surviving for longer periods as plasma densities drop with increasing Saturnian distance, reducing the sputtering rate such that the fraction of a grain (f, by radius) resulting at a distance a_s (Saturn radii) from Enceladus is given as (Horanyi et al. 2008):

$$f = 1.18 \exp\left[-3.34 \arctan(0.2a_s - 0.75)\right]. \tag{17}$$

The sputtering and plasma drag processes result in a change in size distribution of the E ring grains with external distance from Enceladus' orbit (Fig. 13), with no Enceladian grains larger than 0.5 μm expected outside of $9R_S$, as observed by CDA (Srama et al. 2006). Although no changes in size distribution were observed by RPWS (Ye et al. 2016), the small radius range considered there (3–$6R_S$) is one over which the population of the larger grains remains *relatively* constant, with the removal of the larger grains only occurring at approximately $9R_S$ (Horanyi et al. 2008). During their migration and sputtering, it is possible for grains to become "trapped" in a region near Rhea's orbit as their charge changes from negative to positive (Hsu et al. 2011b). Here they can undergo further sputtering, reaching

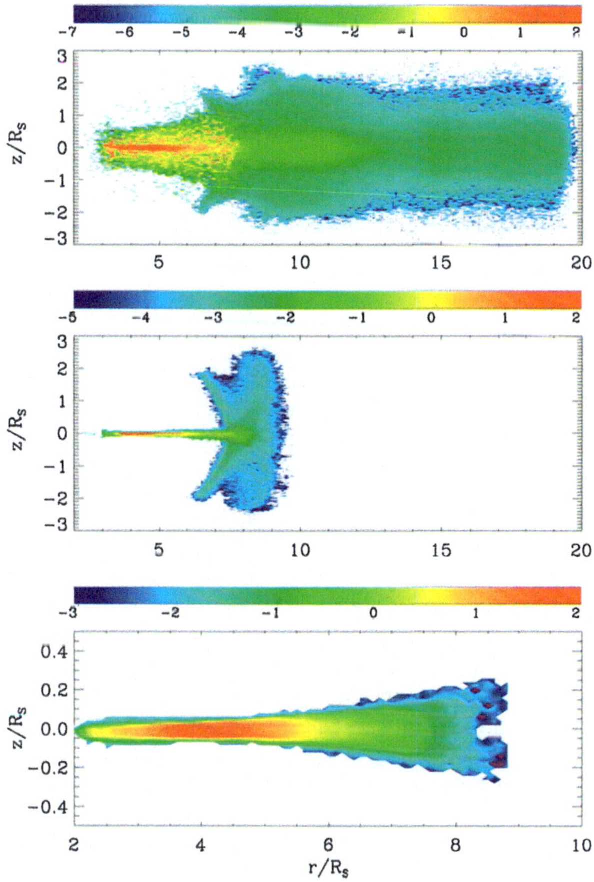

Fig. 13 The azimuthally averaged density distribution of grains with radii: (top) $0.1 < r$ (μm) < 0.5; (middle) $0.5 < r$ (μm) < 1; and (bottom) $1 < r$ (μm) < 3. The largest grains remain confined to the classical E ring, while the smallest particles can reach the orbit of Titan. All grains were released from Enceladus with an initial southward velocity of $100\ \mathrm{m\,s^{-1}}$. The colour scale is logarithmic, and in each case normalised to 100, corresponding to maximum number densities of 7; 1.2; and $0.3\ \mathrm{m^{-3}}$. From Horányi M., Juhasz, A., Morfill, G.E., Geophys. Res. Lett. 35 (4), pp. 1–5, 2008. © 2008 John Wiley and Sons, reprinted with permission

significantly smaller sizes. For Type I and Type II grains (Sect. 4) this may result in complete destruction. However, for Type III grains, with nm-sized silica inclusions (Hsu et al. 2015), a different fate is possible (Hsu et al. 2011b). Preferential sputtering of volatile water ice can release the nanograins previously trapped in the ice matrix, in a region in which the dust equilibrium charge is positive, resulting in positively charged silica nanograins which quickly interact with Saturn's co-rotating, outward pointing E-field (Hsu et al. 2011b; Kempf et al. 2005a), forming silica-dominated Saturnian stream particles (Fig. 6, panel b) (Kempf et al. 2005b; Hsu et al. 2015).

6 Interplanetary Behaviour

With the discovery of tiny nanometric stream particles emitted from the Jovian system (Grün et al. 1993), predictions were made that similar emission may occur from the Saturnian system (Horányi 2000), and which were proved correct by the discovery of Saturnian stream grains (Kempf et al. 2005a). Although the processes involved in accelerating dust out of each planetary system are identical, with similar behaviour in interplanetary space, there are still some differences to be found, in terms of velocities attained during emission and dust trajectories.

6.1 Ionian Dust

Escape from the Jovian system is possible for positively-charged nanograins providing their masses lie in a "sweet spot": not so massive that their motion is dominated by Jupiter's gravity and they remain bound within the system and not so light that their motion is similar to those of ions, with particles trapped gyrating around Jupiter's magnetic field lines.

The upper and lower limits (a_{min}, a_{max}) for the range of particle radii which are able to escape the magnetosphere of a primary are given by Eqs. (18) and (19) (Hsu et al. 2012):

$$a_{\text{min}} \approx \sqrt{\frac{\varepsilon_0 \phi_d B_0}{10\rho_d \Omega}\left(\frac{R_{\text{P}}}{r_0}\right)^3 \left(1 - \sqrt{\frac{GM_{\text{P}}}{\Omega^2 r_0^3}}\right)^{-1}}. \tag{18}$$

$$a_{\text{max}} \approx \sqrt{\frac{6\varepsilon_0 \phi_d}{\rho_d}\frac{\Omega B_0}{GM_{\text{P}}}\left(1 - \frac{r_0}{r_{\text{ms}}}\right)}. \tag{19}$$

Here Ω is the rotation rate of the primary, ϕ_d, m_d and ρ_d are the potential, mass and density of the dust grain, B_0 is the surface magnetic field strength at the planet's equator, r_0 is the radius at which the grain begins to accelerate, R_{P} and M_{P} are the radius and mass of the primary and r_{ms} is the radius of the magnetosphere. G and ε_0 are the gravitational constant and vacuum permittivity respectively.

For Iogenic nanograins, assuming a potential of +5 V, a density of 2200 kg m^{-3}, a starting radius of $r_0 = 6R_J$ and a magnetosphere radius of $100R_J$, Hsu et al. (2012) find the ejected Jovian stream particle size range to be 5.4–156 nm. This is in good agreement with sizes deduced from Cassini CDA measurements of the grains, which found the mean radius to be 6 nm (Postberg et al. 2006). The velocities of the emitted grains as they leave the magnetosphere are approximately given by Eq. (20) (Hsu et al. 2012):

$$v_{\text{ex}} \approx 59.5 \text{ km s}^{-1}\left\{\frac{75.6}{a_{\text{nm}}}\left[\frac{\phi_d}{\rho_{\text{cgs}}}\right]^{\frac{1}{2}}\left(1 - \frac{r_0}{r_{\text{ms}}}\right) - \frac{1}{2}\right\}\left[\frac{R_{\text{P}}}{r_0}\right]^{\frac{1}{2}}. \tag{20}$$

Here the grain size (a_{nm}) and density (ρ_{cgs}) are in nm and g cm^{-3} respectively, with the other values as for Eqs. (19) and (18). For a 10 nm radius grain the ejection speed is approximately 248 km s^{-1}. The values given here are in agreement with those obtained by the reverse trajectory tracking method of Zook et al. (1996), although they differ slightly from results obtained by more complex simulations (Horányi et al. 1997; Hsu et al. 2012; Krüger et al. 2003a,b).

The moving source point (Io[19]), together with Jupiter's tilted magnetosphere (9.6° from rotation axis), and the range of charge-to-mass ratios of the differently-sized grains, produces a complex pattern of emitted grains, Jupiter's "dusty ballerina skirt" (Fig. 14) (Horányi et al. 1993b; Grün et al. 1998). Although grains released on Io's orbital plane tend to experience acceleration along the plane, the tilted magnetic dipole generates a significant out-of-plane Lorentz force component. Dust grains emitted (more rapidly) from the night side of the magnetosphere also have a lower vertical (out of the ecliptic) velocity than those emitted from the day side (Flandes et al. 2011).

Entering interplanetary space the ratio of Lorentz force to (now solar) gravitational force increases by a factor of 1000 (e.g. Hsu et al. 2012) and the Interplanetary Magnetic Field (IMF) begins to strongly affect the particle trajectories (Hamilton 1993; Zook et al. 1996).

[19] Although initially Jupiter's Gossamer ring was postulated as a possible source, Hamilton and Burns (1993).

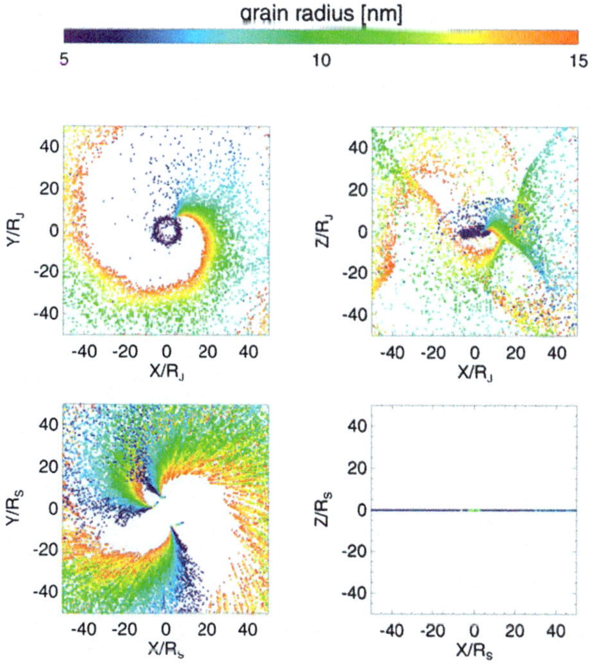

Fig. 14 The upper two panels show the ejection of Io-sourced stream particles from the Jovian system, with trajectories significantly affected by the tilted dipole magnetic field, forming the "dusty ballerina skirt". The lower two panels show the same view, but for Enceladus and Saturn. From Horányi, M., Hartquist, T.W., Haves, O., Mendis, D.A., Morfill, G.E., Reviews of Geophysics 42 (4) pp. RG4002-1-20, 2004. © 2004 John Wiley and Sons, reprinted with permission

On average, the IMF at Jupiter lies along the ecliptic, transported with the solar wind, such that the associated E-field points roughly perpendicular ($\pm z$) to the ecliptic. The acceleration that grains undergo is thus perpendicular to the ecliptic (Hamilton 1993), potentially spreading dust to high latitudes (Krüger et al. 2006a). In particular, interaction with the solar wind and IMF acts to spatially compress or disperse grains, forming "streams" or "bursts" at large Jovian distances from an otherwise near continuous source (Flandes et al. 2011). Dust streams are always preceded by a high IMF event, such as entry or exit from a corotating interaction region (CIR) or coronal mass ejection (CME), generating a compression region (CR) (Flandes et al. 2011). CR intensity also dictates stream duration and intensity (Flandes et al. 2011). The compression, resulting in a spatio-temporal correlation between grain impacts, is responsible for generating the bursts of particles detected by Ulysses, Galileo and Cassini (Grün et al. 1992; Horányi et al. 1993b; Horányi 1996, 2000; Grün et al. 1993, 1994, 1995, 1996b,a; Zook et al. 1996; Krüger et al. 1999; Graps et al. 2000; Krüger and Grün 2002; Krüger et al. 2003a,b, 2006b,a, 2010; Flandes et al. 2011; Hsu et al. 2012), which led Horányi (2000) to theorise that such stream particles may be generated by the Saturnian system.

6.2 Enceladian Dust

E ring grains, having undergone sputtering, release nanometric silica grains, which attain a positive charge in a region approximately 7.5–8R_S from Saturn (Sect. 5). As with Jupiter, interaction of the grains with Saturn's outward-pointing, co-rotating E-field begins to accelerate the particles out of the magnetosphere and into interplanetary space (Horányi 2000; Kempf et al. 2005a; Hsu et al. 2010a, 2011a,b, 2013, 2016). However, unlike Jupiter, Saturn's magnetic dipole is aligned with its rotational axis, resulting in grain acceleration broadly along the equator/ring plane, rather than producing a complex warped "dusty ballerina skirt" (Fig. 14).

Saturnian stream particles are smaller, but emitted from the planetary system at lower velocities than their counterparts are at Jupiter because of Saturn's weaker magnetic field (Horányi 2000; Kempf et al. 2005a; Hsu et al. 2011b). Assuming grain surface potentials and densities of 2.3 V and 2300 kg m^{-3} respectively, and a magnetospheric radius of $25R_S$, Hsu et al. (2012) estimate size ranges for particles capable of ejection from $8R_S$ to be ≥ 0.5–27 nm (Eqs. (18) and (19)). As with the Jovian stream grains, it is possible to estimate the velocity reached by grains as they are accelerated out of the Saturnian magnetosphere (Eq. (21), Hsu et al. 2012):

$$v_{ex} \approx 35.5 \text{ km s}^{-1} \left\{ \frac{23.3}{a_{nm}} \left[\frac{\phi_v}{\rho_{cgs}} \right]^{\frac{1}{2}} \left(1 - \frac{r_0}{r_{ms}} \right) - \frac{1}{2} \right\} \left[\frac{R_P}{r_0} \right]^{\frac{1}{2}}. \tag{21}$$

Here the symbols are defined as in Eq. (20). The resulting emission velocity, for a 3 nm silica grain, starting to accelerate at $8R_S$, is 60 km s^{-1}, significantly slower than for the Jovian stream grains.

After ejection into interplanetary space, the particles interact, via the Lorentz force, with the IMF. With suitable conditions—namely increases in strength of the IMF and IMF direction reversal, it is possible for the ejected stream particles to even reenter Saturn's magnetosphere (Hsu et al. 2013). If reentry does not occur, then the particles spread away from Saturn, interacting further with the IMF in interplanetary space, in the same way as Jovian stream particles.

As at Jupiter, the IMF at Saturn's orbital distance is dominated by its tangential component, with the E-field therefore pointing normal to the ecliptic, either northward or southward depending on IMF sector. Progression through high IMF strength, solar wind compression regions results in temporal-spatial compression of collections of dust grains, forming "bursts" in which the trajectories of the grains are significantly altered from their original line-of-sight trajectories from Saturn, with further acceleration (Hsu et al. 2010b, 2011a, 2012). Entry into low-IMF-strength, rarefaction regions results in no clear clustering of the grains (Hsu et al. 2010b, 2011a, 2012) which continue along line-of-sight trajectories from Saturn. The formation of temporal/spatial clusters of grains by interaction with the IMF gives rise to the illusion of encountering "streams" of grains, whereas a better analogy, especially considering the dynamical link to the solar wind, might be to consider the arrival of grains in gusty "squalls", as IMF conditions change.

7 Europa

Europa (Fig. 15), the smallest of the Galilean moons of Jupiter (Table 1), interacts with Ganymede and Io as part of the Laplace Resonance. The heating power generated within the moon by internal tidal dissipation arising from its forced eccentricity is estimated to be $< 1 \times 10^{13}$ W (Schubert et al. 2010), with 1.52–2.54×10^{11} W provided through the decay of long-lived radionuclides (Hussmann et al. 2010). As with Enceladus, it is likely that serpentinisation reactions occur, although owing to the higher seafloor pressure, only within the upper 25 km of a silicate-rich mantle, producing a heat power of 7.6–12.7×10^7 W (based on Vance et al. 2007's estimates, using radionuclide heating rates from Hussmann et al. 2010).

Fig. 15 Reprocessed (to "realistic" colours) mosaic of images of Europa acquired by the Galileo Solid-State Imaging (SSI) experiment on the spacecraft's first and fourteenth orbits through the Jupiter system, in 1995 and 1998, respectively. Image scale is 1 mile (1.6 kilometres) per pixel. North on Europa is at right. NASA/JPL-Caltech/SETI Institute (PIA19048)

The presence of a subsurface ocean on Europa, indicated by surface geological features (including general geological features, cycloidal cracks etc. (e.g. Carr et al. 1998; Greenberg 2010) and a paucity of craters (Squyres et al. 1983)), magnetospheric interaction due to an induced magnetic field (Khurana et al. 1998; Kivelson 2000) and signs of water vapour density enhancements several kilometres above its surface (Saur et al. 2011; Roth et al. 2014a,b; Sparks et al. 2016), is no longer controversial, although, as with Io and Enceladus, the exact internal structure and composition of Europa is still poorly constrained.

Gravity field data and moments of inertia have been used to generate interior models (e.g. Schubert et al. 1981, 2004), although these are not unique. Europa's moment of inertia of 0.346 ± 0.005 (Schubert et al. 2004) indicates central density enhancement within the moon. The mean density of Europa ($2990 \pm 5\ \mathrm{kg\,m^{-3}}$, Greenberg 2010) also indicates a significant higher density component (silicates and Fe/FeS) to be present within, given the apparent water-ice dominated surface. The exact size of the core (the lack of an intrinsic magnetic field indicates this is probably solid (Kivelson et al. 1979; Khurana et al. 1998; Kivelson 2000)) is dependent on its composition, with FeS-rich cores larger than those of purer metallic Fe (Schubert et al. 2004). Core radii estimates are therefore 280–700 km. The core is believed to be surrounded by a silicate mantle, with again, a compositionally-dependent thickness. With an overlying watery ocean, it is also possible that the top 20–30 km (Sohl et al. 2010) of the mantle is relatively porous and will have undergone some degree of serpentinisation (Sohl et al. 2010), although unlike Enceladus (Sect. 4), it is unknown whether this process still continues. The thicknesses of the ocean and icy crust are also poorly constrained at the present time. Total H_2O thicknesses are estimated to be between 80–170 km (Anderson et al. 1998; Schubert et al. 2009). There are two main schools of thought regarding the state of the ocean and crust: a thin-crust/thick-ocean model (Greenberg et al. 2000; Greenberg 2002) and a thicker crust/thinner ocean model (e.g. Pappalardo et al. 1998; Nimmo et al. 2003). In the former, heat transfer through a $\leq\sim 10$ km thick ice shell is primarily conductive, whereas in the latter, convective heat transfer through the lower, viscoelastic regions of a thicker (~ 25–40 km) ice shell also occurs. This process may also be capable of bringing liquid water closer to the surface via e.g. diapirism (Nimmo and Manga 2002). Typically ocean thicknesses of ~ 150 km (e.g. Sohl et al. 2002) are assumed.

The measured induced magnetic field (Khurana et al. 1998) implies that the subsurface ocean contains a significant proportion of dissolved salts, with a $MgSO_4$-dominated estimated brine composition theorised based on water interaction with chondritic material (Sohl

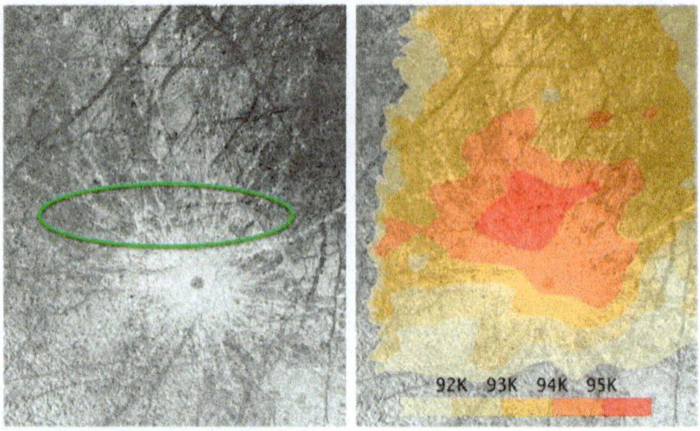

Fig. 16 The region from which the Europa plume seen by Sparks et al. (2016) was estimated to emerge (left), with Galileo thermal imaging showing a coincident warm region shown on the right. Credit: NASA/ESA/W. Sparks (STScI)/USGS Astrogeology Science Center (PIA21444)

et al. 2010). The thinner the ocean, the higher the salt concentration required to generate the observed field.

The previously mentioned indirect water vapour observations provide evidence for cryovolcanism (Saur et al. 2011; Roth et al. 2014b; Sparks et al. 2016, 2017). Saur et al. (2011) detected 130.4 nm and 135.6 nm UV emission surpluses (explained by the dissociation and ionisation of water vapour) at 90°W Europan longitude. Roth et al. (2014b) modelled the hydrogen Lyman-α and oxygen OI 130.4 nm UV line emission observed in 2012 as compatible with two large-scale 200 km high plumes of $\sim 10^{20}$ H_2O m^{-2} column density water vapour, with a spreading angle of 10°, from which they derived a supersonic emission velocity of $\sim 700\ m\,s^{-1}$, requiring $7000\ kg\,s^{-1}$ of emission. Pixel sizes introduced uncertainties of approx. 100 km. The observations of water vapour above the south polar region of Europa, were found to coincide with Europa's orbital apocentre, from which a tidal link was inferred. However, subsequent repeat observations two years later (Roth et al. 2014a), timed to confirm this tidal link failed to do so, with modelling by Rhoden et al. (2015) later indicating that plume activity may instead be most likely at an orbital true anomaly of 120°.

Observations of absorption in the far-UV (FUV) (Sparks et al. 2016), which took place over a similar date range to Roth et al. (2014a)'s unsuccessful plume search, found indicators of plume activity at similar Europan latitudes (but different longitudes than those of Roth et al. 2014b), as well as at lower latitudes (16.5°S). The lower latitude feature was estimated to have a water column density of 1.8×10^{18} H_2O m^{-2} and extend approximately 50 km from the limb. However, as noted by Sparks et al. (2016), the positive FUV observations were nearly temporally coincident with the negative UV observations of Roth et al. (2014a), requiring cautious interpretation as to their reliability. Further Hubble Space Telescope observations by Sparks et al. (2017), again in the FUV, in 2016, found an absorption feature at almost the same low latitude (16.7°) as the 2014 observation, near the crater Pwyll, and coincident (Fig. 16) with enhanced brightness temperatures seen in Galileo Photopolarimeter-Radiometer images from 1998 (Spencer et al. 1999). Although Spencer et al. (1999) consider changes in local surface porosity (and hence thermal inertia) as a possible explanation for the observed thermal anomaly, another option they consider is that of an endogenic heat source, raising the local temperature to 95 K (compared with a base temperature of ≈ 93.5 K). Using these values, Sparks et al. (2017) derive a "hot spot" heat flux of $0.28\ W\,m^{-2}$, implying liquid water at a depth of 1.8–2 km (salinity dependent). Mechanisms for bringing liquid water this close to the surface include crustal thinning over geothermal plumes (O'Brien et al. 2002; Goodman et al. 2004; Lowell and DuBose 2005), diapir-induced melting and transport of liquid salt-rich water (e.g. Fig. 17) (e.g. Head and Pappalardo 1999), or potentially cracking of the ice shell, although this mechanism is not deemed straightforward and

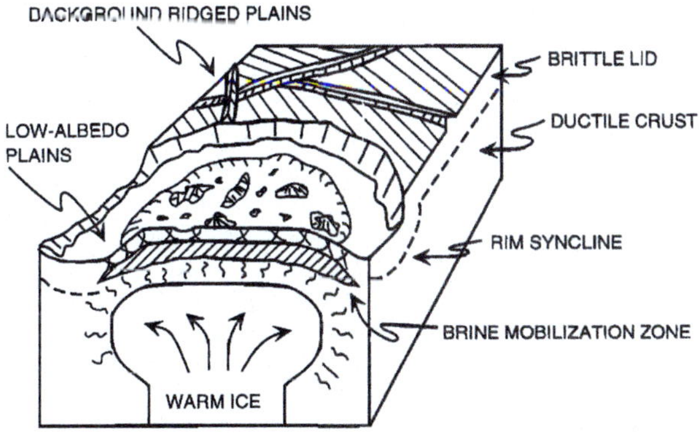

Fig. 17 Schematic showing the geological features on Europa potentially associated with an upwelling warm diapir, including brine mobilisation, from Head, J.W., Pappalardo, R.T., J. Geophys. Res. Planets 104 (E11), pp. 27143–27155, Head and Pappalardo (1999). © 1999 John Wiley and Sons, reprinted with permission

gas exsolvation may also be required (e.g. Crawford and Stevenson 1988). The presence of a hot spot in this region has also been discounted by ALMA observations (Trumbo et al. 2017).

Despite this, to date these observations (Roth et al. 2014b; Sparks et al. 2016, 2017) represent the best—albeit tentative—remote evidence for plume activity, via surface cracks, on Europa. A recent reexamination of Galileo magnetic field and plasma wave observations, originally taken in 1997, has retrospectively discovered that the spacecraft traversed a plume with properties similar to those derived from the remote observations (Jia et al. 2018), further strengthening the case for the existence of Europan plumes.

Ejection velocities for the water vapour in the plumes, based upon an assumed ballistic trajectory and heights of 100–200 km, range between 500–750 $m\,s^{-1}$ (Roth et al. 2014b; Sparks et al. 2017; Lorenz 2016; Berg et al. 2016). These values are similar to those found for the Enceladian plume (Sect. 4), but higher than predicted through a plume-based interpretation of other Europan surface features (Quick et al. 2013). Although the higher ejection velocities are still below Europa's escape velocity (2.025 $km\,s^{-1}$), the heights which could be reached by entrained grains are broadly similar to those reached by the Io plumes (Berg et al. 2016) (Sect. 4). Flight times of 1000 s (Lorenz 2016) through the local plasma environment may allow for charging and extraction of the very smallest grains, as for Io, although this is not yet proven.

Further plume modelling (Teolis et al. 2017a) using an Enceladus-like south polar plume emitting 500 $kg\,s^{-1}$ of water vapour at 1 $km\,s^{-1}$ produced plumes with appreciable gas density at altitudes of up to 1000 km, although particles were not considered. Again considering a south polar source, Southworth et al. (2015) modelled the role of dust grains within a putative plume. Plumes were considered unlikely to directly feed into circumplanetary space, although nano-scale grains and plasma interaction were not discussed in detail. For guaranteed sampling of particles with radii larger than 0.3 μm a flyby distance of 5–100 km was thus recommended, based on a solid grain plume mass flux of 5 $kg\,s^{-1}$.

Without further work on the likelihood of nanoparticle emission from Europa's Hill sphere, it is presently difficult to say whether Europa truly meets our earlier definition of active cosmic dust emission. Given the likely similarities between the Europan and Enceladian oceans, in terms of brininess and the presence of hydrothermal interaction, it is possible that similar inclusions of silica may arise in ejected grains, potentially surviving sputtering in the Jovian magnetosphere. However, in the absence of further proof, it may therefore be that Europa falls into the same category as Triton, that of a dust emitting, but not *cosmic*-dust emitting body.

Table 2 Summary of the lifecycles of cosmic dust emitted from active moons

Process	Cold (e.g. Enceladus)	Hot (e.g. Io)
Heat source	Tidal heating and minor chemical/radionuclide heating	Tidal heating and minor radionuclide heating
Dust formation	Two mechanisms:	Two mechanisms:
	– Gas exsolvation-driven aerosolisation: large salt/organic/silica-rich grains	– Gas exsolvation-driven aerosolisation: large silicate-rich grains
	– Vapour condensation: Small water/water and organics grains, including nanograins	– Vapour condensation: Small salt and volatile-rich grains, including nanograins
Dust ejection	Lofting in vapour/gas plume, with compositional stratification. Majority of large grains (salt-rich) deposited on moon's surface	Lofting in vapour/gas plume, with compositional stratification. Majority of large grains (silicates) deposited on moon's surface
Circumplanetary behaviour	Initial entry into plasma torus, nanograins possibly ejected quickly, larger grains form dust ring, plasma drag increases semi-major axes, sputtering leads to corotating E field interaction and ejection of fraction as stream grains	Initial entry into plasma torus, rapid charge changes after interaction with warmer plasma region leads to corotating E field interaction and ejection of nanograins as stream particles
Interplanetary behaviour	Interaction with CR regions of IMF leads to acceleration, bunching and "squall" formation	Interaction with CR regions of IMF leads to acceleration, bunching and "squall" formation

8 Conclusions

Two main conclusions can be drawn from this review of dust emission by active moons. The first are the, perhaps surprising,[20] similarities between the generation, emission and evolution of cosmic dust generated by Io and Enceladus, despite the significant differences in the temperatures of the volcanism involved (Table 2). Io's high temperature silicate volcanism (Sect. 3.1), powered by intense tidal heating, generates particles of a variety of sizes and compositions via both aerosolisation and condensation processes (Sect. 4.1). A limited sample of these grains, the smallest grains present at the highest altitudes, experience charging conditions which lead to their interaction with the rotating Jovian magnetosphere, extraction from Io's Hill sphere into a plasma torus (Sects. 4.1 and 5.1), and eventual ejection from the Jovian magnetosphere (Sect. 6.1). In interplanetary space, the tiny sub-micron grains interact with the IMF, generating "squalls" of temporally and regionally clustered particles, interpreted as dust streams (Sect. 6.1) and which have velocities high enough to escape the Solar System, potentially contributing to the population of interstellar grains.

Enceladus' low-temperature water cryovolcanism, also powered by tidal heating, generates particles with a range of sizes and compositions (Sect. 4), of which a limited sample (again, the smaller grains) are injected into circumkronian space, Saturn's E ring, (Sect. 5) and gradually evolve until, at sub-micron sizes, they too interact with the planetary mag-

[20] Although partially predicted by Pang et al. (1984) and Horányi (2000).

netic field and are rapidly accelerated out of the Saturnian magnetosphere, subsequently interacting with the IMF and, as with the Ionian grains, forming "squalls" (Sect. 6).

In both cases, the compositions of the grains reflect their formation conditions and subsequent evolution, providing insights into the interiors of their source bodies which would otherwise be unavailable.

This leads to the second conclusion which can be drawn: that spacecraft-based dust detectors, particularly those with the capability of in situ compositional analysis, provide an unforeseen yet critically important opportunity for investigating the interiors of active moons. Namely: Cassini CDA measurements of the sizes and compositions of Iogenic dust grains confirmed predictions of the composition of Io's largest volcanic plumes (e.g. Fegley and Zolotov 2000), whilst CDA measurements of the composition and sizes of Enceladian dust have confirmed the presence of subsurface liquid water (Postberg et al. 2009b, 2011b, 2018) and its present day hydrothermal interaction with the rocky core (Hsu et al. 2015).

The detection of dust grains both within the gas giants' magnetospheres and in interplanetary space has provided a valuable probe into the plasma and magnetic environment of magnetospheres (e.g. Horányi et al. 1993a; Maravilla et al. 1995; Krüger et al. 1998; Hsu et al. 2011b) and the IMF (e.g. Grün et al. 1993; Horányi et al. 1993a; Zook et al. 1996; Hsu et al. 2010b, 2013).

As well as investigating the local dust environment generated by impact gardening around Europa (e.g. Sremčević et al. 2005; Krüger et al. 2003c; Miljković et al. 2012), providing compositional mapping of the surface (Postberg et al. 2011a; Kempf et al. 2012), spacecraft-based dust detectors may also be able to confirm *active* dust emission by Europa through plume fly-throughs (Southworth et al. 2015; Lorenz 2016; Teolis et al. 2017b). Dust detectors such as the SUrface Dust Analyzer, SUDA, onboard the upcoming Europa Clipper, or those on return missions to Enceladus, may therefore enable the internal state and composition of these moons to be further investigated, without the cost and complexity of landers, and complementing other methods (e.g. ion and neutral gas phase mass spectrometry, reflectance spectroscopy) of remote investigation. Despite the wealth of data returned by Galileo and Cassini, outstanding questions concerning dust formation and emission by Io and Enceladus still remain, including:

- Formation: Where in Io's plumes do the nanograins form? Does dust formation occur within brine-filled Tiger Stripes on Enceladus, or within subsurface chambers above an effervescing ocean? Are there plume-to-plume or jet-to-jet size or composition variations in dust from Io or Enceladus?
- Dust size and composition: Do Io nanograins act as condensation nuclei for more volatile (e.g. SO_2) compounds during as they rise in the volcanic plumes? If so, where is this mantle removed? What are the compositions of the high mass organics detected in Enceladian grains? What proportion of grains are salt-rich or organic-rich?

It is possible that other moons, such as Triton, are active and perhaps also emitting dust into circumplanetary space. Future missions to the outer planets may be able to confirm this and perhaps discover new active, cosmic dust emitting, moons. The investigation of cosmic dust emitted from these objects may then provide information about their sources that would otherwise be impossible to determine, as happened for Io and Enceladus.

Acknowledgements The authors thank H. Krüger and an anonymous referee for their helpful and constructive comments, which have improved the manuscript. JH acknowledges funding from Universität Heidelberg. The authors thank ISSI and Andrea Fischer for funding and arranging copyright permissions for many of the figures within this work.

References

O. Abramov, J.R. Spencer, Endogenic heat from Enceladus' south polar fractures: new observations, and models of conductive surface heating. Icarus **199**(1), 189–196 (2009). https://doi.org/10.1016/j.icarus.2008.07.016

J.D. Anderson, G. Schubert, Saturn's gravitational field, internal rotation, and interior structure. Science **317**(5843), 1384–1387 (2007). https://doi.org/10.1126/science.1144835

J.D. Anderson, G. Schubert, R.A. Jacobson, E.L. Lau, W.B. Moore, W.L. Sjogren, Europa's differentiated internal structure: inferences from four Galileo encounters. Science **281**(5385), 2019–2022 (1998). https://doi.org/10.1126/science.281.5385.2019. http://www.sciencemag.org/cgi/doi/10.1126/science.276.5316.1236

J.D. Anderson, R.A. Jacobson, E.L. Lau, W.B. Moore, G. Schubert, Io's gravity field and interior structure. J. Geophys. Res., Planets **106**(E12), 32963–32969 (2001). https://doi.org/10.1029/2000JE001367

R.M. Baland, M. Yseboodt, T.V. Hoolst, Obliquity of the Galilean satellites: the influence of a global internal liquid layer. Icarus **220**(2), 435–448 (2012). https://doi.org/10.1016/j.icarus.2012.05.020

R.M. Baland, M. Yseboodt, T. Van Hoolst, The obliquity of Enceladus. Icarus **268**, 12–31 (2016). https://doi.org/10.1016/j.icarus.2015.11.039. arXiv:1512.0285

D.D. Barbosa, F.V. Coroniti, A. Eviatar, Coulomb thermal properties and stability of the Io plasma torus. Astrophys. J. **274**, 429–442 (1983). https://doi.org/10.1086/161459

M. Běhounková, O. Souček, J. Hron, O. Čadek, Plume activity and tidal deformation on Enceladus influenced by faults and variable ice shell thickness. Astrobiology **17**(9), 941–954 (2017). https://doi.org/10.1089/ast.2016.1629

J.J. Berg, D.B. Goldstein, P.L. Varghese, L.M. Trafton, DSMC simulation of Europa water vapor plumes. Icarus **277**, 370–380 (2016). https://doi.org/10.1016/j.icarus.2016.05.030

C.J. Bierson, F. Nimmo, A test for Io's magma ocean: modeling tidal dissipation with a partially molten mantle. J. Geophys. Res., Planets **121**(11), 2211–2224 (2016). https://doi.org/10.1002/2016JE005005

M.T. Bland, W.B. McKinnon, P.M. Schenk, Constraining the heat flux between Enceladus' tiger stripes: numerical modeling of funiscular plains formation. Icarus **260**, 232–245 (2015). https://doi.org/10.1016/j.icarus.2015.07.016

N.V. Brilliantov, J. Schmidt, F. Spahn, Nucleation and growth of a solid phase expanding into vacuum. Int. J. Mod. Phys. C **18**(04), 676–684 (2007). https://doi.org/10.1142/S0129183107010930

R.H. Brown, R.L. Kirk, T.V. Johnson, L.A. Soderblom, Energy sources for Triton's geyser-like plumes. Science **250**(4979), 431–435 (1990). https://doi.org/10.1126/science.250.4979.431

C. Bu, D.A. Bahr, C.A. Dukes, R.A. Baragiola, The effects of cracking on the surface potential of icy grains in Saturn's E-ring: laboratory studies. Astrophys. J. **825**(2), 106 (2016). https://doi.org/10.3847/0004-637X/825/2/106

O. Čadek, G. Tobie, T. Van Hoolst, M. Massé, G. Choblet, A. Lefèvre, G. Mitri, R.M. Baland, M. Běhounková, O. Bourgeois, A. Trinh, Enceladus's internal ocean and ice shell constrained from Cassini gravity, shape, and libration data. Geophys. Res. Lett. **43**(11), 5653–5660 (2016). https://doi.org/10.1002/2016GL068634

M.H. Carr, M.J. Belton, C.R. Chapman, M.E. Davies, P. Geissler, R. Greenberg, A.S. McEwen, B.R. Tufts, R. Greeley, R. Sullivan, J.W. Head, R.T. Pappalardo, K.P. Klaasen, T.V. Johnson, J. Kaufman, D. Senske, J. Moore, G. Neukum, G. Schubert, J.A. Burns, P. Thomas, J. Veverka, Evidence for a subsurface ocean on Europa. Nature **391**, 363–365 (1998). https://doi.org/10.1038/34857

J.C. Castillo-Rogez, T.B. McCord, Ceres' evolution and present state constrained by shape data. Icarus **205**(2), 443–459 (2010). https://doi.org/10.1016/j.icarus.2009.04.008

J. Castillo-Rogez, T.V. Johnson, M.H. Lee, N.J. Turner, D.L. Matson, J. Lunine, ^{26}Al decay: heat production and a revised age for Iapetus. Icarus **204**(2), 658–662 (2009). https://doi.org/10.1016/j.icarus.2009.07.025

E. Cataldo, L. Wilson, S. Lane, J. Gilbert, A model for large-scale volcanic plumes on Io: implications for eruption rates and interactions between magmas and near-surface volatiles. J. Geophys. Res., Planets **107**(E11), 19 (2002). https://doi.org/10.1029/2001JE001513

G. Choblet, G. Tobie, C. Sotin, M. Běhounková, O. Čadek, F. Postberg, O. Souček, Powering prolonged hydrothermal activity inside Enceladus. Nat. Astron. **1**(12), 841–847 (2017). https://doi.org/10.1038/s41550-017-0289-8

S.A. Collins, Spatial color variations in the volcanic plume at Loki, on Io. J. Geophys. Res. Space Phys. **86**(A10), 8621–8626 (1981). https://doi.org/10.1029/JA086iA10p08621

G.C. Collins, J.C. Goodman, Enceladus' south polar sea. Icarus **189**(1), 72–82 (2007). https://doi.org/10.1016/j.icarus.2007.01.010

G.J. Consolmagno, Io: thermal models and chemical evolution. Icarus **47**, 36–45 (1981)
G.D. Crawford, D.J. Stevenson, Gas-driven water volcanism and the resurfacing of Europa. Icarus **73**(1), 66–79 (1988). https://doi.org/10.1016/0019-1035(88)90085-1
E.N. Crow-Willard, R.T. Pappalardo, Structural mapping of Enceladus and implications for formation of tectonized regions. J. Geophys. Res., Planets **120**(5), 928–950 (2015). https://doi.org/10.1002/2015JE004818
A.G. Davies, *Volcanism on Io—A Comparison with Earth* (Cambridge University Press, Cambridge, 2007)
A.G. Davies, L.P. Keszthelyi, D.A. Williams, C.B. Phillips, A.S. McEwen, R.M.C. Lopes, W.D. Smythe, L.W. Kamp, L.A. Soderblom, R.W. Carlson, Thermal signature, eruption style, and eruption evolution at Pele and Pillan on Io. J. Geophys. Res., Planets **106**(E12), 33079–33103 (2001). https://doi.org/10.1029/2000JE001357
G. De Maria, G. Balducci, M. Guido, V. Piacente, Mass spectrometric investigation of the vaporization process of Apollo 12 lunar samples, in *Proc. Lunar Planet. Sci. Conf.* 2 (MIT Press, Cambridge, 1971), pp. 1367–1380
I. de Pater, C. Laver, A.G. Davies, K. de Kleer, D.A. Williams, R.R. Howell, J.A. Rathbun, J.R. Spencer, Io: eruptions at Pillan, and the time evolution of Pele and Pillan from 1996 to 2015. Icarus **264(August**, 198–212 (2016). https://doi.org/10.1016/j.icarus.2015.09.006. 2007
S.F. Dermott, On the origin of commensurabilities in the Solar System—I. The tidal hypothesis. Mon. Not. R. Astron. Soc. **141**(3), 349–361 (1968). https://doi.org/10.1093/mnras/141.3.349
D. Dhingra, M.M. Hedman, R.N. Clark, P.D. Nicholson, Spatially resolved near infrared observations of Enceladus' tiger stripe eruptions from Cassini VIMS. Icarus **292**, 1–12 (2017). https://doi.org/10.1016/j.icarus.2017.03.002
V. Dikarev, Dynamics of particles in Saturn's E ring: effects of charge variations and the plasma drag force. Astron. Astrophys. **346**, 1011–1019 (1999)
Y. Dong, T.W. Hill, B.D. Teolis, B.A. Magee, J.H. Waite, The water vapor plumes of Enceladus. J. Geophys. Res. Space Phys. **116**(10), 1–13 (2011). https://doi.org/10.1029/2011JA016693
Y. Dong, T.W. Hill, S.Y. Ye, Characteristics of ice grains in the Enceladus plume from Cassini observations. J. Geophys. Res. Space Phys. **120**(2), 915–937 (2015). https://doi.org/10.1002/2014JA020288
M.K. Dougherty, K.K. Khurana, F.M. Neubauer, C.T. Russell, J. Saur, J.S. Leisner, M.E. Burton, Identification of a dynamic atmosphere at Enceladus with the Cassini magnetometer. Science **311**, 1406–1409 (2006). https://doi.org/10.1126/science.1120985
G. Dourneau, Orbital elements of the eight major satellites of Saturn determined from a fit of their theories of motion to observations from the 1886 to 1985. Astron. Astrophys. **267**, 292–299 (1993)
N. Duxbury, R. Brown, The role of an internal heat source for the eruptive plumes on Triton. Icarus **125**(1), 83–93 (1997). https://doi.org/10.1006/icar.1996.5554
I.A.D. Engelhardt, J.E. Wahlund, D.J. Andrews, A.I. Eriksson, S. Ye, W.S. Kurth, D.A. Gurnett, M.W. Morooka, W.M. Farrell, M.K. Dougherty, Plasma regions, charged dust and field-aligned currents near Enceladus. Planet. Space Sci. **117**, 453–469 (2015). https://doi.org/10.1016/j.pss.2015.09.010
B. Fegley, M.Y. Zolotov, Chemistry of sodium, potassium, and chlorine in volcanic gases on Io. Icarus **148**, 193–210 (2000). https://doi.org/10.1006/icar.2000.6490
W.A. Feibelman, Concerning the "D" ring of Saturn. Nature **214**, 793 (1967). https://doi.org/10.1038/214793a0
A. Flandes, Dust escape from Io. Geophys. Res. Lett. **31**(16), L16802 (2004a). https://doi.org/10.1029/2004GL020046
A. Flandes, Dust escape mechanism from Io. Adv. Space Res. **34**(11), 2251–2255 (2004b). https://doi.org/10.1016/j.asr.2003.09.064
A. Flandes, H. Krüger, D.P. Hamilton, J.F. Valdés-Galicia, L. Spilker, R. Caballero, Magnetic field modulated dust streams from Jupiter in interplanetary space. Planet. Space Sci. **59**(13), 1455–1471 (2011). https://doi.org/10.1016/j.pss.2011.05.014
L.A. Frank, W.R. Paterson, K.L. Ackerson, V.M. Vasyliunas, F.V. Coroniti, S.J. Bolton, Plasma observations at Io with the Galileo spacecraft. Science **274**(5286), 394–395 (1996). https://doi.org/10.1126/science.274.5286.394
J. Fuller, J. Luan, E. Quataert, Resonance locking as the source of rapid tidal migration in the Jupiter and Saturn moon systems. Mon. Not. R. Astron. Soc. **458**(4), 3867–3879 (2016). https://doi.org/10.1093/mnras/stw609. arXiv:1601.05804
P. Gao, P. Kopparla, X. Zhang, A.P. Ingersoll, Aggregate particles in the plumes of Enceladus. Icarus **264**, 227–238 (2016). https://doi.org/10.1016/j.icarus.2015.09.030. arXiv:1506.00713
P.E. Geissler, M.T. McMillan, Galileo observations of volcanic plumes on Io. Icarus **197**(2), 505–518 (2008). https://doi.org/10.1016/j.icarus.2008.05.005

P.E. Geissler, A.S. McEwen, L. Keszthelyi, R. Lopes-Gautier, J. Granahan, D.P. Simonelli, Global color variations on Io. Icarus **140**, 265–282 (1999). https://doi.org/10.1006/icar.1999.6128

C.R. Glein, E.L. Shock, Sodium chloride as a geophysical probe of a subsurface ocean on Enceladus. Geophys. Res. Lett. **37**(9), L09204 (2010). https://doi.org/10.1029/2010GL042446

C.R. Glein, J.A. Baross, J.H. Waite, The pH of Enceladus' ocean. Geochim. Cosmochim. Acta **162**, 202–219 (2015). https://doi.org/10.1016/j.gca.2015.04.017. arXiv:1502.01946

J.D. Goguen, B.J. Buratti, R.H. Brown, R.N. Clark, P.D. Nicholson, M.M. Hedman, R.R. Howell, C. Sotin, D.P. Cruikshank, K.H. Baines, K.J. Lawrence, J.R. Spencer, D.G. Blackburn, The temperature and width of an active fissure on Enceladus measured with Cassini VIMS during the 14 April 2012 South Pole flyover. Icarus **226**(1), 1128–1137 (2013). https://doi.org/10.1016/j.icarus.2013.07.012

P. Goldreich, S. Soter, *Q* in the solar system. Icarus **5**, 375–389 (1966). https://doi.org/10.1016/0019-1035(66)90051-0

J.C. Goodman, G.C. Collins, J. Marshall, R.T. Pierrehumbert, Hydrothermal plume dynamics on Europa: implications for chaos formation. J. Geophys. Res. **109**, E03008 (2004). https://doi.org/10.1029/2003JE002073

A. Graps, E. Grun, H. Svedhem, H. Kruger, M. Horanyl, S. Heck a Lammers, Io as a source of the Jovian dust streams. Nature **405**(6782), 48–50 (2000). https://doi.org/10.1038/35011008

A.L. Graps, G.H. Jones, A. Juhasz, M. Horanyi, O. Havnes, The charging of planetary rings. Space Sci. Rev. **137**(1–4), 435–453 (2008). https://doi.org/10.1007/s11214-008-9406-4

R. Greenberg, Tidal-tectonic processes and their implications for the character of Europa's icy crust. Rev. Geophys. **40**(2), 1–34 (2002). https://doi.org/10.1029/2000RG000096

R. Greenberg, The icy Jovian satellites after the Galileo mission. Rep. Prog. Phys. **73**, 6801 (2010). https://doi.org/10.1088/0034-4885/73/3/036801

R. Greenberg, P. Geissler, G. Hoppa, B.R. Tufts, D.D. Durda, R. Pappalardo, J.W. Head, R. Greeley, R. Sullivan, M.H. Carr, Tectonic processes on Europa: tidal stresses, mechanical response, and visible features. Icarus **135**(1), 64–78 (1998). https://doi.org/10.1006/icar.1998.5986

R. Greenberg, P. Geissler, B.R. Tufts, G.V. Hoppa, Habitability of Europa's crust: the role of tidal-tectonic processes. J. Geophys. Res., Planets **105**(E7), 17551–17562 (2000). https://doi.org/10.1029/1999JE001147

E. Grün, H.A. Zook, M. Baguhl, H. Fechtig, M.S. Hanner, J. Kissel, B.A. Lindblad, D. Linkert, G. Linkert, I.B. Mann, Ulysses dust measurements near Jupiter. Science **257**, 1550–1552 (1992). https://doi.org/10.1126/science.11538054

E. Grün, H. Zook, M. Baguhl, A. Balogh, Discovery of Jovian dust streams and interstellar grains by the Ulysses spacecraft. Nature **362**, 428–430 (1993). https://doi.org/10.1038/362428a0

E. Grün, D.P. Hamilton, M. Baguhl, R. Riemann, M. Horanyi, C. Polanskey, Dust streams from comet Shoemaker-Levy 9? Geophys. Res. Lett. **21**(11), 1035–1038 (1994). https://doi.org/10.1029/94GL00701

E. Grün, M. Baguhl, N. Divine, H. Fechtig, D. Hamilton, M. Hanner, J. Kissel, B.A. Lindblad, D. Linkert, G. Linkert, I. Mann, J. McDonnell, G. Morfill, C. Polanskey, R. Riemann, G. Schwehm, N. Siddique, P. Staubach, H. Zook, Two years of Ulysses dust data. Planet. Space Sci. **43**(8), 971–999 (1995). https://doi.org/10.1016/0032-0633(94)00233-H

E. Grün, M. Baguhl, D.P. Hamilton, R. Riemann, H.A. Zook, Constraints from Galileo observations on the origin of Jovian dust streams. Nature **381**(6581), 395–398 (1996a). https://doi.org/10.1038/381395a0

E. Grün, D. Hamilton, R. Riemann, S. Dermott, H. Fechtig, B. Gustafson, M. Hanner, A. Heck, M. Horanyi, J. Kissel, H. Kruger, B. Lindblad, D. Linkert, G. Linkert, I. Mann, J. McDonnell, G. Morfill, C. Polanskey, G. Schwehm, R. Srama, H. Zook, Dust measurements during Galileo's approach to Jupiter and Io encounter. Science **274**, 399–401 (1996b). https://doi.org/10.1126/science.274.5286.399

E. Grün, H. Krüger, A.L. Graps, D.P. Hamilton, A. Heck, G. Linkert, H.A. Zook, S. Dermott, H. Fechtig, B.A. Gustafson, M.S. Hanner, M. Horányi, J. Kissel, B.A. Lindblad, D. Linkert, I. Mann, J.A.M. McDonnell, G.E. Morfill, C. Polanskey, G. Schwehm, R. Srama, Galileo observes electromagnetically coupled dust in the Jovian magnetosphere. J. Geophys. Res., Planets **103**(E9), 20011–20022 (1998). https://doi.org/10.1029/98JE00228

D.P. Hamilton, Motion of dust in a planetary magnetosphere: orbit-averaged equations for oblateness, electromagnetic, and radiation forces with application to Saturn's E-ring. Icarus **101**, 244–264 (1993). https://doi.org/10.1006/icar.1993.1065

D.P. Hamilton, J.A. Burns, Ejection of dust from Jupiter's gossamer ring. Nature **364**(6439), 695–699 (1993). https://doi.org/10.1038/364695a0

C.W. Hamilton, C.D. Beggan, S. Still, M. Beuthe, R.M.C. Lopes, D.A. Williams, J. Radebaugh, W. Wright, Spatial distribution of volcanoes on Io: implications for tidal heating and magma ascent. Earth Planet. Sci. Lett. **361**, 272–286 (2013). https://doi.org/10.1016/j.epsl.2012.10.032

C.J. Hansen, L.W. Esposito, A.I.F. Stewart, J.E. Colwell, A.R. Hendrix, W. Pryor, D.E. Shemansky, R. West, Enceladus' water vapor plume. Science **311**(5766), 1422–1425 (2006). https://doi.org/10.1126/science.1121254

C.J. Hansen, L.W. Esposito, A.I.F. Stewart, B. Meinke, B. Wallis, J.E. Colwell, A.R. Hendrix, K. Larsen, W. Pryor, F. Tian, Water vapour jets inside the plume of gas leaving Enceladus. Nature **456**(7221), 477–479 (2008). https://doi.org/10.1038/nature07542

C.J. Hansen, D.E. Shemansky, L.W. Esposito, A.I.F. Stewart, B.R. Lewis, J.E. Colwell, A.R. Hendrix, R.A. West, J.H. Waite, B. Teolis, B.A. Magee, The composition and structure of the Enceladus plume. Geophys. Res. Lett. **38**(11), 1–5 (2011). https://doi.org/10.1029/2011GL047415

C.J. Hansen, L.W. Esposito, K.M. Aye, J.E. Colwell, A.R. Hendrix, G. Portyankina, D. Shemansky, Investigation of diurnal variability of water vapor in Enceladus' plume by the Cassini ultraviolet imaging spectrograph. Geophys. Res. Lett. **44**(2), 672–677 (2017). https://doi.org/10.1002/2016GL071853

J.W. Head, R.T. Pappalardo, Brine mobilization during lithospheric heating on Europa: implications for formation of chaos terrain, lenticula texture, and color variations. J. Geophys. Res., Planets **104**(E11), 27143–27155 (1999). https://doi.org/10.1029/1999JE001062

M.M. Hedman, P.D. Nicholson, M.R. Showalter, R.H. Brown, B.J. Buratti, R.N. Clark, Spectral observations of the Enceladus plume with Cassini-VIMS. Astrophys. J. **693**(2), 1749–1762 (2009). https://doi.org/10.1088/0004-637X/693/2/1749

M.M. Hedman, J.A. Burns, D.P. Hamilton, M.R. Showalter, The three-dimensional structure of Saturn's E ring. Icarus **217**(1), 322–338 (2012). https://doi.org/10.1016/j.icarus.2011.11.006, arXiv:1111.2568

M.M. Hedman, C.M. Gosmeyer, P.D. Nicholson, C. Sotin, R.H. Brown, R.N. Clark, K.H. Baines, B.J. Buratti, M.R. Showalter, An observed correlation between plume activity and tidal stresses on Enceladus. Nature **500**(7461), 182–184 (2013). https://doi.org/10.1038/nature12371

M. Hedman, D. Dhingra, P. Nicholson, C. Hansen, G. Portyankina, S. Ye, Y. Dong, Spatial variations in the dust-to-gas ratio of Enceladus' plume. Icarus **305**, 123–138 (2018). https://doi.org/10.1016/j.icarus.2018.01.006, arXiv:1801.01567

P. Helfenstein, C.C. Porco, Enceladus' geysers: relation to geological features. Astron. J. **150**(3), 96 (2015). https://doi.org/10.1088/0004-6256/150/3/96

T.W. Hill, M.F. Thomsen, R.L. Tokar, A.J. Coates, G.R. Lewis, D.T. Young, F.J. Crary, R.A. Baragiola, R.E. Johnson, Y. Dong, R.J. Wilson, G.H. Jones, J.E. Wahlund, D.G. Mitchell, M. Horanyi, Charged nanograins in the Enceladus plume. J. Geophys. Res. Space Phys. **117**(5), 1–11 (2012). https://doi.org/10.1029/2011JA017218

J.K. Hillier, S.F. Green, N. McBride, J.P. Schwanethal, F. Postberg, R. Srama, S. Kempf, G. Moragas-Klostermeyer, J.A.M. McDonnell, E. Grun, E. Grün, The composition of Saturn's E ring. Mon. Not. R. Astron. Soc. **377**(4), 1588–1596 (2007). https://doi.org/10.1111/j.1365-2966.2007.11710.x

R. Hodyss, C.D. Parkinson, P.V. Johnson, J.V. Stern, J.D. Goguen, Y.L. Yung, I. Kanik, Methanol on Enceladus. Geophys. Res. Lett. **36**(17), 17103 (2009). https://doi.org/10.1029/2009GL039336

M. Horányi, Charged Dust Dynamics in the Solar System. Annu. Rev. Astron. Astrophys. **34**, 383–418 (1996). https://doi.org/10.1146/annurev.astro.34.1.383

M. Horányi, Dust streams from Jupiter and Saturn. Phys. Plasmas **7**(10), 3847–3849 (2000). https://doi.org/10.1063/1.1288909

M. Horányi, G. Morfill, E. Grün, Mechanism for the acceleration and ejection of dust grains from Jupiter's magnetosphere. Nature **363**, 144–146 (1993a). https://doi.org/10.1038/363144a0

M. Horányi, G. Morfill, E. Grün, The dusty ballerina skirt of Jupiter. J. Geophys. Res. **98**, 245–251 (1993b)

M. Horányi, E. Grün, A. Heck, Modeling the Galileo dust measurements at Jupiter. Geophys. Res. Lett. **24**(17), 2175–2178 (1997). https://doi.org/10.1029/97GL01539

M. Horányi, T.W. Hartquist, O. Havnes, D.A. Mendis, G.E. Morfill, Dusty plasma effects in Saturn's magnetosphere. Rev. Geophys. **42**(4), RG4002 (2004). https://doi.org/10.1029/2004RG000151

M. Horányi, A. Juhasz, G.E. Morfill, Large-scale structure of Saturn's E-ring. Geophys. Res. Lett. **35**(4), 1–5 (2008). https://doi.org/10.1029/2007GL032726

R.R. Howell, R.M.C. Lopes, Morphology, temperature, and eruption dynamics at Pele. Icarus **213**(2), 593–607 (2011). https://doi.org/10.1016/j.icarus.2011.03.008

C.J.A. Howett, J.R. Spencer, J. Pearl, M. Segura, High heat flow from Enceladus' south polar region measured using 10–600 cm^{-1} Cassini/CIRS data. J. Geophys. Res., Planets **116**(3), 1–15 (2011). https://doi.org/10.1029/2010JE003718

H.W. Hsu, S. Kempf, C.M. Jackman, Observation of saturnian stream particles in the interplanetary space. Icarus **206**(2), 653–661 (2010a). https://doi.org/10.1016/j.icarus.2009.06.033

H.W. Hsu, S. Kempf, F. Postberg, R. Srama, C.M. Jackman, G. Moragas-Klostermeyer, S. Helfert, E. Grün, Interaction of the solar wind and stream particles, results from the Cassini dust detector. AIP Conf. Proc. **1216**, 510–513 (2010b). https://doi.org/10.1063/1.3395914

H.W. Hsu, S. Kempf, F. Postberg, M. Trieloff, M. Burton, M. Roy, G. Moragas-Klostermeyer, R. Srama, Cassini dust stream particle measurements during the first three orbits at Saturn. J. Geophys. Res. Space Phys. **116**(8), 1–10 (2011a). https://doi.org/10.1029/2010JA015959

H.W. Hsu, F. Postberg, S. Kempf, M. Trieloff, M. Burton, M. Roy, G. Moragas-Klostermeyer, R. Srama, Stream particles as the probe of the dust-plasma-magnetosphere interaction at Saturn. J. Geophys. Res. Space Phys. **116**(9), 1–23 (2011b). https://doi.org/10.1029/2011JA016488

H.W. Hsu, H. Krüger, F. Postberg, Dynamics, composition, and origin of Jovian and Saturnian dust-stream particles, in *Nanodust in the Solar System: Discoveries and Interpretations* (2012), pp. 77–117. https://doi.org/10.1007/978-3-642-27543-2_5

H.W. Hsu, K.C. Hansen, M. Horányi, S. Kempf, A. Mocker, G. Moragas-Klostermeyer, F. Postberg, R. Srama, B. Zieger, Probing IMF using nanodust measurements from inside Saturn's magnetosphere. Geophys. Res. Lett. **40**(12), 2902–2906 (2013). https://doi.org/10.1002/grl.50604

H.W. Hsu, F. Postberg, Y. Sekine, T. Shibuya, S. Kempf, M. Horányi, A. Juhász, N. Altobelli, K. Suzuki, Y. Masaki, T. Kuwatani, S. Tachibana, S.I. Sirono, G. Moragas-Klostermeyer, R. Srama, Ongoing hydrothermal activities within Enceladus. Nature **519**(7542), 207–210 (2015). https://doi.org/10.1038/nature14262

H.W. Hsu, S. Kempf, S.V. Badman, W.S. Kurth, F. Postberg, R. Srama, Interplanetary magnetic field structure at Saturn inferred from nanodust measurements during the 2013 aurora campaign. Icarus **263**, 10–16 (2016). https://doi.org/10.1016/j.icarus.2015.02.022

T.A. Hurford, P. Helfenstein, J.N. Spitale, Tidal control of jet eruptions on Enceladus as observed by Cassini ISS between 2005 and 2007. Icarus **220**, 896–903 (2012). https://doi.org/10.1016/j.icarus.2012.06.022

H. Hussmann, G. Choblet, V. Lainey, D.L. Matson, C. Sotin, G. Tobie, T. Van Hoolst, Implications of rotation, orbital states, energy sources, and heat transport for internal processes in icy satellites. Space Sci. Rev. **153**(1–4), 317–348 (2010). https://doi.org/10.1007/s11214-010-9636-0

L. Iess, D.J. Stevenson, M. Parisi, D. Hemingway, R.A. Jacobson, J.I. Lunine, F. Nimmo, J.W. Armstrong, S.W. Asmar, M. Ducci, P. Tortora, The gravity field and interior structure of Enceladus. Science **344**(6179), 78–80 (2014). https://doi.org/10.1126/science.1250551

A.P. Ingersoll, S.P. Ewald, Total particulate mass in Enceladus plumes and mass of Saturn's E ring inferred from Cassini ISS images. Icarus **216**(2), 492–506 (2011). https://doi.org/10.1016/j.icarus.2011.09.018

A.P. Ingersoll, S.P. Ewald, Decadal timescale variability of the Enceladus plumes inferred from Cassini images. Icarus **282**, 260–275 (2017). https://doi.org/10.1016/j.icarus.2016.09.018

A.P. Ingersoll, M. Nakajima, Controlled boiling on Enceladus. 2. Model of the liquid-filled cracks. Icarus **272**, 319–326 (2016). https://doi.org/10.1016/j.icarus.2015.12.040

A.P. Ingersoll, A.A. Pankine, Subsurface heat transfer on Enceladus: conditions under which melting occurs. Icarus **206**(2), 594–607 (2010). https://doi.org/10.1016/j.icarus.2009.09.015

W.H. Ip, C.K. Goertz, An interpretation of the dawn–dusk asymmetry of UV emission from the Io plasma torus. Nature **302**(5905), 232–233 (1983). https://doi.org/10.1038/302232a0

R. Jacobson, P. Antreasian, J. Bordi, K. Criddle, R. Ionasescu, J. Jones, R. Mackenzie, M. Meek, D. Parcher, F. Pelletier, W. Owen Jr., D. Roth, I. Roundhill, J. Stauch, The gravity field of the Saturnian system from satellite observations and spacecraft tracking data. Astron. J. **132**(6), 2520–2526 (2006). https://doi.org/10.1086/508812

R. Jaumann, K. Stephan, G.B. Hansen, R.N. Clark, B.J. Buratti, R.H. Brown, K.H. Baines, S.F. Newman, G. Bellucci, G. Filacchione, A. Coradini, D.P. Cruikshank, C.A. Griffith, C.A. Hibbitts, T.B. McCord, R.M. Nelson, P.D. Nicholson, C. Sotin, R. Wagner, Distribution of icy particles across Enceladus' surface as derived from Cassini-VIMS measurements. Icarus **193**(2), 407–419 (2008). https://doi.org/10.1016/j.icarus.2007.09.013

K.L. Jessup, J.R. Spencer, Characterizing Io's Pele, Tvashtar and Pillan plumes: lessons learned from Hubble. Icarus **218**(1), 378–405 (2012). https://doi.org/10.1016/j.icarus.2011.11.013

X. Jia, M.G. Kivelson, K.K. Khurana, W.S. Kurth, Evidence of a plume on Europa from Galileo magnetic and plasma wave signatures. Nat. Astron. **2**, 459–464 (2018). https://doi.org/10.1038/s41550-018-0450-z

T.V. Johnson, G. Morfill, E. Grün, Dust in Jupiter's magnetosphere: an Io source? Geophys. Res. Lett. **7**(5), 305–308 (1980). https://doi.org/10.1029/GL007i005p00305

G.H. Jones, C.S. Arridge, A.J. Coates, G.R. Lewis, S. Kanani, A. Wellbrock, D.T. Young, F.J. Crary, R.L. Tokar, R.J. Wilson, T.W. Hill, R.E. Johnson, D.G. Mitchell, J. Schmidt, S. Kempf, U. Beckmann, C.T. Russell, Y.D. Jia, M.K. Dougherty, J.H. Waite, B.A. Magee, Fine jet structure of electrically charged grains in Enceladus' plume. Geophys. Res. Lett. **36**(16), 1–6 (2009). https://doi.org/10.1029/2009GL038284

A. Juhász, M. Horányi, G.E. Morfill, Signatures of Enceladus in Saturn's E ring. Geophys. Res. Lett. **34**(9), 1–5 (2007). https://doi.org/10.1029/2006GL029120

S. Jurac, R.E. Johnson, J.D. Richardson, Saturn's E ring and production of the neutral torus. Icarus **149**(2), 384–396 (2001). https://doi.org/10.1006/icar.2000.6528

J.S. Kargel, Cryovolcanism on the icy satellites. Earth Moon Planets **67**(1–3), 101–113 (1994). https://doi.org/10.1007/BF00613296

S. Kempf, Interpretation of high rate dust measurements with the Cassini dust detector CDA. Planet. Space Sci. **56**(3–4), 378–385 (2008). https://doi.org/10.1016/j.pss.2007.11.022

S. Kempf, R. Srama, M. Horányi, M. Burton, S. Helfert, G. Moragas-Klostermeyer, M. Roy, E. Grün, High-velocity streams of dust originating from Saturn. Nature **433**(7023), 289–291 (2005a). https://doi.org/10.1038/nature03218

S. Kempf, R. Srama, F. Postberg, M. Burton, S.F. Green, S. Helfert, J.K. Hillier, N. McBride, J.A.M. McDonnell, G. Moragas-Klostermeyer, M. Roy, E. Grün, Composition of saturnian stream particles. Science **307**(5713), 1274–1276 (2005b). https://doi.org/10.1126/science.1106218

S. Kempf, U. Beckmann, R. Srama, M. Horanyi, S. Auer, E. Grün, The electrostatic potential of E ring particles. Planet. Space Sci. **54**(9–10), 999–1006 (2006). https://doi.org/10.1016/j.pss.2006.05.012

S. Kempf, U. Beckmann, G. Moragas-Klostermeyer, F. Postberg, R. Srama, T. Economou, J. Schmidt, F. Spahn, E. Grün, The E ring in the vicinity of Enceladus. I. Spatial distribution and properties of the ring particles. Icarus **193**(2), 420–437 (2008). https://doi.org/10.1016/j.icarus.2007.06.027

S. Kempf, U. Beckmann, J. Schmidt, How the Enceladus dust plume feeds Saturn's E ring. Icarus **206**(2), 446–457 (2010). https://doi.org/10.1016/j.icarus.2009.09.016

S. Kempf, R. Srama, E. Grün, A. Mocker, F. Postberg, J.K. Hillier, M. Horányi, Z. Sternovsky, B. Abel, A. Beinsen, R. Thissen, J. Schmidt, F. Spahn, N. Altobelli, Linear high resolution dust mass spectrometer for a mission to the Galilean satellites. Planet. Space Sci. **65**(1), 10–20 (2012). https://doi.org/10.1016/j.pss.2011.12.019

L. Keszthelyi, A. McEwen, Thermal models for basaltic volcanism on Io. Geophys. Res. Lett. **24**(20), 2463–2466 (1997). https://doi.org/10.1029/97GL01368

L. Keszthelyi, W.L. Jaeger, E.P. Turtle, M. Milazzo, J. Radebaugh, A post-Galileo view of Io's interior. Icarus **169**(1), 271–286 (2004). https://doi.org/10.1016/j.icarus.2004.01.005

L. Keszthelyi, W. Jaeger, M. Milazzo, J. Radebaugh, A.G. Davies, K.L. Mitchell, New estimates for Io eruption temperatures: implications for the interior. Icarus **192**(2), 491–502 (2007). https://doi.org/10.1016/j.icarus.2007.07.008

K.K. Khurana, M.G. Kivelson, D.J. Stevenson, G. Schubert, C.T. Russell, R.J. Walker, C. Polanskey, Induced magnetic fields as evidence for subsurface oceans in Europa and Callisto. Nature **395**(6704), 777–780 (1998). https://doi.org/10.1038/27394

K.K. Khurana, X. Jia, M.G. Kivelson, F. Nimmo, G. Schubert, C.T. Russell, Evidence of a global magma ocean in Io's interior. Science **332**(6034), 1186–1189 (2011). https://doi.org/10.1126/science.1201425

S.W. Kieffer, R. Lopes-Gautier, A. McEwen, W. Smythe, L. Keszthelyi, R.W. Carlson, Prometheus: Io's wandering plume. Science **288**(5469), 1204–1208 (2000). https://doi.org/10.1126/science.288.5469.1204

M.R. Kirchoff, W.B. McKinnon, P.M. Schenk, Global distribution of volcanic centers and mountains on Io: Control by asthenospheric heating and implications for mountain formation. Earth Planet. Sci. Lett. **301**(1–2), 22–30 (2011). https://doi.org/10.1016/j.epsl.2010.11.018

E.S. Kite, A.M. Rubin, Sustained eruptions on Enceladus explained by turbulent dissipation in tiger stripes. Proc. Natl. Acad. Sci. USA **113**(15), 3972–3975 (2016). https://doi.org/10.1073/pnas.1520507113

M.G. Kivelson, Galileo magnetometer measurements: a stronger case for a subsurface ocean at Europa. Science **289**(5483), 1340–1343 (2000). https://doi.org/10.1126/science.289.5483.1340

M.G. Kivelson, J.A. Slavin, D.J. Southwood, Magnetospheres of the Galilean satellites. Science **205**(4405), 491–493 (1979)

H. Krüger, E. Grün, Dust en-route to Jupiter and the Galilean satellites, in *COSPAR Colloquia Series*, vol. 15 (2002), pp. 144–159. https://doi.org/10.1016/S0964-2749(02)80336-5. arXiv:astro-ph/0205478

H. Krüger, E. Grün, A. Graps, S. Lammers, Observations of electromagnetically coupled dust in the Jovian magnetosphere. Astrophys. Space Sci. **264**(1–4), 247–256 (1998). https://doi.org/10.1023/A:1002474912055

H. Krüger, E. Grun, A. Heck, S. Lammers, Analysis of the sensor characteristics of the Galileo dust detector with collimated Jovian dust stream particles. Planet. Space Sci. **47**(8–9), 1015–1028 (1999). https://doi.org/10.1016/S0032-0633(99)00027-6

H. Krüger, P. Geissler, M. Horányi, A.L. Graps, S. Kempf, R. Srama, G. Moragas-Klostermeyer, R. Moissl, T.V. Johnson, E. Grün, Jovian dust streams: a monitor of Io's volcanic plume activity. Geophys. Res. Lett. **30**(21), 2101 (2003a). https://doi.org/10.1029/2003gl017827

H. Krüger, M. Horányi, E. Grün, Jovian dust streams: probes of the Io plasma torus. Geophys. Res. Lett. **30**(2), 10–13 (2003b). https://doi.org/10.1029/2002GL015920. arXiv:astro-ph/0211061

H. Krüger, A.V. Krivov, M. Sremčević, E. Grün, Impact-generated dust clouds surrounding the Galilean moons. Icarus **164**(1), 170–187 (2003c). https://doi.org/10.1016/S0019-1035(03)00127-1. arXiv:astro-ph/0304381

H. Krüger, N. Altobelli, B. Anweiler, S.F. Dermott, V. Dikarev, A.L. Graps, E. Grün, B.A. Gustafson, D.P. Hamilton, M.S. Hanner, M. Horányi, J. Kissel, M. Landgraf, B.A. Lindblad, D. Linkert, G. Linkert, I. Mann, J.A.M. McDonnell, G.E. Morfill, C. Polanskey, G. Schwehm, R. Srama, H.A. Zook, Five years of Ulysses dust data: 2000–2004. Planet. Space Sci. **54**(9–10), 932–956 (2006a). https://doi.org/10.1016/j.pss.2006.04.015

H. Krüger, A.L. Graps, D.P. Hamilton, A. Flandes, R.J. Forsyth, M. Horányi, E. Grün, Ulysses jovian latitude scan of high-velocity dust streams originating from the jovian system. Planet. Space Sci. **54**(9–10), 919–931 (2006b). https://doi.org/10.1016/j.pss.2006.05.010

H. Krüger, V. Dikarev, B. Anweiler, S.F. Dermott, A.L. Graps, E. Grün, B.A. Gustafson, D.P. Hamilton, M.S. Hanner, M. Horányi, J. Kissel, D. Linkert, G. Linkert, I. Mann, J.A.M. McDonnell, G.E. Morfill, C. Polanskey, G. Schwehm, R. Srama, Three years of Ulysses dust data: 2005 to 2007. Planet. Space Sci. **58**(7–8), 951–964 (2010). https://doi.org/10.1016/j.pss.2009.11.002. arXiv:astro-ph/9809133

L.A. Morabito, S.P. Synnott, P.N. Kupferman, S.A. Collins, Discovery of currently active extraterrestrial volcanism. Science **204**(4396), 972 (1979). https://doi.org/10.1126/science.204.4396.972

V. Lainey, J.E. Arlot, O. Karatekin, T. Van Hoolst, Strong tidal dissipation in Io and Jupiter from astrometric observations. Nature **459**(7249), 957–959 (2009). https://doi.org/10.1038/nature08108

V. Lainey, Ö. Karatekin, J. Desmars, S. Charnoz, J.-E. Arlot, N. Emelyanov, C. Le Poncin-Lafitte, S. Mathis, F. Remus, G. Tobie, J.-P. Zahn, Strong tidal dissipation in Saturn and constraints on Enceladus' thermal state from astrometry. Astrophys. J. **752**(1), 14–33 (2012). https://doi.org/10.1088/0004-637X/752/1/14

V. Lainey, R.A. Jacobson, R. Tajeddine, N.J. Cooper, C. Murray, V. Robert, G. Tobie, T. Guillot, S. Mathis, F. Remus, J. Desmars, J.-E. Arlot, J.-P. De Cuyper, V. Dehant, D. Pascu, W. Thuillot, C. Le Poncin-Lafitte, J.-P. Zahn, New constraints on Saturn's interior from Cassini astrometric data. Icarus **281**, 286–296 (2017). https://doi.org/10.1016/j.icarus.2016.07.014. arXiv:1510.05870

H.M. Lamadrid, J.D. Rimstidt, E.M. Schwarzenbach, F. Klein, S. Ulrich, A. Dolocan, R.J. Bodnar, Effect of water activity on rates of serpentinization of olivine. Nat. Commun. **8**(May), 1–9 (2017). https://doi.org/10.1038/ncomms16107

D. Laufer, A. Bar-Nun, I. Pat-El, R. Jacovi, Experimental studies of ice grain ejection by massive gas flow from ice and implications to comets, Triton and Mars. Icarus **222**(1), 73–80 (2013). https://doi.org/10.1016/j.icarus.2012.10.030

A. Le Gall, C. Leyrat, M.A. Janssen, G. Choblet, G. Tobie, O. Bourgeois, A. Lucas, C. Sotin, C. Howett, R. Kirk, R.D. Lorenz, R.D. West, A. Stolzenbach, M. Massé, A.H. Hayes, L. Bonnefoy, G. Veyssière, F. Paganelli, Thermally anomalous features in the subsurface of Enceladus's south polar terrain. Nat. Astron. **1**, 0063 (2017). https://doi.org/10.1038/s41550-017-0063

E. Lellouch, G. Paubert, J.I. Moses, N.M. Schneider, D.F. Strobel, Volcanically emitted sodium chloride as a source for Io's neutral clouds and plasma torus. Nature **421**(6918), 45–47 (2003). https://doi.org/10.1038/nature01292

G. Leone, L. Wilson, A.G. Davies, The geothermal gradient of Io: consequences for lithosphere structure and volcanic eruptive activity. Icarus **211**(1), 623–635 (2011). https://doi.org/10.1016/j.icarus.2010.10.016

R.M. Lopes, J.R. Spencer, *Io After Galileo: A New View of Jupiter's Volcanic Moon* (Springer, Berlin, 2007). https://doi.org/10.1007/978-3-540-48841-5

R.M.C. Lopes, R.L. Kirk, K.L. Mitchell, A. LeGall, J.W. Barnes, A. Hayes, J. Kargel, L. Wye, J. Radebaugh, E.R. Stofan, M.A. Janssen, C.D. Neish, S.D. Wall, C.A. Wood, J.I. Lunine, M.J. Malaska, Cryovolcanism on Titan: new results from Cassini radar and VIMS. J. Geophys. Res., Planets **118**(3), 416–435 (2013). https://doi.org/10.1002/jgre.20062

R.D. Lorenz, Europa ocean sampling by plume flythrough: astrobiological expectations. Icarus **267**, 217–219 (2016). https://doi.org/10.1016/j.icarus.2015.12.018

A.E.H. Love, *Treatise on Mathematical Theory of Elasticity*, 4th edn. (1944). https://archive.org/details/atreatiseonmath01lovegoog

R.P. Lowell, M. DuBose, Hydrothermal systems on Europa. Geophys. Res. Lett. **32**(January), 4–7 (2005). https://doi.org/10.1029/2005GL022375

V.V. Makarov, M. Efroimsky, Tidal dissipation in a homogeneous spherical body. II. Three examples: Mercury, Io, and Kepler-10b. Astrophys. J. **795**(7), 1–10 (2014). https://doi.org/10.1088/0004-637X/795/1/7

B. Malvoisin, N. Brantut, M.A. Kaczmarek, Control of serpentinisation rate by reaction-induced cracking. Earth Planet. Sci. Lett. **476**, 143–152 (2017). https://doi.org/10.1016/j.epsl.2017.07.042

D. Maravilla, K. Flammer, D. Mendis, On the injection of fine dust from the Jovian magnetosphere. Astrophys. J. **438**, 968–974 (1995). http://adsabs.harvard.edu/abs/1995ApJ...438..968M

H.R. Martens, A.P. Ingersoll, S.P. Ewald, P. Helfenstein, B. Giese, Spatial distribution of ice blocks on Enceladus and implications for their origin and emplacement. Icarus **245**, 162–176 (2015). https://doi.org/10.1016/j.icarus.2014.09.035

E.S. Martin, The distribution and characterization of strike-slip faults on Enceladus. Geophys. Res. Lett. **43**(6), 2456–2464 (2016). https://doi.org/10.1002/2016GL067805

D.L. Matson, J.C. Castillo-Rogez, A.G. Davies, T.V. Johnson, Enceladus: a hypothesis for bringing both heat and chemicals to the surface. Icarus **221**(1), 53–62 (2012). https://doi.org/10.1016/j.icarus.2012.05.031

I. Matsuyama, F. Nimmo, Tectonic patterns on reoriented and despun planetary bodies. Icarus **195**(1), 459–473 (2008). https://doi.org/10.1016/j.icarus.2007.12.003

N. McBride, J. Hillier, S. Green, R. Srama, S. Kempf, F. Postberg, G. Moragas-Klostermeyer, J. McDonnell, E. Grün, Cassini cosmic dust analyser: composition of dust at Saturn, in *ESA—Workshop on Dust in Planetary Systems SP-643* (2007), pp. 107–110

W.J. McDoniel, D.B. Goldstein, P.L. Varghese, L.M. Trafton, Three-dimensional simulation of gas and dust in Io's Pele plume. Icarus **257**, 251–274 (2015). https://doi.org/10.1016/j.icarus.2015.03.019

W.J. McDoniel, D.B. Goldstein, P.L. Varghese, L.M. Trafton, The interaction of Io's plumes and sublimation atmosphere. Icarus **294**, 81–97 (2017). https://doi.org/10.1016/j.icarus.2017.04.021

A.S. McEwen, L.A. Soderblom, Two classes of volcanic plumes on Io. Icarus **55**(2), 191–217 (1983). https://doi.org/10.1016/0019-1035(83)90075-1

W.B. McKinnon, Effect of Enceladus's rapid synchronous spin on interpretation of Cassini gravity. Geophys. Res. Lett. **42**(7), 2137–2143 (2015). https://doi.org/10.1002/2015GL063384

P. Meier, H. Kriegel, U. Motschmann, J. Schmidt, F. Spahn, T.W. Hill, Y. Dong, G.H. Jones, A model of the spatial and size distribution of Enceladus' dust plume. Planet. Space Sci. **104**, 216–233 (2014). https://doi.org/10.1016/j.pss.2014.09.016

D.A. Mendis, W.I. Axford, Satellites and magnetospheres of the outer planets. Annu. Rev. Earth Planet. Sci. **2**(1), 419–474 (1974). https://doi.org/10.1146/annurev.ea.02.050174.002223

N. Meyer-Vernet, On the charge of nanograins in cold environments and Enceladus dust. Icarus **226**(1), 583–590 (2013). https://doi.org/10.1016/j.icarus.2013.06.014. arXiv:1306.2228

K. Miljković, J.K. Hillier, N.J. Mason, J.C. Zarnecki, Models of dust around Europa and Ganymede. Planet. Space Sci. **70**(1), 20–27 (2012). https://doi.org/10.1016/j.pss.2012.06.006. arXiv:1206.2833v1

W.B. Moore, G. Schubert, J.D. Anderson, J.R. Spencer, The interior of Io, in *Io After Galileo: A New View of Jupiter's Volcanic Moon* (2007), pp. 89–108. https://doi.org/10.1007/978-3-540-48841-5_5

J.M. Moore, W.B. McKinnon, J.R. Spencer, A.D. Howard, P.M. Schenk, R.A. Beyer, F. Nimmo, K.N. Singer, O.M. Umurhan, O.L. White, S.A. Stern, K. Ennico, C.B. Olkin, H.A. Weaver, L.A. Young, R.P. Binzel, M.W. Buie, B.J. Buratti, A.F. Cheng, D.P. Cruikshank, W.M. Grundy, I.R. Linscott, H.J. Reitsema, D.C. Reuter, M.R. Showalter, V.J. Bray, C.L. Chavez, C.J.A. Howett, T.R. Lauer, C.M. Lisse, A.H. Parker, S.B. Porter, S.J. Robbins, K. Runyon, T. Stryk, H.B. Throop, C.C.C. Tsang, A.J. Verbiscer, A.M. Zangari, A.L. Chaikin, D.E. Wilhelms, F. Bagenal, G.R. Gladstone, T. Andert, J. Andrews, M. Banks, B. Bauer, J. Bauman, O.S. Barnouin, P. Bedini, K. Beisser, S. Bhaskaran, E. Birath, M. Bird, D.J. Bogan, A. Bowman, M. Brozovic, C. Bryan, M.R. Buckley, S.S. Bushman, A. Calloway, B. Carcich, S. Conard, C.A. Conrad, J.C. Cook, O.S. Custodio, C.M.D. Ore, C. Deboy, Z.J.B. Dischner, P. Dumont, A.M. Earle, H.A. Elliott, J. Ercol, C.M. Ernst, T. Finley, S.H. Flanigan, G. Fountain, M.J. Freeze, T. Greathouse, J.L. Green, Y. Guo, M. Hahn, D.P. Hamilton, S.A. Hamilton, J. Hanley, A. Harch, H.M. Hart, C.B. Hersman, A. Hill, M.E. Hill, D.P. Hinson, M.E. Holdridge, M. Horanyi, C. Jackman, R.A. Jacobson, D.E. Jennings, J.A. Kammer, H.K. Kang, D.E. Kaufmann, P. Kollmann, S.M. Krimigis, D. Kusnierkiewicz, J.E. Lee, K.L. Lindstrom, A.W. Lunsford, V.A. Mallder, N. Martin, D.J. McComas, R.L. McNutt, D. Mehoke, T. Mehoke, E.D. Melin, M. Mutchler, D. Nelson, J.I. Nunez, A. Ocampo, W.M. Owen, M. Paetzold, B. Page, J.W. Parker, F. Pelletier, J. Peterson, N. Pinkine, M. Piquette, S. Protopapa, J. Redfern, J.H. Roberts, G. Rogers, D. Rose, K.D. Retherford, M.G. Ryschkewitsch, E. Schindhelm, B. Sepan, M. Soluri, D. Stanbridge, A.J. Steffl, D.F. Strobel, M.E. Summers, J.R. Szalay, M. Tapley, A. Taylor, H. Taylor, G.L. Tyler, M.H. Versteeg, M. Vincent, R. Webbert, S. Weidner, G.E. Weigle, K. Whittenburg, B.G. Williams, K. Williams, S. Williams, W.W. Woods, E. Zirnstein, The geology of Pluto and Charon through the eyes of New Horizons. Science **351**, 1284–1293 (2016). https://doi.org/10.1126/science.aad7055

G. Morfill, E. Grün, T. Johnson, Dust in Jupiter's magnetosphere: physical processes. Planet. Space Sci. **28**(12), 1087–1100 (1980). https://doi.org/10.1016/0032-0633(80)90067-7

S. Mostefaoui, G.W. Lugmair, P. Hoppe, ^{60}Fe: a heat source for planetary differentiation from a nearby supernova explosion. Astrophys. J. **625**(1), 271–277 (2005). https://doi.org/10.1086/429555, http://adsabs.harvard.edu/abs/2005ApJ...625..271M

A.L. Nahm, S.A. Kattenhorn, A unified nomenclature for tectonic structures on the surface of Enceladus. Icarus **258**, 67–81 (2015). https://doi.org/10.1016/j.icarus.2015.06.009

M. Nakajima, A.P. Ingersoll, Controlled boiling on Enceladus. 1. Model of the vapor-driven jets. Icarus **272**, 309–318 (2016). https://doi.org/10.1016/j.icarus.2016.02.027

M. Neveu, S.J. Desch, J.C. Castillo-Rogez, Aqueous geochemistry in icy world interiors: equilibrium fluid, rock, and gas compositions, and fate of antifreezes and radionuclides. Geochim. Cosmochim. Acta **212**, 324–371 (2017). https://doi.org/10.1016/j.gca.2017.06.023

R.C. Newton, C.E. Manning, Solubilities of corundum, wollastonite and quartz in H_2O–NaCl solutions at 800 °C and 10 kbar: interaction of simple minerals with brines at high pressure and temperature. Geochim. Cosmochim. Acta **70**(22), 5571–5582 (2006). https://doi.org/10.1016/j.gca.2006.08.012

P.D. Nicholson, M.R. Showalter, L. Dones, R.G. French, S.M. Larson, J.J. Lissauer, C.A. McGhee, P. Seitzer, B. Sicardy, G.E. Danielson, Observations of Saturn's ring-plane crossings in August and November 1995. Science **272**(5261), 509–515 (1996). https://doi.org/10.1126/science.272.5261.509

F. Nimmo, M. Manga, Causes, characteristics and consequences of convective diapirism on Europa. Geophys. Res. Lett. **29**(23), 24:1–24:4 (2002). https://doi.org/10.1029/2002GL015754

F. Nimmo, R.T. Pappalardo, Diapir-induced reorientation of Saturn's moon Enceladus. Nature **441**(7093), 614–616 (2006). https://doi.org/10.1038/nature04821

F. Nimmo, B. Giese, R.T. Pappalardo, Estimates of Europa's ice shell thickness from elastically-supported topography. Geophys. Res. Lett. **30**(5), 1233 (2003). https://doi.org/10.1029/2002GL016660

F. Nimmo, C. Porco, C. Mitchell, Tidally modulated eruptions on Enceladus: Cassini ISS observations and models. Astron. J. **148**(3), 46 (2014). https://doi.org/10.1088/0004-6256/148/3/46

D. O'Brien, P.E. Geissler, R. Greenberg, A melt-through model for chaos formation on Europa. Icarus **156**(1), 152–161 (2002). https://doi.org/10.1006/icar.2001.6777

G.W. Ojakangas, D.J. Stevenson, Episodic volcanism of tidally heated satellites with application to Io. Icarus **66**(2), 341–358 (1986). https://doi.org/10.1016/0019-1035(86)90163-6

K.M. Ostdiek, T.S. Anderson, W.K. Bauder, M.R. Bowers, A.M. Clark, P. Collon, W. Lu, A.D. Nelson, D. Robertson, M. Skulski, R. Dressler, D. Schumann, J.P. Greene, W. Kutschera, M. Paul, Activity measurement of ^{60}Fe through the decay of ^{60m}Co and confirmation of its half-life. Phys. Rev. C **95**(5), 055809 (2017). https://doi.org/10.1103/PhysRevC.95.055809

K.D. Pang, C.C. Voge, J.W. Rhoads, J.M. Ajello, The E ring of Saturn and satellite Enceladus. J. Geophys. Res. **89**(B11), 9459–9470 (1984). https://doi.org/10.1029/JB089iB11p09459

R.T. Pappalardo, J.W. Head, R. Greeley, R.J. Sullivan, C. Pilcher, G. Schubert, W.B. Moore, M.H. Carr, J.M. Moore, M.J.S. Belton, D.L. Goldsby, Geological evidence for solid-state convection in Europa's ice shell. Nature **391**(6665), 365–368 (1998). https://doi.org/10.1038/34862

R.T. Pappalardo, M.J.S. Belton, H.H. Breneman, M.H. Carr, C.R. Chapman, G.C. Collins, T. Denk, S. Fagents, P.E. Geissler, B. Giese, R. Greeley, R. Greenberg, J.W. Head, P. Helfenstein, G. Hoppa, S.D. Kadel, K.P. Klaasen, J.E. Klemaszewski, K. Magee, A.S. McEwen, J.M. Moore, W.B. Moore, G. Neukum, C.B. Phillips, L.M. Prockter, G. Schubert, D.A. Senske, R.J. Sullivan, B.R. Tufts, E.P. Turtle, R. Wagner, K.K. Williams, Does Europa have a subsurface ocean? Evaluation of the geological evidence. J. Geophys. Res., Planets **104**(E10), 24015–24055 (1999). https://doi.org/10.1029/1998JE000628

R.T. Pappalardo, W.B. McKinnon, K.K. Khurana, *Europa* (University of Arizona Press, Tucson, 2009). https://doi.org/10.2307/j.ctt1xp3wdw

D.A. Patthoff, S.A. Kattenhorn, A fracture history on Enceladus provides evidence for a global ocean. Geophys. Res. Lett. **38**(18), 1–6 (2011). https://doi.org/10.1029/2011GL048387

S.J. Peale, Origin and evolution of the natural satellites. Annu. Rev. Astron. Astrophys. **37**(1), 533–602 (1999). https://doi.org/10.1146/annurev.astro.37.1.533

S.J. Peale, P. Cassen, R.T. Reynolds, Melting of Io by tidal dissipation. Science **203**(4383), 892–894 (1979). https://doi.org/10.1126/science.203.4383.892

C.C. Porco, P. Helfenstein, P.C. Thomas, A.P. Ingersoll, J. Wisdom, R. West, G. Neukum, T. Denk, R. Wagner, T. Roatsch, S. Kieffer, E. Turtle, A. McEwen, T.V. Johnson, J. Rathbun, J. Veverka, D. Wilson, J. Perry, J. Spitale, A. Brahic, J.A. Burns, A.D. Delgenio, L. Dones, C.D. Murray, S. Squyres, Cassini observes the active south pole of Enceladus. Science **311**(5766), 1393–1401 (2006). https://doi.org/10.1126/science.1123013

C. Porco, D. DiNino, F. Nimmo, How the geysers, tidal stresses, and thermal emission across the south polar terrain of Enceladus are related. Astron. J. **148**(3), 45 (2014). https://doi.org/10.1088/0004-6256/148/3/45

C.C. Porco, F. Nimmo, D. DiNino, Enceladus' 101 geysers: phantoms? Hardly, in *European Planetary Science Congress 2015* (2015)

F. Postberg, S. Kempf, R. Srama, S.F. Green, J.K. Hillier, N. McBride, E. Grün, Composition of jovian dust stream particles. Icarus **183**(1), 122–134 (2006). https://doi.org/10.1016/j.icarus.2006.02.001

F. Postberg, S. Kempf, J.K. Hillier, R. Srama, S.F. Green, N. McBride, E. Grün, The E-ring in the vicinity of Enceladus. II. Probing the moon's interior—the composition of E-ring particles. Icarus **193**(2), 438–454 (2008). https://doi.org/10.1016/j.icarus.2007.09.001

F. Postberg, S. Kempf, D. Rost, T. Stephan, R. Srama, M. Trieloff, A. Mocker, M. Goerlich, Discriminating contamination from particle components in spectra of Cassini's dust detector CDA. Planet. Space Sci. **57**(12), 1359–1374 (2009a). https://doi.org/10.1016/j.pss.2009.06.027
F. Postberg, S. Kempf, J. Schmidt, N. Brilliantov, B. a, B. Abel, U. Buck, R. Srama, Sodium salts in E-ring ice grains from an ocean below the surface of Enceladus. Nature **459**(June), 1–4 (2009b). https://doi.org/10.1038/nature08046
F. Postberg, E. Grün, M. Horanyi, S. Kempf, H. Krüger, J. Schmidt, F. Spahn, R. Srama, Z. Sternovsky, M. Trieloff, Compositional mapping of planetary moons by mass spectrometry of dust ejecta. Planet. Space Sci. **59**(14), 1815–1825 (2011a). https://doi.org/10.1016/j.pss.2011.05.001
F. Postberg, J. Schmidt, J. Hillier, S. Kempf, R. Srama, A salt-water reservoir as the source of a compositionally stratified plume on Enceladus. Nature **474**(7353), 620–622 (2011b). https://doi.org/10.1038/nature10175
F. Postberg, N. Khawaja, B. Abel, G. Choblet, C.R. Glein, M.S. Gudipati, B.L. Henderson, H.W. Hsu, S. Kempf, F. Klenner, G. Moragas-Klostermeyer, B. Magee, L. Nlle, M. Perry, R. Reviol, J. Schmidt, R. Srama, F. Stolz, G. Tobie, M. Trieloff, J.H. Waite, Macromolecular organic compounds from the depths of Enceladus. Nature **558**(7711), 564–568 (2018). https://doi.org/10.1038/s41586-018-0246-4
L.C. Quick, O.S. Barnouin, L.M. Prockter, G.W. Patterson, Constraints on the detection of cryovolcanic plumes on Europa. Planet. Space Sci. **86**, 1–9 (2013). https://doi.org/10.1016/j.pss.2013.06.028
A.R. Rhoden, T.A. Hurford, L. Roth, K. Retherford, Linking Europa's plume activity to tides, tectonics, and liquid water. Icarus **253**, 169–178 (2015). https://doi.org/10.1016/j.icarus.2015.02.023. arXiv:1011.1669v3
A.R. Rhoden, W. Henning, T.A. Hurford, D.A. Patthoff, R. Tajeddine, The implications of tides on the Mimas ocean hypothesis. J. Geophys. Res., Planets **122**(2), 400–410 (2017). https://doi.org/10.1002/2016JE005097
J.H. Roberts, The fluffy core of Enceladus. Icarus **258**, 54–66 (2015). https://doi.org/10.1016/j.icarus.2015.05.033
M.N. Ross, G. Schubert, Tidally forced viscous heating in a partially molten Io. Icarus **64**(3), 391–400 (1985). https://doi.org/10.1016/0019-1035(85)90063-6
L. Roth, K.D. Retherford, J. Saur, D.F. Strobel, P.D. Feldman, M.A. McGrath, F. Nimmo, Orbital apocenter is not a sufficient condition for HST/STIS detection of Europa's water vapor aurora. Proc. Natl. Acad. Sci. USA **111**(48), E5123–E5132 (2014a). https://doi.org/10.1073/pnas.1416671111
L. Roth, J. Saur, K.D. Retherford, D.F. Strobel, P.D. Feldman, M.A. McGrath, F. Nimmo, Transient water vapor at Europa's south pole. Science **343**(6167), 171–174 (2014b). https://doi.org/10.1126/science.1247051
L. Roth, J. Saur, K.D. Retherford, A. Blöcker, D.F. Strobel, P.D. Feldman, Constraints on Io's interior from auroral spot oscillations. J. Geophys. Res. Space Phys. **122**(2), 1903–1927 (2017). https://doi.org/10.1002/2016JA023701
J. Saur, P.D. Feldman, L. Roth, F. Nimmo, D.F. Strobel, K.D. Retherford, M.A. McGrath, N. Schilling, J.C. Gérard, D. Grodent, Hubble Space Telescope/Advanced Camera for Surveys observations of Europa's atmospheric ultraviolet emission at eastern elongation. Astrophys. J. **738**, 153–166 (2011). https://doi.org/10.1088/0004-637X/738/2/153
L. Schaefer, B. Fegley, Alkali and halogen chemistry in volcanic gases on Io. Icarus **173**(2), 454–468 (2005). https://doi.org/10.1016/j.icarus.2004.08.015
P. Schenk, H. Hargitai, R. Wilson, A. Mcewen, P. Thomas, The mountains of Io: global and geological perspectives from Voyager and Galileo. J. Geophys. Res. **106**(25), 33201–33222 (2001). https://doi.org/10.1029/2000JE001408
P. Schenk, J. Schmidt, O. White, The snows of Enceladus. EPSC Abstracts 6:EPSC-DPS2011-1358-1-2 (2011)
P.M. Schenk, R.N. Clark, C.J.A. Howett, A.J. Verbiscer, J.H. Waite (eds.), *Enceladus and the Icy Moons of Saturn* (University of Arizona Press, Tucson, 2018)
P. Schippers, M. Blanc, N. André, I. Dandouras, G.R. Lewis, L.K. Gilbert, A.M. Persoon, N. Krupp, D.A. Gurnett, A.J. Coates, S.M. Krimigis, D.T. Young, M.K. Dougherty, Multi-instrument analysis of electron populations in Saturn's magnetosphere. J. Geophys. Res. Space Phys. **113**(A7), A07208 (2008). https://doi.org/10.1029/2008JA013098
J. Schmidt, N. Brilliantov, F. Spahn, S. Kempf, Slow dust in Enceladus' plume from condensation and wall collisions in tiger stripe fractures. Nature **451**(7179), 685–688 (2008). https://doi.org/10.1038/nature06491
N.M. Schneider, M.H. Burger, E.L. Schaller, M.E. Brown, R.E. Johnson, J.S. Kargel, M.K. Dougherty, A. Na, No sodium in the vapour plumes of Enceladus. Nature **459**(7250), 1102–1104 (2009). https://doi.org/10.1038/nature08070

G. Schubert, D.J. Stevenson, K. Ellsworth, Internal structures of the Galilean satellites. Icarus **47**(1), 46–59 (1981). https://doi.org/10.1016/0019-1035(81)90090-7

G. Schubert, D.J. Andersen, T. Spohn, W.B. Mckinnon, Interior composition, structure and dynamics of the Galilean satellites, in *Jupiter the Planet, Satellites and Magnetosphere* (2004), pp. 281–306

G. Schubert, F. Sohl, H. Hussmann, Interior of Europa, in *Europa*, ed. by R. Pappalardo, W.B. McKinnon, K.K. Khurana (University of Arizona Press, Tucson, 2009), pp. 353–367

G. Schubert, H. Hussmann, V. Lainey, D.L. Matson, W.B. McKinnon, F. Sohl, C. Sotin, G. Tobie, D. Turrini, T. Van Hoolst, Evolution of icy satellites. Space Sci. Rev. **153**(1–4), 447–484 (2010). https://doi.org/10.1007/s11214-010-9635-1

M. Segatz, T. Spohn, M.N. Ross, G. Schubert, Tidal dissipation, surface heat flow, and figure of viscoelastic models of Io. Icarus **75**(2), 187–206 (1988). https://doi.org/10.1016/0019-1035(88)90001-2

P.K. Seidelmann, B.A. Archinal, M.F. A'Hearn, A. Conrad, G.J. Consolmagno, D. Hestroffer, J.L. Hilton, G.A. Krasinsky, G. Neumann, J. Oberst, P. Stooke, E.F. Tedesco, D.J. Tholen, P.C. Thomas, I.P. Williams, Report of the IAU/IAG Working Group on cartographic coordinates and rotational elements: 2006. Celest. Mech. Dyn. Astron. **98**(3), 155–180 (2007). https://doi.org/10.1007/s10569-007-9072-y

Y. Sekine, T. Shibuya, F. Postberg, H.W. Hsu, K. Suzuki, Y. Masaki, T. Kuwatani, M. Mori, P.K. Hong, M. Yoshizaki, S. Tachibana, S. Si, High-temperature water–rock interactions and hydrothermal environments in the chondrite-like core of Enceladus. Nat. Commun. **6**, 8604 (2015). https://doi.org/10.1038/ncomms9604

D. Shoji, H. Hussmann, Frequency-dependent tidal dissipation in a viscoelastic Saturnian core and expansion of Mimas' semi-major axis. Astron. Astrophys. **599**, L10 (2017). https://doi.org/10.1051/0004-6361/201630230

M.R. Showalter, J.N. Cuzzi, S.M. Larson, Structure and particle properties of Saturn's E ring. Icarus **94**(2), 451–473 (1991). https://doi.org/10.1016/0019-1035(91)90241-K

A.P. Showman, I. Mosqueira, J.W. Head, On the resurfacing of Ganymede by liquid water volcanism. Icarus **172**, 625–640 (2004). https://doi.org/10.1016/j.icarus.2004.07.011

B.A. Smith, E.M. Shoemaker, S.W. Kieffer, A.F. Cook, The role of SO_2 in volcanism on Io. Nature **280**, 738–743 (1979). https://doi.org/10.1038/280738a0

B.A. Smith, L.A. Soderblom, D. Banfield, C. Barnet, A.T. Basilevsky, R.F. Beebe, K. Bollinger, J.M. Boyce, A. Brahic, G.A. Briggs, R.H. Brown, C. Chyba, S.A. Collins, T. Colvin, A.F. Cook II, D. Crisp, S.K. Croft, D. Cruikshank, J.N. Cuzzi, G.E. Danielson, M.E. Davies, E. De Jong, L. Dones, D. Godfrey, J. Goguen, I. Grenier, V.R. Haemmerle, H. Hammel, C.J. Hansen, C.P. Helfenstein, C. Howell, G.E. Hunt, A.P. Ingersoll, T.V. Johnson, J. Kargel, R. Kirk, D.I. Kuehn, S. Limaye, H. Masursky, A. McEwen, D. Morrison, T. Owen, W. Owen, J.B. Pollack, C.C. Porco, K. Rages, P. Rogers, D. Rudy, C. Sagan, J. Schwartz, E.M. Shoemaker, M. Showalter, B. Sicardy, D. Simonelli, J. Spencer, L.A. Sromovsky, C. Stoker, R.G. Strom, V.E. Suomi, S.P. Synott, R.J. Terrile, P. Thomas, W.R. Thompson, A. Verbiscer, J. Veverka, Voyager 2 at Neptune—imaging science results. Science **246**(4936), 1422–1449 (1989). https://doi.org/10.1126/science.246.4936.1422

L.A. Soderblom, S.W. Kieffer, T.L. Becker, R.H. Brown, A.F. Cook, C.J. Hansen, T.V. Johnson, R.L. Kirk, E.M. Shoemaker, Triton's geyser-like plumes: discovery and basic characterization. Science **250**(4979), 410–415 (1990). https://doi.org/10.1126/science.250.4979.410

F. Sohl, T. Spohn, D. Breuer, K. Nagel, Implications from Galileo observations on the interior structure and chemistry of the Galilean satellites. Icarus **157**(1), 104–119 (2002). https://doi.org/10.1006/icar.2002.6828

F. Sohl, M. Choukroun, J. Kargel, J. Kimura, R. Pappalardo, S. Vance, M. Zolotov, Subsurface water oceans on icy satellites: chemical composition and exchange processes. Space Sci. Rev. **153**(1–4), 485–510 (2010). https://doi.org/10.1007/s11214-010-9646-y

O. Souček, J. Hron, M. Běhounková, O. Čadek, Effect of the tiger stripes on the deformation of Saturn's moon Enceladus. Geophys. Res. Lett. **43**(14), 7417–7423 (2016). https://doi.org/10.1002/2016GL069415

B.S. Southworth, S. Kempf, J. Schmidt, Modeling Europa's dust plumes. Geophys. Res. Lett. **42**(24), 10541–10548 (2015). https://doi.org/10.1002/2015GL066502

F. Spahn, N. Albers, M. Hörning, S. Kempf, A.V. Krivov, M. Makuch, J. Schmidt, M. Seiß, M. Sremčević, E ring dust sources: implications from Cassini's dust measurements. Planet. Space Sci. **54**(9–10), 1024–1032 (2006a). https://doi.org/10.1016/j.pss.2006.05.022

F. Spahn, J. Schmidt, N. Albers, M. Hörning, M. Makuch, M. Seiß, S. Kempf, R. Srama, V. Dikarev, S. Helfert, G. Moragas-Klostermeyer, A. Krivov, M. Sremčević, A.J. Tuzzolino, T. Economou, E. Grün, Cassini dust measurements at Enceladus and implications for the origin of the E ring. Science **311**(5766), 1416–1418 (2006b). https://doi.org/10.1126/science.1121375

W.B. Sparks, K.P. Hand, M.A. McGrath, E. Bergeron, M. Cracraft, S.E. Deustua, Probing for evidence of plumes on Europa with HST/STIS. Astrophys. J. **829**(2), 121–132 (2016). https://doi.org/10.3847/0004-637X/829/2/121

W.B. Sparks, B.E. Schmidt, M.A. Mcgrath, K.P. Hand, J.R. Spencer, M. Cracraft, S.E. Deustua, Active cryovolcanism on Europa? Astrophys. J. Lett. **839**(2), L18–L23 (2017). https://doi.org/10.3847/2041-8213/aa67f8. arXiv:1704.04283

J. Spencer, C. Howett, Enceladus heat flow from high spatial resolution thermal emission observations. EPSC Abstracts **8**, 1–2 (2013). http://meetingorganizer.copernicus.org/EPSC2013/EPSC2013-840-1.pdf

J.R. Spencer, L.K. Tamppari, T.Z. Martin, L.D. Travis, Temperatures on Europa from Galileo photopolarimeter-radiometer: nighttime thermal anomalies. Science **284**(5419), 1514–1516 (1999). https://doi.org/10.1126/science.284.5419.1514

J.R. Spencer, J. Pearl, M. Segura, F.M. Flasar, A. Mamoutkine, P. Romani, B.J. Buratti, A.R. Hendrix, L. Spilker, R.M.C. Lopes, Cassini encounters Enceladus: background and the discovery of a south polar hot spot. Science **311**(5766), 1401–1405 (2006). https://doi.org/10.1126/science.1121661

J.R. Spencer, S.A. Stern, A.F. Cheng, H.A. Weaver, D.C. Reuter, K. Retherford, A. Lunsford, J.M. Moore, O. Abramov, R.M.C. Lopes, J.E. Perry, L. Kamp, M. Showalter, K.L. Jessup, F. Marchis, P.M. Schenk, C. Dumas, Io volcanism seen by New Horizons: a major eruption of the Tvashtar volcano. Science **318**(5848), 240–243 (2007). https://doi.org/10.1126/science.1147621

J.N. Spitale, C.C. Porco, Association of the jets of Enceladus with the warmest regions on its south-polar fractures. Nature **449**(7163), 695–697 (2007). https://doi.org/10.1038/nature06217. arXiv:1507.02142v2

J.N. Spitale, T.A. Hurford, A.R. Rhoden, E.E. Berkson, S.S. Platts, Curtain eruptions from Enceladus' south-polar terrain. Nature **521**(7550), 57–60 (2015). https://doi.org/10.1038/nature14368

T. Spohn, G. Schubert, Oceans in the icy Galilean satellites of Jupiter? Icarus **161**(2), 456–467 (2003). https://doi.org/10.1016/S0019-1035(02)00048-9

S.W. Squyres, R.T. Reynolds, P.M. Cassen, S.J. Peale, Liquid water and active resurfacing on Europa. Nature **301**(5897), 225–226 (1983). https://doi.org/10.1038/301225a0

R. Srama, S. Kempf, G. Moragas-Klostermeyer, S. Helfert, T.J. Ahrens, N. Altobelli, S. Auer, U. Beckmann, J.G. Bradley, M. Burton, V.V. Dikarev, T. Economou, H. Fechtig, S.F. Green, M. Grande, O. Havnes, J.K. Hillier, M. Horanyi, E. Igenbergs, E.K. Jessberger, T.V. Johnson, H. Krüger, G. Matt, N. McBride, A. Mocker, P. Lamy, D. Linkert, G. Linkert, F. Lura, J.A.M. McDonnell, D. Möhlmann, G.E. Morfill, F. Postberg, M. Roy, G.H. Schwehm, F. Spahn, J. Svestka, V. Tschernjawski, A.J. Tuzzolino, R. Wäsch, E. Grün, In situ dust measurements in the inner Saturnian system. Planet. Space Sci. **54**(9–10), 967–987 (2006). https://doi.org/10.1016/j.pss.2006.05.021

M. Sremčević, A.V. Krivov, H. Krüger, F. Spahn, Impact-generated dust clouds around planetary satellites: model versus Galileo data. Planet. Space Sci. **53**(6), 625–641 (2005). https://doi.org/10.1016/j.pss.2004.10.001

R.G. Strom, N.M. Schneider, R.J. Terrile, A.F. Cook, C. Hansen, Volcanic eruptions on Io. J. Geophys. Res. Space Phys. **86**(A10), 8593–8620 (1981). https://doi.org/10.1029/JA086iA10p08593

P.J. Tackley, G. Schubert, G.A. Glatzmaier, P. Schenk, J.T. Ratcliff, J.P. Matas, Three-dimensional simulations of mantle convection in Io. Icarus **149**(1), 79–93 (2001). https://doi.org/10.1006/icar.2000.6536

R. Tajeddine, K.M. Soderlund, P.C. Thomas, P. Helfenstein, M.M. Hedman, J.A. Burns, P.M. Schenk, True polar wander of Enceladus from topographic data. Icarus **295**, 46–60 (2017). https://doi.org/10.1016/j.icarus.2017.04.019

B.D. Teolis, M.E. Perry, B.A. Magee, J. Westlake, J.H. Waite, Detection and measurement of ice grains and gas distribution in the Enceladus plume by Cassini's Ion Neutral Mass Spectrometer. J. Geophys. Res. Space Phys. **115**(9), 1–12 (2010). https://doi.org/10.1029/2009JA015192

B.D. Teolis, D.Y. Wyrick, A. Bouquet, B.A. Magee, J.H. Waite, Plume and surface feature structure and compositional effects on Europa's global exosphere: preliminary Europa mission predictions. Icarus **284**, 18–29 (2017a). https://doi.org/10.1016/j.icarus.2016.10.027

B.D. Teolis, D.Y. Wyrick, A. Bouquet, B.A. Magee, J.H. Waite, Plume and surface feature structure and compositional effects on Europa's global exosphere: preliminary Europa mission predictions. Icarus **284**, 18–29 (2017b). https://doi.org/10.1016/j.icarus.2016.10.027

P.C. Thomas, Sizes, shapes, and derived properties of the saturnian satellites after the Cassini nominal mission. Icarus **208**(1), 395–401 (2010). https://doi.org/10.1016/j.icarus.2010.01.025

P. Thomas, J. Burns, P. Helfenstein, S. Squyres, J. Veverka, C. Porco, E. Turtle, A. McEwen, T. Denk, B. Giese, Shapes of the saturnian icy satellites and their significance. Icarus **190**(2), 573–584 (2007). https://doi.org/10.1016/j.icarus.2007.03.012

P.C. Thomas, R. Tajeddine, M.S. Tiscareno, J.A. Burns, J. Joseph, T.J. Loredo, P. Helfenstein, C. Porco, Enceladus's measured physical libration requires a global subsurface ocean. Icarus **264**, 37–47 (2016). https://doi.org/10.1016/j.icarus.2015.08.037. arXiv:1509.07555

F. Tian, A. Stewart, O. Toon, K. Larsen, L. Esposito, Monte Carlo simulations of the water vapor plumes on Enceladus. Icarus **188**(1), 154–161 (2007). https://doi.org/10.1016/j.icarus.2006.11.010

G. Tobie, A. Mocquet, C. Sotin, Tidal dissipation within large icy satellites: applications to Europa and Titan. Icarus **177**(2), 534–549 (2005). https://doi.org/10.1016/j.icarus.2005.04.006

R.L. Tokar, The interaction of the atmosphere of Enceladus with Saturn's plasma. Science **311**(5766), 1409–1412 (2006). https://doi.org/10.1126/science.1121061

B.J. Travis, G. Schubert, Keeping Enceladus warm. Icarus **250**, 32–42 (2015). https://doi.org/10.1016/j.icarus.2014.11.017

S.K. Trumbo, M.E. Brown, B.J. Butler, ALMA thermal observations of a proposed plume source region on Europa. Astron. J. **154**(4), 148 (2017). https://doi.org/10.3847/1538-3881/aa8769. arXiv:1708.07922

B.M. Tutolo, A.J. Luhmann, N.J. Tosca, W.E. Seyfried, Serpentinization as a reactive transport process: the brucite silicification reaction. Earth Planet. Sci. Lett. **484**, 385–395 (2018). https://doi.org/10.1016/j.epsl.2017.12.029

R.H. Tyler, W.G. Henning, C.W. Hamilton, Tidal heating in a magma ocean within Jupiter's moon Io. Astrophys. J. Suppl. Ser. **218**(2), 22 (2015). https://doi.org/10.1088/0067-0049/218/2/22

T. Van Hoolst, R.M. Baland, A. Trinh, The diurnal libration and interior structure of Enceladus. Icarus **277**, 311–318 (2016). https://doi.org/10.1016/j.icarus.2016.05.025

W.R. Van Schmus, Natural radioactivity of the crust and mantle, in *Global Earth Physics: A Handbook of Physical Constants* (1995), pp. 283–291

S. Vance, J. Harnmeijer, J. Kimura, H. Hussmann, B. Demartin, J.M. Brown, Hydrothermal systems in small ocean planets. Astrobiology **7**(6), 987–1005 (2007). https://doi.org/10.1089/ast.2007.0075

P. Varga, B. Süle, E. Illés-Almár, On the tidal heating of Enceladus. J. Geodyn. **48**(3–5), 247–252 (2009). https://doi.org/10.1016/j.jog.2009.09.031

G.J. Veeder, D.L. Matson, T.V. Johnson, A.G. Davies, D.L. Blaney, The polar contribution to the heat flow of Io. Icarus **169**(1), 264–270 (2004). https://doi.org/10.1016/j.icarus.2003.11.016

A.J. Verbiscer, M.F. Skrutskie, D.P. Hamilton, Saturn's largest ring. Nature **461**(7267), 1098–1100 (2009). https://doi.org/10.1038/nature08515

J.E. Wahlund, M. André, A. Eriksson, M. Lundberg, M. Morooka, M. Shafiq, T. Averkamp, D. Gurnett, G. Hospodarsky, W. Kurth, K. Jacobsen, A. Pedersen, W. Farrell, S. Ratynskaia, N. Piskunov, Detection of dusty plasma near the E-ring of Saturn. Planet. Space Sci. **57**(14–15), 1795–1806 (2009). https://doi.org/10.1016/j.pss.2009.03.011

J.H. Waite, M.R. Combi, W.H. Ip, T.E. Cravens, R.L. McNutt, W. Kasprzak, R. Yelle, J. Luhmann, H. Niemann, D. Gell, B. Magee, G. Fletcher, J. Lunine, W.L. Tseng, Cassini ion and neutral mass spectrometer: Enceladus plume composition and structure. Science **311**(5766), 1419–1422 (2006). https://doi.org/10.1126/science.1121290

J.H. Waite, W.S. Lewis, B.A. Magee, J.I. Lunine, W.B. McKinnon, C.R. Glein, O. Mousis, D.T. Young, T. Brockwell, J. Westlake, M.J. Nguyen, B.D. Teolis, H.B. Niemann, R.L. McNutt, M. Perry, W.H. Ip, Liquid water on Enceladus from observations of ammonia and ^{40}Ar in the plume. Nature **460**(7254), 487–490 (2009). https://doi.org/10.1038/nature08153

J.H. Waite, C.R. Glein, R.S. Perryman, B.D. Teolis, B.A. Magee, G. Miller, J. Grimes, M.E. Perry, K.E. Miller, A. Bouquet, J.I. Lunine, T. Brockwell, S.J. Bolton, Cassini finds molecular hydrogen in the Enceladus plume: evidence for hydrothermal processes. Science **356**(6334), 155–159 (2017). https://doi.org/10.1126/science.aai8703

W.W. Wegner, W.G. Ernst, Experimentally determined hydration and dehydration reaction rates in the system $MgO\text{-}SiO_2\text{-}H_2O$ (1983)

R.J. Wilson, R.L. Tokar, M.G. Henderson, T.W. Hill, M.F. Thomsen, D.H. Pontius, Cassini plasma spectrometer thermal ion measurements in Saturn's inner magnetosphere. J. Geophys. Res. Space Phys. **113**(A12), A12218 (2008). https://doi.org/10.1029/2008JA013486

S.Y. Ye, D. Gurnett, W. Kurth, T. Averkamp, S. Kempf, H.W. Hsu, R. Srama, E. Grün, Properties of dust particles near Saturn inferred from voltage pulses induced by dust impacts on Cassini spacecraft. J. Geophys. Res. Space Phys. **119**, 6294–6312 (2014). https://doi.org/10.1002/2014JA020024

S.Y. Ye, D.A. Gurnett, W.S. Kurth, In-situ measurements of Saturn's dusty rings based on dust impact signals detected by Cassini RPWS. Icarus **279**, 51–61 (2016). https://doi.org/10.1016/j.icarus.2016.05.006

S.K. Yeoh, Z. Li, D.B. Goldstein, P.L. Varghese, D.A. Levin, L.M. Trafton, Constraining the Enceladus plume using numerical simulation and Cassini data. Icarus **281**, 357–378 (2017). https://doi.org/10.1016/j.icarus.2016.08.028

J. Zhang, D.B. Goldstein, P.L. Varghese, N.E. Gimelshein, S.F. Gimelshein, D.A. Levin, Simulation of gas dynamics and radiation in volcanic plumes on Io. Icarus **163**(1), 182–197 (2003). https://doi.org/10.1016/S0019-1035(03)00050-2

J. Zhang, D.B. Goldstein, P.L. Varghese, L. Trafton, C. Moore, K. Miki, Numerical modeling of ionian volcanic plumes with entrained particulates. Icarus **172**(2), 479–502 (2004). https://doi.org/10.1016/j.icarus.2004.06.016

M.Y. Zolotov, An oceanic composition on early and today's Enceladus. Geophys. Res. Lett. **34**(23), 1–5 (2007). https://doi.org/10.1029/2007GL031234

M.Y. Zolotov, B. Fegley, Oxidation state of volcanic gases and the interior of Io. Icarus **141**(1), 40–52 (1999). https://doi.org/10.1006/icar.1999.6164
M.Y. Zolotov, B. Fegley, Eruption conditions of Pele volcano on Io inferred from chemistry of its volcanic plume. Geophys. Res. Lett. **27**(17), 2789–2792 (2000). https://doi.org/10.1029/2000GL011608
H.A. Zook, E. Grun, M. Baguhl, D.P. Hamilton, G. Linkert, J.C. Liou, R. Forsyth, J.L. Phillips, Solar wind magnetic field bending of Jovian dust trajectories. Science **274**(5292), 1501–1503 (1996). https://doi.org/10.1126/science.274.5292.1501

Space Sci Rev (2019) 215:11
https://doi.org/10.1007/s11214-018-0577-3

Circumplanetary Dust Populations

Frank Spahn[1] · Manuel Sachse[1] · Martin Seiß[1] · Hsiang-Wen Hsu[2] · Sascha Kempf[2] · Mihály Horányi[2]

Received: 25 April 2018 / Accepted: 31 December 2018 / Published online: 28 January 2019

Abstract We summarize the current state of observations of circumplanetary dust populations, including both dilute and dense rings and tori around the giant planets, ejecta clouds engulfing airless moons, and rings around smaller planetary bodies throughout the Solar System. We also discuss the theoretical models that enable these observations to be understood in terms of the sources, sinks and transport of various dust populations. The dynamics and resulting transport of the particles can be quite complex, due to the fact that their motion is influenced by neutral and plasma drag, radiation pressure, and electromagnetic forces—all in addition to gravity. The relative importance of these forces depends on the environment, as well as the makeup and size of the particles. Possible dust sources include the generation of ejecta particles by impacts, active volcanoes and geysers, and the capture of exogenous particles. Possible dust sinks include collisions with moons, rings, or the central planet, erosion due to sublimation and sputtering, even ejection and escape from the circumplanetary environment.

Keywords Circumplanetary dust · Planetary rings and tori · Dust sources and sinks · Dust dynamics

1 Introduction

Dust, being particles in a size range from about tens of nanometers to tens of micrometers, is one of the most abundant constituents of the universe. Our current knowledge suggests that there was no dust in the universe when the first massive stars and galaxies formed after the "big bang" from just the primitive elements hydrogen and helium.

Cosmic Dust from the Laboratory to the Stars
Edited by Rafael Rodrigo, Jürgen Blum, Hsiang-Wen Hsu, Detlef Koschny, Anny-Chantal Levasseur-Regourd, Jesús Martín-Pintado, Veerle Sterken and Andrew Westphal

✉ F. Spahn
fspahn@uni-potsdam.de

1 Institut für Physik und Astronomie, Universität Potsdam, Potsdam, Germany

2 Department of Physics, University of Colorado Boulder, Boulder, CO, USA

This had changed when these first-generation stars terminated in gigantic cosmic fireworks—supernovae, leaving a considerable amount of dust and heavy elements that comprise the basic building blocks of new star generations and a wealth of certain cosmic structures—including life. However, the dust populations observed today are not these primordial products of later stars, but are instead generated by erosion of larger objects lacking an atmosphere (see the review by Szalay et al. 2018, this issue) like planets, satellites, and asteroids—as well as geophysical activities of satellites (e.g. volcanism on the Galilean moon Io or "ice geysers" on the Saturnian moon Enceladus, see the review by Hillier et al. 2018) or comets (outgassing or breakup).

After their creation, the dust particles are subject to various forces, first of all the gravity of its source moon, the planet, other satellites, and the star (Sun). In addition, the smaller the dust particle, the more important non-gravitational forces become. The size of a dust-grain therefore determines its dynamical "fate" and decides when and where it will meet its final destination. For instance, a micron-sized particle (e.g. with a radius of $s = 0.5\ \mu\text{m}$) ejected from one of the Galilean satellites of Jupiter quickly develops (after several orbital periods) large eccentricities due to Solar radiation, so that it likely impacts the planet after a relatively short period of time (months to years). Much smaller grains (nanometers) are dominated by Jupiter's strong electromagnetic field, so that they can leave the planet's Hill sphere as so-called "stream particles". Much larger particles (tens of micrometers up to sub-millimeters) remain close to their parent moon and likely re-collide with it. This means, depending on the source, the mass $m(s)$ and size s of the dust grain, its dynamics decisively determines its lifetime τ_{life} in circumplanetary space. The distribution of lifetimes of a dust particle ensemble in turn determines its phase-space distribution for given dust sources (initial distributions). The longer a dust grain remains in a certain dust population (rings, tori, or dust clouds around atmosphereless satellites), the larger its lifetime τ_{life}, and thus, its contribution to the dust population. The existence of sinks and a finite lifetime τ_{life} also means that material is steadily removed from the system, so that the sources have to replenish the material at the same rate in order to guaranty an observable dust-distribution in a dynamic equilibrium.

Numerical and analytical methods to quantify the dust phase-space distribution and related quantities for certain dust populations are presented in Sect. 2. This sketch of a theoretical description of dust populations is followed by Sect. 3, which describes different planetary dust environments, including dust clouds embracing planetary satellites, as well as faint rings and tori along the orbits of their parent bodies. In the first instance this concerns the extended dust populations of the four giant planets. Special attention is paid to Saturn motivated by the stunning results of the Cassini mission, having been in orbit around Saturn for more than a decade. Additionally, we also address dust about the dwarf-planets Pluto-Charon, the putative tori about our neighbor Mars and very special of "tiny rings" around the centaurs Chiron and Chariklo, and the dwarf-planet Haumea.

2 Theoretical Modeling of Dust Populations

In this section, we sketch the theoretical description of such circumplanetary dust populations as faint and diffuse planetary rings. However, the approach can also be used for other particle systems in space.

The basic variable for a general description of particle systems (dust) in orbit around a central body (planet, star) is the distribution function $\tilde{n}(\mathbf{X}, s)$ in the phase space $\mathbf{X} = (\mathbf{r}, \mathbf{v})$

of a particle with a radius[1] s defining the particle number

$$dN = \tilde{n}(\mathbf{X}, s)d^6\mathbf{X} \tag{1}$$

expected in the phase space volume $d^6\mathbf{X}$. The evolution of this distribution function is described by the kinetic equation

$$\begin{aligned}\frac{D\tilde{n}}{Dt} = \frac{\partial\tilde{n}}{\partial t} + \dot{\mathbf{X}}\cdot\nabla_{\mathbf{X}}\tilde{n} &= \frac{\partial\tilde{n}}{\partial t} + \dot{\mathbf{r}}\cdot\nabla\tilde{n} + \dot{\mathbf{v}}\cdot\nabla_{\mathbf{v}}\tilde{n}\\ &= q_+ - q_- + \left\langle\frac{\partial\tilde{n}}{\partial t}\right\rangle_{\text{coll}},\end{aligned} \tag{2}$$

where the right hand side of Eq. (2) contains sources $q_+(\mathbf{X}, t; s)$ and sinks $q_-(\mathbf{X}, t; s)$ of particles of the size s. The last term stands for collisional interactions between the particles, whose form usually assumes only two-body interactions are relevant.

The collisional term is only important for dense rings, where the collisional frequency $\omega_c > \Omega$ is larger than the orbital mean motion Ω—but can safely be neglected for tenuous dust rings. Dense rings are composed of constituents with sizes between centimeters up to a few meters and their mean free paths are in the same order of magnitude. This causes the collision term $\langle\dot{\tilde{n}}\rangle_{\text{coll}}$ to dominate (Stewart et al. 1984). However, also the sources q_+ and sinks q_- are of importance because aggregation and fragmentation[2] of ring aggregates also need to be taken into account (Spahn et al. 2004). Recent Cassini images during the final F ring orbit phase strongly indicate that mutual gravity in combination with contact forces may cause an ongoing balance between aggregation and dissolution of ring aggregates of subkilometre size. It should be noted that for dense rings, both sources q_+ and sinks q_- are actually metamorphoses between different states (sizes s and phase space states $\mathbf{X}$) of the ring material (accretion vs fragmentation) at a certain distance from the center of the planet. These quantities do not address processes like genesis of the ring matter, in contrast to the case of diffuse rings described below. The formation of dense planetary rings is still an open question—and the theories of structure formation in dense planetary rings are based upon the assumption of a given ring mass (Charnoz et al. 2018), created by a yet unknown formation process (catastrophic disruption), which is steadily modified by collisions under the action of a Keplerian shear caused by the planet's gravity (Brilliantov et al. 2015).

After this brief excursion to the dynamics of dense rings, we will focus on circumplanetary dust populations—the main subject of this review. The main difference between the diffuse rings and their dense counterparts is an enormously larger mean free path $\lambda \propto s^{-2}n^{-1}$, where particle number density is defined by

$$n(\mathbf{r}, t, s) = \int d^3\mathbf{v}\tilde{n}(\mathbf{r}, \mathbf{v}, t; s). \tag{3}$$

Both, the sizes s of the dust particles—in the range of nanometers up to submillimetres—and the density n are orders of magnitudes smaller than for dense rings (whose constituents range from centimeters up to tens of meters). For instance, in the dense main rings of Saturn tons of icy ring aggregates jostle quite close to the ring plane in vertical (thickness) range of less than 10 meters. Assuming the majority of aggregates in dense rings are centimeter-sized (a power-law size distribution has been observed in dense rings and also theoretically

[1] Spherical shape of the grains is assumed, for simplicity.

[2] Both processes act as sources as well as sinks.

founded, see Brilliantov et al. 2015) gives a number density of the order $n \sim 10^3\ \mathrm{m}^{-3}$. By contrast, the densest part of the E ring of Saturn has been found near its main dust source Enceladus with $n \sim 1\ \mathrm{m}^{-3}$ (Ye et al. 2014b), which together with the small sizes of the E ring ice particles of $s \approx 1\ \mu\mathrm{m}$ gives 6 up to 7 orders of magnitude difference in the mean free path λ. In other words, the particles in diffuse rings practically never collide with another particle during their lifetime, giving $\langle \dot{\tilde{n}} \rangle_{\mathrm{coll}} \to 0$.

Consequently, for dilute rings, the dust sources $q_+(\mathbf{X}; s)$ and sinks $q_-(\mathbf{X}; s)$ (assumed to be steady in time t) are the only important terms in the right-hand side of Eq. (2). Recalling that $\dot{\mathbf{v}} = \mathbf{F}/m(s)$ characterise the specific particle forces, one finds that the evolution of $\tilde{n}$ is driven by the dust sources q_+, the sinks q_-, and the dynamics

$$m(s)\ddot{\mathbf{r}} = \mathbf{F}_{GP} + \sum_i \mathbf{F}_{GS_i} + \mathbf{F}_{EM} + \mathbf{F}_{RF} + \mathbf{F}_{PD}. \tag{4}$$

The forces on the right hand side of Eq. (4) are the gravity of the planet $\mathbf{F}_{GP}$ and its satellites $\mathbf{F}_{GS_i}$, followed by the Lorentz force $\mathbf{F}_{EM}$, radiation forces $\mathbf{F}_{RF}$ (caused by the Sun and containing radiation pressure and Poynting-Robertson drag), and the plasma drag $\mathbf{F}_{PD}$.

The first three terms on rhs are conservative, i.e. they do not exert work (a Hamiltonian exists) so that the dynamics does not change the phase space volume set by the sources (Liouville theorem). In this case, the kinetics is quantified by a continuity equation for the probability density $\tilde{n}$

$$\frac{\partial \tilde{n}}{\partial t} + \nabla_{\mathbf{X}} \cdot [\dot{\mathbf{X}}\tilde{n}] = q_+ - q_-, \tag{5}$$

meaning that the "probability fluid" is incompressible, i.e. $\nabla_{\mathbf{X}} \cdot \dot{\mathbf{X}} = 0$, confirming the conservation of the phase space volume under dynamical changes. Integrated over the phase space volume of the dust population

$$\mathrm{d}N = \tilde{n}\mathrm{d}^6\mathbf{X}; \ \Rightarrow N = \int_{\mathcal{B}} \mathrm{d}^6\mathbf{X}\tilde{n} \tag{6}$$

$$N^{+/-} = \int_{\mathcal{B}} \mathrm{d}^6\mathbf{X} q_{+/-}, \tag{7}$$

gives the number, source and sinks of dust particles of the kind s, respectively. Integrating Eq. (5) yields the total particle number of a certain size s

$$\frac{\mathrm{dN(s)}}{\mathrm{dt}} = N^+ - N^-, \tag{8}$$

where integration of divergence-term of the balance (5) (generalized Gauß-theorem) vanishes because the distribution function takes $\tilde{n} \to 0$ at the phase space boundary due to normalization requirements.

For the investigations reported in this review we consider stationary states, Eq. (5) equals zero i.e. $\dot{\tilde{n}} = 0$, i.e. the divergence of mass fluxes should vanish $\nabla_{\mathbf{X}} \cdot (\dot{\mathbf{X}}\tilde{n}) \to 0$ and the sources balance according to $q_+ = q_-$ at "quasi"-equilibrium. This together with the constancy of the phase space volume of Hamiltonian particle ensembles guarantees that the number of particles with radius s once set at the sources (initial state vector $\mathbf{X}_0$) is constant, quantified by

$$\tilde{n}(\mathbf{X}; s) = \tilde{n}\big[\mathbf{X}_0(\mathbf{X}); s\big]\left|\frac{\partial(\mathbf{X}_0)}{\partial(\mathbf{X})}\right|. \tag{9}$$

The last factor on the right hand side is the Jacobi determinant mediated by the dependence of the state **X** on the initial state $\mathbf{X}_0$ at the sources. This relation has successfully been used to quantify the impact-generated ejecta clouds embracing atmosphereless satellites elaborated by Krivov et al. (2003); Sremčević et al. (2003; see Sect. 2.1.1). In these cases, conservative forces can safely be assumed because the lifetimes of the particles in the vicinity of their sources are comparatively short (hours to days), so that dissipative forces (drag, radiation) do not play a role.

For circumplanetary dust populations, dust rings and -tori, this changes because the lifetimes of dust particles are on average much longer (thousands of orbital periods, about 100 years and more). During that time span dissipative forces, like plasma and Poynting-Robertson drag, have a considerable effect, especially for smaller, submicrometre-sized grains. In such cases the phase space trajectories cannot be computed analytically based on the initial state. Consequently, numerical simulations are necessary to characterize the Jacobian in Eq. (9), and to quantify the continuation of the density in the phase space.

In the following three subsections the creation (sources), transport (dynamics) and the removal (sinks) of dust will be discussed.

2.1 "Birth" of Dusty Grains: Dust Source Processes

Both exogenous as well as endogenous processes contribute as the sources of circumplanetary dust. Here we discuss three dust production processes most relevant for planetary environments: impact ejecta production, volcanism on active moons, and interplanetary particle capture.

2.1.1 Ejecta by Micrometeoroid Impacts—Impact-Ejection Process

A comprehensive description of the impact-ejection and the dust distribution around airless bodies is provided in the review by Szalay et al. (2018, this issue). Here we focus on the particles that escape the gravity of their parent bodies and add to the planetary dust environment as arcs (Hedman et al. 2009; Sun et al. 2017) or rings (Krivov et al. 2002a).

Generally, the amount of material ejected into circumplanetary space depends on the surface properties of the target as well as its gravity. For the process of impact-ejection, the mass flux of unbound ejecta is given by the mass production rate M^+ times the fraction of ejecta launched with speeds larger than the target's escape velocity $F(> v_{\mathrm{esc}})$

$$M^+_{\mathrm{esc}} = M^+ F(> v_{\mathrm{esc}}), \tag{10}$$

where the mass production rate scales linearly with the target's cross sectional area, $M^+ \sim R_{\mathrm{M}}^2$, while the fraction of unbound ejecta depends on the relation between the minimal ejection speed and the target's escape velocity, which in turn is proportional to the radius of the body, $v_{\mathrm{esc}} \sim R_{\mathrm{M}}$.

The mass flux of the ejected particles Eq. (10) is distributed over the ejection speeds v according to a power-law, consistent with impact experiments (Hartmann 1985):

$$f(v) = \frac{\gamma}{u_{\min}}\left(\frac{v}{u_{\min}}\right)^{-\gamma-1}\Theta(v - u_{\min}), \tag{11}$$

with the normalization

$$\int_0^\infty f(v)\mathrm{d}v = 1, \tag{12}$$

where $\Theta(x)$ is the unit step function, which is unity for $x \geq 0$ and zero otherwise. The index γ depends on properties of the target material and ranges from $\gamma = 1$ for (highly) porous to $\gamma = 2$ for nonporous materials (Krivov et al. 2003). The minimal speed $u_{\min}$—ensuring the convergence of the integral Eq. (12)—is dominated by the impact physics, in particular energy balance (Krivov et al. 2003). Assuming that the kinetic energy of the ejecta is a few percent to a few tens of percent of the kinetic energy of the impactor (Asada 1985; Hartmann 1985), typical values for $u_{\min}$ are a few 10 m/s (Krivov et al. 2003). These values are similar to the escape speeds of small planetary satellites with sizes of several 10 km, but about two orders of magnitude smaller than the escape speeds of the largest satellites in the Solar System.

With the above distribution Eq. (11), Eq. (10) gives for the mass flux of escaping ejecta

$$M^+_{\mathrm{esc}} = F_{\mathrm{imp}} Y \pi \left[\frac{u_{\min}}{\sqrt{\frac{8}{3}\pi G\rho}}\right]^\gamma R_{\mathrm{M}}^{2-\gamma} \quad \text{for } u_{\min} \leq v_{\mathrm{esc}} \tag{13}$$

and

$$M^+_{\mathrm{esc}} = F_{\mathrm{imp}} Y \pi R_{\mathrm{M}}^2 \quad \text{for } u_{\min} > v_{\mathrm{esc}}, \tag{14}$$

where the first equation applies to large bodies and the second equation to smaller ones with the transition at an object radius of $R_{\mathrm{M}} \approx 100$ km. Below that, the majority of particles escape the target's gravity and the amount of ejected material scales only with the target surface area. In Eqs. (13) and (14) the quantities F_{imp}, Y, ρ and G denote the impactor flux, the ejecta yield, the body's bulk density, and the gravitational constant, respectively. The yield is defined as the ratio of the total mass ejected by an impactor to its own mass, which strongly depends on the impact speed v_{imp} as well as the impactor mass m_{imp} (or size s_{imp}) and the composition of the target surface (silicate fraction x in %). We use an empirical relation for the yield from Koschny and Grün (2001), which reads (in SI units)

$$Y = 2.85 \times 10^{-8} (0.015)^{x/100} \rho_{\mathrm{t}} m_{\mathrm{imp}}^{0.23} v_{\mathrm{imp}}^{2.46}, \tag{15}$$

where

$$\rho_{\mathrm{t}} = \left(\frac{1 - x/100}{930} + \frac{x/100}{2800}\right)^{-1} \tag{16}$$

is the mean mass density of the material at the target surface assuming a linear mixing model with mass densities $\rho_{\mathrm{ice}} = 930\ \mathrm{kg/m^3}$ for ice at a temperature of 100 K and $\rho_{\mathrm{sil}} = 2800\ \mathrm{kg/m^3}$ for silicate.

Recently, Sachse et al. (2015) have generalized the impact-ejection model by allowing for a correlation between the size and the speed of the ejecta. This model upgrade is motivated by the fact that most of the processes (shock waves, drag etc.) mediate the momentum transfer (force $\mathbf{F}$) via the particle's surface ($\propto s^2$), but the acceleration $\mathbf{a}$ is dominated by the dust grain's mass $m(s) \propto s^3$ via Newton's second law $\mathbf{a} = \mathbf{F}/m \propto s^{-1}$, explaining why heavier particles get less accelerated.

The mass flux of unbound impact-ejecta is shown in Fig. 1 as a function of the radius of their parent body (solid lines), for the smallest (highly dissipative surfaces: $\gamma = 1.0$,

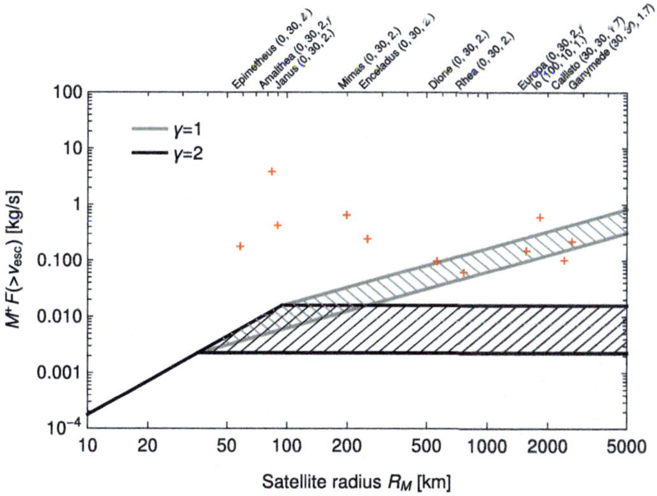

Fig. 1 Mass flux of unbound ejecta generated by impacts of micro-meteoroids onto planetary satellites of different sizes. The solid lines are Eqs. (13) and (14) evaluated for fixed impactor properties (impactor radius $s_{\mathrm{imp}} = 10\ \mu\mathrm{m}$, speed $v_{\mathrm{imp}}^{\infty} = 10$ km/s, projectile flux $F_{\mathrm{imp}}^{\infty} = 10^{-15}\ \mathrm{kg/m^2/s}$) and surface properties (pure ice), and for typical mean mass densities of planetary satellites between 0.5 $\mathrm{g/cm^3}$ (upper bound) and 3.5 $\mathrm{g/cm^3}$ (lower bound of hatched area). The red crosses are the mass fluxes of the moons given on the upper abscissa also considering the effect of the planet's gravitational focusing and the moons' surface properties. The fluxes for individual moons are systematically above the general trends primarily because of gravitational focusing. For the impactor properties at Jupiter, we use the values from Krivov et al. (2003), and for Saturn from Kempf et al. (2018a). The surface properties and related impact-ejection parameters of the moon are listed after the moons' names (silicate fraction in %, fraction of the kinetic energy of the impactor that goes into the kinetic energy of the ejecta in %, slope of cumulative speed distribution γ)

gray lines) and the largest slope (pure crystalline icy surfaces: $\gamma = 2.0$, black lines) of the ejecta speed distribution, and for typical mean mass densities of planetary satellites between 0.5 $\mathrm{g/cm^3}$ (upper bound) and 3.5 $\mathrm{g/cm^3}$ (lower bound of hatched area). To eliminate the influence of other parameters, we have used fixed impactor properties (projectile radius $s_{\mathrm{imp}} = 10\ \mu\mathrm{m}$, impact speed $v_{\mathrm{imp}}^{\infty} = 10$ km/s, projectile flux $F_{\mathrm{imp}}^{\infty} = 10^{-15}\ \mathrm{kg/m^2/s}$) and surface properties (pure ice).[3] In reality, each moon orbits at a distinct distance from its planet and has a characteristic surface. Therefore we have calculated the mass fluxes for a selection of moons (red crosses) also considering the effect of the planet's gravitational focusing on the flux $F_{\mathrm{imp}}^{\infty}$ (Spahn et al. 2006a) and the moons' surface properties on the impactor properties and the yield.

In general, large moons eject large amounts of material, but most of it falls back to their surfaces; while smaller moons eject less material, but most, if not all, escapes. Considering only the size of the moons, the amount of escaping material can vary by more than three orders of magnitude, where the larger moons are generally more efficient sources of dust. This, however, may change when the gravitational focusing by the planet and the moon's surface properties are also considered. For example, the four small inner satellites of Jupiter (one of them, Amalthea, is shown in Fig. 1) can be more efficient dust sources than the much larger, but more distant Galilean moons. This demonstrates that the mere extent of a body is not the only factor that determines its efficiency as a dust source.

[3] These dynamical properties determine the mass of ejected material (yield).

Dust Clouds Around Moons Given the velocity distribution Eq. (11) of the ejecta, for the dust clouds embracing a source moon of mass M_M, radius R_M, and at a distance r from the planet. For such dust clouds that fraction of dust with velocities below its escape speed, $u < v_{esc} \propto R_M$, is essentially relevant. These grains move along short-term ballistic trajectories inside the Hill scale $h \approx a_M \sqrt[3]{M_M/(3M_P)}$ (mass of the planet: M_P) before they fall back onto the surface. Thus the motion is well approximated by the Keplerian two-body problem—i.e. the gravity of the source satellite dominates the motion of the particle. The ballistic trajectories are sections of a Kepler ellipse determined by the integrals of motion—the energy $E(\mathbf{X}_0)$, and angular momentum $\mathbf{L}(\mathbf{X}_0)$ at the site of the initial impact $\mathbf{X}_0$ of the moon's surface ($\mathbf{X}_0 = (\mathbf{r}_0, \mathbf{v}_0)$—phase space coordinates of the grain of size s). These two constants of motion are sufficient to describe the phase space density distribution of the dust $\tilde{n}(\mathbf{X})$ embracing the source moon by using the Jacobi transformation Eq. (9), while grains falling back are steadily replaced by fresh ejecta.

A rough estimate of the stratification of the ejected dust at large distances R from the center of the source moon can be derived under the following assumptions: Given an isotropic projectile influx F_{imp} onto the surface of the source satellite, a yield Y (Koschny and Grün 2001), an ejecta distribution Eq. (11), and spherical symmetry of the resulting dust cloud, an integration of the phase space density over the ejection velocities $\mathbf{v}$ yields the spatial density for grains of a size s (Krivov et al. 2003)

$$n(R; s) = \int \mathrm{d}^3\mathbf{v}\tilde{n}(\mathbf{X}) = CR^{-5/2}. \tag{17}$$

However, this trend holds only far from the surface of the source moon and for bound ejecta, i.e. dust grains which will fall back on the surface. A general expression of the density $n(R; s)$ including also escaping ejecta at all ranges of R, i.e. also at low altitudes, are more involved and can be found in Krivov et al. (2003).

In a next step Sremčević et al. (2003) have waived the spherical symmetry in favor of the "windshield" effect, which is caused by Keplerian motion of the source satellite around its planet, where the symmetry breaking is expressed in the pre-factor C of the stratification Eq. (17). This radial density trend has been confirmed by the analysis of data obtained with the dust detector (DDS) of the Galileo spacecraft at the Galilean moons (Krüger et al. 2003b; Sremčević et al. 2005).

Some grains of the dust cloud around the moon could "leak" to form an arc or a complete ring along its orbit (e.g., Sect. 3.1.1). Dust clouds or dust halos are of particular interest, simply because they provide current samples of the moon surfaces/ring material to an orbiting spacecraft. In situ sampling of such dust halos around the Galilean satellites and the Moon by the Galileo and LADEE missions, respectively, provided constraints on the dust creation mechanism as well as the impactor flux (Horányi et al. 2015, see also chapter by Szalay et al. 2018, this issue).

Surprisingly, except for the dust arcs or tori near Pallene, Anthe and Methone (see Sect. 3.1.1; Sun et al., 2017), the Cosmic Dust Analyzer (CDA) of the Cassini spacecraft could not detect such impact-generated dust clouds around the Saturn's major icy moons, likely because of the elevated dust background from the extensive E ring.

2.1.2 Endogenic Dust Production: Volcanism on Io and Enceladus

Two volcanic moons, the Jovian moon Io and the Saturnian moon Enceladus, are active dust sources located inside the giant planets' magnetospheres. Particles with sizes of tens of

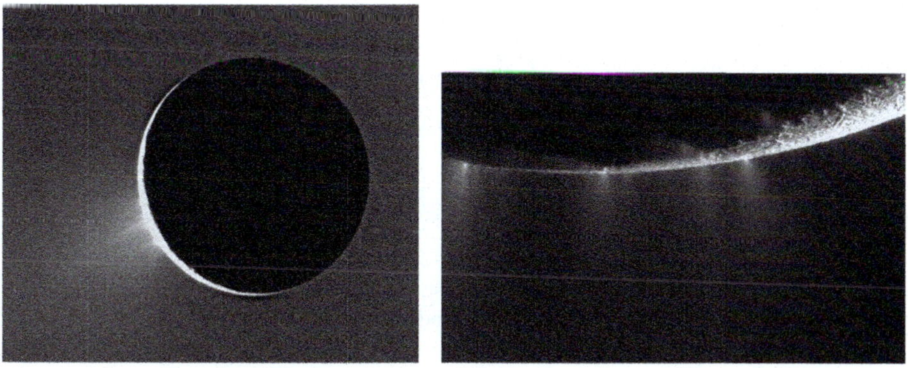

a) b)

Fig. 2 A plume of water vapor and icy grains towering over Enceladus' south pole as seen by the Cassini spacecraft (credit: NASA/JPL, PIA07758, PIA11688)

nanometer formed from the volcanic plumes of Io are dynamically coupled to the magnetosphere and eventually ejected into interplanetary space (Grün et al. 1998; Graps et al. 2000). In contrast, the dust emission from Enceladus' cryovolcanism discovered by the Cassini mission (Dougherty et al. 2006) is driven by the outward flowing water vapor supplied by the subsurface liquid sources. Our discussion below will be focused on the dust emission from Enceladus, as it shows resemblances to the aforementioned impact-ejection. Readers interested in the Io dust emission are referred to the review by Hillier et al. (2018) as well as Sect. 2.3.2 addressing the criterion for charged nanograin ejection from the magnetosphere (so-called "stream particles").

Enceladus is a medium-sized icy moon of Saturn with a radius of about 250 km located in the brightest part of the E-ring (Sect. 3.1.1). The pure-water-ice surface composition and the mean density of 1.6 $\mathrm{g\,cm^{-3}}$ imply that Enceladus is a differentiated body. In between its rocky core (radius 200 km) and the ice crust (5–30 km thick) lies a subsurface global ocean (Postberg et al. 2009; Iess et al. 2014; Thomas et al. 2016; Le Gall et al. 2017). A plume composed mainly of water vapor and ice grains emerges from the four warm surface features called "Tiger Stripes" located in its geologically young South Polar Terrain (Porco et al. 2006; Spahn et al. 2006b; Spencer et al. 2006; Waite et al. 2006; Spitale et al. 2015). Roughly 150–300 kg/s of water vapor is emitted from the plume (Hansen et al. 2006). Through curtains and jets, tens of kg/s of ice grains launched by the outflowing water vapor form the dusty plume as seen in Fig. 2. Most of them fall back to the moon, with $< 10\%$ capable of entering into the E ring (Juhász and Horányi 2004; Spahn et al. 2006b; Schmidt et al. 2008; Kempf et al. 2010; Mitchell et al. 2015). The plume activity dominates the ejecta production by at least a factor of several, making it the major source of Saturn's E ring (Spahn et al. 2006b; Kempf et al. 2010).

A grain formation model considering nucleation at narrow nozzles in the vents connecting the surface to the subsurface waters successfully explain several plume observations. In this model, the size-dependent speed distribution f_c of ice grains at the outlet of the channel can be parameterized as (Schmidt et al. 2008)

$$f_c(s, v) = \left(1 + \frac{s}{s_c}\right)\frac{s}{s_c}\frac{v}{v_{\mathrm{gas}}^2}\left(1 - \frac{v}{v_{\mathrm{gas}}}\right)^{\frac{s}{s_c}-1}, \tag{18}$$

with the normalization

$$\int_0^{v_{gas}} f(s,v)\mathrm{d}v = 1 \tag{19}$$

and the critical radius[4]

$$s_c = \frac{\rho_{gas}}{\rho_g}\sqrt{\frac{8k_B T_{gas}}{\pi m_0}}\left[1+\frac{\pi}{8}(1-\beta_{abs})\right]\frac{L_{col}}{v_{gas}}. \tag{20}$$

Here v_{gas} is the gas speed at the channel outlet, which constitutes an upper bound for the speed of the particles as they are (re-)accelerated by the gas (after collisions with the walls) and cannot move faster. The square-root in Eq. (20) is the mean thermal speed of the gas molecules of a mass m_0. The critical radius s_c further depends on parameters of the gas (density ρ_{gas}, temperature T_{gas}, speed v_{gas}), the bulk density of the grains ρ_g, and the characteristic length between successive wall-collisions L_{col}. This length is proportional to the diameter of the channel rather than its length and therefore the speed of the particles does not depend on the thickness of the ice crust. Small particles (with radius $s < s_c$) are (re-)accelerated very efficiently by the gas and obtain speeds close to the gas speed. Larger particles ($s > s_c$) are on average much slower than the gas because of their large mass-to-surface ratio. The condensation coefficient β_{abs} quantifies the adsorption of water molecules by growing grains and ranges from 0.1 to 1 (Shaw and Lamb 1999; Batista et al. 2005).

Three composition types are known for ice grains originated from Enceladus (Postberg et al. 2009, 2011a). All types are dominated by water ice but differ in the amount of salt and other minor constituents. Types I and II both show traces of salt (salinity 10^{-8}–10^{-6}) and they are labeled "salt-poor". In addition to water ice, Type II particles contain also large amounts of organic compounds and/or silicates, which are assumed to originate from the rocky core (Zolotov 2007). Types III particles have a significantly higher salinity than the other types (10^{-3}–10^{-2}) and they are referred to as "salt-rich". Particles with an intermediate salinity are rare.

The different composition of the various particle types suggests that they are formed by different mechanisms. Type I particles are assumed to be formed from the salt-poor vapor phase[5] by homogeneous nucleation. By contrast, Type II and Type III particles are assumed to be formed by heterogeneous nucleation. Bursting bubbles of rising volatile gases are thought to eject salty water droplets into the vapor phase, which serve as nucleation seeds where water molecules can deposit. In this case, Type II particles are thought to be formed from a thin organic-rich film on top of the oceanic water table, whereas Type III particles are thought to be formed from the salty ocean water.

Cassini measurements in the plume (Postberg et al. 2011a) show that directly above the south pole most grains mass are salt-rich (> 99% by mass, corresponding about 70% by number). This suggests that salt-rich particles are generally larger (> half a micron) than salt-poor ones and therefore they are ejected with lower speeds (see Eq. 18). The size difference is consistent with the idea that they started as frozen water droplets with additional growth by condensation as they ascend.

Another place where internal geological activity could be a source of dust particles is Jupiter's moon Europa. The structure of a hypothetic dusty plume on Europa, assuming a

[4]For triple point conditions of water and wall-collision lengths $L_{Col} \approx 10^{-1}$ m the critical radii take submicrometers (Schmidt et al. 2008).

[5]The salt is not fully dissolved in the water, leading to a small salt concentration in the vapor phase (Postberg et al. 2009).

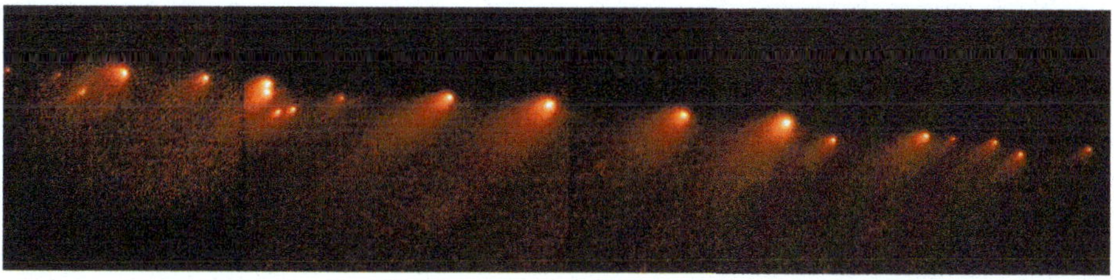

Fig. 3 The comet Shoemaker-Levy 9 (SL9), formed a 'string of pearls' of 21 pieces after its earlier close encounter with Jupiter resulting in its tidal disruption in 1992. During this event copious amounts of dust have been released in Jupiter's magnetosphere (credit: H. Weaver (JHU), T. Smith (STScI), NASA)

similar grain emission mechanism, has been carried out (Southworth et al. 2015), as motivated by the recent putative detection of intermittent, stochastic plume activities (Roth et al. 2014; Sparks et al. 2016, 2017) and the associated surface thermal anomaly site (McGrath and Sparks 2017; Sparks et al. 2017). Because of Europa's much stronger gravity (an escape speed of ~2000 m/s, compared to ~240 m/s at Enceladus), *in situ* detection of such a dusty plume requires a much lower flyby altitude of 5 to 100 km (Southworth et al. 2015).

2.1.3 Other Sources: Particle Capture

Capturing exogenous dust particles into orbits around planets requires mechanisms that efficiently dissipate their energy and angular momentum. Three body interactions and atmospheric drag are traditionally recognized processes (Jewitt and Haghighipour 2007). However, the interaction of small charged dust particles with the electric and magnetic fields in planetary magnetospheres (Axford and Mendis 1974; Colwell and Horányi 1996) can be a much more efficient way to capture dust grains into circumplanetary orbits (Horanyi 1996).

At Jupiter, captured interplanetary and interstellar dust grains have been suggested to form the diffuse faint ring detected by the Galileo Dust Detector System (Grün et al. 1997). This ring is comprised of particles with radii 0.5 to 1.5 μm on both retrograde and prograde orbits in a 4:1 ratio, with semimajor axes $3 < a < 20\ R_J$ Jovian radii,[6] eccentricities $0.1 < e < 0.3$, and inclinations $i < 20°$ or $i > 160°$. The size range, the orbital characteristics, and the density of this cloud are consistent with numerical trajectory integrations of dust particles of both interplanetary and interstellar origin entering Jupiter's magnetosphere (Colwell et al. 1998).

In addition to the continual influx of dust particles, there are additional intermittent enhancements of dust influx due to comets (Hill and Mendis 1979, 1980). One such relatively recent rare event was the tidal disruption of comet Shoemaker-Levy-9 (SL9) in 1992, breaking apart into 21 fragments forming a 'string of pearls' (Fig. 3), eventually all impacting Jupiter in 1994 (Weaver et al. 1993). Numerically integrating the orbital evolution of small dust particles copiously generated from the break up of SL9 it has been predicted that in about a decade following the break up, a small fraction of the dust will settle into orbits well inside the magnetosphere (Fig. 4). The forming ring is expected to be comprised of dust particles with radii in the range of 1.5 to 2.5 μm following retrograde orbits in the radial range of 4.5 to 6 R_J with a peak optical depth in the range of $10^{-8} < \tau_{max} < 2 \times 10^{-6}$ (Horányi 1994).

[6] Jupiter radius $R_J = 71492$ km.

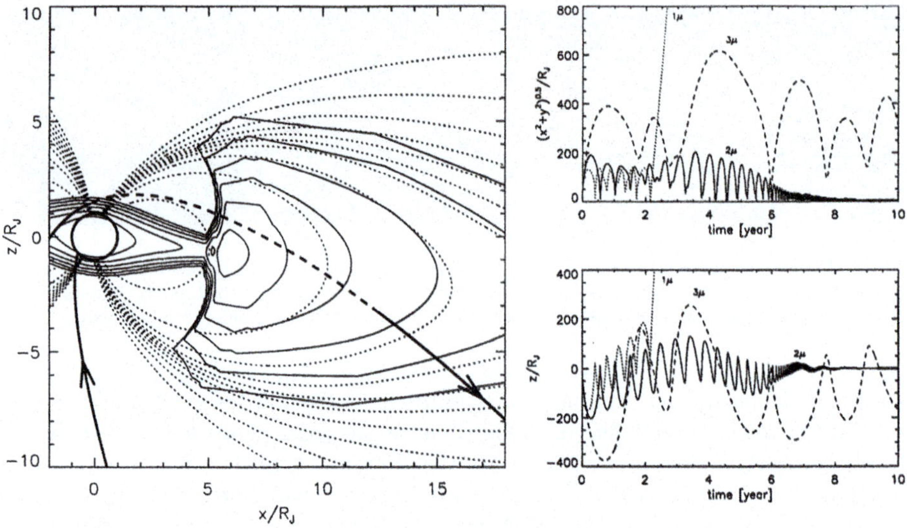

Fig. 4 *Left*: The trajectory of SL9 (thick continuous and dashed line) in a jovicentric coordinate system (z is along the rotational axis and x is in the equatorial plane of Jupiter pointing towards the Sun's position on July 20, 1994). Dust production was assumed to continue during the first 12 h following the break up, along the trajectory of SL9 (dashed section). The arrows show the direction of the motion of the SL9 fragments. The snapshot of plasma density contours (thin lines) mark the values of $n_p = 10^{-2}, 10^{-1}, 10^0, 10^1, 10^2, 10^3$ cm^{-3} and the magnetic field lines shown (dotted lines) pierce the magnetic equator at $L = 4, 8, \ldots, R_J$. *Right*: The trajectory of dust grains with radii of 1, 2, and 3 μm released at the break up of SL9 (t = 0). The top panel shows their distance in the equatorial plane of Jupiter; the bottom panel shows their position above/below that plane (Horányi 1994)

Magnetospheric dust capture is also likely to be operating at Saturn, however the lifetime of the captured particles there is expected to be vanishingly short due to collisions with E ring particles (Mitchell et al. 2004, 2005)

2.2 "Life" of Particles: Dust Dynamics

The mobility of cosmic dust makes it a unique tool to study the planetary environment and their source(s). Understanding the dynamical evolution of cosmic dust is an essential part of the so-called "Dust Astronomy" concept. Proposed by Grün et al. (2001), the Dust Astronomy concept envisions planetary scientists using dust grains very much like astronomers using photons to study the environment that cannot be probed by other means. Studying the dynamical transportation and distribution processes of dust grains from their parent bodies allows us to identify structures of dusty rings and the subtle forces operating behind them, as well as to probe the composition of their source bodies remotely.

The most important parameter regarding the dust dynamics is their masses (resp. sizes and shapes). Celestial objects much larger than typical dust particles, such as the planets and the satellites, strictly follow the rule of gravity. However, for dust with sizes ranging from a few micrometers down to nanometers the dynamics is far more involved and complex, other forces become comparable or even dominate the gravity of the planet. Additionally, erosion processes, e.g. plasma sputtering, cause the grain size to vary throughout its lifetime, crucially affecting their dynamical evolution. In other words, it is the microscopic scale of cosmic dust grains that leads to their intriguing dynamical behavior. As a result of these dynamical processes, grains can be put into unstable orbits, i.e., either colliding with other bodies or leaving the system, or survive and accumulate to form dusty rings around planets.

The lifetime of circumplanetary dust is in the range of hundreds to thousands of years. This relatively short dynamical "life" of individual dust particles indicates a non-primordial origin of the circumplanetary dust-populations, and requires their steady material replenishment from the aforementioned dust sources (see Sect. 2.1).

2.2.1 *Dust-Relevant Forces: Gravity & Perturbations*

Dust grains around planets are subject to a wealth of forces in addition to the everywhere-dominating gravity of the planet and its satellites (and sometimes even the Sun). Their dynamical fate is largely determined by the ratio of these forces, which is dominated by the mass (size) of the dust grain. Generally speaking, larger grains (i.e., $\geq$ a few micrometers) behave quite similar to planetary moons, with their dynamics basically governed by gravity. This explains why such larger dust particles will closely follow the orbits of their source bodies. In contrast to grains of interplanetary or interstellar origin, the majority of such larger, circumplanetary grains are expected to move in prograde, Keplerian orbits. This means that the location of larger dust particles provides a first-order indication for their putative source bodies.

However, the less massive (smaller) a grain, the more important radiation pressure and Lorentz forces become. The radiation pressure force depends on the photon scattering properties of the particles, which is a function of the radiation's wavelength and the particle's size. The radiation pressure is a directional force and its amplitude is often expressed by a so-called β value—the ratio between the radiation pressure and the Sun's gravity as a function grain size (Burns et al. 1979). The maximum values of β range from 0.5 to 1.5, depending on the material properties and peak at sizes around 0.2 micron. Grains with $\beta > 1$ are essentially blown away from the Sun, forming the so-called β meteorites in interplanetary space moving in hyperbolic orbits. Similarly, solar radiation pressure is a major perturbation force for circumplanetary dust.

One example of the action of the radiation on circumplanetary dust particles is the "heliotropic" dusty ringlet within Saturn's Cassini Division (Hedman et al. 2010). It is composed of grains ~20 micrometers across with orbits that are not symmetric with respect to the planet but instead have their orbital apsidal lines shifted towards the Sun. This heliotropicity is caused by the solar radiation that modifies the grain's orbital eccentricity (and inclination) in conjunction with an orbital precession caused by planetary oblateness. The situation for sub-micron grains becomes even more complex as the amplitudes of radiation pressure, the Lorentz force (which also causes orbital precession), and gravity all become comparable.

Furthermore, drag forces may also influence considerably the dynamics of sub-micron particles. Drag is caused by the momentum transfer from the neutral gas, plasma ions, or the non-direct component of the solar radiation pressure (i.e., the Poynting-Robertson drag, caused by the motion of the grain relative to the incident light). The drag forces result in either loss or gain of the grain's orbital energy, corresponding to a reduction or an increment of the semi-major axis, respectively. Drag forces become more effective for the smaller grains because of their larger surface-to-mass ratio. The effects and the relative importance of drag forces depend on individual environment and thus vary planet by planet. For example, grains in Uranus' inner most ring ζ may experience significant gas drag from Uranus' extensive hydrogen corona (Uranus' gas exosphere), while grains in Saturn's E ring gradually gain orbital energy from the co-rotating water group ions of the magnetospheric plasma (direct plasma drag) and their orbital semi-major axes expand outwards.

2.2.2 *Charging Processes and Dust–Plasma Interactions*

Their high surface-to-mass ratio makes dust grains especially susceptible to interactions with the magnetospheric plasma environment. In the case of circumplanetary dust, we focus on the following three aspects: grain charging, sputtering erosion, and dusty plasma effects.

Grains in space are typically charged by collecting plasma electrons and ions, emitting electrons due to the bombardment of energetic plasma particles and the Solar UV photons. The Lorentz force acting on grains depends on its charge-to-mass ratio. Thus, the grain dynamics is essentially coupled to the plasma conditions through the charging processes in presence of a planetary electro-magnetic field. The Orbital Motion Limited theory is often used to calculate the charging currents for an electrostatic isolated dust grain, i.e., a grain immersed in an homogeneous plasma with a plasma Debye length (λ_D), $s \ll \lambda_D < d$, where d is the inter-grain distance. A grain's charge $q(s)$ and its polarity are determined by the sum of all charging currents, which can be written as

$$\dot{q}_d = J_e + J_i + J_{ph} + J_{\mathrm{sec}}, \tag{21}$$

where $J_{e,i}$ represent the collection currents of ambient plasma electrons and ions, respectively, J_{ph} is the grain photoemission caused by Solar ultraviolet radiation, and J_{sec} is the secondary electron emission caused by energetic electron populations (Horanyi 1996).

To first order, the amount of charge resting on a dust grain (usually expressed in the surface potential) is controlled by the electron temperature. The charging time is determined by the plasma density, the plasma electron temperature, and by the size of the particle when considering grain sizes $\ll 1$ µm (see Horanyi, 1996 for more details).

The properties of thermal plasma in the Jovian and Saturnian magnetospheres are mainly functions of the distance from the planet. As the magnetospheric plasma is supplied by the ionization of neutrals provided from the source moons (i.e., Io and Enceladus, see the review by Hillier et al. 2018), the plasma density is the highest (2000 and 100 cm^{-3}, respectively) while the plasma temperature is the lowest near the orbits of the source moons. The outcome of such configurations is that the grains are charged negatively in the vicinity of the source moons, due to the dominance of electron collection current, and become charged positively further away, mainly because of the secondary and photo- electron emission. The grain charging conditions in the dusty rings around Uranus and Neptune are less clear because the lack of in situ measurements and also the more complicated magnetospheric configurations.

Plasma sputtering is an erosion process of solid material caused by impacts of energetic ions and it constitutes an important grain loss mechanism, which depends on both, the grain material and the ion energy. This is discussed further in Sect. 2.3.3.

In most space environments the amount of electric charge carried by grains is negligible compared to the ambient plasma. However, in dust-rich regions the grain charge density may become significant to make the charged grains an integral part of the plasma, forming the so–called dusty plasma species. The dust-population must be considered being an extra, complex plasma species whose mass and charge are poorly known.

One indication of a dusty plasma is the so-called electron bite-out phenomenon. Because the electron collection current mostly dominates other charging processes, grains are expected to become negatively charged and serve as a major electron sink under dusty plasma conditions. Laboratory and in situ space measurements show that the presence of abundant charged grains is the cause of the reduction of free plasma electrons in comparison to plasma ions (Barkan et al. 1994; Wahlund et al. 2009). It has been suggested that the presence of abundant sub-micron and nano- dust populations causes a substantial electron depletion in the plume of Enceladus (Morooka et al. 2011; Jones et al. 2009; Yaroshenko et al. 2009; Hill et al. 2012; Meyer-Vernet 2013). Partly due to the complexity of data interpretation, the amplitudes of the electron depletion derived from different methods vary, ranging from 50% (Kriegel et al. 2014) to more than 99% (Morooka et al. 2011) of plasma electrons being attached to grains. However, it is evident that the reduction of free electrons leaves a measurable signature in the magnetic field structure, adding to the already complex interactions between Enceladus' plume and the magnetosphere (Simon et al. 2011; Kriegel et al. 2014; Ye et al. 2014b; Dong et al. 2015; Meier et al. 2015; Yaroshenko and Lühr 2016).

2.3 "Final Destination" of Dust: Dust Sinks

The "birth-life-death"-cycle is concluded when the dust particles are removed from the planetary system by one of the sinks. Potential sinks for circumplanetary dust particles are collisions with larger bodies (planet, its main dense rings, and moons), escape from the planetary system, interactions with other particles, plasma sputtering, and possibly mutual dust particle collisions.

The relative importance of these sinks depends on the governing dynamics in the planetary system and thus is also size-dependent. For instance, plasma sputtering is more important for micron-sized E ring ice grains in Saturn's magnetosphere, but not as much, compared to other sinks, in Jupiter's main rings. In the following, we discuss each of the above mentioned sinks in more detail.

2.3.1 Impact with the Planet, Its Dense Rings & Its Satellites

In most cases the dominant sink for circumplanetary dust particles is re-impacting either their parent body, or the planet, or (in the case of Saturn) dense rings.

The collision probability between two orbiting objects has been studied by several authors (Öpik 1951; Wetherill 1967; Kessler 1981). The basic idea is to apply the particle-in-a-box model, where the bodies are distributed homogeneously in the interaction volume and can move freely between collisions. Similar to the kinetic theory of gases, with the relative speed Δv between a dust grain and another body (target), the flux density $F = n\Delta v$ quantifies the number of impacts per unit cross-sectional area and per unit time. The spatial density of the dust particles $n = \Delta t/(T\Delta V)$ scales with the fraction of time $\Delta t/T$ of their orbital period the particle spends in the interaction volume ΔV. Then the number of collisions with a body of cross-sectional area σ in the time period t is $P = f\sigma t$, so that the collision probability P_{coll} per orbital period $t = T$ reads

$$P_{\text{coll}} = \frac{2\Delta v\sigma\Delta t}{\Delta V}, \tag{22}$$

where the factor two considers two crossings per orbit (when the particle is moving outwards and when it is moving inwards).

For the collision probability per orbital period T of a dust particle with a planetary satellite on a (nearly) circular orbit, the evaluation of the above equation gives (Kessler 1981; Spahn and Wiebicke 1989)

$$P_{\text{coll}} = \frac{\Delta v\sigma T}{2\pi^3 a_{\text{M}} a \sin i\sqrt{(a_{\text{M}} - r_{\text{min}})(r_{\text{max}} - a_{\text{M}})}} \tag{23}$$

for $r_{\text{min}} < a_{\text{M}} < r_{\text{max}}$ and $P = 0$ otherwise. $r_{\text{min/max}}$ are the pericenter and apocenter of the particle orbit, respectively, a and a_{M} are the semi-major axes of the particle and the moon, respectively, and i is the inclination between both orbits.

The collisional cross-section σ of the moon differs from its geometric cross-section πR_{M}^2 because the moon's gravity is not considered in this geometric model. It has to be corrected for the effect of gravitational focusing by the moon, i.e. by the Safronov factor (Safronov 1972)

$$\sigma = \pi R_{\mathrm{M}}^2 \left(1 + \frac{v_{\mathrm{esc}}^2}{\Delta v^2}\right), \tag{24}$$

where the expression in the brackets is close to unity for small moons or encounters with high relative velocities.

A collision with the planet occurs if the pericenter of the particle orbit lies inside the planet, i.e.

$$P_{\mathrm{coll}} = 1 \quad \text{if} \quad r_{\min} < R_{\mathrm{p}}. \tag{25}$$

The collision probability for encounters with multiple bodies is additive

$$P_{\mathrm{tot}} = \sum_i P_{\mathrm{coll}}^{(i)} \tag{26}$$

and the lifetime is given by

$$\tau_{\mathrm{life}}^{(\mathrm{coll})} = \frac{T}{P_{\mathrm{tot}}}. \tag{27}$$

2.3.2 *Escape from the Planetary System ("Streams")*

An object can potentially leave a planetary system when its apocenter exceeds the planet's Hill sphere, i.e. the probability of escape is

$$P_{\mathrm{esc}} = 1 \quad \text{if} \quad r_{\max} > R_{\mathrm{h}}, \tag{28}$$

where

$$R_{\mathrm{h}} = a_{\mathrm{P}} \left[\frac{M_{\mathrm{P}}}{3(M_{\odot} + M_{\mathrm{P}})}\right]^{1/3}, \tag{29}$$

is the Hill sphere of the planet, e.g. for Jupiter, $R_{\mathrm{h}} \approx 740\ R_{\mathrm{J}}$. M_{P} and a_{P} are the mass and the semi-major axis of the planet, respectively, and $M_{\odot}$ is the mass of the Sun.

For example, nano-sized particles have a very large charge-to-mass ratio, so that they are primarily governed by electromagnetic forces. Depending on the charge of the particles and the orientation of the magnetic field, the grains are forced on highly elliptic orbits, potentially satisfying condition (28), or on hyperbolic orbits, in which case they are guaranteed to escape the system. In the Jovian system, positively charged grains experience an effective acceleration that is radially repelling, so that they are rapidly expelled from the system. Dust streams originating from the Jovian system were first detected by the spacecraft Ulysses (Grün et al. 1993) and later by Galileo (Grün et al. 1996). With the energy gained from the corotating electric field, grains with sufficient charge–to–mass ratio and the "correct" charge polarity, depending on the magnetic field orientation, could overcome the gravity of the host planet and escape the magnetosphere (Horányi et al. 1993).

A measure for the strength of the planet's magnetic field, is the L parameter, defined as the ratio of the electric component of the Lorentz force relative to the planet's gravity

(Hamilton 1993)

$$L = \frac{q_d B_0 R_p^3 \Omega_p}{G M_p m_d} \tag{30}$$

where q_d and m_d are the grain charge and mass, respectively. B_0, R_P, and Ω_P are the surface magnetic field strength at equator, the radius, and the rotation rate of the host planet, respectively.

The lifetime of electromagnetically dominated ($L > 1/2$) dust particles has been estimated by Liu et al. (2016)

$$\tau_{\mathrm{life}}^{(\mathrm{em})} \simeq \frac{R_h}{2}\sqrt{\frac{a_M(2L-1)}{G M_p (L-1)^2}}, \tag{31}$$

which evaluates to a few tenths or hundredths of a year.

Considering the energy conservation (Kempf et al. 2005; Hsu et al. 2012), the upper limit of grain radius ($s_{\max}$) for ejection by the outward pointing co-rotational electric field is

$$s_{\max} = \sqrt{\frac{6\varepsilon_0 \phi_d}{\rho_d}\frac{\Omega_p B_0}{G M_P}\left(1 - \frac{r_0}{r_{ms}}\right)}, \tag{32}$$

where ϵ_0 is the vacuum permittivity, and ρ_d and $\phi_d = q_d/4\pi\epsilon_0 s$ are the grain bulk density and surface potential, respectively. r_0 is the starting distance measured in planet radii and r_{ms} is the size of the magnetosphere, i.e., where the acceleration occurs.

The nanodust dynamics in interplanetary space is governed by the interplanetary magnetic field originating from the Sun, which provides a direct constraint on their dynamical properties (Zook et al. 1996; Hsu et al. 2010). The volcanic plumes of Io are identified as the major sources of the Jovian stream particles (Graps 2001). The smaller, slower Saturnian counter part is supplied from Saturn's E ring (ultimately Enceladus) and potentially the main rings (Kempf et al. 2005; Horányi 2000; Hsu et al. 2011, 2015). More details can be found in the review by Hsu et al. (2012).

2.3.3 *Plasma Sputtering & Particle-Particle Interactions*

Plasma Sputtering Energetic ions bombarding the surface of a dust particle can liberate ions and neutrals from the grain. This sputtering process leads to the steady erosion of the particle, with a mass loss rate

$$dM/dt = 4\pi s^2 \langle m \rangle \int F(E) Y(E) dE, \tag{33}$$

where M is the mass of the grain, s is its radius, $F(E)$ is the flux of bombarding plasma ions with energy E, and $Y(E)$ is the yield, the number of liberated ions (with average mass $\langle m \rangle$) in response to a single incoming ion impact. The yield is independent of the particle size s as long as the characteristic penetration distance of the bombarding ions $\lambda \gg s$. There is a significant increase in the yield Y for $\lambda \lesssim s$ (Jurac et al. 1998).

Equation (33) indicates that for $\lambda \le s$, the radius of the dust particles decreases at a size-dependent rate

$$\frac{ds}{dt} = -\frac{s}{\tau_{\mathrm{life}}^{(\mathrm{sp})}}, \tag{34}$$

where $\tau_{\text{life}}^{(sp)}$ is the characteristic lifetime against sputtering. For the conditions in Saturn's E-ring, for example, using the energy dependent yield $Y(E)$ in this plasma environment (Jurac et al. 2001a,b) the ice grain lifetime is $\tau_{\text{life}}^{(\text{sp})} = 50$ years for $s = 1$ μm near Enceladus, increasing to $\tau_{\text{life}}^{(\text{sp})} = 600$ years near Titan (Horányi et al. 2008).

Particle Collisions Another potential sink are mutual collisions between dust grains resulting in restitution, aggregation or fragmentation of both collision partners (Spahn et al. 2004). The latter two cases—aggregation and fragmentation—generally reduce the number of grains of that mass range which the collision partners originally belong to. Even in the stationary state, where source- and sink particle rates balance as well as size- and particle number densities will not change, material is steadily eroded to nano- or molecular scales and finally removed from the system.

The outcome of a collision depends on the masses m_α; m_β and the relative ("thermal") speed $c(m_\alpha, m_\beta)$ between the collision partners[7]—quantities that can be derived from observations of circumplanetary dust rings.

The vertical thickness provides a measure for the velocity dispersion $c \approx \langle H \rangle \Omega$ of the dust ring, where Ω is the Keplerian mean motion (resp. Keplerian shear $2\partial u_\varphi/\partial r = 2\partial(r\Omega)/\partial r = -\Omega$). The observed thickness of most of the dust-rings in the Solar System point to high thermal speeds at collisions resulting in fragmentation of the collision partners. The kinetics of such processes is rather involved (Spahn et al. 2004) mainly because of the poor knowledge of the debris distribution $\mathcal{P}(m, \mathbf{v}|m_\alpha, m_\beta; \mathbf{v}_\beta - \mathbf{v}_\alpha)$.

Fortunately, such a definite description of a collision is not needed for the estimates in this review. In order to estimate an upper bound for the collisional lifetime of the particles, we only consider destructive impacts and assume that their frequency is sufficiently high. In this way we neglect bouncing or aggregating collisions, and thus, we intentionally overrate fragmenting collisions as sinks for this calculation. Additionally, we assume that the dust ring consists of identical particles of radius $s_0 = 1$ μm: $\hat{n}(s) \propto \delta(s - s_0)$, which crucially simplifies the calculations.[8] These simplifications suffice to provide an order-of-magnitude estimate of the fragmentation sink, so that the particle removal rate can be written as

$$\dot{n} \approx -n\omega_\text{c}(n) \approx -n \underbrace{\left[\pi s_0^2 v_\text{th} n\right]}_{\text{collision frequency } \omega_c} \propto -n^2, \tag{35}$$

with the particle number density n. The analytical solution of this conservative, non-linear decay model is

$$n(t) = \frac{n_0}{1 + \omega_\text{c} t}, \tag{36}$$

with the starting density $n_0 = n(0)$ and the corresponding initial collision frequency $\omega_\text{c} = \pi s_0^2 v_\text{th} n_0$. The half-value period is thus found as $t_{1/2} = \omega_\text{c}^{-1}$, i.e. the number density has dropped by 50% after the time $t_{1/2}$, which provides a characteristic timescale of the efficiency of a fragmentation sink. It has to be noted that the decay occurs much slower than in

[7]The velocity dispersion c^2 is second central moment of the velocity distribution $f(v)$, i.e. c could be considered as the averaged modulus of the thermal speed.

[8]Power-law size distributions $n(s) \propto s^{-q}$ are often observed, but for slopes $q > 2$ optical properties are dominated by the smallest particles, alternatively, narrow distributions, as the E ring dust, support this simplification.

the case for a constant ω_c giving $n(t) = n_0 \exp(-\omega_c t)$. Nevertheless, the initial half-value period $t_{1/2}$ are very similar in both cases!

In order to estimate the timescale ω_c^{-1}, we will use the observable optical depth $\tau \approx \pi s_0^2 n_S$ with the projected (ring-plane) surface number density $n_S \approx n\langle H\rangle$ and recalling the ring thickness given by $\langle H\rangle \approx v_{th}/\Omega$, one may write[9]

$$\omega_c \approx \tau\Omega = 2\pi\frac{\tau}{T}. \tag{37}$$

For diffuse rings the optical depth $\tau(r)$ is much more variable than the orbital period $T(r)$ so that it dominates the influence of collisions as a sink.

For dust rings with an optical depth of $\sim 10^{-6}$, inter-particle collisions are only important for grains with lifetimes longer than 10^5 orbital periods (>100 years). Although we have favored fragmenting collisions in our sink evaluation, we come to the conclusion that mutual collisions between ring particles are practically negligible for the most dust populations. Potential exceptions for several dusty rings will be discussed later in this paper, like Saturn's F ring and/or the dusty arcs of Neptune.

3 Observations of Dust Populations

3.1 Introduction

Since the era of deep space missions to the outer Solar System and space telescopes, we have been aware of planetary rings surrounding all giant planets from Jupiter to Neptune. There are two types of planetary rings:

- *Dense* rings are constituted of macroscopic aggregates with sizes ranging from centimeters to a few meters, which collide more than a dozen times per orbit (depending on the optical depth, τ). These physical contacts are highly dissipative, causing dense rings to be hydrodynamic granular systems. One of the most remarkable properties of dense planetary rings is their extreme flatness—for example, Saturn's rings show a lateral width up to hundred thousand kilometers whereas they are only a few meters thick!
- *Diffuse* rings consist mainly of micrometer- or even sub-micrometer-sized dust particles, which are strongly affected by a suite of forces apart from gravity. The resulting much higher mobility of the dust particles, compared to the aggregates of dissipative dense rings, suggests much smoother ring edges and a larger extent of dusty rings in any direction—unless being otherwise confined or disturbed by nearby moons like Saturn's F ring or Uranus' λ ring. Such circumplanetary dust populations are the subject of this review.

In the following subsections, we will sketch the dust environments of the planets of the Solar System including the Earth, Mars, the dwarf planet Pluto, and even the properties of the miniature versions of rings girdling two Centaurs and a Kuiper-belt object are summarized.

[9]The index "0" is dropped in following because real dust rings are steadily sustained by sources so that the quantities n and τ are assumed stationary!

Fig. 5 "The Lord of the Rings"—Saturn and its gorgeous rings occult the Sun leaving behind a glowing shine of all rings. This high phase angle, i.e., the angle between the observer and the light source, mosaic highlights the optically thin dusty rings, where the grain sizes are comparable to the wavelength of visible light. The insert figure shows the section with enhanced contrast to better display the faint details. The diffuse rings visible here are, sorted by the Saturn-centric distance, Janus/Epimetheus ring (2.5 R_S), G ring ($\sim$2.8 R_S), Pallene ring (3.5 R_S), and E ring ($\sim$4 R_S). The optically thick Main rings (A, B, and C) are viewed from the unlit side and therefore the apparent brightness is inversely proportional to the rings' optical depth. Source: CICLOPS

3.1.1 The "Lord of the Rings"—Saturn

The giant planet Saturn hosts the widest variety of rings currently known:

There are at first its beautiful, extremely flat, dense main rings, the A and B rings, separated by the Cassini division, which has similar properties to the C ring (see Fig. 5).

These dense rings are radially embedded in comparatively dilute rings. Saturn's innermost ring D (Fig. 6), is a faint, dusty ring located interior to the C ring, between 66900 to 74500 km from Saturn's center. Further outwards from the main rings, there are the dusty F ring, G ring, and the faint, but vertically as well as laterally extended E ring, the latter being sustained by the geysers of the small icy moon Enceladus.

The outermost "ring" of Saturn is the Phoebe ring—an extremely faint, but vastly extended dust complex.

In the following we will discuss all faint dust rings of Saturn separately (Table 1).

Saturn's D Ring Saturn's innermost ring D (Fig. 6), is a faint, dusty ring located interior to the C ring (<74500 km from Saturn's center) and likely extends all the way down to the planet's cloud decks. Its first detection is reported by Guérin (1973), yet most of our understanding of the D ring is based on the Voyager and Cassini observations. The ring is characterized by narrow ringlets and broad, wave-like patterns (Showalter 1996). Overall, the D ring grain size ranges from 1 to 100 micrometers, with a trend of increasing grain size with the radial distance (Hedman et al. 2007a).

D ring structures show measurable changes over less than 3 decades in between the missions to Saturn (Hedman et al. 2007a). Three embedded ringlets, D73, D72, and D68, named by the proxies of their radial distances in 1000 km, display radial changes, either in position (e.g., D72 shifted by roughly 100 km (Hedman et al. 2007a)) or in width. The innermost D68

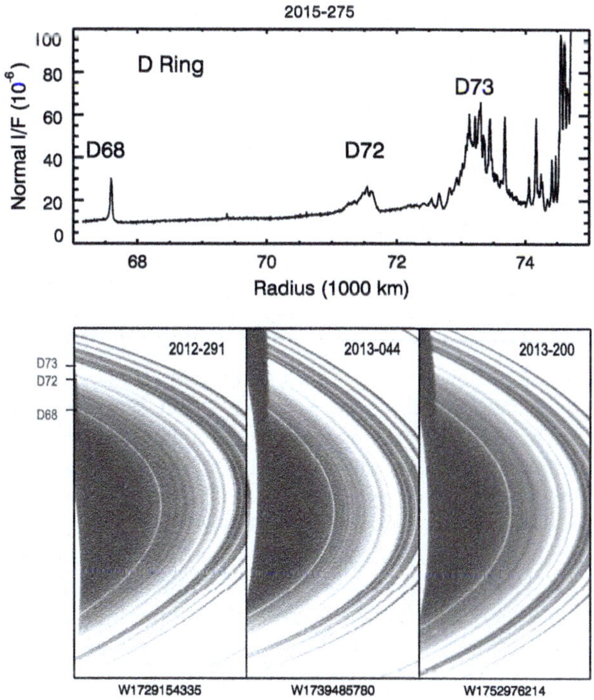

Fig. 6 High phase angle view of Saturn's D ring by Cassini. *Top*. Brightness profile of Saturn's D ring taken at a high phase angle (136°) in 2015. Adopted from Hedman et al. (2018). *Bottom*. Structural changes in the region in between D68 and D72 ringlets are clearly visible comparing these high-phase-angle images (∼ 175°) taken less than a year apart. Adopted from Hedman and Showalter (2016)

Table 1 Dusty rings of Saturn

Ring	Radius [km]	Width [km]	Optical depth
D	66900–74510	7500	10^{-4} to 10^{-3}
F	140180	30–500	10^{-3} to 1
Janus-Epimetheus	149000–154000	5000	?
G	166000–175000	9000	$\sim 10^{-6}$
Methone (arc)	194440 ± 20	?	$\sim 10^{-6}$
Anthe (arc)	197700	?	$\sim 10^{-6}$
Pallene	212000	2500	$\sim 10^{-6}$
E	180000–1206000	1000000	$\sim 10^{-6}$
Phoebe	∼4000000–13000000	9000000	$\sim 10^{-8}$

ringlet displays longitudinal brightness variations and its precession is consistent with the perturbations from Saturn's high order gravity field (Hedman et al. 2014). Periodic patterns in the outer D ring (73200–74000 km, i.e., outside D73 ringlet) are interpreted as a vertical corrugation, whose gradual decreasing wavelength allows the timing of the disturbance to be traced back to 1984, likely an impact event associated with the perturbation seen in the C ring (Hedman et al. 2007a, 2015). Significant changes of the dust sheet between D72 and D68 ringlets are also seen in Cassini (late 2011) and likely also Voyager (1979) images, which may indicate a periodic, systematic change of the D ring environment associated with the solar activity (Hedman and Showalter 2016). The highly dynamic D ring provides an unique window to examine sporadic events in the past as well as the physical conditions in the close vicinity of Saturn.

During the Cassini Grand Finale mission, micron-sized grains at the inner edge of the D ring were directly sampled, revealing a dust sheet peaked 4 km below the ring plane with a scale height of 300 km at a few thousand km altitude (Hsu et al. 2018). The registered dust density is about 10^{-5} m^{-3} (Hsu et al. 2018), orders of magnitude lower than that during the earlier Ring Grazing orbits (ring plane crossing distance of $\sim 2.5\ R_S$)[10] and that of other diffuse rings.

Saturn's E Ring Thanks to the Cassini mission, Saturn's E ring is the most studied diffuse ring by far. It has been covered in great detail by recent reviews (Horányi et al. 2009; Hedman et al. 2018; Kempf et al. 2018b). Therefore, we refer interested readers to those papers and will only address issues relevant to the discussions presented in this review.

The E ring is the second largest planetary ring in the solar system, encompassing the icy satellites Mimas ($r_M = 3.07\ R_S$), Enceladus ($r_E = 3.95\ R_S$), Tethys ($r_{Te} = 4.88\ R_S$), Dione ($r_D = 6.25\ R_S$), Rhea ($r_R = 8.73 R_S$), and Titan ($r_{Ti} = 20.25\ R_S$). The ring was spotted first in photographic observations obtained during the 1966–1967 ring plane crossing by Feibelman (1967). CCD observations during the next ring plane crossing in 1981 revealed that the ring's edge-on brightness peaks near Enceladus' mean orbital distance and that the ring extends at least until 9 R_S (Baum et al. 1981). Perhaps the most striking finding was, however, the unusual blue color of the ring (Larson et al. 1981), implying a narrow grain size range centered between 0.3 μm and 3 μm (Nicholson et al. 1996).

The evidence from these early observations indicated a close relation between Enceladus and the E ring (Showalter et al. 1991; Nicholson et al. 1996; de Pater et al. 1996), and many works prior to the Cassini mission focused on the dynamic evolution of E ring grains in order to explain its existence and unique properties. Ejecta with a broad size distribution produced from hypervelocity impacts on the embedded icy moons was considered as the main production mechanism, which, in retrospect, was not correct. Processes sensitive to grain sizes were studied in order to explain the narrow grain size distribution of the E ring, as inferred from its blueness, which include the secular orbital evolution under the influences of various forces, electrostatic charging, and plasma sputtering (Horányi et al. 1992; Hamilton 1993; Hamilton and Burns 1994; Hamilton and Krivov 1996; Dikarev 1999; Juhász and Horányi 2002; Horányi et al. 2008). While it is clear now that the narrow grain size distribution largely stems from the aforementioned relationship between particle size and ejection speed of grains emitted from Enceladus' plume (Schmidt et al. 2008; Postberg et al. 2011b; Kempf et al. 2010), the theoretical modeling of these processes nonetheless correctly illustrate the long–term evolution of E ring particles and lays out the foundation of modern understanding of dusty rings.

The Cassini mission has revolutionized our understanding of the E ring and its sources. The most important finding was that the ring particle population is to a large extent supplied by the geological activity of Enceladus (Spahn et al. 2006b; Porco et al. 2006). The fact that ongoing endogenous activities of a 500 km-diameter-icy moon serve as a major source of Saturn's E ring and the magnetospheric neutral and plasma was not expected based on our prior understanding of minor bodies in the outer Solar System. Another important finding was that the E ring is much larger than perceived before Cassini. Measurements by Cassini CDA revealed that ring extents at least to Titan's orbit with its density decreasing smoothly outside the Enceladus orbit (Srama et al. 2006).

The differential size distribution of the ring particles ≥ 900 nm inside Rhea's orbit scales like $\hat{n}(s) \sim s^q$ with slopes q between -5 and -4 (Kempf et al. 2008). Cassini RPWS reported for grains < 10 μm slopes of -4 at Enceladus and of -3 at 3 R_S (Ye et al. 2014a). The

[10] Saturn radius $R_S = 60268$ km.

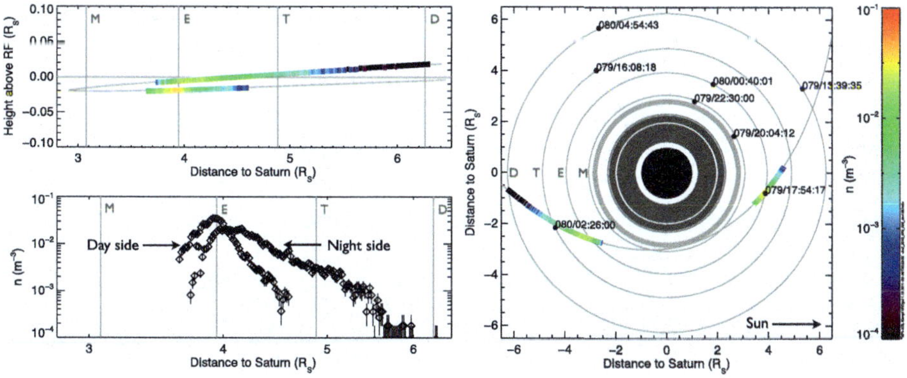

Fig. 7 Radial density profile of Saturn's E ring for grains ≥ 0.9 μm obtained during Cassini's almost equatorial traversal of the inner E ring in orbit 128 (lower left plot). On the night side the ring flares out while on the day side the ring is strongly "compressed" towards Saturn. Because the traversal took less than a day, the radial profiles are due to the same dust population. (from Kempf et al. 2012)

radial density profile of the ring can be roughly described as a pair of power laws centered at the densest point of the ring (Showalter et al. 1991), which is located 0.05 R_S outside the Enceladus orbit (Kempf et al. 2008). The empirical ring profile for grains ≥ 900 nm derived from CDA measurements during the first two years of the Cassini mission roughly reproduces Cassini data sets obtained outside of Mimas' and inside of Dione's orbit:[11]

$$n(r,z) = n_0 \exp\left(-\frac{(z-z_0(r))^2}{2\sigma(r)^2}\right)\begin{cases}(r/r_c)^{+50} & r \leq r_c,\\ (r/r_c)^{-20} & r > r_c\end{cases} \tag{38}$$

and

$$\sigma(r) = \sigma_c + (r - r_c)\begin{cases}\frac{\sigma_i-\sigma_c}{r_i-r_c} & r \leq r_c,\\ \frac{\sigma_0-\sigma_c}{r_0-r_c} & r > r_c\end{cases}$$

with

$$z_0(r) = \begin{cases}z_0(r_i)\frac{r-r_c}{r_i-r_c} & r \leq r_c,\\ 0 & r > r_c\end{cases} \tag{39}$$

(Kempf et al. 2008). The best fit parameters are $\sigma_i = 2293$ km, $\sigma_c = 1826$ km, $r_c = 3.98\ R_S$, $\sigma_0 = 2336$ km, and $z_0(r_i) = -1220$ km.

The ring exhibits a pronounced day-night asymmetry (Fig. 7, Kempf et al. 2012, Hedman et al. 2012), while its vertical structure is affected by seasonal effects (Kempf et al. 2008; Hedman et al. 2012; Ye et al. 2016). It is worth noting that the mechanism responsible for the day–night asymmetry is not yet understood.

The surface of E ring grains are continually bombarded by ambient plasma particles and photons, causing a charge transfer between the ring particles and their environment. Cassini CDA allowed for the first time a direct measurement of the electrostatic charge of the ring

[11] cylindrical coordinates (r, z) are used here.

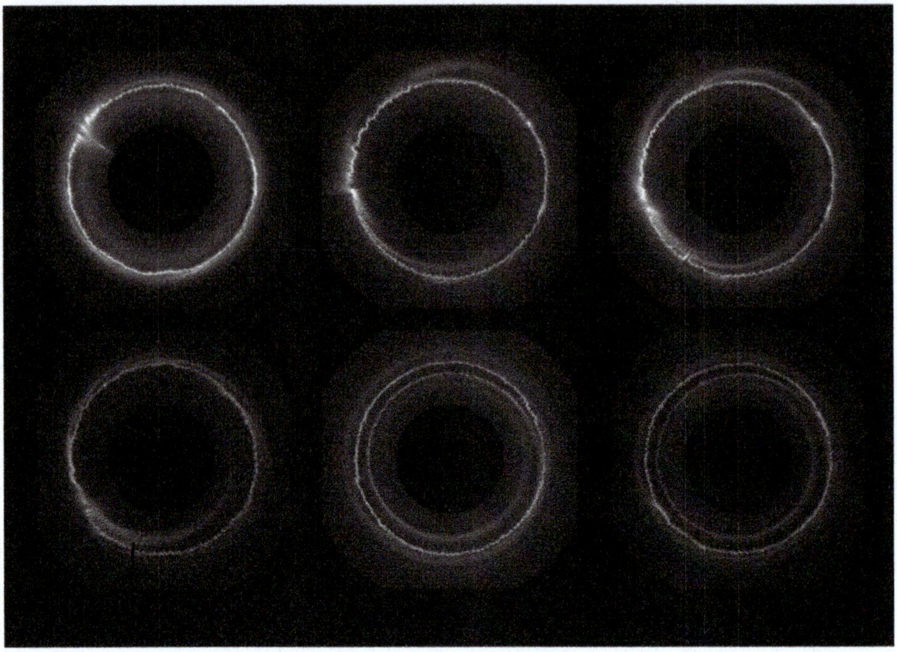

Fig. 8 Six 360° full panorama views of the F ring at different moments of time. The radial scales are largely exaggerated compared to the azimuthal ones (credit: JPL/NASA)

particle population. Particles inside Rhea's orbit have been found to be negatively charged with a potential of ~ -2 V, while the charging of grains outside Rhea's orbit is dominated by photoionization, resulting in positive grain potentials of about +3 V (Kempf et al. 2006).

In addition to the ring dynamics, compositional studies of E ring ice grains play an important role in understanding their source, i.e., the plume and the internal properties of Enceladus. While the majority of E ring grains are composed of pure water ice (Hillier et al. 2007), the presence of a minor salt-rich ice grain population (Postberg et al. 2009) and the nanometer-sized silica (SiO_2) particles (Hsu et al. 2015; Sekine et al. 2015) suggest an alkaline subsurface ocean with ongoing hydrothermal activities underneath the icy crust of Enceladus (Iess et al. 2014). These findings significantly contribute to our understanding of Enceladus as one of the most habitable worlds in the Solar System. We refer the readers interested in the composition of E ring grains to another recent review article by Postberg et al. (2018).

Saturn's F Ring The F ring, located at the Roche Limit of Saturn, marks a region of a transition between the existence of stationary dense planetary rings (closer to Saturn) and the wealth of icy Kronian moons. The main reason is the tidal force $\propto R_M \Omega^2(r)$ tending to disrupt a gravitationally bound body (moon) of a size R_M orbiting at a distance r to the planet. The Roche limit denotes the distance from the planet that separates regions far from the planet where extended, selfgravitating bodies can exist and regions closer in where such objects are disrupted by tidal forces.

The F ring consists of a broad dusty envelope of a radial width between 300 and 500 km with a narrow, bright central core about radially 50 km wide (Murray and French 2018). This picture is well illustrated by the 6 "panorama" snapshots of the F ring in Fig. 8, which

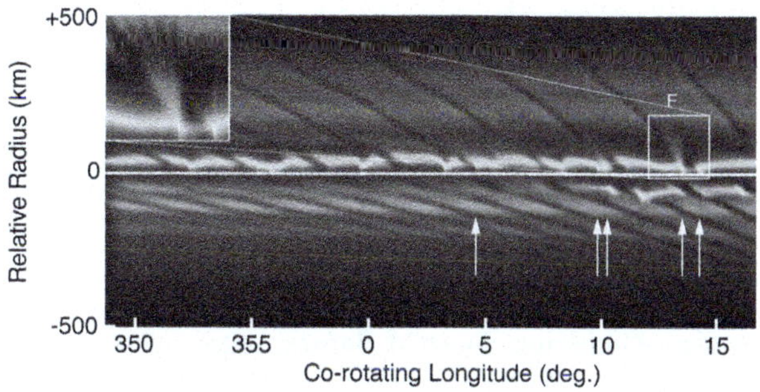

Fig. 9 "Fans" (upper left corner insert) have been observed by the Cassini-cameras, which originate from mass concentrations at the bright strand, triggered by the periodic perturbations by the ring "shepherd" Prometheus (credit: NASA/JPL)

shows both its core and envelope, as well as time variations of these structures. The fuzzy appearance of the bright strands of the core (optical depth: $\tau \approx 1$) point to frequent disruptive collisional events between macroscopic bodies, which might be short lived moonlet sized aggregates, temporary rubble piles or even longer lasting, stable moonlets. Such collisions produce a great amount of dust that locally changes the brightness of the main strand. Occasionally dust spirals originating from such collisions make their way through the outskirts of the F ring.

Larger objects have been spotted orbiting among the bright strands of the F ring core during the Cassini mission. Most of them are only of temporary character, i.e. they could not be tracked for an extended period of time, unlike the largest propeller moonlets in the A ring of Saturn. Nevertheless, these accreted objects must have survived several orbital periods so that they can gravitationally influence the neighboring ring material, causing the kinematic wake like features, similar to those Prometheus carves in the F ring, that can survive many orbits producing the so-called *fans* as shown in the Fig. 9. The fans are similar to the kinematic waves decorating the moonlet's propellers or the edges of the wider gaps swept free by Pan and Daphnis. However, unlike the moonlets populating the A ring, these temporary density-concentrations cannot create propeller-like gaps in the F ring. This is because the velocity dispersion c^2 of the surrounding dusty ring material is much larger than in the dense rings, further assisted by the manifold of non-gravitational perturbations the dust particles are exposed to. These effects strongly amplify the dispersion which, in turn, prevent a complete formation of a propeller-gap by the gravitational scattering at the massive but temporary aggregate. This tendency might be supported by the fact that massive bodies (moonlets, aggregates) in the F ring travel on chaotic orbits (Winter et al. 2007) which might further hinder a propeller formation.

After the long term observations of the Cassini mission at Saturn, the following picture of the F ring emerges: a ≈ 500 km wide dusty envelop of an optical depth of $\tau \approx (1-2) \times 10^{-3}$ hosts a narrow, bright core with optical depths reaching $\tau \approx 1$ in the central strand.

The latter accommodates larger (possibly up to km sized) aggregates/rubble piles and also very likely more resistant moonlets which are likely covered by more or less thick dust layers.

We suggest that this relatively large optical depth $\tau \approx O(1)$ of the central strand is probably sustained mainly by the dust- and debris layers shaken off of the moonlets or large aggregates in mutual collisions. In order for these impacts to be frequent enough to supply all the visible dust, the mutual impact rates between the moonlets/temporary aggregates need to be sufficiently high. This requires an optical depth the moonlets of about $\tau_{\text{moonlets}} \subset (10^{-3}, 0.1)$ giving a collision frequency $\omega_{\text{moonlets}} \approx 3\Omega\tau_{\text{moonlets}}$ to sustain the central strand's high optical depth. The moonlet's collision frequency ω_{moonlets} is, on the one hand, concluded from the lifetime (about a few months) of larger clumps or temporary moonlets (Murray and

French 2018), but also, on the other hand, by the high optical depth of the central strand and related short dust particle lifetimes in the F ring. These larger aggregates or moonlets are concentrated mainly in that radial narrow area where the gravitational torques exerted by the shepherds Prometheus and Pandora cancel and where the gravitating bodies concentrate (Cuzzi et al. 2014). On the other hand, without the steady stirring by the shepherds, another satellite would likely form. In this sense Pandora and Prometheus guarantee the existence of this lively, exciting ring.

Impact-Generated Dust at Saturn One interesting observational fact, obtained from measurements by the Cassini-spacecraft, is the lack of impact-generated dust-clouds embracing the satellites embedded in the faint E ring of Saturn. Such dusty "cocoons" surrounding atmosphereless moons (see review by Szalay et al. 2018, this issue), considered evidence for ongoing micrometeoroid erosion, have been clearly detected with Galileo's dust detector (DDS) at Jupiter's Galilean satellites (Krivov et al. 2003; Sremčević et al. 2005). Similar dust "atmospheres" were expected at the numerous Kronian moons (Spahn et al. 2006a), but they were not detected by the Cassini-experiments. Exceptions are the dust haloes embracing Saturn's main rings, the Phoebe ring, Saturn's G ring, as well as three tiny satellites at the inner edge of the E ring. The latter ring, fed by the geysers of the ice-moon Enceladus, possibly obscures the otherwise very faint dust clouds around the E ring moons.

Saturn's G Ring The G ring of Saturn is a very faint dust ring ($\tau \sim 10^{-6}$) that is located about halfway between the denser F ring and the inner edge of the thick and wide E ring. It extends from 166000 km (2.75 R_S) to 175000 km (2.90R_S), having a radial width of about 9000 km (Hedman et al. 2007b). Its spectral color appears to be neutral to slightly red, suggesting a size distribution power-law q between 1.5 to 3.5 (Throop and Esposito 1998). Near the ring's inner edge at 168496 km (2.80 R_S), there is a noticeably brighter, 250 km-wide arc that extends over one sixth of the ring's circumference and is centered on the icy, sub-kilometer moonlet Aegaeon. It is believed that the arc is composed of larger icy particles up to a few meters in diameter, while the rest of the ring consists of smaller dust particles generated by micrometeoroid impacts which is subsequently dragged outward by plasma-drag. The arc itself is longitudinally confined by the 7:6 eccentricity corotation resonance with Mimas (Hedman et al. 2007b).

Arcs and Rings near Anthe, Methone, Pallene The three tiny moons—Anthe, Methone, Pallene, which were discovered during the Cassini mission (Porco et al. 2007)—are located between Mimas and Enceladus. The innermost of these three moons is Methone with a diameter of 3.2 km and a semi-major axis of 194440 km (3.23 R_S) followed by Anthe which has a diameter of about 2 km and a semi-major axis of 197700 km (3.28 R_S). The outermost moon Pallene has a diameter of 4.4 km and a semi-major axis of 212280 km (3.52 R_S; Porco et al. 2007). Based on the Cassini imaging data, Hedman et al. (2009) described arcs of material at the orbital positions of Anthe and Methone. The orbit of Pallene is completely surrounded by a dust-torus. The earliest evidence for an arc of Methone came from two charged particle absorption signatures measured with the Magnetospheric Imaging Instrument aboard the Cassini spacecraft (Roussos et al. 2008).

Pallene Ring The ring of Pallene has been observed with the cameras at rather high phase angles (> 175°), indicating the predominance of μm-sized particles (Hedman et al. 2009). Furthermore, the ring has a radial full width at half maximum (henceforth FWHM) of 2500 km and there is no evidence of larger-scale variations in the ring in azimuth. The torus

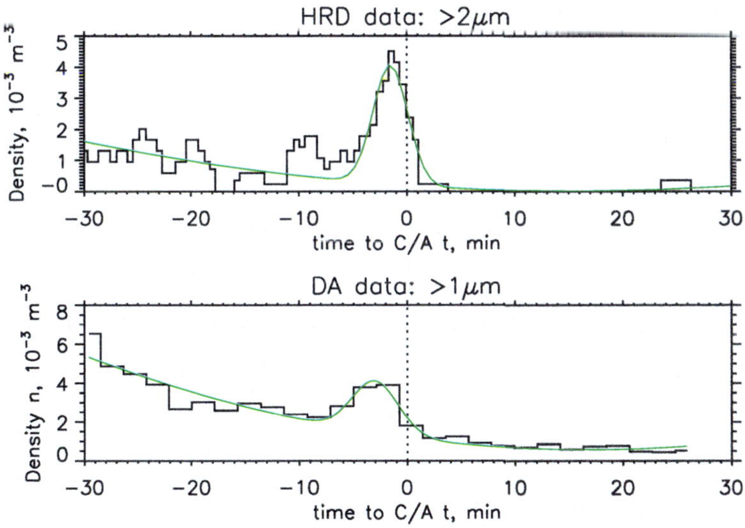

Fig. 10 Comparison between the measured densities by the HRD and DA instruments of the CDA. The offsets between the density peaks at $t \approx -2$ (top panel) respectively $t \approx -3$ (bottom panel) and the closes approach (dotted vertical line at $t = 0$) if caused by an outward displacement of the Pallene ring likely driven by plasma drag

follows Pallene's orbit and its vertical thickness is smaller than the vertical excursions of the source moon.

Cassini passed the orbit of Pallene (from July 2004 until the end of 2015) 15 times at a distance of 500 km or closer. During these passings Pallene's ring has been probed in-situ with the Cosmic Dust Analyser (CDA, Srama et al. 2004) aboard the spacecraft Cassini, which confirmed a dust torus of micrometer-sized particles along the orbit of Pallene. The Dust Analyzer (DA) unit of the CDA provided only useful data during 6 flybys (1 in 2010, 4 in 2011 and 1 in 2015) including 5 certain torus detections. The High Rate Detector (HRD) unit of the CDA recorded a single useful profile during the 2010 flyby (Seiß et al. 2014).

The cross-section of the torus measured by CDA/HRD can be fitted with a two-dimensional Gaussian distribution, yielding a radial and vertical full width at half maximum of 2300 km and 270 km, respectively, and a maximum particle density of $n = 2.7 \times 10^{-3}\ \mathrm{m}^{-3}$. Additionally, the data clearly show an enhancement of larger particles in the torus compared to the background E-ring particle sizes.

The radial mean position of the torus is shifted radially outwards by around 1100 km. This suggests a systematic enhancement of the semi-major axes (a gain of orbital energy) of the dust particles (in comparison to Pallene) which is possibly caused by plasma drag.

During the 2010 flyby the HRD succeeded to scan the Pallene ring which allows for a direct comparison with the DA-data (see Fig. 10). The interesting fact is that the DA has recorded particles larger than 1 μm, while the HRD is sensitive to particles larger than 2 μm, meaning a comparison of both data-sets yields information about different parts of the size distribution.

Larger particles are found in a narrower torus centered closer to Pallene's orbit than small grains. The amount of large and small particles is about the same level considering the uncertainty in the fits. In contrast to this, the E ring provides 15 times more 1 μm particles than 2 μm particles. So, we conclude that the majority of particles forming Pallene's torus are larger than 2 μm particles, while typical E-ring particles ($s \approx 1$ μm) form the background of sub-micrometer dust.

Anthe and Methone Arcs The brightest images of the arcs of the moons Anthe and Methone are taken by the Cassini cameras at low phase angles. Thus, these arcs host much larger particles (10–100 μm) compared to those forming the Pallene ring (Hedman et al. 2009). This means that these arcs are dominated by gravitational forces. Indeed the grains are confined by co-rotational eccentricity resonances with the moon Mimas at $\sim$3.08 R_S. These arcs are radially narrower than the Pallene ring—also pointing to smaller thermal random-walk speeds, and to larger grains in the arcs.

The Anthe arc is shaped by the 11:10 Mimas resonance which the moon also occupies. It has an azimuthal length of 20° while the co-rotation resonance would restrict the length to 32°. Further, the Anthe arc shows time variable substructures in the images. For example two distinct strands separated by 200 km ahead of Anthe (0 to 5°), could be observed in some of the images. The moon seems shifted forward at the end of 2007 and beginning of 2008 with respect to the arc, but this is partially explained by the libration of Anthe around the gyro-center of the resonance with an amplitude of 7°.

Methone's arc is much shorter, only 10° wide, but the arc particles are orbiting in a 15:14 Mimas resonance that restricts the particles to a region smaller than 24°. Also Methone appeared shifted with respect to the arc caused by a libration inside the resonance with an amplitude of 5°.

The arcs are sustained by the impact-ejection, resulting mainly in escaping ejecta from the surfaces of the nearby moons Anthe and Methone (Krivov et al. 2003). Small moons are supposed to be relatively efficient in ejecting dust to space because particles can escape easily due to the small gravity of these moons (Burns et al. 1999). For instance, the escape speeds of these three moons are well below 1 m/s, making them effective ejecta sources (see Sect. 2.1).

Numerical integrations have been performed to verify this hypothesis (Sun et al. 2017), where micrometer-sized dust was released from the surfaces of the moons, including the gravity of Mimas and Saturn, Lorentz force, solar radiation pressure, and plasma drag. The latter is important for the shape of the arcs and for the lifetime of the particles. The plasma drag restricts the arcs to an azimuthal region of 11 and 15° for the moons Methone and Anthe, respectively, which is shorter than the theoretical co-rotation length but better fits the observations.

Small particles ($<$ 5 μm for Methone, $<$ 10 μm for Anthe) do not stay in the arcs and escape in course of a few orbits due to the plasma drag, while larger ones stay much longer in the co-rotational region and have lifetimes of about 100 yrs before they escape or re-impact the source moons. This explains why the particles appeared much larger than those in the Pallene ring. The simulations also confirmed that the radial width of the arcs is smaller than the Pallene ring (Sun et al. 2017).

Dust-Halo around Saturn's Dense Rings A dust halo around Saturn's Main rings has long been speculated as one of the pathways for ring–planet interactions (Northrop and Hill 1982, 1983a,b; Connerney 1986; Northrop and Connerney 1987). The rings' large surface–to–mass ratio means that they are prone to the erosion by infalling exogenous material. The incoming material (usually non water-ice) not only contaminates the main rings but also provides additional angular momentum, driving the so–called ballistic transport process that modifies the ring structure and redistributes the non-icy ring constituents (Cuzzi and Durisen 1990; Cuzzi and Estrada 1998; Estrada et al. 2015). Nanometer-sized ejecta produced from

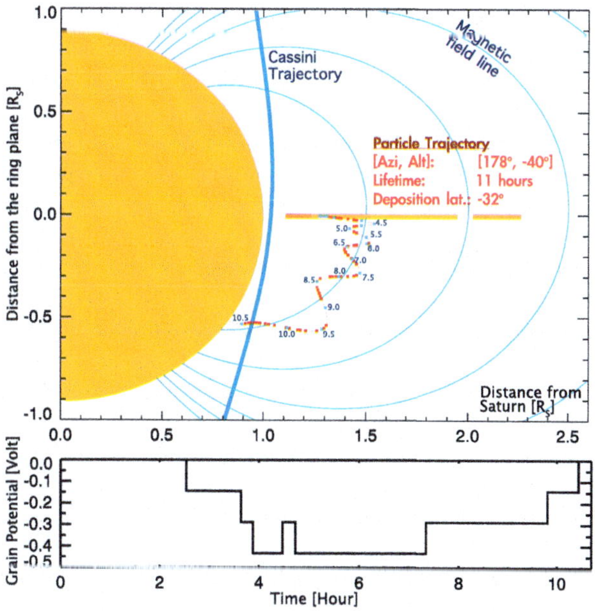

Fig. 11 *Top.* A side view of a simulated trajectory of a charged grain 10 nm in radius ejected from Saturn's C ring that falls into the southern hemisphere after about 11 hours. The grain is loosely bound to the magnetic field line with its gyro radii in the order of 0.1 Saturn radii (∼6000 km) changing with the grain charge-to-mass ratio and local magnetic field strength. The blue ticks mark half hour time interval along the grain trajectory. *Bottom.* The grain charge history. Each "jump" marks the change of one elementary charge

the rings may fall into Saturn following a complex dynamical evolution. In the proximity of Saturn, the magnetic field is so strong that it dominates the motion of charged nanograins (Jontof-Hutter and Hamilton 2012a). Because of the northward offset of Saturn's magnetic dipole, charged grains are preferably transported to and deposited on the southern hemisphere (Connerney 1986; Liu and Ip 2014; Ip et al. 2016), as shown in Fig. 11.

The nanodust population was observed during the Cassini Grand Finale mission. During these highly eccentric polar orbits, the Cassini spacecraft repeatedly traversed the region in-between the cloud tops of Saturn and the innermost D ring. This region was found to be dominated by grains with radii of a few tens of nanometers within $\pm 0.3\ R_S$ around the ring plane (Hsu et al. 2018). The grain mass density, however, is still dominated by larger, less abundant micron-sized grains concentrated at the ring plane. The nanograin detection profile is consistent with fast, tiny ejecta produced from the main rings, with a characteristic enhancement in the southern hemisphere. The profile is also consistent with the ionospheric H_3^+ infrared emission pattern, indicating that the mass depositing to the planet from the infalling charged nanograins is the cause of local enhancement of the H_3^+ in the ionosphere, i.e., the so-called "ring rain" effect (O'Donoghue et al. 2013; Moore et al. 2015). High ejection speeds (a few 100s m/s and up to several km/s) are required to reproduce the measured profiles, which implies an inverse ejecta size–speed relation, i.e., the smaller the ejecta, the faster the ejection speed (e.g., Sect. 2.1.1). It is worth noting that these fast, tiny ejecta may not be significant regarding the ballistic transport, but could play a substantial role in ring-planet interactions because of the high ejection speeds and the "dynamical assistance" from the electromagnetic forces. Similarly, measurements from the energetic particle Imager and the neutral gas spectrometer instruments onboard Cassini during the Grand Finale orbits also indicate the infall of nanograins, likely originating from the D ring, into the atmosphere of Saturn (Mitchell et al. 2018; Waite et al. 2018). These measurements illustrate the ring mass loss through the transport of microscopic particles as well as their interactions with the atmosphere, which may in fact be the common fate of the rings around giant planets.

Spokes In addition to the continually generated ejecta particles forming a permanently present dust halo above the main rings, there are also intermittently appearing radial features over the B-ring, the so-called spokes. The exact triggering mechanism, and subsequent evolution of the cloud of small dust particles comprising a spoke still remains poorly understood, but most likely are related to the plasma environment and the magnetic fields piercing the rings (Horányi et al. 2004). Spokes, what little we know, have been discussed in text books as a prime example of complex dusty plasma phenomena (Bliokh et al. 1995; Shukla and Mamun 2002). Spokes were recognized in images taken by Voyagers 1 and 2 (Smith et al. 1981, 1982), but they were perhaps noticed even earlier in ground-based observations (Robinson 1980). The key physical characteristics of spokes based on the Voyager (Porco and Danielson 1982; Grün et al. 1983, 1984, 1992a) and Cassini (Mitchell et al. 2013) data can be summarized as follows: they are generally most common near the dawn ring ansa, and seem to form primarily in that region. Spokes develop on a timescale of minutes, and can become more intense over a period of a few hours. Appearing at radial distances that are near to, or straddle, the kronosynchronous orbit. They move around the ring nearly co-rotating with the planet. The radial termini of new spokes expand radially at $\simeq 0.5$ km/s in both the inward and outward directions. The proposed spoke formation theories that arguably have been the most promising are those that rely on: meteoroid impacts onto the rings (Goertz and Morfill 1983), field-aligned electron beams originating from the auroral regions of Saturn (Hill and Mendis 1982), atmospheric lightning related high-energy electron beams (Jones et al. 2006; Horányi et al. 2010), or dusty plasma waves (Yaroshenko et al. 2008). The most recent set of observations from Cassini imaging (Mitchell et al. 2013) are yet to be fully harvested to identify the missing pieces for understanding spokes.

Phoebe's Ring Phoebe is the largest of Saturn's irregular satellites (mean diameter 213 km), orbiting the planet on a distant (semi-major axis $a = 215\ R_S$), eccentric ($e = 0.16$), retrograde and inclined ($i = 152°$) orbit. Based on its unusually low albedo and its surface reflectance spectra, combined with its orbital characteristics, it has been suggested to be a captured object originating from the Kuiper Belt (Johnson and Lunine 2005). Just like all other bodies in the Solar System, Phoebe is also being continually bombarded by interplanetary and interstellar dust particles generating secondary ejecta dust. The orbital evolution of the escaping fraction of these particles is dominated by the gravity of Saturn, with perturbations by solar gravity and radiation pressure. It was suggested early on that the unusual brightness asymmetry between the leading and trailing hemispheres of the moon Iapetus ($a = 59\ R_S$) is due to the deposition of dark dust, possibly with organic compounds, originating from Phoebe (Mendis and Axford 1974; Soter 1974; Mendis and Axford 2008; Strazzulla 1986).

The escaping fraction of the secondary ejecta particles disperse from Phoebe into a large torus that was first noticed in the mid-infrared (wavelength $\lambda = 24$ and 70 μm) images taken by the Spitzer Space Telescope (Verbiscer et al. 2009). This is the largest planetary ring structure in our Solar System, spanning a radial distance of approximately 128 to 260 R_S, with a vertical extent of at $\sim 40\ R_S$ (matching the vertical excursions of Phoebe above/below the equatorial plane of Saturn), and normal optical depth $\tau \sim 2 \times 10^{-8}$. Followup optical observations by the Cassini (Tamayo et al. 2014), and mid-infrared images ($\lambda = 22$ μm) by WISE (Hamilton et al. 2015) missions indicate that ring spans an even bigger region, from 80 to 270 R_S. The ring has been suggested to be comprised dominantly of particles with radii < 20 μm in an unusually steep power-law size distribution with index > 4 (Tamayo et al. 2016). Curiously, the Cassini CDA found no indication of this ring during the final approaching phase to Saturn in 2004 (Kempf et al. 2011).

Table 2 Properties of the Jovian rings (Ockert-Bell et al. 1999; Showalter et al. 1987; Throop et al. 2004; Krivov et al. 2002a,b)

Name	Radius [km]	Width [km]	Thickness [km]	Optical depth	Dust fraction
Inner region					
Halo ring	92000–122500	30500	12500	$\sim 1 \times 10^{-6}$	100%
Main ring	122500–129000	6500	30–300	$\sim 6 \times 10^{-6}$	25%
Amalthea ring	129000–182000	53000	2000	$\sim 1 \times 10^{-7}$	100%
Thebe ring	129000–226000	97000	8400	$\sim 3 \times 10^{-8}$	100%
Mid region					
Galilean "ring"	226000–3500000	~3000000	~10000–100000	$\sim 1 \times 10^{-9}$	100%
Outer region					
Outer "ring"	> 3500000	?	?	?	100%

3.1.2 The Rings of Jupiter

The Jovian system can be divided into three regions. The *inner region* comprises the planet's main ring system and the four small inner moons Metis (1.8 R_J[12]), Adrastea (1.8 R_J), Amalthea (2.5 R_J), and Thebe (3.1 R_J), which have radii of several tens of kilometers. The *mid region* between about 5 and 30 R_J contains the four large Galilean moons Io (5.9R_J), Europa (9.4 R_J), Ganymede (15.0 R_J), and Callisto (26.3 R_J), which besides Earth's Moon and Saturn's moon Titan are the largest moons in the Solar System. The *outer region* between about 30 and 300 R_J is populated by a large number (> 60) of very small irregular satellites, which are mostly only a few kilometers in size and on highly inclined or retrograde orbits. The physical and orbital properties of the Jovian rings are summarized in Table 2. We also refer the interested readers to another recent review article by de Pater et al. (2018b).

Inner Region The main rings of Jupiter consist primarily of dust and they are much fainter than the dense rings of Saturn. They were first observed in 1979 by the spacecraft Voyager 1 (Smith et al. 1979) during its flyby at the Jovian system. The rings have been investigated in more detail by the Galileo spacecraft (Ockert-Bell et al. 1999) while it was orbiting Jupiter from 1995–2003, and by the New Horizons spacecraft during its flyby at Jupiter in 2007. The main rings have also been observed by the Hubble Space Telescope (Meier et al. 1999) and by large telescopes on Earth (de Pater et al. 1999).

The main ring system extends from 92000 km (1.29 R_J) to 226000 km (3.11 R_J, Ockert-Bell et al. 1999) and it has four components: a vertically thick inner torus called the "halo ring"; a narrow and relatively flat, but relatively bright "main ring"; and two wider and thicker, but fainter outer "gossamer rings" that embrace each other. The source of all of these rings is most likely the ejection of material by impacts of fast micro-meteoroids on the four inner moons and possibly a larger number of smaller, not yet detected bodies embedded in the rings.

The brightest component of the main ring is located between 122500 km (1.72 R_J) and 129000 km (1.81 R_J, Ockert Bell et al. 1999), thus having a radial width of about 6500 km. Its outer edge coincides with the orbit of the second innermost satellite Adrastea, which

[12] $R_J = 71492$ km is the equatorial radius of Jupiter.

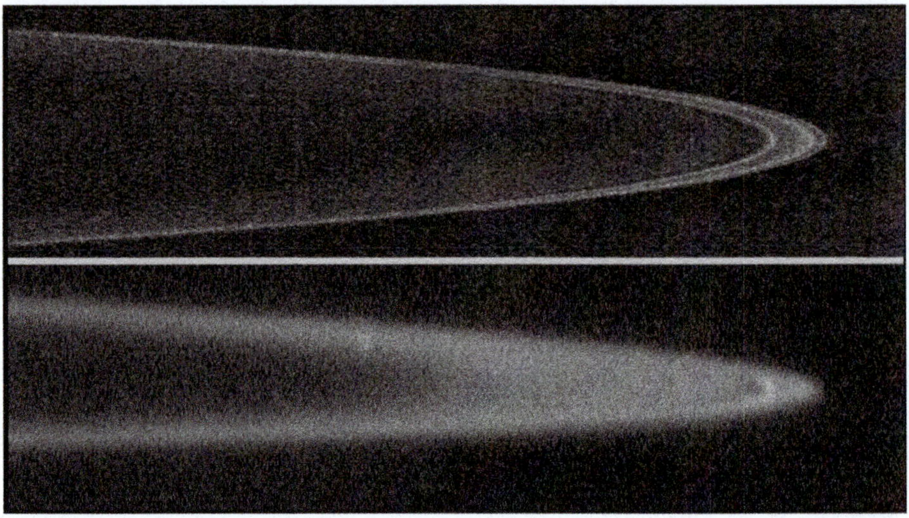

Fig. 12 Jupiter's main rings in back-scattered (above) and forward-scattered (below) light as seen by the New Horizons spacecraft in 2007. The main ring is the brightest component of the main ring system and supplied by impact-ejection from the four inner moons of Jupiter (Metis, Adrastea, Amalthea, and Thebe) and possibly a larger number of smaller bodies. The gap at Metis' orbit is more pronounced in back-scattered than in forward-scattered light (credit: NASA/JPL, PIA09342)

plays the role of a shepherd moon that keeps the ring confined. The ring has a fine substructure, whose appearance depends on the viewing geometry, namely whether it is viewed in forward- (emphasizing small dust particles) or backward-scattered light (emphasizing larger embedded bodies).

In forward-scattered light, the brightness is fairly constant inside the orbit of Adrastea (see Fig. 12) with a maximum near the ring's center at 126000 km, from where it slowly decreases to the ring's inner edge. Additionally, there is a small gap in the brightness inwards the orbit of Metis called the "Metis notch". The main ring is much brighter in forward-scattered than in backward-scattered light. In backward-scattered light, the ring appears to consist of two different parts: a wide and fainter inner part inside the orbit of Metis and a narrow and brighter outer part around the orbits of Metis and Adrastea. The outer part shows two major ringlets separated by a gap that appears more pronounced than in forward-scattered light (Fig. 12). The main ring is relatively thin ($\sim$30 km) for a faint ring in backward-scattered light, but much thicker ($\sim$300 km) in forward-scattered light (Showalter et al. 1987). In 2007, the New Horizons spacecraft performed a search for new small moons in the main ring. No new moons were found, but the spacecraft detected seven small, not yet explained clumps of ring particles inside the orbit of Adrastea (Showalter et al. 2007).

Interestingly, vertical corrugation spirals in the main ring have been identified in the Galileo imaging data by Showalter et al. (2011) which date back to the capture of then comet Shoemaker-Levy in 1994. Since that original event the spirals wind up at a rate dominated by Jupiter's gravitational field—an effect which has been confirmed by the imaging data of the spacecraft New Horizons in 2007 (Showalter et al. 2007, 2011).

The optical properties of the main ring can be explained by a combination of large amounts of small dust particles with sizes of $s \sim 0.1$–10 μm and a population of larger bodies with $s \sim 1$ mm–1 km. The data is best-fit when the small particles are distributed by a broken power law $n(s) \sim s^{-q}$, where the index is $q_1 = 2.0 \pm 0.2$ for particles with

$s < 15 \pm 0.3$ μm and $q_2 = 5 \pm 1$ for larger ones (Brooks et al. 2004), and by using non-spherical rather than spherical dust grains (Throop et al. 2004). For the non-spherical case, the optical depth is 4.7×10^{-6} for the dust particles and 1.3×10^{-6} for the larger bodies. The size of the largest bodies was limited by the search of the New Horizons spacecraft to less than 0.5 km (Showalter et al. 2007).

The halo ring is the innermost and vertically thickest component of Jupiter's main ring system. Its outer edge coincides with the inner edge of the main ring at 122500 km (1.72 R_J), from where the ring becomes gradually thicker towards Jupiter. Its optical depth rapidly decreases at 100000 km (de Pater et al. 1999), but the ring extends further inward to 92000 km (1.29 R_J, Ockert-Bell et al. 1999). This makes a radial width of the dust halo of about 30000 km and a height of at least 10000 km above and below the ring plane, so that it resembles a thick torus rather than a flat ring. However, the brightness rapidly decreases further away from the ring plane, so that most of the material is concentrated within only a few hundred kilometers (Burns et al. 2004). In contrast to the main ring, it has no clear internal structure and its appearance depends only weakly on the viewing geometry. The halo ring appears brightest in forward-scattered light, probably because the ring is mainly composed of micrometer-sized particles, which scatter visible light preferentially in the forward direction. While the halo's optical depth ($\sim 1 \times 10^{-6}$) is similar to that of the main ring, the halo ring is much more vertically extended, and so the number density of particles is much lower in the halo ring.

The Gossamer ring is a very faint ring that has two components. They extend from the main ring to the outer two of the four inner moons of Jupiter, Amalthea and Thebe, respectively, which are most likely their sources. The rings embrace each other and are much thicker than the main ring. The Amalthea ring is the inner of the gossamer rings. It ranges from 129000 km (1.81 R_J) to the orbit of Amalthea at 182000 km (2.55 R_J, de Pater et al. 1999) and has a vertical thickness (FWHM) of about 2000 km. The particle size distribution is similar to the main ring with a power law index of $q = 2.0 \pm 0.5$, but the optical depth ($\sim 1 \times 10^{-7}$) is about one order of magnitude lower than that of the main ring. The Thebe ring is the outer of the gossamer rings and extends to 226000 km (3.11 R_J, de Pater et al. 1999), but additional material called the "Thebe extension" is visible beyond the orbit of Thebe (Burns et al. 1999), extending up to 280000 km (3.75 R_J, Krüger et al. 2009). The Thebe ring has a vertical thickness (FWHM) of about 8000 km. In forward-scattered light, both rings are brightest near their top and bottom edges and gradually become brighter towards Jupiter. The Thebe ring appears to be about three times fainter than the Amalthea ring, which in turn appears to be about 30 times fainter than the main ring. The Thebe ring's optical depth ($\sim 3 \times 10^{-8}$) is about three times lower than that of the Amalthea ring.

Mid Region Aside from the main ring system, there are other dust populations further away from the planet. Dust clouds around the Galilean moons formed by ejecta from impacts of interplanetary meteoroids have been detected in various flybys by the spacecraft Galileo (Krüger et al. 1999, 2003b). A small fraction of the ejected particles can escape the gravity of their parent bodies and forms a dust ring between their orbits. In contrast to the dense rings of Saturn or the main rings of Jupiter, the "Galilean ring" is a very faint, but laterally and vertically extended ring. It reaches from within the outer edge of the main ring system (3.1 R_J) to far beyond the orbit of Callisto (26.3 R_J), thus being one of the widest known planetary dust rings with a radial width of several 100000 km. The ring's maximal density of $n \sim 100\ \mathrm{km}^{-3}$ is near Europa's orbit and decreases further outwards. Numerical simulations of micron-sized dust particles ejected from the Galilean moons (Krivov et al. 2002a; Sachse 2018) reproduce the observed density and show that the particles have inclinations of up to 45°. This corresponds to an optical depth of $\tau \sim 10^{-9}$ near Europa's orbit, which means that the ring is not detectable by remote observations. The three-dimensional structure of the Galilean ring is most likely similar to the structure of Saturn's E ring. The lifetimes of different-sized particles differ by up to three orders of magnitude (~10–10000 yr), where the larger particles have generally larger lifetimes (Liu et al. 2016; Sachse 2018). This implies that the size distribution in the Galilean ring is flatter than the initial size distribution of the ejecta.

In addition to the continuous ejecta production, an endogenic dust production within the Jupiter system comes from the volcanic activities on the inner most Galilean moon Io at 5.9 R_J. Grains of a broad size range form from condensation of ascending plume gas. The tiniest ash particles of roughly 10 nm radii, once acquired sufficient electric charges, could be accelerated by Jupiter's magnetosphere and ejected into interplanetary space with speeds comparable to the Solar wind (Horányi et al. 1993; Graps et al. 2000). These so-called "stream particles" were discovered during the first Ulysses flyby of Jupiter in 1992 by the on board dust detector (Grün et al. 1993). The detection patterns as registered by the Galileo Dust Detection System (DDS, Grün et al. 1992b) suggest that they are modulated by the Jovian magnetic field (Grün et al. 1998), the Io Plasma Torus (IPT), as well as the temporal variation of Io's volcanic activities (Krüger et al. 2003a). Cassini CDA performed the first composition measurement of the Jovian stream particles during its Jupiter flyby in 2000/2001, which shows that these ash particles are mainly composed of sodium chlorite (NaCl), with trace amounts of silicates and refractory sulfuric compounds (Postberg et al. 2006). See review by Hillier et al. (2018) and review by Hsu et al. (2012).

Outer Region Dust in the outer region of the Jovian system has been detected by the DDS-detector aboard the spacecraft Galileo over the course of several orbital revolutions about Jupiter (Krivov et al. 2002b). The density of micron-sized dust particles is nearly constant at $n \sim 10\ \text{km}^{-3}$ between 50 and 350 R_J and sharply drops down outside 350 R_J. This density is much larger than the background density generated by the *direct flux* of interplanetary meteoroids and consistent with the density expected for dust ejected from the small irregular moons by *impacts* of interplanetary meteoroids. The size distribution of the particles is rather unusual as grains smaller than 1 μm are depleted. This is because strong perturbations by solar radiation and solar gravity force the small particles to rapidly collide with the planet, while lager particles are affected to a lesser extent.

3.1.3 The Rings of Uranus & Neptune

Uranian Rings While the ring system of Uranus does not look as spectacular as that of Saturn at first sight, it is as complex and diverse as a ring system could be. Uranus' rings were first "seen" in 1977 by stellar occultation (Elliot et al. 1977). Modern high–speed photometry provided km–level resolution power sufficient to study the structure of these narrow, dark rings in detail (Elliot 1979). Follow–up observations, including the 1986 Voyager 2 flyby (Smith et al. 1986) and the Earth-bound observations during the 2006–2007 ring plane crossing event (Showalter and Lissauer 2006; de Pater et al. 2006, 2007), provided new discoveries and further constraints about this distant ring system.

The main rings of Uranus are packed with narrow rings and small satellites (radius $\leq$100 km) within a distance of two Uranian radii,[13] as shown in Fig. 13. The narrow rings (rings 6, 5, 4, α, β, η, γ, δ, and ϵ) are optically thick and are dominated by macroscopic

[13] 1 $R_U = 25559$ km.

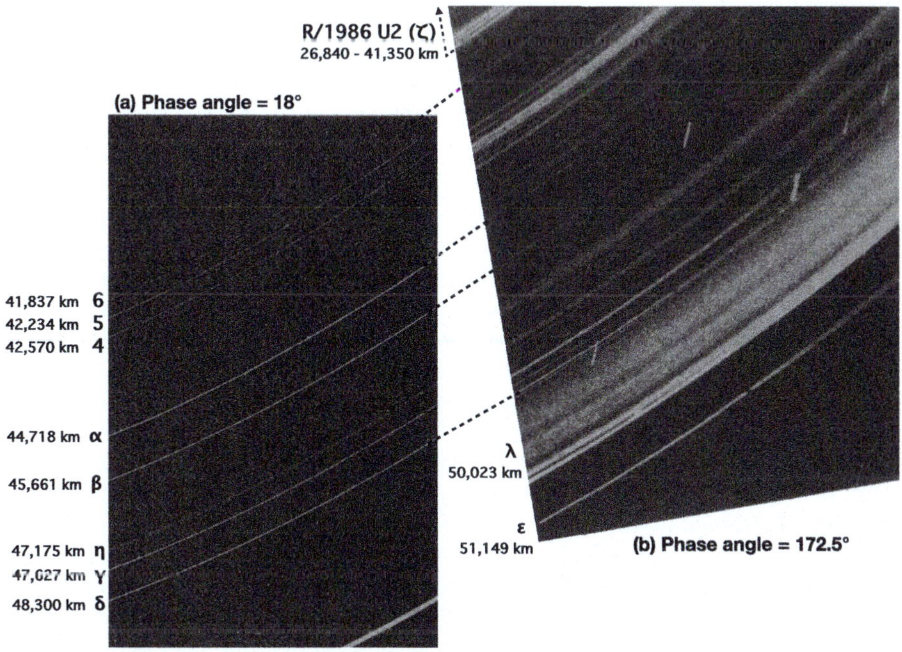

Fig. 13 Rings of Uranus viewed by Voyager 2 at different phase angles. Panel (**a**) shows the backscattered light from the main rings of Uranus. Panel (**b**) is the only high phase angle image of Uranus' rings. Several distinct dust bands co-exist with the rings appeared in the low-phase angle image. Dusty rings, e.g., λ and ζ rings, are only visible at high phase angle. The marked dash lines are only visual guides. The large eccentricity of the ring ϵ ($e \sim 0.08$) is apparent by comparing these two images. The feature marked ring ζ in (**b**) only represents the outer edge of a broader ring, extending inwards to ~35000 km (de Pater et al. 2007) (credit: JPL/Caltech/NASA)

particles, with an optical depth ranging from 0.1 to 1 and a radial width from ≤ 10 to 100 km. The overall low albedo (~0.04, Sinton (1977), de Kleer et al. (2013)) and flat color spectra indicate their main constituent to be irradiated hydrocarbon (Lanzerotti et al. 1987).

In addition to the optically thick rings, numerous confined dust bands across the main ring region were seen in the only high phase angle image taken by Voyager 2 (Fig. 13 and Table 3). The presence of a dust component is also evident in the recent ground–based observation (de Pater et al. 2006; Dunn et al. 2010), thanks to the nearly edge-on view from Earth during 2006–2007. Some of the bands are not associated with any narrow ring described above. These bands are composed of particles efficient in forward–scattering, suggesting these are micron-sized dust particles. Ring λ is predominantly composed of particles ~ µm in size as it shows the highest contrast between high and low phase images. The analysis of its low phase angle images taken by Voyager 2 showed a five–cycle periodic variation in brightness, possibly resulting from the interactions with the moon Cordelia and another unknown moon (Showalter 1995).

Because of the short dynamical lifetime ($\leq 10^3$ years) caused by the drag force from the extended hydrogen exosphere of Uranus (Broadfoot et al. 1986), it was proposed that the observed dust bands are continuously supplied from impact ejecta produced within the rings (Esposito and Colwell 1989; Colwell and Esposito 1990b). The band structure therefore is a manifestation of various ring-shaping processes, including orbital resonances with satellites, as well as electromagnetic and gas drag forces.

Table 3 Dusty rings of Uranus

Ring	Radius [km]	Width [km]	Optical depth
ζ/R1986U2	26840–41350	14510	$\sim 10^{-3}$
λ	50023	1–2	~ 0.1
ν	66100–69900	3800	5×10^{-5}
μ	86000–103000	17000	8×10^{-5}

Keck observations during 2006–2007, when the rings of Uranus appeared edge–on to the observers on Earth, showed significant changes in the ring structure compared to the Voyager era. The brightness peak of the inner most ring ζ, presumably associated with the ring called R/1986 U2, moved outward by a couple thousands of km. This may indicate a large-scale dust migration occurred in between the Voyager 2 flyby in 1986 and 2006 (de Pater et al. 2007, 2013).

Two outer diffuse rings were discovered during the same period (Showalter and Lissauer 2006; de Pater et al. 2007). The inner one, named ring ν, is located between the orbits of the moons Portia and Rosalind with peak intensity at 67300 km (2.63 R_U). Its spectrally red color resembles Saturn's G ring, indicating a broad particle size distribution (de Pater et al. 2007). The outer ring μ, is spatially wider with its brightest part coinciding with the orbit of the tiny moon Mab ($\sim$25 km) at 97700 km (3.82 R_U) (Showalter and Lissauer 2006). Its peculiar blue spectral color is similar to Saturn's E ring (de Pater et al. 2004, 2006), indicating that processes other than impactor–ejecta may play an important role in shaping the grain size distribution (Sfair and Giuliatti Winter 2009, 2012; Jontof-Hutter and Hamilton 2012a,b; Hedman et al. 2018). So far the origins of these two diffuse rings are not yet clear.

Dust impacts inferred from the Voyager 2 plasma instruments, Plasma Wave Spectrometer (PWS) (Meyer-Vernet et al. 1986) and Planetary Radio Astronomy (PRA) (Gurnett et al. 1986), during the ring plane crossing (RPX) at 115401 km (4.52 R_U) may correlate with the newly discovered μ ring, or the material drifting inwards from outer moons, e.g., Miranda. The derived dust number density is in the range of 10^5 to 10^6 km^{-3}, comparable to the density Voyager 1 registered during its Saturn RPX at 6.1 R_S in Saturn's E ring with $\tau \sim 10^{-8} - 10^{-7}$ (Meyer-Vernet et al. 1996).

Further outward the major moons of Uranus—Miranda (5.05 R_U), Ariel (7.46 R_U), Umbriel (10.4 R_U), Titania (17.0 R_U), and Oberon (22.8 R_U)—are located. All moons except Umbriel exhibit endogenic resurfacing events as indicated by the Voyager 2 imaging data. The outer four moons show a color dichotomy—the leading side is 2–23% redder than the trailing side. In addition, the overall color becomes redder with increasing distance to Uranus (Buratti and Mosher 1991). Recent modeling work suggests that this color asymmetry could result from the infall of dust generated as impact ejecta from the outer captured, irregular satellites (Tamayo et al. 2013).

Neptune's Ring System The ring system of Neptune[14] consists of 3 narrow rings and 2 broad rings with dust content of up to 80%, much higher in comparison to the main rings of Saturn and Uranus. The optical depth of Neptune's rings is larger than that of Jupiter's rings, yet the albedo of the ring material is very low at about 5%, similar to the adjacent ring moons (Smith et al. 1989; Ferrari and Brahic 1994). Neptune's rings are named after astronomers

[14] Neptune's radius $R_N = 24764$ km.

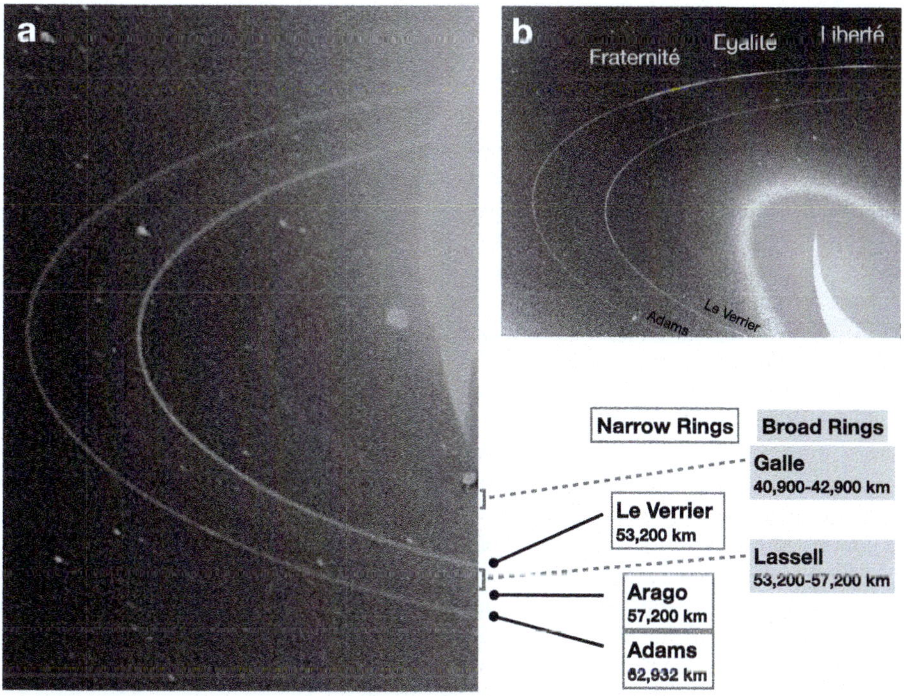

Fig. 14 Neptune's rings and ring arcs viewed by Voyager 2 at a high phase angle of 135°. Panel (**a**) shows the five rings of Neptune marked with their name and radii. Panel (**b**) shows three of the Adams ring arcs. The orbital motion direction is clockwise (credit: JPL/Caltech/NASA)

Table 4 Rings of Neptune

Ring	Radius [km]	Width [km]	Optical depth
Galle	40900–42900	2000	$\sim 10^{-4}$
Le Verrier	53200 ± 20	113	$\sim 10^{-3}$
Lassell	53200–57200	4000	$\sim 10^{-4}$
Arago	57200	<100	?
Adams	62932	15–50	~ 0.01

who contributed to our knowledge about the planet. With increasing planetcentric distance, they are Galle, Le Verrier, Lassell, Arago, and Adams (Fig. 14 and Table 4).

One intriguing topic that attracted a great deal of attention was the discovery of ring arcs in the outermost Adams ring. Early searches for Neptune's rings through ground-based occultations did not confirm their existence as the blockage confusingly at most only appeared at one side of the planet, which contradicted the collective experience of a planetary ring at the time. It was realized that the ring must be partial in the form of arcs based on the simultaneous observations at multiple locations in 1984 (Hubbard et al. 1986). Four bright arcs of Adams ring were then identified during the Voyager 2 flyby in 1989 (Smith et al. 1989) and named Fraternité, Egalité, Liberté, and Courage.

Observations indicate that the brighter arcs, corresponding to an optical depth of the order of 0.01, are locations with high concentration of dust supplied from collisional fragmentation of larger, unresolvable ring particles (Colwell and Esposito 1990a; Dumas et al. 2002; de Pater et al. 2005). However, the mechanism leading to the confinement of four ring arcs is still not fully understood. Confinement by resonances with the moon Galatea, which lies inside the Adams ring (Porco 1991; Namouni and Porco 2002) or by small co-orbital moons yet to be discovered (Lissauer 1985; Renner et al. 2014) are the most plausible scenarios. Confirmation of these two hypotheses requires close–up observations to provide constraints, either on the eccentricity of Galatea (Namouni and Porco 2002) or the presence of ring moons below the Voyager detection limit (about 7 km) (Renner et al. 2014). More recent observations show that the arcs have changed since the discovery in 1989. The arcs Liberté and Courage nearly disappeared (Dumas et al. 1999; de Pater et al. 2005; Showalter et al. 2005). The trailing arcs Fraternité and Egalité persist but the latter became relatively fainter in 2007 (Renner et al. 2014).

The Voyager PRA and PWS instruments were sensitive to impact events by grains with radii larger than ~10 μm. Data from these two instruments during the Neptune flyby indicate a warped dust disk around the geometric equator and a much more tenuous halo population at high latitudes (Gurnett et al. 1991). The dust number density in the equatorial disk is estimated to be around $10^2\ m^{-3}$ and about two orders of magnitude lower in the halo (Gurnett and Kurth 1995). The warping of the dust disk likely reflects the gravitational perturbation of the grains' Laplace plane from the inclined, massive moon Triton (Gurnett and Kurth 1995; Porco et al. 1995). Electromagnetic forces are likely responsible of transporting grains eroded from the rings and moons in the equatorial plane into inclined orbits, forming the halo population (Gurnett and Kurth 1995).

The rings of the ice giant planets are characterized by densely packed narrow rings and small ring moons. We refer the interested readers to recent review articles by Nicholson et al. (2018) for rings of Uranus and de Pater et al. (2018a) for rings of Neptune. Such arrangement can be well explained by the ring–moon interactions, i.e., the exchange of angular momentum and collisional fragmentation from exogenous hypervelocity impacts as well as mild (~m/s) inter–particle collisions (Murray and Thompson 1990; Colwell and Esposito 1990b,a). Their variabilities, as indicated by scarce but illustrative observations, reinforce the idea that the rings as we know them are snapshots of a constantly changing landscape and a manifestation of the shaping processes at work rather than a testimony of eternity (see also Esposito 2014). Moons are shattered forming rings, which eventually are ground down and lost as dust. Encompassed by diffuse rings and dust discs and halo, these two systems are rich in dust. Future in situ measurements of microscopic relics will help to reconstruct the violent history witnessed by the ice giant systems as well as the minor bodies in the outer Solar System.

3.1.4 Dust-Populations at Mars and Pluto-Charon

Here we address putative circumplanetary dust-populations where an observational confirmation is still missing: the dust populations at Mars and around Pluto-Charon.

The Putative Dust-Tori Around Mars Erosion of atmosphereless Solar System bodies by micrometeoroids is the major source of dust in the Solar System (see Sect. 2.1.1). This inspired Soter (1971) to numerically simulate the dynamics of millimeter-sized ejecta from the Martian satellites Phobos and Deimos that produce low-optical depth debris tori around their orbits. This first dynamical study has been refined by all relevant perturbation forces—i.e. Mars' oblateness (J2), radiation pressure, Lorentz force—supplemented by steadily improved dust production models (see Sect. 2.1.1) by various authors (Juhász and Horányi,

1995; Krivov and Hamilton, 1997; and references therein). As a result of these model upgrades, Krivov et al. (1996) have classified the Martian dust environment into 3 populations (apart from short-living millimeter down to sub-millimeter tori covering the orbits of Phobos and Deimos):

- *Population I:* contains dust of sizes 10 μm–100 μm remaining thousands of years in orbit close to both sources, forming the dominant dust component at Mars;
- *Population II:* cover particles between 1 and 10 μm which have much shorter lifetimes through loss to Mars owing to radiation pressure and the J2 gravitational force component (oblateness) of the planet;
- *Population III:* consists of sub-micrometer and smaller dust particles which suffer strong force fluctuations (related to even shorter lifetimes) and form an extended, faint dust component surrounding the other two, more prominent, populations;

An interesting structural addendum has been predicted by Makuch et al. (2005): The smallest part (10 μm–15 μm) of the Population I dust generated at Deimos has lifetimes sufficiently long so that they migrate towards Mars because of the Poynting-Robertson drag. This inward creeping causes a rather contracted torus—more symmetric than the rest of Population I belt—located between the orbits of either Martian moons (Makuch et al. 2005). All the theoretical studies have yielded rather low optical depths of those tori, $\tau \approx 10^{-7}...10^{-9}$ (e.g. Juhász and Horányi, 1995; Krivov and Hamilton, 1997; Showalter et al., 2006), so that a detection, even in situ, will be a great experimental challenge.

Nevertheless, these theoretical predictions seemed to be partly confirmed by the magnetometer of the *Phobos* 2 spacecraft, which showed unusual fluctuations of the **B** field when crossing the orbit of the satellite Phobos. One explanation could be the Population I dust torus embracing the orbit of the moon Phobos (Riedler et al. 1989) constituting an obstacle for the Solar wind and possibly causing the related magnetic field fluctuations. Thus, Baumgärtel et al. (1996) investigated the influence of a torus of charged dust on the flow of the Solar wind around Mars. They found that only an unusually narrow and flat dust ring would be compatible with the observations of the magnetometer of Phobos 2. This, however, contradicts the modeling of such ejecta-generated dust configurations, so that the authors rule out the option of a dust torus. Great hope of a detection of the putative dust tori was set in the Japanese space mission *Nozomi*. Unfortunately, during the traverse from Earth to Mars (which had required a couple of gravity assists at Earth), the spacecraft has been hit by a Sun-storm caused by a Solar eruption end of 2002, which finally lead to the loss of the spacecraft.

Another analysis of magnetic field measurements from the *Mars Global Surveyor* (Øieroset et al. 2010) shows neither the sign of existence of Mars' dusty tori nor significant activities from the Marian moons.

The Putative Dust in the Pluto-System With the advent of the New Horizon mission, the scientific interest had been rising in the Pluto-Charon system. In this context Weaver et al. (2006) detected two new, rather small satellites of Pluto by processing *Hubble Space Telescope* (HST) images: the moons Nix (size ≈ 42 km × 36 km; semi-major axis $a \approx$ 48675 km) and Hydra (size ≈ ? km; semi-major axis $a \approx$ 64780 km). Given the fact that small moons are efficient dust sources, Showalter et al. (2011) started a search for impact-ejection generated dust rings or -tori in the HST-data which, unfortunately, did not reveal dust populations but another small satellite: Kerberos (size ≈ 5km . . . 8 km) that is located between the former two (semi-major axis $a \approx$ 59000 km). A further more sophisticated HST observing program yielded yet another small satellite Styx (size ≈ 7km × 5 km) located

Table 5 Parameters of Chariklo, Chiron & Haumea

Central body	a [AU]	e	i [degrees]	Mean radius [km] semi-axes
Chariklo	15.822	0.172	23.4	124 ± 9
Chiron	13.684	0.382	6.9	90
Haumea	43.355	0.189	28.204	1161 ± 30
				852 ± 4
				513 ± 16

closer to the double dwarf-planets Pluto-Charon (semi-major axis $a \approx 42000 \pm 2000$ km)—but again no hints for any dust population. All these observations have yielded an upper detection limit of putative dust rings, -arc, or -tori of Pluto-Charon of about $\tau \approx 10^{-8}$ to 10^{-7} (Lauer et al. 2017; Sicardy et al. 2017).

Two possible explanations for the non-detection of dust at Pluto are discussed currently:

(i) The dynamical fate of the tiny moons and related dust at the edge of chaos (Showalter and Hamilton 2015) might explain the rather low dust particle number densities and optical depths τ of putative dust populations at Pluto. Although the recently detected small moons constitute quite efficient dust sources, Solar radiation pressure combined with the relatively low gravity of the double-dwarf-planet Pluto-Charon cause a rather unstable dynamical situation of the ejecta (Porter and Stern 2015), reducing their lifetime and justifying rather low optical depths of the resulting rings or tori.
(ii) Analyses of imaging data of the *New Horizon* mission concerning satellite crater records point to a lack of impacting projectiles of sizes less than < 1 km (Lauer et al. 2017). This deficit of a part of the projectile family would reduce considerably the dust production rate—possibly explaining why Pluto-dust-rings could not be detected yet.

Similarly to the case of the putative Mars-dust tori, at Pluto we face a situation where a prediction could not be confirmed with observations—neither with the HST nor in situ with the *New Horizon* spacecraft.

3.1.5 Rings Around Centaurs Chiron, Chariklo, and the Dwarf-Planet Haumea

Recently, tiny-versions of planetary rings have been detected (Braga-Ribas et al. 2014; Ortiz et al. 2015, 2017) by using stellar occultations scanning the vicinities of the Centaurs (10199) Chariklo and (2060) Chiron, as well as the trans-pluto dwarf planet Haumea.

Currently it is not clear whether these "exotic" rings contain a considerable dust fraction. We argue that widths, shapes and configurations of these rings point to some similarities with, e.g. the rings Uranus or Neptune—rings which contain both, dense, narrow, sharp-edged ringlets embedded in broader dusty components (see Sect. 3.1.3). This and also their scientific interest to some extent warrant their mention in this review.

In Table 5, the positions, or respectively, the orbital elements (semi-major axis a, eccentricity e, inclination i) as well as information about the sizes and—in the case of the dwarf-planet Haumea—the shape (semi-major axes of an ellipsoid) of the certain central bodies are summarized.

Haumea's mass $m = (4.006 \pm 0.040) \times 10^{21}$ kg can be determined from the orbits of its two known satellites, Hi'iaka and Namaka, while from its elliptical shape (see Table 5) the density $\varrho = (1.885 \pm 80) \times 10^3$ kg/m^3 can be concluded. The masses of the two Centaurs are not known.

Table 6 Parameters of the ring systems

Central body	Ring part	a or r [km]	e	W [km]	τ
Chariklo	Ca1	390.6 ± 3.3	0.45	6...7.2	0.3–0.45
	Ca2	404.8 ± 3.3	0.45	3.4...3.6	<0.1
Chiron	–	324 ± 10	–	3 & 7	0.7...1
Haumea	–	2287	–	≈70	0.5

Table 6 summarizes the parameters of the three ring systems: The orbital elements (a, e, i; or if not known, the distance r from the central body), the width W of the rings and their optical depths τ.

Both Centaurs are girdled by quite tiny rings—that one of the trans-Plutonian dwarf-planet Haumea is one order of magnitude larger—but being still orders of magnitude smaller than the rings of the gas-giants. Ortiz et al. (2017) have found that the ring systems of Chariklo and Chiron appear very similar. Unfortunately, detailed data of the positions and widths of the pair of Chiron's rings is not tabulated there. Thus, in Table 6, we present the two values of 3 and 7 kilometers for the two Chiron rings together with 14 km separation between them, as mentioned in the original papers.

3.1.6 Rings Around the Earth

For completeness we finish this section by briefly discussing the dust environment of the Earth. While we do not have a natural ring system, the continual debris production from disintegrating and colliding artificial satellites represents a growing hazard. The dynamics of small debris in the near Earth environment has been investigated near the beginning of the space age, as the potential of generating an artificial ring system about the Earth by the degradation of the rapidly growing number of Earth orbiting satellites was immediately recognized. The survey of perturbations on the orbital motion of small particles due to gravity, drag, and radiation pressure (Colombo et al. 1966a,b; Shapiro et al. 1966; Lautman et al. 1966) were carried over for studies of natural planetary ring systems. The effects of electrostatic charging on the lifetime of space debris around the Earth (Horányi et al. 1988; Juhász and Horányi 1997) also turned out to be relevant to study the dynamics of charged ring particles at Jupiter and Saturn.

The Moon, as all other moons orbiting other planets, is also a source of particles through its continual bombardment, but due to its large distance to Earth, perturbations by solar gravity, radiation pressure, and electromagnetic effects prevent most small escaping dust particles from getting near the Earth (Colombo et al. 1966b; Juhász and Horányi 1999). Since the Moon is a relatively large object, the majority of the ejecta particles remain on bound ballistic orbits and return to the surface, forming a permanently present but spatially and temporarily variable ejecta cloud (see the review by Szalay et al. 2018, this issue).

4 Discussion and Conclusion

In this review circumplanetary dust populations in the Solar System are presented together with a general theoretical description. All these dust rings have the following common "life-cycle": (a) the creation of dust at certain sources ("birth"); (b) the subsequent dynamical "life" of these generated grains which is terminated at their (c) sinks ("death").

Despite of the diversities of all the presented circumplanetary dust complexes—caused by the different source processes, different planetary environments, e.g., certain distances from the Sun and the resulting variety of dynamical particle paths—a widespread reason for dust creation is the cosmic erosion caused by hypervelocity impacts of meteoroids—the impact-ejection (see Sect. 2.1.1). This degradation of the atmosphereless cosmic bodies, bears various information about the projectiles—interplanetary or interstellar micrometeoroids—as well as the target material—the surfaces of planets, moons, and minor bodies such as asteroids. Following the concept of dust-astronomy (see Sect. 2.2) with dust-detectors/analyzers aboard interplanetary space vehicles (as Galileo- or Cassini spacecraft) the composition, mass, velocity of a detected dust grain can be obtained. Their origins can then be retraced with dynamical calculations leading us to the putative source—moons, rings or other airless bodies. Furthermore, the flux of particles stemming from the same source carry indirect information about the projectiles and parameters of the related impact-ejection process. In this way we are able to remotely probe alien, distant worlds—as has been successfully demonstrated with the spacecrafts Galileo and Cassini at Jovian- and Saturnian system, respectively.

If we compare the gradual erosion of airless surfaces by the impact-ejection to the geology concept of *uniformitarianism*, the analogs to the *catastrophism* would be the (cryo-)volcanic activities from Io and Enceladus (and possibly Europa) as well as large collision events, e.g., Shoemaker–Levy 9 at Jupiter. With the supply from these processes, the circumplanetary dust environments reach the present landscape due to the persistent sculpture by various forces and interactions with the moons, rings, and magnetospheres. Study of circumplanetary dust, despite being only a snapshot, can provide a window to look back to the history to understand the evolution and the origin of the planetary systems in our, and likely many other, extra-solar systems.

This review presented observations and theoretical models of a large variety of dust populations in various circumplanetary environments throughout the Solar System. Despite the large differences in the characteristics of these environments and dust populations, we recognize several universal processes that commonly shape the sources, transport, and sinks of dust particles in all of these environments. The possible sources of dust particles include ejecta generation by the continual bombardment of surfaces by interplanetary meteoroids, active volcanoes and geysers, and the capture of interplanetary dust particles, for example. The dynamics, and hence the transport, of newly generated particles involves a variety of possible forces, including drag, radiation pressure, electromagnetic forces, and gravity, for example. In addition, in regions with sufficiently large densities, dust-dust collisions, and self-gravity can sculpt the spatial distribution of dust particles.

Our recognition of these universal processes is based on the combination of both remote sensing and in situ observations. Especially the combination of imaging observations of large-scale structures, and the in situ observations of dust particles, complemented by theoretical modeling, proved to be very successful in characterizing circumplanetary dust populations. Recent dust measurements from instruments onboard Galileo at Jupiter, and especially onboard Cassini at Saturn led to the recognition that these measurements are critical in the analysis and interpretation of a number of different observations, including magnetospheric plasmas and fields. The Cassini CDA instrument, capable of measuring the mass, charge, velocity and chemical makeup of impacting dust particles, was critical in the discovery of Enceladus' subsurface ocean and its hydrothermal activity.

The processes responsible for sustaining and shaping circumplanetary dust environments today are the very same processes that shaped the evolution of our early Solar System, and are active in all extra solar dust disks. Hence, the emerging understanding of the workings of the easily accessible dust environments, e.g. planetary rings, asteroids, the putative dust clouds of Phobos and Deimos, even our own Moon, can greatly improve our understanding of the history of our own Solar System.

Acknowledgements We thank an anonymous referee for his considerable advise in making the paper more convincing. We acknowledge the support by the NASA/ESA Cassini-Huygens mission and by ISSI. The work of M. Seiß and M. Sachse has been funded by the DLR (German Space Agency) with the project (50OH1401).

References

N. Asada, Fine fragments in high-velocity impact experiments. J. Geophys. Res. **90**, 12445 (1985). https://doi.org/10.1029/JB090iB14p12445

W.I. Axford, D.A. Mendis, Satellites and magnetospheres of the outer planets. Annu. Rev. Earth Planet. Sci. **2**, 419 (1974). https://doi.org/10.1146/annurev.ea.02.050174.002223

A. Barkan, N. D'angelo, R.L. Merlino, Charging of dust grains in a plasma. Phys. Rev. Lett. **73**, 3093–3096 (1994). https://doi.org/10.1103/PhysRevLett.73.3093

E.R. Batista, P. Ayotte, A. Bilić, B.D. Kay, H. Jónsson, What determines the sticking probability of water molecules on ice? Phys. Rev. Lett. **95**(22), 223201 (2005). https://doi.org/10.1103/PhysRevLett.95.223201

W.A. Baum, T. Kreidl, J.A. Westphal, G.E. Danielson, P.K. Seidelmann, D. Pascu, D.G. Currie, Saturn's E ring. Icarus **47**, 84–96 (1981). https://doi.org/10.1016/0019-1035(81)90093-2

K. Baumgärtel, K. Sauer, A. Bogdanov, E. Dubinin, M. Dougherty, "Phobos events": signatures of solar wind dust interaction. Planet. Space Sci. **44**(6), 589–601 (1996)

P. Bliokh, V. Sinitsin, V. Yaroshenko, *Dusty and Self-Gravitational Plasmas in Space*. Astrophysics and Space Science Library (Kluwer, Dordrecht, 1995)

F. Braga-Ribas, B. Sicardy, J.L. Ortiz, C. Snodgrass, F. Roques, R. Vieira-Martins, J.I.B. Camargo, M. Assafin, R. Duffard, E. Jehin, J. Pollock, R. Leiva, M. Emilio, D.I. Machado, C. Colazo, E. Lellouch, J. Skottfelt, M. Gillon, N. Ligier, L. Maquet, G. Benedetti-Rossi, A.R. Gomes, P. Kervella, H. Monteiro, R. Sfair, M. El Moutamid, G. Tancredi, J. Spagnotto, A. Maury, N. Morales, R. Gil-Hutton, S. Roland, A. Ceretta, S.-H. Gu, X.-B. Wang, K. Harpsøe, M. Rabus, J. Manfroid, C. Opitom, L. Vanzi, L. Mehret, L. Lorenzini, E.M. Schneiter, R. Melia, J. Lecacheux, F. Colas, F. Vachier, T. Widemann, L. Almenares, R.G. Sandness, F. Char, V. Perez, P. Lemos, N. Martinez, U.G. Jørgensen, M. Dominik, F. Roig, D.E. Reichart, A.P. Lacluyze, J.B. Haislip, K.M. Ivarsen, J.P. Moore, N.R. Frank, D.G. Lambas, A ring system detected around the Centaur (10199) Chariklo. Nature **508**, 72–75 (2014). https://doi.org/10.1038/nature13155

N.V. Brilliantov, P.L. Krapivsky, A. Bodrova, F. Spahn, H. Hayakawa, V. Stadnichuk, J. Schmidt, Size distribution of particles in Saturn's rings from aggregation and fragmentation. Proc. Natl. Acad. Sci. **112**(31), 9536–9541 (2015)

A.L. Broadfoot, F. Herbert, J.B. Holberg, D.M. Hunten, S. Kumar, B.R. Sandel, D.E. Shemansky, G.R. Smith, R.V. Yelle, D.F. Strobel, H.W. Moos, T.M. Donahue, S.K. Atreya, J.L. Bertaux, J.E. Blamont, J.C. Mcconnell, A.J. Dessler, S. Linick, R. Springer, Ultraviolet spectrometer observations of Uranus. Science **233**, 74–79 (1986). https://doi.org/10.1126/science.233.4759.74

S.M. Brooks, L.W. Esposito, M.R. Showalter, H.B. Throop, The size distribution of Jupiter's main ring from Galileo imaging and spectroscopy. Icarus **170**, 35–57 (2004). https://doi.org/10.1016/j.icarus.2004.03.003

B.J. Buratti, J.A. Mosher, Comparative global albedo and color maps of the Uranian satellites. Icarus **90**, 1–13 (1991). https://doi.org/10.1016/0019-1035(91)90064-Z

J.A. Burns, P.L. Lamy, S. Soter, Radiation forces on small particles in the solar system. Icarus **40**, 1–48 (1979). https://doi.org/10.1016/0019-1035(79)90050-2

J.A. Burns, M.R. Showalter, D.P. Hamilton, P.D. Nicholson, I. de Pater, M.E. Ockert-Bell, P.C. Thomas, The formation of Jupiter's faint rings. Science **284**, 1146 (1999). https://doi.org/10.1126/science.284.5417.1146

J.A. Burns, D.P. Simonelli, M.R. Showalter, D.P. Hamilton, C.D. Porco, H. Throop, L.W. Esposito, Jupiter's ring-moon system, in *Jupiter. The Planet, Satellites and Magnetosphere*, ed. by F. Bagenal, T.E. Dowling, W.B. McKinnon (2004), pp. 241–262

S. Charnoz, R.M. Canup, A. Crida, L. Dones, Planetary ring systems, in *The Origin of Planetary Ring Systems*, ed. by M.S. Tiscareno, C.D. Murray (Cambridge University Press, New York, 2018), pp. 517–538
G. Colombo, D.A. Lautman, I.I. Shapiro, The Earth's dust belt: fact or fiction? 2, Gravitational focusing and Jacobi capture. J. Geophys. Res. **71**, 5705 (1966a)
G. Colombo, I.I. Shapiro, D.A. Lautman, The Earth's dust belt: fact or fiction? 3, Lunar ejecta. J. Geophys. Res. **71**, 5719 (1966b). https://doi.org/10.1029/JZ071i023p05719
J.E. Colwell, L.W. Esposito, A model of dust production in the Neptune ring system. Geophys. Res. Lett. **17**, 1741–1744 (1990a). https://doi.org/10.1029/GL017i010p01741
J.E. Colwell, L.W. Esposito, A numerical model of the Uranian dust rings. Icarus **86**, 530–560 (1990b). https://doi.org/10.1016/0019-1035(90)90232-X
J.E. Colwell, M. Horányi, Magnetospheric effects on micrometeoroid fluxes. J. Geophys. Res. **101**, 2169–2176 (1996). https://doi.org/10.1029/95JE03103
J.E. Colwell, M. Horányi, E. Gren, Capture of interplanetary and interstellar dust by the Jovian magnetosphere. Science **280**, 88 (1998). https://doi.org/10.1126/science.280.5360.88
J.E.P. Connerney, Magnetic connection for Saturn's rings and atmosphere. Geophys. Res. Lett. **13**, 773–776 (1986). https://doi.org/10.1029/GL013i008p00773
J.N. Cuzzi, R.H. Durisen, Bombardment of planetary rings by meteoroids—general formulation and effects of Oort cloud projectiles. Icarus **84**, 467–501 (1990). https://doi.org/10.1016/0019-1035(90)90049-F
J.N. Cuzzi, P.R. Estrada, Compositional evolution of Saturn's rings due to meteoroid bombardment. Icarus **132**, 1–35 (1998). https://doi.org/10.1006/icar.1997.5863
J.N. Cuzzi, A.D. Whizin, R.C. Hogan, A.R. Dobrovolskis, L. Dones, M.R. Showalter, J.E. Colwell, J.D. Scargle, Saturn's F ring core: calm in the midst of chaos. Icarus **232**, 157–175 (2014). https://doi.org/10.1016/j.icarus.2013.12.027
K. de Kleer, I. de Pater, M. Ádámkovics, H. Hammel, Near-infrared spectra of the Uranian ring system. Icarus **226**, 1038–1044 (2013). https://doi.org/10.1016/j.icarus.2013.07.016
I. de Pater, M.R. Showalter, J.J. Lissauer, J.R. Graham, Keck infrared observations of Saturn's E and G rings during Earth's 1995 ring plane crossings. Icarus **121**, 195–198 (1996)
I. de Pater, M.R. Showalter, J.A. Burns, P.D. Nicholson, M.C. Liu, D.P. Hamilton, J.R. Graham, Keck infrared observations of Jupiter's ring system near Earth's 1997 ring plane crossing. Icarus **138**, 214–223 (1999). https://doi.org/10.1006/icar.1998.6068
I. de Pater, S.C. Martin, M.R. Showalter, Keck near-infrared observations of Saturn's E and G rings during Earth's ring plane crossing in August 1995. Icarus **172**, 446–454 (2004). https://doi.org/10.1016/j.icarus.2004.07.012
I. de Pater, S.G. Gibbard, E. Chiang, H.B. Hammel, B. Macintosh, F. Marchis, S.C. Martin, H.G. Roe, M. Showalter, The dynamic Neptunian ring arcs: evidence for a gradual disappearance of Liberté and resonant jump of courage. Icarus **174**, 263–272 (2005). https://doi.org/10.1016/j.icarus.2004.10.020
I. de Pater, H.B. Hammel, S.G. Gibbard, M.R. Showalter, New dust belts of Uranus: one ring, two ring, red ring, blue ring. Science **312**, 92–94 (2006). https://doi.org/10.1126/science.1125110
I. de Pater, H.B. Hammel, M.R. Showalter, M.A. van Dam, The dark side of the rings of Uranus. Science **317**, 1888 (2007). https://doi.org/10.1126/science.1148103
I. de Pater, D.E. Dunn, D.M. Stam, M.R. Showalter, H.B. Hammel, M. Min, M. Hartung, S.G. Gibbard, M.A. van Dam, K. Matthews, Keck and VLT AO observations and models of the Uranian rings during the 2007 ring plane crossings. Icarus **226**, 1399–1424 (2013). https://doi.org/10.1016/j.icarus.2013.08.001
I. de Pater, D.P. Hamilton, M.R. Showalter, H.B. Throop, J.A. Burns, The rings of Jupiter, in *Planetary Ring Systems Properties, Structure, and Evolution*, ed. by M.S. Tiscareno, C.D.E. Murray. (Cambridge University Press, New York, 2018a), pp. 125–134. https://doi.org/10.1017/9781316286791.006. Chap. 6
I. de Pater, S. Renner, M.R. Showalter, B. Sicardy, The rings of Neptune, in *Planetary Ring Systems Properties, Structure, and Evolution*, ed. by M.S. Tiscareno, C.D.E. Murray. (Cambridge University Press, New York, 2018b), pp. 112–124. https://doi.org/10.1017/9781316286791.005. Chap. 5
V.V. Dikarev, Dynamics of particles in Saturn's E ring: effects of charge variations and the plasma drag force. Astron. Astrophys. **346**, 1011–1019 (1999)
Y. Dong, T.W. Hill, S.-Y. Ye, Characteristics of ice grains in the Enceladus plume from Cassini observations. J. Geophys. Res. Space Phys. **120**, 915–937 (2015). https://doi.org/10.1002/2014JA020288
M.K. Dougherty, K.K. Khurana, F.M. Neubauer, C.T. Russell, J. Saur, J.S. Leisner, M.E. Burton, Identification of a dynamic atmosphere at Enceladus with the Cassini magnetometer. Science **311**, 1406–1409 (2006). https://doi.org/10.1126/science.1120985
C. Dumas, R.J. Terrile, B.A. Smith, G. Schneider, E.E. Becklin, Stability of Neptune's ring arcs in question. Nature **400**, 733–735 (1999). https://doi.org/10.1038/23414
C. Dumas, R.J. Terrile, B.A. Smith, G. Schneider, Astrometry and near-infrared photometry of Neptune's inner satellites and ring arcs. Astron. J. **123**, 1776–1783 (2002). https://doi.org/10.1086/339022

D.E. Dunn, I. de Pater, D. Stam, Modeling the Uranian rings at 2.2 μm: comparison with Keck AO data from July 2004. Icarus **208**, 927–937 (2010). https://doi.org/10.1016/j.icarus.2010.03.027
J.L. Elliot, Stellar occultation studies of the Solar System. Annu. Rev. Astron. Astrophys. **17**, 445–475 (1979). https://doi.org/10.1146/annurev.aa.17.090179.002305
J.L. Elliot, E. Dunham, D. Mink, The rings of Uranus. Nature **267**, 328–330 (1977). https://doi.org/10.1038/267328a0
L. Esposito, *Planetary Rings* (2014)
L.W. Esposito, J.E. Colwell, Creation of the Uranus rings and dust bands. Nature **339**, 605–607 (1989). https://doi.org/10.1038/339605a0
P.R. Estrada, R.H. Durisen, J.N. Cuzzi, D.A. Morgan, Combined structural and compositional evolution of planetary rings due to micrometeoroid impacts and ballistic transport. Icarus **252**, 415–439 (2015). https://doi.org/10.1016/j.icarus.2015.02.005
W.A. Feibelman, Concerning the "D" ring of Saturn. Nature **214**, 793–794 (1967)
C. Ferrari, A. Brahic, Azimuthal brightness asymmetries in planetary rings. 1: Neptune's arcs and narrow rings. Icarus **111**, 193–210 (1994). https://doi.org/10.1006/icar.1994.1140
C.K. Goertz, G.E. Morfill, A model for the formation of spokes in Saturn's ring. Icarus **53**, 219–229 (1983)
A.L. Graps, Io revealed in the Jovian dust streams. Ph.D. Thesis, Ruprecht-Karls-Universität, Heidelberg (2001)
A.L. Graps, E. Grün, H. Svedhem, H. Krüger, M. Horányi, A. Heck, S. Lammers, Io as a source of the Jovian dust streams. Nature **405**, 48–50 (2000)
E. Grün, G.E. Morfill, R.J. Terrile, T.V. Johnson, G. Schwehm, The evolution of spokes in Saturn's B ring. Icarus **54**, 227–252 (1983). https://doi.org/10.1016/0019-1035(83)90194-X
E. Grün, G.W. Garneau, R.J. Terrile, T.V. Johnson, G.E. Morfill, Kinematics of Saturn's spokes. Adv. Space Res. **4**, 143–148 (1984). https://doi.org/10.1016/0273-1177(84)90019-X
E. Grün, C.K. Goertz, G.E. Morfill, O. Havnes, Statistics of Saturn's spokes. Icarus **99**(1), 191–201 (1992a)
E. Grün, H. Fechtig, M.S. Hanner, J. Kissel, B.-A. Lindblad, D. Linkert, D. Maas, G.E. Morfill, H.A. Zook, The Galileo dust detector. Space Sci. Rev. **60**, 317–340 (1992b)
E. Grün, H.A. Zook, M. Baguhl, A. Balogh, S.J. Bame, H. Fechtig, R. Forsyth, M.S. Hanner, M. Horanyi, J. Kissel, B.-A. Lindblad, D. Linkert, G. Linkert, I. Mann, J.A.M. McDonnell, G.E. Morfill, J.L. Phillips, C. Polanskey, G. Schwehm, N. Siddique, P. Staubach, J. Svestka, A. Taylor, Discovery of Jovian dust streams and interstellar grains by the ULYSSES spacecraft. Nature **362**, 428–430 (1993). https://doi.org/10.1038/362428a0
E. Grün, M. Baguhl, D.P. Hamilton, R. Riemann, H.A. Zook, S. Dermott, H. Fechtig, B.A. Gustafson, M.S. Hanner, M. Horányi, K.K. Khurana, J. Kissel, M. Kivelson, B.A. Lindblad, D. Linkert, G. Linkert, I. Mann, J.A.M. McDonnell, G.E. Morfill, C. Polanskey, G. Schwehm, R. Srama, Constraints from Galileo observations on the origin of Jovian dust streams. Nature **381**, 395–398 (1996). https://doi.org/10.1038/381395a0
E. Grün, P. Staubach, M. Baguhl, D.P. Hamilton, H.A. Zook, S. Dermott, B.A. Gustafson, H. Fechtig, J. Kissel, D. Linkert, G. Linkert, R. Srama, M.S. Hanner, C. Polanskey, M. Horányi, B.A. Lindblad, I. Mann, J.A.M. McDonnell, G.E. Morfill, G. Schwehm, South-North and radial traverses through the interplanetary dust cloud. Icarus **129**, 270–288 (1997)
E. Grün, H. Krüger, A.L. Graps, D.P. Hamilton, A. Heck, G. Linkert, S. Dermott, H. Fechtig, B.A. Gustafson, M.S. Hanner, M. Horányi, J. Kissel, B.A. Lindblad, D. Linkert, I. Mann, J.A.M. McDonnel, G.E. Morfill, C. Polanskey, G. Schwehm, R. Srama, Galileo observes electromagnetically coupled dust in the Jovian magnetosphere. J. Geophys. Res. **103**, 20011–20022 (1998)
E. Grün, S. Kempf, H. Krüger, M. Landgraf, R. Srama, Dust astronomy: a new approach to the study of interstellar dust, in *Meteoroids 2001 Conference*, ed. by B. Warmbein. ESA Special Publication, vol. 495 (2001), pp. 651–662
P. Guérin, Les Anneaux de Saturne en 1969. Etude Morphologique et Photométrique. I. Obtention et Dépouillement des Photographies. Icarus **19**, 202–211 (1973). https://doi.org/10.1016/0019-1035(73)90123-1
D.A. Gurnett, W.S. Kurth, Plasma waves and related phenomena in the magnetosphere of Neptune, in *Neptune and Triton*, ed. by D.P. Cruikshank (Arizona University Press, Tucson, 1995), pp. 389–423
D.A. Gurnett, W.S. Kurth, F.L. Scarf, R.L. Poynter, First plasma wave observations of Uranus. Science **233**, 106–109 (1986). https://doi.org/10.1126/science.233.4759.106
D.A. Gurnett, W.S. Kurth, L.J. Granroth, S.C. Allendorf, R.L. Poynter, Micron-sized particles detected near Neptune by the Voyager 2 plasma wave instrument. J. Geophys. Res. **96**, 19177 (1991)
D.P. Hamilton, Motion of dust in a planetary magnetosphere—orbit-averaged equations for oblateness, electromagnetic, and radiation forces with application to Saturn's E ring. Icarus **101**, 244–264 (1993). https://doi.org/10.1006/icar.1993.1022
D.P. Hamilton, J.A. Burns, Origin of Saturn's E ring: selfsustained—naturally. Science **264**, 550–553 (1994)

D.P. Hamilton, A.V. Krivov, Circumplanetary dust dynamics: effects of solar gravity, radiation pressure, planetary oblateness, and electromagnetism. Icarus **123**, 503–523 (1996). https://doi.org/10.1006/icar.1996.0175

D.P. Hamilton, M.F. Skrutskie, A.J. Verbiscer, F.J. Masci, Small particles dominate Saturn's Phoebe ring to surprisingly large distances. Nature **522**, 185–187 (2015). https://doi.org/10.1038/nature14476

C.J. Hansen, L. Esposito, A.I.F. Stewart, J. Colwell, A. Hendrix, W. Pryor, D. Shemansky, R. West, Enceladus' water vapor plume. Science **311**, 1422–1425 (2006). https://doi.org/10.1126/science.1121254

W.K. Hartmann, Impact experiments. I—Ejecta velocity distributions and related results from regolith targets. Icarus **63**, 69–98 (1985). https://doi.org/10.1016/0019-1035(85)90021-1

M.M. Hedman, M.R. Showalter, A new pattern in Saturn's D ring created in late 2011. Icarus **279**, 155–165 (2016). https://doi.org/10.1016/j.icarus.2015.09.017

M.M. Hedman, J.A. Burns, M.R. Showalter, C.C. Porco, P.D. Nicholson, A.S. Bosh, M.S. Tiscareno, R.H. Brown, B.J. Buratti, K.H. Baines, R. Clark, Saturn's dynamic D ring. Icarus **188**, 89–107 (2007a). https://doi.org/10.1016/j.icarus.2006.11.017

M.M. Hedman, J.A. Burns, M.S. Tiscareno, C.C. Porco, G.H. Jones, E. Roussos, N. Krupp, C. Paranicas, S. Kempf, The source of Saturn's G ring. Science **317**, 653 (2007b). https://doi.org/10.1126/science.1143964

M.M. Hedman, C.D. Murray, N.J. Cooper, M.S. Tiscareno, K. Beurle, M.W. Evans, J.A. Burns, Three tenuous rings/arcs for three tiny moons. Icarus **199**, 378–386 (2009). https://doi.org/10.1016/j.icarus.2008.11.001

M.M. Hedman, J.A. Burt, J.A. Burns, M.S. Tiscareno, The shape and dynamics of a heliotropic dusty ringlet in the Cassini division. Icarus **210**, 284–297 (2010). https://doi.org/10.1016/j.icarus.2010.06.017

M.M. Hedman, J.A. Burns, D.P. Hamilton, M.R. Showalter, The three-dimensional structure of Saturn's E ring. Icarus **217**, 322–338 (2012). https://doi.org/10.1016/j.icarus.2011.11.006

M.M. Hedman, J.A. Burt, J.A. Burns, M.R. Showalter, Non-circular features in Saturn's D ring: D68. Icarus **233**, 147–162 (2014). https://doi.org/10.1016/j.icarus.2014.01.022

M.M. Hedman, J.A. Burns, M.R. Showalter, Corrugations and eccentric spirals in Saturn's D ring: new insights into what happened at Saturn in 1983. Icarus **248**, 137–161 (2015). https://doi.org/10.1016/j.icarus.2014.10.021

M.M. Hedman, F. Postberg, D.P. Hamilton, S. Renner, H.-W. Hsu, Dusty rings, in *Planetary Ring Systems. Properties, Structure, and Evolution*, ed. by M.S. Tiscareno, C.D. Murray (Cambridge Univ. Press, New York, 2018), pp. 308–337

J.R. Hill, D.A. Mendis, Charged dust in the outer planetary magnetospheres: physical and dynamical processes. Moon Planets **21**, 3–16 (1979)

J.R. Hill, D.A. Mendis, On the origin of striae in cometary dust tails. Astrophys. J. **242**, 395–401 (1980). https://doi.org/10.1086/158472

J.R. Hill, D.A. Mendis, The dynamical evolution of the Saturnian ring spokes. J. Geophys. Res. **87**, 7413–7420 (1982). https://doi.org/10.1029/JA087iA09p07413

T.W. Hill, M.F. Thomsen, R.L. Tokar, A.J. Coates, G.R. Lewis, D.T. Young, F.J. Crary, R.A. Baragiola, R.E. Johnson, Y. Dong, R.J. Wilson, G.H. Jones, J.-E. Wahlund, D.G. Mitchell, M. Horányi, Charged nanograins in the Enceladus plume. J. Geophys. Res. Space Phys. **117**, 05209 (2012). https://doi.org/10.1029/2011JA017218

J.K. Hillier, S.F. Green, N. McBride, J.P. Schwanethal, F. Postberg, R. Srama, S. Kempf, G. Moragas-Klostermeyer, J.A.M. McDonnell, E. Grün, The composition of Saturn's E ring. Mon. Not. R. Astron. Soc. **377**, 1588–1596 (2007). https://doi.org/10.1111/j.1365-2966.2007.11710.x

J.K. Hillier, J. Schmidt, H.-W. Hsu, F. Postberg, Dust emission by active moons. Space Sci. Rev. **214**, 51–96 (2018). https://doi.org/10.1007/s11214-018-0539-9

M. Horányi, New Jovian ring? Geophys. Res. Lett. **21**, 1039–1042 (1994). https://doi.org/10.1029/94GL01326

M. Horanyi, Charged dust dynamics in the solar system. Annu. Rev. Astron. Astrophys. **34**, 383–418 (1996). https://doi.org/10.1146/annurev.astro.34.1.383

M. Horányi, Dust streams from Jupiter and Saturn. Phys. Plasmas **7**(10), 3847–3850 (2000)

M. Horányi, H.L.F. Houpis, D.A. Mendis, Charged dust in the Earth's magnetosphere. I—Physical and dynamical processes. Astrophys. Space Sci. **144**, 215–229 (1988). https://doi.org/10.1007/BF00793182

M. Horányi, J.A. Burns, D.P. Hamilton, The dynamics of Saturn's E ring particles. Icarus **97**, 248–259 (1992). https://doi.org/10.1016/0019-1035(92)90131-P

M. Horányi, G. Morfill, E. Grün, Mechanism for the acceleration and ejection of dust grains from Jupiter's magnetosphere. Nature **363**, 144–146 (1993)

M. Horányi, T.W. Hartquist, O. Havnes, D.A. Mendis, G.E. Morfill, Dusty plasma effects in Saturn's magnetosphere. Rev. Geophys. **42**, 4002 (2004). https://doi.org/10.1029/2004RG000151

M. Horányi, A. Juhász, G.E. Morfill, Large-scale structure of Saturn's E-ring. Geophys. Res. Lett. **35**, 04203 (2008). https://doi.org/10.1029/2007GL032726

M. Horányi, J.A. Burns, M.M. Hedman, G.H. Jones, S. Kempf, Diffuse rings, in *Saturn from Cassini-Huygens*, ed. by M.K. Dougherty, L.W. Esposito, S. Krimigis (Springer, Berlin, 2009), p. 511. https://doi.org/10.1007/978-1-4020-9217-6_16

M. Horányi, G.E. Morfill, T.E. Cravens, Spokes in Saturn's B ring: could lightning be the cause? IEEE Trans. Plasma Sci. **38**, 874–879 (2010). https://doi.org/10.1109/TPS.2009.2034455

M. Horányi, J.R. Szalay, S. Kempf, J. Schmidt, E. Grün, R. Srama, Z. Sternovsky, A permanent, asymmetric dust cloud around the Moon. Nature **522**, 324–326 (2015). https://doi.org/10.1038/nature14479

H. Hsu, S. Kempf, C.M. Jackman, Observation of Saturnian stream particles in the interplanetary space. Icarus **206**, 653–661 (2010). https://doi.org/10.1016/j.icarus.2009.06.033

H.-W. Hsu, S. Kempf, F. Postberg, M. Trieloff, M. Burton, M. Roy, G. Moragas-Klostermeyer, R. Srama, Cassini dust stream particle measurements during the first three orbits at Saturn. J. Geophys. Res. Space Phys. **116**, 08213 (2011). https://doi.org/10.1029/2010JA015959

H.-W. Hsu, H. Krüger, F. Postberg, Dynamics, composition, and origin of Jovian and Saturnian dust-stream particles, in *Nanodust in the Solar System: Discoveries and Interpretations*, ed. by I. Mann, N. Meyer-Vernet, A. Czechowski. Astrophysics and Space Science Library, vol. 385 (2012), p. 77. https://doi.org/10.1007/978-3-642-27543-2_5

H.-W. Hsu, F. Postberg, Y. Sekine, T. Shibuya, S. Kempf, M. Horányi, A. Juhász, N. Altobelli, K. Suzuki, Y. Masaki, T. Kuwatani, S. Tachibana, S.-I. Sirono, G. Moragas-Klostermeyer, R. Srama, Ongoing hydrothermal activities within Enceladus. Nature **519**, 207–210 (2015). https://doi.org/10.1038/nature14262

H.-W. Hsu, J. Schmidt, S. Kempf, F. Postberg, G. Moragas-Klostermeyer, M. Seiß, H. Hoffmann, M. Burton, S. Ye, W.S. Kurth, M. Horányi, N. Khawaja, F. Spahn, D. Schirdewahn, J. O'Donoghue, L. Moore, J. Cuzzi, G.H. Jones, R. Srama, In situ collection of dust grains falling from Saturn's rings into its atmosphere. Science **362**, 3185 (2018). https://doi.org/10.1126/science.aat3185

W.B. Hubbard, A. Brahic, B. Sicardy, L.-R. Elicer, F. Roques, F. Vilas, Occultation detection of a Neptunian ring-like arc. Nature **319**, 636–640 (1986). https://doi.org/10.1038/319636a0

L. Iess, D.J. Stevenson, M. Parisi, D. Hemingway, R.A. Jacobson, J.I. Lunine, F. Nimmo, J.W. Armstrong, S.W. Asmar, M. Ducci, P. Tortora, The gravity field and interior structure of Enceladus. Science **344**, 78–80 (2014). https://doi.org/10.1126/science.1250551

W.-H. Ip, C.-M. Liu, K.-C. Pan, Transport and electro-dynamical coupling of nano-grains ejected from the Saturnian rings and their possible ionospheric signatures. Icarus **276**, 163–169 (2016). https://doi.org/10.1016/j.icarus.2016.04.004

D. Jewitt, N. Haghighipour, Irregular satellites of the planets: products of capture in the early solar system. Annu. Rev. Astron. Astrophys. **45**, 261–295 (2007). https://doi.org/10.1146/annurev.astro.44.051905.092459

T.V. Johnson, J.I. Lunine, Saturn's moon Phoebe as a captured body from the outer Solar System. Nature **435**, 69–71 (2005). https://doi.org/10.1038/nature03384

G.H. Jones, E. Roussos, N. Krupp, C. Paranicas, J. Woch, A. Lagg, D.G. Mitchell, S.M. Krimigis, M.K. Dougherty, Enceladus' varying imprint on the magnetosphere of Saturn. Science **311**, 1412–1415 (2006). https://doi.org/10.1126/science.1121011

G.H. Jones, C.S. Arridge, A.J. Coates, G.R. Lewis, S. Kanani, A. Wellbrock, D.T. Young, F.J. Crary, R.L. Tokar, R.J. Wilson, T.W. Hill, R.E. Johnson, D.G. Mitchell, J. Schmidt, S. Kempf, U. Beckmann, C.T. Russell, Y.D. Jia, M.K. Dougherty, J.H. Waite, B.A. Magee, Fine jet structure of electrically charged grains in Enceladus' plume. Geophys. Res. Lett. **36**, 16204 (2009). https://doi.org/10.1029/2009GL038284

D. Jontof-Hutter, D.P. Hamilton, The fate of sub-micron circumplanetary dust grains I: aligned dipolar magnetic fields. Icarus **218**, 420–432 (2012a). https://doi.org/10.1016/j.icarus.2011.09.033

D. Jontof-Hutter, D.P. Hamilton, The fate of sub-micron circumplanetary dust grains II: multipolar fields. Icarus **220**, 487–502 (2012b). https://doi.org/10.1016/j.icarus.2012.04.032

A. Juhász, M. Horányi, Dust torus around Mars. J. Geophys. Res., Planets **100**(E2), 3277–3284 (1995)

A. Juhász, M. Horányi, Dynamics of charged space debris in the Earth's plasma environment. J. Geophys. Res. **102**, 7237–7246 (1997). https://doi.org/10.1029/96JA03672

A. Juhász, M. Horányi, Magnetospheric screening of cosmic dust. J. Geophys. Res. **104**, 12577–12584 (1999). https://doi.org/10.1029/1999JA900091

A. Juhász, M. Horányi, Saturn's E ring: a dynamical approach. J. Geophys. Res. Space Phys. **107**, 1–10 (2002). https://doi.org/10.1029/2001JA000182

A. Juhász, M. Horányi, Seasonal variations in Saturn's E-ring. Geophys. Res. Lett. **311**, 19703 (2004). https://doi.org/10.1029/2004GL020999

S. Jurac, R.E. Johnson, B. Donn, M. Carlo, Calculations of the sputtering of grains: enhanced sputtering of small grains. Astrophys. J. **503**, 247–252 (1998). https://doi.org/10.1086/305994

S. Jurac, R.E. Johnson, J.D. Richardson, Saturn's E ring and production of the neutral torus. Icarus **149**, 384–396 (2001a). https://doi.org/10.1006/icar.2000.6528

S. Jurac, R.E. Johnson, J.D. Richardson, C. Paranicas, Satellite sputtering in Saturn's magnetosphere. Planet. Space Sci. **49**, 319–326 (2001b). https://doi.org/10.1016/S0032-0633(00)00153-7

S. Kempf, R. Srama, M. Horányi, M. Burton, S. Helfert, G. Moragas-Klostermeyer, M. Roy, E. Grün, High-velocity streams of dust originating from Saturn. Nature **433**, 289–291 (2005)

S. Kempf, U. Beckmann, R. Srama, M. Horányi, S. Auer, E. Grün, The electrostatic potential of E ring particles. Planet. Space Sci. **54**, 999–1006 (2006). https://doi.org/10.1016/j.pss.2006.05.012

S. Kempf, U. Beckmann, G. Moragas-Klostermeyer, F. Postberg, R. Srama, T. Economou, J. Schmidt, F. Spahn, E. Grün, The E ring in the vicinity of Enceladus I: spatial distribution and properties of the ring particles. Icarus **193**, 420–437 (2008). https://doi.org/10.1016/j.icarus.2007.06.027

S. Kempf, U. Beckmann, J. Schmidt, How the Enceladus dust plume feeds Saturn's E ring. Icarus **206**, 446–457 (2010). https://doi.org/10.1016/j.icarus.2009.09.016

S. Kempf, M. Horanyi, A. Juhasz, R. Srama, G. Moragas-Klostermeyer, The Phoebe dust ring as seen as by the Cassini dust detector CDA, in *AGU Fall Meeting Abstracts*, 13-1673 (2011)

S. Kempf, M. Horanyi, A. Juhasz, A. Cruz, R. Srama, F. Postberg, F. Spahn, J. Schmidt, The 3-dimensional structure of Saturn's E ring inferred from Cassini CDA observations, in *European Planetary Science Congress* (2012), p. 701

S. Kempf, N. Altobelli, J.N. Cuzzi, P.R. Estrada, R. Srama, Nature (submitted, 2018a)

S. Kempf, M. Horányi, H. Hsu, T.W. Hill, A. Juhasz, H.T. Smith, Saturn's diffuse E ring and its connection with Enceladus, in *Enceladus and the Icy Moons of Saturn* (University Arizona Press, Tucson, 2018b)

D.J. Kessler, Derivation of the collision probability between orbiting objects the lifetimes of Jupiter's outer moons. Icarus **48**, 39–48 (1981). https://doi.org/10.1016/0019-1035(81)90151-2

D. Koschny, E. Grün, Impacts into ice-silicate mixtures: crater morphologies, volumes, depth-to-diameter ratios and yield. Icarus **154**, 391–401 (2001)

H. Kriegel, S. Simon, P. Meier, U. Motschmann, J. Saur, A. Wennmacher, D.F. Strobel, M.K. Dougherty, Ion densities and magnetic signatures of dust pickup at Enceladus. J. Geophys. Res. Space Phys. **119**, 2740–2774 (2014). https://doi.org/10.1002/2013JA019440

A.V. Krivov, D.P. Hamilton, Martian dust belts: waiting for discovery. Icarus **128**, 335–353 (1997)

A.V. Krivov, L.L. Sokolov, V.V. Dikarev, Dynamics of Mars-orbiting dust: effects of light pressure and planetary oblateness. Celest. Mech. Dyn. Astron. **63**, 313–339 (1996). https://doi.org/10.1007/BF00692293

A.V. Krivov, H. Krüger, E. Grün, K.-U. Thiessenhusen, D.P. Hamilton, A tenuous dust ring of Jupiter formed by escaping ejecta from the Galilean satellites. J. Geophys. Res., Planets **107**, 5002 (2002a). https://doi.org/10.1029/2000JE001434

A.V. Krivov, I. Wardinski, F. Spahn, H. Krüger, E. Grün, Dust on the outskirts of the Jovian system. Icarus **157**, 436–455 (2002b). https://doi.org/10.1006/icar.2002.6848

A.V. Krivov, M. Sremčević, F. Spahn, V.V. Dikarev, K.V. Kholshevnikov, Impact-generated dust clouds around planetary satellites: spherically symmetric case. Planet. Space Sci. **51**, 251–269 (2003). https://doi.org/10.1016/S0032-0633(02)00147-2

H. Krüger, A.V. Krivov, D.P. Hamilton, E. Grün, Detection of an impact-generated dust cloud around Ganymede. Nature **399**, 558–560 (1999). https://doi.org/10.1038/21136

H. Krüger, M. Horányi, E. Grün, Jovian dust streams: probes of the Io plasma torus. Geophys. Res. Lett. **30**(2), 1058 (2003a). https://doi.org/10.1029/2002GL015920

H. Krüger, A.V. Krivov, M. Sremčević, E. Grün, Impact-generated dust clouds surrounding the Galilean moons. Icarus **164**, 170–187 (2003b). https://doi.org/10.1016/S0019-1035(03)00127-1

H. Krüger, D.P. Hamilton, R. Moissl, E. Grün, Galileo in-situ dust measurements in Jupiter's gossamer rings. Icarus **203**, 198–213 (2009). https://doi.org/10.1016/j.icarus.2009.03.040

L.J. Lanzerotti, W.L. Brown, C.G. Maclennan, A.F. Cheng, S.M. Krimigis, Effects of charged particles on the surfaces of the satellites of Uranus. J. Geophys. Res. **92**, 14949–14957 (1987). https://doi.org/10.1029/JA092iA13p14949

S.M. Larson, J.W. Fountain, B.A. Smith, H.J. Reitsema, Observations of the Saturn E ring and a new satellite. Icarus **47**, 288–290 (1981). https://doi.org/10.1016/0019-1035(81)90173-1

T.R. Lauer et al., The New Horizons and Hubble Space Telescope search for rings, dust, and debris in the Pluto-Charon System (2017). arXiv:1709.07981 [astro-ph.EP]

D.A. Lautman, I.I. Shapiro, G. Colombo, The Earth's dust belt: fact or fiction? 4, Sunlight-pressure air-drag capture. J. Geophys. Res. **71**, 5733 (1966). https://doi.org/10.1029/JZ071i023p05733

A. Le Gall, C. Leyrat, M.A. Janssen, G. Choblet, G. Tobie, O. Bourgeois, A. Lucas, C. Sotin, C. Howett, R. Kirk, R.D. Lorenz, R.D. West, A. Stolzenbach, M. Massé, A.H. Hayes, L. Bonnefoy, G. Veyssière,

F. Paganelli, Thermally anomalous features in the subsurface of Enceladus's South polar terrain. Nat. Astron. **1**, 0063 (2017). https://doi.org/10.1038/s41550-017-0063

J.J. Lissauer, Shepherding model for Neptune's arc ring. Nature **318**, 544 (1985). https://doi.org/10.1038/318544a0

C.-M. Liu, W.-H. Ip, A new pathway of Saturnian ring-ionosphere coupling via charged nanograins. Astrophys. J. **786**, 34 (2014). https://doi.org/10.1088/0004-637X/786/1/34

X. Liu, M. Sachse, F. Spahn, J. Schmidt, Dynamics and distribution of Jovian dust ejected from the Galilean satellites. J. Geophys. Res., Planets **121**, 1141–1173 (2016). https://doi.org/10.1002/2016JE004999

M. Makuch, A.V. Krivov, F. Spahn, Long-term dynamical evolution of dust ejecta from Deimos. Planet. Space Sci. **53**, 357–369 (2005)

M.A. McGrath, W.B. Sparks, Galileo ionosphere profile coincident with repeat plume detection location at Europa. Res. Notes Am. Astron. Soc. **1**, 14 (2017). https://doi.org/10.3847/2515-5172/aa988e

R. Meier, B.A. Smith, T.C. Owen, E.E. Becklin, R.J. Terrile, Near infrared photometry of the Jovian ring and Adrastea. Icarus **141**, 253–262 (1999)

P. Meier, U. Motschmann, J. Schmidt, F. Spahn, T.W. Hill, Y. Dong, G.H. Jones, H. Kriegel, Modeling the total dust production of Enceladus from stochastic charge equilibrium and simulations. Planet. Space Sci. **119**, 208–221 (2015). https://doi.org/10.1016/j.pss.2015.10.002

D.A. Mendis, W.I. Axford, Satellites and magnetospheres of the outer planets. Annu. Rev. Earth Planet. Sci. **2, 419. 2(1)**, 419–474 (1974)

D.A. Mendis, W.I. Axford, Revisiting Iapetus following recent Cassini observations. J. Geophys. Res. Space Phys. **113**, 11217 (2008). https://doi.org/10.1029/2008JA013532

N. Meyer-Vernet, On the charge of nanograins in cold environments and Enceladus dust. Icarus **226**, 583–590 (2013). https://doi.org/10.1016/j.icarus.2013.06.014

N. Meyer-Vernet, M.G. Aubier, B.M. Pedersen, Voyager 2 at Uranus—grain impacts in the ring plane. Geophys. Res. Lett. **13**, 617–620 (1986). https://doi.org/10.1029/GL013i007p00617

N. Meyer-Vernet, A. Lecacheux, B.M. Pedersen, Constraints on Saturn's E ring from the Voyager 1 radio astronomy instrument. Icarus **123**, 113–128 (1996). https://doi.org/10.1006/icar.1996.0145

C.J. Mitchell, J.E. Colwell, M. Horányi, Dust capture by the Saturnian magnetosphere. IEEE Trans. Plasma Sci. **32**, 598–600 (2004). https://doi.org/10.1109/TPS.2004.826100

C.J. Mitchell, J.E. Colwell, M. Horányi, Tenuous ring formation by the capture of interplanetary dust at Saturn. J. Geophys. Res. Space Phys. **110**, 9218 (2005). https://doi.org/10.1029/2004JA010577

C.J. Mitchell, C.C. Porco, H.L. Dones, J.N. Spitale, The behavior of spokes in Saturn's B ring. Icarus **225**, 446–474 (2013). https://doi.org/10.1016/j.icarus.2013.02.011

C.J. Mitchell, C.C. Porco, J.W. Weiss, Tracking the Geysers of Enceladus into Saturn's E ring. Astron. J. **149**, 156 (2015). https://doi.org/10.1088/0004-6256/149/5/156

D.G. Mitchell, M.E. Perry, D.C. Hamilton, J.H. Westlake, P. Kollmann, H.T. Smith, J.F. Carbary, J.H. Waite, R. Perryman, H.-W. Hsu, J.-E. Wahlund, M.W. Morooka, L.Z. Hadid, A.M. Persoon, W.S. Kurth, Dust grains fall from Saturn's D-ring into its equatorial upper atmosphere. Science **362**, 2236 (2018). https://doi.org/10.1126/science.aat2236

L. Moore, J. O'Donoghue, I. Müller-Wodarg, M. Galand, M. Mendillo, Saturn ring rain: model estimates of water influx into Saturn's atmosphere. Icarus **245**, 355–366 (2015). https://doi.org/10.1016/j.icarus.2014.08.041

M.W. Morooka, J.-E. Wahlund, A.I. Eriksson, W.M. Farrell, D.A. Gurnett, W.S. Kurth, A.M. Persoon, M. Shafiq, M. André, M.K.G. Holmberg, Dusty plasma in the vicinity of Enceladus. J. Geophys. Res. Space Phys. **116**, 12221 (2011). https://doi.org/10.1029/2011JA017038

C.D. Murray, R.S. French, Planetary ring systems, in *The F ring of Saturn*, ed. by M.S. Tiscareno, C.D. Murray (Cambridge University Press, New York, 2018), pp. 338–362

C.D. Murray, R.P. Thompson, Orbits of shepherd satellites deduced from the structure of the rings of Uranus Nature **348**, 499–502 (1990). https://doi.org/10.1038/348499a0

F. Namouni, C. Porco, The confinement of Neptune's ring arcs by the moon Galatea. Nature **417**, 45–47 (2002). https://doi.org/10.1038/417045a

P.D. Nicholson, M.R. Showalter, L. Dones, Observations of Saturn's ring-plane crossing in August and November. Science **272**, 509–516 (1996)

P.D. Nicholson, I.D. Pater, R.G. French, M.R. Showalter, The rings of Uranus, in *Planetary Ring Systems Properties, Structure, and Evolution*, ed. by M.S. Tiscareno, C.D.E. Murray. (Cambridge University Press, New York, 2018), pp. 93–111. https://doi.org/10.1017/9781316286791.004. Chap. 4

T.G. Northrop, J.E.P. Connerney, A micrometeorite erosion model and the age of Saturn's rings. Icarus **70**, 124–137 (1987). https://doi.org/10.1016/0019-1035(87)90079-0

T.G. Northrop, J.R. Hill, Stability of negatively charged dust grains in Saturn's ring plane. J. Geophys. Res. **87**, 6045–6051 (1982). https://doi.org/10.1029/JA087iA08p06045

T.G. Northrop, J.R. Hill, The adiabatic motion of charged dust grains in rotating magnetospheres. J. Geophys. Res. **88**, 1–11 (1983a). https://doi.org/10.1029/JA088iA01p00001
T.G. Northrop, J.R. Hill, The inner edge of Saturn's B ring. J. Geophys. Res. **88**, 6102–6108 (1983b). https://doi.org/10.1029/JA088iA08p06102
M.E. Ockert-Bell, J.A. Burns, I.J. Daubar, P.C. Thomas, J. Veverka, M.J.S. Belton, K.P. Klaasen, The structure of Jupiter's ring system as revealed by the Galileo imaging experiment. Icarus **138**, 188–213 (1999). https://doi.org/10.1006/icar.1998.6072
J. O'Donoghue, T.S. Stallard, H. Melin, G.H. Jones, S.W.H. Cowley, S. Miller, K.H. Baines, J.S.D. Blake, The domination of Saturn's low-latitude ionosphere by ring 'rain'. Nature **496**, 193–195 (2013). https://doi.org/10.1038/nature12049
M. Øieroset, D.A. Brain, E. Simpson, D.L. Mitchell, T.D. Phan, J.S. Halekas, R.P. Lin, M.H. Acuña, Search for Phobos and Deimos gas/dust tori using in situ observations from Mars Global Surveyor MAG/ER. Icarus **206**, 189–198 (2010). https://doi.org/10.1016/j.icarus.2009.07.017
E.J. Öpik, Collision probabilities with the planets and the distribution of interplanetary matter. Proc. R. Ir. Acad., A Math. Phys. Sci. **54**, 165–199 (1951)
J.L. Ortiz, R. Duffard, N. Pinilla-Alonso, A. Alvarez-Candal, P. Santos-Sanz, N. Morales, E. Fernández-Valenzuela, J. Licandro, A. Campo Bagatin, A. Thirouin, Possible ring material around centaur (2060) Chiron. Astron. Astrophys. **576**, 18 (2015). https://doi.org/10.1051/0004-6361/201424461
J.L. Ortiz, P. Santos-Sanz, B. Sicardy, G. Benedetti-Rossi, D. Bérard, N. Morales, R. Duffard, F. Braga-Ribas, U. Hopp, C. Ries, V. Nascimbeni, F. Marzari, V. Granata, A. Pál, C. Kiss, T. Pribulla, R. Komžík, K. Hornoch, P. Pravec, P. Bacci, M. Maestripieri, L. Nerli, L. Mazzei, M. Bachini, F. Martinelli, G. Succi, F. Ciabattari, H. Mikuz, A. Carbognani, B. Gaehrken, S. Mottola, S. Hellmich, F.L. Rommel, E. Fernández-Valenzuela, A.C. Bagatin, S. Cikota, A. Cikota, J. Lecacheux, R. Vieira-Martins, J.I.B. Camargo, M. Assafin, F. Colas, R. Behrend, J. Desmars, E. Meza, A. Alvarez-Candal, W. Beisker, A.R. Gomes-Junior, B.E. Morgado, F. Roques, F. Vachier, J. Berthier, T.G. Mueller, J.M. Madiedo, O. Unsalan, E. Sonbas, N. Karaman, O. Erece, D.T. Koseoglu, T. Ozisik, S. Kalkan, Y. Guney, M.S. Niaei, O. Satir, C. Yesilyaprak, C. Puskullu, A. Kabas, O. Demircan, J. Alikakos, V. Charmandaris, G. Leto, J. Ohlert, J.M. Christille, R. Szakáts, A.T. Farkas, E. Varga-Verebélyi, G. Marton, A. Marciniak, P. Bartczak, T. Santana-Ros, M. Butkiewicz-Bąk, G. Dudziński, V. Alí-Lagoa, K. Gazeas, L. Tzouganatos, N. Paschalis, V. Tsamis, A. Sánchez-Lavega, S. Pérez-Hoyos, R. Hueso, J.C. Guirado, V. Peris, R. Iglesias-Marzoa, The size, shape, density and ring of the dwarf planet Haumea from a stellar occultation. Nature **550**, 219–223 (2017). https://doi.org/10.1038/nature24051
C.C. Porco, An explanation for Neptune's ring arcs. Science **253**, 995–1001 (1991). https://doi.org/10.1126/science.253.5023.995
C.A. Porco, G.E. Danielson, The periodic variation of spokes in Saturn's rings. Astron. J. **87**, 826–833 (1982). https://doi.org/10.1086/113162
C.C. Porco, P.D. Nicholson, J.N. Cuzzi, J.J. Lissauer, L.W. Esposito, Neptune's ring system, in *Neptune and Triton*, Arizona, Tucson ed. by D.P. Cruikshank (1995), pp. 703–804
C.C. Porco, P. Helfenstein, P.C. Thomas, A.P. Ingersoll, J. Wisdom, R. West, G. Neukum, T. Denk, R. Wagner, T. Roatsch, S. Kieffer, E. Turtle, A. McEwen, T.V. Johnson, J. Rathbun, J. Veverka, D. Wilson, J. Perry, J. Spitale, A. Brahic, J.A. Burns, A.D. Del Genio, L. Dones, C.D. Murray, S. Squyres, Cassini observes the active South pole of Enceladus. Science **311**, 1393–1401 (2006). https://doi.org/10.1126/science.1123013
C.C. Porco, P.C. Thomas, J.W. Weiss, D.C. Richardson, Saturn's small inner satellites: clues to their origins. Science **318**, 1602 (2007). https://doi.org/10.1126/science.1143977
S.B. Porter, A.S. Stern, Orbits of potential Pluto satellites and rings between Charon and Hydra (2015). arXiv:1505.05933 [astro-ph.EP]
F. Postberg, S. Kempf, R. Srama, S.F. Green, J.K. Hillier, N. McBride, E. Grün, Composition of Jovian dust stream particles. Icarus **183**, 122–134 (2006). https://doi.org/10.1016/j.icarus.2006.02.001
F. Postberg, S. Kempf, J. Schmidt, N. Brilliantov, A. Beinsen, B. Abel, U. Buck, R. Srama, Sodium salts in E-ring ice grains from an ocean below the surface of Enceladus. Nature **459**, 1098–1101 (2009). https://doi.org/10.1038/nature08046
F. Postberg, J. Schmidt, J. Hillier, S. Kempf, R. Srama, A salt-water reservoir as the source of a compositionally stratified plume on Enceladus. Nature **474**, 620–622 (2011a). https://doi.org/10.1038/nature10175
F. Postberg, E. Grün, M. Horányi, S. Kempf, H. Krüger, R. Srama, Z. Sternovsky, M. Trieloff, Compositional mapping of planetary moons by mass spectrometry of dust ejecta. Planet. Space Sci. (2011b). https://doi.org/10.1016/j.pss.2011.05.001
F. Postberg, R.N. Clark, C.J. Hansen, A.J. Coates, C.M. Dale Ore, F. Scipioni, M.M. Hedman, J.H. Waite, Plume and surface composition of Enceladus, in *Enceladus and the Icy Moons of Saturn* (University of Arizona Press, Tucson, 2018)

S. Renner, B. Sicardy, D. Souami, B. Carry, C. Dumas, Neptune's ring arcs: VLT/NACO near-infrared observations and a model to explain their stability. Astron. Astrophys. **563**, 133 (2014). https://doi.org/10.1051/0004-6361/201321910

W. Riedler, K. Schwingenschuh, D. Möhlmann, V.N. Oraevskii, E. Eroshenko, J. Slavin, Magnetic fields near Mars: first results. Nature **341**, 604–607 (1989)

L.J. Robinson, Closing in on Saturn. Sky Telesc. **60**, 481 (1980)

L. Roth, J. Saur, K.D. Retherford, D.F. Strobel, P.D. Feldman, M.A. McGrath, F. Nimmo, Transient water vapor at Europa's South pole. Science **343**, 171–174 (2014). https://doi.org/10.1126/science.1247051

E. Roussos, G.H. Jones, N. Krupp, C. Paranicas, D.G. Mitchell, S.M. Krimigis, J. Woch, A. Lagg, K. Khurana, Energetic electron signatures of Saturn's smaller moons: evidence of an arc of material at Methone. Icarus **193**, 455–464 (2008). https://doi.org/10.1016/j.icarus.2007.03.034

M. Sachse, A planetary dust ring generated by impact-ejection from the Galilean satellites. Icarus **303**, 166–180 (2018). https://doi.org/10.1016/j.icarus.2017.10.011

M. Sachse, J. Schmidt, S. Kempf, F. Spahn, Correlation between speed and size for ejecta from hypervelocity impacts. J. Geophys. Res., Planets **120**, 1847–1858 (2015). https://doi.org/10.1002/2015JE004844

V.S. Safronov, Evolution of the protoplanetary cloud and formation of the earth and planets (1972)

J. Schmidt, N. Brilliantov, F. Spahn, S. Kempf, Slow dust in Enceladus' plume from condensation and wall collisions in tiger stripe fractures. Nature **451**, 685–688 (2008). https://doi.org/10.1038/nature06491

M. Seiß, R. Srama, K.-L. Sun, M. Seiler, G. Moragas-Klostermeyer, S. Kempf, F. Spahn, Pallene dust torus observations by the cosmic dust analyzer, in *European Planetary Science Congress*, vol. 9 (2014), p. 375

Y. Sekine, T. Shibuya, F. Postberg, H. Hsu, K. Suzuki, Y. Masaki, T. Kuwatani, M. Mori, P.K. Hong, M. Yoshizaki, S. Tachibana, S.-I. Sirono, High-temperature water–rock interactions and hydrothermal environments in the chondrite-like high-temperature water-rock interactions and hydrothermal environments in the chondrite-like core of Enceladus. Nat. Commun. **6**, 8604 (2015)

R. Sfair, S.M. Giuliatti Winter, Orbital evolution of the μ and ν dust ring particles of Uranus. Astron. Astrophys. **505**, 845–852 (2009). https://doi.org/10.1051/0004-6361/200911886

R. Sfair, S.M. Giuliatti Winter, The role of Mab as a source for the μ ring of Uranus. Astron. Astrophys. **543**, 17 (2012). https://doi.org/10.1051/0004-6361/201117346

I.I. Shapiro, D.A. Lautman, G. Colombo, The Earth's dust belt: fact or fiction?: 1. Forces perturbing dust particle motion. J. Geophys. Res. **71**, 5695–5704 (1966). https://doi.org/10.1029/JZ071i023p05695

R.A. Shaw, D. Lamb, Experimental determination of the thermal accommodation and condensation coefficients of water. J. Chem. Phys. **111**, 10659–10663 (1999). https://doi.org/10.1063/1.480419

M.R. Showalter, Arcs and clumps in the Uranian λ ring. Science **267**, 490–493 (1995). https://doi.org/10.1126/science.267.5197.490

M.R. Showalter, Saturn's D ring in the Voyager images. Icarus **124**, 677–689 (1996). https://doi.org/10.1006/icar.1996.0241

M.R. Showalter, D.P. Hamilton, Resonant interactions and chaotic rotation of Pluto's small moons. Nature **522**, 45–49 (2015). https://doi.org/10.1038/nature14469

M.R. Showalter, J.J. Lissauer, The second ring-moon system of Uranus: discovery and dynamics. Science **311**, 973–977 (2006). https://doi.org/10.1126/science.1122882

M.R. Showalter, J.A. Burns, J.N. Cuzzi, J.B. Pollack, Jupiter's ring system—new results on structure and particle properties. Icarus **69**, 458–498 (1987)

M.R. Showalter, J.N. Cuzzi, S.M. Larson, Structure and particle properties of Saturn's E ring. Icarus **94**, 451–473 (1991)

M.R. Showalter, J.A. Burns, I. de Pater, D.P. Hamilton, J.J. Lissauer, G. Verbanac, Updates on the dusty rings of Jupiter, Uranus and Neptune, in *Dust in Planetary Systems*. LPI Contributions, vol. 1280 (2005), p. 130

M.R. Showalter, D.P. Hamilton, P.D. Nicholson, A deep search for Martian dust rings and inner moons using the Hubble Space. Planet. Space Sci. **54**, 844–854 (2006). https://doi.org/10.1016/j.pss.2006.05.009

M.R. Showalter, A.F. Cheng, H.A. Weaver, S.A. Stern, J.R. Spencer, H.B. Throop, E.M. Birath, D. Rose, J.M. Moore, Clump detections and limits on moons in Jupiter's ring system. Science **318**, 232 (2007). https://doi.org/10.1126/science.1147647

M.R. Showalter, M.M. Hedman, J.A. Burns, The impact of comet Shoemaker-Levy 9 sends ripples through the rings of Jupiter. Science **332**, 711 (2011). https://doi.org/10.1126/science.1202241

P.K. Shukla, A.A. Mamun, *Introduction to the Physics of Dusty Plasmas*. IOP Series in Plasma Physics (Bristol, Institute of Physics, 2002)

B. Sicardy et al., Rings beyond the Giant Planets (2017). arXiv:1612.03321 [astro-ph.EP]

S. Simon, J. Saur, H. Kriegel, F.M. Neubauer, U. Motschmann, M.K. Dougherty, Influence of negatively charged plume grains and hemisphere coupling currents on the structure of Enceladus' Alfvén wings:

analytical modeling of Cassini magnetometer observations. J. Geophys. Res. Space Phys. **116**, 04221 (2011). https://doi.org/10.1029/2010JA016338

W.M. Sinton, Uranus—the rings are black. Science **198**, 503 (1977). https://doi.org/10.1126/science.198.4316.503

B.A. Smith, L.A. Soderblom, T.V. Johnson, A.P. Ingersoll, S.A. Collins, E.M. Shoemaker, G.E. Hunt, H. Masursky, M.H. Carr, M.E. Davies, A.F. Cook, J.M. Boyce, T. Owen, G.E. Danielson, C. Sagan, R.F. Beebe, J. Veverka, J.F. McCauley, R.G. Strom, D. Morrison, G.A. Briggs, V.E. Suomi, The Jupiter system through the eyes of Voyager 1. Science **204**, 951–957 (1979). https://doi.org/10.1126/science.204.4396.951

B.A. Smith, L. Soderblom, R.F. Beebe, J.M. Boyce, G. Briggs, A. Bunker, S.A. Collins, C. Hansen, T.V. Johnson, J.L. Mitchell, R.J. Terrile, M.H. Carr, A.F. Cook, J.N. Cuzzi, J.B. Pollack, G.E. Danielson, A.P. Ingersoll, M.E. Davies, G.E. Hunt, H. Masursky, E.M. Shoemaker, D. Morrison, T. Owen, C. Sagan, J. Veverka, R. Strom, V.E. Suomi, Encounter with Saturn—Voyager 1 imaging science results. Science **212**, 163–191 (1981)

B.A. Smith, L. Soderblom, R.M. Batson, P.M. Bridges, J.L. Inge, H. Masursky, E. Shoemaker, R.F. Beebe, J. Boyce, G. Briggs, A. Bunker, S.A. Collins, C. Hansen, T.V. Johnson, J.L. Mitchell, R.J. Terrile, A.F. Cook, J.N. Cuzzi, J.B. Pollack, G.E. Danielson, A.P. Ingersoll, M.E. Davies, G.E. Hunt, D. Morrison, T. Owen, C. Sagan, J. Veverka, R. Strom, V.E. Suomi, A new look at Saturn system: the Voyager 2 images. Science **215**, 503–537 (1982)

B.A. Smith, L.A. Soderblom, R. Beebe, D. Bliss, R.H. Brown, S.A. Collins, J.M. Boyce, G.A. Briggs, A. Brahic, J.N. Cuzzi, D. Morrison, Voyager 2 in the Uranian system—imaging science results. Science **233**, 43–64 (1986). https://doi.org/10.1126/science.233.4759.43

B.A. Smith, L.A. Soderblom, D. Banfield, C. Barnet, A.T. Basilevksy, R.F. Beebe, K. Bollinger, J.M. Boyce, A. Brahic, G.A. Briggs, R.H. Brown, C. Chyba, S.A. Collins, T. Colvin, A.F. Cook, D. Crisp, S.K. Croft, D. Cruikshank, J.N. Cuzzi, G.E. Danielson, M.E. Davies, E. de Jong, L. Dones, D. Godfrey, J. Goguen, I. Grenier, V.R. Haemmerle, H. Hammel, C.J. Hansen, C.P. Helfenstein, C. Howell, G.E. Hunt, A.P. Ingersoll, T.V. Johnson, J. Kargel, R. Kirk, D.I. Kuehn, S. Limaye, H. Masursky, A. McEwen, D. Morrison, T. Owen, W. Owen, J.B. Pollack, C.C. Porco, K. Rages, P. Rogers, D. Rudy, C. Sagan, J. Schwartz, E.M. Shoemaker, M. Showalter, B. Sicardy, D. Simonelli, J. Spencer, L.A. Sromovsky, C. Stoker, R.G. Strom, V.E. Suomi, S.P. Synott, R.J. Terrile, P. Thomas, W.R. Thompson, A. Verbiscer, J. Veverka, Voyager 2 at Neptune: imaging science results. Science **246**, 1422–1449 (1989). https://doi.org/10.1126/science.246.4936.1422

S. Soter, The dust belts of Mars. Technical report, Center for Radiophysics and Space Research (1971)

S. Soter, Brightness asymmetry of Iapetus, Paper presented at IAU Coll. No. 28, Cornell Univ. (Aug. 1974)

B.S. Southworth, S. Kempf, J. Schmidt, Modeling Europa's dust plumes. Geophys. Res. Lett. **42**, 10 (2015). https://doi.org/10.1002/2015GL066502

F. Spahn, H.-J. Wiebicke, Long-term gravitational influence of moonlets in planetary rings. Icarus **77**, 124–134 (1989)

F. Spahn, N. Albers, M. Sremčević, C. Thornton, Kinetic description of coagulation and fragmentation in dilute granular particle ensembles. Europhys. Lett. **67**, 545–551 (2004)

F. Spahn, N. Albers, M. Hörning, S. Kempf, A.V. Krivov, M. Makuch, J. Schmidt, M. Seiß, M. Sremčević, E ring dust sources: implications from Cassini's dust measurements. Planet. Space Sci. **54**, 1024–1032 (2006a). https://doi.org/10.1016/j.pss.2006.05.022

F. Spahn, J. Schmidt, N. Albers, M. Hörning, M. Makuch, M. Seiß, S. Kempf, R. Srama, V. Dikarev, S. Helfert, G. Moragas-Klostermeyer, A.V. Krivov, M. Sremčević, A.J. Tuzzolino, T. Economou, E. Grün, Cassini dust measurements at Enceladus and implications for the origin of the E ring. Science **311**, 1416–1418 (2006b). https://doi.org/10.1126/science.1121375

W.B. Sparks, K.P. Hand, M.A. McGrath, E. Bergeron, M. Cracraft, S.E. Deustua, Probing for evidence of plumes on Europa with HST/STIS. Astrophys. J. **829**, 121 (2016). https://doi.org/10.3847/0004-637X/829/2/121

W.B. Sparks, B.E. Schmidt, M.A. McGrath, K.P. Hand, J.R. Spencer, M. Cracraft, S.E. Deustua, Active cryovolcanism on Europa? Astrophys. J. Lett. **839**, 18 (2017). https://doi.org/10.3847/2041-8213/aa67f8

J.R. Spencer, J.C. Pearl, M. Segura, F.M. Flasar, A. Mamoutkine, P. Romani, B.J. Buratti, A.R. Hendrix, L.J. Spilker, R.M.C. Lopes, Cassini encounters Enceladus: background and the discovery of a South polar hot spot. Science **311**, 1401–1405 (2006). https://doi.org/10.1126/science.1121661

J.N. Spitale, T.A. Hurford, A.R. Rhoden, E.E. Berkson, S.S. Platts, Curtain eruptions from Enceladus' South-polar terrain. Nature **521**, 57–60 (2015). https://doi.org/10.1038/nature14368

R. Srama, T.J. Ahrens, N. Altobelli, S. Auer, J.G. Bradley, M. Burton, V.V. Dikarev, T. Economou, H. Fechtig, M. Görlich, M. Grande, A. Graps, E. Grün, O. Havnes, S. Helfert, M. Horanyi, E. Igenbergs, E.K. Jessberger, T.V. Johnson, S. Kempf, A.V. Krivov, H. Krüger, A. Mocker-Ahlreep, G. Moragas-Klostermeyer, P. Lamy, M. Landgraf, D. Linkert, G. Linkert, F. Lura, J.A.M. McDonnell, D. Möhlmann, G.E. Morfill,

M. Müller, M. Roy, G. Schäfer, G. Schlotzhauer, G.H. Schwehm, F. Spahn, M. Stübig, J. Svestka, V. Tschernjawski, A.J. Tuzzolino, R. Wäsch, H.A. Zook, The Cassini cosmic dust analyzer. Space Sci. Rev. **114**, 465–518 (2004). https://doi.org/10.1007/s11214-004-1435-z

R. Srama, S. Kempf, G. Moragas-Klostermeyer, S. Helfert, T.J. Ahrens, N. Altobelli, S. Auer, U. Beckmann, J.G. Bradley, M. Burton, V.V. Dikarev, T. Economou, H. Fechtig, S.F. Green, M. Grande, O. Havnes, J.K. Hillier, M. Horányi, E. Igenbergs, E.K. Jessberger, T.V. Johnson, H. Krüger, G. Matt, N. McBride, A. Mocker, P. Lamy, D. Linkert, G. Linkert, F. Lura, J.A.M. McDonnell, D. Möhlmann, G.E. Morfill, F. Postberg, M. Roy, G.H. Schwehm, F. Spahn, J. Svestka, V. Tschernjawski, A.J. Tuzzolino, R. Wäsch, E. Grün, In situ dust measurements in the inner Saturnian system. Planet. Space Sci. **54**, 967–987 (2006). https://doi.org/10.1016/j.pss.2006.05.021

M. Sremčević, A.V. Krivov, F. Spahn, Impact-generated dust clouds around planetary satellites: asymmetry effects. Planet. Space Sci. **51**, 455–471 (2003). https://doi.org/10.1016/S0032-0633(03)00050-3

M. Sremčević, A.V. Krivov, H. Krüger, F. Spahn, Impact-generated dust clouds around planetary satellites: model versus Galileo data. Planet. Space Sci. **53**, 625–641 (2005). https://doi.org/10.1016/j.pss.2004.10.001

G.R. Stewart, D.N.C. Lin, P. Bodenheimer, Collision-induced transport processes in planetary rings, in *Planetary Rings*, ed. by R. Greenberg, A. Brahic (University Arizona Press, Tucson, 1984), pp. 447–512

G. Strazzulla, Organic material from Phoebe to Iapetus. Icarus **66**, 397–400 (1986). https://doi.org/10.1016/0019-1035(86)90167-3

K.-L. Sun, M. Seiß, M.M. Hedman, F. Spahn, Dust in the arcs of Methone and Anthe. Icarus **284**, 206–215 (2017). https://doi.org/10.1016/j.icarus.2016.11.009

J.R. Szalay, A.R. Poppe, J. Agarwal, D. Britt, I. Belskaya, M. Horányi, T. Nakamura, M. Sachse, F. Spahn, Dust phenomena relating to airless bodies. Space Sci. Rev. **214**, 98 (2018). https://doi.org/10.1007/s11214-018-0527-0

D. Tamayo, J.A. Burns, D.P. Hamilton, Chaotic dust dynamics and implications for the hemispherical color asymmetries of the Uranian satellites. Icarus **226**, 655–662 (2013). https://doi.org/10.1016/j.icarus.2013.06.018

D. Tamayo, M.M. Hedman, J.A. Burns, First observations of the Phoebe ring in optical light. Icarus **233**, 1–8 (2014). https://doi.org/10.1016/j.icarus.2014.01.021

D. Tamayo, S.R. Markham, M.M. Hedman, J.A. Burns, D.P. Hamilton, Radial profiles of the Phoebe ring: a vast debris disk around Saturn. Icarus **275**, 117–131 (2016). https://doi.org/10.1016/j.icarus.2016.04.009

P.C. Thomas, R. Tajeddine, M.S. Tiscareno, J.A. Burns, J. Joseph, T.J. Loredo, P. Helfenstein, C. Porco, Enceladus's measured physical libration requires a global subsurface ocean. Icarus **264**, 37–47 (2016). https://doi.org/10.1016/j.icarus.2015.08.037

H.B. Throop, L.W. Esposito, G ring particle sizes derived from ring plane crossing observations. Icarus **131**, 152–166 (1998)

H.B. Throop, C.C. Porco, R.A. West, J.A. Burns, M.R. Showalter, P.D. Nicholson, The Jovian rings: new results derived from Cassini, Galileo, Voyager, and Earth-based observations. Icarus **172**, 59–77 (2004). https://doi.org/10.1016/j.icarus.2003.12.020

A.J. Verbiscer, M.F. Skrutskie, D.P. Hamilton, Saturn's largest ring. Nature **461**, 1098–1100 (2009). https://doi.org/10.1038/nature08515

J.-E. Wahlund, M. André, A.I.E. Eriksson, M. Lundberg, M.W. Morooka, M. Shafiq, T.F. Averkamp, D.A. Gurnett, G.B. Hospodarsky, W.S. Kurth, K.S. Jacobsen, A. Pedersen, W. Farrell, S. Ratynskaia, N. Piskunov, Detection of dusty plasma near the E-ring of Saturn. Planet. Space Sci. **57**, 1795–1806 (2009). https://doi.org/10.1016/j.pss.2009.03.011

J.H. Waite, M.R. Combi, W.-H. Ip, T.E. Cravens, R.L. McNutt, W. Kasprzak, R. Yelle, J. Luhmann, H. Niemann, D. Gell, B. Magee, G. Fletcher, J. Lunine, W.-L. Tseng, Cassini ion and neutral mass spectrometer: Enceladus plume composition and structure. Science **311**, 1419–1422 (2006). https://doi.org/10.1126/science.1121290

J.H. Waite, R.S. Perryman, M.E. Perry, K.E. Miller, J. Bell, T.E. Cravens, C.R. Glein, J. Grimes, M. Hedman, J. Cuzzi, T. Brockwell, B. Teolis, L. Moore, D.G. Mitchell, A. Persoon, W.S. Kurth, J.-E. Wahlund, M Morooka, L.Z. Hadid, S. Chocron, J. Walker, A. Nagy, R. Yelle, S. Ledvina, R. Johnson, W. Tseng, O.J. Tucker, W.-H. Ip, Chemical interactions between Saturn's atmosphere and its rings. Science **362**, 2382 (2018). https://doi.org/10.1126/science.aat2382

H.A. Weaver, P.D. Feldman, M.F. A'Hearn, C. Arpigny, R.A. Brown, E.F. Helin, D.H. Levy, B.G. Marsden, K.J. Meech, S.M. Larson, K.S. Noll, J.V. Scotti, Z. Sekanina, C.S. Shoemaker, E.M. Shoemaker, T.E. Smith, A.D. Storrs, D.K. Yeomans, B. Zellner, HST observations of comet Shoemaker-Levy (1993e), in *AAS/Division for Planetary Sciences Meeting Abstracts #25*. Bulletin of the American Astronomical Society, vol. 25 (1993), p. 1042

H.A. Weaver, S.A. Stern, M.J. Mutchler, A.J. Steffl, M.W. Buie, W.J. Merline, J.R. Spencer, E.F. Young, L.A. Young, Discovery of two new satellites of Pluto. Nature **439**, 943–945 (2006). https://doi.org/10.1038/nature04547

G.W. Wetherill, Collisions in the asteroid belt. J. Geophys. Res. **72**, 2429 (1967). https://doi.org/10.1029/JZ072i009p02429

O.C. Winter, D.C. Mourão, S.M. Giuliatti Winter, F. Spahn, C. da Cruz, Moonlets wandering on a leash-ring. Mon. Not. R. Astron. Soc. **380**, 54–57 (2007)

V.V. Yaroshenko, H. Lühr, Electrical conductivity of the dusty plasma in the Enceladus plume. Icarus **278**, 79–87 (2016). https://doi.org/10.1016/j.icarus.2016.05.033

V.V. Yaroshenko, M. Horányi, G.E. Morfill, The wave mechanism of spoke formation in Saturn's rings, in *Multifacets of Dusty Plasmas, Fifths International Conference on the Physics of Dusty Plasmas*, vol. 1041 (AIP, New York, 2008), pp. 215–216

V.V. Yaroshenko, S. Ratynskaia, J. Olson, N. Brenning, J.-E. Wahlund, M. Morooka, W.S. Kurth, D.A. Gurnett, G.E. Morfill, Characteristics of charged dust inferred from the Cassini RPWS measurements in the vicinity of Enceladus. Planet. Space Sci. **57**, 1807–1812 (2009). https://doi.org/10.1016/j.pss.2009.03.002

S.-Y. Ye, D.A. Gurnett, W.S. Kurth, T.F. Averkamp, M. Morooka, S. Sakai, J.-E. Wahlund, Electron density inside Enceladus plume inferred from plasma oscillations excited by dust impacts. J. Geophys. Res. Space Phys. **119**, 3373–3380 (2014a). https://doi.org/10.1002/2014JA019861

S.-Y. Ye, D.A. Gurnett, W.S. Kurth, T.F. Averkamp, S. Kempf, H.-W. Hsu, R. Srama, E. Grün, Properties of dust particles near Saturn inferred from voltage pulses induced by dust impacts on Cassini spacecraft. J. Geophys. Res. Space Phys. **119**, 6294–6312 (2014b). https://doi.org/10.1002/2014JA020024

S.-Y. Ye, D.A. Gurnett, W.S. Kurth, In-situ measurements of Saturn's dusty rings based on dust impact signals detected by Cassini RPWS. Icarus **279**, 51–61 (2016). https://doi.org/10.1016/j.icarus.2016.05.006

M.Y. Zolotov, An oceanic composition on early and today's Enceladus. Geophys. Res. Lett. **34**, 23203 (2007). https://doi.org/10.1029/2007GL031234

H.A. Zook, E. Grun, M. Baguhl, D.P. Hamilton, G. Linkert, J.-C. Liou, R. Forsyth, J.L. Phillips, Solar wind magnetic field bending of Jovian dust trajectories. Science **274**, 1501–1503 (1996). https://doi.org/10.1126/science.274.5292.1501

The manufacturer's authorised representative in the EU is Springer Nature Customer Service Centre GmbH, Europaplatz 3, 69115 Heidelberg, Germany. If you have any concerns regarding our products, please contact ProductSafety@springernature.com

Printed and bound by CPI Group (UK) Ltd, Croydon, CR0 4YY
28/09/2026
02224092-0008